IET MATERIALS, CIRCUITS AND DEVICES SERIES 64

Phase-Locked Frequency Generation and Clocking

Other volumes in this series

Volume 2 **Analogue IC Design: The current-mode approach** C. Toumazou, F.J. Lidgey and D.G. Haigh (Editors)
Volume 3 **Analogue–Digital ASICs: Circuit techniques, design tools and applications** R.S. Soin, F. Maloberti and J. France (Editors)
Volume 4 **Algorithmic and Knowledge-Based CAD for VLSI** G.E. Taylor and G. Russell (Editors)
Volume 5 **Switched Currents: An analogue technique for digital technology** C. Toumazou, J.B.C. Hughes and N.C. Battersby (Editors)
Volume 6 **High-Frequency Circuit Engineering** F. Nibler *et al.*
Volume 8 **Low-Power High-Frequency Microelectronics: A unified approach** G. Machado (Editor)
Volume 9 **VLSI Testing: Digital and mixed analogue/digital techniques** S.L. Hurst
Volume 10 **Distributed Feedback Semiconductor Lasers** J.E. Carroll, J.E.A. Whiteaway and R.G.S. Plumb
Volume 11 **Selected Topics in Advanced Solid State and Fibre Optic Sensors** S.M. Vaezi-Nejad (Editor)
Volume 12 **Strained Silicon Heterostructures: Materials and devices** C.K. Maiti, N.B. Chakrabarti and S.K. Ray
Volume 13 **RFIC and MMIC Design and Technology** I.D. Robertson and S. Lucyzyn (Editors)
Volume 14 **Design of High Frequency Integrated Analogue Filters** Y. Sun (Editor)
Volume 15 **Foundations of Digital Signal Processing: Theory, algorithms and hardware design** P. Gaydecki
Volume 16 **Wireless Communications Circuits and Systems** Y. Sun (Editor)
Volume 17 **The Switching Function: Analysis of power electronic circuits** C. Marouchos
Volume 18 **System on Chip: Next generation electronics** B. Al-Hashimi (Editor)
Volume 19 **Test and Diagnosis of Analogue, Mixed-Signal and RF Integrated Circuits: The system on chip approach** Y. Sun (Editor)
Volume 20 **Low Power and Low Voltage Circuit Design with the FGMOS Transistor** E. Rodriguez-Villegas
Volume 21 **Technology Computer Aided Design for Si, SiGe and GaAs Integrated Circuits** C.K. Maiti and G.A. Armstrong
Volume 22 **Nanotechnologies** M. Wautelet *et al.*
Volume 23 **Understandable Electric Circuits** M. Wang
Volume 24 **Fundamentals of Electromagnetic Levitation: Engineering sustainability through efficiency** A.J. Sangster
Volume 25 **Optical MEMS for Chemical Analysis and Biomedicine** H. Jiang (Editor)
Volume 26 **High Speed Data Converters** A.M.A. Ali
Volume 27 **Nano-Scaled Semiconductor Devices** E.A. Gutiérrez-D (Editor)
Volume 28 **Security and Privacy for Big Data, Cloud Computing and Applications** L. Wang, W. Ren, K.R. Choo and F. Xhafa (Editors)
Volume 29 **Nano-CMOS and Post-CMOS Electronics: Devices and modelling** Saraju P. Mohanty and Ashok Srivastava
Volume 30 **Nano-CMOS and Post-CMOS Electronics: Circuits and design** Saraju P. Mohanty and Ashok Srivastava
Volume 32 **Oscillator Circuits: Frontiers in design, analysis and applications** Y. Nishio (Editor)
Volume 33 **High Frequency MOSFET Gate Drivers** Z. Zhang and Y. Liu
Volume 34 **RF and Microwave Module Level Design and Integration** M. Almalkawi
Volume 35 **Design of Terahertz CMOS Integrated Circuits for High-Speed Wireless Communication** M. Fujishima and S. Amakawa
Volume 38 **System Design with Memristor Technologies** L. Guckert and E.E. Swartzlander Jr.
Volume 39 **Functionality-Enhanced Devices: An alternative to Moore's law** P.-E. Gaillardon (Editor)
Volume 40 **Digitally Enhanced Mixed Signal Systems** C. Jabbour, P. Desgreys and D. Dallett (Editors)

Volume 43 **Negative Group Delay Devices: From concepts to applications** B. Ravelo (Editor)
Volume 45 **Characterisation and Control of Defects in Semiconductors** F. Tuomisto (Editor)
Volume 47 **Understandable Electric Circuits: Key concepts, 2nd Edition** M. Wang
Volume 51 **Modelling Methodologies in Analogue Integrated Circuit Design** G. Dundar and M.B. Yelten (Editors)
Volume 53 **VLSI Architectures for Future Video Coding** M. Martina (Editor)
Volume 54 **Advances in High-Power Fiber and Diode Laser Engineering** Ivan Divliansky (Editor)
Volume 58 **Magnetorheological Materials and their Applications** S. Choi and W. Li (Editors)
Volume 60 **IP Core Protection and Hardware-Assisted Security for Consumer Electronics** A. Sengupta and S. Mohanty
Volume 67 **Frontiers in Securing IP Cores; Forensic detective control and obfuscation techniques** A Sengupta
Volume 68 **High Quality Liquid Crystal Displays and Smart Devices: Vol. 1 and Vol. 2** S. Ishihara, S. Kobayashi and Y. Ukai (Editors)
Volume 69 **Fibre Bragg Gratings in Harsh and Space Environments: Principles and applications** B. Aïssa, E I. Haddad, R.V. Kruzelecky, W.R. Jamroz
Volume 70 **Self-Healing Materials: From fundamental concepts to advanced space and electronics applications, 2nd Edition** B. Aïssa, E I. Haddad, R.V. Kruzelecky, W.R. Jamroz
Volume 71 **Radio Frequency and Microwave Power Amplifiers: Vol. 1 and Vol. 2** A. Grebennikov (Editor)
Volume 73 **VLSI and Post-CMOS Electronics Volume 1: VLSI and post-CMOS electronics and Volume 2: Materials, devices and interconnects** R. Dhiman and R. Chandel (Editors)

Phase-Locked Frequency Generation and Clocking

Architectures and circuits for modern wireless and wireline systems

Edited by
Woogeun Rhee

The Institution of Engineering and Technology

Published by The Institution of Engineering and Technology, London, United Kingdom

The Institution of Engineering and Technology is registered as a Charity in England & Wales (no. 211014) and Scotland (no. SC038698).

First published 2020

The Institution of Engineering and Technology
Michael Faraday House
Six Hills Way, Stevenage
Herts, SG1 2AY, United Kingdom

www.theiet.org

British Library Cataloguing in Publication Data
A catalogue record for this product is available from the British Library

ISBN 978-1-78561-885-7 (hardback)
ISBN 978-1-78561-886-4 (PDF)

Typeset in India by MPS Limited

Contents

About the editor

Woogeun Rhee is a Professor in the Institute of Microelectronics and the Department of Microelectronics and Nanoelectronics at Tsinghua University, Beijing, China. He has over 20 years of professional career in integrated circuit design with nearly 10 years in industry and 13 years in academia. From 1997 to 2001, he was with Conexant Systems, Newport Beach, CA, where he developed a low-power low-cost fractional-N frequency synthesizer product. From 2001 to 2006, he was with IBM Thomas J. Watson Research Center, Yorktown Heights, NY and worked on clocking circuits for high-speed I/O serial links, including low-jitter phase-locked loops, clock-and-data recovery circuits, and on-chip testability circuits. He has published >150 IEEE publications and currently holds 24 US patents. He is an IEEE Distinguished Lecturer (2016–2017) and an Ex-Officio Administrative Committee (AdCom) member of the Solid-State Circuits Society (2020).

Part I

Basic architectures and system perspectives

Chapter 1
Evolution of monolithic phase-locked loops

Woogeun Rhee[1]

The phase-locked loop (PLL) is a key building block in both wireline and wireless systems. In the wireline system, low-jitter clock generation and clock-and-data recovery (CDR) circuits are critical in high data rate I/O links. In the wireless system, the delta-sigma ($\Delta\Sigma$) fractional-N frequency synthesizer plays an important role in modern transceivers not only as a local oscillator but also as a phase modulator with direct-digital modulation. Due to various applications, diversified PLL architectures and circuit techniques have been proposed in consideration of performance, power, and cost, making it difficult for circuit designers to choose the right design solution. Hence, it would be somewhat helpful for circuit designers to gain a historical view of the PLL integrated circuit. This chapter reviews major innovations in the development of early-stage complementary-metal-oxide-semiconductor (CMOS) PLLs. In the first section, a brief overview of the development of PLL integrated circuits (ICs) is introduced. Then, major innovations in the history of CMOS PLLs are described in the following sections by considering architecture, circuit, and application aspects.

1.1 Overview

Even though the PLL system was developed in the 1930s, back then it was considered a complex system with limited use in a few coherent communication systems such as a satellite system until the advent of the monolithic PLL. One of the most popular commercial applications of the PLL IC was the TV system in which the PLL was considered a versatile building block not only for horizontal and vertical synchronizations in screen display but also for sub-carrier recovery and phase synchronization in color signal demodulation. Nowadays, the PLL IC is widely adopted in various applications including frequency synthesis, CDR, frequency modulation, and demodulation.

Due to the limited speed performance of CMOS technology, monolithic PLLs were implemented with either bipolar or BiCMOS technology in the early days. In

[1]Department of Microelectronics and Nanoelectronics, Institute of Microelectronics, Tsinghua University, Beijing, China

the late 1970s, PLLs having additional system features such as frequency-shift keying (FSK) demodulation (Figure 1.1(a)) and the clock-recovery PLL with an extended capture range (Figure 1.1(b)) were implemented with Bipolar technology [1,2]. The development of the CMOS PLL came along with the advance of digital systems. A fully integrated CMOS PLL appeared in the late 1980s in the area of chip-to-chip communication and CDR systems [3–7]. The invention of a phase/frequency detector (PFD) and a charge pump significantly elevated the PLL architecture [8,9]. A voltage-controlled oscillator (VCO) based on a ring oscillator instead of a relaxation oscillator was proposed for high-speed clock generation. A PLL based on a voltage-controlled delay line (VCDL) instead of a VCO was also proposed [10], which was later named as a delay-locked loop (DLL). The use of the bang-bang phase detector (BBPD) was proposed for robust and high-speed operation at the cost of the nonlinear bandwidth control [10,11].

In the 1990s, the fundamentals of product-level PLL architectures were developed in both wireline and wireless systems. A differential ring VCO was

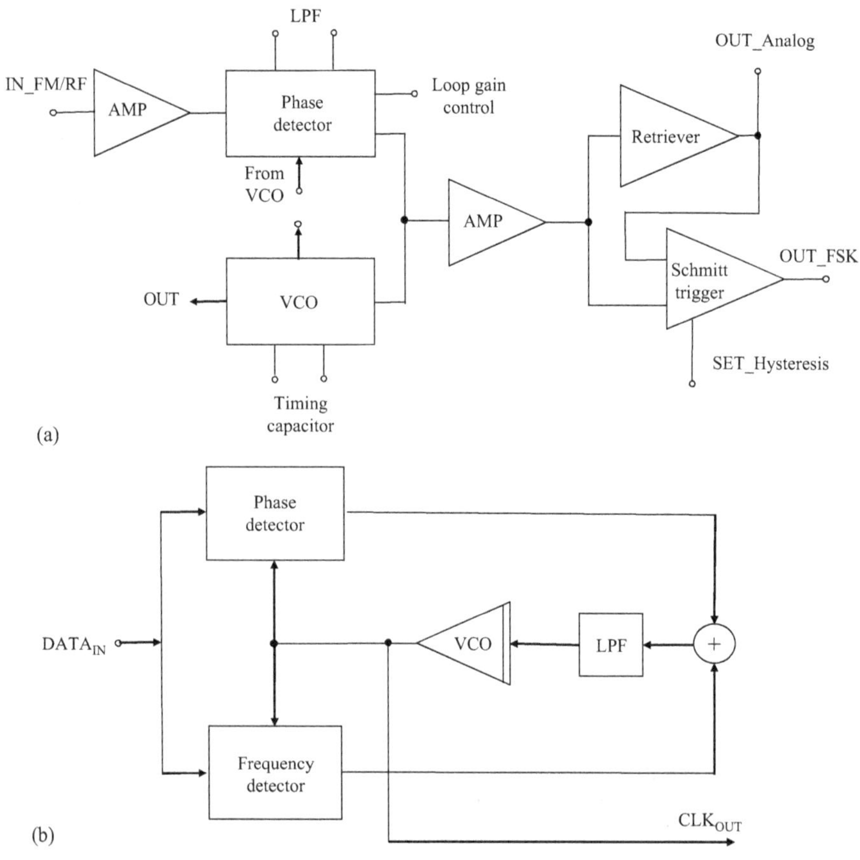

Figure 1.1 Monolithic PLLs in Bipolar: (a) PLL with FSK demodulation [1] and (b) phase and frequency-locked loop for clock recovery [2]

proposed not only to enable robust clock generation with high power supply rejection (PSR) but also to achieve flexible multi-phase generation [12]. Multi-phase data sampling by utilizing the ring VCO alleviates the clock speed requirement for CDR systems. The DLL architecture that is able to track a frequency offset to a certain degree was further developed, which eliminated the need for multiple VCOs in multiple I/O links [13]. In wireless systems, a fractional-N frequency synthesizer with $\Delta\Sigma$ modulation greatly relaxed design tradeoffs of the conventional integer-N frequency synthesizer [14,15]. The $\Delta\Sigma$ fractional-N frequency synthesizer also enabled direct frequency/phase modulation, building a basis toward digital-intensive transmitter systems [16,17].

Thanks to numerous design innovations along with the advance of CMOS technology, PLL architectures have brought more and more diversity since the 2000s. A fully on-chip inductor capacitor (LC) VCO was realized in radio frequency (RF) transceivers for cellular applications. The $\Delta\Sigma$ fractional-N PLLs have been extensively used for both wireless and wireline applications such as two-point modulation, frequency-modulated continuous waveform signal generation, and spread-spectrum clock generation. One of the most significant innovations in the 2000s was the digital-intensive PLL (DPLL)* [18], which overcame the leakage current problem of the on-chip loop filter, while offering technology scaling and digital calibration features. Below is a brief summary of the development of monolithic PLLs.

- 1970s: Beginning of monolithic design
  - Monolithic PLL with non-CMOS technologies
  - Elevated PLL architectures with PFD invention
- 1980s: CMOS PLLs for synchronization, clock generation and clock recovery
  - Use of ring VCO
  - PLL for I/O clock de-skewing
  - PLL with variable delay line (later named as DLL)
  - Bang-bang PLL for robust and high-speed clocking
- 1990s: PLL fundamentals for wireline and wireless
  - Differential ring VCOs and high PSR PLLs
  - Multi-phase data sampling with ring VCO
  - Phase-rotator-based semidigital DLL
  - $\Delta\Sigma$ fractional-N PLL for frequency synthesis and modulation
  - On-chip LC VCOs for single-chip transceivers
- 2000s: Boom in PLL architectures
  - Diversified and fragmented PLL architectures
  - PLL with on-chip LC VCO for wireless and high-speed wireline
  - Advanced PLL-based frequency modulation for wireless and wireline
  - Digital-intensive PLLs

*To avoid confusion with *true* digital PLL that has a counter-based numerically controlled oscillator often employed in the TV system, the term *digital-intensive PLL* (DPLL) instead of the *all-digital PLL* is used in this chapter.

1.2 Architectures and circuits

In this section, PLL innovations that have given a significant impact on products from the late 1980s to the early 2000s are covered. The PLL architecture in the 2000s is much more fragmented and diversified. Some of the state-of-the-art architectures from the 2000s will be covered in the following chapters of this book.

1.2.1 PLL with PFD and charge pump

A charge pump PLL (CP-PLL) with the invention of a phase/frequency detector (PFD) greatly simplified the PLL design and paved the way for the modern on-chip clock generation [8,9]. The use of the charge pump made it possible to realize a type-2 feedback system with passive devices only, i.e. resistors and capacitors. Designing a low noise and a wide bandwidth PLL became easier than ever without requiring an active integrator. Moreover, the use of the PFD substantially relaxed the complex large-signal behavior of the PLL, as it provides frequency-sensitive phase detection when the PLL is out-of-lock-in range. The frequency detection feature of the PFD was also utilized as part of a frequency-acquisition loop in the CDR system.

One drawback of the PFD is the existence of a dead zone. Unlike other phase detectors that provide direct current (DC) information based on a balanced duty cycle at the output, the PFD generates a pulse whose width is proportional to a phase error. When the phase error is close to zero, the PFD having a very narrow pulse width has difficulty in tuning on a charge pump in the following stage [19]. The dead-zone problem was overcome by having a minimum common-mode delay for both the up and down outputs as illustrated in Figure 1.2 [20]. The CP-PLL with the PFD has become a standard PLL topology for most clock generation and frequency synthesis applications. Nowadays, a time-to-digital converter (TDC) in the DPLL and hybrid phase detectors also receive high attention for low-noise clock generation systems.

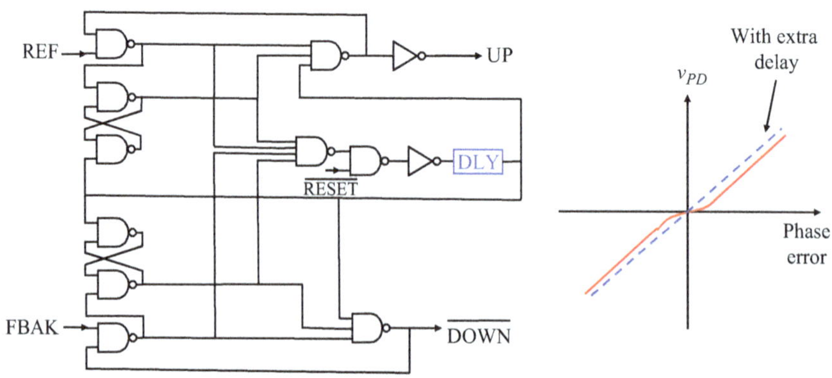

Figure 1.2 Dead-zone-free PFD with extra delay cells [20]

1.2.2 PLL with ring VCO

As the clock frequency increases, clock skew between an on-chip clock and an off-chip clock becomes critical in chip-to-chip communication. The first fully integrated CMOS PLL was designed for clock de-skewing purposes [3]. Figure 1.3 shows a block diagram of the PLL that reduces the clock skew between an external clock and an internal clock. A VCO based on a current-starved ring oscillator was proposed to overcome the limited speed and tuning range of the conventional relaxation oscillator. Since then, on-chip clock de-skewing based on the PLL has been a viable solution to enhance the data rate, making the PLL a key building block in high-speed serial I/O links. The design of a wide bandwidth PLL became popular as a high open-loop gain of the PLL could significantly improve the phase noise performance after having the low-Q ring VCO locked in phase with a stable reference source.

Even though the ring VCO was highly attractive for an on-chip PLL design, poor PSR of the ring VCO made it difficult for it to be useful in an actual very-large-scale integration system until a differential ring VCO was developed [12]. Figure 1.4(a) shows a schematic of the differential ring VCO. A linear-mode P-type metal-oxide-semiconductor (PMOS) transistor is used as a load resistor, so that resistance can be varied by a gate voltage. By having a voltage-controlled resistor with a replica bias circuit, a voltage swing at the output is nearly constant regardless of the tail current value. If a passive resistor is used instead of the linear-mode PMOS transistor, the voltage swing increases linearly with the tail current, resulting in a narrow tuning range since the oscillation frequency of the differential ring VCO is mainly determined by the ratio of the tail current to the voltage swing for a

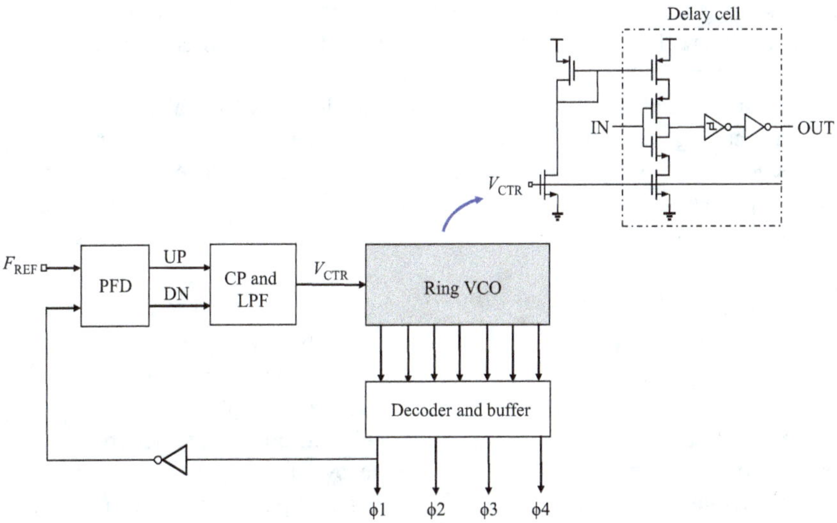

Figure 1.3 CMOS PLL with ring VCO [3]

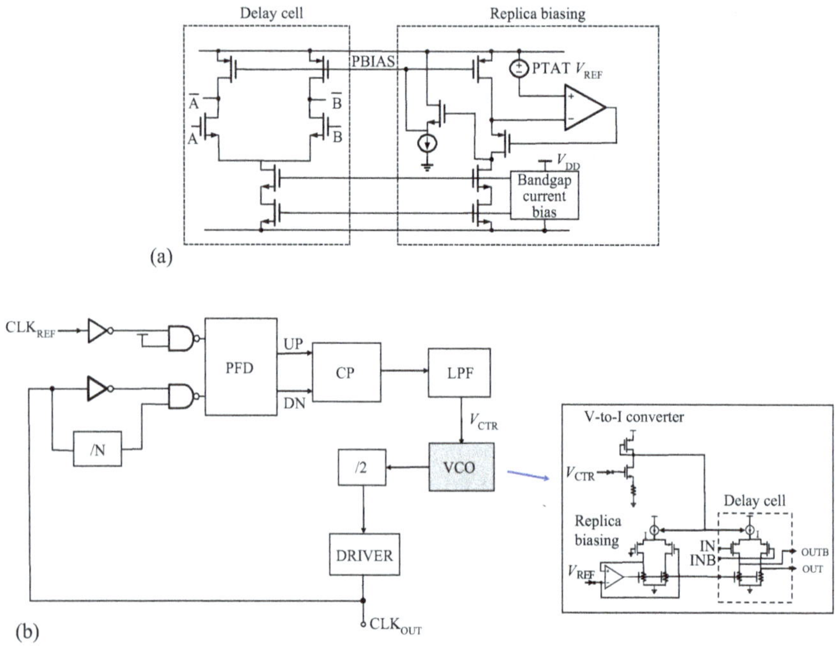

Figure 1.4 (a) Differential ring VCO [12] and (b) clock generator for microprocessors [21]

given load capacitance. To further improve the PSR performance of the ring VCO, a PMOS input pair with NMOS load transistors was proposed [21]. In this way, the output swing of the ring VCO is ground referenced, resulting in higher PSR as shown in Figure 1.4(b). A product-level PLL with the proposed differential ring VCO was successfully launched in microprocessors.

The differential ring VCO not only provides high PSR but also enables even-number phase generation, e.g. quadrature phase generation. Compared to the single-ended ring VCO, the differential ring VCO is able to generate 2^k number of multiple phases, which is highly useful for flexible data slicing with lower clock frequency than the data rate in wireline systems [12].

1.2.3 PLL with dual-path control

A control path of the type-2 PLL can be split into two separate paths; an integral path and a proportional-gain path. Since the integral path stores DC information, the proportional-gain path does not have to induce a static phase error to provide a DC voltage to the VCO. As a result, the type-2 PLL does not suffer from a static phase error problem when there is a frequency jump at the PLL input. The proportional-gain path is analogous to a small-signal path, while the integral path being analogous to a large-signal path. In the design of an integrated PLL, having a separate control for the proportional-gain path and the integral path offers a flexible

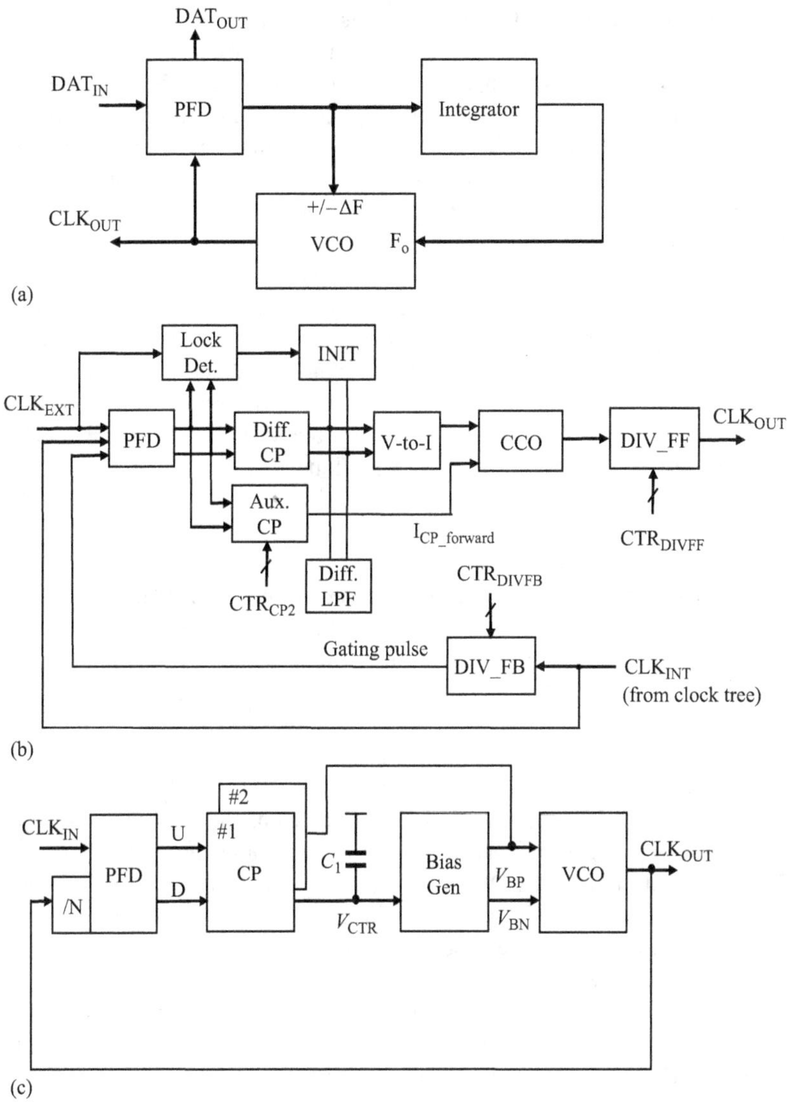

Figure 1.5 (a) Dual-path CDR [22], (b) resistor-less PLL [23], and (c) self-biased PLL [24]

way of designing a robust PLL at the circuit level. The PLL having a dual charge pump was proposed to have a flexible gain control of the proportional-gain path and the integral path [20].

A PLL shown in Figure 1.5(a) is considered one of the earliest dual-path PLLs implemented with Bipolar technology [22]. The differential ring VCO contains an extra bang-bang delay cell that is directly controlled by the output of a phase/frequency logic block. Other delay cells are continuously tuned by an integrator. Since there exists a bypass path of the integrator from the phase detector to the VCO, a zero of the system is automatically created, stabilizing the type-2 feedback system. A similar concept can be applied to CP-PLLs. Figure 1.5(b) shows a PLL without having a resistor in the loop filter [23]. In the early days, the standard digital CMOS process did not offer a linear poly resistor, only having an *N*-well resistor whose temperature coefficient was as large as 5,000 ppm. For that reason, the *N*-well resistor in a loop filter caused a large variation of the loop bandwidth. By having an auxiliary charge pump that provides a feed-forward path over the integral path, a stable type-2 PLL could be realized without requiring passive resistors. The technique of the resistor-less architecture was further extended to implement a *self-biased* PLL where the same kinds of bias currents were used for both the charge pump and the current-controlled oscillator (CCO) as depicted in Figure 1.5(c), so that the gain variations of the charge pump and the VCO could be correlated over process, voltage, and temperature (PVT) variations [24].

1.2.4 Delay-locked loop

As the VCO is essentially a voltage-to-frequency converter, jitter propagation in the phase domain due to a frequency jump is inevitable in the PLL. Accordingly, jitter accumulation caused by a supply voltage jump is a big concern when a PLL with a ring VCO is placed in microprocessors or other digital systems where billions of transistors generate switching noise. A phase-locked circuit that employs a VCDL instead of the VCO as shown in Figure 1.6(a) was proposed to improve noise immunity, ease loop stabilization, and enable dynamically adjustable clock period [10]. Later, the VCDL-based PLL was named as the DLL that significantly improved the clocking performance in the chip-to-chip communication (Figure 1.6(b)) [25,26]. The application of the DLL, however, was limited to clock or data synchronization since the input of the VCDL could be saturated with a frequency offset. Therefore, the DLL could not be adopted for clocking systems that require frequency multiplication or frequency tracking.

Two major architectures have been proposed to realize DLL-based frequency multiplication. The first approach used an edge combiner that combined multiple phases of the DLL to generate high frequency as illustrated in Figure 1.7(a) [27]. Since the DLL used the low-frequency crystal oscillator input and generated multiple delayed edges, excellent phase noise performance could be achieved. The main drawback of this architecture lied in the difficulty of phase matching and glitch-less edge combining with high-frequency operation. As a second approach, a multiplying DLL (MDLL) shown in Figure 1.7(b) was developed to enable frequency multiplication [28]. Compared to the PLL, the MDLL exhibited superior jitter performance by flushing away the accumulated jitter with periodic synchronization to a clean reference clock. In some sense, the MDLL can also be classified as an injection-locked PLL since it contains a gated ring VCO rather than the VCDL.

Other than the DLL-based frequency multiplication, the DLL-based frequency tracking is another important aspect for the CDR system that needs to track

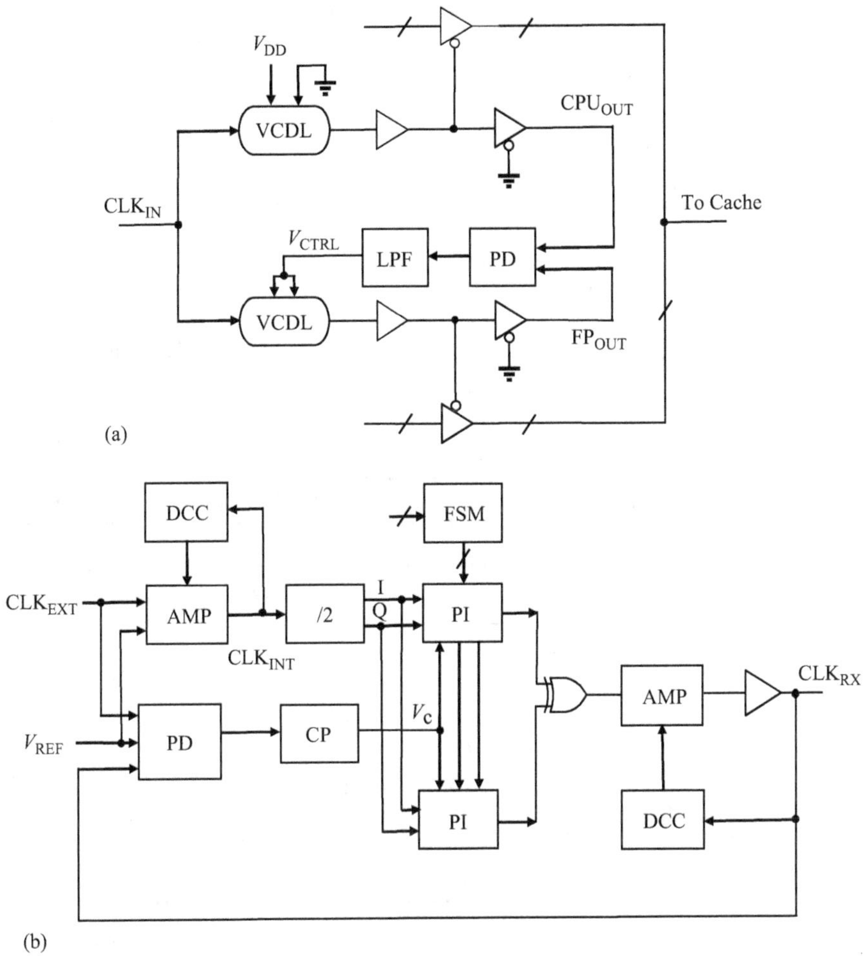

Figure 1.6 (a) PLL with variable delay line [10] and (b) DLL for 500 Mb/s I/O [26]

frequency offset to a certain degree. For multiple I/O links, using conventional CDRs with multiple VCOs is problematic since each VCO with a slight frequency difference can cause a crosstalk problem. A semidigital DLL with a phase-interpolator (PI) shown in Figure 1.8(a) was proposed to achieve frequency tracking based on the phase rotation of digitized phases [25]. Multiple reference phases for phase interpolation was generated by a PLL with a multi-phase ring VCO. Later, a dual DLL architecture shown in Figure 1.8(b) was developed [13], so that multiple DLLs with a single PLL could be used for multiple links. The PI generates more phases with fine resolution, and the digitized phases are rotated when the phase movement exceeds 360°. This kind of a phase-rotator-based DLL became a dominant CDR architecture for low-cost multiple I/O links. Note that the basic

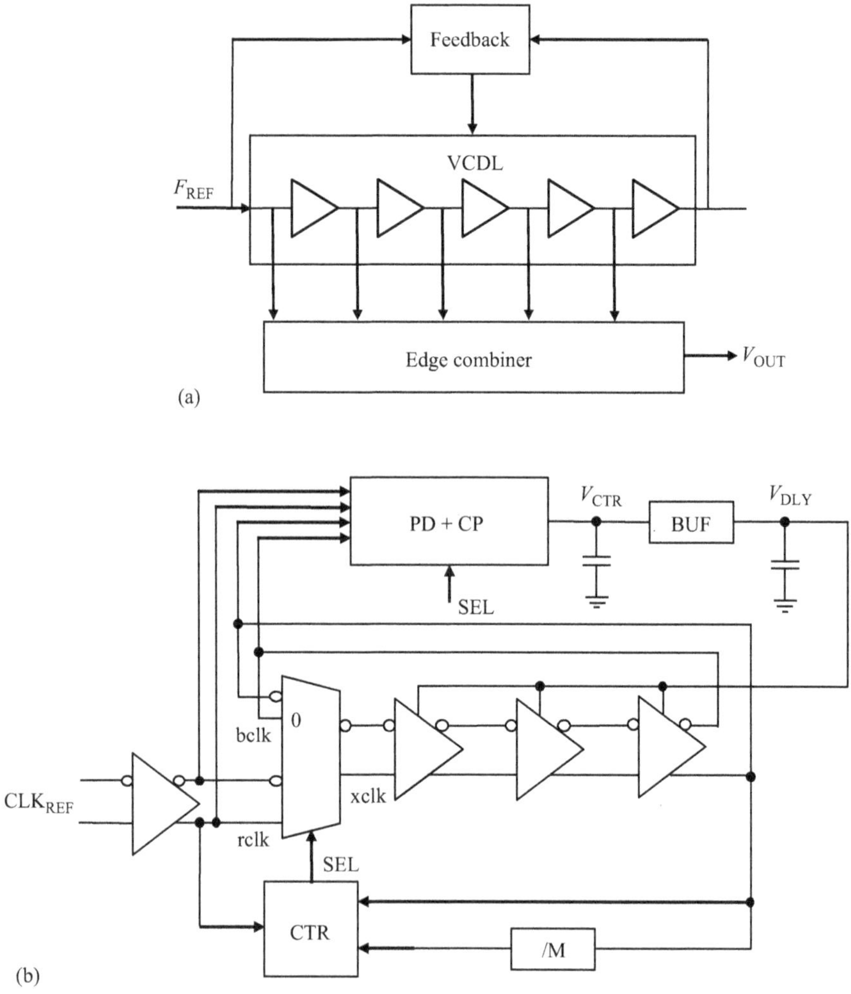

Figure 1.7 DLL-based frequency multiplier: (a) DLL with edge combiner [27] and (b) MDLL [28]

principle of the frequency tracking based on the phase rotator is analogous to that of the PI-based fractional divider as illustrated in Figure 1.9 [29,30].

1.2.5 Bang-bang PLL

Since the non-return-to-zero (NRZ) data does not contain clock information, a linear phase detector (PD) in the CDR PLL needs to generate a narrow pulse at the transition edge of the NRZ data to provide RZ-like information. At the same time, phase error information needs to be contained in the pulse by having the pulse width proportional to the phase error. Such a linear PD not only suffers from a

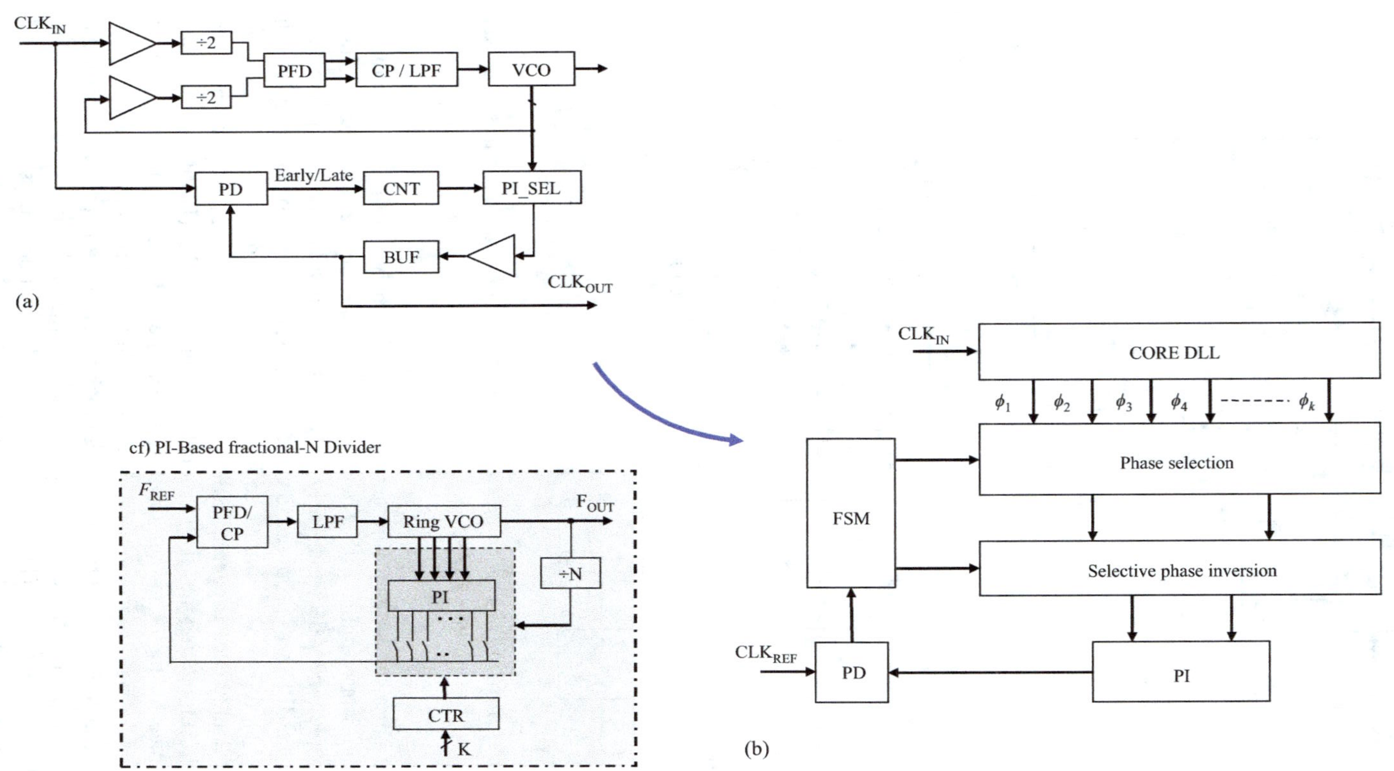

Figure 1.8 DLL-based CDR with frequency tracking: (a) cascaded PLL and semidigital DLL [25] and (b) dual semidigital DLL [13]

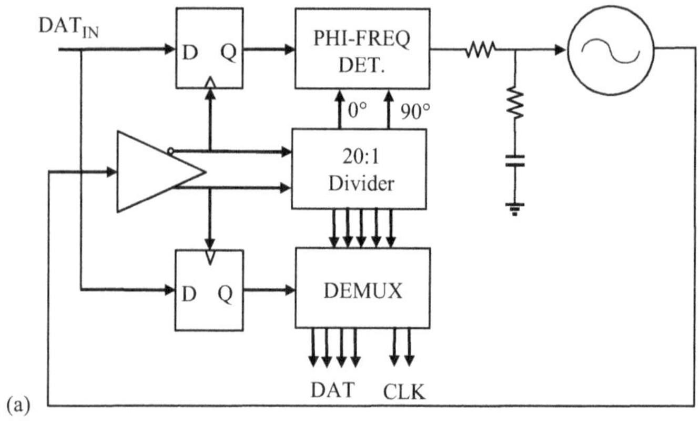

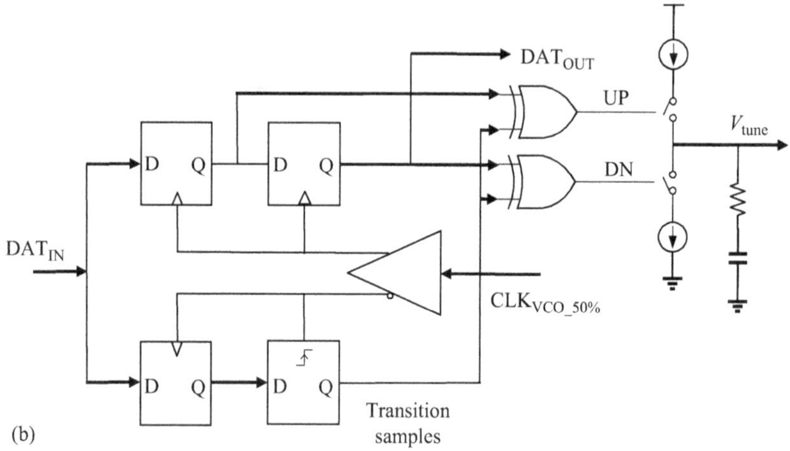

Figure 1.9 (a) BBPD to reduce mismatch between clock and data paths [11] and (b) Alexander PD [31]

speed limit but also causes a substantial phase offset between the clock-recovery path and the data retiming path when the data rate gets close to the technology limit. A simple binary-decision (early/late) PD based on a D-type flip-flop (DFF), also called a BBPD, was proposed to achieve high-speed operation and reduce path mismatch [11], followed by improved BBPDs for the NRZ data, e.g. an Alexander PD [31], a double-edge triggered BBPD [32], a ternary PD [33], and so on. Since then, the PLL employing the BBPD, namely, a bang-bang PLL (BBPLL) has been a popular choice for high data rate CDR systems. The BBPLL architecture has also been adopted for synchronization and clock generation circuits. For instance, the first design of the DLL employed the BBPD for clock synchronization since the DLL did not suffer from jitter accumulation [10]. Nowadays, the digital-intensive PLL using the BBPD, namely, a 1-bit TDC, also receives high attention because of the design complexity of the linear TDC.

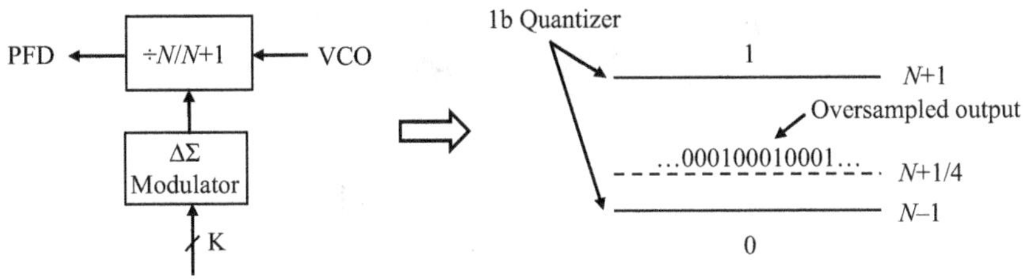

Figure 1.10 ΔΣ fractional-N frequency synthesizer [14,15]

1.2.6 Delta-sigma fractional-N PLL

The fractional-*N* frequency synthesizer is a PLL-based synthesizer that synthesizes frequencies with a frequency resolution finer than phase detector frequency. The fractional-*N* method originally comes from the *Digiphase* technique [34], which is later called as a fractional-*N* technique in the commercial version [35]. Since the phase detector frequency is higher than the frequency resolution, the frequency division ratio is substantially reduced, offering several advantages over integer-*N* frequency synthesizers [36]. Firstly, the in-band phase noise requirement of PLL circuits is greatly relaxed. Secondly, reference spur performance is not so sensitive to the leakage current and to the charge pump current mismatch. Thirdly, the fractional-*N* technique offers agile frequency switching. The main problem of the conventional fractional-*N* frequency synthesizer is unwanted spur generation, which is caused by periodic operation of the dual-modulus divider. By employing the ΔΣ modulation technique, pseudo-random modulation can be achieved to suppress fractional spurs, while pushing the modulation noise away from carrier frequency [14,15]. The desired fractional division ratio is similar to the DC input of an oversampling analog-to-digital converter, and the integer divider is analogous to the one bit quantizer as illustrated in Figure 1.10.

The development of the ΔΣ fractional-*N* PLL is considered one of the most significant innovations in the area of frequency synthesis. Not only has the ΔΣ fractional-*N* PLL improved the performance of frequency synthesizers, but it has also elevated the architecture of modern transceivers. By having the frequency resolution fully determined by the reference frequency and the number of modulation bits, the ΔΣ fractional-*N* PLL could directly accommodate modulated digital data at the input of the ΔΣ modulator, thus achieving direct-digital frequency or phase modulation [16,17]. To overcome the bandwidth limitation of the PLL, modulated data with a pre-emphasis technique were proposed as depicted in Figure 1.11 [16].

1.2.7 Digital-intensive PLL

One notable architecture among modern PLLs is the digital-intensive PLL [18] (Figure 1.12). High attention toward the DPLL in the early days was mainly associated with technology issues. As CMOS technology entered the era of sub-0.1 μm process, the loop filter of the conventional PLL suffered a lot from gate leakage current, process variation, and poor scalability problems. Consisting of a TDC, a digital loop filter, and

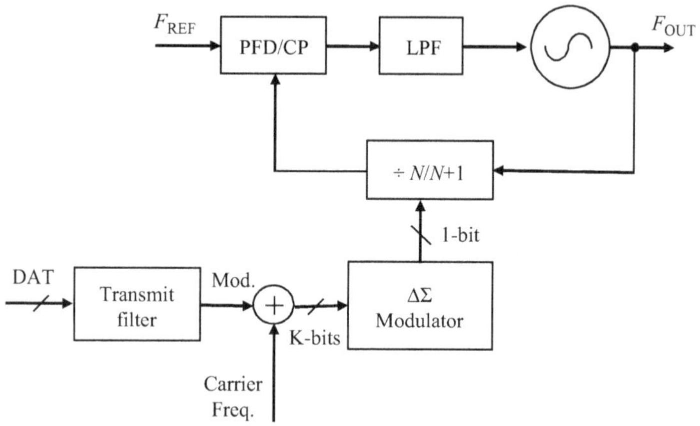

Figure 1.11 Direct digital phase/frequency modulation with pre-emphasis [16]

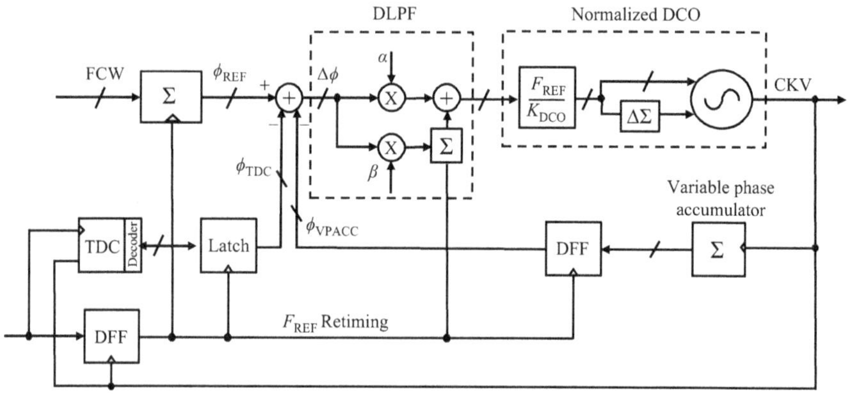

Figure 1.12 Digital-intensive PLL [18]

a digitally controlled oscillator (DCO), the DPLL offers the following merits over the traditional CP-PLL: (1) avoiding leakage current problem of the loop filter; (2) high integration and good area scalability with advanced CMOS technology; (3) easy for design migration from old technology to new technology; (4) fast frequency acquisition and linear modulation with DCO calibration; and (5) low in-band phase noise performance without the charge pump and analog bias current. Adopting the DPLL architecture, however, does not necessarily guarantee superior performance to conventional PLLs in spite of the fancy term *digital-intensive* or *all-digital*. In practice, the main bottleneck for the design of high-performance DPLLs lies in the implementation of a high-resolution TDC. Poor resolution or nonlinearity of the TDC results in degraded in-band phase noise or fractional spur performance. Accordingly, a complex TDC or a high-resolution digital-to-time converter is often required for the design of high-performance DPLLs even with the state-of-the-art CMOS technology.

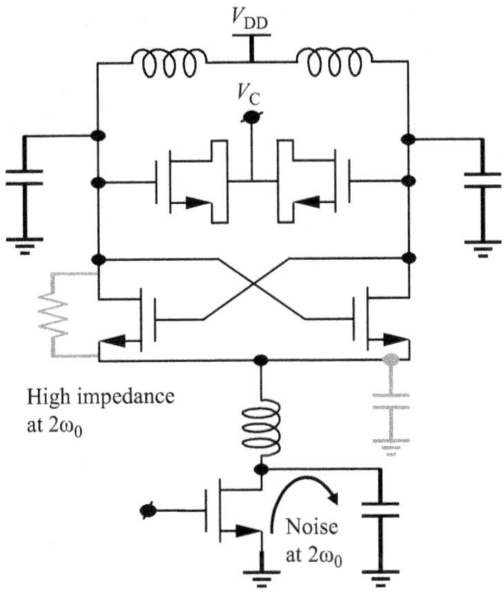

Figure 1.13 Noise filtering technique in LC VCO [48]

1.2.8 LC VCO

As the performance of CMOS PLLs had rapidly been improved with architecture and circuit innovations, one of the ultimate goals of RF engineers was to integrate an LC VCO in RF transceivers. In the early days, the quality factor of an on-chip inductor was not high enough to be useful in Bipolar or CMOS VCOs for cellular phone applications [37,38]. For instance, the global system for mobile communications (GSM) radio not only requires low in-band phase noise for a <1 rms-degree phase error but also aims for the out-of-band phase noise of –165 dBc/Hz at 20-MHz offset frequency to eliminate an external band-pass filter at the transmitter output. A bondwire inductor was proposed as a remedy to realize an integrated LC VCO with in-package solution [39], but inductance variation of the bondwire was not tolerable for mass production and the in-package integration was not considered a cost-effective solution. The development of the accumulation-mode MOS varactor [40], the electromechanically tunable varactor [41,42], good inductor modeling [43–45], and modern study of on-chip VCOs [46,47] made it possible to implement on-chip differential LC VCOs for several wireless standards. As to the GSM standard, a novel filtering technique as shown in Figure 1.13 [48] made a significant contribution to lower the noise factor of differential LC oscillators. Nowadays, LC VCOs are also employed in >6-Gb/s wireline systems to achieve sub-picosecond rms jitter performance.

1.2.9 Frequency divider

In the conventional dual-modulus divider, we employ a prescaler followed by cascaded toggle flip-flops (TFFs) and synchronous counters. Because of the required feedback time for dual-modulus control, the 4/5 prescaler instead of the 2/3 prescaler is mostly used. The use of the 4/5 prescaler requires three synchronous DFFs, which gives a large capacitive loading to the VCO buffer. Since the basic operation principle of the prescaler is based on a pulse swallowing method [49], a similar operation can be achieved when phase selection is performed with the phase resolution equivalent to the VCO period. Figure 1.14(a) shows the basic concept of a phase-selection prescaler [50]. As shown at the bottom diagram, a phase control circuit delays the phase by 90° whenever /5 is done. The 90° delay at the output of the /4 is equivalent to one period delay of the VCO clock. Hence, this operation is similar to the 4/5 prescaler whose operation is based on the pulse swallowing technique. The benefit of using this topology is that it reduces the capacitive loading to the VCO buffer since the VCO is followed by two cascaded TFFs.

The frequency divider based on the standard topology consisting of a prescaler, a main counter, and an auxiliary counter has a drawback of large minimum division ratio for seamless division values [49]. For fractional-*N* frequency synthesizers where a high reference frequency is used, lower frequency division values are needed. Figure 1.14(b) shows a ($2^k + M$) divider topology whose minimum division ratio is 2^k by having cascaded 2/3 prescalers [16]. However, having the cascaded 2/3 prescalers gives a harsh timing margin for modulus control, resulting in higher power consumption than that of the standard topology.

An injection-locked frequency divider (ILFD) was developed to overcome the speed limit of the broadband static frequency dividers. The ILFD can be classified as an analog frequency divider or a regenerative frequency divider since the input frequency is a multiple of the free-running frequency of the oscillator. Figure 1.14(c) shows an example of a differential ILFD performing divide-by-two operation [50]. Since the common-source connection of differential input pair oscillates at twice the output frequency, the common-source node is used as an injection node for a divide-by-two operation. The ILFD has become the natural choice for millimeter-wave frequency synthesizers these days. The main drawbacks of the ILFD are narrow locking range and high sensitivity to PVT variations.

1.3 Application aspects

In this section, we will give an overview of early-stage CMOS PLLs by considering four main applications; synchronization, clock generation, clock-and-data recovery, and frequency synthesis/modulation. Frequency demodulation is also an important role of the PLL as it was done in the early monolithic PLL [1], but the digital modulation based on the fractional-*N* PLL becomes a key feature of the modern PLL. The architectures and circuits discussed in the previous section will be reviewed based on applications aspects.

1.3.1 Synchronization

As clock frequency increases, clock skew between an internal clock and an external clock becomes critical in chip-to-chip communication, e.g. data communication

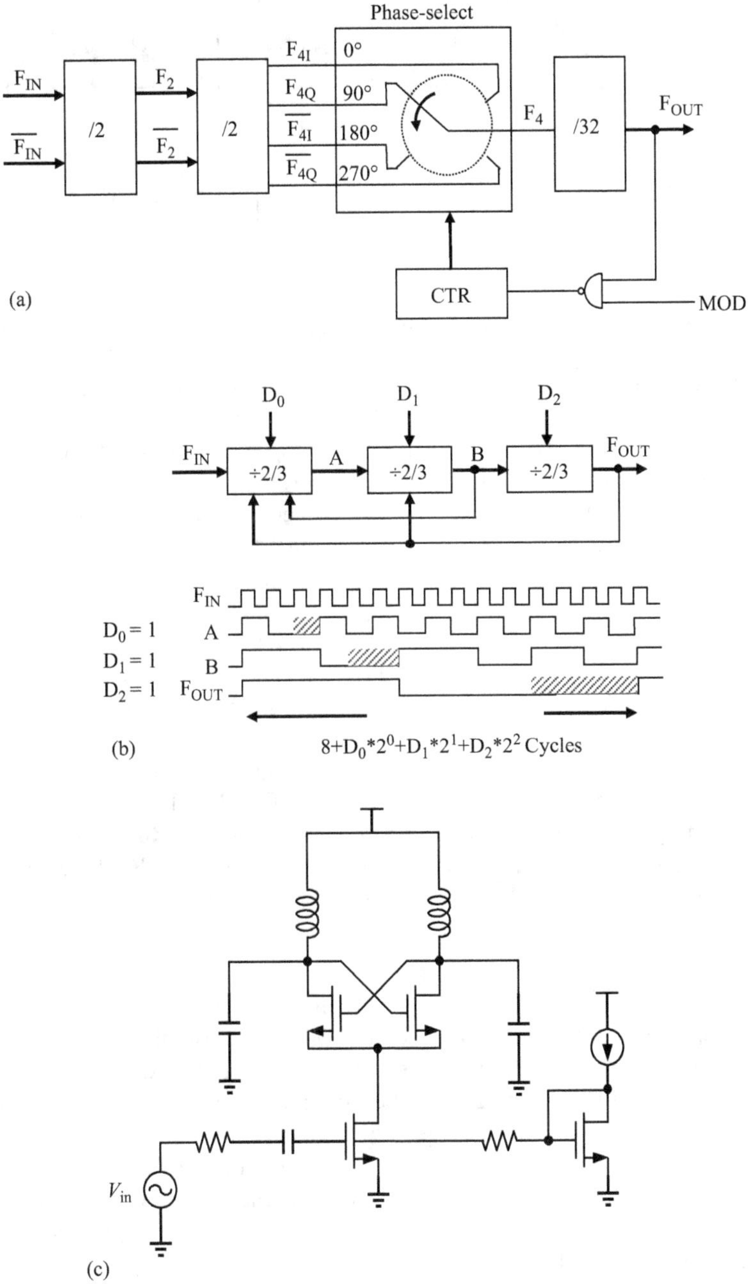

Figure 1.14 Customized frequency dividers: (a) phase-selection prescaler [50], (b) ($2^k + M$) divider [16], and (c) injection-locked divider [51]

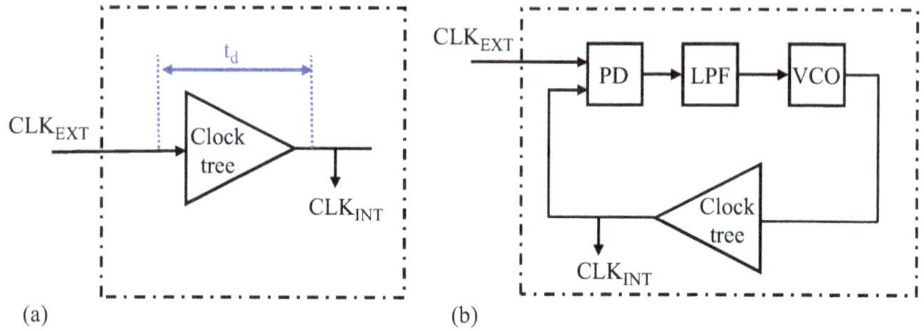

Figure 1.15 PLL for clock de-skewing: (a) open-loop based and (b) PLL based

between the microprocessor and memory chips. The clock skew is mainly caused by a large clock tree that buffers an external clock to drive a large number of internal nodes. As depicted in Figure 1.15, the propagation delay of the clock tree varies greatly over PVT variations and is difficult to be matched with simulation. Hence, the clock skew could substantially degrade the data throughput for high-speed communication. A CMOS PLL with a ring VCO was proposed to mitigate the clock skew problem [3]. By including the clock tree in a feedback path of the PLL and taking an output of a single gate in the clock tree as an input of the phase detector, the output of the clock tree could be synchronized to the external clock. In this way, all internal clocks in different chips could be synchronized to a single external clock, resulting in an improved phase margin in data slicing.

One major problem of using the PLL in digital systems was that the VCO was highly sensitive to supply noise. The DLL was developed to improve noise immunity by employing the VCDL instead of the VCO [10]. The DLL also enabled clock stretching as the phase error was not propagated. A 500-Mb/s memory system was demonstrated by employing the DLL for the I/O interface [26]. Since then, the use of the PLL or the DLL became essential for the high data rate I/O interface, building a basis for high-speed serial I/O links.

1.3.2 Clock generation

The development of the ring VCO made the CMOS PLL useful for high-frequency clock generation, while the PFD and the charge pump paved a way for the modern PLL design. The use of the ring VCO became more realistic after a fully differential ring VCO was developed [12]. A PLL clock generator with a wide tuning range of 5-to-110 MHz for microprocessors was demonstrated [21]. As clock speed rapidly increases, the clock jitter becomes the bottleneck to determine the overall performance. The importance of the clock jitter in high-speed digital systems is analogous to the ground bouncing issue in high-frequency analog/RF circuits as illustrated in Figure 1.16.

Besides speed and noise performance, robust clock generation is another important factor. Since most PLL clock generators are implemented with standard CMOS technology, loop parameter variations over process and temperate are substantial. In the early days, only the *N*-well resistor was available in standard

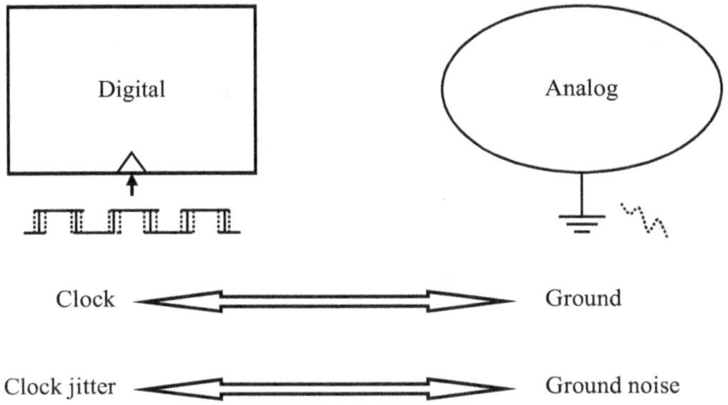

Figure 1.16 Jitter in the digital system in comparison with ground noise

digital CMOS. A resistor-less PLL clock generator with an auxiliary charge pump was proposed [23]. Moreover, the PLL clock generator that used the same kind of bias currents for both the charge pump and the CCO was proposed to implement a self-biased architecture [24].

Since one of the biggest concerns was the jitter propagation of the ring VCO due to supply voltage noise, DLL-based architectures were proposed for digital clock generation. A hot-standby PLL with a reconfigurable delay line (RDL) was designed to achieve fast settling as well as good immunity [52]. During an unlocked state, the RDL is configured as a ring VCO for frequency multiplication. When the PLL enters a lock-in range, the RDL is reconfigured to a VCDL, so that the PLL becomes a first-order DLL for fast settling and low jitter. The MDLL was also developed to enable frequency multiplication [28]. Compared to the hot-standby PLL that uses the RDL, the MDLL employs a gated ring VCO whose phase is periodically locked by a clean reference clock, so that the accumulated jitter of the ring VCO is flushed away at each reference clock period.

As CMOS technology advances, the gate leakage current of the MOS capacitor becomes a serious problem as the MOS capacitor is used as a large integration capacitor in the loop filter (LPF). In standard CMOS, the metal-to-metal capacitor is not available. Even if it is available, the small density of the metal-to-metal capacitor significantly increases the LPF area. The DPLL has received high attention since it overcomes the technology issues with digital LPF and reconfigurable features. Designing a high-performance TDC is critical for low-noise DPLLs.

1.3.3 Clock-and-data recovery

The PLL is a unique device that can perform as an auto-tracking high-Q band-pass filter with wide tunability. Such a property has been employed for the CDR system where a clean clock is extracted from noisy data and used for data retiming as depicted in Figure 1.17. Several high-speed CMOS PLLs for the CDR system appeared nearly at the same time as other clock generation PLLs [4–6]. In the CDR circuit, designing a high-speed phase detector (PD) that directly takes an NRZ

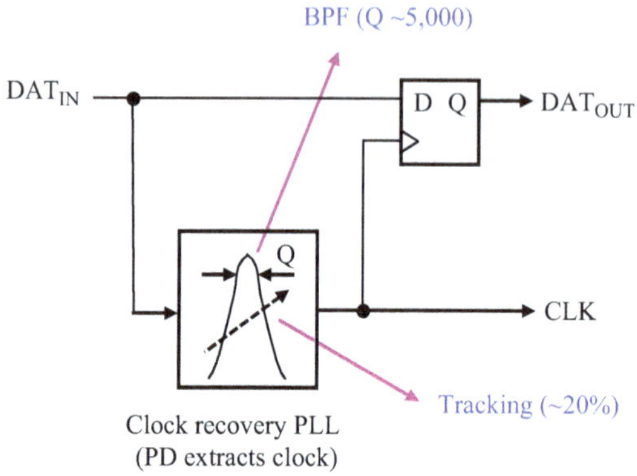

Figure 1.17 PLL-based CDR

signal is challenging. The Hogge PD was proposed to detect an edge of the NRZ data and generate an error voltage linearly proportional to the phase difference between the data edge and the VCO phase [53]. To overcome the speed limit of the linear PD, nonlinear PDs such as an Alexander PD [31], a BBPD [11,54], or a double-edge triggered BBPD [32] were proposed.

In addition to circuit innovations, several important CDR architectures have been developed. A quasi-PLL CDR architecture that utilized oversampled data to perform phase tracking and data decision with a digital logic block instead of the conventional PD was developed [12]. In this architecture, a 32-phase differential ring VCO was designed to realize a multi-phase sampler as shown in Figure 1.18(a). An effective sample rate of about 1 Gb/s was achieved with a 30-MHz ring VCO.

The CDR system based on the type-2 PLL exhibits gain peaking near the bandwidth in the jitter transfer function, resulting in a jitter accumulation problem for CDR repeaters. This kind of jitter peaking should be tightly controlled for the synchronous optical network (SONET) where a single clock handles data retiming across the entire network with repeaters. For that reason, a heavily overdamped PLL had to be designed to have the jitter peaking of less than 0.1 dB. This serious jitter-peaking issue was nicely solved by combining the loop dynamics of both the type-2 PLL and the DLL [55]. A delay- and phase-locked loop (D/PLL) architecture was proposed to realize a jitter-peaking-free CDR by phase shifting the input data by the DLL. A system zero is formed in the feedback system by overlapping the loop dynamics of both the type-2 PLL and the DLL as illustrated in Figure 1.18(b).

The PLL-based CDR inherently offers frequency tacking and can further extend the capture range with the help of a frequency-acquisition aid circuit that extracts frequency information from non-periodic NRZ data [2]. However, the drawback of the PLL-based CDR system is the jitter generation and accumulation problem mainly caused by a VCO especially when the CDR circuit is embedded in the I/O link of a large digital system. Even though the DLL shows high immunity against the supply noise, it suffers from the out-of-lock problem when there is a

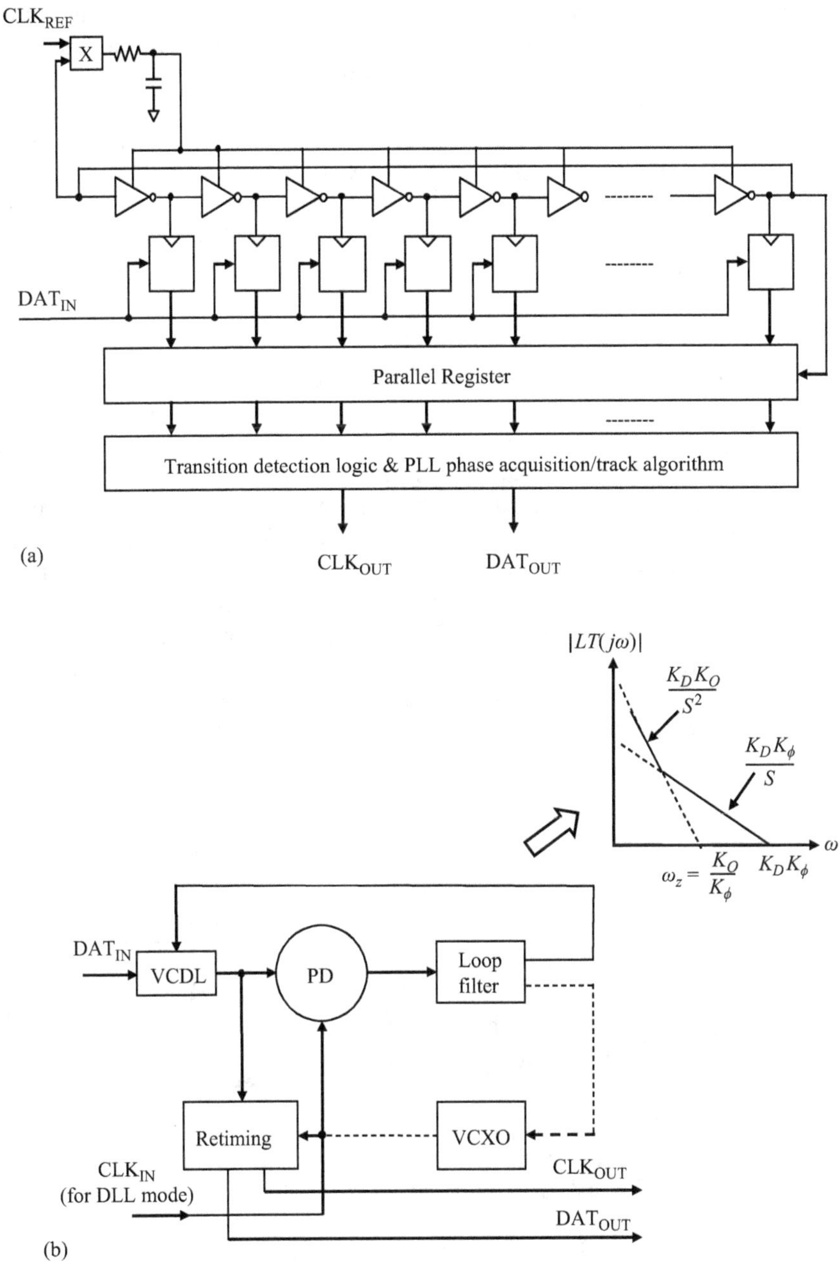

Figure 1.18 (a) Hybrid CDR [9] and (b) D/PLL CDR [55]

frequency offset between a transmitter and a receiver in wireline communication systems. A semidigital DLL architecture was developed to achieve frequency tracking with discrete phases [13]. By rotating the discrete phases generated by a main DLL, infinite phase accumulation was achieved. In [25], a cascaded PLL-DLL architecture was proposed earlier. The second-stage DLL performed a similar function, but the frequency tracking was mainly achieved by the first-stage PLL. After the development of the semidigital dual DLL, low-cost yet robust CDR systems could be built. Specifically in multiple I/O links, only a single PLL is used for clock multiplication, while the CDR is performed by multiple DLLs.

1.3.4 Frequency synthesis and modulation

Unlike the digital clock generator whose ultimate goal is to minimize peak-to-peak jitter in the time domain regardless of random jitter or deterministic jitter, the frequency synthesizer aims for high spectral purity. Accordingly, phase noise and spur performances in frequency domain are important in both in-band and out-of-band regions. For instance, high out-of-band phase noise can cause a reciprocal mixing problem, while poor in-band phase noise can increase integrated phase error and degrade the bit-error-rate performance as illustrated in Figure 1.19. As a feedback system, the PLL-based frequency synthesizer has fundamental design tradeoffs such as phase noise, spur, and settling time. These tradeoffs become more critical when the frequency synthesizer uses a high division ratio, that is, the ratio of the output frequency to the reference frequency. The matching and noise requirements of the charge pump becomes quite challenging with high division ratios [56].

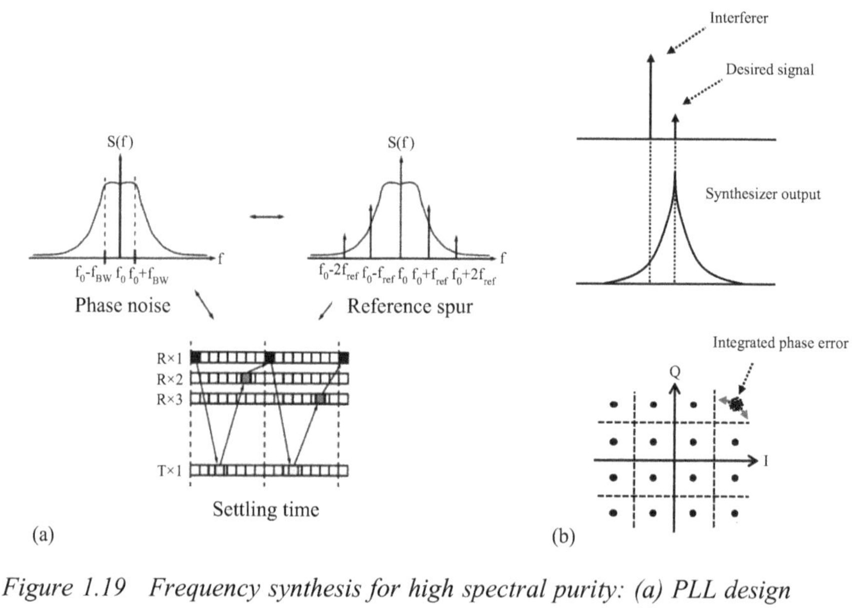

Figure 1.19 Frequency synthesis for high spectral purity: (a) PLL design tradeoffs and (b) effects on transceiver system performance

Offering fine frequency resolution with high phase detection frequency, the fractional-N frequency synthesizer significantly improves the performance with a much lower division ratio. However, the use of the fractional-N frequency synthesizer originally based on the *Digiphase* technique [34,35] had been limited due to the fractional spur problem until the $\Delta\Sigma$ fractional-N frequency synthesizer was developed [14,15]. The operation principle of the $\Delta\Sigma$ modulated frequency divider is fully digital based since arbitrary frequency resolution can be obtained by simply increasing the number of bits of the digital modulator.

The $\Delta\Sigma$ fractional-N frequency synthesizer generates output frequency with fine resolution by giving a (static) frequency control word to the $\Delta\Sigma$ modulator. If the frequency control word of the $\Delta\Sigma$ modulator is modulated, the output frequency of the $\Delta\Sigma$ fractional-N frequency synthesizer will be modulated. Hence, direct-digital frequency/phase modulation can be achieved by the $\Delta\Sigma$ fractional-N frequency synthesizer [16,17]. However, the modulation rate is limited by a PLL bandwidth. To overcome the bandwidth problem, a pre-emphasis method in the digital domain before the $\Delta\Sigma$ modulation was proposed [16]. Modulation quality depends on the transfer function matching between the digital filter and the analog PLL. The two-point modulation method that had been proposed to realize an all-pass transfer function by injecting modulation information to both the input and the output of the VCO [57] became an attractive solution to build a robust transmitter system with the $\Delta\Sigma$ fractional-N frequency synthesizer [58,59]. The performance of the two-point modulation, however, is extremely sensitive to the nonlinearity of the VCO. After the DPLL having the DCO was developed, the two-point modulation method became a viable solution for frequency modulation in modern transmitters. Some examples can be found in later chapters of this book.

1.4 Summary

In this chapter, a historical view of the monolithic PLL development from the late 1980s to the early 2000s is presented. Fundamental innovations that gave a significant impact on the product development of CMOS PLL ICs during the early stages are reviewed. Also, an overview of PLL development by considering system and application aspects is given. As the PLL architectures have been diversified and fragmented since the 2000s, addressing fundamental innovations in recent PLL architectures would be challenging without considering the product-level impact. Numerous state-of-the-art architectures from recent years are introduced in the following chapters. The purpose of this chapter is to help readers gain an insight into the trends of the PLL IC design in the early stage, while emphasizing that the fundamental concept and principle of the PLL would not change much regardless of the architecture evolution.

Acknowledgments

The author would like to thank Bowen Wang and Jiahao Zhao in the Institute of Microelectronics at Tsinghua University for a lot of help drawing the figures.

References

[1] E. N. Multhi, "A monolithic phase-locked loop with post detection processor," *IEEE Journal of Solid-State Circuits*, vol. SC-14, pp. 155–161, Feb. 1979.

[2] R. R. Cordell, J. B. Forney, C. N. Dunn, and W. G. Garrett, "A 50 MHz phase- and frequency-locked loop," *IEEE Journal of Solid-State Circuits*, vol. SC-14, pp. 1003–1010, Dec. 1979.

[3] D.-K. Jeong, G. Borriello, D. A. Hodges, and R. H. Katz, "Design of PLL-based clock generation circuits," *IEEE Journal of Solid-State Circuits*, vol. SC-22, pp. 255–261, Apr. 1987.

[4] R. H. Leonowich and J. M. Steininger, "A 45-MHz CMOS phase/frequency-locked loop timing recovery circuit," in *Dig. Tech. Papers Int. Solid State Circuits Conf. (ISSCC)*, Feb. 1988, pp. 14–15.

[5] S. Y. Sun, "An analog PLL-based clock and data recovery circuit with high input jitter tolerance," *IEEE Journal of Solid-State Circuits*, vol. 24, pp. 325–330, Apr. 1989.

[6] R. Baumert, P. Metz, M. Pedersen, R. Pritchett, and J. Young, "A monolithic 50–200 MHz CMOS clock recovery and retiming circuit," in *Proc. IEEE Custom Integrated Circuits Conf. (CICC)*, May 1989, pp. 14.5.1–14.5.4.

[7] K. M. Ware, H. S. Lee, and C. G. Sodini, "A 200-MHz CMOS PLL with dual phase detectors," *IEEE Journal of Solid-State Circuits*, vol. 24, pp. 1560–1568, Dec. 1989.

[8] J. I. Brown, "A digital phase and frequency sensitive detector," *IEEE Proceeding Letters*, pp. 717–178, Apr. 1971.

[9] F. M. Gardner, "Charge-pump phase-lock loops," *IEEE Transactions on Communications,* vol. COM-28, pp. 1849–1858, Nov. 1980.

[10] M. Johnson and E. Hudson, "A variable delay line PLL for CPU-coprocessor synchronization," *IEEE Journal of Solid-State Circuits*, vol. 23, pp. 1218–1223, Oct. 1988.

[11] R. C. Walker, T. Hornak, C. Yen, and K. H. Springer, "A chipset for gigabit rate data communication," in *Proc. Bipolar Circuits Technology Meeting*, pp. 288–290, Sept. 1989.

[12] B. Kim, D. N. Helman, and P. R. Gray, "A 30-MHz hybrid analogdigital clock recovery circuit in 2-pm CMOS," *IEEE Journal of Solid-State Circuits*, vol. 25, pp. 1385–1394, Dec. 1990.

[13] S. Sidiropoulos and M. Horowitz, "A semidigital dual delay-locked loop", *IEEE Journal of Solid-State Circuits*, vol. 32, pp. 1683–1692, Nov. 1997.

[14] B. Miller and R. J. Conley, "A multiple modulator fractional divider," in *Proc. IEEE Frequency Control Symp.*, Mar. 1990, pp. 559–568.

[15] T. A. Riley, M. A. Copeland, and T. A. Kwasniewski, "Delta-sigma modulation in fractional-N frequency synthesis," *IEEE J. Solid-State Circuits*, vol. 28, pp. 553–559, May 1993.

[16] M. H. Perrott, T. L. Tewksbury, III, and C. G. Sodini, "A 27-mW CMOS fractional-N synthesizer using digital compensation for 2.5-Mb/s GFSK modulation," *IEEE J. Solid-State Circuits*, vol. 32, pp. 2048–2060, Dec. 1997.

[17] N. Filiol, T. Riley, C. Plett, and M. Copeland, "An agile ISM band frequency synthesizer with built-in GMSK data modulation," *IEEE Journal of Solid-State Circuits*, vol. 33, pp. 998–1008, July 1998.

[18] R. Staszewski, J. L. Wallberg, S. Rezeq *et al.*, "All-digital PLL and transmitter for mobile phones," *IEEE Journal of Solid-State Circuits*, vol. 40, pp. 2469–2482, Dec. 2005.

[19] R. F. Bitting and W. P. Repasky, "A 30-128MHz frequency synthesizer standard cell," in *Proc. IEEE Custom Integrated Circuits Conf. (CICC)*, May 1992, pp. 24.1.1–24.1.6.

[20] D. Mijuskovic, M. Bayer, T. Chomicz *et al.*, "Cell-based fully integrated CMOS frequency synthesizers," *IEEE Journal of Solid-State Circuits*, vol. 29, pp. 271–279, Mar. 1994.

[21] I. A. Young, J. K. Greason, and K. L. Wong, "A PLL clock generator with 5 to 110 MHz of lock range for microprocessors," *IEEE J. Solid-State Circuits*, vol. 27, pp. 1599–1607, Nov. 1992.

[22] B. Lai and R. C. Walker, "A monolithic 622 Mb/s clock extraction data retiming circuit," in *Dig. Tech. Papers Int. Solid State Circuits Conf. (ISSCC)*, Feb. 1991, pp. 144–145.

[23] I. Novof, J. Austin, R. Kelkar, D. Strayer, and S. Wyatt, "Fully-integrated CMOS phase-locked loop with 15 to 240 MHz locking range and ±50 ps jitter," *IEEE J. Solid-State Circuits*, vol. 30, pp. 1259–1266, Nov. 1995.

[24] J. Maneatis, "Low-jitter process-independent DLL and PLL based on self-biased techniques," *IEEE Journal of Solid-State Circuits*, vol. 31, pp. 1723–1732, Nov. 1996.

[25] M. Horowitz, A. Chan, J. Cobrunson *et al.*, "PLL design for a 500 MB/s interface," in *Dig. Tech. Papers Int. Solid State Circuits Conf. (ISSCC)*, Feb. 1993, pp. 160–161.

[26] T. Lee, K. Donnelly, J. Ho *et al.*, "A 2.5 V CMOS delay-locked loop for an 18 Mbit, 500 MB/s DRAM," *IEEE Journal Solid-State Circuits*, vol. 29, pp. 1491–1496, Dec. 1994.

[27] G. Chien and P. R. Gray, "A 900-MHz local oscillator using a DLL-based frequency multiplier technique for PCS applications," *IEEE Journal Solid-State Circuits*, vol. 35, pp. 1996–1999, Dec. 2000.

[28] R. Farjad-Rad, W. Dally, H.-T. Ng *et al.*, "A low-power multiplying DLL for low-jitter multigigahertz clock generation in highly integrated digital chips," *IEEE Journal Solid-State Circuits*, vol. 37, pp. 1804–1812, Dec. 2002.

[29] T. Weigandt and S. Mehta, "A phase interpolation technique for fractional-N frequency synthesis," in *1997 Website of RF Integrated Circuit Design Group* (http://kabuki.eecs.berkeley.edu/rf/), University of California, Berkeley, 1997.

[30] W. Rhee, "Design of low-jitter 1-GHz phase-locked loops for digital clock generation," in *Proc. IEEE Int. Symp. on Circuit and Systems (ISCAS)*, May 1999, pp. 520–523.

[31] J. D. H. Alexander, "Clock recovery from random binary signals," *Electronics Letters*, vol. 11, no. 22, pp. 541–542, Oct. 1975.

[32] M. Soyer, “A monolithic 2.3-Gb/s 100-mW clock and data recovery circuit in silicon bipolar technology,” *IEEE Journal of Solid-State Circuits*, vol. 28, pp. 1310–1313, Dec. 1993.

[33] N. Ishihara and Y. Akazawa, “A monolithic 156 Mb/s clock and data recovery PLL circuit using the sample-and-hold technique,” *IEEE Journal of Solid-State Circuits*, vol. 28, pp. 1310–1313, Dec. 1993.

[34] G. C. Gillette, “Digiphase synthesizer,” in *Proc. IEEE Frequency Control Symp.*, 1969, pp. 201–210.

[35] J. Gibbs and R. Temple, “Frequency domain yields its data to phase-locked synthesizer,” *Electronics*, pp. 107–113, Apr. 1978.

[36] W. Rhee, *Multi-Bit Modulation Technique for Fractional-N Frequency Synthesizers*, Ph.D. Thesis, University of Illinois, Urbana-Champaign, Aug. 2000.

[37] N. M. Nguyen and R. G. Meyer, “A 1.8-GHz monolithic LC voltage controlled oscillator,” *IEEE Journal of Solid-State Circuits*, vol. 27, pp. 444–450, Mar. 1992.

[38] P. Basedau and Q. Huang, “A 1GHz, 1.5V monolighic LC oscillator in 1-μm CMOS,” in *Proc. IEEE Eur. Solid-State Circuits Conf. (ESSCIRC)*, Sept. 1994, pp. 172–175.

[39] J. Craninckx and M. Steyeart, “A 1.8 GHz CMOS low-phase noise voltage-controlled oscillator with prescaler,” *IEEE Journal of Solid-State Circuits*, vol. 30, pp. 1474–1482, Dec. 1995.

[40] F. Svelto, S. Deantoni, and R. Castello, “A 1.3 GHz low-phase noise fully tunable CMOS LC VCO,” *IEEE Journal of Solid-State Circuits*, vol. 35, pp. 356–361, Mar. 2000.

[41] D. Young and B. Boser, “A micromachined based RF low noise voltage controlled oscillator,” in *Proc. IEEE Custom Integrated Circuits Conference (CICC)*, May 1997, pp. 431–434.

[42] A. Dec and K. Suyama, “A 1.9-GHz CMOS VCO with micromachined electromechanically tunable capacitors,” *IEEE Journal Solid-State Circuits*, vol. 35, pp. 1231–1237, Aug. 2000.

[43] J. R. Long and M. A. Copeland, “Modeling of monolithic inductors and transformers for silicon RFIC design,” in *Proc. IEEE MTT-S Int. Symp. Tech. Wireless Appl.*, Feb. 1995, pp. 129–134.

[44] M. Soyuer, J. N. Burghartz, K. A. Jenkins, S. Ponnapalli, J. F. Ewen, and W. E. Pence, “Multi-level monolithic inductors in silicon technology,” *Electron. Lett.*, vol. 31, no. 5, pp. 359–360, Mar. 1995.

[45] C. P. Yue and S. S. Wong, “On-chip spiral inductors with patterned ground shields for Si-based RF ICs,” *IEEE Journal of Solid-State Circuits*, vol. 33, pp. 743–752, May 1998.

[46] C. Samori, A. L. Lacaita, F. Villa, and F. Zappa, “Spectrum folding and phase noise in LC tuned oscillators,” *IEEE Trans. Circuits Syst. II*, vol. 45, pp. 781–790, July 1998.

[47] A. Hajimiri and T. H. Lee, “A general theory of phase noise in electrical oscillators,” *IEEE Journal of Solid-State Circuits*, vol. 33, pp. 179–194, Feb. 1998.

[48] E. Hegazi, H. Sjöland, and A. A. Abidi, "A filtering technique to lower *LC* oscillator phase noise" *IEEE Journal of Solid-State Circuits*, vol. 12, pp. 1921–1930, Dec. 2001.
[49] W. F. Egan, *Frequency Synthesis by Phase Lock*, 2nd ed. New York: Wiley Interscience, 2000.
[50] J. Craninckx and M. Steyaert, "A 1.75-GHz/3-V dual modulus divide-by-128/129 prescaler in 0.7 μm CMOS," *IEEE Journal of Solid-State Circuits*, vol. 31, pp. 890–897, July 1996.
[51] R. Rategh and T. H. Lee, "Superharmonic injection-locked frequency dividers," *IEEE Journal of Solid-State Circuits*, vol. 34, pp. 813–821, June 1999.
[52] M. Mizuno, K. Furuta, T. Andoh *et al.*, "A 0.18 μm CMOS hot-standby phase-locked loop using a noise-immune adaptive-gain voltage-controlled oscillator," in *Dig. Tech. Papers Int. Solid State Circuits Conf. (ISSCC)*, Feb. 1995, pp. 268–269.
[53] C. R. Hogge, Jr., "A self correcting clock recovery circuit," *IEEE Transactions on Electron Devices*, vol. ED-32, no. 12, pp. 2704–2706, Dec. 1985.
[54] R. C. Walker, J. Wu, C. Stout *et al.*, "A 2-chip 1.5Gb/s bus-oriented serial link interface," in *Dig. Tech. Papers Int. Solid State Circuits Conf. (ISSCC)*, Feb. 1992, pp. 226–227.
[55] T. H. Lee and J. F. Bulzacchelli, "A 155 MHz clock recovery delay- and phase-locked loop," *IEEE Journal of Solid State Circuits*, vol. 27, pp. 1736–1746, Dec. 1992.
[56] W. Rhee, "Design of high-performance CMOS charge pumps in phase-locked loops," in *Proc. IEEE Int. Symp. on Circuit and Systems (ISCAS)*, May 1999, pp. 545–548.
[57] M. J. Underhill and R. I. H. Scott, "Wideband frequency modulation of frequency synthesisers," *Electronics Letters*, vol. 15, no. 13, pp. 541–542, June 1979.
[58] C. Diirdodt, M. Friedrich, C. Grewing *et al.*, "The first very low-IF RX, 2-point modulation TX CMOS system on chip Bluetooth solution," in *IEEE RFIC Symp. Dig.*, May 2001, pp. 99–102.
[59] D. Theil, C. Diirdodt, A. Hanke *et al.*, "A fully integrated CMOS frequency synthesizer for Bluetooth," in *IEEE RFIC Symp. Dig.*, May 2001, pp. 103–106.

Chapter 2
Fractional-N frequency synthesis

Sudhakar Pamarti[1]

Fractional-N phase-locked loops (PLLs), particularly delta-sigma fractional-N PLLs, are some of the most popular integrated frequency synthesizers. The fractional-N PLL is essentially a variant of the integer-N PLL. Just like the integer-N PLL, it employs negative feedback to synchronize a local oscillator to a periodic reference signal. The synchronization ensures that the controlled oscillator's frequency is a multiple of the reference frequency, f_{REF}. However, unlike the integer-N PLL which ensures that the local oscillator frequency is exactly N times the reference frequency, f_{REF}, the fractional-N PLL ensures that it is $(N + \alpha)$ times the reference frequency, where α is a fractional value. In effect, the fractional-N PLL offers phase synchronization and fine grain frequency synthesis [1,2].

The fine grain frequency synthesis is very useful in generating local oscillator (LO) signals in wireless communication transceivers where finely spaced channels are a norm. For example, a state-of-the-art WiMAX frequency synthesizer has 36-Hz frequency resolution, −130-dBc/Hz phase noise at the 3-MHz offset, lower than −75-dBc reference spur, and less than 11-μs settling time [3]. In addition, in some cases, by changing α in time, the fractional-N PLL is used to realize phase/frequency modulation [1,4–6], spread-spectrum clocking [7], and frequency hopping [8].

2.1 Basics

2.1.1 Fractional-N operation

The block diagram of a typical fractional-N PLL is shown in Figure 2.1. Just like in an integer-N PLL, negative feedback is used to adjust the control voltage, $v_{CTRL}(t)$, and hence the oscillation frequency, f_{PLL}, of a voltage-controlled oscillator (VCO). A phase-frequency detector (PFD) compares arrival times of rising edges of the frequency divider output, $v_{DIV}(t)$, to those of a low-noise periodic reference signal, $v_{REF}(t)$, of frequency f_{REF}; a charge pump (CP) dumps/draws proportionate charge into/from a low-pass loop filter to adjust $v_{CTRL}(t)$.

In an integer-N PLL, in steady state, the frequency divider generates one rising edge for every N rising edges of $v_{PLL}(t)$. In steady state, the rising edges of $v_{REF}(t)$

[1]Electrical and Computer Engineering Department, University of California, Los Angeles, CA, USA

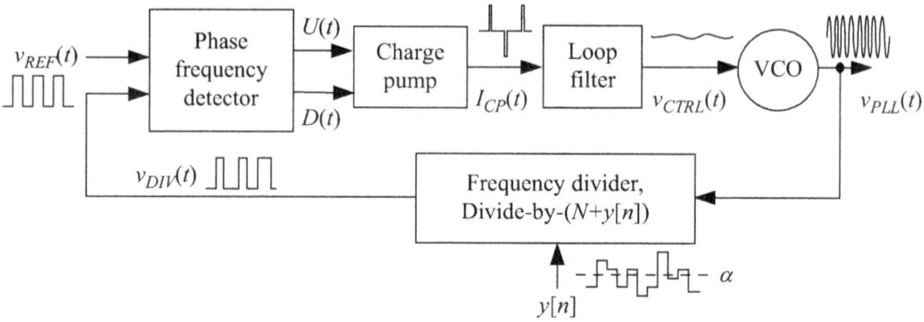

Figure 2.1 Simplified block diagram of an example fractional-N PLL. The main difference with a conventional integer-N PLL is that the frequency division ratio is potentially changed every cycle, according to a sequence of small integers, y[n], that average to a fraction, α. As a result, the PLL's output frequency tunes to $(N + \alpha)f_{REF}$

and $v_{DIV}(t)$ are aligned except for possibly a constant offset; consequently, $f_{PLL} = N \times f_{REF}$. The desired channel is synthesized by the choice of N but with a frequency resolution limited to f_{REF}.

In contrast, in a fractional-N PLL, the frequency division modulus is changed once every reference period according to a sequence of small integers, $y[n]$. Consequently, $v_{REF}(t)$ and $v_{DIV}(t)$ are in phase synchronization only in an average sense. The PLL still locks, but now $f_{PLL} = (N + avg(y[n])) \times f_{REF} = (N + \alpha) \times f_{REF}$, i.e., f_{PLL} is a fractional multiple of f_{REF}; hence, the name fractional-N PLL. More importantly, the desired channel is synthesized by the choice of both N and α and the frequency resolution is no longer limited to f_{REF}. In fact, arbitrarily fine resolution frequency resolution is achievable.

2.1.2 Excess phase noise

Note that the frequency divider never divides exactly by $(N + \alpha)$ but divides by $(N + y[n])$. As a result, even in steady state, $v_{DIV}(t)$ has the same frequency as $v_{REF}(t)$ only in an average sense. As a result, in each period of the reference, $v_{DIV}(t)$ and $v_{REF}(t)$ invariably exhibit instantaneous phase errors. The PLL suppresses these phase errors outside of its bandwidth but residual errors result in excess phase noise at the PLL's output, compared to integer-N operation.

This behavior is most easily quantified by breaking the negative feedback loop in steady-state operation and observing the evolution of phase errors between $v_{REF}(t)$ and $v_{DIV}(t)$ in time. As illustrated in Figure 2.2, it can be readily shown that the phase detector effectively sees instantaneous timing errors of:

$$\tau[n] = T_{PLL} \sum_{k=0}^{n-1} (y[k] - \alpha) \tag{2.1}$$

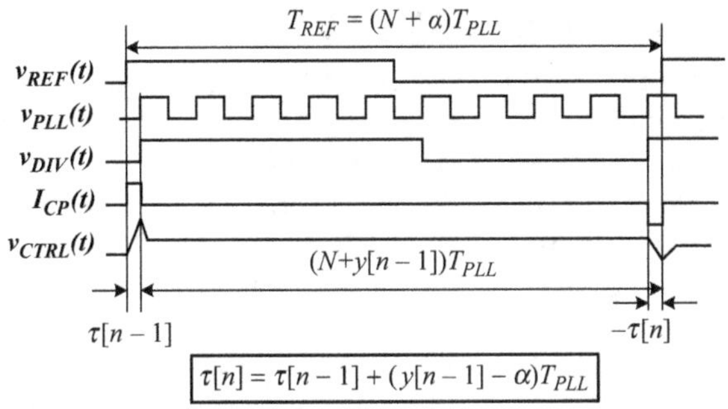

Figure 2.2 Timing diagram showing fractional-N operation in steady state

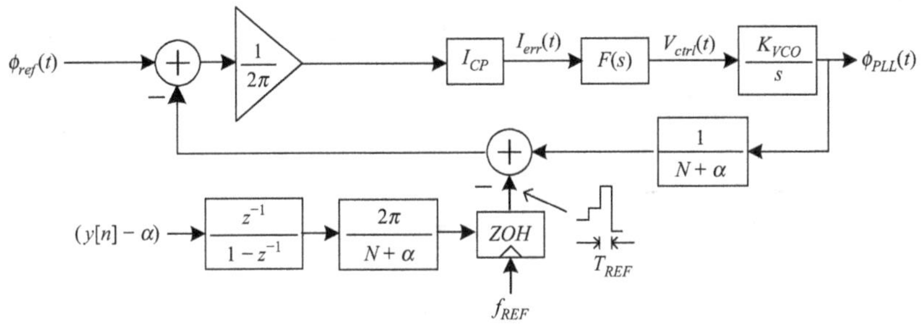

Figure 2.3 A simplified model for excess phase noise caused by the fractional-N operation

where T_{PLL} is one PLL period and instantaneous phase errors of

$$\phi[n] = 2\pi \frac{\tau[n]}{T_{REF}} = \frac{2\pi}{N+\alpha} \sum_{k=0}^{n-1} (y[k] - \alpha) \tag{2.2}$$

where T_{REF} is one reference period and it is assumed that the fractional operation starts at $n = 0$. Accordingly, the excess phase noise contributed by these instantaneous phase errors can be modeled as shown in Figure 2.3, where "ZOH" is a "zeroth-order hold" block [9].

Intuitively, dividing the PLL output by $(N + y[n])$ instead of $(N + \alpha)$ results in instantaneous frequency errors between $v_{REF}(t)$ and $v_{DIV}(t)$, proportional to $(y[n] - \alpha)$; these are converted (accumulated) into phase errors that are subsequently acted upon by the PLL. Readers will recognize in Figure 2.3 that, effectively, the zeroth-order-held phase errors are indeed processed by the continuous-time model of the PLL.

Much of the work in fractional-N PLLs has focused on dealing with the excess phase noise resulting from fractional-N operation: how to choose $y[n]$ effectively to reduce it, how to design the PLL to minimize it in relation to other PLL noise sources, etc. These questions are addressed in more detail in the rest of this chapter.

2.1.3 Delta-sigma fractional-N PLLs

In the original fractional-*N* PLL, the desired fractional value, α, is digitally accumulated and the 1-bit carry out is used as $y[n]$ [5,10], as shown in Figure 2.4. It can be readily seen that $y[n]$ averages to α. However, the accumulator-based choice presents a big problem: $y[n]$ is periodic with a period that depends on α, as illustrated in the figure, for the case of alpha $= 0.75$. As a result, the phase errors are also strongly periodic and manifest as strong undesirable tones called *fractional spurs* in the power spectral density (PSD) of $v_{PLL}(t)$. The fractional spurs occur at frequency offsets that are integer multiples of αf_{REF} away from the center frequency.

Modern fractional-*N* PLLs employ a digital delta-sigma modulator (DDSM) to generate $y[n]$ and are referred to as *delta-sigma fractional-N PLLs*. Figure 2.5 shows an example second-order DDSM [11]. The DDSM generates $y[n]$ using a digital quantizer ensconced within a negative feedback loop set by the digital filters, $F(z)$ and $G(z)$. The DDSM input, typically represented as 2's complement

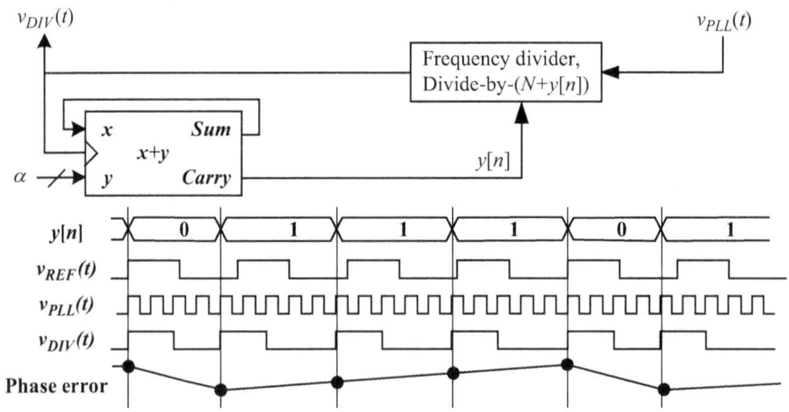

Figure 2.4 *Original fractional-N PLLs used a digital accumulator to generate the varying frequency modulus, N + y[n]. The resulting phase error is periodic causing strong spurious tones in the phase noise PSD*

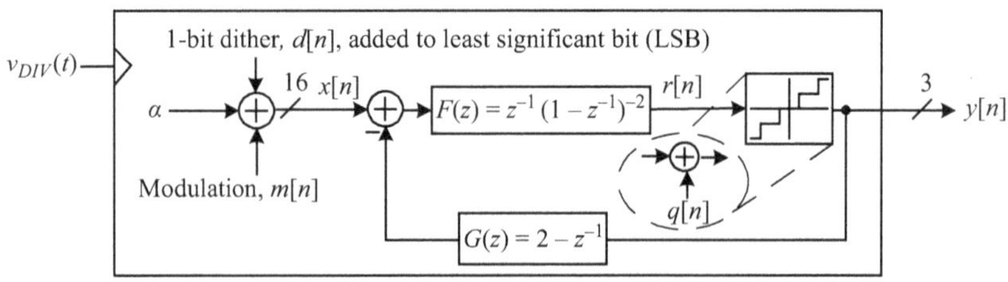

Figure 2.5 *Block diagram of an example second-order digital delta-sigma modulator: it employs a digital quantizer within a negative feedback loop*

numbers with m-bit fractional values, includes the desired fraction, α, and an optional, slowly varying sequence of small fractions, $m[n]$, called the "modulation sequence." The optional 1-bit random "dither sequence," $d[n]$, is discussed later. Note that, as in the original fractional-N PLL, the DDSM is clocked by the frequency divider output, $v_{DIV}(t)$.

The digital quantizer itself quantizes numbers with m fractional bits to nearby integers. While it can be realized in multiple ways, e.g., rounding, truncation, etc., the popular "mid-tread" form is illustrated in Figure 2.6 for the special case of quantizing 3-bit fractional values. Note that, in some cases, the quantizer output, $y[n]$, may saturate, as illustrated.

Momentarily ignoring the modulation sequence, $m[n]$, the central idea is that $y[n]$ still averages to α, but the DDSM ensures that $(y[n] - \alpha)$ has a high-pass shaped PSD, as described below. In other words, the instantaneous frequency (and phase) errors seen by the PLL in response to the fractional-N operation have mostly high-frequency content. Since the PLL attenuates such high-frequency frequency/phase errors, excess phase noise resulting from a delta-sigma fractional-N operation can be very low.

2.1.3.1 Model of excess phase in a delta-sigma fractional-*N* PLL

The digital quantizer is typically, and correctly, modeled as an additive "quantizer error" source, $q[n]$. While the additive model is correct, the properties of the quantizer error depend on a lot of factors discussed in detail later. For now, note that, while it is not true in general, $q[n]$ is assumed to be an *i.i.d.* (independent, identically distributed) sequence whose samples are uniformly distributed over one quantization step, i.e., $|q[n]| \leq 0.5$. Accordingly, for a reasonably large number of fractional bits, m, $q[n]$ is assumed to be a white sequence with a variance of 1/12.

Modeling the quantizer as an additive source allows treating the DDSM as a simple two-input ($x[n]$ and $q[n]$) linear time-invariant system. Specifically, as

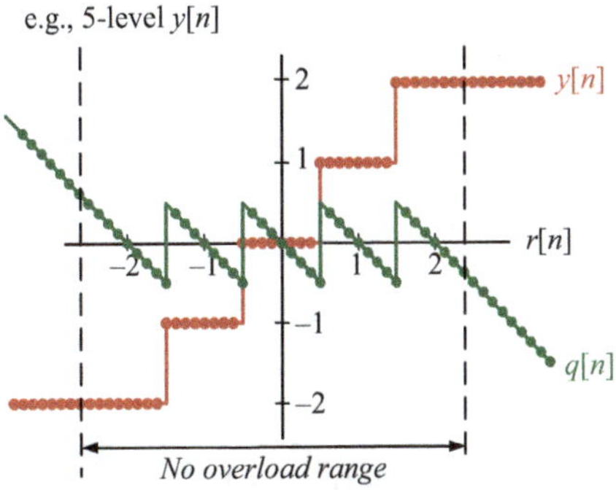

Figure 2.6 Example "mid-tread" digital quantizer that quantizes numbers with 3-bit fractional values to integers. In some cases, the output of the quantizer may saturate

shown in Figure 2.7, the DDSM can be described using so-called signal and noise transfer functions, $STF(z)$ and $NTF(z)$,

$$Y(z) = STF(z)X(z) + NTF(z)Q(z) \tag{2.3}$$

where Z-transforms of $x[n]$, $q[n]$, and $y[n]$ are assumed to exist and are represented by $X(z)$, $Q(z)$, and $Y(z)$, respectively, and

$$STF(z) = \frac{F(z)}{1 + F(z)G(z)} \text{ and } NTF(z) = \frac{1}{1 + F(z)G(z)}. \tag{2.4}$$

Note that the filtered quantizer error, represented by the $NTF(z)Q(z)$ term, is often referred to as "quantization noise" or "shaped quantization noise" and will be henceforth denoted by $e[n]$.

Typically, $F(z)$ and $G(z)$ are chosen such that noise transfer function, $NTF(z)$, is a high-pass filter, whereas the signal transfer function, $STF(z)$, is just a delay. For example, in the case of the second-order DDSM, $F(z) = z^{-1}(1 - z^{-1})^{-2}$, $G(z) = 2 - z^{-1}$, resulting in $NTF(z) = (1 - z^{-1})^2$ corresponding to a second-order high-pass filter, and $STF(z) = z^{-1}$. Similarly, an Lth-order DDSM, i.e., a DDSM with $NTF(z) = (1-z^{-1})^L$ and $STF(z) = z^{-L}$ can be realized using $F(z) = z^{-1}(1-z^{-1})^{-L}$ and $G(z) = z[1 - (1 - z^{-1})^L]$.

With the DDSM model described above, a model for the excess phase of the delta-sigma fractional-N PLL can be readily derived and is shown in Figure 2.8. Both $m[n]$ and $q[n]$ contribute to instantaneous frequency (and hence, phase) errors between $v_{REF}(t)$ and $v_{DIV}(t)$, which are, in turn, suppressed by the low-pass filter, $A_{\phi}(j2\pi f)$, representing the PLL. Note that $A_{\phi}(j2\pi f)$ is the phase transfer function of the PLL, normalized to unity gain at dc.

While the contributions of both $m[n]$ and $q[n]$ are low-pass filtered by the PLL, there is a crucial difference: in a typical DDSM, the quantizer error, $q[n]$, is high-pass filtered, while the input, $x[n]$, is simply delayed. As a result, the quantization noise, $e[n]$, is mostly suppressed potentially resulting in very low residual phase noise at the PLL output. Furthermore, the PLL output's instantaneous frequency

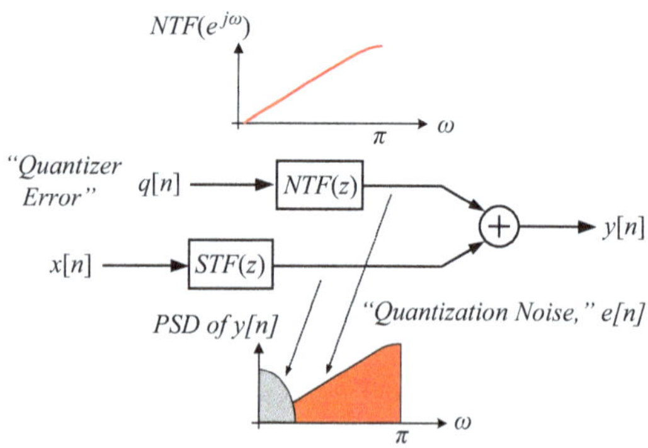

Figure 2.7 A linear model of the digital delta-sigma modulator; the noise transfer function, NTF(z), is typically chosen to be a high-pass filter

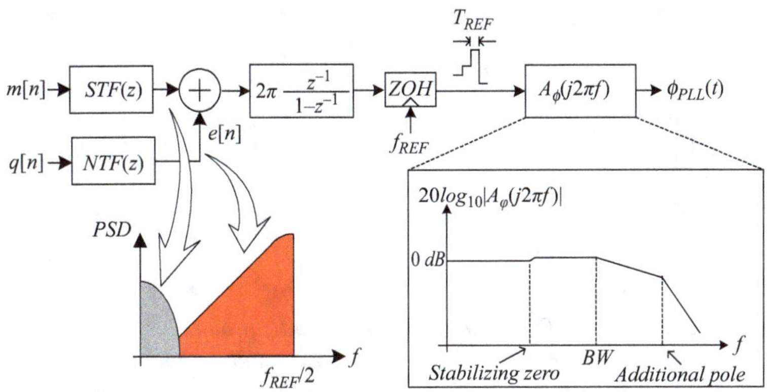

Figure 2.8 Model of excess phase in the delta-sigma fractional-N PLL; $m[n] \times f_{REF}$ is the desired frequency modulation, while $q[n]$ is the DDSM quantizer error. Note that the shaped DDSM quantization noise, $e[n]$, is high-pass shaped and is mostly suppressed by a PLL with a low enough bandwidth

now follows $m[n] \times f_{REF}$, at least within the PLL's bandwidth, enabling several interesting applications described below.

Continuing to assume that the quantizer error, $q[n]$, is a white noise sequence with a variance of 1/12, the phase noise resulting from quantization in an Lth-order DDSM delta-sigma fractional-N PLL can be shown to be:

$$S_{\phi,DDSM}(f) = \frac{\pi^2}{f_{REF}} \cdot \left| 2\sin\left(\frac{\pi f}{f_{ref}}\right) \right|^{2(L-1)} \cdot \left| A_{\varphi}(j2\pi f) \right|^2 \frac{\text{rad}^2}{\text{Hz}} \tag{2.5}$$

where $A_{\phi}(j2\pi f)$ is the transfer function of the PLL normalized to its dc gain.

It is very important to note that when non-linear blocks, such as the digital quantizer, are wrapped in negative feedback, as in the DDSM, complicated behavior ensues. Specifically, even for constant inputs (i.e., $x[n] = \alpha$), both $y[n]$ and $q[n]$ become periodic with a period that depends on α. The DDSM is said to exhibit so-called "limit cycles." The 1-bit random sequence called "dither" or "least significant bit (LSB) dither," added to the LSB of the DDSM input, breaks up such limit cycles in most cases [12]. The quantizer error sequence, $q[n]$, has been shown to indeed become an *i.i.d.* sequence whose samples are uniformly distributed over one quantization step, i.e., $|q[n]| \leq 0.5$, as assumed so far. Later sections discuss limit cycles and dithering in more detail.

Figure 2.9 shows simulated PSDs of the shaped quantization noise, $e[n]$, PLL output, and PLL phase noise for an example second-order delta-sigma fractional-N PLL with $f_{REF} = 48$ MHz, $N = 51$, $\alpha = 0.081875$, and approximately 500 kHz bandwidth. One-bit LSB dither was employed. Note the smooth PSD of $e[n]$ with a 40 dB/decade slope corresponding to a second-order NTF ($L = 2$). Note also the PSD of the PLL's phase noise which increases at 20 dB/decade till the bandwidth

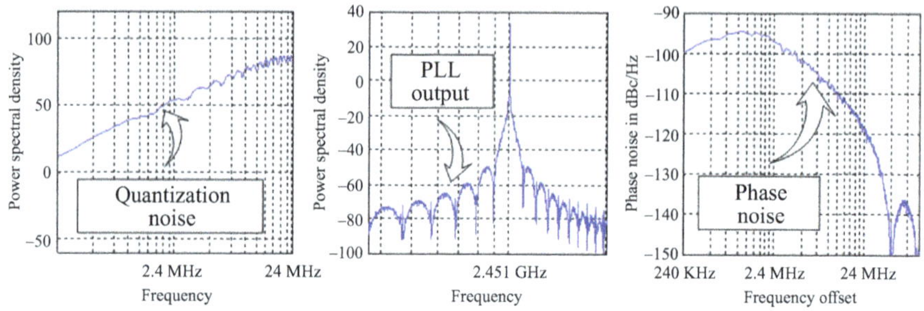

Figure 2.9 Simulated PSD of DDSM shaped quantization noise, e[n], PLL output, and the PLL phase noise for an example delta-sigma fractional-N PLL illustrating the high-pass shaping of the DDSM's quantization error and subsequent low-pass filtering by the PLL. The example PLL has f_{ref} = 48 MHz, N = 51, α = 0.081875, and approximately 500 kHz bandwidth; all other sources of noise in the PLL are suppressed

of the PLL, beyond which, it is suppressed by the PLL. Note that the choice of an appropriate $NTF(z)$ for use in a delta-sigma fractional-N PLL is discussed in more detail in the next section.

2.1.4 Applications

The delta-sigma fractional-N PLL has many uses. Since α can be any fraction, arbitrarily fine grain frequency resolution can be achieved simply by increasing the number of fractional bits, m, in the DDSM's input word. For example, while an integer-N PLL with f_{ref} = 50 MHz has a frequency synthesis resolution of 50 MHz, the fractional-N version of the same PLL using a DDSM with m = 20 bits has a frequency resolution of just 50 Hz! The fine resolution makes the delta-sigma fractional-N PLL the preferred choice for LO synthesis applications.

Furthermore, the PLL bandwidth, which is typically limited to less than f_{REF}/10 for stability considerations [13], can be set to a high value because frequency resolution is not constrained by f_{REF} anymore. A high PLL bandwidth offers several attendant benefits. It enables faster switching from one frequency to another which is beneficial for many communication standards. A high PLL bandwidth also suppresses VCO phase noise much better, which is an important factor in integrated solutions with noisy ring oscillators or even LC oscillators that suffer from poor quality of passive components. In scenarios in which the PLL is followed directly by a powerful power amplifier, the wider bandwidth provides improved immunity from VCO pulling.

Perhaps most importantly, with enough PLL bandwidth, wide bandwidth angle modulation becomes possible. As described earlier, the instantaneous frequency of the delta-sigma fractional-N PLL tracks $m[n] \times f_{REF}$ within the bandwidth of the PLL. Appropriate frequency modulation can be achieved simply by setting $m[n]$. Since the reference is typically derived from a stable crystal oscillator, this approach can control the frequency modulation index very precisely. Note that even phase modulation is achieved by suitably setting $m[n]$.*

Whether it is frequency or phase modulation, the delta-sigma fractional-N PLL enables direct modulation transmitters that avoid conventional quadrature mixer-based frequency up-conversion approaches and save hardware and power consumption [1,2,7,8,14]. Another popular application of the delta-sigma fractional-*N* PLL is in spread-spectrum clock generation: $m[n]$ is chosen to impart the desired triangular or saw-tooth wave-like modulation of a clock frequency (spreading). Many radar and imaging applications often employ wide bandwidth frequency modulated signals, e.g., frequency chirp-based radars, and can benefit from delta-sigma fractional-*N* PLLs. Less common, but promising applications include new efficient transmitter and power amplifier architectures such as switching, polar, and out-phasing architectures that require wide bandwidth phase-modulated signals [15–17].

2.1.5 Challenges

Designing good fractional-*N* PLLs presents multiple challenges. First, the phase noise resulting from fractional-*N* operation, while small, can still be a significant problem in many applications. This is especially true when wide PLL bandwidth is desired to realize one of the aforementioned benefits. This is often referred to as the "bandwidth—quantization noise tradeoff": while higher PLL bandwidth allows higher bandwidth modulation, it lets more quantization noise through as well, and vice versa. A few important approaches to address this tradeoff are described in later sections.

Second, the fractional-*N* operation results in much larger instantaneous phase errors than in an integer-*N* PLL. It can be shown that, due to fractional-*N* operation alone, in a second-order delta-sigma fractional-*N* PLL, $v_{REF}(t)$ and $v_{DIV}(t)$ can differ by one full period of PLL output, T_{PLL}. This increases to $2T_{PLL}$ for a third-order delta-sigma fractional-*N* PLL. Such large phase errors imply that more noise, especially from the charge pump circuitry appears in the PLL's output. As a result, fractional-*N* PLLs, even low bandwidth ones which suppress most of the quantization noise, are known to exhibit up to 10 dB higher in-band noise floor than their integer-*N* counterparts.

Third, non-ideal operation of the circuit components of the PLL can, and often does, result in strong spikes in the PSD of the PLL output. It is important to note that these spikes in the PSD typically occur at harmonics of αf_{REF}, i.e., cause fractional spurs. Fractional spurs present a difficult challenge in most applications. For a charge pump PLL, the most important non-idealities are charge pump current mismatch, modulus-dependent divider delays, undesired coupling between parts of the PLL, etc. A later section describes these in greater detail. The sources of the fractional spurs may change if a different phase detector and/or a loop filter were chosen but invariably, fractional spurs are observed in the presence of circuit

*For phase modulation, $m[n]$ should follow the derivate of the desired phase modulation, approximately.

non-linearity. Typically, many, but not all, fractional spur contributions are suppressed outside the PLL's bandwidth. However, close-in fractional spurs remain a particularly challenging problem, especially in wireless communication applications. The remainder of the chapter summarizes several efforts to reduce phase noise and fractional spurs even for wide bandwidth PLLs.

2.2 Digital delta-sigma modulator design

2.2.1 Modulator order

The impact of DDSM quantization error, $q[n]$, on the fractional-N PLL's phase noise can be reduced by employing higher-order noise transfer functions, e.g., $NTF(z) = (1 - z^{-1})^3$. As shown in Figure 2.10, increasing the DDSM order suppresses the quantization noise power for frequencies ω less than $\pi/3$, i.e., in continuous frequency terms, for f less than $f_{REF}/6$. Of course, quantization noise power increases at higher frequencies, but if the PLL bandwidth is less than $f_{REF}/6$, much of this appears in the stop-band of the PLL. Of course, additional poles in the PLL's loop filter may be needed to suppress the higher out-of-band quantization noise. This is an excellent approach to achieving very low quantization-induced phase noise in fractional-N PLLs with low bandwidths, much less than $f_{REF}/10$. However, if the application requires higher loop bandwidth, the additional poles in the loop filter can degrade the stability (phase margin) of a Type-II PLL. As a result, typical fractional-N PLLs are confined to second- or third-order DDSMs, with fourth-order DDSMs employed rarely, but in low bandwidth cases only.

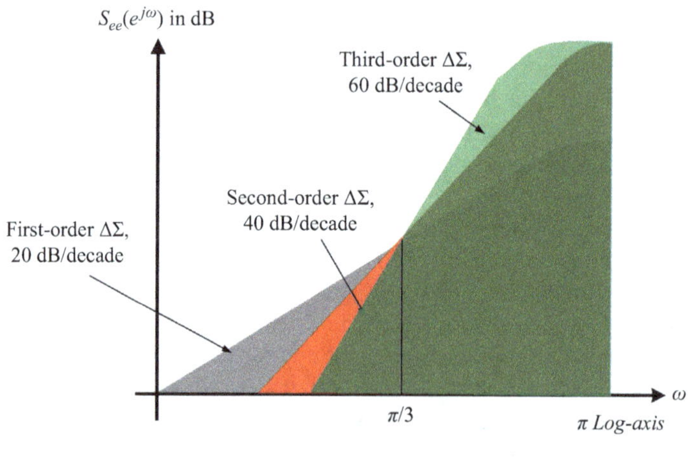

Figure 2.10 A sketch of the PSD of shaped quantization noise, e[n], in DDSMs of first, second, and fourth order. Increasing the DDSM order reduces close-in phase noise for frequencies ω less than $\pi/3$, i.e., for continuous frequencies of f less than $f_{REF}/6$

Two contrasting approaches have been proposed to address this problem. One approach introduces poles in the $NTF(z)$ to keep out-of-band quantization noise manageable [18]. However, this usually comes with the price of higher in-band noise potentially negating the rationale of using a higher-order $NTF(z)$ in the first place. Furthermore, the limit cycle behavior of $q[n]$ in such DDSMs is not well studied; and, as described below, the efficacy of the dither in breaking up limit cycles in such DDSMs is unclear, at best.

The second approach employs Type-I PLLs. Since Type-I PLLs have one less pole at dc, additional loop filters can be readily employed to realize sharp PLL transfer functions while maintaining excellent loop stability. However, Type-I PLLs are known to exhibit large phase errors in steady state. Combined with the larger instantaneous phase errors, the fractional-N PLL can become more sensitive to errors in the phase detector, frequency divider, and the charge pump circuitry, resulting in potentially stronger fractional spurs.

2.2.2 DDSM architecture

DDSMs can be implemented in multiple ways. For example, apart from the "single-loop" architecture shown in Figure 2.5, the second-order DDSM can be realized in at least three other ways, shown in Figure 2.11: (a) a multi-loop structure, (b) an "error-feedback" structure, or (c) a cascade of lower order DDSMs (called the 1-1

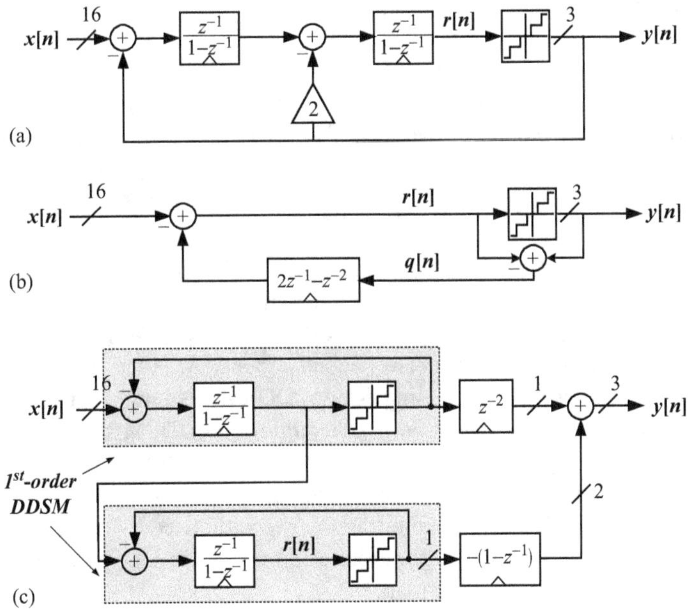

Figure 2.11 Different realizations of the second-order DDSM: (a) single-loop architecture, (b) error-feedback structure, (c) cascade of two first-order DDSMs or a 1-1 MASH architecture. Another structure is shown in Figure 2.5

MASH DDSM). Higher-order DDSMs can be realized in even more different ways. Note that all the example second-order DDSMs have the same $NTF(z)$ and $STF(z)$ except, perhaps, for some inconsequential difference in sample delays.

However, for use in a fractional-N PLL, the statistical properties of the quantizer error, $q[n]$, are just as important as $STF(z)$ and $NTF(z)$, which have been derived using the linear, additive error source model of the quantizer.

First consider the single-loop, multi-loop, and error-feedback structures—all of which employ a single digital quantizer. Assuming that the digital quantizers do not saturate, it becomes apparent that the feedback paths, which carry integer-valued signals, do not affect the properties of the quantizer error, $q[n]$. It is easy to see that the forward path, from $x[n]$ to $r[n]$, the input of the digital quantizer, is the same in all these second-order single quantizer DDSMs[†]: comprising two digital accumulators.

The situation in the MASH DDSM is a little different: the overall quantization noise is determined entirely by the quantization error, $q[n]$, from the final stage since the quantization error from the preceding stage is canceled out. Ignoring all quantizer outputs since they are integer-valued, the path from $x[n]$ to the final quantizer's input is again comprised of two digital accumulators. In effect, as long as the quantizer does not saturate, all the four second-order DDSMs with the same $STF(z)$ and $NTF(z)$ are equivalent even in regard to the statistics of $q[n]$. Similar results can be proved for higher-order DDSMs as well.

It is important to address a common misconception here. Researchers often, but incorrectly, cite stability concerns with higher-order single-loop DDSMs and instead prefer a MASH DDSM. As argued above, the MASH and the single-loop DDSMs are equivalent here. The real difference is in whether the quantizer saturates or overloads in some manner or not. Consider the single-loop and MASH second-order DDSM examples with $NTF(z) = (1 - z^{-1})^2$. Ignoring dither and modulation, in either case,

$$y[n] = \alpha + q[n] - 2q[n-1] + q[n-2] \tag{2.6}$$

If $q[n]$ were a white *i.i.d.* sequence such that $|q[n]| \leq 0.5$ as is typically assumed, the PSD of $(y[n]–\alpha)$ would possess the second-order high-pass shape dictated by $NTF(z)$. However, this requires that $|y[n]| \leq |\alpha| + 2$, e.g., if $\alpha = 0.3$, then $y[n]$ should have four levels. The second-order MASH has at least four-level output by construction; a 2-bit single-loop DDSM also meets the condition.[‡] However, if a 1-bit quantizer is used in the single-loop DDSM, as many designers do, then $q[n]$ is no longer white, bounded in magnitude by 0.5, or independent of α. This condition, referred to as "quantizer overload," results in strong fractional spurs and in the PSD of $(y[n] - \alpha)$ not exhibiting the shape dictated by $NTF(z)$, for a large range of values of α. The latter condition is referred to as DDSM instability. It is preferable to avoid quantizer overload.[§]

[†]Except perhaps for some difference in delays, which is inconsequential.

[‡]Technically, if modulation, $m[n]$, and/or dither, $d[n]$, are considered, five levels would be needed for all the second-order DDSMs discussed here.

[§]Note that avoiding overload is not necessary for this discussion but is sufficient.

However, as mentioned before, note that a large range of $y[n]$ results in large instantaneous phase errors between $v_{REF}(t)$ and $v_{DIV}(t)$ and an attendant increase in the contribution of CP phase noise. Designers can compromise between the usable input range of the DDSM, the range of $y[n]$, and achievable quantization noise shaping by employing poles in the $NTF(z)$ [11,18]. Interested readers are referred to [11] for a more detailed discussion of delta-sigma modulator design and stability, albeit in the context of analog modulators.

2.2.3 Dithering

As mentioned before, without dithering, DDSMs exhibit limit cycle behavior, which, in turn, results in strong *fractional spurs* in the PSD of the PLL phase noise [11]. The 1-bit dither, $d[n]$, added to the LSB of the DDSM input has been theoretically shown to break up limit cycles and render the quantization error, $q[n]$, *i.i.d.* and uniformly distributed for a fairly large class of DDSMs [12,19]. Without getting too mathematical, the theoretical conditions under which LSB dither is effective are as follows:

1. The DDSM employs two or more integrators between the dither and the relevant digital quantizer.
2. The quantizer does not overload, i.e., $y[n]$ is allowed enough levels to accommodate a desired input range and noise shaping.

While seemingly restrictive, these conditions are fairly general and cover most of the multi-bit DDSMs employed including second and higher-order DDSMs of single-loop, multi-loop, error-feedback, and MASH structures. The dither itself causes additional frequency modulation (FM) noise which is very low and can be further suppressed in some cases by high-pass filtering prior to addition to the DDSM input [19].

Note that, in practice, the dither sequence, $d[n]$, is usually a pseudorandom sequence generated by a linear feedback shift register (LFSR). Unfortunately, the pseudorandom sequence generated by an LFSR only approximates a true random *i.i.d.* 1-bit sequence. In reality, it is periodic with the sequence length. Consequently, $y[n]$ does exhibit periodicity, but the period can be made arbitrarily long by using LFSRs of the desired length. For example, a 30th-order LFSR can generate a pseudorandom sequence that repeats only after a trillion samples; with $f_{REF} = 50$ MHz, the sequence repeats only after 20 seconds!

Researchers have suggested alternatives to dithering with varying degrees of success and with attendant advantages and disadvantages. For example, researchers have suggested adding dither to the input of the quantizer, rather than at the DDSM input. Such dither can break the limit cycles and render $q[n]$ *i.i.d.* provided the dither itself is *i.i.d.*, and its samples are uniformly distributed over an entire quantization step, i.e., over [–0.5, 0.5]. This represents a lot of random noise being introduced into the DDSM. It increases the PLL's DDSM-induced phase noise floor by as much as 3 dB and could need a substantial amount of digital hardware to generate.

A popular alternative approach sets specially chosen integrator initial conditions [20]. This approach requires extensive simulations to pick initial conditions specific to various values of α. More recently, limit cycle period extension approaches [21] have been reported. They do not eliminate the fractional spurs but make them independent of α and move them to very low offset frequencies where they may be tolerated. However, they do not guarantee that $q[n]$ is *i.i.d.* or that its samples are uniformly distributed over [–0.5, 0.5] as typically assumed for arbitrary inputs, limiting their use cases.

2.3 Circuit design issues

While the components of a fractional-*N* PLL are essentially the same as in an integer-*N* PLL, they do differ significantly in the performance required of them. In particular, non-ideal circuit behavior and inevitable matching errors, e.g., PFD and/or CP non-linearity and modulus-dependent divider delays (MDDD) cause strong fractional spurs and quantization noise "fold-in" in fractional-*N* PLLs [1,22,23]. These effects are aggravated in wide bandwidth fractional-*N* PLLs. In the following, issues pertaining to the design of charge pump-based fractional-*N* PLLs are discussed. While fractional-*N* PLLs employing other kinds of phase detectors and loop filters have been reported, the charge pump-based fractional-*N* PLL remains popular and serves as a good example.

2.3.1 PFD and CP

Ideally, every reference period, the PFD/CP combination dumps into the loop filter, an amount of charge that is linearly proportional to the phase (or time) difference between the PFD's inputs. This is similar to what happens in an integer-*N* PLL but with one important difference. Specifically, owing to the fractional-*N* operation, even in steady state, the PFD/CP circuits process much larger instantaneous phase errors, $\phi[n]$. This was quantified earlier in Section 2.1.2. The large instantaneous phase errors have two important repercussions.

2.3.1.1 Increased charge pump noise contribution

Larger phase errors imply that the charge pump transistors stay ON for longer durations each reference period, thereby dumping more noise charge into the loop filter than in an integer-*N* PLL. Ignoring modulation, $m[n]$, and assuming that $q[n]$ is *i.i.d.* and $|q[n]| \leq 0.5$, it can be shown that, for a second-order DDSM, the largest magnitude timing error between $v_{REF}(t)$ and $v_{DIV}(t)$ due to fractional-*N* operation alone, in steady state, is:

$$|\tau[n]|_{max} = T_{PLL}\left|\sum_{k=0}^{n-1}(y[k]-\alpha)\right| = T_{PLL}|q[n]-q[n-1]| \leq T_{PLL} \tag{2.7}$$

This means that, in the worst case, the charge pump dumps noise charge for a duration of $(T_{PLL} + T_D)$, where T_{PLL} is one period of the PLL output and T_D is the so-called dead-zone duration (i.e., when both up and down charge pumps are ON [24]). Proceeding similarly, it can be shown that for an Lth-order DDSM ($L \geq 2$),

$$|\tau[n]|_{max} \leq 2^{L-2}\, T_{PLL} \tag{2.8}$$

and accordingly, the charge pump can dump noise charge for up to $(2^{L-2}T_{PLL} + T_D)$ seconds. In contrast, in an ideal integer-N PLL, where phase errors are negligibly small and are caused essentially by phase noise or any phase offsets, the charge pump dumps noise charge for about T_D seconds only. Consequently, the charge pump's noise contribution can be $(1 + 2^{L-2}T_{PLL}/T_D)$ times higher. This can raise the in-band noise floor by as much as 10 dB in a fractional-N PLLs compared to an integer-N PLLs. For example, in a third-order DDSM fractional-N, 2.5 GHz PLL with $T_D = 200$ ps, the noise can be 6× higher than the corresponding integer-N PLL.[¶] Designers need to take this additional noise contribution into account while designing charge pumps for fractional-N PLLs.

2.3.1.2 PFD/CP non-linearity

Most forms of non-linearity acting on the instantaneous phase errors from fractional-N operation, $\phi[n]$, are known to result in

1. Strong fractional spurs at harmonics of αf_{REF}.
2. High-pass shaped DDSM quantization noise "folding in" and raising in-band noise floor.

This is true even in a well-dithered DDSM in which $q[n]$ is *i.i.d.*, and hence, $\phi[n]$ has a smooth, spur-free high-pass PSD. This might seem surprising but is expected. The dithering makes $q[n]$ only pairwise independent, i.e., $q[n]$ becomes independent of $q[m]$ for all distinct integers m and n. Unfortunately, the pairwise independence only renders $q[n]$ white and spur-free but does not completely eliminate underlying periodic evolution of the DDSM with the sample index, n.

Several sources of errors in both the PFD and the CP can result in non-linearity in the PFD/CP transfer characteristic, i.e., charge dumped into the loop filter, $Q[n]$, vs timing error between PFD inputs, $\phi[n]$, as illustrated in Figure 2.12(a) [25,26]. For small $|\phi[n]|$, the PFD may not respond at all resulting in the so-called PFD *dead zone*. Even if the PFD responds linearly, the CP may not fully turn ON in time resulting in an effective slope that is smaller than the nominal CP current, I_{CP}. Both problems are alleviated by using a sufficiently large CP dead zone, T_D [24], but this comes at the expense of higher CP noise contribution.

In the charge pump, transistor size mismatches and finite and mismatched pump output impedances result in different slopes ($I_{CP+} \neq I_{CP-}$) for positive and negative phase errors, $\phi[n]$. The current mismatch adds a non-linear error charge proportional to $0.5 \times (I_{CP+}-I_{CP-})|\phi[n]|$, every reference period. The current matching is poor over corners and various values of α, because the CP current sources are composed of dissimilar transistors (pMOS vs nMOS) with finite, mismatched output impedances. Using amplifier-based negative feedback to match currents is a popular solution [27] but has amplifier bandwidth and headroom limitations. Using a type-I PLL loop [3] or, more commonly, adding a DC offset current to the CP [4,22,28] avoids the non-linearity by operating away from

[¶]While the white noise component of the charge pump can increase by $10 \log_{10} 6 = 7.7$ dB, charge pump flicker noise, which is correlated, can increase by $20 \log_{10} 6 = 15.5$ dB.

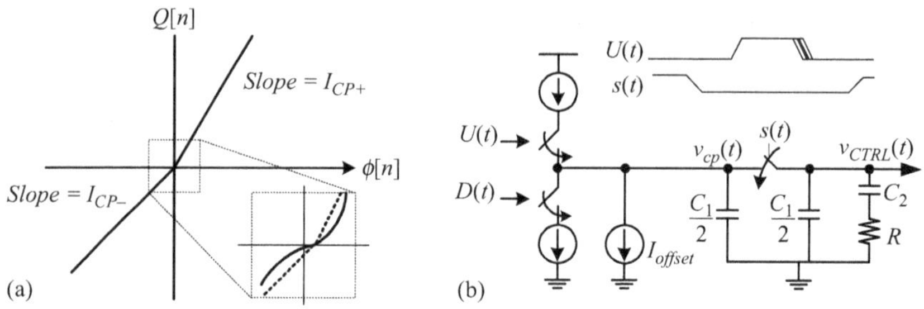

Figure 2.12 (a) Sketch of an example PFD/CP transfer curve. The non-linearity causes fractional spurs and noise folding in-band. (b) The non-linearity from the PFD/CP can be addressed in multiple ways: by adding an offset current, I_{offset}, so that the PLL operates with a non-zero phase offset, i.e., mean($\phi[n]$) $\neq$ 0, and/or by employing a switched capacitor loop filter

$\phi[n] = 0$. However, this increases CP phase noise contribution and results in a strong reference spur.

A sampled loop filter [22], such as the one shown in Figure 2.12(b), reduces the reference spur, but it brings its own specific problems such as clock feed-through and charge injection. The pedestal-based CP linearization technique [1] dynamically cancels the non-linear error charge by keeping one half of both the I_{CP+} and I_{CP-} current sources ON for $(T_{ped} - \tau[n])$ seconds longer, each reference period, where $\tau[n] = 2\pi\phi[n]/(N + \alpha)$ is the timing error between $v_{REF}(t)$ and $v_{DIV}(t)$. It offers at least 9 dB spur suppression but is limited by matching between like transistors, e.g., nMOS vs nMOS. Spur suppression can be further improved by randomizing the chosen pump halves [29].

Other sources of non-linearity include dynamic errors such as ringing on the $v_{CTRL}(t)$ node due to bonding wire inductance when the loop filter is off-chip or PFD delay modulation caused by its own switching transients [22]. Careful capacitive decoupling or regulation of the PFD supply or using a sampled loop filter may resolve dynamic errors [22].

2.3.2 Multi-modulus frequency dividers

Multi-modulus frequency dividers are commonly implemented using dual-modulus prescalers [24] or a chain of 2/3 counters [30]. Either way, the delay through the frequency divider varies with the modulus, $N + y[n]$, primarily because of the critical path through the divider changes with $y[n]$. These MDDDs non-linearly distort the PSD of $y[n]$ resulting in both fractional spurs and quantization noise fold-in.

Synchronizing the divider output to the VCO signal by a flip-flop can reduce MDDD [1]. However, care must be taken to avoid meta-stability; the attendant circuitry could result in high power consumption [1,22,31]. Note that if the divider is realized as a chain of 2/3 cells, the "mod" signal in the first (high frequency) 2/3 cell stage can be used as a synchronized divider output. A simple way of eliminating the MDDD is illustrated in Figure 2.13 in simplified form [32]. The divider modulus, $N + y[n]$, is counted in two steps: first, a variable number of cycles, $N_1 + y[n]$, are counted followed by a fixed number of cycles, $N - N_1$, as shown. Although the divider output suffers from MDDD when counting $N_1 + y[n]$ cycles, it does not suffer from any significant modulus-dependent delays when counting $N - N_1$. Adding a divide-by-2 circuit at the divider output picks up a clean rising edge that is not contaminated by MDDD.

Using a "successive requantizer" instead of a DDSM to generate $y[n]$ eliminates the fractional spurs caused by both PFD/CP non-linearity and MDDD using purely digital techniques [22]. As shown in Figure 2.14, a high-precision digital representation of α is quantized using a cascade of single-bit quantization blocks. The single quantization block works by adding a random error sample, $s_{d,n}$, and truncating the LSB of the result. A Markov chain generates $s_{d,n}$ with statistical properties that guarantee that both the PFD/CP non-linearity and MDDD do not cause fractional spurs. Readers are referred to [22,33] for details of the finite state machine that implements the Markov chain. Very low in-band fractional spurs

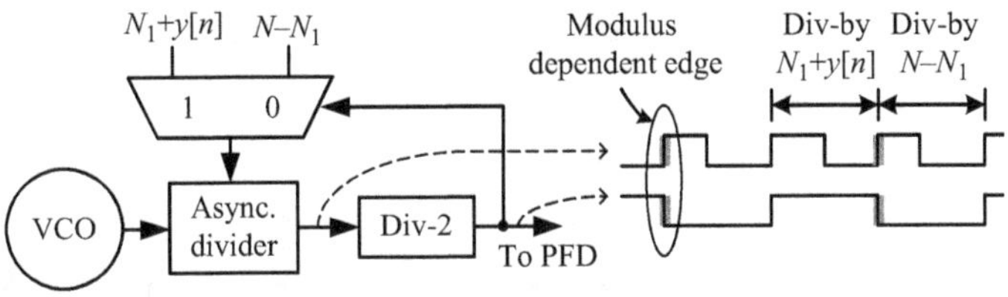

Figure 2.13 Suppressing modulus-dependent divider delays using a two-step division process

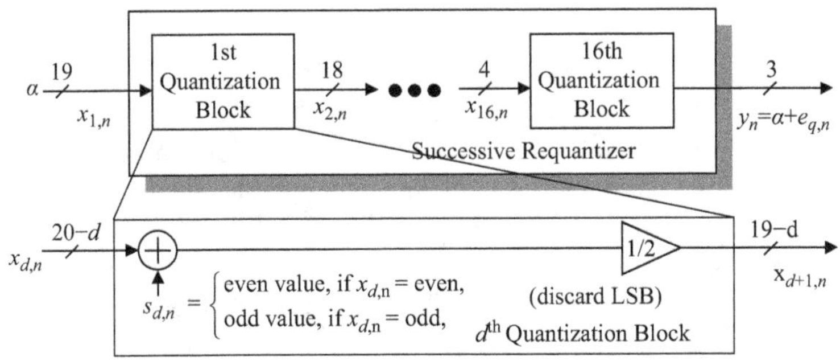

Figure 2.14 A simplified block diagram of the successive requantizer-based fractional-N PLL. The fractional-N modulus sequence, y[n], is generated such that both modulus-dependent divider delays and most PFD/CP non-linearity acting on the instantaneous phase errors, $\phi[n]$, do not generate fractional spurs

(<–64 dBc) have been reported [22]. However, the successive requantizer adds more quantization noise than a conventional DDSM, which is suppressed by phase noise cancellation techniques described in the next section.

2.4 Wideband fractional-*N* PLLs

Wide bandwidth fractional-*N* PLLs offer several distinct benefits: better immunity from VCO pulling, lower VCO phase noise contribution, and direct angle modulation including spread-spectrum clocking. However, as mentioned earlier, wider PLL bandwidth implies that much of the DDSM quantization noise remains unsuppressed resulting in poor phase noise.

The earliest attempts to enable wide bandwidth fractional-*N* PLL-based frequency synthesis did not resort to wide PLL bandwidths. Instead, the high-frequency content of the modulation signal, $m[n]$, which would be suppressed by a narrowband PLL, is either pre-emphasized digitally before the DDSM [6] or applied directly to a second VCO control port (called *two-point modulation*) [34]. However, these approaches require precise knowledge of the PLL's transfer function to maintain modulation accuracy. Several elaborate calibration techniques that estimate the dynamics of the PLL have been reported [35], but it is unclear if these techniques can scale to modulation bandwidths beyond a few MHz.

2.4.1 Phase noise cancellation

Extending the fractional-*N* PLL bandwidth enables direct in-loop modulation and relaxes the requirements of the aforementioned pre-emphasis and two-point modulation techniques. Phase noise cancellation techniques have been reported to cancel the quantization noise that would have been otherwise suppressed by a narrow PLL bandwidth [1,3,4,22,31,32,36,37]. The central idea is that the noise charge dumped into the loop filter during each reference period, $I_{CP}\tau[n]$, is nominally known:

$$Q_{noise}[n] = I_{CP}\tau[n] = I_{CP}T_{PLL}\sum_{k=0}^{n-1} e[k] \tag{2.9}$$

Since the shaped quantization noise, $e[n]$, is known or can be digitally computed using signals from the DDSM, an equal but opposite amount of cancellation charge can be dumped into the loop filter to nullify $Q_{noise}[n]$. A simple way is to use a digital-to-analog current converter (I-DAC) that operates for a fixed duration, T_{DAC}, and sources/sinks a variable amount of current, $i_{DAC}[n]$, each reference period:

$$Q_{cancel}[n] = T_{DAC}i_{DAC}[n] = T_{DAC}I_{DAC}\sum_{k=0}^{n-1} e[k] \tag{2.10}$$

An example application of the phase noise cancellation system [1] is shown in Figure 2.15. Note that the shaped quantization, $e[n]$, is computed by subtracting (appropriately delayed) DDSM input from its output. Since $e[n]$ is a high-resolution

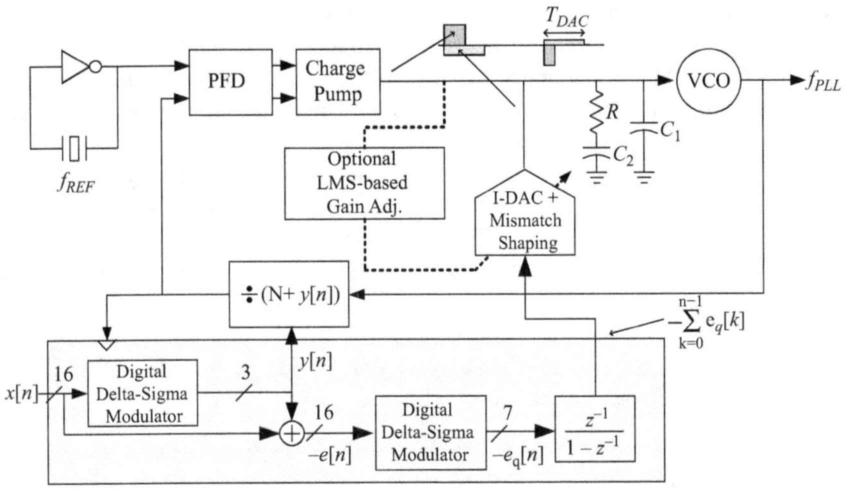

Figure 2.15 A simplified block diagram of an example delta-sigma fractional-N PLL with phase noise cancellation

sequence, it is quantized down to a 7-bit sequence using another DDSM, and the quantized version, $e_q[n]$, is integrated, and the I-DAC pumps in a proportionate amount of cancellation charge for a fixed duration, T_{DAC}. While conceptually simple, the main challenges in phase noise cancellation are in (a) ensuring good matching between the main (CP) and the cancellation (I-DAC) paths and (b) handing the high dynamic range of the cancellation path.

2.4.1.1 Handling high dynamic range of cancellation path

In Figure 2.15, a DDSM ($L > 2$) quantizes 16-bit $e[n]$ to 7-bit $e_q[n]$ such that the error of this "requantization" has a high-pass shaped PSD. Since the PLL will effectively low-pass filter, any charge introduced into the loop filter, $e_q[n]$, would be greatly suppressed as well. Furthermore, errors in the I-DAC will cause fractional spurs and noise folding in. To avoid this, dynamic element matching (DEM) techniques can be employed in the DAC. Several variations in the type of the DAC and the choice of the DEM algorithm employed in the DAC have been reported [3,4]. Similarly, researchers have interchanged the order of the integration of $e[n]$ and its requantization to a lower resolution sequence or used an MASH architecture in the main DDSM to reduce digital hardware requirements. However, these are not discussed here for the sake of brevity.

2.4.1.2 Matching the main and cancellation paths

Since the quantization noise and the compensation charge go through two very different paths, the gain and propagation delay mismatches between the paths limit cancellation effectiveness [38]. Certain simple techniques can readily achieve up to 20 dB of cancellation [1]. First, making T_{DAC} an integer multiple of T_{PLL} ensures that the path gains are matched even as the fractional value, α, is changed. Second, making the I-DAC's unit elements be scaled replicas of the CP greatly improves matching. Third, the propagation delay mismatch between the two paths may be alleviated by clocking PFD and DAC using the same reference signal. Adequate DAC pulse width and properly positioned DAC pulses can further suppress quantization noise and lower reference spurs [38].

Even better phase noise cancellation can be achieved by employing adaptive gain adjustment techniques as indicated by the optional block in Figure 2.15. The idea is to monitor the loop filter voltage, $v_{CTRL}(t)$, for any residual variations that are dependent on $e[n]$, and accordingly adjust the I-DAC gain using a least mean square algorithm [31,32,36]. With such adaptive gain calibration, the DDSM-induced quantization noise has been suppressed by up to 32 dB [31]. Such cancellation can suppress the DDSM-induced phase noise enough to approach the noise levels of a comparable integer-N PLL.

2.4.2 Multi-phase fractional-N PLLs

The central idea behind the multi-phase fractional-N PLL is to use an M-phase VCO to enable fractional frequency division, i.e., division by $(N + v[n]/M)$ rather than $(N + y[n])$ in the PLL's feedback path, where $y[n]$ and $v[n]$ are integer sequences that average to α and $M\alpha$, respectively. Accordingly, the instantaneous frequency (and hence, phase) errors are M times smaller, thereby lowering the quantization noise by $20 \times \log_{10} M$-dB [39].

To divide by $(N + v[n]/M)$, for every divider cycle, the frequency divider switches its input from VCO phase $\theta_{ph[n-1]}$ to $\theta_{ph[n]}$, where $ph[n] = (ph[n-1] + v[n]) \bmod M$. The phase switch ensures that the divider counts an extra $v[n]/M$ cycles of the VCO, while the modulo-operation accounts for wrapping around one VCO cycle. For example, for an eight-phase VCO, i.e., $M = 8$, to divide by $(N + 3/8)$, the divider switches its input phases in the order, $\ldots \theta_0, \theta_3, \theta_6, \theta_1, \theta_4, \theta_7, \theta_2, \theta_5, \theta_0, \ldots$ and so on. A simplified block diagram is shown in Figure 2.16. Momentarily assuming that the optional dither filter, $G(z)$, is zero, it can be seen that the "phase select logic" chooses the appropriate phase. Note that the wrap-around is equivalent to the divider counting an extra VCO cycle, i.e., increasing $y[n]$.

While the quantization noise is greatly reduced, the fractional division approach suffers from two main challenges. First, low-noise multi-phase VCOs are difficult to design: ring oscillator-based VCOs can generate multiple phases readily but suffer from much higher phase noise than LC oscillator-based VCOs. Second, mismatches among the multiple VCO phases cause strong fractional spurs even when a properly dithered DDSM chooses $v[n]$. Phase error calibration techniques have been suggested [40] but calibrating sub-picosecond errors over operational corners is very challenging.

The fractional spurs can be suppressed to an extent by randomizing the phase selection. As shown in Figure 2.16, the optional finite impulse response (FIR) dither filter, $G(z)$, in conjunction with the 1-bit pseudorandom sequence, introduces a large dither, $\eta[n]$, that randomizes the phase selection. It has been theoretically guaranteed that such randomization eliminates fractional spurs caused by

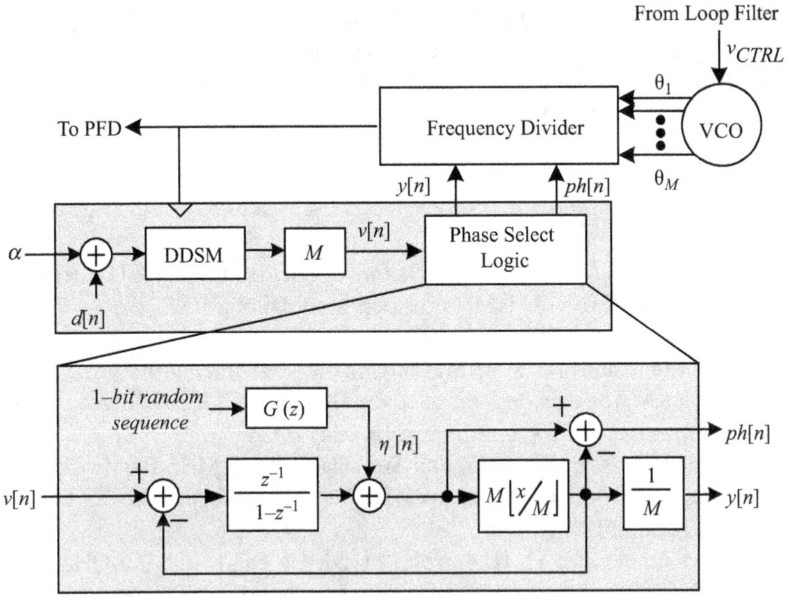

Figure 2.16 Simplified block diagram of a fractional-N PLL based on a multi-phase VCO and fractional frequency division. The VCO generates M multiple phases among which the frequency divider switches, once every reference period, to count an additional v[n]/M VCO cycles. The phase select logic block keeps track of the phases to be selected including any wrap-around the M VCO cycles. An optional 1-bit random sequence and a special dither filter, G(z), randomize the phase selection to suppress fractional spurs resulting from phase mismatches

memoryless phase mismatches if at least some coefficients of the FIR dither filter, $G(z)$, have amplitudes of 1, 2, 4, ..., $M/2$ [41]. In spite of the constraints, $G(z)$ can be chosen to aggressively shape dither-induced phase noise so that it is suppressed by the PLL. However, instantaneous phase errors seen by the PLL are large with the attendant disadvantages of higher charge pump noise contribution and susceptibility to CP non-linearity that were discussed earlier.

2.5 Conclusion

Fractional-*N* PLLs, particularly delta-sigma fractional-*N* PLLs, are some of the most popular integrated frequency synthesizers. They offer very fine grain frequency synthesis and find use in a wide range of applications such as LO synthesis, phase/frequency modulation, spread-spectrum clocking, frequency hopping, etc. While very useful, fractional-*N* PLLs present interesting technical challenges in the form of excess phase noise and fractional spurs. However, with the help of clever quantization noise shaping techniques and careful circuit design, phase noise and spurious tone performance close to integer-*N* PLLs can be achieved without sacrificing their attendant benefits.

References

[1] S. Pamarti, L. Jansson, and I. Galton, "A wideband 2.4-GHz delta-sigma fractional-*N* PLL with 1-Mb/s in-loop modulation," *IEEE Journal of Solid State Circuits*, vol. 39, no. 1, 2004, pp. 49–62.

[2] B. De Muer and M. S. J. Steyaert, "A CMOS monolithic ΔΣ-controlled fractional-*N* frequency synthesizer for DCS-1800," *IEEE Journal of Solid-State Circuits*, vol. 37, no. 7, pp. 835–844, 2002.

[3] H. Hedayati, B. Bakkaloglu, and W. Khalil, "A 1 MHz-Bandwidth Type-I ΔΣ Fractional-*N* Synthesizer for WiMAX Applications," *ISSCC Dig. Tech. Papers*, pp. 390–391, 2009.

[4] S. E. Meninger and M. H. Perrott, "A 1-MHz bandwidth 3.6-GHz 0.18-μm CMOS fractional-*N* synthesizer utilizing a hybrid PFD/DAC structure for reduced broadband phase noise," *IEEE Journal of Solid-State Circuits*, vol. 41, no. 4, pp. 966–980, 2006.

[5] T. A. D. Riley and M. A. Copeland, "A simplified continuous phase modulator technique," *IEEE Transactions on Circuits and Systems II – Analog and Digital Signal Processing*, vol. 41, no. 5, 1994, pp. 321–328.

[6] M. H. Perrott, T. L. Tewksbury III, and C. G. Sodini, "A 27-mW CMOS fractional-*N* synthesizer using digital compensation for 2.5-Mb/s GFSK modulation," *IEEE Journal of Solid-State Circuits*, vol. 32, no. 12, pp. 2048–2060, 1997.

[7] H.-R. Lee, O. Kim, G. Ahn, and D.-K. Jeong, "A low-jitter 5000 ppm spread spectrum clock generator for multi-channel SATA transceiver in 0.18μm CMOS," in *Proc. IEEE Int. Solid-State Circuits Conf.*, Feb. 2005, pp. 162–163.

[8] K. F. Tsang and C. M. Yuen, "A low-voltage fast switching frequency synthesizer for FH-SS applications," *IEEE Journal of Solid-State Circuits*, vol. 53, no. 12, pp. 1431–1435, 2006.

[9] M. Perrott, M. Trott, and C. Sodini, "A modeling approach for Σ-Δ fractional-*N* frequency synthesizers allowing straightforward noise analysis," *IEEE Journal of Solid-State Circuits*, vol. 37, no. 8, pp. 1028–1038, 2002.

[10] C. A. Kingsford-Smith, "Device for synthesizing frequencies which are rational multiples of a fundamental frequency," U. S. Patent 3928813, Dec. 23, 1975.

[11] S. R. Norsworthy, R. Schreier, and G. C. Temes, *Delta-Sigma Data Converters – Theory, Design, and Simulation*, New York: IEEE Press, 1997.

[12] S. Pamarti, J. Welz, and I. Galton, "Statistics of the quantization noise in 1-bit dithered single-quantizer digital delta-sigma modulators," *IEEE Transactions on Circuits and Systems I: Regular Papers*, vol. 54, no. 3, 2007, pp. 492–503.

[13] F.M. Gardner, *Phaselock Techniques*, Second Ed., New York: McGraw-Hill, 1993.

[14] M. A. Ferriss and M. P. Flynn, "A 14 mW fractional-*N* PLL modulator with a digital phase detector and frequency switching scheme," *IEEE Journal of Solid-State Circuits*, vol. 43, no. 11, pp. 2464–2471, 2008.

[15] J. Groe, "Polar transmitters for wireless communications," *IEEE Communications Magazine*, vol. 45, no. 9, pp. 58–63, 2007.

[16] M. E. Heidari, M. Lee, and A. A. Abidi, "All-digital outphasing modulator for a software-defined transmitter," *IEEE Journal Solid-State Circuits*, vol. 44, no. 4, pp. 1260–1271, 2009.

[17] C. D. Presti, F. Carrara, A. Scuderi, P. M. Asbeck, and G. Palmisano, "A 25 dBm digitally modulated CMOS power amplifier for WCDMA/EDGE/OFDM with adaptive digital predistortion and efficient power control," *IEEE Journal of Solid State Circuits*, vol. 44, no. 7, pp. 1883–1896, 2009.

[18] W. Rhee, B.-S. Song, and A. Ali, "A 1.1-GHz CMOS fractional-*N* frequency synthesizer with a 3-b third-order ΔΣ modulator," *IEEE Journal of Solid-State Circuits*, vol. 35, no. 10, pp. 1453–1460, 2000.

[19] S. Pamarti and I. Galton, "LSB dithering in MASH delta-sigma D/A converters," *IEEE Transactions on Circuit Systems – I: Regular Papers*, vol. 54, no. 4, pp. 779–790, 2007.

[20] M. Kozak and I. Kale, "Rigorous analysis of delta-sigma modulators for fractional-N PLL frequency synthesis," *IEEE Transactions on Circuits and Systems – I: Regular Papers*, vol. 51, no. 6, 2004, pp. 1148–1162.

[21] K. Hosseini and P. M. Kennedy, "Maximum sequence length MASH digital delta–sigma modulators," *IEEE Transactions on Circuits and Systems I: Regular Papers*, vol. 54, no. 12, pp. 2628–2638, 2007.

[22] K. Wang, A. Swaminathan, and I. Galton, "Spurious-tone suppression techniques applied to a wide-bandwidth 2.4 GHz fractional-*N* PLL," *IEEE Journal of Solid-State Circuits*, vol. 43, no. 12, pp. 2787–2797, 2008.

[23] T. A. D. Riley, N. M. Filiol, Q. Du, and J. Kostamovaara, "Techniques for in-band phase noise reduction in ΔΣ synthesizers," *IEEE Transactions on Circuits and Systems II: Analog and Digital Signal Processing*, vol. 50, no. 11, 2003, pp. 794–803.

[24] B. Razavi, *RF Microelectronics*, Englewood Cliffs, NJ: Prentice-Hall, 1998.

[25] H. Hedayati, B. Bakkaloglu, and W. Khalil, "Closed-loop nonlinear modeling of wideband ΣΔ fractional-*N* frequency synthesizers," *IEEE Transactions on Microwave Theory and Techniques*, vol. 54, no. 10, pp. 3654–3663, 2006.

[26] H. Arora, N. Klemmer, J. C. Morizio, and P. D. Wolf, "Enhanced phase noise modeling of fractional-*N* frequency synthesizers," *IEEE Transactions on Circuits and Systems I: Regular Papers*, vol. 52, no. 2, pp. 379–395, 2005.

[27] J. S. Lee, M. S. Keel, S. I. Lim, and S. Kim, "Charge pump with perfect current matching characteristics in phase-locked loops," *Electronic Letters*, vol. 36, no. 23, pp. 1907–1908, 2000.

[28] T.-H. Lin, C.-L. Ti, and Y.-H. Liu, "Dynamic current-matching charge pump and gated-offset linearization technique for delta-sigma fractional-*N* PLLs,"

IEEE Transactions on Circuits and Systems I: Regular Papers, vol. 56, no. 5, 2009, pp. 877–885.

[29] P.-E. Su and S. Pamarti, "Mismatch shaping techniques to linearize charge pump errors in fractional-*N* PLLs," *Transactions on Circuits and Systems I: Regular Papers*, vol. 57, no. 6, June 2010, pp. 1221–1230.

[30] C. Vaucher, I. Ferencic, M. Locher, S. Sedvallson, U. Voegeli, and Z. Wang, "A family of low-power truly modular programmable dividers in standard 0.35-μm CMOS technology," *IEEE Journal of Solid-State Circuits*, vol. 35, no. 7, pp. 1039–1045, 2000.

[31] A. Swaminathan, K. J. Wang, and I. Galton, "A wide-bandwidth 2.4 GHz ISM band fractional-*N* PLL with adaptive phase noise cancellation," *IEEE Journal of Solid-State Circuits*, vol. 42, no. 12, pp. 2639–2650, 2007.

[32] C.-M. Hsu, M. Z. Straayer, and M. H. Perrott, "A low-noise wide-BW 3.6-GHz digital ΔΣ fractional-*N* frequency synthesizer with a noise-shaping time-to-digital converter and quantization noise cancellation," *IEEE Journal of Solid State Circuits*, vol. 43, no. 12, pp. 2776–2786, 2008.

[33] A. Swaminathan, A. Panigada, E. Marsy, and I. Galton, "A digital requantizer with shaped requantization noise that remains well behaved after nonlinear distortion," *IEEE Transactions on Signal Processing*, vol. 55, no. 11, pp. 5382–5394, 2007.

[34] W. W. Si, D. Weber, S. Abdollahi-Alibeik *et al.*, "A single-chip CMOS bluetooth v2.1 radio SoC," *IEEE Journal of Solid-State Circuits*, vol. 43, no. 12, pp. 2896–2904, 2008.

[35] D. R. McMahill and C. G. Sodini, "A 2.5-Mb/s GFSK 5.0-Mb/s 4-FSK automatically calibrated Σ-Δ frequency synthesizer," *IEEE Journal of Solid-State Circuits*, vol. 37, no. 1, pp. 18–26, 2002.

[36] M. Gupta and B.-S. Song, "A 1.8GHz spur-cancelled fractional-*N* frequency synthesizer wit LMS-based DAC gain calibration," *IEEE Journal of Solid-State Circuits*, vol. 41, no. 12, pp. 2842–2851, 2006.

[37] E. Temporiti, G. Albasini, I. Bietti, R. Castello, and M. Colombo, "A 700-kHz bandwidth ΣΔ fractional synthesizer with spurs compensation and linearization techniques for WCDMA applications," *IEEE Journal of Solid-State Circuits*, vol. 39, no. 9, pp. 1446–1454, 2004.

[38] S. Pamarti and I. Galton, "Phase-noise cancellation design tradeoffs in delta-sigma fractional-N PLLs," *IEEE Transactions on Circuits and Systems II: Analog and Digital Signal Processing*," vol. 50, pp. 829, 2003.

[39] C.-H. Heng and B.-S. Song, "A 1.8-GHz CMOS fractional-N frequency synthesizer with randomized multiphase VCO," *IEEE Journal of Solid State Circuits*, vol. 38, no. 6, pp. 848–854, 2003.

[40] C.-H. Park, O. Kim, and B. Kim, "A 1.8-GHz self-calibrated phase locked loop with precise I/Q matching," *IEEE Journal of Solid State Circuits*, vol. 36, no. 5, pp. 777–786, 2001.

[41] S. Pamarti and S. Delshadpour, "A spur elimination technique for phase interpolation-based fractional-*N* PLLs," *IEEE Transactions on Circuits and Systems I*, vol. 55, no. 6, 2008, pp. 1639–1647.

Chapter 3

Clock data recovery: a system perspective

Fulvio Spagna[1]

In recent years, the serial IO (SIO) links have experienced a significant increase in data rates. Recent published data show SIO links being developed with data rates in excess of 64 Gbps and many companies are working on SIOs supporting data rates in excess of 100 Gbps. This growth in data rates reflects the ever-increasing demand in system bandwidth originating in a variety of very diverse applications from data center to autonomous driving, from base stations to Internet of Things. As the data rate increases, the SIO designer is faced with numerous challenges in trying to optimize the SIO performance, power consumption, area, and reliability. Advanced signal processing techniques are, without doubt, among the tools at the designer disposal to overcome these challenges but their effectiveness can be greatly enhanced by the proper selection of clock and data recovery (CDR) architecture.

3.1 Basic communication system

The block diagram in Figure 3.1 details a baseband PAM communication system. Its components are from left to right:

- A transmitter (TX), which includes the following:
  - A line encoder, which converts the bit sequence b_k into a symbol sequence a_k that belongs to the alphabet $\{\pm 1, \pm 3, \ldots, \pm L - 1\}$. The case $L = 2$ (PAM2) is probably the most common choice for data rates below 56 Gbps. The case $L = 4$ (PAM4) represents the most common alternative for higher data rates. Most of the examples and considerations in this chapter are taken from PAM2 implementations but similar arguments can be made for other alphabet choices.
  - A filter, which, if present, may provide a certain amount of equalization.
  - An output driver, which, in general, will have some intrinsic bandwidth limitation and converts the symbol sequence into a continuous time waveform.

  In many cases, the transmitter has the ability to equalize the signal. As an example, the PCIe standard [1] assumes a three-tap filter FIR filter with

[1]INTEL, Santa Clara CA, USA

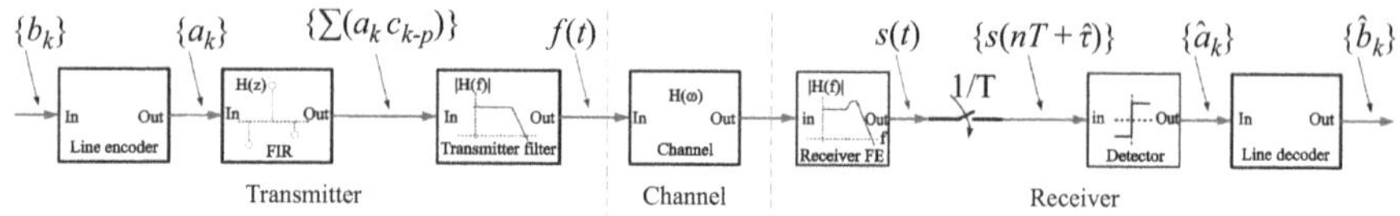

Figure 3.1 Baseband PAM communication system

well-defined ranges for the precursor and postcursor. The set of possible precursor and postcursor values defines a transmitter equalization map which the designer can take advantage of in the design of the receiver.

- A channel, supposed to be linear and time invariant, by means of which the transmitted signal reaches the receiver.

  The channel is completely characterized by its impulse response $h(t)$. As the insertion loss increases, the impulse response spreads out in time causing the transmitted bits to interfere with each other (inter symbol interference, ISI).
- A receiver (RX), which includes
  - an input front end that amplifies and equalizes the incoming signal;
  - a sampler, followed by a detector and a line decoder by means of which the original bit sequence b_k is recovered.

To facilitate interoperability between components designed by different manufacturers, the standards usually define some minimum receiver capabilities in the form, for example, of a reference receiver, which may include a continuous time linear equalizer (CTLE) as well as a certain number of decision feedback equalizer (DFE) taps.

The overall system response is completely characterized by the baseband pulse response $g(t)$ obtained by convolving the transmitter pulse response with the channel and the receiver front-end impulse responses. The signal at the input of the sampler can be written as:

$$s(t) = \sum_{k=-m}^{q} a_k g(t - kT - \tau) + n(t) \tag{3.1}$$

which contains a first term that consists of the summation of a finite number of copies of the system baseband response $g(t)$ shifted in time, each one associated with a symbol a_k, and a second term that represents a zero-mean noise component representing impairments present in the system such as thermal noise and crosstalk. Here, the variable T represents the symbol period (also referred to as unit interval or UI) while the variable τ represents the delay, in general unknown, from the transmitter to the sampler input. Since it is sufficient to know the delay value τ modulo T, in the following it will be assumed that $\{-T/2 \leq \tau \leq +T/2\}$. The indices in the summation term in (3.1) include m postcursor terms and q precursor term which capture, respectively, the ISI contributed by the last m bits and the ISI of the next q bits. In cases where the $f^{-0.5}$ loss dominates at low frequency (skin effect), the channel impulse pulse response is characterized by a long tail and the number of postcursor terms in (3.1) greatly exceeds the number of precursor terms. Nevertheless, as the data rate increases the leading edge of the pulse response may extend over several UIs and its effect cannot be ignored.

Because the symbol sequence is being transmitted at a known symbol rate, $1/T$, it is reasonable to attempt to recover the transmitted symbol sequence on the receiver side by sampling the continuous time signal $s(t)$ at the same rate $1/T$ to produce a sampled version of the received signal, s_p:

$$s_p = s(pT + \hat{\tau}) = \sum_{j=-q}^{+m} a_{p-j} g(jT + \varepsilon) + n(pT + \hat{\tau}) \tag{3.2}$$

where the variable $\hat{\tau}$ represents an estimate of the delay τ and $\varepsilon = (\tau - \hat{\tau})$ represents the estimate error.

To facilitate the recovery of the transmitted signal sequence, it is common, in high insertion loss situations, to include in the receiver equalizer a DFE. The DFE operation [2] is based on the idea that knowledge of the most recent r detected symbols can be used to cancel the ISI contributed by these symbols so that (3.2) can be rewritten as:

$$s_p = \sum_{j=-q}^{+m} a_{p-j} g(jT + \varepsilon) - \sum_{j=1}^{r} \hat{s}_{p-j} \hat{g}(jT + \varepsilon) + n(pT + \hat{\tau}) \tag{3.2a}$$

where $(r < m)$, symbol $\hat{s_p}$ represents the (detected) bit associated with the previous samples, and symbol $\hat{g}$ represents the pulse response postcursor terms estimate. Provided that both the recovered bits and the ISI estimates are correct, the previous expression reduces to:

$$s_p = \sum_{j=(r+1)}^{+m} a_{p-j} g(jT + \varepsilon) + a_p g(0 + \epsilon) + \sum_{j=-q}^{-1} a_{p-j} g(jT + \varepsilon) + n(pT + \hat{\tau}) \tag{3.2b}$$

which in addition to any precursor term that may be present contains only the postcursor ISI terms that have not been canceled by the DFE. It should be noted that, due to the discrete time nature of these expressions, (3.2a) and (3.2b) are only valid at the sampling time where the ISI cancellation is designed to take place. The DFE removes ISI without enhancing noise as in the linear equalization case; however, an incorrect past decision will increase the chance of producing long burst of errors. In cases where the SIO protocol mandates use of error-correcting codes (ECC), the burst of errors may still represent a problem if the length of the error burst exceeds the correction capabilities of the ECC. To mitigate this problem, the designer needs to carefully partition the receiver equalization between TX, CTLE, and DFE so as to limit the length of the DFE and the magnitude of the tap values.

The average value of the received signal is referred to as the *cursor amplitude* and can be written as:

$$\langle s_p \mid a_p = 1\rangle = g(0+\varepsilon) + \sum_{j\neq 0} g(jT)\langle a_{p-j}\rangle + \langle n(pT+\tau)\rangle \tag{3.3}$$

which, under the assumptions that symbols $a_n(n \neq p)$ assume with equal probability the values $\{-1, +1\}$ and that the noise term has zero mean, reduces to $g(0+\varepsilon)$.

3.2 Clock data recovery

In the absence of other criteria, it is often convenient to set the time origin so that $t = 0$ corresponds to the peak of the pulse response $g(t)$ which then coincides with the cursor position when $\varepsilon = 0$. The recovery of the original symbol sequence a_k requires estimating the system delay so that the timing error $\varepsilon = 0$ and removing, through a suitable combination of transmitter and receiver equalization, the ISI introduced by the channel so that:

$$g(jT) = \begin{cases} 0, & j \neq 0 \\ 1, & j = 0 \end{cases} \tag{3.4}$$

and:

$$s_p = a_p g(0) + n(pT+\tau) \tag{3.5}$$

To accomplish the first task, the receiver needs to have some means to evaluate, at each sampling time jT, the difference ε_j between the delay τ and its current estimate $\hat{\tau}_j$. The task of the CDR circuit is to provide such an estimate and to force the error to zero by continuously adjusting the sampling position. As an example, if the desired sampling position were to coincide with the peak of the baseband pulse response, a very reasonable estimate of the timing error could be given by the baseband pulse response derivative. In addition to providing a well-defined equilibrium point in correspondence of the desired sampling point, the sign of the pulse response derivative indicates whether the next sampling event should be advanced or retarded while its magnitude determines the sampling error magnitude. In such an example, the CDR task would be to reconstruct the pulse response derivative starting from the sampled signal.

As the determination of the timing error is a discrete time process, it is convenient to capture the discrete time nature of the CDR from the start. Figure 3.2 shows the block diagram of a digital phase-locked loop (DPLL) as used in high-speed SIO CDR applications. Here, φ_{in} represents the phase information embedded in the input data stream, φ_{vco} represents the phase of the sampling clock as generated, for example, by a digitally controlled oscillator (DCO) or a phase interpolator (PI), φ_{err} is a function of the difference between the input signal and the sampling clock phases and, finally, *phUpdate* represents the output of the loop filter, i.e. the

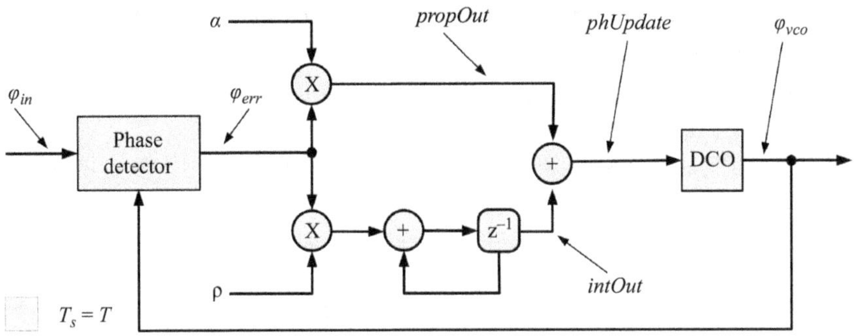

Figure 3.2 Discrete-time CDR block diagram; in this example the loop update rate ($1/T_s$) is equal to the symbol rate ($1/T$)

correction to the sampling clock phase. The kind of representation portrayed in Figure 3.2 is commonly referred to as a phase domain model and it is a useful representation to study the loop dynamics and loop performance abstracting away the interaction between the timing recovery and the other portion of the system such as the receiver equalization or the effects of non-idealities present in the actual components. Within these bounds, the phase domain model offers a unified simulation environment where, for example, the performance of a linear phase detector (PD) can directly be compared, with minimal effort, against the performance of a non-linear or quasi-linear PD. Likewise, bandwidth and acquisition characteristics can similarly be explored vis-à-vis latency, quantization and transition density effects.

The loop filter implementation shown in the block diagram is an example of a proportional-integral filter. The parameters α and ρ, which allow control of the loop dynamics, define, respectively, the (total) gain of the *proportional path* (so-called because its output, *propOut*, is *proportional* to the phase error) and the (total) gain of the *integral path* (so-called because its output, *intOut*, is the accumulated version of the phase error). In the following, the phase error will be expressed in UI so that the parameters α and ρ are dimensionless quantities.

3.2.1 Phase detector

Up to this point, the PD has been functionally defined by saying that its output is a function of the difference between the phase of the input signal and the phase of the sampling clock. This functional relationship, also referred to as the CDR timing function, can be, broadly speaking, categorized as:

- Zero crossing timing error detectors (also referred to as Alexander phase detectors) and
- Baud rate timing error detectors (also referred to as Mueller and Müller phase detector).

Additionally, depending on how the timing function information is processed, we can define the phase detector implementation type to be:

- bang-bang (or early/late)—if the phase detector generates an output based only on the *sign* of the timing function and
- linear—if the phase detector generates an output which is a linear (or approximately linear) function of the timing function.

In both the Alexander and Mueller-Muller phase detector cases, the implementation type can be *bang-bang* or *linear*.

The PD choice ultimately depends on the receiver architecture, the signal coding, and system integration considerations related to SerDes power and area. Whatever the final choice may be, it is desirable to optimize loop latency and update rates to provide the right trade-off between power efficiency and performance. If the receiver architecture is based on an analog-to-digital converter (A/D) then the availability of the received sample magnitude makes having a linear PD operating at an update rate equal to the symbol rate a possible option as this will add little incremental complexity. Because of the (quasi) linear nature of the system, the CDR loop can then be analyzed using the classical tools of control theory which can be used to optimize its performance. If, on the other side, the receiver architecture is based on a one-bit quantizer (sampler), only the sign of the received sample is available, the CDR loop is irremediably non-linear [3] and the system definition and analysis must largely rely on system simulation.

3.2.2 Closed-loop response

Under the assumption that φ_{err} is a linear function of $\varphi_{in} - \varphi_{vco}$, the closed-loop gain for the system in Figure 3.2 can be written as:

$$H_{cl}(z, \alpha, \rho) = \frac{\Phi_{vco}(z)}{\Phi_{in}(z)} = \frac{(\alpha(z-1) + \rho)}{(z-1)^2 + (\alpha(z-1) + \rho)} \tag{3.6}$$

where $z = e^{-sT_s}$ and $1/T_s$ is the rate at which the loop is updated. The transfer function has the form of a low-pass filter with one zero, at $z_0 = 1 - \rho/\alpha$, and two poles located at:

$$z_{p1,p2} = \frac{1}{2}\left(2 - \alpha \pm \alpha\sqrt{1 - \frac{4\rho}{\alpha^2}}\right) \tag{3.7}$$

which are real and distinct if the discriminant $1 - 4\rho/\alpha^2$ is positive, real and coincident if the discriminant is zero and become complex if the discriminant is negative. For stability reasons, the poles need to be located inside the unit circle. The closed-loop gain at DC ($z = 1$) is one, while the closed-loop gain at Nyquist ($z = -1$) is $(-2\alpha + \rho)/(4 - 2\alpha + \rho)$.

To a first order, parameter α sets the bandwidth of the DPLL while the ratio $4\rho/\alpha^2$ sets the peaking; in order to maintain the same peaking for different bandwidths, the integral gain needs to change quadratically with the proportional gain (Figure 3.3).

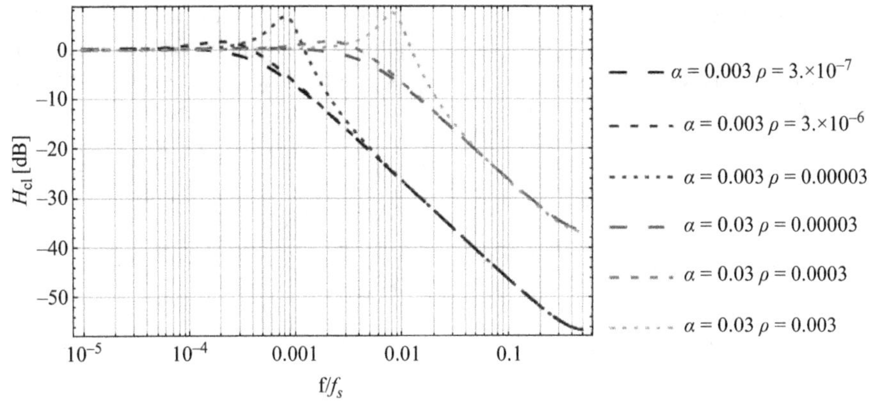

Figure 3.3 Digital PLL CDR: closed-loop response

The case where $\rho = 0$ (first-order loop) reduces to:

$$H_{\rm cl}(z,\alpha) = \frac{\Phi_{vco}(z)}{\Phi_{in}(z)} = \frac{\alpha}{z+\alpha-1} \tag{3.8}$$

which, for the case $(\alpha) = 1$, degenerates to a simple delay (a situation that has no counterpart in an analog phase locked loop) [3]. Whether or not the CDR loop should be first order or second order is largely determined by platform and architecture considerations. In cases where the PLLs on both sides of the link are locked to the same reference clock (common clock configuration), a first-order loop may, in some cases, be preferable. When this is not the case, or when the receiver sampling clock is generated by a DCO, a second-order loop is required to absorb the frequency differences between the two sides of the link.

The assumption that φ_{err} is a linear function of $\varphi_{in} - \varphi_{vco}$ represents a convenient abstraction which in practice is rarely met. Even in the so-called linear case, the PD linearity may in fact depend on factors (such as the amplitude of the input signal and the magnitude of the phase error) which may restrict the range over which the PD operation is truly linear. The case in which the phase detector output produces an output equal to the sign of the phase error (bang-bang phase detector) is however clearly non-linear. In this case, barring an attempt to linearize the phase detector operation, the designer will typically resort to simulation to gain some insight on the CDR operation.

3.2.3 Latency considerations

With the exception of the integral path accumulator and DCO accumulator, the block diagram of Figure 3.2 ignored the effect of delays. Such an approach is valid only at very low data rates as the phase detector, the integral accumulator and the clock distribution within the receiver will introduce latency, in the forms of transport delay or pipelining delays, to carry out the logical operations at speed.

The effects that these delays have on the loop operation may be discussed considering a case in which the *phase error* incurs a delay of l clock cycles as it is processed through the phase detector, the loop filter (proportional path), the *Vco* accumulator, the clock tree, etc. and assuming, for completeness, an additional processing delay of q clock cycles in the loop filter integral path. Under these conditions, the (linear) system closed-loop response can be written as:

$$H_{\text{cl}}(z,\alpha,\rho,l,q) = \frac{(\alpha z^q(z-1)+\rho)}{z^{l+q}(z-1)^2+(\alpha z^q(z-1)+\rho)} \tag{3.9}$$

and, for the first-order case:

$$H_{\text{cl}}(z,\alpha,0,l,0) = \frac{\alpha}{z^l(z-1)+\alpha} \tag{3.10}$$

The delays l and q introduce a number of zeros and poles in the CDR transfer function which affect the stability of the system limiting the maximum possible gain. As shown in Figure 3.4, the latency impact becomes more evident as the target bandwidth becomes a higher fraction of the loop update rate f_s. At the same time, the additional delay q turns out to be less important that delay l (the output of the integral path is, after all, the accumulation of all the previous phase error estimates) suggesting that in designing the CDR optimization of the term l should take priority over q.

Since the timing error estimation is predicated on the detection of the incoming symbols, there are cases where the detection process itself introduces a significant delay which is then reflected in the latency of the timing error estimation. In these cases, the designer may elect, for the purpose of generating a timing error estimate,

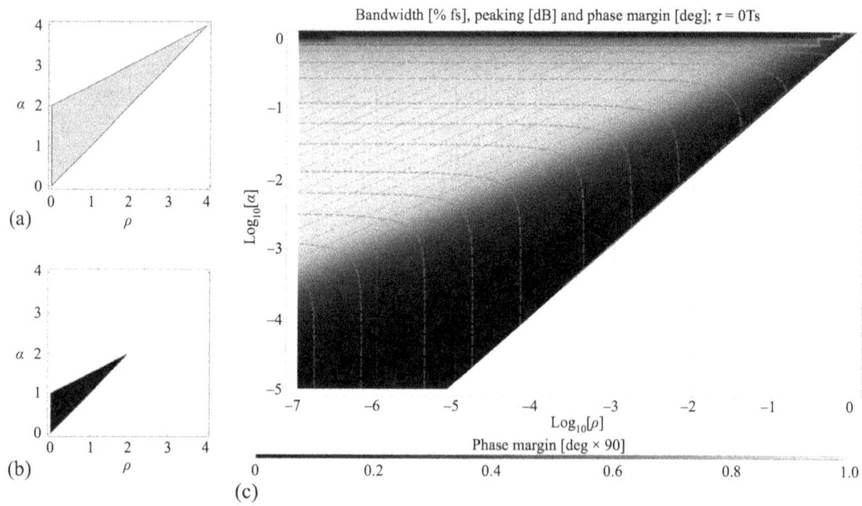

Figure 3.4 Delay impact on DPLL transfer function: (a) stability region with no delay; (b) stability region reduction due to non zero delay; (c) relationship between loop bandwidht, peaking and phase margin for the zero delay case

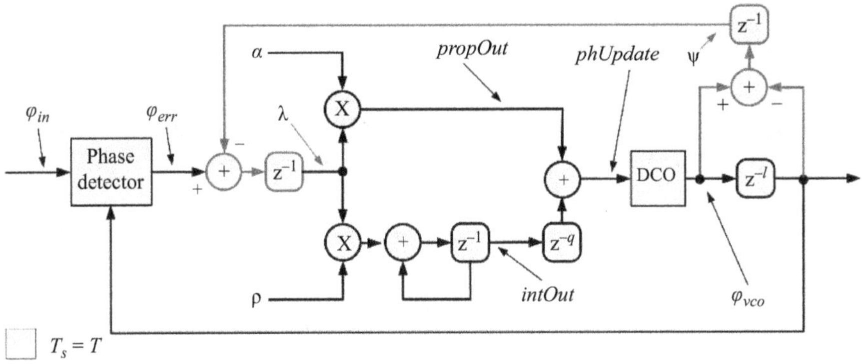

Figure 3.5 Latency compensated digital PLL CDR

to process the samples through a different path with shorter latency perhaps at the cost of a noisier estimate. Alternatively, if the PD is linear, the problem can be addressed by precomputing the effects that past updates, not yet absorbed by the sampling clock phase because of latency, have on the current phase error estimation. A control loop based on this idea is known as a Smith predictor and an example applied to DPLL design is shown in Figure 3.5 which is characterized by the transfer function:

$$H_{cl}(z,\alpha,\rho) = z^{-l}\frac{(\alpha z^q(z-1)+\rho)}{z^q\left((z-1)^2+(\alpha(z-1)+\rho)\right)} \tag{3.11}$$

indicating that, apart from a delay term which can be factored out, the effect of the *l-clock cycle delay* has no bearing on the loop dynamics. In the end, the effectiveness of this approach is limited by the accuracy of the analog components in the loop (e.g. the phase detector linearity) but, within these limitations, the latency effects can be largely controlled if the bulk of the delay is digital in nature. A more complete discussion which includes the effects of the compensation network delay may be found in [4].

3.2.4 Update rate

Setting the sampling period T_s to be equal to the symbol period T makes it possible (at least in principle) to design a CDR system with very wide bandwidth. In most cases, the logic complexity in the phase detector implementation and in the integral path accumulator are such that trying to operate the loop with an update rate equal to the symbol rate may not be possible or comes at a significant cost in power, area and latency. Because the typical bandwidth required by most SIO standards is a (small) fraction of the symbol rate, the loop update rate can become another parameter by which the designer can optimize the system.

Figure 3.6 shows a possible alternative to the block diagram in Figure 3.2. Here, the phase detector early/late outputs are accumulated and down-sampled with

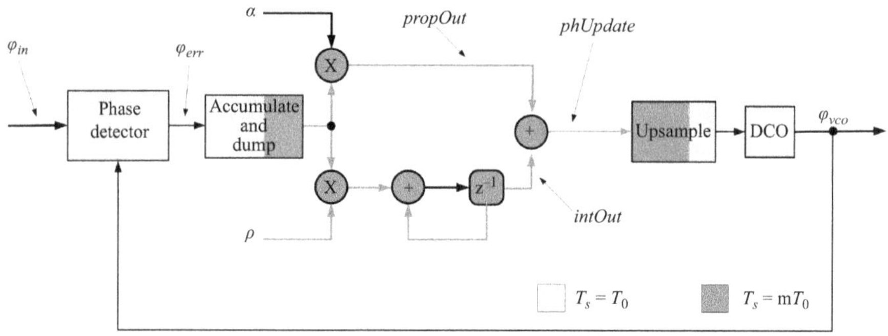

Figure 3.6 Multi-rate CDR

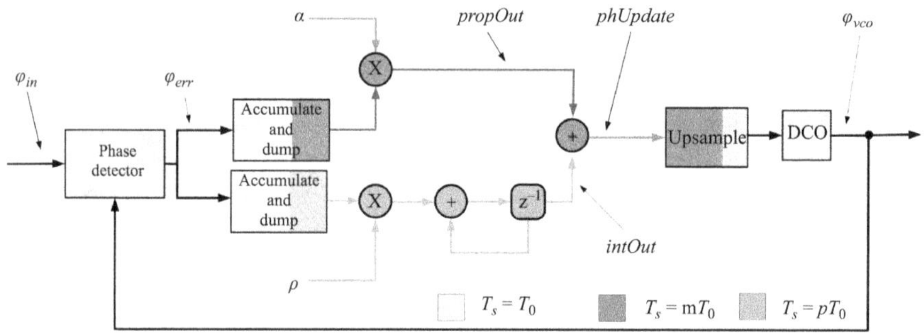

Figure 3.7 Multi-rate CDR: distinct update rates for the proportional and integral paths

an n:1 ratio so that the loop digital core (i.e. the filter) is now operating at the lower rate $1/(nT)$ and the output of the loop filter can be thought of as if processed by a zero-hold sampler in front of the DCO. In the case of a linear PD, the accumulation of n distinct phase error estimates may be processed to produce an average phase error, in the case of a bang-bang phase detector, the accumulation of n distinct early/late may be processed by a sequential filter (e.g. majority vote filter). One obvious consequence of this type of architecture is that the time in between corrections is now n-times longer so that the phase error can grow n times larger than in the original case. In addition, because the desired bandwidth is now a larger fraction of the loop update rate which, as indicated in the previous section, makes it more sensitive to pipelining delays, this arrangement is effective only to the extent that the overall latency remains a low fraction of the update rate.

As noted in the previous section, the latencies of the proportional and integral paths have a very different impact on the system stability. This observation can be exploited to further refine the design by maximizing the update rate on the proportional path while reducing the update rate in the integral path and minimizing the power and high-speed design complexity associated with this path (Figure 3.7).

In doing so, the update rate and latency on the integral path become less critical and can generally be easily handled by standard logic synthesis while the phase detector and proportional path remain custom logic.

The analysis of a multi-rate system can become quite involved. The reader is referred to [3,5] for a more complete treatment of the topic.

3.2.5 Jitter tolerance

The designer needs to have metrics by which to compare the different tradeoffs that arise in architecting the CDR loop. One such metric is the ability of the sampling clock to track the phase of the input signal. This can be quantified by writing an expression for the normalized phase error:

$$\frac{\Phi_{\text{in}}(z) - \Phi_{vco}(z)}{\Phi_{in}(z)} = \frac{z^{l+q}(z-1)^2}{z^{l+q}(z-1)^2 + z^q\alpha(z-1) + \rho} \tag{3.12}$$

For frequencies within the CDR bandwidth, $\Phi_{vco}(z) \sim \Phi_{in}(z)$ and the normalized error approaches zero. Conversely, for frequencies outside of the CDR bandwidth,

$\Phi_{vco}(z) \rightarrow 0$ and the normalized error goes to one.

The reciprocal of the normalized phase error defines the maximum input phase magnitude that the CDR can track for a certain phase error. Since the ability to track the input phase has sense only in a bit error-free context, the maximum tolerable error is $\pm 0.5\ UI = 1.0\ UI_{pp}$, which puts an upper limit to the input phase magnitude that the system can handle [6]. Excess latency or peaking in the transfer function translates into a loss of jitter tolerance margin and should be avoided (Figure 3.8).

Because the jitter tolerance is defined without assuming any specific PD or CDR architecture, this metric is adopted by many SIO standards as one of main criteria by which to define the receiver performance. To this end, most SIO standards specify a jitter tolerance mask which defines the minimum magnitude of a sinusoidal phase modulation which must be tolerated at a specified BER (typically 1E − 12). Because this mask is to be measured in the presence of impairments of

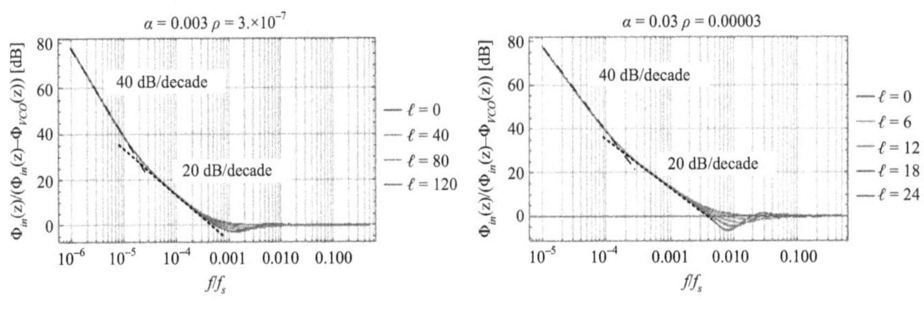

Figure 3.8 Jitter tolerance profile

various type (DCO noise, cross-talk, receiver noise, residual equalization error, pulse response asymmetries, and spread spectrum modulation, just to name a few) the minimum high-frequency jitter tolerance required is well below the ideal $\pm 0.5UI_{pk}$ (a minimum value of $0.1UI_{pp}$ being very typical).

Because the jitter tolerance is such a useful metric to establish the receiver margin, it is not uncommon to provide an on-die jitter tolerance capability that can be used *in situ* to assess the link robustness in the presence of jitter. This functionality, sometime referred to as jitter injection margining (JIM), needs to provide modulation frequencies well in excess of the CDR bandwidth and phase modulation depths of several UIs to properly characterize the high-frequency and low-frequency tracking capabilities.

3.3 Timing function selection

The elimination of all the ISI terms (3.4) represents a desirable objective but one that, except in very trivial cases, it is not one practically realizable in actual implementation. One criterion for reliable detection is to require that the cursor amplitude be larger than the residual equalization error (peak distortion minimization):

$$|g(0)| - \sum_{j\neq 0} \left|a_{p-j} g(jT)\right| > 0 \tag{3.13}$$

The following sections show the timing functions of two types of phase detectors commonly used in high-speed SIO under the assumption of incomplete ISI cancellation.

3.3.1 Alexander phase detector

The Alexander phase detector (or zero crossing time detector) extracts the timing information from the zero crossings of the signal $s(t)$. Because a waveform zero crossings is associated with a data transition, it is implicit in the operation of the Alexander phase detector the fact that timing information can only be extracted in the presence of a data transitions and that the signal needs to be sampled twice per UI, once at the center of the symbol interval (to recover that transmitted data) and once at some later time (edge sample).

Figure 3.9 illustrates the idea upon which the phase detector operation is based. In the presence of a transition, that is when two consecutive data samples (dark dots) have opposite values, the value of the edge samples (light dots) determines whether or not the sampling is aligned with the incoming data. In the first waveform, the edge sample, in the presence of transitions assumes a "zero" value, meaning that, on the average, it will appear with equal probability as $+1$ or a -1 to the receiver sampler indicating proper alignment. In the middle waveform, the edge sample sign is always equal to the sign of the first data sample, indicating that

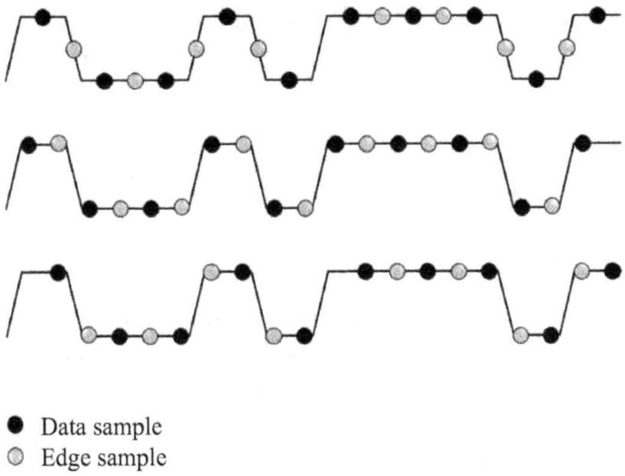

Figure 3.9 Alexander phase detector operation

sampling clock is *early* with respect to the incoming signal. In the bottom waveform, the edge sample sign is always equal to the sign of the second data sample, indicating that sampling clock is *late* with respect to the incoming signal.

The signal sampled by the data clock can be written as:

$$s_p(t) = \underbrace{\sum_{j=1}^{m} a_{p-j} g(jT + t_0)}_{postcursor} + \underbrace{a_p g(0 + t_0)}_{cursor} + \underbrace{\sum_{j=-q}^{-1} a_{p-j} g(jT + t_0)}_{precursor} \tag{3.14}$$

which identifies the ISI components related to the pre- and postcursor portions of the pulse response, $g(t)$.

An equivalent expression for the edge samples can be written as:

$$\begin{aligned} s_{p+\frac{1}{2}} = {} & \underbrace{\sum_{j=1}^{m} a_{p-j} g\left(\left(j + \frac{1}{2}\right)T + t_0\right)}_{postcursor} \pm \underbrace{g\left(t_0 + \frac{1}{2}T\right)}_{current\ bit} \mp \underbrace{g\left(t_0 - \frac{1}{2}T\right)}_{next\ bit} \\ & + \underbrace{\sum_{j=-q}^{-2} a_{p-j} g\left(\left(j + \frac{1}{2}\right)T + t_0\right)}_{precursor} \end{aligned} \tag{3.15}$$

where the signs of the *current bit* and the *next bit* have been chosen to indicate a data transition, i.e. $a_p a_{p+1} = -1$. Since, by definition, the *edge sample* should be zero at a zero crossing the previous expression simplifies to:

$$0 = g\left(t_0 + \frac{1}{2}T\right) - g\left(t_0 - \frac{1}{2}T\right) + \underbrace{\sum_{j \neq p, j \neq p+1} a_{p-j} g((j + \epsilon)T + t_0)}_{ISI\ terms} \tag{3.16}$$

and, under the usual assumption that each one of the other bits has equal probability to assume -1 or $+1$ value, this reduces to:

$$g\left(t_0 + \frac{1}{2}T\right) = g\left(t_0 - \frac{1}{2}T\right) \tag{3.17}$$

which indicates that the sampling point t_0 (or timing reference point) is identified by the condition that pulse response value at $t_0 + 0.5T$ equals the pulse response value at $t_0 - 0.5T$. If the pulse response $g(t)$ did exhibit even symmetry, this would effectively be stating that the sampling occurs at the peak of the pulse response. The condition of even symmetry, however, is not generally met, especially at high data rate and/or with high insertion loss interconnect where, as indicated earlier, the pulse response is asymmetrically spread out over many UIs. However, by judicious choice of transmitter and receiver equalization, the pulse response may approximate even symmetry in which case the sampling point turns out to be very close to the peak of the pulse response.

Figure 3.10(a) shows the Alexander timing function for three different equalized pulse responses associated with different levels of ISI. Because of equalization, the pulse responses are reasonably symmetrical and the corresponding timing functions cross zero in correspondence of points at or very nearly at the peak of the pulse response. The slope of the timing function around zero is of course a measure of the phase detector gain while the sign indicates whether the sampling phase should be advanced or retarded. The timing function depicted in this figure has been generated for an isolated pulse response so in a sense what is plotted should be considered the timing function average value (which is consistent with the assumption in (3.16) that the ISI term has zero average).

Because the Alexander timing error requires data and edge samples, careful consideration must be given to provide the necessary equalization of both data and

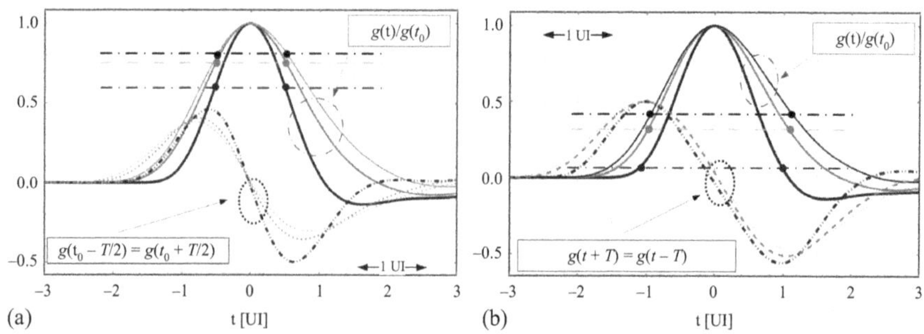

Figure 3.10 Timing functions for different pulse responses exhibiting different degrees of insertion loss: (a) Alexander phase detector and (b) Mueller-Müller phase detector

edge samples. For a bang-bang architecture based on a 1-bit quantizer, some possibilities worth noting are:

- No DFE, all the equalization is provided by a CTLE common to both data and edge samplers [7,8].
- Shared path for edge and data samplers. A DFE provides equalization for the edge samples and the data samples equalization is provided by the CTLE [9].
- Split paths for the edge and data samplers. The CTLE, which is common to both paths, is tuned to provide optimum equalization for the edge samples, and a DFE corrects only the data samples.
- Split paths for the edge and data samplers. The CTLE is common to both paths and separate DFE engines provide the final equalization for the edge and data samplers.

The first option is, from a hardware standpoint, the most economical but it is also the most limiting one being, in general, appropriate only at low data rates. The challenge, from a design point of view, lies in building a filter which provides equalization for data and edge samples, has enough adjustability to track out environmental changes without disrupting the data recovery, and can be tuned to provide a different amount of equalization and gain depending on the nature of the interconnect.

The last option is, on the other side, the most expensive from a hardware point of view, as it requires independent DFEs for the data and edge interleaves. This is also the most versatile as samples and edges are now truly independent and the DFE ability to continuous adapt can be used to track environmental changes simplifying the CTLE requirements.

The option next to the last trades off analog complexity in the CTLE for a more power efficient design but its effectiveness depends on the CTLE ability effectively remove the ISI from the edge samples. The remaining two options are more efficient than the first two but introduce an interaction between the data (edge) DFE feedback and edge (data) samplers that is difficult to control at high data rates.

While the presence of an edge sampler adds to the equalization complexity, a nice property of the Alexander PD is that the timing estimation relies only on the data and edge samples. The gain and DFE loops, which rely on error signals generated by comparing the analog sample against a predefined target voltage (the cursor $g(0)$), do not participate directly in the phase error estimate and, because they need to track environmental induced changes, can be refreshed at a slower rate which can translate into beneficial power savings.

3.3.2 Mueller-Müller phase detector

As in the Alexander Phase detector case, the Mueller-Müller phase detector extracts the timing information directly from the samples of the signal $s(t)$ but it does so by sampling the incoming signal once per UI. As a consequence, the receiver requires a *single clock phase* to generate data samples and timing information. This property is particularly convenient in receiver architectures where the input signal is sampled by an A/D (as the cost of an edge sampling clock would be prohibitive).

As indicated earlier, the Alexander phase detector timing function represents a discrete time *approximation* of the pulse response derivative based on samples of the signal $s(t)$ spaced 1 UI apart. As it turns out, it is possible to generate, through an appropriate linear combination of samples s_n, an approximation of the pulse response derivative based on samples spaced 2 UI apart [10]:

$$f_n(t) = s_n(t)\hat{a}_{n-1} - s_{n-1}(t)\hat{a}_n \tag{3.18}$$

where $\hat{a}_n$ represents the sampler output (decision). Under the assumptions that the decisions are correct, $\hat{a}_n \equiv a_n$ (3.18) reduces to

$$f_n(t) = s_n\, a_{n-1} - s_{n-1}\, a_n \tag{3.19}$$

which after some manipulations reduces to:

$$f(t) = \underbrace{g(t+T)}_{postcursor} - \underbrace{g(t-T)}_{precursor} \tag{3.20}$$

This is referred to as a Mueller-Müller type A timing function. In a bang-bang implementation where the analog sample is not available, s_n is replaced by the *sign*, η, of the difference $s_n - g(0)$:

$$f_n(t) = \eta_n a_{n-1} - \eta_{n-1} a_n \tag{3.21}$$

Figure 3.10(b) shows the timing function plotted for the same three pulse responses used in the previous example; note that even symmetry on a 2UI scale is a little more difficult to achieve so the error in the zero crossing position relative to the pulse response peak is a bit larger than in the Alexander timing function case. In addition, because the slope of the timing function is adversely impacted by a soft slope in the pulse response leading edge, it is necessary to provide sufficient pre-emphasis in the transmitter so that the CDR operation is not limited by it. Once again, the timing function is computed directly from the isolated pulse response and represents the timing function average value.

Because the Mueller-Muller timing function exploits amplitude differences created by the ISI to generate an estimate of the timing error, the case of a low ISI (or excess bandwidth) system may become problematic because the only place where significant amplitude differences are observable is near a data transition (Figure 3.11). This is not a desirable situation as it implies that the sampling point will also be at the edge of the UI. A possible way to get around this problem is to add a controlled amount of ISI in the receiver. An example of a solution of this kind [11] is to integrate the signal before sampling it; the integration operation

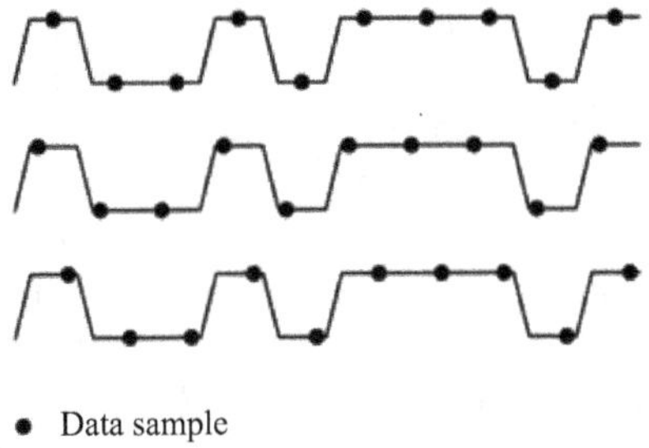

Figure 3.11 Ambiguity in the phase detector response in a low ISI situation

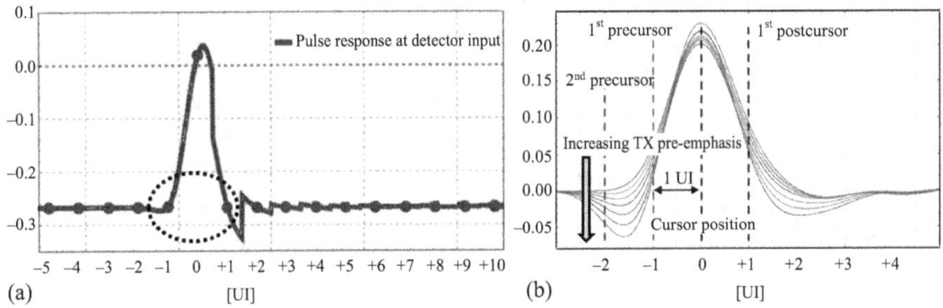

Figure 3.12 (a) Zero first precursor criteria and (b) transmitter pre-emphasis impact on pulse response

introduces additional ISI at Nyquist but it also introduces a relationship between the sample amplitude and the sampling phase which can be used to estimate the timing error. As a side benefit, this approach provides a more power efficient solution than one based on a non-integrating topology since the capacitance on the sampler node is exploited for the purpose of integrating the signal.

Earlier (3.2b) it was pointed out that a common technique to reduce the impact of the ISI is to add a DFE to the receiver. The interaction of the Mueller-Muller timing function with the zero-forcing action of the DFE causes a shift in the sampling position defined by (3.21). This modification to the original timing function is sometimes referred to as the zero first precursor criteria because the sampling point is now defined by:

$$g(t - T) = 0 \tag{3.22}$$

under the assumption that the first postcursor ISI has been completely canceled out [Figure 3.12(a)]. As a result of this new timing function, the sampling point generally shifts to the left of the baseband pulse response peak and this, in turn, causes the DFE tap correction to increase (stressing the linearity of the analog front end) and the cursor value to decrease (increasing the analog front-end gain requirements). For this reason, the ability to maintain the sampling position close to the peak of the baseband pulse response is predicated on the ability to create a negative going overshoot on the leading edge of the pulse response approximately 1 UI before the pulse response peak. This could be done, for example, by increasing the transmitter pre-emphasis but such a strategy is limited by the fact that the transmitter cursor amplitude decreases (which requires more gain in the receiver) and a new ISI term, associated with a second precursor [Figure 3.12(b)], appears.

In observing the equalized pulse response when a DFE zero forces the first postcursor ISI to zero [as in Figure 3.12(a)], one cannot fail to observe that the pulse response around the first postcursor has a very nice shape that could actually meet the requirements previously outlined for a timing function. This is, in a nutshell, the idea for the type-B Mueller-Müller timing function (zero forcing of g_1). This approach does introduce a potential conflict if both PD and DFE try to control the value of the first postcursor correction. A later section will present one possible way to get around this problem.

The two versions of this Baud rate PD work best in different cases. The type A algorithm works best when the pulse response exhibits even symmetry as in the case of lower ISI systems; type B, on the other hand, does not rely on the symmetry of the pulse response and so it is more suitable to situations where the ISI is higher and the pulse distortion is more severe. In either case, the interaction between the timing function and the overall equalization solution needs to be understood: in the presence of the DFE, type A requires a zero crossing on the precursor side with a sharply defined slope, while type B requires a mechanism by which the PD and equalization of the first tap precursor can operate jointly.

Finally, in both cases, the timing estimate processed by a Baud rate PD will have no useful timing information in the presence of a constant 1T clock pattern. In general, this is not a limiting factor as, in most cases, the signal is scrambled but may become a concern in some cases where the protocol relies on a clock pattern for in-band signaling (e.g. skip ordered sets in PCIe) since these cannot be eliminated by scrambling.

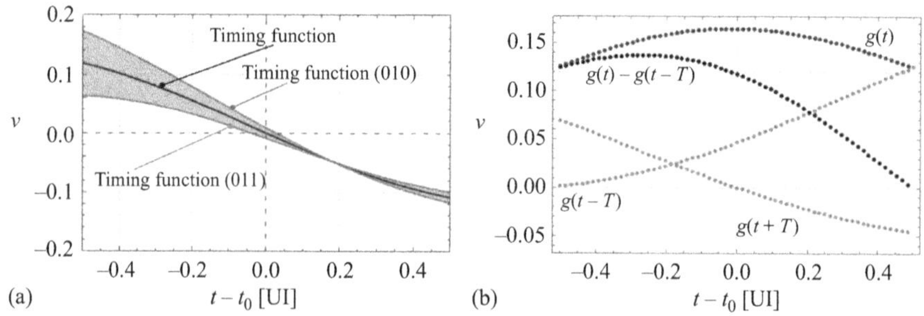

Figure 3.13 Alexander PD: (a) average timing function vs 010 and 110 timing functions and (b) pulse response decomposition

3.3.3 Timing function analysis

In describing the timing functions for the Alexander and Baud rate PDs, it was mentioned that the equations represent the average value of the timing functions. If the ISI is not completely canceled, the interaction with neighboring pulses may modify these functions.

The average timing function can be compared against the timing function associated with a specific repeating pattern, for example, a 010 or 110 sequence. In the example shown below, Figure 3.13(a), the area shaded in green represents a measure of the ISI impact due to these two patterns; the 110 timing function crosses zero later than the average timing function and the 010 crosses earlier suggesting that, in the presence of long repeating patterns, the sampling position may shift away from its average value.

In addition, the pulse response can be decomposed in its constituent parts [Figure 3.13(b)] to help identify how the various components contribute to the overall timing function.

3.4 Eye diagram

One of the classical ways [12] to assess the receiver performance is based on the concept of the eye diagram (also referred to as the eye pattern) as shown in Figure 3.14. The *eye diagram* of a signal is built by folding the signal on a time segment of length equal to the symbol period. While this operation is very feasible in a simulator environment, it is considerably more difficult to do in hardware.

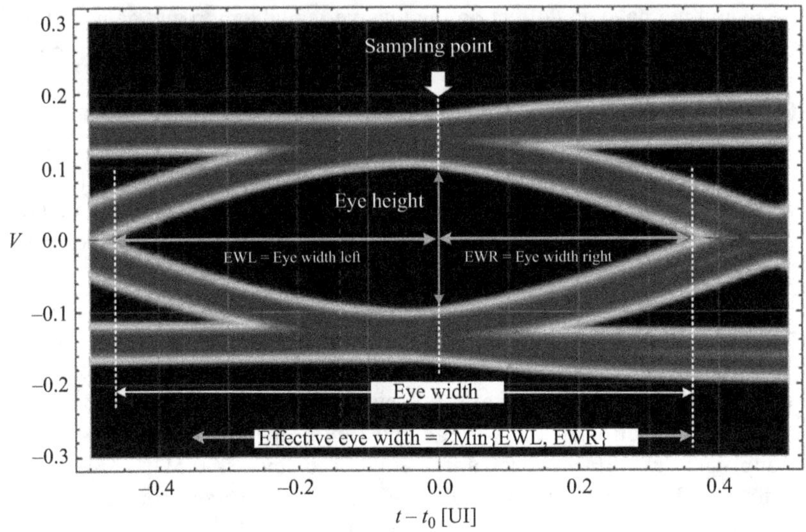

Figure 3.14 Statistical eye diagram

3.4.1 Statistical eye construction

A discussion on the means by which an eye diagram can be constructed in hardware should probably start by agreeing on what information should be conveyed by the eye diagram. There are perhaps different ways to answer this question but the one of interest here is related to the receiver ability to tolerate noise (vertical margin) and jitter (horizontal margin) (Figure 3.14). It is clear that to satisfy this requirement the designer needs to provide two additional degrees of freedom to independently change the horizontal and vertical sampling positions. This can be accomplished by having an auxiliary sampler (or samplers), connected to the input of the data sampler, with an independent threshold control and a mechanism to advance/retard its clock with respect to the nominal sampling position. This approach sometimes referred to as non-destructive eye diagram generation extracts the desired information without affecting the operation of the normal data path but comes at a considerable hardware cost as it is embedded in the high-speed portion of the signal path. Destructive methods, in contrast, are typically less costly from a hardware point of view as they rely on manipulating the threshold of the existing data sampler and/or applying a defined amount of phase modulation to the recovered clock at a frequency higher than the CDR (using, for example, the JIM hardware referred to in a previous section) to detect errors in the link.

The vertical margin at the nominal sampling time can be measured with an auxiliary sampler or with the data sampler as discussed above but a direct measurement of the horizontal margin is more ambiguous. If the horizontal margin is defined as the receiver ability to tolerate high-frequency phase modulation, it can be argued that in these conditions the only thing that should change is the sampling position. In other words, because the modulation is faster than what the CDR or gain and equalization loops can track, it should be assumed that sampling position change would not have any effect on the receiver equalization and gain. In light of these considerations, the arrangement in which the auxiliary sampler clock is advanced/retarded with respect to the nominal sampling position would seem to satisfy these requirements as it observes the signal at the input of the sampler without disturbing either the timing, gain or equalization loops. One important exception is the case in which the receiver, in addition to a CTLE, includes a DFE. In this case, as seen in (3.2b), the auxiliary sampler sees a combination of a (valid) forward signal and a DFE feedback signal which, however, is only meant to be valid at the sampling point.

If the purpose of the test were to examine the setup and hold behavior of the sampler, the real-time sampler input signal showing the interaction of the DFE feedback signal with the forward signal would be the right signal to observe. Extracting margin information from such signal appears more ambiguous. In this case, a better indication of the voltage and timing margin is obtained by artificially applying the DFE feedback at the edge of the eye (+ 0.5UI). This removes the distortion that a time-varying DFE feedback signal would have on the eye diagram and allows one to compare the effects that high frequency timing jitter would have on the receiver. Unfortunately, this is not something that can be directly done in hardware so in the end this remains an approach that can only be used in simulation. Figure 3.15 compares the results obtained in a case where

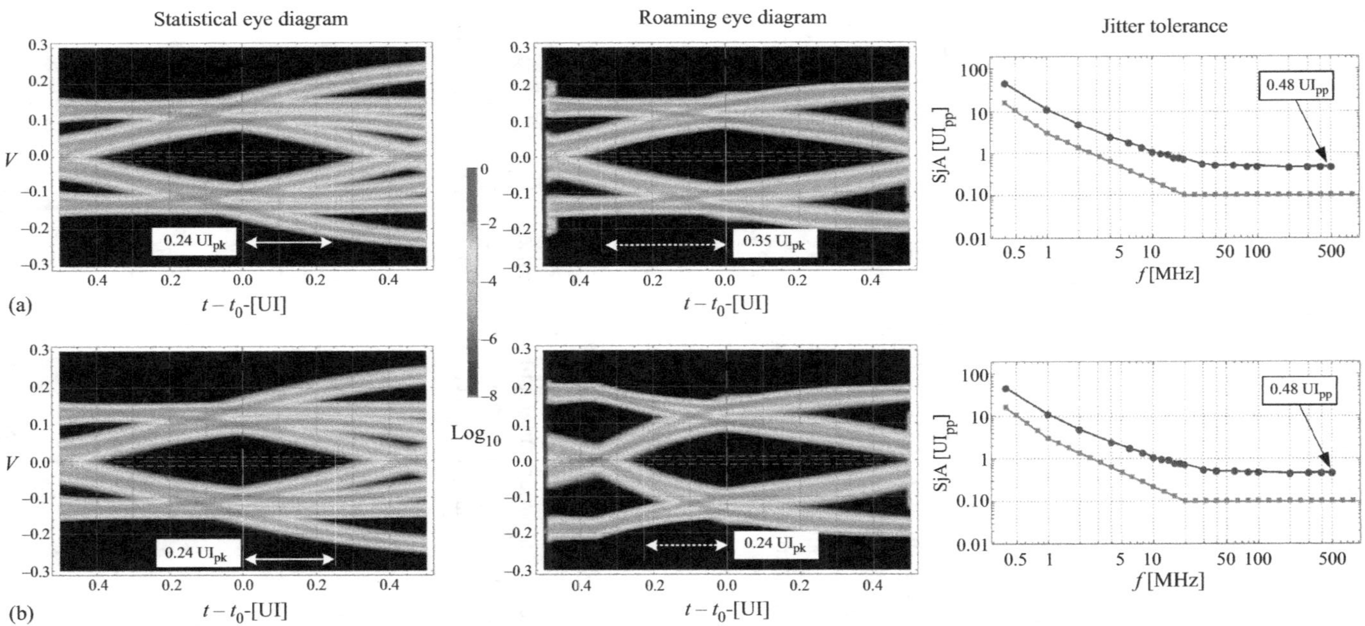

Figure 3.15 Statistical eye, roaming eye, and jitter tolerance plot: (a) fast DFE tap1 feedback path and (b) slow DFE tap 1 feedback path

the DFE feedback is fast (i.e. a small fraction of a UI) and when the DFE feedback is slow (i.e. a large fraction of a UI). The eye diagram on the left (statistical eye) has the DFE feedback kept constant and applied at the edge of the eye, the eye diagram in the middle (roaming eye) is what is observed in real time at the input of the sampler. The jitter tolerance results show how the horizontal margin is not well represented by the case where the application of dfe feedback is not placed at the edges of the eye.

3.4.2 Tri-bit eye diagram

While an eye diagram can be built from a data set of any length, the ability to predict the system bit error rate (BER) requires the generation of a *statistical eye diagram* capturing a large amount of data. A fair amount of information is, however, obtainable with simple manipulations of the system pulse response $g(t)$ and comparatively little computational effort to generate an eye diagram based on all possible three-bit sequences (tri-bit eye diagram).

The tri-bit eye diagram is particularly useful in establishing, by visual inspection, the interaction between the precursor, cursor, and postcursor components of the baseband pulse response. In Figure 3.16, the sum of the precursor and postcursor terms is clearly visible as the distance between the 111 trajectory and the 010 trajectory. Likewise, the difference between the postcursor and the precursor is visible as the distance between the 011 and 110 trajectories. This simple example shows how we can conclude, by visual inspection, that not only we have a significant residual ISI term due to the un-equalized precursor and postcursor but also that these terms are approximately equal.

To be clear, the tri-bit eye diagram does not contain the same information as a statistical eye but is a useful tool to quickly assess the structure of the statistical eye diagram associated with a particular pulse response and sampling point choice. Because the tri-bit eye is generated from a pulse response analysis and represents an idealized version of what we expect the eye to looks like, superimposing the

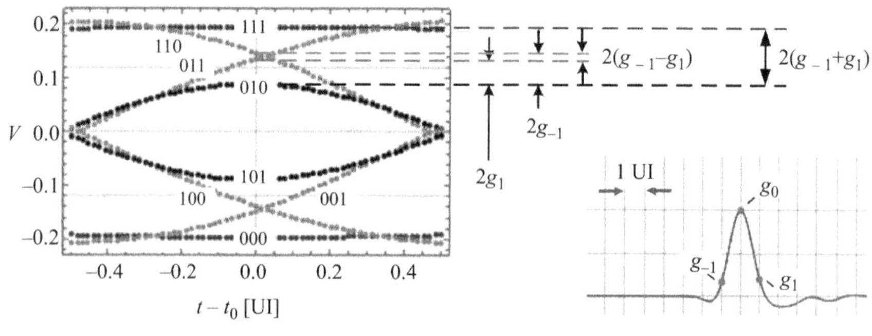

Figure 3.16 Tri-bit eye diagram

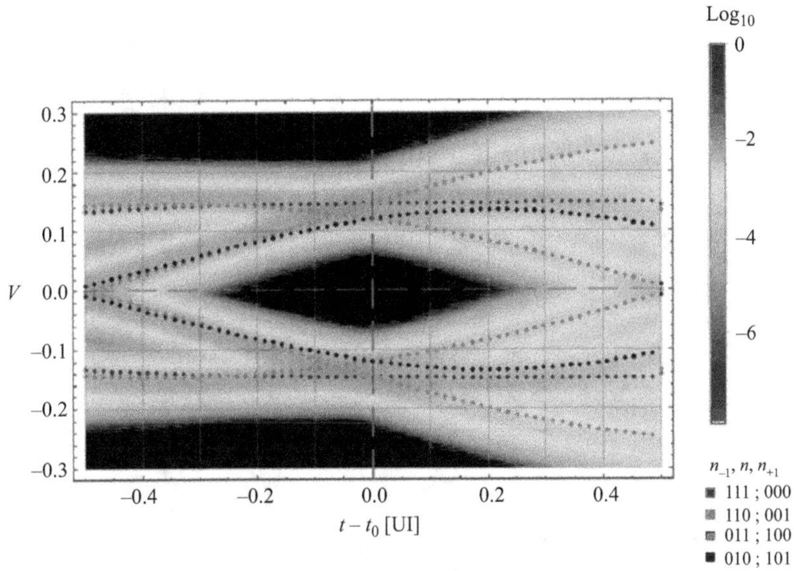

Figure 3.17 The tri-bit eye and its relation to the statistical eye

tri-bit eye on the statistical eye can show differences which may point out effects as gain shortcoming, excess jitter or asymmetries caused by non-idealities in the circuit. In that sense, the tri-bit eye becomes a diagnostic tool which the designer can use to gauge to what extent the actual performance differs from the design target. The example in Figure 3.17 shows a reasonable match between the tri-bit eye and the statistical eye indicating that to a first-order sampling position, gain and equalization are working as expected.

3.5 An optimization example

In addition to the jitter tolerance, the high-speed SIO performance can be characterized by means of margin maps [13]. A margin map visualizes relevant system metrics as a function of the possible transmitter equalization settings. Since robust link operation is synonymous with a larger solution space (meaning good bit error rate (BER) over a large portion of the equalization map), the margin map is an effective visualization technique to quantify the robustness of the serial IO design.

As an example, Figure 3.18 shows, for a low loss channel, the eye height (vertical margining) and effective eye width metrics reported for the TX equalization region defined by the PCIe Express standard [1]. The margin map on the left shows that the eye height is larger than 140 mV_{pp} in the region framed by

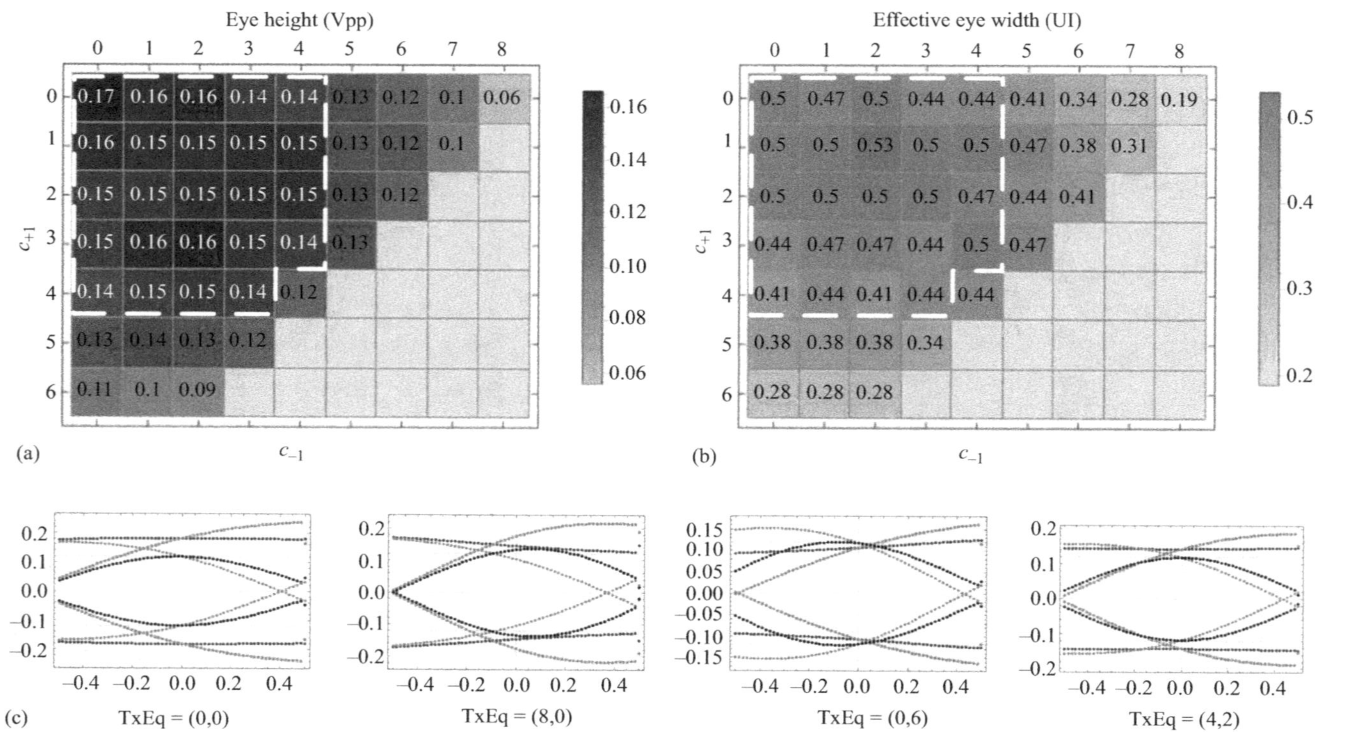

Figure 3.18 Margin maps for a representative low ISI scenario; (a) eye height; (b) effective eye width; and (c) 3-bit eye diagrams for selected TX equalization values

the white dashed line and decreases outside in the peripheral region of the map. On the right map, the effective eye width is greater than 40% of a UI in the same region and degrades moving away from the central region of the map. Examination of the tri-bit eye diagrams at different points of the margin map reveals how these metrics are affected by the TX equalization. In the upper left corner, which corresponds to no equalization on the transmitter side (TxEq = {0,0}), the tri-bit eye indicates the presence of a large precursor component and some horizontal axis encroachment on the right side of the eye. As the transmitter de-emphasis increases, moving to the right on the x axis (TxEq = {8,0}), the eye encroachment on the right side increases as the amount of TX significantly over-equalize the link. Conversely, moving downward along the y-axis, i.e. increasing the transmitter pre-emphasis (TxEq = {0,6}), it is the left side of the tri-bit eye that is affected as the pulse response leading edge becomes sharper in response to the increasing transmitter pre-emphasis. Increasing pre-emphasis and de-emphasis by similar amounts (TxEq = {4,2}), help maintaining a reasonable symmetry in the pulse response a fact which is clearly reflected in the resulting more symmetrical tri-bit eye diagram.

As the interconnect length increases, a significant reduction in the solution space and in the vertical margin can be observed [Figure 3.19(a)] clearly showing the limits of a CTLE-only based equalization. Addition of an 8-taps DFE [Figure 3.19(b)] improves the best eye height and eye width but the solution space remains smaller than in the low ISI case. To identify the factors that limit the design performance it is useful to look at the residual equalization error for one of the low margin points at the periphery of the map [Figure 3.19(c)]. Here, the dashed line represents the cursor position, the terms to its left represent the precursor ISI and the terms to its right represent the postcursor ISI not corrected by the DFE. While the postcursor ISI components are a small fraction of the cursor amplitude, the precursor terms, in particular, the first precursor term (30% of the cursor in this case) stands out as the dominant ISI error term. At the same time, the margin map shows that the transmitter pre-emphasis not really effective in reducing the effect of the precursor ISI. As indicated earlier, the reason is that while an increase in the TX pre-emphasis reduces the first precursor ISI term, it does so by increasing the ISI term associated with the second precursor [Figure 3.12(b)].

The tri-bit eye diagrams [Figure 3.19(d)] offer a clue on how to solve the first precursor ISI problem. Since the precursor magnitude is related to the difference between the 011 and 010 trajectories, a reduction in the precursor ISI magnitude could be achieved by moving the sampling point to the left of its nominal position (as determined by the Alexander or Baud Rate PD). Applying such timing offset delta would not only greatly reduce the impact of the precursor ISI but also benefit the eye symmetry which was originally limited by the 110/001 trajectories, and, as a result of the sampling point shift to the right, would now become more open and symmetric.

A possible technique for controlling the amount by which the sampling point needs to shift can be glimpsed by looking at the tri-bit eye diagram in conjunction with the pulse response components in an interval centered around the sampling

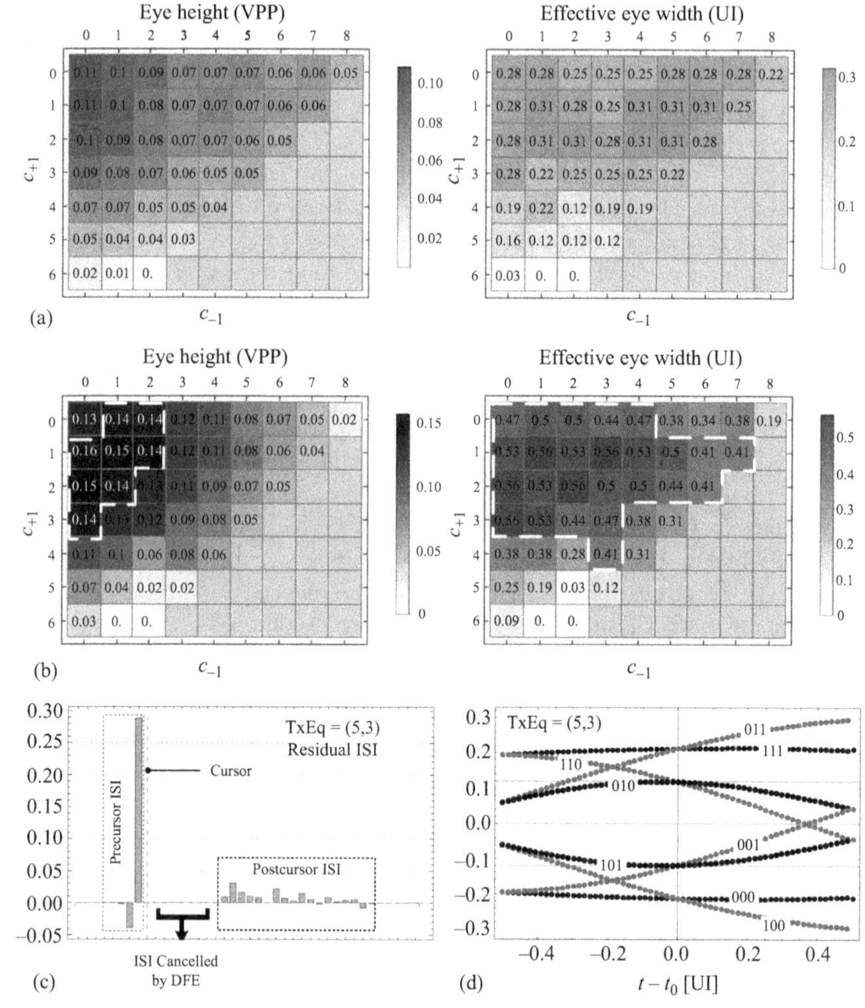

Figure 3.19 Margin maps for a representative high ISI scenario: (a) CTLE equalizer; (b) CTLE + DFE equalizer; (c) residual ISI for TxEq = {5,3}; and (d) tri-bit eye diagram for TxEq = {5,3}

point [Figure 3.20(a) and (b)]. As the sampling point is moved to the to the left, the difference between the cursor "g(t)" and precursor "g($t - T$)" traces increases to reach a maximum as shown in Figure 3.20(b) after which the cursor minus precursor difference decreases. In terms of ISI impact, this suggests that operating at the point where the cursor minus precursor difference is maximum, would very effectively minimize the impact of the precursor so this seems a very reasonable optimization metric [14]. In fact, comparing the residual equalization error before

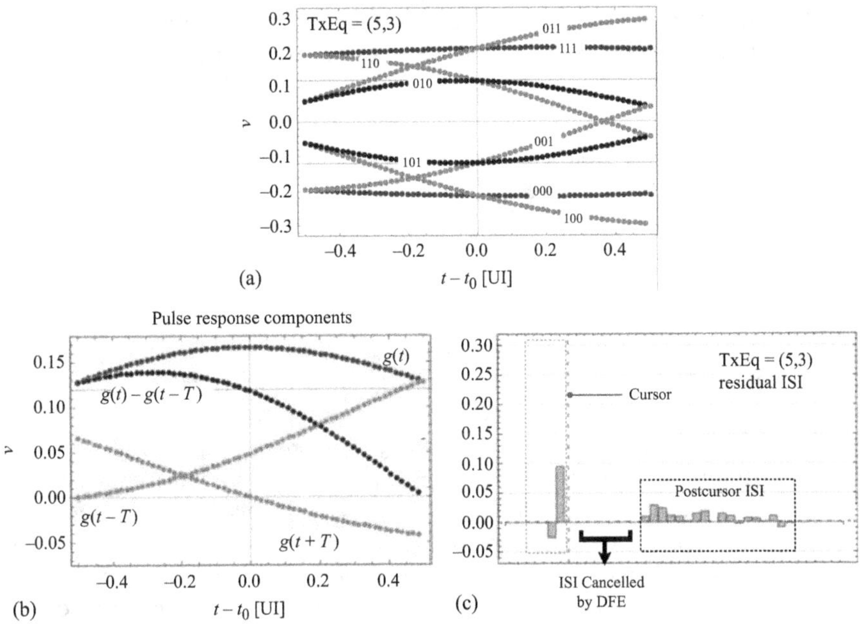

Figure 3.20 (a) Tri-bit eye; (b) timing function decomposition; and (c) residual ISI

and after applying the sampling point offset, shows a drastic reduction in the value of the precursor ISI [Figure 3.20(c)]. However, because the sampling point has shifted to the left of its original position, this is accompanied by a small increase in the postcursor ISI for the terms that are beyond the DFE reach.

The idea of minimizing the cursor precursor difference can be used with either an Alexander or Baud rate phase detector. In the type-B variant of the Baud rate timing function, the minimization of the cursor precursor difference can be used to control the magnitude of the DFE first postcursor, while the PD drives the sampling point so that the first postcursor is zero (zero forcing criteria). With this choice, the potential conflict arising from both PD and DFE trying to determine the value of the first postcursor correction is removed. In an Alexander PD implementation, the same objective is obtained by shifting the edge clock with respect to the data clock. Because the Alexander PD requires the additional sampling at the edge, the same analysis needs to be repeated to assess the impact of residual equalization error on the edge samples. When the amount of ISI on the edge path becomes a factor limiting the receiver BER, it will be necessary to add a DFE to the edge path.

The previous example indicates that under the conditions of residual equalization error, the system performance may not be optimal when sampling at the peak of the pulse response (i.e. when $\varepsilon = 0$). The peak distortion minimization equation (3.13) should then be generalized to comprehend that the phase error signal is the difference between the current sampling point and a reference sampling point t_0 the definition of which becomes another tool in the hands of the designer:

$$|g(t_0)| - \sum_{j \neq 0} |a_{p-j} g(t_0 + jT)| > 0 \tag{3.23}$$

3.6 Conclusion

With shrinking unit intervals (UI), the optimal horizontal centering of the reference sampling point becomes an important consideration for the choice of the phase detector and associated timing recovery method. Optimization of the timing recovery and link equalization can be greatly enhanced by the proper choice of timing function which, in conjunction, with proper equalization techniques will greatly improve the system performance. The material in this chapter has presented an overview of some of the considerations involved in the choice of phase detectors for good horizontal centering as well as good dynamic tracking characteristics and has introduced some analysis techniques to make quantitative comparisons to enable architectural decisions.

References

[1] PCI Express Base Specification Revision 4.0, Version 1.0, http://www.pcisig.com
[2] C. A. Belfiore and J. H. Park Jr, "Decision Feedback Equalization", Proc. IEEE, Vol. 67, No. 8, Aug. 1979.
[3] F. M. Gardner, "Phaselock Techniques", 3rd Edition, Wiley Interscience, Hoboken, NJ, 2005.
[4] F. Spagna. "Phase Locked Loop Using Delay Compensation Techniques", 2000 Fifth IEEE Symposium on Computers and Communications (ISCC), 2000; pp. 417–423.
[5] R. E. Crochiere and L. R. Rabiner, "Multirate Digital Signal Processing", Prentice Hall, Englewood Cliffs, NJ, 1983.
[6] N. Da Dalt and A. Sheikholeslami, "Understanding Jitter and Phase Noise", Cambridge University Press, Cambridge, 2018.
[7] L. Chen, F. Spagna, P. Marzolf *et al.* "A 90 nm 1-4.25-Gb/s Multi Data Rate Receiver for High Speed Serial Links", 2006 IEEE Asian Solid-State Circuits Conference, pp. 391–394, 2006.
[8] S. Agarwal and V. S. Rao. "A 5-Gb/s Adaptive Ctle with Eye-Monitoring for Multi-Drop Bus Applications", 2014 IEEE 57th International Midwest Symposium on Circuits and Systems (MWSCAS), pp. 410–413, 2014.
[9] C. Holdenried, R. Bespalko, M. v. Ierssel *et al.* "Performance of Edge Tap Decision Feedback Equalization methods for Wireline receivers", 2014

IEEE 12th International New Circuits and Systems Conference (NEWCAS) , pp. 5–8, 2014.

[10] K. Mueller and M. Muller, "Timing Recovery in Digital Synchronous Data Receivers", *IEEE Transactions on Communications*, Vol. COM-24, No. 5, May 1976–.

[11] F. Spagna, L. Chen, M. Deshpande *et al.*, "A 78 mW 11.8-Gb/s Serial Link Transceiver with Adaptive RX Equalization and Baud-rate CDR in 32 nm CMOS", ISSCC Dig. Tech Papers, pp. 366–367, 2010.

[12] R. W. Lucky, J. Salz, and E. J. Weldon, "Principles of Data Communication", New York, McGraw-Hill, 1968; pp. 59–82.

[13] F. Spagna, "Clock and Data Recovery Systems", 2018 IEEE Custom Integrated Circuits Conference (CICC).

[14] Y. Lu and E. Alon, "Design Techniques for a 66 Gb/s 46 mW 3-Tap Decision Feedback Equalizer in 65 nm CMOS", IEEE Journal of Solid-State Circuits, Vol. 48, No. 12, Dec. 2013.

Chapter 4

Silicon-based THz frequency synthesizers with wide locking range

Payam Heydari[1]

Terahertz (THz) band (0.3–3 THz), extending from the highest millimeter-wave (mm-wave) frequency range to the lowest infrared frequency, has drawn great attention. Recent research in the THz region has been expanding to a wide range of applications such as spectroscopy, imaging, and ultrahigh-data-rate wireless communications [1–6]. For imaging applications, due to the short wavelength and non-ionized characteristic of the THz wave, it provides a non-invasive and high spatial resolution image detection, which fits the requirement of security surveillance including stand-off imaging for the detection of hidden objects [7,8]. The SNR and the resolution can be improved by utilizing a synchronized carrier for time integration. For wireless communication applications, the availability of very wide unused bandwidth in the THz spectrum makes it possible to build up a wireless data link with data rates in excess of tens of gigabits-per-second by using a modulation scheme such as amplitude-shift keying or on-off keying [9–13]. Due to large atmospheric attenuation of THz waves, one promising application of this spectrum is in the area of scientific satellite communication [14,15].

An accurate and stable THz source proves to be essential for coherent data link, active sensing, and frequency metrology. Recently, silicon-based signal sources at a low-THz range (e.g., 200 and 300 GHz) using coupled VCOs and harmonic generation have been reported [16–18]. Chiang *et al.* [19] present a push–push VCO in which the frequency variation is achieved using inductive tuning. The lower loss of an inductive-tuning mechanism compared to the traditional varactor tuning makes this design amenable to the THz frequency range. Pfeiffer *et al.* [20] present an SiGe 530-GHz array-based radiator comprising 16 radiating elements each with its own ring-antenna and a balanced triple-push oscillator with no tuning capability. Free-running signal sources exhibit severe frequency fluctuation and are vulnerable to temperature and process variations. Moreover, the VCOs reported in prior work exhibit a limited tuning range, not often sufficient to tackle these variations and meet bandwidth requirements of the systems they are designed for. For example, in a radar system, both bandwidth and frequency of operation

[1]Department of Electrical Engineering and Computer Science, University of California, Irvine, CA

contribute to the resolution, in that, the former affects the cross-range resolution, while the latter contributes to the range resolution [21]. Consequently, the need for precise oscillation frequency with wide tuning range and low close-in phase-noise calls for closed-loop source topologies. Millimeter-wave phase-locked loops (PLLs) incorporating push–push VCOs have been demonstrated up to 164 GHz [22] in silicon technologies. Seo *et al.* presented a 300-GHz PLL with a 0.12% locking range and a divider ratio of 10 in a III–IV technology [23]. Due to the low divider ratio, the PLL in this work demands a high-frequency input reference signal which could be generated from another PLL, resulting in a high phase-noise and system complexity.

In this chapter, we present a 300-GHz frequency synthesizer incorporating a triple-push VCO with a Colpitts-based active varactor (CAV) and a frequency divider with three-phase injection [24]. The synthesizer, implemented in 90-nm SiGe Bipolar complementary-metal-oxide-semiconductor (BiCMOS) with a f_T/f_{max} of 240/315 GHz, achieves 7.9% of locking range (280.32–303.36 GHz) and generates −14 dBm of output power at 290 GHz. Based on the measurement result, the frequency-scaled phase noise of the 294.9-GHz signal is −77.8 dBc/Hz (−82.5 dBc/Hz) at the 100-kHz (1 MHz) offset [24].

4.1 Architecture of the THz frequency synthesizer

In a THz frequency synthesizer, the VCO and the first stage of the divider chain following the VCO are two of the most important building blocks, which dominate the PLL performance including locking range, phase noise, and output power. As the operation frequency approaches the device f_{max}, transistors barely provide sufficient power gain or negative resistance even with a proper feedback loop. The transistor's gain limitations along with sever loss of passive components such as capacitors, varactors, and inductors make it difficult to design frequency dividers or fundamental oscillators at THz frequencies. A PLL operating at a lower frequency followed by a frequency tripler (e.g., a harmonic-based or an injection-locked tripler) alleviates some of the challenges, as was extensively discussed in [25]. While the harmonic-based tripler (HBT) is amenable to high frequencies close to f_{max}, it consumes additional DC power and degrades the overall PLL phase noise. The total phase noise, S_{Total} ($\Delta\omega$), of a lower-frequency PLL followed by the HBT is expressed as [25]:

$$S_{Total}(\Delta\omega) = 3^2 S_{Fund}(\Delta\omega) + S_{HT}(\Delta\omega) + S_{MN}(\Delta\omega) \tag{4.1}$$

where S_{HT} ($\Delta\omega$) is the phase noise of the low-frequency PLL, S_{HT} ($\Delta\omega$) and S_{MN} ($\Delta\omega$) represent the phase-noise contributions from the harmonic transistors (the transistors generating third harmonic signals) and the output matching network of the HBT, respectively. The tripler's noise contributions, S_{HT} ($\Delta\omega$) and S_{MN} ($\Delta\omega$), cause a deviation from the ideal value, that is, the HBT's input phase-noise S_{Fund} $(\Delta\omega)|_{dB}$ plus 20 $\log_{10}(3)$.

Figure 4.1 shows the architecture of the 300-GHz frequency synthesizer composed of a triple-push VCO, a three-phase injection-locked divider (÷4) followed by a ÷256 divider chain, a phase-frequency detector/charge pump (PFD/CP) with a tunable current (I_{CP}: 150–300 μA), and a 2-bit programable third-order loop filter [24]. The input reference is fed by a 96-MHz crystal oscillator. The VCO's third harmonic signal will be the synthesizer's output ($3f_0 = 300$ GHz) and its three-phase fundamental signals (f_0) are fed to the divider. The use of the triple-push VCO results in the injection of lower-frequency signal f_0 to the divider chain, thereby relaxing the divider design requirement and lowering its DC power consumption. Although one can clearly implement a higher harmonic VCO and further decrease the fundamental frequency, it is noteworthy that the VCO's harmonic output power will be limited by the transistor nonlinearity. Since no additional frequency multiplier follows the synthesizer in this design, the noise contribution of the harmonic generation transistors, S_{HT} $(\Delta\omega)$, in (4.1) will no longer exist, resulting in lower phase noise.

By using the triple-push VCO, the first-stage divider within the synthesizer operates at around 100 GHz. An injection-locked frequency divider (ILFD) is commonly used at mm-wave frequencies [26]. Nonetheless, an ILFD exhibits a small locking range, which limits the frequency range of the synthesizer. One of the solutions to improve the locking range is to use multi-phase injection [27]. The availability of three-phase fundamental frequency signals in the triple-push VCO allows us to consider an ILFD with three-phase injection. The three-phase injection-locked divider exhibits a wide locking range, which encompasses the VCO's tuning range. The synthesizer locking range, therefore, follows the VCO's frequency-tuning range. The ÷256 divider chain uses emitter coupled logic (ECL)- and current mode logic (CML)-type topologies for their wide bandwidth characteristics. The PFD is implemented based on a conventional structure similar to the one presented in [28]. By sending an "up" and a "down" current pulse to the CP

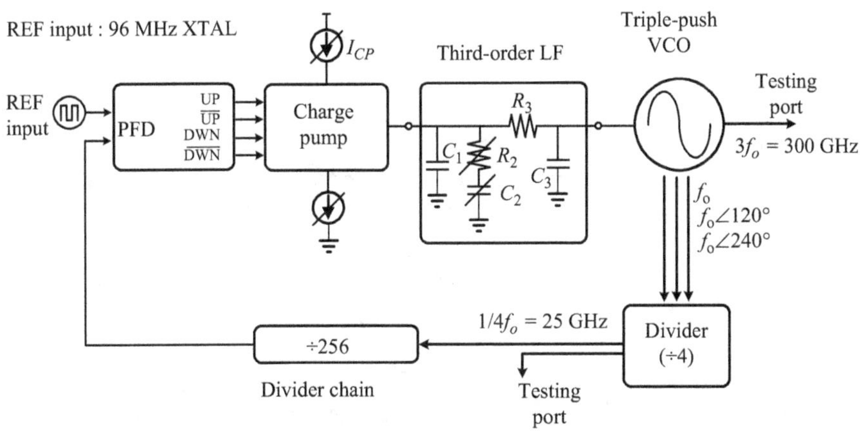

Figure 4.1 The architecture of the 300-GHz frequency synthesizer [24]

during each reference cycle, the PFD dead zone is eliminated and the loop gain for the small phase error is improved.

An on-chip third-order loop filter is utilized, in which C_2 produces the first pole and together with R_2 it is used to generate a zero for the loop stability. C_1 is used to smoothen the control voltage ripples. R_3 and C_3 further suppress reference spurs and high-frequency noise. The loop filter's component values are $C_1 = 800$ fF, $R_2 = 13$ kΩ, $C_2 = 22$ pF, $R_3 = 21$ kΩ, and $C_3 = 300$ fF.

4.2 Triple-push VCO with CAV

An alternative way of producing a high-frequency signal, especially when it is close to the transistor's f_{max}, is to make use of the oscillator's harmonics. To retain the output power, minimizing the resonator's loss at both the harmonic and fundamental frequencies is essential. For example, if the resonator's loss is minimized only at the oscillator's fundamental frequency and increased at its harmonics, the fundamental signal swing will be improved and the transistors will thus generate a higher intrinsic harmonic power. This harmonic power, once passing through the resonator, will be degraded significantly. In addition, the frequency-tuning mechanism, introduced at the fundamental frequency, exhibits ever-increasing loss at the harmonic frequency, thereby further degrading the harmonic power. Besides, the relative input–output phase difference ϕ and gain A of a single transistor, defined in Figure 4.2, affect the output power that this transistor can generate [29]. If a transistor operates in its optimum condition, expressed by (4.7), it generates the maximum output power, as will be described later. In our VCO design, all these considerations are taken into account in order to improve the output power, and subsequently, a better phase noise if the VCO operates in the current-limited region [30], which is usually the case for THz VCOs.

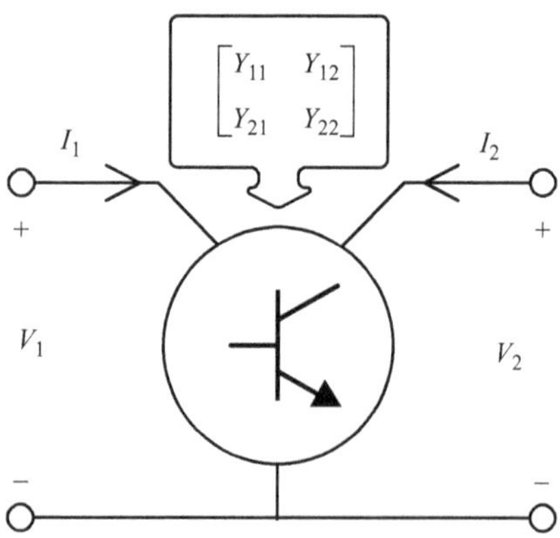

Figure 4.2 A two-port Y-parameter representation of a single transistor

Starting from the transistor's input–output phase and gain conditions, shown in Figure 4.2, is a two-port Y-parameter representation of the transistor. The net power that flows into the transistor can be expressed as [31]:

$$P_N = V_1^* I_1 + V_2^* I_2 \tag{4.2}$$

where * denotes the complex conjugate. Representing I_1 and I_2 with respect to V_1 and V_2, the real power $\mathrm{Re}[-P_N]$ that flows out of the transistor is expressed as:

$$\mathrm{Re}[-P_N] = |V_1|^2\left[-A\left|Y_{12}+Y_{21}^*\right|\cos(\angle(Y_{12}+Y_{21}^*)+\phi_{BC})\right] - |V_1|^2(G_{11}+A^2G_{22}) \tag{4.3}$$

where

$$A = \frac{|V_2|}{|V_1|}, \phi_{BC} = \angle\frac{|V_2|}{|V_1|} \tag{4.4}$$

The real power is composed of the power dissipated (P_R) and the power generated (P_G) inside the transistor, defined as:

$$P_R = |V_1|^2\left[(G_{11} + A^2G_{22})\right] \tag{4.5}$$

$$P_G = -A\,|V_1|^2\left|Y_{12} + Y_{21}^*\right|\cos(\angle(Y_{12} + Y_{21}^*) + \phi_{BC}) \tag{4.6}$$

In Eq. (4.6), P_G reaches its maximum absolute value if:

$$\phi_{BC} = \phi_{BC,opt} = (2k+1) - \angle(Y_{12} + Y_{21}^*) \tag{4.7}$$

where k is an integer. Under a given bias condition, there exists an optimum phase difference between the base and the collector, $\phi_{BC,opt}$, that a transistor generates the maximum power. Figure 4.3 shows the simulated $\phi_{BC,opt}$ versus frequency for a transistor with an aspect ratio of L_E/L_B = 4 μm/0.09 μm and biased at its

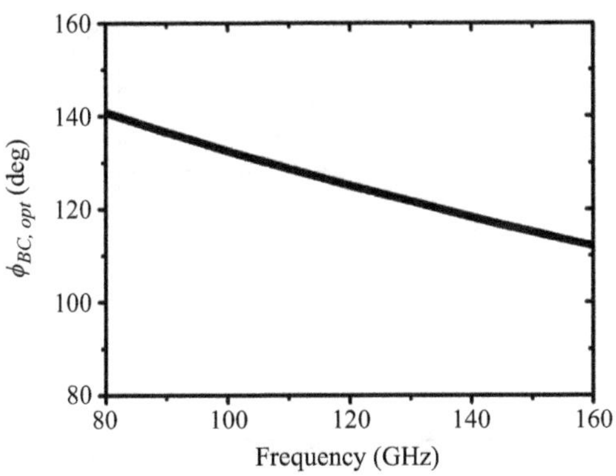

Figure 4.3 The simulated $\phi_{BC,opt}$ versus frequency for a transistor with an aspect ratio of L_E/L_B = 4 μm/0.09 μm and biased at its maximum f_{max}

maximum f_{max}. The optimum phase condition varies from 140° to 110° across the frequency range from 80 to 160 GHz. In the VCO design, the transistors are designed to satisfy this phase condition so as to achieve substantially better VCO performance.

Based on the net power analysis of the single transistor, a systematic approach of designing a 300-GHz oscillator is now described. A harmonic ring oscillator offers the ability to achieve $\phi_{BC,opt}$ by adding extra phase shift between stages. Figure 4.4(a) shows an example of a two-stage ring (i.e., cross-coupled) oscillator, where the second harmonic signal $2f_0$ is retrieved from a common output terminal. Referring to Figure 4.3, $\phi_{BC,opt}$ at a fundamental oscillation frequency of 150 GHz is 115°. Because of the 180° constant phase difference between two stages, an extra phase shift of 65°, realized using a t-line, is required to achieve $\phi_{BC} = \phi_{BC,opt}$. A larger phase shift requires a longer t-line with inevitably larger in-series loss in the signal path. This loss degrades the fundamental voltage swing inside the ring which, in turn, degrades the second harmonic power. Consequently, ring oscillators requiring larger extra phase shifts to achieve $\phi_{BC,opt}$ will produce a smaller harmonic power. Another example using a three-stage ring is shown in Figure 4.4(b). By virtue of its design, the three-stage ring oscillator requires a small phase shift of 10° to satisfy the optimum phase condition. In addition, compared to the two-stage counterpart, more branches participate in the harmonic power generation, resulting in a higher output power. The three-stage ring oscillator is thus chosen as the VCO core.

The conventional approach for frequency tuning in a three-stage ring VCO, shown in Figure 4.5, employs a varactor in parallel with a buffer at the base of each ring transistor. The input parasitic capacitor of the buffer, however, decreases the oscillation frequency. Decreasing the loading inductor, L_C, boosts up the oscillation frequency at the expense of lowering the gain of each stage. Both the loss of the buffer and the varactor degrade the base voltage swing of the ring transistor, especially the severe varactor loss at high frequency becomes a bottleneck.

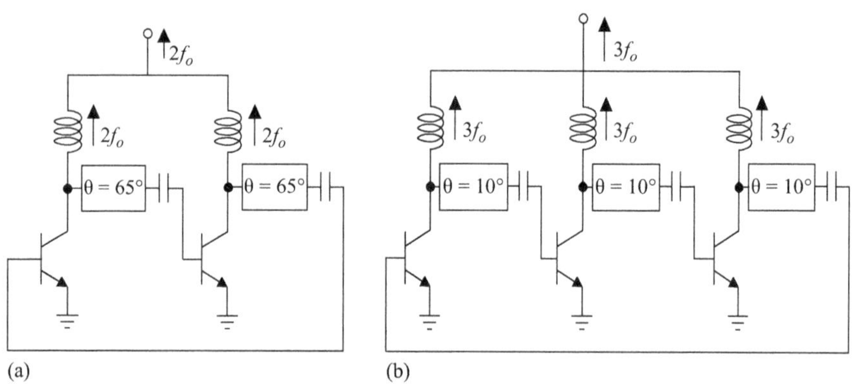

Figure 4.4 (a) A two-stage ring oscillator with a phase shift of 65° between stages. (b) A three-stage ring oscillator with a phase shift of 10°

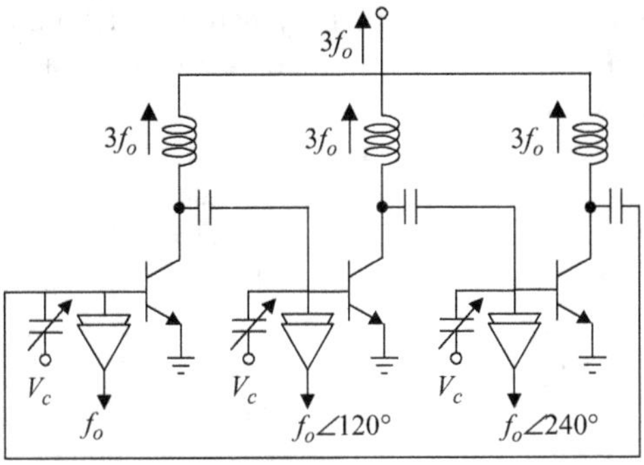

Figure 4.5 A conventional three-stage ring VCO employing a varactor in parallel with a buffer at the base terminals of the transistors

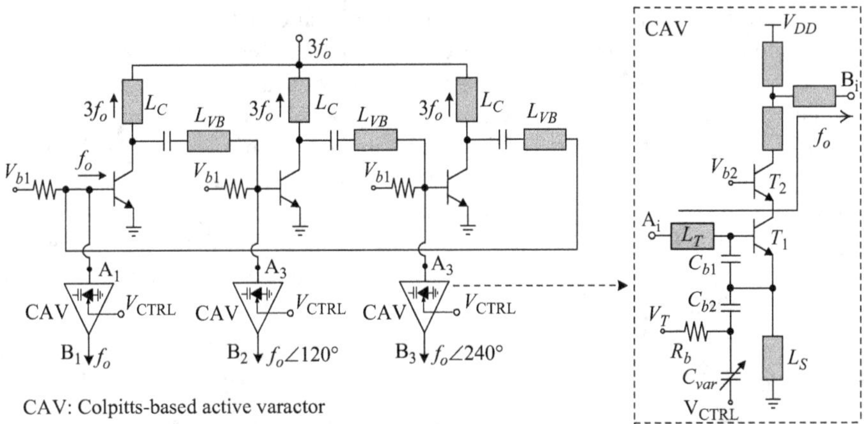

Figure 4.6 The schematic of the triple-push VCO with CAV [24]

Therefore, a new frequency-tuning circuitry called the CAV is introduced [24]. The schematic of the VCO with CAV is shown in Figure 4.6. The VCO's f_0 is traveling along the VCO three-stage ring with a 120° phase difference between stages. Its third harmonic signals generated from the transistors nonlinearity will be in-phase and collected at the common-mode output $3f_0$. The t-line L_{VB} is used to introduce 10° phase shift for transistor's optimum phase condition. In a conventional Colpitts oscillator, the Colpitts cell is an essential part of the oscillator that enables sustainable oscillation by providing a loss compensation mechanism for the resonator. On the other hand, the CAV in this design merely realizes extremely low-loss tunable capacitance and isolates the varactor loss from the VCO ring. In addition, it buffers the VCO's f_0 to the divider and thus avoids loading the VCO ring. In the CAV schematic, node A_i $(1 \leq i \leq 3)$ is connected to each stage of the VCO ring and output node B_i $(1 \leq i \leq 3)$ is connected to the divider. Inside the

CAV, the cascode stage $T_1 - T_2$ with its T-junction matching network buffers the VCO's f_0 and feeds the signal to the divider. The bypass capacitor C_{b2}, the large resistor R_B, and bias voltage V_T are used to level-shift V_{CTRL} within the voltage range required by the charge pump. The t-line L_T models the short interconnect between the CAV and the VCO ring in the actual layout. The three CAVs are placed in close proximity of the VCO ring so that the L_T effect on the oscillation frequency and tuning range becomes negligible.

The simulation results of the CAV including the tunable capacitor C_{tune} and the parallel conductance Re[Y_P] seen from node A_i ($1 \leq i \leq 3$) are shown in Figure 4.7. C_{tune} monotonically decreases from 42 to 26 fF with V_{CTRL} varying from 0 to 2 V for frequency tuning. The negative conductance Re[Y_P] across the tuning range compensates for the large varactor loss and also helps the VCO startup condition. The simulated fundamental voltage swings of the conventional and the CAV tuning ring VCOs, respectively, with the same tuning range and oscillation frequency are shown in Figure 4.8(a) and (b). The VCO achieves more than three times higher base voltage swing than the conventional counterpart.

In addition to providing 10° phase shift, L_{VB} also enhances the $3f_0$ power collection by blocking $3f_0$ from flowing to the next stage as shown in Figure 4.9(a). The current source I_{3f0} represents the third harmonic current generated from a transistor and C_p is the input parasitic capacitor of the next stage. At the third harmonic frequency, L_{VB} partially resonates with C_P, resulting in high impedance. Therefore, the large portion of the third harmonic current will flow into the load, I_{out}. The plot in Figure 4.9(b) is the current I_{out} normalized to I_0 (the current flowing to the load without L_{VB}). As L_{VB} increases to 40 μm, I_{out} is improved by as much as 1.9 times.

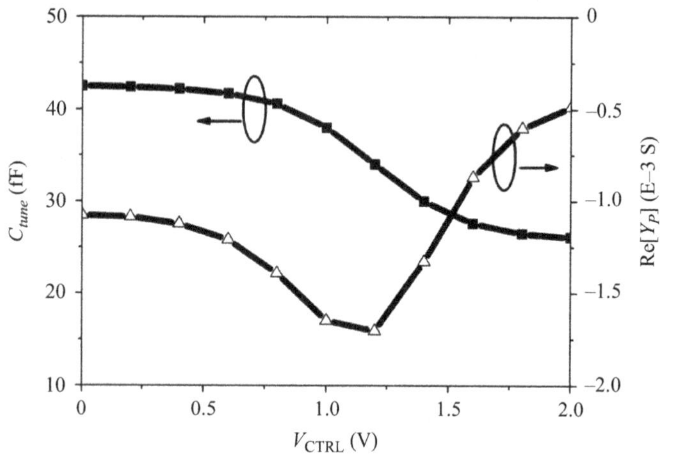

Figure 4.7 The simulated tunable capacitor C_{tune} and the parallel conductance Re[Y_P] of the CAV (seen from node A_i of Figure 4.6)

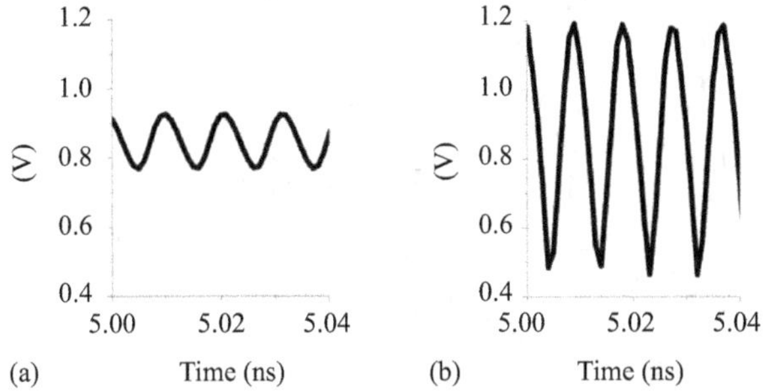

Figure 4.8 *The simulated fundamental voltage swings for (a) a three-stage ring VCO with a conventional varactor tuning as shown in Figure 4.5 and (b) a three-stage ring VCO with CAV tuning*

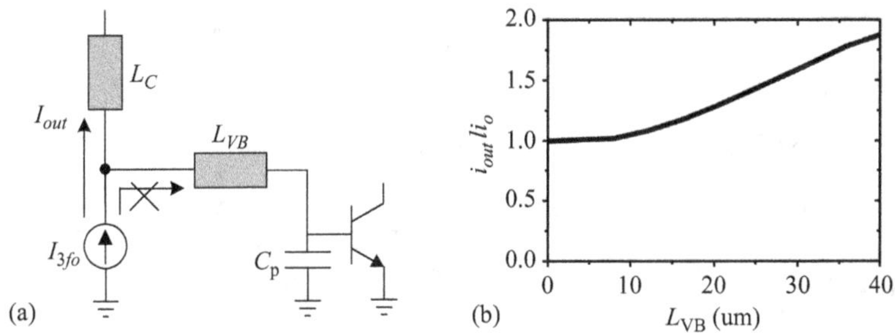

Figure 4.9 *(a) The third harmonic power enhancement by using L_{VB} to partially resonate with C_p. (b) The simulation of the normalized third harmonic current flowing to the load versus the length of the t-line L_{VB}*

4.3 Three-phase injection locked divider (÷ 4)

The conceptual loop of the three-phase injection-locked divider [24] is shown in Figure 4.10. Each stage is composed of an amplifier, a mixer, and a low-pass filter (LPF). The free-running frequency of the loop is designed close to $1/4f_0$, and each stage contributes 120° phase difference between its input and output. The three third harmonic signals ($3/4f_0$) generated from the amplifiers are all naturally in-phase. The input three-phase signals coming from the VCO's three-phase output are mixed with $3/4f_0$ signals. The mixers' outputs ($1/4f_0$, $1/4f_0\angle120°$, $1/4f_0\angle240°$) are injected back to the loop and are added constructively with the loop's signal. In the steady state, the loop is locked to the injected signals and divide-by-4 operation is performed. The same conceptual loop can readily perform divide-by-5 by simply swapping the last two input signals, as depicted in Figure 4.11. Assuming that the free-running frequency of the loop is close to $1/5f_0$, the three-phase input signals are mixed with the loop's fourth harmonic ($4/5f_0$). The loop will be locked to the mixer's output signals ($1/5f_0$, $1/5f_0\angle120°$, and $1/5f_0\angle240°$).

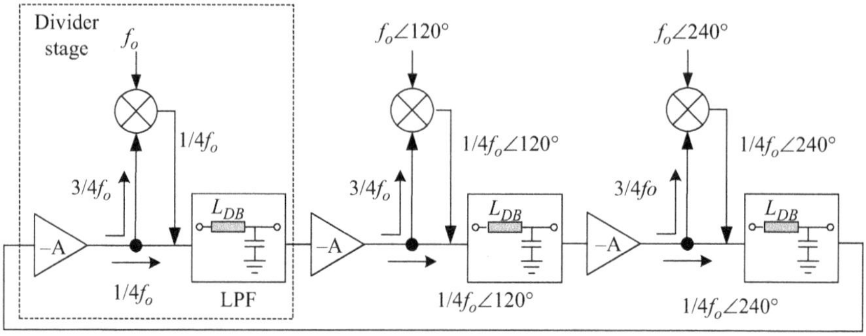

Figure 4.10 The conceptual loop of the three-phase injection-locked divider (divide-by-4) [24]

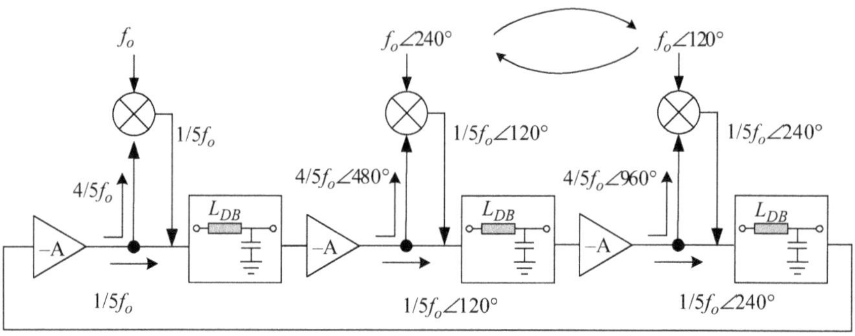

Figure 4.11 By swapping the last input signals, the same conceptual loop performs divide-by-5

Compared with the single injection, the multi-phase injection achieves a wider locking range and requires a lower input power. For a general case of the N-stage divider with M-phase injection, in the phasor domain, each injection current, I_{in}, introduces a phase change ϕ to the loop signal (I_{osc}), and each LPF needs to compensate for a phase change θ in order to satisfy Barkhausen criteria. The phase change θ of each LPF is expressed as:

$$\theta = \frac{M}{N}\phi \tag{4.8}$$

If I_{in} is much smaller than I_{osc}, the normalized single-side locking range is then expressed as [32]:

$$\frac{\Delta f}{f_a} \leq \frac{M}{N} \cdot \frac{1 + \tan^2 \dfrac{\pi}{N}}{\tan \dfrac{\pi}{N}} \cdot \left| \frac{I_{in}}{I_{osc}} \right| \tag{4.9}$$

where f_a is the free-running frequency of the N-stage divider. From (4.9), it is seen that (a) the locking range is proportional to the number of injected phases (M); (b) Under the same locking range, the injection current I_{in} is lowered by M times; (c) With higher mixer's efficiency, the injection current is increased, resulting in a wider locking range. Figure 4.12 demonstrates the special case when $N = M = 3$. With three-phase injections, the phase change of each LPF θ is equal to the phase change (ϕ) introduced by the injection current. The overall phase shift of each LPF thus becomes $-\pi/3 + \phi$.

Figure 4.13 shows the circuit implementation of this divider. The amplifying stages (T_{a2}–T_{c2}) with the t-lines of L_{DM} and L_{DB} form the divider's three-stage ring and transistors T_{a1}–T_{c1} act as the three mixing cells. The three-phase input signals

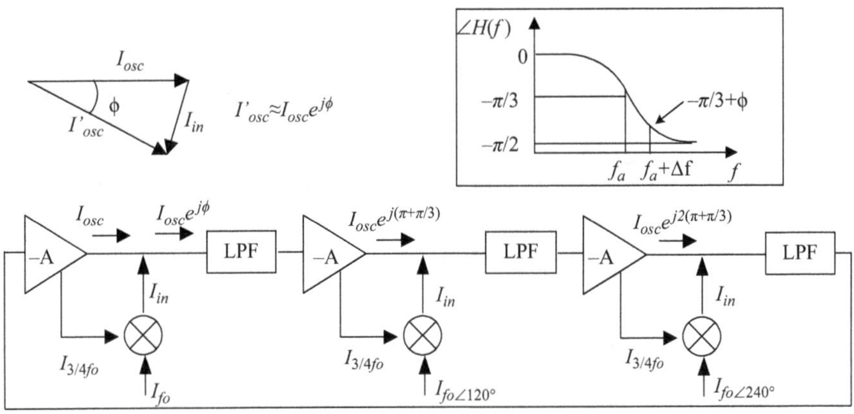

Figure 4.12 *The three-stage divider with three-phase injection ($N = M = 3$), where the phase change of each LPF is equal to the phase change ϕ introduced by the injection current I_{in}*

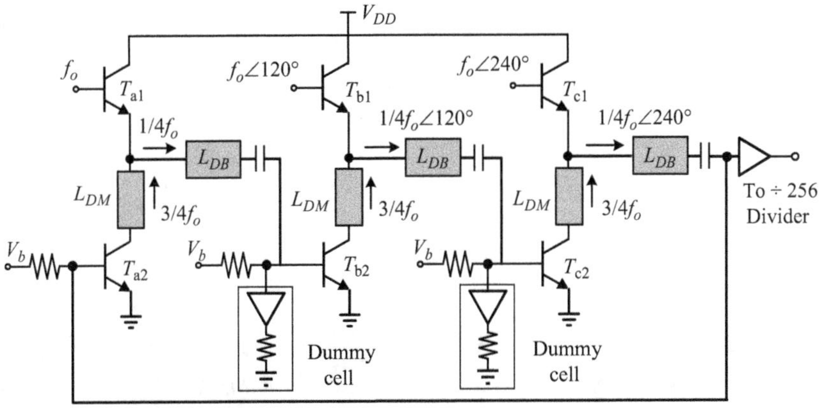

Figure 4.13 *The schematic of the three-phase injection-locked divider (÷ 4) [24]*

are fed to the base terminals of the mixing cells and mixed with the loop's third harmonic signals. The mixer's outputs ($1/4f_0$, $1/4f_0\angle 120°$, and $1/4f_0\angle 240°$) flow back to the loop at three injection points. Similar to the VCO of Figure 4.6, L_{DB} blocks the loop's third harmonic signal $3/4f_0$ to flow to the next stage and forces it to flow into the mixer. This enhances the mixing efficiency and thus further increases the locking range. A buffer is used to feed the divider's output $1/4f_0$ to the rest of the divider chain ($\div 256$) and two dummy buffers are added to provide symmetric parasitic loading to other stages.

4.4 Integration and layout of VCO and divider

At high frequencies, any interconnect or cross-over introduces parasitics that cause severe signal distortion. This distortion will adversely affect the performance of the synthesizer's VCO and dividers such as phase noise, output power, and input sensitivity. The structural similarity of the VCO and the three-phase injection divider helps us co-design/co-optimize these blocks and develop a compact layout. The integration of the VCO and the divider within the synthesizer is shown in Figure 4.14, where the inner blue circle indicates the VCO ring. The VCO's f_0 travels along the inner blue ring and its third harmonic signals are collected coherently at the center's $3f_0$ pad. The t-line's length realizing L_C determines the perimeter of the VCO ring and the length of L_{VB}. The t-line L_{VB}, used to introduce 10° phase shift for the transistor optimum condition, forms the ring's perimeter. By narrowing L_C and widening L_{VB} to proper values, the L_{VB}'s t-line is fitted into

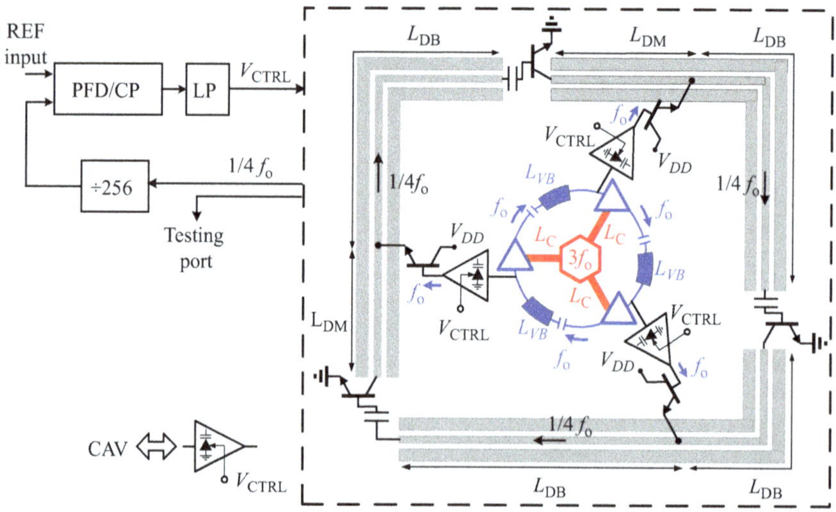

Figure 4.14 The integration of the VCO and divider within the synthesizer where the inner blue circle and the outer rectangle indicate the VCO's and the divider's three-stage rings, respectively [24]

the VCO ring. Three CAVs are placed in the layout with respect to divider's three-phase inputs such that no extra t-line or signal cross-over is required. The outer rectangle realizes the divider's loop, which is locked to the VCO ring and performs divide-by-4. The divider's output $1/4f_0$ is then fed to the synthesizer divide-by-256 and forms a close-loop through PFD/CP and LF. Two testing ports ($3f_0$ and $1/4f_0$) are used to measure the synthesizer's output spectra and its locking behavior.

4.5 Measurement

Three individual circuits including a divider, a VCO + divider, and a 300-GHz frequency synthesizer were fabricated in 90-nm SiGe BiCMOS with six metal layers and f_T/f_{max} of 260 GHz/315 GHz. Their die micrographs are shown in Figure 4.15(a) and (b). The synthesizer chip area is 1.6 × 1.6 mm^2 including DC and RF pads. All the t-lines were implemented using coplanar waveguides (CPWs) with a 2.8 μm thick-top-metal (M6) as the signal line and M3 as the shielding ground. The CPWs and the interconnection lines were carefully characterized using planar 3-D electromagnetic simulations.

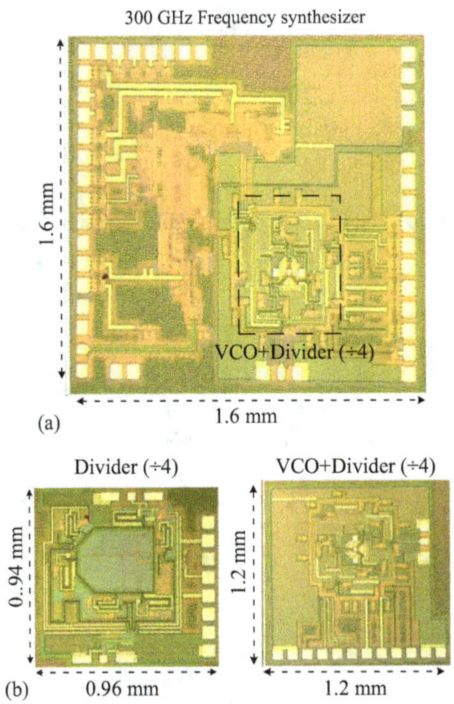

Figure 4.15 The die micrographs of (a) the 300-GHz frequency synthesizer and (b) the standalone divider (÷ 4) and the VCO+divider [24]

In order to generate a three-phase signal for measuring the divider's input sensitivity, an on-chip three-way power divider and two on-chip phase shifters were used to provide input to the standalone divider. In the divider measurement setup shown in Figure 4.16, an external W-band signal from a multiplier (×6) was applied to the divider input using a WR-10 GSG waveguide probe, and the divider output was connected to the spectrum analyzer (SA). With the measured divider's output spectrum and the input frequency, the divider's locking to the input signal was verified. By sweeping the input frequency and power, the divider's input sensitivity was measured. Figure 4.17 shows the plot of the input sensitivity versus input frequency for two bias settings ($V_b = 1.35$ V and $V_b = 1.48$ V). With the two bias settings, the measured overall locking range is 91.9–101.8 GHz and the divider's average DC power consumption is 48.4 mW. At $V_b = 1.35$ V, the divider's free-running frequency is 99.5 GHz. At the higher bias voltage (i.e., $V_b = 1.48$ V), the free-running frequency decreases to 95.5 GHz due to the larger C_π's of the

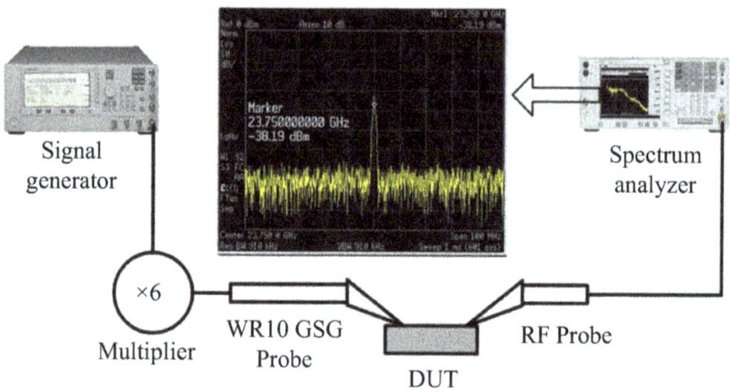

Figure 4.16 The standalone divider measurement setup

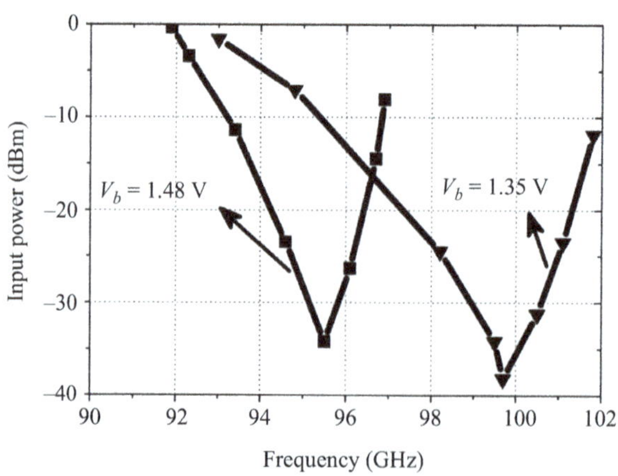

Figure 4.17 The measured divider's input sensitivity with two bias settings

divider's amplifying transistors. Moreover, under this bias condition, the amplifying transistors generate smaller third harmonic signals, resulting in locking-range degradation.

The VCO + divider chip was used to measure the VCO free-running performance and verify the divider's functionality with its input fed by the VCO. Figure 4.18 shows the measurement setup. For the VCO's output power measurement, the VCO output was connected to a WR-3 waveguide GSG probe followed by the calorimetry power sensor. The waveguide taper was used to transform the waveguide interface from WR-3 to WR-10. The measured power was calibrated by the loss of the signal path including the probe and the taper. For the VCO's oscillation frequency and tuning range measurement, the VCO's output signal was down-converted to the SA by using the harmonic mixer. Meanwhile, with the divider's measured output spectrum, the divider locking characteristic was validated over the whole VCO tuning range. Since the divider is locked to the VCO and the conversion loss of the harmonic mixer is high (>70 dB), the VCO phase noise is measured at the divider output. This phase-noise measurement is close to the direct measurement of the VCO's output. This is because that the output phase-noise profile of an injection-locked divider is similar to a first-order PLL, where the input noise is low-pass-filtered, while the divider noise is high-pass-filtered and is thus negligible [33].

The measured VCO's oscillation frequency and output power is shown in Figure 4.19. The oscillation frequency varies from 280 to 303.36 GHz (i.e., 8% of tuning range) with V_{CTRL} varying from 0 to 2.6 V. The measured output power is −14 dBm with less than 0.5 dB variation. Figure 4.20 shows the measured phase-noise profile at the divider's output where the input VCO frequency is 100.4 GHz. The phase noise at 1-MHz offset is −101.9 dBc/Hz, and the corresponding VCO phase noise is −80.3 dBc/Hz (with a factor of 20 $\log_{10}$ 4). The VCO's DC power consumption is 105.6 mW.

The synthesizer measurement setup is similar to that of the VCO where the synthesizer's $3f_0$ output was down-converted to the SA and its $1/4f_0$ was connected

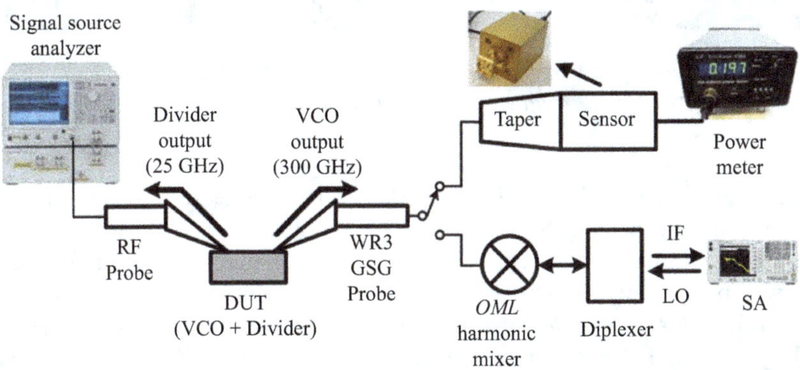

Figure 4.18 The VCO + divider measurement setup

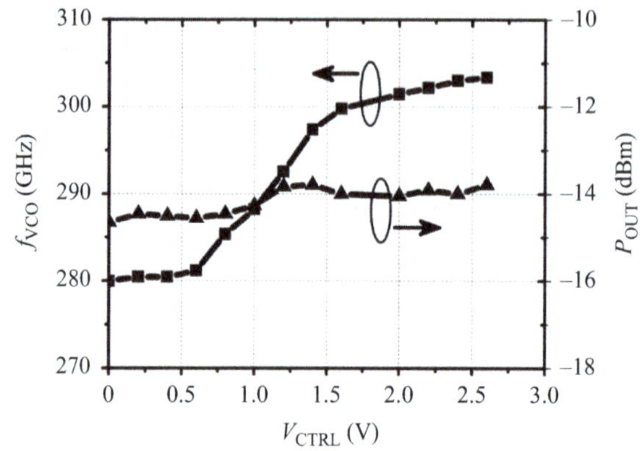

Figure 4.19 The measured VCO's oscillation frequency and output power versus V_{CTRL}

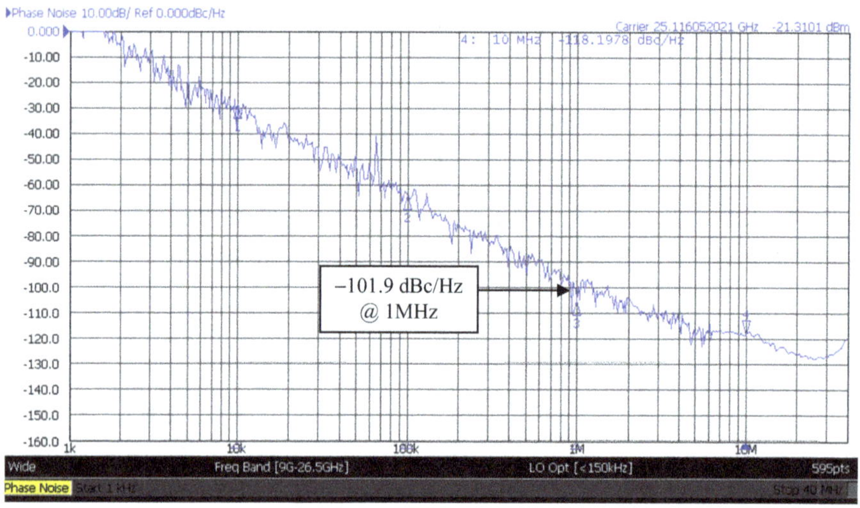

Figure 4.20 The measured phase-noise profile at the divider's output where the input VCO frequency is 100.4 GHz

to the signal source analyzer. The synthesizer input was fed using a crystal oscillator or a signal generator (SG) for measuring the synthesizer's locking range. The measured synthesizer's output spectrum is shown in Figure 4.21 when the synthesizer was locked at 300.8 GHz. Figure 4.22 shows the measured phase-noise profile at the $1/4f_0$ output port when the synthesizer was locked to a 95-MHz reference feeding by the SG. The profile depicts a close-loop locking behavior, in which the frequency response for input noise is low-pass-filtered and VCO noise is high-pass-filtered. The synthesizer's locking range was measured by sweeping the input

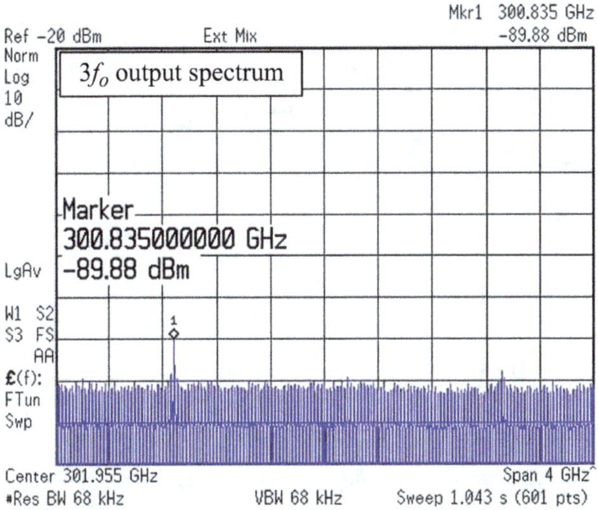

Figure 4.21 The measured synthesizer's output spectrum

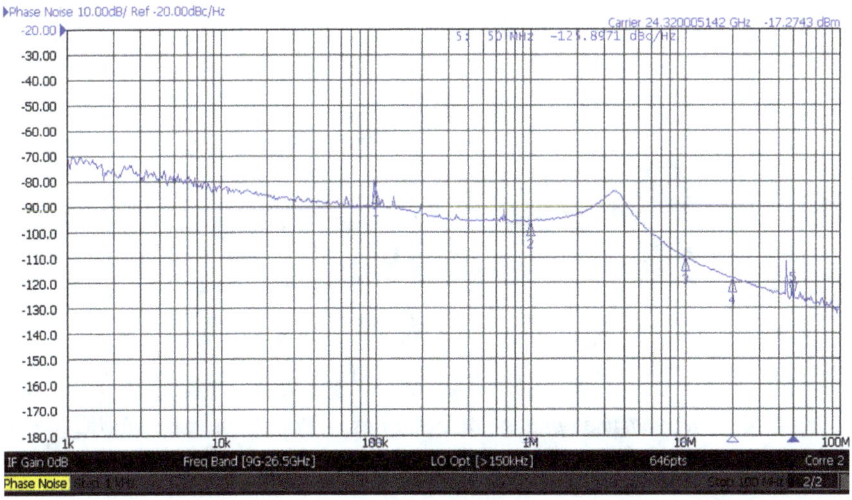

Figure 4.22 The measured phase-noise profile at the synthesizer's $1/4f_0$ output when the synthesizer was locked to a 95-MHz reference feeding by the SG

frequency from the SG and measuring the output spectra. The measurement of the synthesizer's scale-up phase noise versus its output frequency is shown in Figure 4.23. The locking range varies from 280 to 303 GHz, which closely follows the VCO's tuning range. The minimum phase noise at the 1-MHz offset is −75.4 dBc/Hz when the synthesizer's output frequency is 296 GHz. With the input

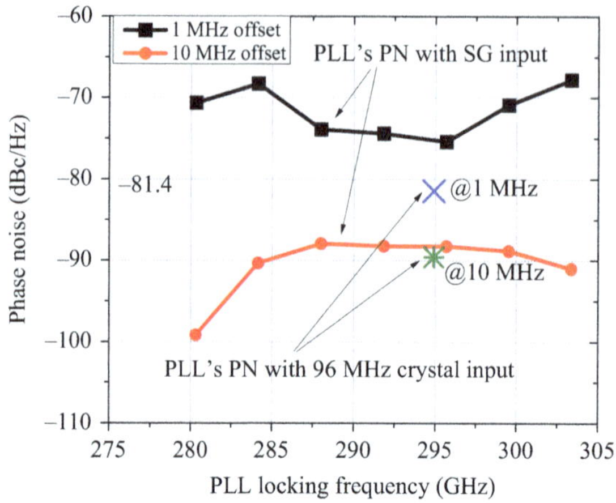

Figure 4.23 The measurement of the synthesizer's scale-up phase noise versus its output frequency with the synthesizer's input reference from an SG and a crystal oscillator

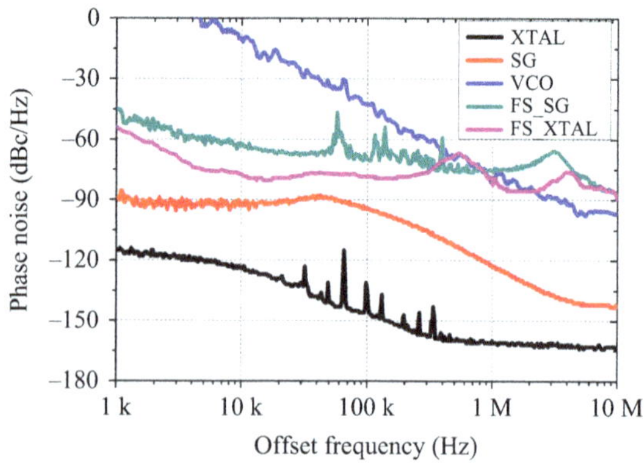

Figure 4.24 The measured phase-noise profiles of the synthesizer, the VCO, and [24]

signal fed from a 96-MHz crystal oscillator (i.e., the synthesizer's output frequency is 294.9 GHz), the measured phase noise is −82.5 dBc/Hz (−89.6 dBc/Hz) at the 1-MHz (10 MHz) offset. Figure 4.24 shows the measured phase-noise profiles at 294.9 GHz for the free-running VCO, the synthesizer with its input from the SG (labeled as FS_SG) and from the crystal oscillator (labeled as FS_XTAL), input signal from the SG (labeled as SG) and from the crystal oscillator (labeled as XTAL). In a comparison between the two plots of FS_SG and FS_XTAL, the in-band phase noise follows the input noise, and thus a better input signal results in a lower phase noise. Note that one of the important advantages using a synthesizer over a VCO is that, because of the close-loop locking behavior, the phase noise within the synthesizer's loop bandwidth is significantly reduced (e.g., −45 and −77.8 dBc/Hz at the 100-kHz offset for the VCO and the synthesizer, respectively). Table 4.1 shows the measured performance summaries of the divider, VCO, and synthesizer.

Tables 4.2 and 4.3 provide the performance comparison with the prior work for the VCO and the 300-GHz synthesizer, respectively. In the VCO's comparison table, the VCO in this design demonstrates the widest tuning range and a better phase noise with a less DC power consumption. Note that for the oscillators reported in [34], the frequency tuning is achieved by changing the supply voltage and thus suffer from considerable output power variation.

In the synthesizer comparison table, Seo *et al.* [23] report the highest synthesizer in III–IV technology with an f_{max} of 800 GHz, and Shahramian *et al.* [22]

Table 4.1 Performance summary

Three-phase injection-locked divider		**Triple-push VCO**		**300-GHz frequency synthesizer**	
Frequency (GHz)	93–101.8	Frequency (GHz)	280–303.36	Frequency (GHz)	280.32–303.36
Divider ratio	4	Tuning range	8%	Divider ratio	1,024
Locking range	10.2%	Output power (dBm)	−14	Locking range	7.9%
Input power (dBm)	<0	PN @ 1 MHz offset (dBc/Hz)	−80.28	PN @ 100 kHz offset (dBc/Hz)	−77.8 @ 294.9 GHz
Supply (V)	2	Supply (V)	1.8	PN @ 1 MHz offset (dBc/Hz)	−82.5 @ 294.9 GHz
DC power (mW)	48.4	DC power (mW)	105.6	DC power (mW)	376

Table 4.2 VCO comparison

	This work	**[17]**	**[18]**	**[34]**
Frequency (GHz)	290	290	324	288
Tuning range (%)	8	4.5	1.2	1.4*
PN @ 1 MHz offset (dBc/Hz)	−80.28	−78	−78	−87
Output power (dBm)	−14	−1.2	−46	−1.5
DC power (mW)	105.6	325	12	275
FOM_T†	−153.4	−154	−111.4	−153.2
Technology	90-nm SiGe BiCMOS	65-nm CMOS	90-nm CMOS	65-nm CMOS

*Tuning achieved by changing supply voltage.

†$FOM_T = PN - 20\log\left(\frac{f_o}{\Delta f}\cdot\frac{FTR}{10}\right) + 10\log\left(\frac{P_{DISS}}{1\text{ mW}}\right) - P_{out}$

Table 4.3 Synthesizer comparison

	This work [24]	[23]	[22]
Frequency (GHz)	280.32–303.36 (third)	300.76–301.12 (fund.)	160–169 (second)
Divider ratio	1,024	10	128
Locking range	7.9%	0.12%	5.5%
PN @ 100-kHz/ 1-MHz offset (dBc/Hz)	−77.8/−82.5	−78/−85	−75/−78
DC power (P_D)	376 mW	301.6 mW	1,250 mW
FOM_T* @ 100-kHz/1-MHz offset	−179.4/ −163.9 dBc/Hz	−144.4/ −131.4 dBc/Hz	−163.1/ −146.1 dBc/Hz
Technology (f_{max})	90-nm SiGe BiCMOS (315 GHz)	InP HBT (800 GHz)	130-nm SiGe BiCMOS (280 GHz)

$*FOM_T = PN - 20\log\left(\frac{f_o}{\Delta f}\cdot\frac{Locking\ Range}{10}\right) + 10\log\left(\frac{P_D}{1\ \mathrm{mW}}\right)$

present the highest in silicon using a technology similar to our work. Our synthesizer in [24] demonstrates a higher operation frequency and a divider ratio, a wider locking range (60 times wider than that in [23]), and a comparable phase noise. The use of a crystal oscillator instead of an SG makes our synthesizer a practical and complete system. The synthesizer achieves a 20 dB better figure-of-merit (FOM_T) among the other work.

4.6 Conclusion

A fully integrated THz frequency synthesizer including a triple-push VCO with a CAV and a three-phase injection-locked divider was implemented in 90-nm SiGe BiCMOS. By using the VCO's third-order harmonics as the output signal and injecting its fundamental to the divider chain, the synthesizer's operation frequency was designed to become close to the transistor f_{max}, while relaxing the divider's operation frequency. Meanwhile, the three-phase injection improved the divider's locking range to cover the VCO's tuning range. This synthesizer thus achieved a wide locking range which follows the VCO's tuning range. The CAV provided frequency tunability and varactor isolation to the VCO and acted as a buffer between the VCO and the divider. In addition to the similar structure of the VCO and the divider, an efficient circuit co-design and a compact layout were achieved.

References

[1] P. H. Siegel, "Terahertz technology," *IEEE Trans. Microw. Theory Tech.*, vol. 50, no. 3, pp. 910–928, 2002.

[2] H.-J. Song and T. Nagatsuma, "Present and future of terahertz communications," *IEEE Trans. Terahertz Sci. Technol.*, vol. 1, no. 1, pp. 256–263, 2011.

[3] J. Wiltse, "History of millimeter and submillimeter waves," *IEEE Trans. Microw. Theory Tech.*, vol. 32, no. 9, pp. 1118–1127, 1984.
[4] K. B. Cooper, R. J. Dengler, N. Llombart *et al.*, "THz imaging radar for standoff personnel screening," *IEEE Trans. Terahertz Sci. Technol.*, vol. 1, no. 1, pp. 169–182, 2011.
[5] E. Pickwell and V. P. Wallace, "Biomedical applications of terahertz technology," *J. Phys. D: Appl. Phys.*, vol. 39, no. 17, pp. R301–R310, 2006.
[6] P. Siegel, "Terahertz technology in biology and medicine," *IEEE Trans. Microw. Theory Tech.*, vol. 52, no. 10, pp. 2438–2447, 2004.
[7] V. Krozer, T. Löffler, J. Dall *et al.*, "Terahertz imaging systems with aperture synthesis techniques," *IEEE Trans. Microw. Theory Techn.*, vol. 58, no. 7, pp. 2027–2039, 2010.
[8] R. Han, Y. Zhang, Y. Kim *et al.*, "Active terahertz imaging using Schottky diodes in CMOS: Array and 860 GHz pixel," *IEEE J. Solid-State Circuits*, vol. 48, no. 10, pp. 2296–2308, 2013.
[9] C. Wang, B. Lu, C. Lin *et al.*, "0.34 THz wireless link based on high-order modulation for future wireless local area network applications," *IEEE Trans. Terahertz Sci. Technol.*, vol. 4, no. 1, pp. 75–85, 2014.
[10] J.-D. Park, S. Kang, and A. M. Niknejad, "A 0.38 THZ fully integrated transceiver utilizing a quadrature push-push harmonic circuitry in SiGe BiCMOS," *IEEE J. Solid-State Circuits*, vol. 47, no. 10, pp. 2344–2354, 2012.
[11] H. J. Song, K. Ajito, Y. Muramoto, A. Wakatsuki, T. Nagatsuma, and N. Kukutsu, "24 Gbit/s data transmission in 300 GHz band for future terahertz communications," *Electron. Lett.*, vol. 48, pp. 953–954, 2012.
[12] H.-J. Song, J.-Y. Kim, K. Ajito, M. Yaita, and N. Kukutsu, "Fully integrated ASK receiver MMIC for terahertz communications at 300 GHz," *IEEE Trans. Terahertz Sci. Technol.*, vol. 3, pp. 445–452, 2013.
[13] Z. Wang, P.-Y. Chiang, P. Nazari, C.-C. Wang, Z. Chen, and P. Heydari, "A CMOS 210 GHz fundamental transceiver with OOK modulation," *IEEE J. Solid-State Circuits*, vol. 49, no. 3, pp. 564–580, 2014.
[14] S. U. Hwu, K. B. deSilva, and C. T. Jih, "Terahertz (THz) wireless systems for space applications," in *IEEE Sensors Applications Symposium (SAS)*, 2013, pp. 171–175.
[15] J. Park, S. Kang, S. Thyagarajan, E. Alon, and A. Niknejad, "A 260 GHz fully integrated CMOS transceiver for wireless chip-to-chip communication," in *Proc. Symp. VLSI Circuits*, Jun. 2012, pp. 48–49.
[16] K. Sengupta and A. Hajimiri, "Distributed active radiation for THz signal generation," in *IEEE ISSCC Dig. Tech. Papers*, 2011, pp. 288–289.
[17] Y. Tousi, O. Momeni, and E. Afshari, "A 283-to-296 GHz VCO with 0.76 mW peak output power in 65 nm CMOS," in *IEEE ISSCC Dig. Tech. Papers*, Feb. 2012, pp. 258–259.
[18] D. Huang, T. R. LaRocca, M.-C. F. Chang *et al.*, "Terahertz CMOS frequency generator using linear superposition technique," *IEEE J. Solid-State Circuits*, vol. 43, no. 12, pp. 2730–2738, 2008.

[19] P.-Y. Chiang, O. Momeni, and P. Heydari, "A 200 GHz inductively tuned VCO with −7 dBm output power in 130 nm SiGe BiCMOS," *IEEE Trans. Microw. Theory Tech.*, vol. 61, no. 10, pp. 3666–3673, 2013.

[20] U. R. Pfeiffer, Y. Zhao, J. Grzyb *et al.*, "A 0.53 THz Reconfigurable source array with up to 1 mW radiated power for terahertz imaging applications in 0.13 μm SiGe BiCMOS," in *IEEE ISSCC Dig. Tech. Papers*, Feb. 2014, pp. 226–257.

[21] T. Bryllert, V. Drakinskiy, K. B. Cooper, and J. Stake, "Integrated 200–240 GHz FMCW radar transceiver module," *IEEE Trans. Microw. Theory Tech.*, vol. 61, no. 10, pp. 3808–3815, 2013.

[22] S. Shahramian, A. Hart, A. Tomkins *et al.*, "Design of a dual W- and D-band PLL," *J. Solid-State Circuits*, vol. 46, no. 5, pp. 1011–1022, 2011.

[23] M. Seo, M. Urteaga, M. Rodwell, M.-J. Choe *et al.*, "A 300 GHz PLL in an InP HBT technology," *IEEE MTT-S Int. Microw. Symp. Dig.*, pp. 1–4, 2011.

[24] P.-Y. Chiang, Z. Wang, O. Momeni, and P. Heydari, "A silicon-based THz frequency synthesizer with wide locking range," *IEEE J. Solid-State Circuits*, vol. 49, no. 12, pp. 2951–2963, 2014.

[25] C.-C. Wang, Z. Chen, and P. Heydari, "W-band silicon-based frequency synthesizers using injection-locked and harmonic triplers," *IEEE Trans. Microw. Theory Techn.*, vol. 60, no. 5, pp. 1307–1320, 2012.

[26] V. Jain, B. Javid, and P. Heydari, "A BiCMOS dual-band millimeter-wave frequency synthesizer for automotive radars," *IEEE J. Solid-State Circuits*, vol. 44, no. 8, pp. 2100–2113, 2009.

[27] A. Mirzaei, M. Heidari, R. Bagheri, S. Chehrazi, and A. Abidi, "Injection-locked frequency dividers based on ring oscillators with optimum injection for wide lock range," in *Proc. IEEE Symp. VLSI Circuits*, Sep. 2006, pp. 174–175.

[28] D. Mijuskovic, M. Bayer, T. Chomicz *et al.*, "Cell-based fully integrated CMOS frequency synthesizers," *IEEE J. Solid-State Circuits*, vol. 29, no. 3, pp. 271–279, 1994.

[29] R. Spence, Linear Active Networks. New York: Wiley-Interscience, 1970.

[30] A. Hajimiri and T. H. Lee, "A general theory of phase-noise in electrical oscillators," *IEEE J. Solid-State Circuits*, vol. 33, no. 2, pp. 179–194, 1998.

[31] O. Momeni and E. Afshari, "High power terahertz and millimeter-wave oscillator design: A systematic approach," *IEEE J. Solid-State Circuits*, vol. 46, no. 3, pp. 583–597, 2011.

[32] J. Chien and L. Lu, "Analysis and design of wideband injection-locked ring oscillators with multiple-input injection," *IEEE J. Solid-State Circuits*, vol. 42, no. 9, pp. 1906–1915, 2007.

[33] S. Verma, H. Rategh, and T. Lee, "A unified model for injection-locked frequency dividers," *IEEE J. Solid-State Circuits*, vol. 38, pp. 1015–1027, 2003.

[34] J. Grzyb, Y. Zhao, and U. R. Pfeiffer, "A 288 GHz lens-integrated balanced triple-push source in a 65nm CMOS technology," *IEEE J. Solid-State Circuits*, vol. 48, pp. 1751–1761, 2013.

Part II

Digital-intensive phase-locked loops

Chapter 5
Time-to-digital converters

Fa Foster Dai[1] *and Hechen Wang*[1,2]

5.1 Introduction

Integrated circuits, especially analog circuits, are highly sensitive to process, voltage, and temperature (PVT) variations. The information processed by analog circuits is often embedded in the amplitude of the waveforms, which requires circuits with high precision and high linearity. On the other hand, analog signal processing in time domain, such as time-to-digital converters (TDCs) and digital-to-time converters (DTCs), was not widely adopted until the late 1990s when semiconductor technology advanced to the sub-micrometer region [1].

Similar to digital circuits, time domain signal processing takes advantages of the transistor dimension scaling to achieve improved transition speed as well as power and area efficiency. In the deep sub-micrometer domain, the benefit of time signal processing finally emerged. The process scaling results in the reduction of the gate delay and the improvement of temporal resolution, which is exactly the opposite of the degraded amplitude resolution due to the reduced supply voltage for small feature-size devices. The resolution of the time interval measurement using the TDC reaches the pico-seconds (ps) or even femto-seconds (fs) region [2]. As a result, digital phase-locked loops (DPLLs) have adopted TDCs or DTCs for rigorous time measurement or phase shifting [3]. Furthermore, many precision analog blocks can be enhanced using time signal processing. Alternate ADC architectures using TDCs for fine quantization have been reported [4].

TDC together with DTC plays a critical role in time domain signal processing as depicted in Figure 5.1. While ADCs and DACs perform the conversions between amplitude domain and digital domain, TDCs and DTCs convert the time information to digital data and vice versa [5]. TDCs and DTCs are often used in DPLL frequency synthesizers where they serve as phase detectors [6]. With the aid of the TDC, the DPLL gains an extra ability to adaptively adjust loop parameters, such as the loop bandwidth and the locking time [7]. Furthermore, with highly linear TDC and DCO, direct phase modulation through the DPLL can be implemented with adaptively tuned loop bandwidth [8].

[1]Department of Electrical and Computer Engineering, Auburn University, Auburn, AL, USA
[2]Wireless Communication Research, Intel Labs, Intel Corporation, Hillsboro, OR, USA

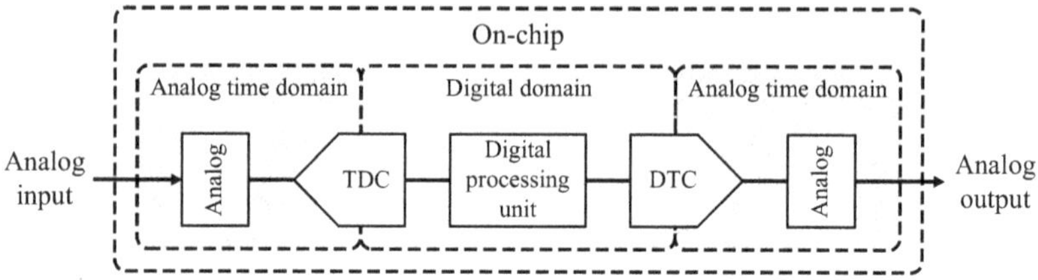

Figure 5.1 Conceptual drawing of a time domain signal processing system

While TDC and DTC are the key building blocks for DPLLs, the TDC is also used in the time interval measurement, and the DTC is used for phase shifting or as a time delay element. In addition, the TDC is used to digitize the received waveform envelope in a super-regenerative terahertz receiver [9] or a wireless polar receiver [10]. The time-based signal process can find applications in emerging areas such as internet-of-things (IoT) and edge inference artificial intelligence networks. Planted into sensor nodes in the network, TDC and DTC assist the network to plan the activation timing and duration, which directly affect the lifespan and accuracy of each node in the network [11].

5.2 Time–digital conversion basis

The first TDC prototype was invented for the measurement of the physics experiment in 1942 by Bruno Rossi. Due to the technology limitation, it used capacitors for time interval to voltage conversion. The corresponding voltage is then converted to digital bits using an ADC. The appearance of silicon-based integrated TDC was postponed till the late 1980s due to excessive timing requirements of the technology process [1,12]. While some well-established concepts such as the Vernier mechanism by Pierre Vernier (1584–1638) [13] and time stretching or amplifying [14] for dividing the time interval into measurable pieces are still valid, the implementation of TDCs advanced tremendously during the past decade. The old devices using vacuum tubes are now implemented in deep sub-micro complementary-metal-oxide-semiconductor (CMOS) technologies.

5.2.1 Time–digital conversion key parameters

Both TDC and DTC are data converters that transform information between analog domain and digital domain. Generally, most of the performance parameters can be directly borrowed from ADC and DAC characterizations, although a few subtle things must be clarified.

5.2.1.1 Basic structural parameters

Resolution

Resolution is the most elementary parameter. It represents the finest detectability that the converter can reach, which is in time unit of picoseconds with a modern

COMS process technology. A finite resolution causes quantization noise, which directly affects DPLL's in-band phase noise, one of the crucial performance indicators in DPLL designs.

Number of bits (detectable range)
Number of bits (NoB) is a parameter to determine the maximum time interval that a converter can reach. The NoB sometimes can also be expressed as a detectable range. An appropriate NoB balances DPLL's locking time, acquisition range, and power consumption. As the resolution of a TDC has been preset to meet the in-band phase noise requirement, the detectable range can be calculated as resolution $\times\ 2^{\mathrm{NoB}}$. In contrast, the NoB of an ADC implies ADC's resolution, since its full-scale input range is normally limited by its supply voltage, e.g., 1 V. Consequently, the larger the NoB is, the finer resolution the ADC can achieve.

5.2.1.2 Performance specifications

As a data converter, a linear conversion is desired. However, the measured transfer function deviates from the ideal one due to PVT variations, device mismatches, and other sources of inaccuracy. The performance parameters characterize the nonlinearity of the converter.

The linearity specifications generally include two categories: DC accuracy and dynamic performances. In an integer-*N* DPLL, the input to the TDC is relatively time invariant at the lock state. In the acquisition phase of an integer-*N* DPLL, the linearity of the TDC is not a big concern as long as its transfer function is monotonic. To the extreme, a highly nonlinear bang-bang TDC can be used to phase locking effectively. On the other hand, the TDC must quantize a time-varying dynamic signal in a fractional-*N* DPLL or in a PLL with direct digital phase modulation. The specifications that the designer need to focus on are thus application driven.

DC accuracy
The parameters specifying converter's DC accuracy are gain error, offset error, differential nonlinearity (DNL), and integral nonlinearity (INL). These four specifications provide a complete description of converters' DC performance. TDC and DTC perform the opposite functions with similar characteristics. We use TDC as an example to illustrate their performance specifications.

Transfer function: The transfer function or the transfer curve of a converter describes the relationship between the input time interval and the output digital code. The output of the plot is a sequence of 2^{NoB} codes, where NoB represents TDC's detectable range.

Gain error: Gain error describes the difference between the ideal code transition and the actual measured transition from code zero to the highest code. This can also be observed as a change in the slope of the transfer function shown in Figure 5.2.

Offset error: An ideal transfer function intersects the origin of the plot. The first code transition boundary occurs at 1/2 least-significant bit (LSB). When the offset error exists, a left or right shifting of the entire transfer function can be observed, as shown in Figure 5.2.

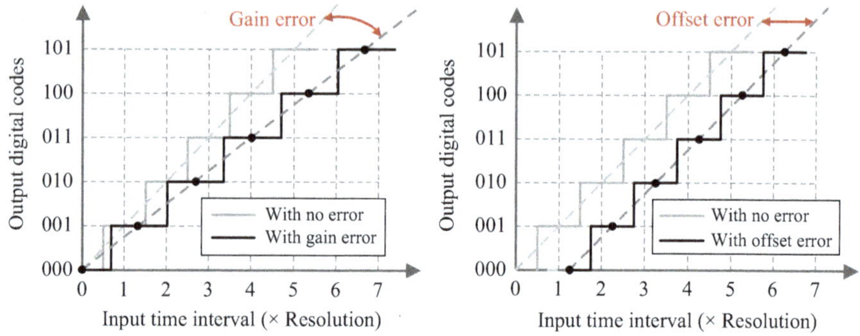

Figure 5.2 Gain error and offset error in the TDC transfer function

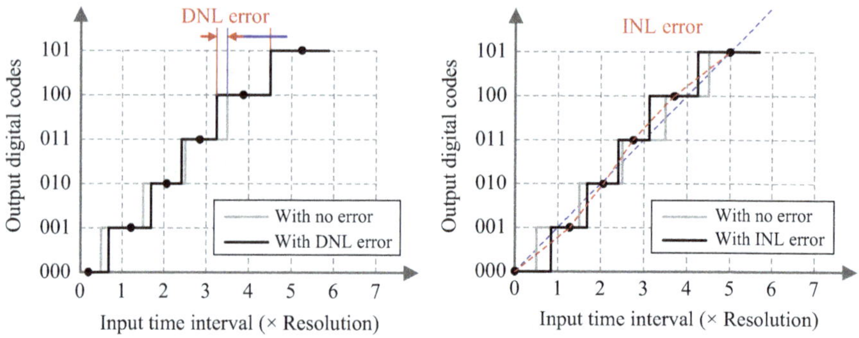

Figure 5.3 DNL and INL errors in the TDC transfer function

DNL: Under ideal conditions, each transition step on a TDC's transfer function curve should be uniform in width, which is equal to the TDC's resolution. Deviation of each step size from the TDC's ideal resolution is specified as its DNL, expressed in LSBs. This can be observed as unevenly distributed transition steps in the transfer curve. In Figure 5.3, the selected output code width is shown as larger than the previous code's step size. This difference is DNL error and calculated as follows:

$$DNL = \frac{(T_{n+1} - T_n)}{Resolution_{TDC}}$$

INL: The INL is the deviation of a TDC's transfer function from the ideal straight line, which connects the highest and lowest data points and can be treated as an ideal transfer curve with infinite resolution. The INL error is depicted as the difference between the ideal time interval lengths at which code transitions occur and the actual input time interval, as shown in Figure 5.3. DNL is related to INL in the sense that the INL error at any given time interval in the transfer function is the total summation of all DNL errors of the previous TDC codes.

Dynamic performance

Data converter's dynamic performance is normally analyzed in the frequency domain and is typically calculated by performing fast Fourier transform (FFT) on the output code sequence of the TDC within a given time window. In Figure 5.4, the fundamental tone is the input signal frequency. This is the signal measured with the TDC. The rest of the spectral energy is considered as unwanted signals, including all kinds of noises and harmonic distortions. Signal distortion is mainly caused by the nonlinearity in the conversion transfer function curve. Therefore, INL and DNL will affect the dynamic performance of a TDC as well.

Signal-to-noise ratio: The signal-to-noise ratio (SNR) is the ratio of the input signal power to the noise power (excluding harmonic distortion), expressed in decibels (dB):

$$\mathrm{SNR(dB)} = 10\log_{10}\left(\frac{P_{signal}}{P_{noise}}\right)$$

where P_{signal} is the power of the signal and P_{noise} is the power of noise. It is a comparison between the noise within the band of interest with respect to the input signal power. Theoretically, the noise contributor in the TDC is only quantization noise, which is determined by TDC's resolution.

The temporal quantum size A_q is TDC's resolution or one LSB, namely,

$$A_q = \frac{\text{Detectable Range}}{2^{NoB}} = 1\ \text{LSB}$$

Finite temporal resolution introduces a quantization error between the analog input time interval and its digital output code representation. Assuming that the quantization error is evenly distributed over the converted code from $-1/2$ LSB to $+1/2$ LSB, the expectation value of the error amplitude can be expressed as:

$$E\{\epsilon^2\} = \frac{1}{A_q}\int_{-(1/2)A_q}^{+(1/2)A_q} \epsilon^2 d\epsilon = \frac{1}{A_q}\left[\frac{\epsilon^2}{3}\right]_{-(1/2)A_q}^{+(1/2)A_q} = \frac{{A_q}^2}{12}$$

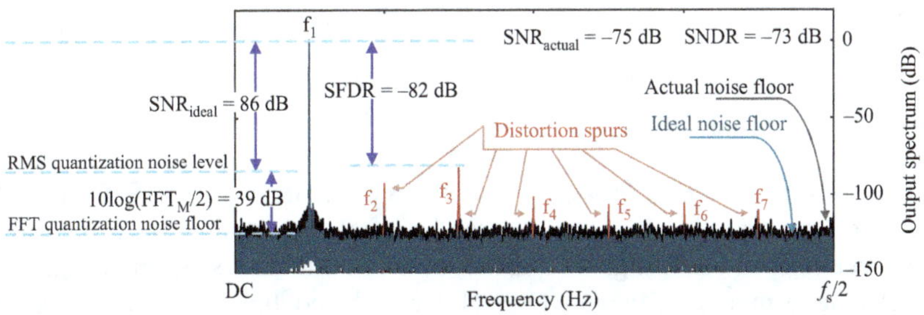

Figure 5.4 Spectrum plot for a 14-bit data converter (16,384-point FFT)

This quantization noise is uncorrelated in the entire frequency band. By using this result, the maximum theoretical SNR can be calculated. The root-mean-square (rms) value of a full-scale input time interval is given by:

$$A_{rms} = \frac{2^{NoB} A_q}{2\sqrt{2}}$$

Thus, the theoretical SNR can be written as:

$$\mathrm{SNR(dB)} = 10\log_{10}\left(\frac{A_{rms}{}^2}{E\{\epsilon^2\}}\right) = 10\log_{10}\left(\frac{3}{2}2^{2NoB}\right) = 6.02NoB + 1.76$$

Signal-to-noise and distortion ratio: Nonlinearity in the data converter results in harmonic distortion when analyzed in the frequency domain. Such distortion is observed as spurious tones or spurs in the spectrum at harmonics of the measured signal as illustrated in Figure 5.4.

Signal-to-noise and distortion ratio (SNDR or SINAD) offers a more comprehensive picture by including the noise and harmonic distortion into the analysis. It can be calculated using the following equation:

$$\mathrm{SNDR(dB)} = 10\log_{10}\left[\frac{A_1{}^2}{A_2{}^2 + A_3{}^2 + \cdots + A_n{}^2 + A_{noise}{}^2}\right]$$

where A_1 is the amplitude of the fundamental tone and A_n is the amplitude of the n-th harmonic tone. For an M-point FFT result, if the fundamental tone is in frequency bin m (with amplitude of A_m), the SNDR can be obtained from the FFT result as:

$$\mathrm{SNDR(dB)} = 10\log_{10}\left[A_m{}^2\left(\sum_{k=1}^{m-1} A_k{}^2 + \sum_{k=m+1}^{M/2} A_k{}^2\right)^{-1}\right]$$

To avoid any spectral leakage around the fundamental tone, often several bins around the fundamental frequency are ignored. Effective-number-of-bits (ENoB) is simply the SNDR expressed in bits rather than decibels by solving the "ideal SNR" as:

$$\mathrm{ENoB} = \frac{\mathrm{SNDR} - 1.76\ \mathrm{dB}}{6.02\ \mathrm{dB/bit}}$$

Spurious-free dynamic range: Spurious-free dynamic range (SFDR) is the difference between the power of the fundamental tone and its peak harmonic component or peak of the highest spurious tone in dB numbers. Spurious tones are generated at harmonics of the input signal due to the nonlinear conversion transfer function or at sub-harmonics of the sampling frequency due to mismatch or clock on-chip coupling.

5.3 Time–digital converter architectures

Similar to ADCs, various types of TDC architectures have been developed with a variety of schemes to balance resolution, detectable range, linearity, conversion speed, bandwidth, design complexity, power consumption, and area in order to fulfill a broad range of application-driven requirements.

5.3.1 Counter-based TDC

In the time when deep sub-micrometer CMOS technologies were not available, two types of TDCs, counter-based TDC and ADC-based TDC, were used. If high resolution is not required, usually larger than 1 ns, a counter-based TDC is adequate for the conversion. This type of the TDC is simply a high-frequency counter that increments every clock cycle. The implementation includes two basic building blocks: a frequency generator and a counter. The counter is controlled by an enabler, which is triggered by the input signal. The current contents of the counter represent the input time interval. With this approach, the measurement is an integer multiple of the clock cycles. Thus, the measurement resolution is limited to the clock period. For instance, a 1-ns temporal resolution requires a clock frequency equals to 1 GHz. Finer resolution can only be achieved with faster clock and unfortunately higher power consumption. Also, the accuracy of the measurement depends on the purity of the generated frequency, namely, its jitter or phase noise.

A PLL frequency synthesizer is commonly adopted to generate the counter clock in the giga-hertz range with good spectral purity. Although pico-second resolution is realizable, the power consumption of such a PLL and the associated counter operating at such a high-frequency disqualifies the approach for sub-nanosecond TDC applications.

5.3.2 Analog-to-digital conversion-based TDC

ADC-based time interval measurement is another approach during the CMOS technological underdevelopment period. By utilizing advanced techniques available in ADC designs, such as sigma-delta modulation and successive-approximation register, this type of TDC achieves a much finer resolution even with large feature-size technologies [15,16].

This type of TDC consists of two major building blocks, an integrator and an ADC. The integrator is normally built with a capacitor and a constant current source. The capacitor is charged by the current source during the measured time interval and converts the time interval into voltage domain. The converted voltage is then digitized by an ADC. A basic block diagram is given in Figure 5.5.

Although pico-second level resolution is achievable owing to well-developed ADC techniques, the performance of this type of TDC is limited by several practical issues. Primarily, the time-to-voltage conversion must meet the linearity requirement of the overall TDC. The integrator used in the TDC is similar to a charge pump, in which a constant current source is connected to a holding capacitor during the measurement time interval. However, the output resistance of the current source is limited,

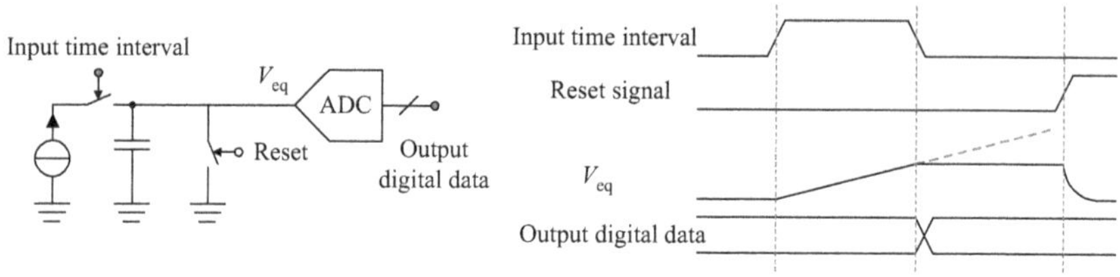

Figure 5.5 Block diagram of the basic ADC-based time-to-digital converter

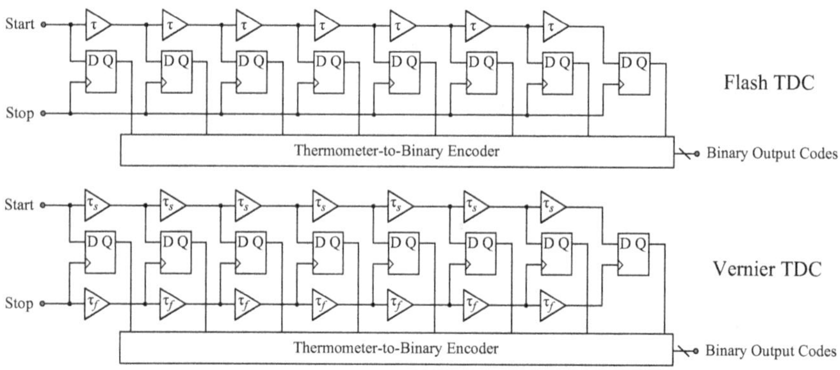

Figure 5.6 Block diagrams of a flash TDC and a Vernier delay line TDC

especially in deep sub-micrometer technology, which leads to a degraded linearity performance. This issue could be partially neutralized by employing an operational amplifier (op-amp)-based active integrator. However, the congenital deficiency of the op-amp bandwidth greatly reduces the conversion speed of the TDC.

5.3.3 Flash (single delay line) TDC

In order to achieve the resolution lower than the clock period, uniformly distributed sub-period clock phases can be generated using delay lines, as shown in Figure 5.6. The resolution of a delay line TDC is, therefore, determined by the propagation delay of a single delay cell [17]. The delay cells in the delay line are usually built with CMOS inverters and its resolution is typically limited to a few tens of pico-seconds with current technology. The TDC detectable range is the product of the number of delay stages and its resolution. The flash TDC structure is widely adopted in various applications due to its simplicity and low latency as the output is immediately available by when the stop signal arrives. Similar to the flash ADC, the numbers of the delay cells and the comparators grow exponentially when the number of bits increases. Additionally, mismatches between the delay cells and the comparators directly affect the linearity performance of the TDC. The variation in the propagation delay accumulates along the delay line. Thus, the measurement error accumulates and reaches its maximum when the signal propagates to the end of the delay line.

5.3.4 Vernier delay line TDC

The resolution that a flash TDC can reach is limited by the delay of each stage, which is normally in the range of a few tens of picoseconds in advanced technology. However, the DPLL's in-band phase noise requires the TDC with resolution finer than 10 ps. To further improve the resolution, Vernier topology is therefore adopted, which reveals a way to achieve sub-gate-delay resolution [12,13].

As shown in Figure 5.6, a Vernier TDC consists of two delay lines with a slight difference in their time delays. The start (lead) signal is fed to the line with a larger delay, i.e., the slow delay line, and the stop (lag) signal is fed to the fast delay line. During the measurement, the propagation of the stop signal is relatively faster and will catch up with the start signal eventually after passing a certain number of delay cells. The resolution of the Vernier TDC is thus the delay difference between the unit delays in the slow line and the fast line. Therefore, a fine resolution independent of the gate delay can be implemented.

Compared to the flash TDC, the latency in the Vernier TDC is equal to the time that two signals catch up with each other. Consequently, for a given range, the length of the delay lines in the Vernier TDC is the quotient of the maximum input time interval divided by the TDC resolution. As a result, the length of the delay lines is considerably larger than the flash TDC. Essentially, the Vernier delay line TDC trades the resolution with the penalty of increased hardware and power consumption as well as the reduced conversion rate.

5.3.5 Vernier ring TDC

The Vernier ring TDC architecture, shown in Figure 5.7, is a variant of the Vernier TDC that places the delay cells in a ring format such that the delay cells can be

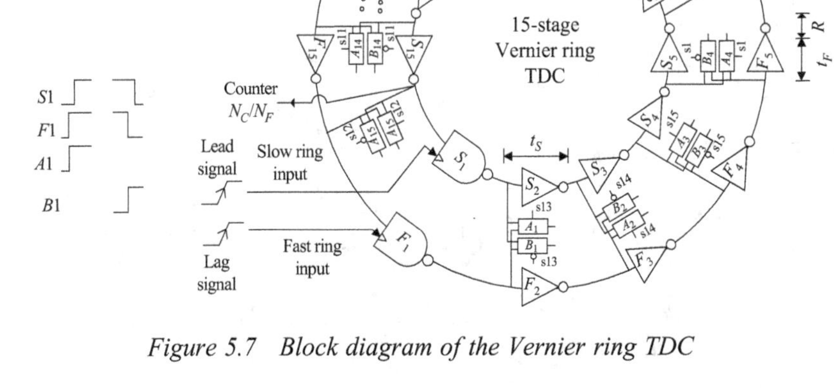

Figure 5.7 Block diagram of the Vernier ring TDC

reused for measuring large time intervals [18]. When the leading edge arrives, it will be fed into the slow ring. Prior to the arrival of the lag edge, the converter uses only the slow delay chain, resulting in a single delay line TDC. When the lag edge comes, it will be fed into the fast ring and the converter automatically changes to the Vernier mode. When the lag edge catches the leading edge, comparison stops. The reuse of the Vernier delay cells in a ring configuration achieves fine resolution and large detectable range simultaneously with a small area. The only drawback is its conversion speed.

5.3.6 Oversampling gated ring oscillator (GRO) TDC

Counter-based TDC can also achieve a sub-gate-delay resolution with the aid of a ring oscillator with multi-phase outputs. The ring oscillator intrinsically contains multi-phase outputs evenly distributed in one oscillation cycle. The temporal difference between each phase becomes the resolution of the quantizer. As shown in Figure 5.8, the GRO TDC includes a ring oscillator, a counter, and a register. When the rising edge arrives, the counter starts to count and when the falling edge comes, the counting stops. The quantization error in previous measurement cycle, denoted as $qe[k-1]$ in the plot, is the residue of the measurement. The gated ring structure preserves the phase of the waveform when the counting is stopped and starts the oscillation waveform at the same phase preserved from the last measurement cycle. This process effectively moves the quantization error of $qe[k-1]$ to the beginning of the current measurement, namely, the current measurement cycle starts $qe[k-1]$ earlier. As a result, there are no quantization errors between the measurements cycles. In another word, the quantization error is first-order canceled, achieving the first-order noise shaping effect [19].

With the first-order noise shaping and oversampling, the quantization noise at low frequencies is shaped to the high-frequency band. Since the high-frequency components are filtered in the system such as a DPLL, the GRO TDC is able to achieve a temporal resolution even below 1 ps. However, it is important to point out that the improved resolution is only achievable for repeated measurements over

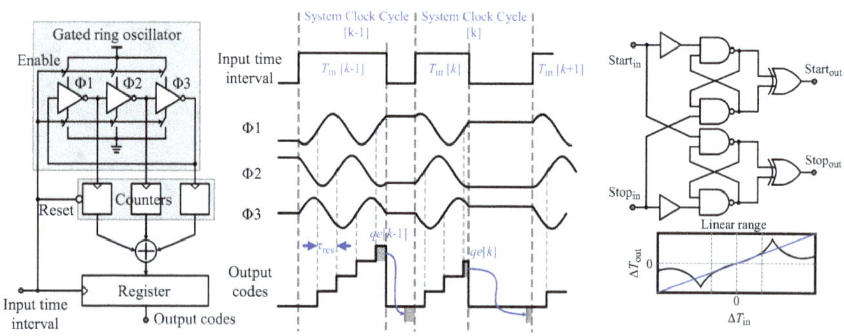

Figure 5.8 Block diagram of GRO TDC. SR-latch-based time amplifier and its time domain transfer curve

time such as the case of a PLL. The fine resolution is not realizable for a single shot measurement. Consequently, the actual conversion bandwidth of this GRO TDC is limited to only a few megahertz, similar to the sigma-delta ADCs. Another drawback of the GRO TDC is the uncompensated quantization errors due to its leakage currents. The preserved phases between each state are stored as charge on parasitic capacitors in the delay cells. The discharge during the stop time is inevitable, especially in deep sub-micrometer technologies [20].

5.3.7 Time amplifier-based TDC

An alternate approach to accomplish the sub-gate-delay resolution is to amplify the time interval using a time amplifier [14]. Many techniques developed for ADC can be adopted to the TDC designs. Voltage amplification is commonly seen in pipelined ADC designs. The residue voltage from the previous stage is amplified before sent to the next stage. Due to the amplification, the requirement on the comparators can be relaxed. However, the linearity of the converter is highly dependent on the amplifiers between the stages.

While a linear amplifier can be relatively easily implemented in the voltage domain, it is very challenge to build a highly linear time amplifier. An SR-latch-based time amplifier is depicted in Figure 5.8. Although fine resolution can be achieved with pipelined TDC topology, the detectable range is limited due to the narrow dynamic range of the time amplifier.

5.4 Advanced time-to-digital conversion techniques

In DPLL designs, the TDC directly affects the in-band phase noise and fractional spurious level [21]. The in-band phase noise of a DPLL is related to the TDC quantized resolution as

$$\mathcal{L} = \frac{(2\pi)^2}{12}\left(\frac{\tau_{TDC}}{T_{DCO}}\right)^2 \frac{1}{f_{ref}}$$

where τ_{TDC} is the TDC resolution, T_{DCO} is the oscillation period of the digitally controlled oscillator (DCO) output, and f_{ref} is the reference frequency of the DPLL [22]. In order to lower in-band phase noise, the TDC resolution has been reduced to around 1-ps level according to the recently reported data [3,23]. However, improving TDC's linearity performance faces increasing challenges particularly for high-resolution TDC. The TDC nonlinearity not only jeopardizes PLL's in-band phase noise but also leads to a deteriorated fractional spur level in fractional-N PLLs. When operating in a fractional-N mode, a multi-modulus divider toggles the division ratio between N and $N + 1$ and generates a gradually increased phase difference between reference signal and divided feedback signal [24]. Quantized by the TDC, a staircase ramp signal is generated at TDC output, as shown in Figure 5.9. Directly feeding this signal into DCO through the loop filter may lead to an unacceptable spur level or unstable loop dynamics. In order to reduce the

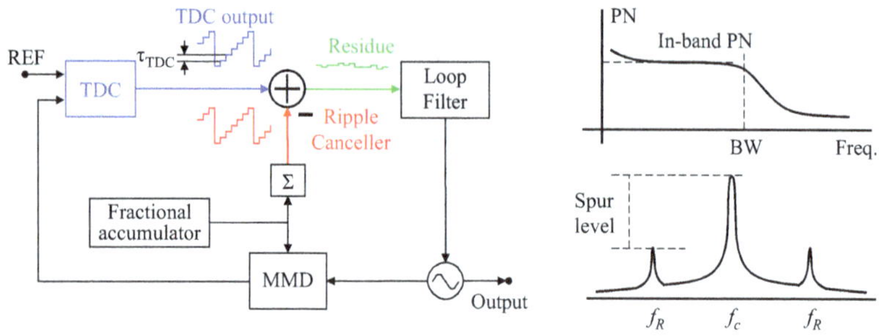

Figure 5.9 Illustration of in-band phase noise and fractional spur level related to TDC performance in a DPLL with a fractional-N ripple canceler

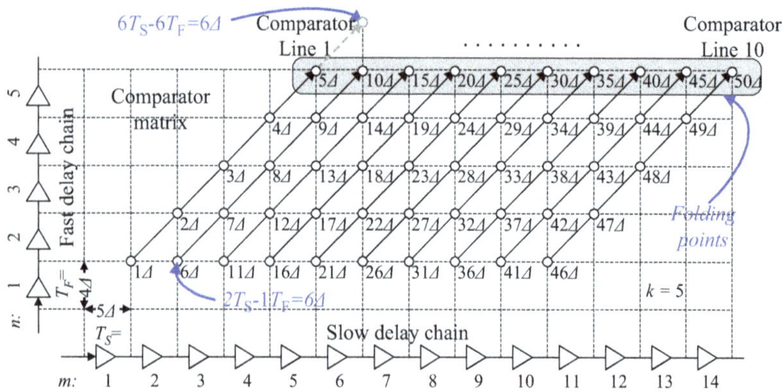

Figure 5.10 Illustration of a conventional 2-D Vernier TDC topology

fractional spurious tones, a digital ripple cancellation technique is often employed. As illustrated in Figure 5.9, an ideal staircase ramp is generated following the variation of the loop division ratio in fractional mode. This ideal staircase ramp is subtracted from the TDC raw output to cancel its staircase signal while extracting the needed DC component. As a result, a less rippled TDC output for DCO tuning is generated, leading to an improved fractional spur performance. However, this spur suppression technique is highly sensitive to TDC's linearity.

5.4.1 Linearity issues associated with 2-D Vernier TDC

To achieve reasonable in-band noise and fractional spur performances in DPLL designs, TDCs are required to have sub-gate delay resolution according to in-band phase noise analysis given above, while its detection range should cover at least one DCO oscillation cycle. Considering all these constraints, the 2-D Vernier TDC is a preferred candidate architecture [25]. A Vernier TDC with a 2-D comparator array, as illustrated in Figure 5.10, is evolved from a prior-art Vernier TDC with a 1-D comparator line. 2-D Vernier TDC has reduced a number of delay elements and a much higher conversion rate compared with other types of Vernier TDCs. However, the linearity of a 2-D Vernier TDC is more sensitive to delay variations compared to 1-D Vernier TDCs [21].

With a closer look of Figure 5.10, the TDC consists of two delay chains with two different unit delays τ_S and τ_F, respectively. The resolution is defined by the delay difference, for instance, $\tau_S - \tau_F = 5\Delta - 4\Delta = \Delta$ in this case. The 2-D Vernier TDC breaks the comparator line into multiple sections and forms a 2-D comparator array instead of forming a long single comparator line. The 2-D Vernier TDC uses less delay stages to cover the same detection range. However, each of the segmented comparator line contains k comparators, e.g., k is equal to 5 in this case. The folding points cycled by the gray box in Figure 5.10 indicate the comparison signal's transition locations into next comparator lines. The extended sixth comparison point in the first comparator line is equal to $6\tau_S - 6\tau_F = 6\Delta$, while the first comparison point in the second line is equal to $2\tau_S - 1\tau_F = 6\Delta$. In order to ensure a smooth transition between each comparator lines, the two comparisons should produce an identical delay response to the input signals, namely, $6\tau_S - 6\tau_F = 2\tau_S - 1\tau_F$. In general, a linear conversion using 2-D comparator array topology requires that:

$$k(\tau_S - \tau_F) = \tau_S$$

This condition demands precisely matched delays in both delay chains against the PVT variations. A small delay deviation can lead to large periodic nonlinearity. Mismatches introduce slope errors, gaps, or overlaps between the comparator lines, producing periodic errors in both differential nonlinearity points of the 2-D comparator array topology. To illustrate the problem, a 4% delay mismatch is assumed in the simulation. Figure 5.11 presents the simulated TDC transfer curve, DNL and INL. These plots illustrate that a small delay mismatch could lead to a large nonlinearity in the 2-D Vernier TDC. Indeed, the number of periodic cycles in the nonlinearity plots corresponds to the number of comparator lines with their peaks located at the folding points of the comparator array.

A nonlinear TDC transfer curve can lead to a high fractional spur level in a fractional-N DPLL. A 2-D Vernier TDC-based DPLL with a digital ripple canceler was presented in [21]. Figure 5.11 gives the measured TDC output in its fractional-N operation. The signal has a periodic cycle of $1/f_\Delta = 1/(F \cdot f_{ref})$, where f_Δ is the closest fractional spur offset frequency and F is the fractionality. In this case, the DPLL is running with a division ratio of $29 + 1/64$ and a reference frequency of 80 MHz. The synthesized frequency is centered at 2.32125 GHz. The closest spur is located at $f_\Delta = 1/64 \times 80$ MHz $= 1.25$ MHz. The 2-D folding point errors can be clearly seen from the measured waveform given in Figure 5.11. Subtracted from the ideal staircase waveform generated by the fractional accumulator, the measured TDC residue error and its filtered version are plotted in Figure 5.11. Note that the residue error represents the TDC's INL. Even with the digital ripple canceling technique, the filtered TDC residue error still shows nonlinearity associated with the folding point errors caused by the TDC 2-D comparator array, which will be addressed later. The fractional spurs are affected by TDC nonlinearity and can be

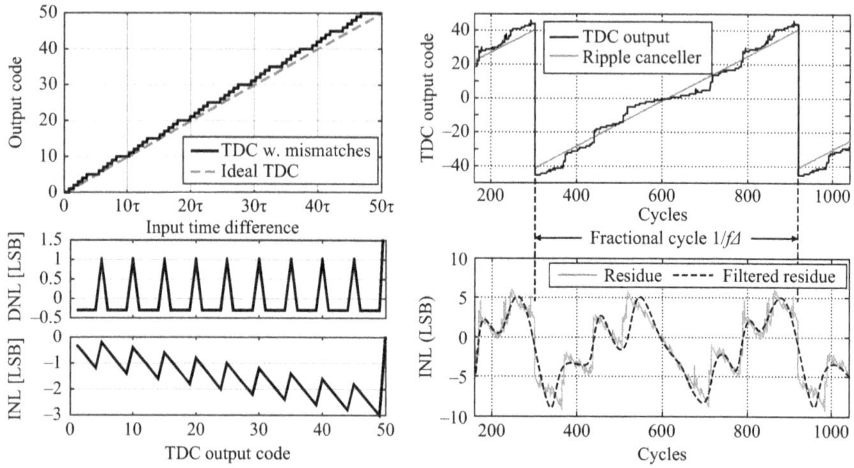

Figure 5.11 Simulated TDC transfer curve, DNL, and INL with 4% delay mismatch. Measured TDC outputs after the fractional-N ripple canceler, showing TDC nonlinearity caused by periodic folding errors of the 2-D comparator array

analyzed taking Fourier transform of the DCO control signal. The smoothed residue error in time domain is mapped into the frequency domain as multiple tones located at f_Δ, $2f_\Delta$, $3f_\Delta$, etc. The DCO output in the loop is modulated by the filtered residue error signal. The frequency components of this filtered control signal will be up-converted to DCO's output, showing as fractional spurs.

5.4.2 Spiral comparator array for 2-dimensional Vernier TDC

Previous discussion reveals that 2-D Vernier TDC peak nonlinearity appears at folding comparator line folding locations. By following a saw-tooth arrangement in the traditional 2-D array, the last comparator in n-th comparator line faces much larger accumulated delay mismatches comparing with the first comparator in the $(n + 1)$-th comparator line, leading to a discontinuous transfer curve. To break this trend, we thus propose to configure the 2-D comparator array in a spiral arrangement as shown in Figure 5.12 [25]. In this arrangement, instead of folding the comparator line in a saw-tooth form in one direction, we rearrange the comparator path in a spiral shape. Referring to Figure 5.12, the comparison points start with climbing up along the comparator line from node "0" in a way similar to the conventional 2-D Vernier TDC. When reaching the folding point, the comparison folds back counter-clockwise to the left side and continuous downwards, as shown in Figure 5.12. The separated two sides of the comparator array have opposite comparison mechanisms: on the right-hand side, it satisfies that $m\tau_S - n\tau_F$ and defined as the positive plane, while, on the left-hand side, it meets that $n\tau_F - m\tau_S$ and defined as the negative plane, where n and m are the unit delay index of the fast

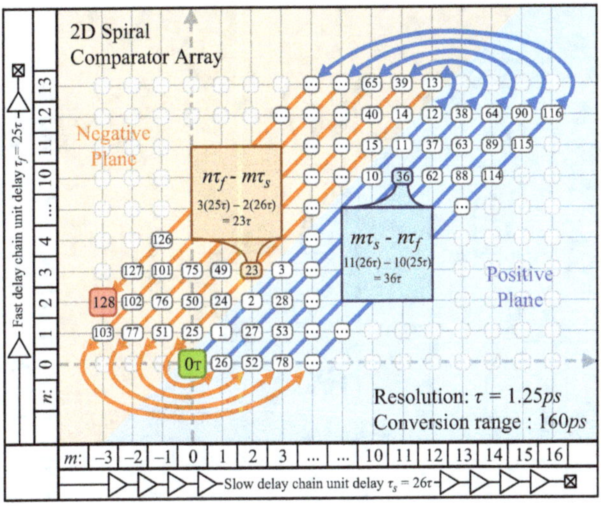

Figure 5.12 Proposed Vernier TDC with a 2-D spiral comparator array

and slow delay chains, respectively. As a result, the mismatches along the comparison path are partially compensated, resulting in an improved linearity performance.

To compare different comparator arrangements, we use reduced numbers of delay cells to illustrate the proposed spiral arrangement in comparison with other two conventional 2-D arrangements shown in Figure 5.13. Scheme "1" shows the proposed 2-D spiral comparator array and is used as the benchmark for evaluation. Schemes "2" and "3" provide two options that achieve the same resolution of "1τ" and the conversion range of "20τ" by using conventional arrangements. Scheme "2" uses the same amount of unit delay cells as that of Scheme "1." However, in order to maintain the "1τ" resolution, the temporal delay of its unit delay cells has to be reduced to 5τ comparing to 10τ in Scheme "1." Thus, Scheme "2" is more sensitive to delay mismatches and parasitic effects. Scheme "3" is built with the same temporal unit delay as that of Scheme "1"; yet, its comparator line length is doubled to fulfill the 2-D TDC linear requirement given in the previous equation, resulting in degraded nonlinearity and increased power consumption due to longer delay lines.

From topological point of view, the comparison path in the 2-D spiral Vernier TDC forms a spiral shape. The comparison starts from the center of the comparator array and gradually fans out to the outer lines, alternately across the positive plane on the right and the negative plane on the left when the input time interval increases. If there are mismatches, the errors accumulated on positive and negative planes will partially cancel with each other. In contrast, in the conventional 2-D comparator arrays, the comparison path follows a saw-tooth shape, moving in one direction on the positive plane, which accumulates mismatch errors.

To further analyze the nonlinearity of the above three schemes, a theoretical model is built based on the theory presented in [26]. Two kinds of errors, absolute delay error $\varepsilon_{\mathrm{Absolute}}$ and local delay error $\varepsilon_{\mathrm{Local}}$, are added, where $\varepsilon_{\mathrm{Absolute}}$ is the

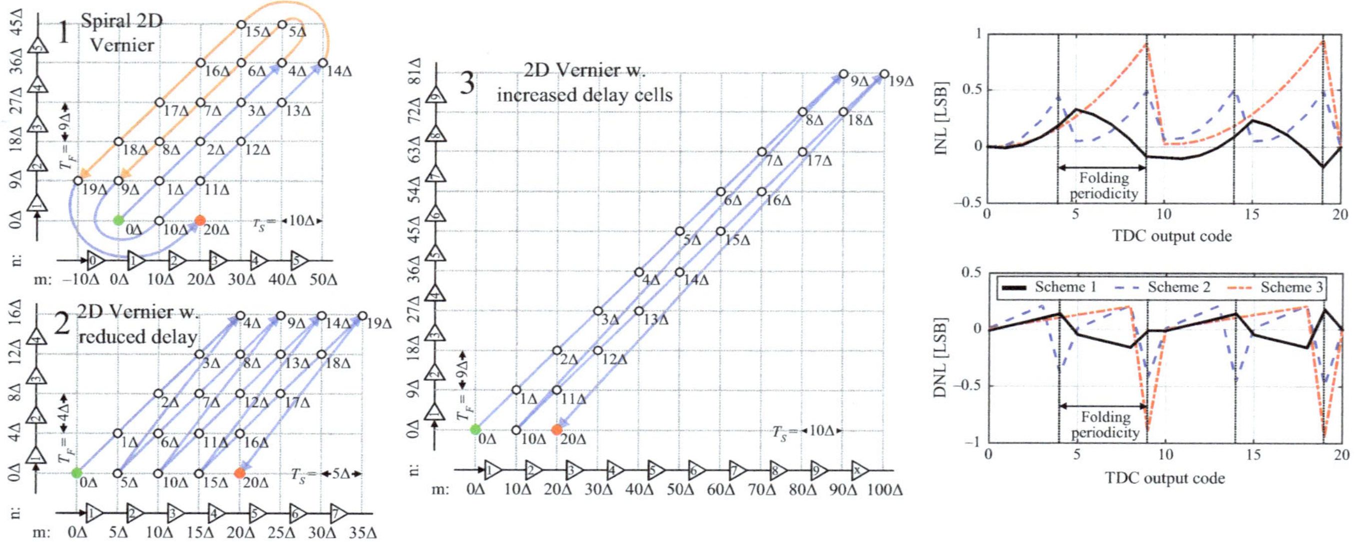

Figure 5.13 Comparisons among proposed 2-D spiral comparator arrangement (Scheme 1) and conventional 2-D comparator arrangements (Schemes 2 and 3), indicating a better linearity achieved by using the spiral comparator array formation with less delay elements

fixed delay error applied to all the unit delay cells in the delay chains and ε_{Local} is a gradually increased delay error along the delay chains, which models the unevenly distributed on-chip doping level. The simulated TDC INL in Figure 5.13 reveals that Scheme "1" has two opposite INL slopes alternatively appears along the TDC detectable range. As a result, its INL is bounded around zero. The INL slope of Scheme "2" is the largest due to the reduced temporal delay of unit delay cells. Scheme "3" ends up with the largest INL due to its longest delay lines along one direction. Moreover, the transitions between comparator lines are much smoother in a spiral arrangement compared with traditional saw-tooth cases, which end up with a much improved DNL as shown in Figure 5.13. Among all these three options, the spiral scheme achieves the best INL and DNL performance and has the least number of unit delay cells, which indicates less mismatches and lower power.

5.4.3 Linearization improvement techniques

5.4.3.1 Delay interpolation of unit delay cells

Vernier TDCs' nonlinearity mainly comes from the temporal delay errors of the delay units. Minimizing the delay error is the prerequisite for improving its linearity. The unit delay cell in the delay chain comprises a pair of cascaded inverters as shown in Figure 5.14. To reduce mismatch, both fast and slow delay chains employ identical unit delay cells. In this design, the unit delay is tunable from 19 to 43 ps with seven digitally controlled bits to obtain digital calibration compatibility and meet tuning requirements against PVT variations. The seven delay tuning bits are constructed with six pairs of NMOS and PMOS transistors sized with binary weights. The first and second LSBs of the tuning bits share the same transistor pair. A pair of keep-life NMOS and PMOS transistors is connected in parallel with delay tuning transistors. The median value of the delay tuning range can be varied by adjusting the size ratio between keep-life transistors and the tuning transistors. Overall, each tunable delay cell is a 7-bit DTC with quantization errors due to its digitized tuning steps. Moreover, the quantization granularity is not evenly distributed due to the intrinsic nonlinearity of the MOSFETs. Indeed, the transfer curve of the delay cell follows an exponential curve approximately, as shown in Figure 5.14. For instance, to achieve a 32-ps time delay, the closest reachable delay in the DTC is 31.9 ps as shown in Figure 5.14. This 0.1-ps time difference introduces a 0.3% delay error that leads to an INL of more than 1.5 LSB according to simulations.

To deal with this issue, we propose to interpolate the precise delay amount by using $\Delta\Sigma$ modulation. The delay interpolation with $\Delta\Sigma$ noise shaping is illustrated in Figure 5.15, where the unit delay cells can be digitally tuned to four adjacent reachable discrete delays. To obtain the desired interpolated delay of 32 ps, a second-order $\Delta\Sigma$ output sequence is used to sequentially select these four delays as the timing diagram shown in the bottom of Figure 5.15, where the time-average value of the temporal delay amount is 32 ps. The static delays corresponding to the four-digital delay-controlled-words (DCWs) vary among four reachable levels. Controlled by the second-order $\Delta\Sigma$ modulator, the spectrum of the delay sequence demonstrates the 40-dB/dec noise shaping effect. The loading on the pseudosupply

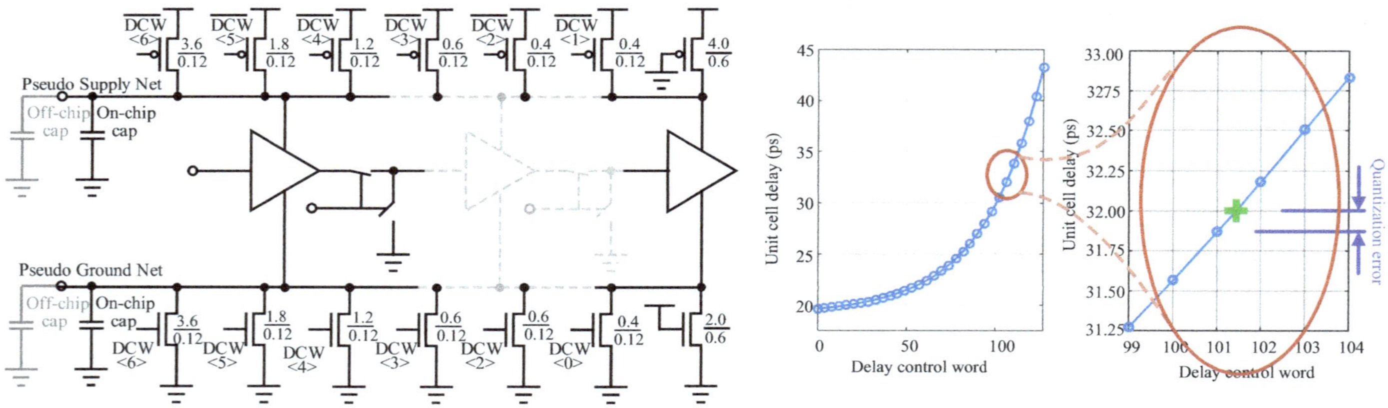

Figure 5.14   Seven-bit digitally controlled tunable unit delay cell circuit diagram and its digitally controlled tunable unit delay cell transfer curve

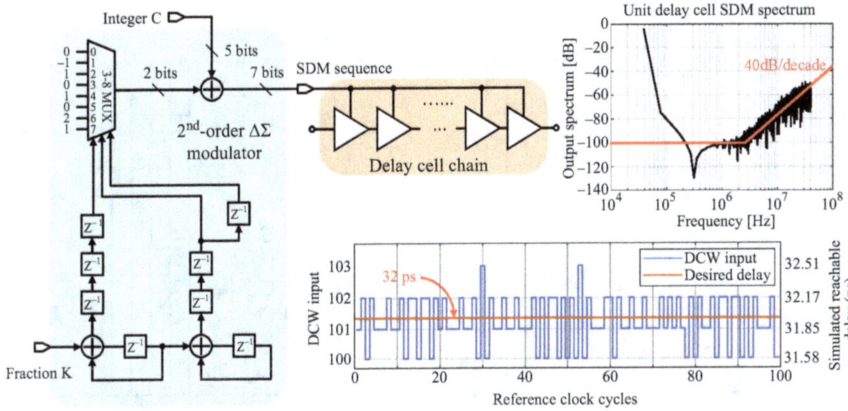

Figure 5.15 A second-order ΔΣ modulator used for DTC delay interpolation with quantization error noise shaping

and ground lines of the delay cell (see Figure 5.14) naturally provides a low-pass filter with about 2-MHz bandwidth that helps removing the shaped high-frequency noise, leading to a smooth time-averaged interpolation delay value. Note that this delay tuning bandwidth is determined by the loading of pseudosupply and ground lines and is not related to the TDC conversion bandwidth since the signal bandwidth of the delay cells are related to the inverter speed only.

The architecture of the second-order ΔΣ modulator is also shown in Figure 5.15 [27]. In our case, the integer value C is set to be 101. The fractional value K is determined by the distances between two adjacent quantized delay steps as well as the accumulator size, which has 10 bits in this design. Additionally, the adjustable fractional value K is used to compensate the mismatch between the seven–digital-controlled switches and the nonlinearity of the DTC transfer curve in this design.

5.4.3.2 2-D comparator array folding error randomization

As discussed in Section 1.4.1, 2-D Vernier TDCs suffer from periodic nonlinearity due to the transition errors at the folding points between different comparator lines in a 2-D array (see Figure 5.10). Even with delay calibration and the spiral comparator arrangement, the delay mismatch between the delay lines still cannot be eliminated completely. As a result, the folding errors are inevitable and periodic INL peaks at the folding locations are still present, which results in a high fractional spur level at the DPLL output.

Comparator folding locations are fixed in hardware once the delay chains and comparator parameters are chosen. To reduce the nonlinearity, we propose to randomize the folding locations using multiple comparator configurations. If there are multiple sets of comparator line folding locations that can satisfy the equation, we can choose different folding points in each comparison cycle, leading to a reconfigurable comparator array architecture that randomizes the mismatch errors. Figure 5.16 illustrates four valid configurations of a spiral 2-D comparator array, in

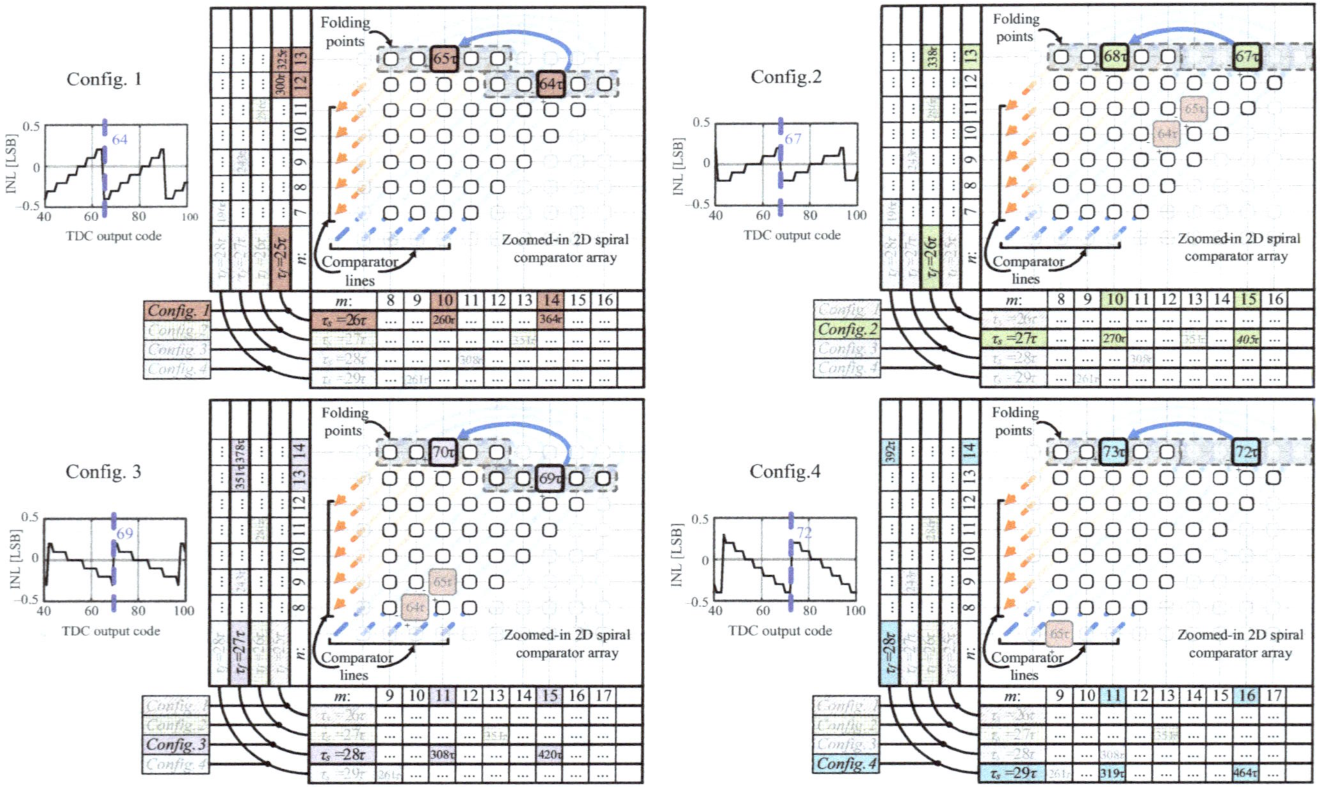

Figure 5.16 Four sets of delay settings used for folding error randomization and the resultant INLs. Also shown are four sets of spiral comparator array configurations that meet the delay requirement and their corresponding folding point locations

which "Config. 1" is the first arrangement with delay $\tau_f = 25\tau$, $\tau_s = 26\tau$. The enlarged squares labeled with "64τ" and "65τ" indicate one of comparator line folding locations in Config. 1, where the maximum periodic error occurs. In "Configs. 2–4" with different delay settings, the "64τ" nodes are moved to different locations and the corresponding folding points in the simulated INL curves are shifted to TDC output codes 67, 69, and 72, respectively, while the time resolution settings of "1τ" among all configurations are kept the same. The randomization block diagram is shown in Figure 5.17. With the tunable delay cell and reconfigurable comparator array, only one set of hardware is required. A ΔΣ output sequence generated by a second-order ΔΣ modulator preloads one of the four configuration settings to the TDC at each reference cycle and selects their corresponding output by controlling a 4-1 multiplexer.

A delay variation increases with the length of the delay chain. A signal goes through a longer delay chain and has a larger delay variation, as illustrated in Figure 5.18. Moreover, due to the layout geometry, the comparison points close to the end of the transfer curve, which correspond to a large input time interval, face larger delay deviation from layout mismatches and parasitics. When combined with the spiral comparator array architecture, the randomization level is automatically adapted to input signals' time interval. As shown in Figure 5.18, comparison point P4 located in the first comparator line has the same location among four different configurations. In other words, it experiences no ΔΣ randomization effect. The other three comparison points P_{34}, P_{94}, and P_{124}, all have four different locations in four configurations. The second point among the four points is selected to be the default "0" point of a second ΔΣ output sequence. Then, the other three points are assigned to "−1," "+1," and "+2," correspondingly. The distance between each randomization point increases when comparison points move away from its nominal location.

The four configurations have the same resolution, namely, they have the same amount of the quantization error. The difference in their nonlinearity characteristics

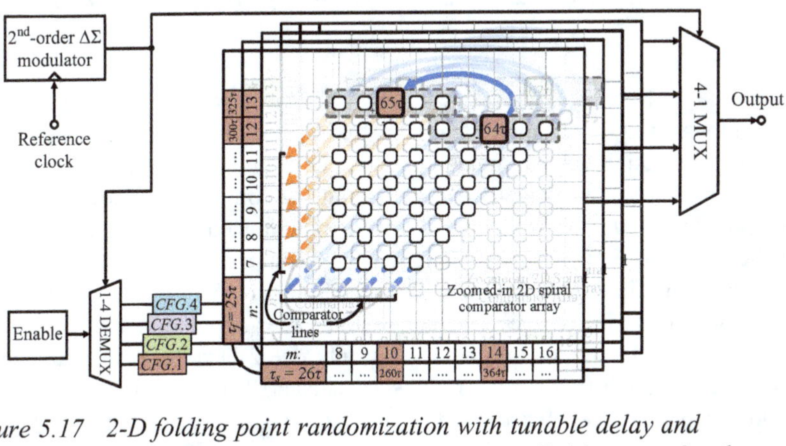

Figure 5.17 2-D folding point randomization with tunable delay and reconfigurable comparator array controlled by second-order ΔΣ modulators

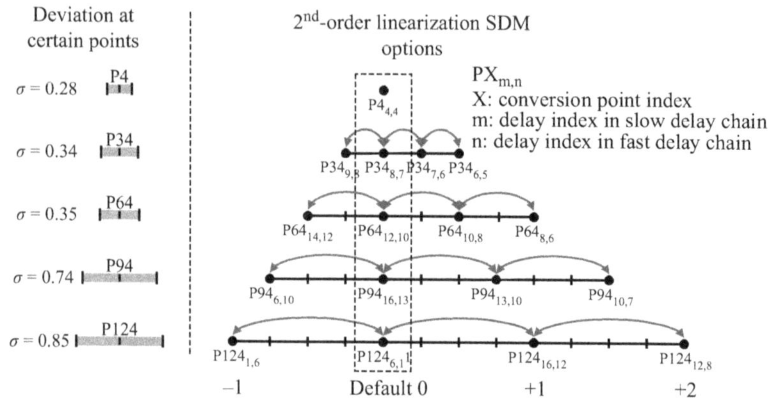

Figure 5.18 Delay variations and second-order $\Delta\Sigma$ randomization points

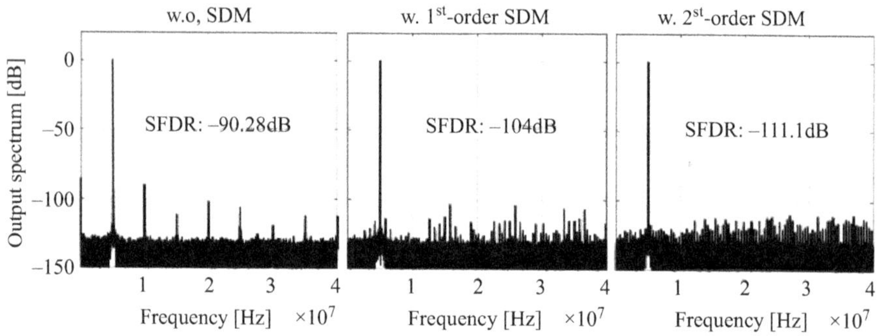

Figure 5.19 Simulated TDC output spectrum without, with the first order, and the second-order $\Delta\Sigma$ modulators for folding error randomization

lies upon the folding locations, namely, where the error peaks. Selecting different configurations using a $\Delta\Sigma$ modulator, the folding locations can be randomized, while the quantization noise remains the same. This randomization technique does not have a noise shaping effect, nor does it limit the conversion bandwidth. In the process, the spur cause by TDC nonlinearity is randomized, while its resolution is untouched. Figure 5.19 gives a simulated TDC output spectrum for the cases without, with the first order, and with the second-order $\Delta\Sigma$ modulations. A 20-dB SFDR improvement is achieved with the second-order $\Delta\Sigma$ modulation.

The reconfigurable structure comprises only one 2-D spiral comparator array in hardware, although four configurations are needed. Therefore, power consumption and area penalty are minimal for the proposed linearization technique. The four configurations are always available in the 2-D spiral comparator array and one of the four valid configurations is selected based on the output sequence of a $\Delta\Sigma$ modulator at the beginning of each comparison cycle. The outputs are further processed by a thermometer to binary encoders to produce the final TDC output.

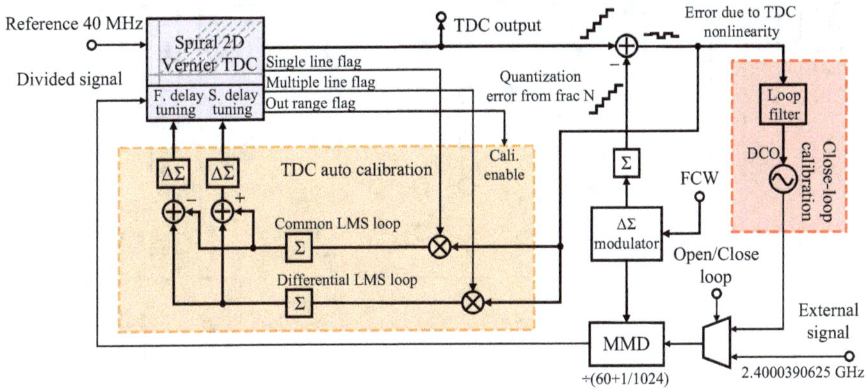

Figure 5.20 Automatic 2-D Vernier TDC closed–open-loop delay calibrations

5.4.3.3 TDC delay calibration

Calibration is another linearization technique. A closed-loop automatic digital calibration technique based on the least-mean-square (LMS) algorithm is developed in [28]. This work leverages the technique with additional open-loop calibration capability for TDC stand-alone applications. The block diagram of the TDC calibration circuit is shown in Figure 5.20. The calibration is accomplished with a 40-MHz reference clock. The loop's output frequency is set to a certain fractional number with a minimal fractional part such as 60 + 1/1024, shown in Figure 5.20. With a small fractional number, the quantization error generated by the factional-*N* accumulator forms a staircase ramp waveform with fine step size that can be used to sweep the TDC input time interval over one DCO cycle. The corresponding TDC output was further subtracted from the calculated ideal ramp signal, creating an error signal corresponding to TDC nonlinearity that is used to automatically adjust the TDC delays with the optimization goal of minimizing this error. Two LMS loops are designed to collect the differential and common error signals used for fast and slow delay calibrations. Similarly, the open-loop calibration uses an external signal to provide a fractional frequency the same as that used in the closed-loop calibration. However, although the frequency is pulled close to the desired value, the phase error can still be unknown in an open-loop operation. A large phase error could saturate the TDC and fail the calibration algorithm. Thus, an out-range-flag is generated from the TDC to indicate whether the input phase error is out of the TDC detection range or not. This flag is used to validate the input used for automatic calibration. A pair of optimized DCWs for slow and fast delays is obtained during the calibration and set as pair "0." The second-order ΔΣ modulator will select four adjacent DCWs to form the pair "−1," "1," and "2" for precise delay interpolation.

5.5 TDC implementation examples

5.5.1 TDC architecture and circuit implementation

Figure 5.21 presents the block diagram of the proposed TDC system including the proposed spiral comparator arrangement and two ΔΣ modulation-based TDC

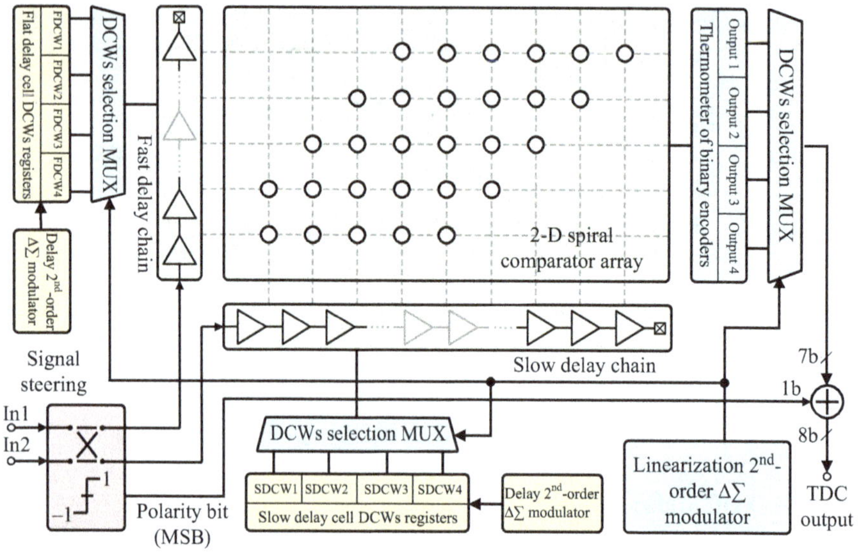

Figure 5.21 Block diagram of the proposed reconfigurable 2-D spiral Vernier TDC with second-order ΔΣ linearization

linearization techniques. In summary, the 2-D spiral comparator array improves both linearity and detection range of the TDC. A ΔΣ modulator is employed in delay interpolation to minimize the quantization errors introduced by digitally tuned delay cells. The folding point errors commonly seen in 2-D comparator arrays are randomized by using a reconfigurable comparator array controlled by the output of another ΔΣ modulator. The 2-D spiral Vernier TDC produces seven output bits. A steering module detects lead/lag or polarity information and outputs the most-significant bit, forming the eighth bit of the TDC. To ensure that there is no dead zone around the zero-crossing point, the same comparator is used in the steering module with its decision standard derivation around 0.2 ps based on Monte Carlo simulations. This decision error is smaller than the quantization error, i.e., half of LSB (0.625 ps in the design).

5.5.2 Experiment results of the proposed TDC

The proposed TDC was fabricated in a 45-nm CMOS SOI technology. The measured full-range transfer curves of the TDC with and without the second-order ΔΣ modulator are given in Figure 5.22. The TDC covers a conversion range from −160 to 160 ps, namely, 8 bits output with a 1.25-ps resolution. Sinusoidal modulated delay signals are generated with an arbitrary waveform generator and are fed into the TDC to perform a spectrum measurement. Figure 5.22 gives the measured TDC output spectrum results with inputs equal to 1.01 and 32.7 MHz under three different configurations: (i) without ΔΣ modulation, (ii) with the first-order ΔΣ modulation, and (iii) with the second-order ΔΣ modulation. A 10-dB SFDR improvement is achieved with the second-order ΔΣ modulation.

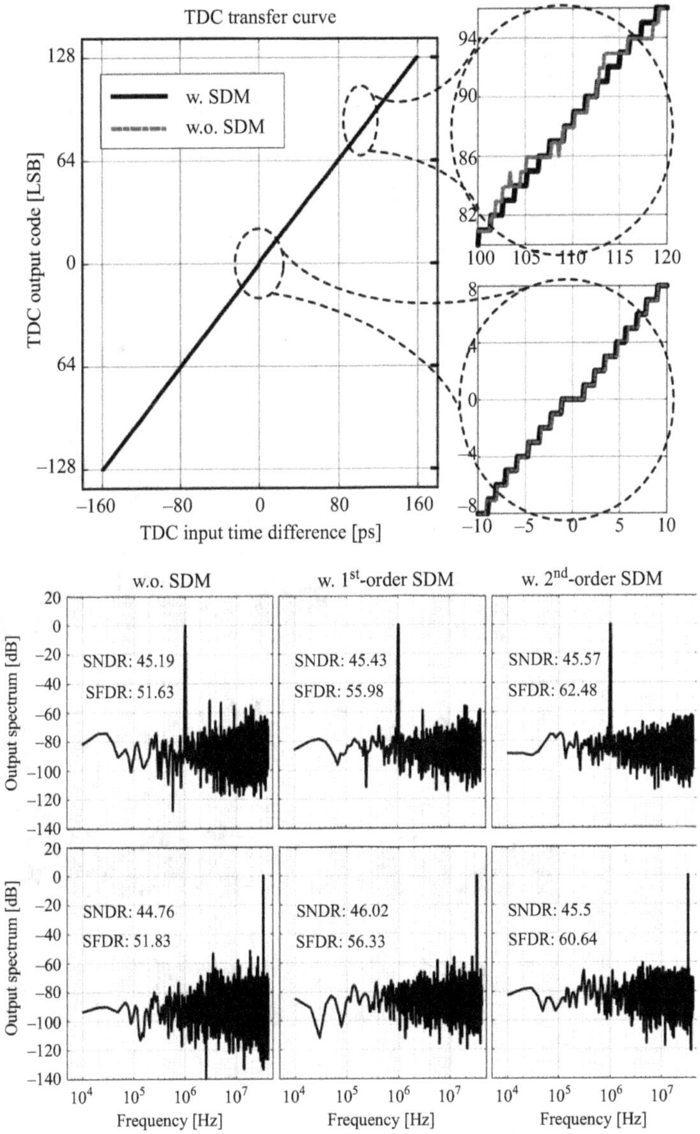

Figure 5.22 Measured transfer curves and output power spectrum density with 1.01- and 32.7-MHz input signals under three configurations: (i) without $\Delta\Sigma$ modulation, (ii) with the first-order $\Delta\Sigma$ modulation, and (iii) with the second-order $\Delta\Sigma$ modulation

Measured histogram plots of the TDC output codes with a ramp input signal sweeping the entire detectable range of the TDC under different settings are presented in Figure 5.23, illustrating the efficacy of each proposed TDC linearization techniques. Without calibration, the TDC transfer curve is extremely nonlinear with missing codes

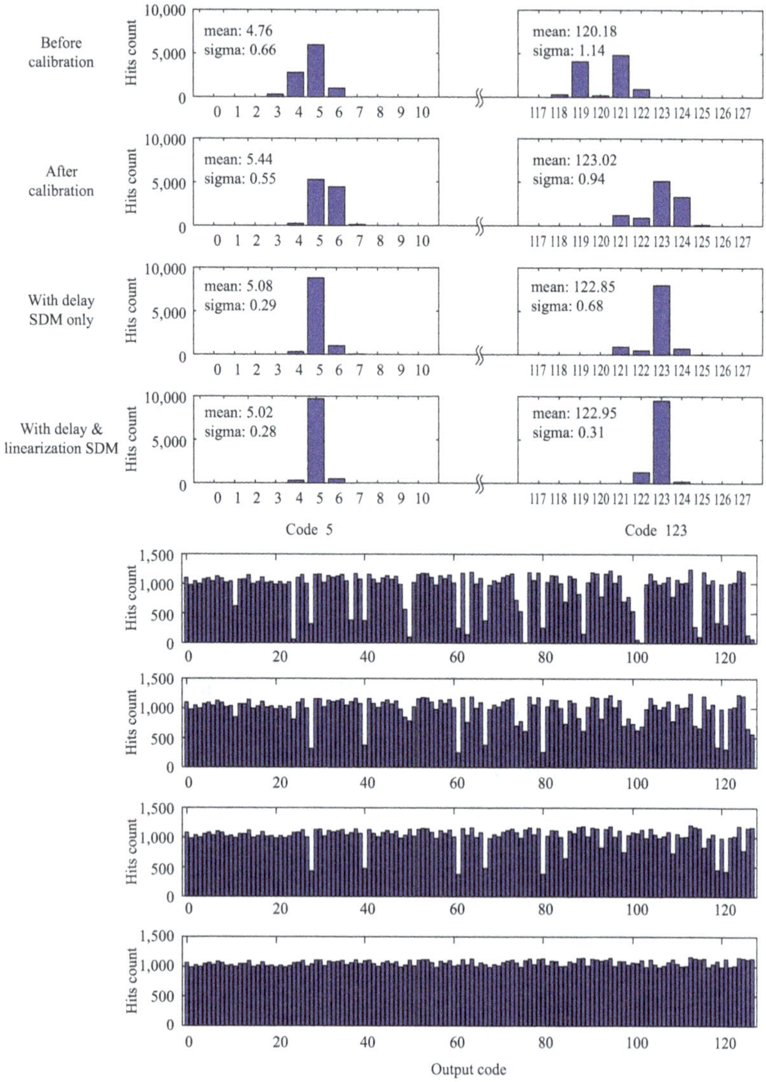

Figure 5.23 Single-shot precision measurement with 10,000 tests measured for four different input time intervals located at codes 5, 64, 99, and 123. Measured histogram plots of TDC output codes with a ramp input signal sweeping the entire detectable range under different settings

or code gaps due to the 2-D folding errors, layout mismatches, and unexpected parasitic effects. The 2-D folding error effect has been greatly reduced by automatic delay calibration and delay interpolation using $\Delta\Sigma$ modulators. The folding error residues together with layout mismatches and unexpected parasitic effects are further eliminated by the second-order $\Delta\Sigma$ randomization for the folding locations. The last histogram plot with evenly distributed code hits indicates a highly linear transfer curve.

Two different time interval signals are further illustrated with single-shot precision measurements, as shown in Figure 5.23, where 10,000 tests are measured for four different input time intervals located at codes 5 and 123, respectively.

For comparison, linearity performances are measured with the second-order, first-order, and no ΔΣ modulations, as shown in Figure 5.24. Periodic errors as large as 1.27 LSB are observed in the measured INL without ΔΣ modulation, showing dominant nonlinearity associated with the folding point errors of a 2-D comparator array. With the second-order ΔΣ randomization, the measured INL and DNL have much smaller errors of 0.34 LSB and 0.25 LSB, respectively.

This proposed TDC is designed to cover 8 bits with a resolution of 1.25 ps. The effective number of bits (ENoB) is 7.58 bits, and the effective resolution is 1.67 ps. In the measurement, the TDC consumes 0.3 mW under a conversion rate of 80 MS/s and a 1-V power supply when the TDC input is fed with a staircase sweeping ramp signal similar to a fractional-N mode operation. As shown in the die photo of Figure 5.25, the 2-D Vernier TDC core occupies an area of 0.03 mm^2.

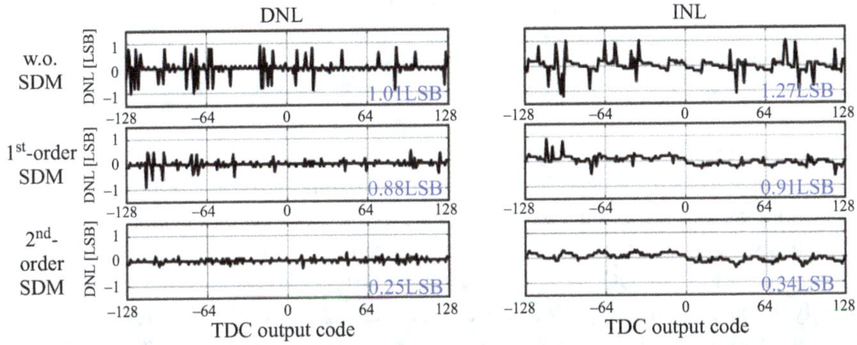

Figure 5.24 Measured TDC DNL and INL without ΔΣ modulation, with the first-order ΔΣ modulation, and with the second-order ΔΣ modulation

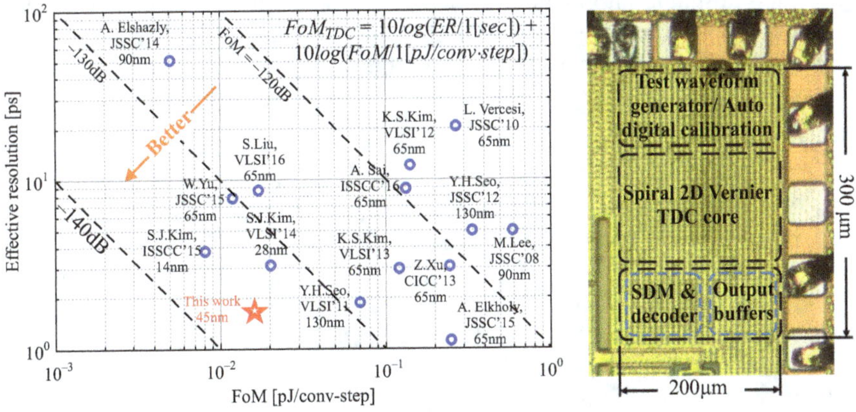

Figure 5.25 Performance summary and comparison with prior-art TDC designs. Die photograph of the TDC prototype chip

Table 5.1 Performance summary and comparison with recently reported TDCs

	JSSC'15 [32]	ISSCC'15 [33]	ISSCC'16 [16]	JSSC'18 [34]
Topology	1–3 MASH	Stochastic	SS-ADC	2D spiral Vernier
Process (nm)	65	14	65	45
NoB	11	10	6.1	8
ENoB[1]	9.42	8.28	5.76	7.58
Resolution	2.64	1.17 ps	6 ps	1.25 ps
ER[2]	7.89	3.85 ps	7.60 ps	1.67 ps
Speed (MS/s)	150	100	40	80
DNL (LSB)/(ps)	–	0.8/0.94	–	0.25/0.31
INL (LSB)/(ps)	2.0	2.3/2.7	0.27/1.6	0.34/0.4
Power (mW)	3.52	0.78	0.36	0.07–0.69
Area (mm^2)	0.03	0.036	0.022	0.04
FoM[3]	0.012	0.008	0.131	0.016
FoM_{TDC}[4]	−130.2	−135.1	−120.0	−135.7

[1]$ENoB = NoB - \log_2(\text{INL} + 1)$.
[2]Effective resolution (ER) = Resolution $\times 2^{(NOB - ENOB)}$.
[3]$FoM = \text{Power}/(2^{NOB} \times F_S)$ [pJ/conv-step].
[4]$FoM_{TDC} = 20\log(ER/1\ [\text{s}]) + 10\log(FoM_{ADC}/1[pJ/\text{conv-step}])$.

Other auxiliary circuits occupy another 0.03-mm^2 space. The performance summary and comparison with prior-art TDC designs are listed in Table 5.1. FoM is based on a well-accepted data converter figure of merit (FoM) evaluation criterion that takes the power consumption, detectable range, and conversion rate into consideration [29]. For TDC designs, effective resolution is an important factor that directly impacts DPLL's performance. Considering both effective resolution and FoM, we summarized the performances of recently reported state-of-the-art TDC designs and presented the comparison in Figure 5.25, demonstrating a very competitive TDC design among the state-of-the-art with excellent linearity performance. The presented TDC design provides precise time measurement and digitization of timing information up to 1.25-ps resolution, which supports a wide variety of applications, including DPLL, direct digital modulator, time-based communication transceivers [30,31], and millimeter-wave imaging radars [31].

References

[1] Y. Arai and T. Baba, "A CMOS time to digital converter VLSI for high-energy physics," *1988 Symposium on VLSI Circuits*, Tokyo, 1988, pp. 121–122.

[2] B. Shen, T. Tokairin, M. Okada, M. Kitsunezuka, T. Maeda, and M. Fukaishi, "An 8.5 mW, 0.07 mm^2 ADPLL in 28 nm CMOS with sub-ps resolution TDC and < 230 fs RMS jitter," *2013 Symposium on VLSI Circuits*, Kyoto, 2013, pp. C192–C193.

[3] C. W. Yao, W. F. Loke, R. Ni *et al.*, "24.8 A 14nm fractional-*N* digital PLL with $0.14ps_{rms}$ jitter and −78dBc fractional spur for cellular RFICs," *2017*

IEEE Int. Solid-State Circuits Conf. (ISSCC), San Francisco, CA, 2017, pp. 422–423.
[4] Z. Su, H. Wang, H. Zhao, Z. Chen, Y. Wang, and F. Dai, "A 280MS/s 12b SAR-assisted hybrid ADC with time domain sub-range quantizer in 45nm CMOS," *2019 IEEE Custom Integrated Circuits Conference (CICC)*, Austin, 2019, pp. 1–4.
[5] Z. Su, H. Wang, H. Zhao, X. Liu, and F. Dai, "An 8-bit 80-Ms/S fully self-timed SAR ADC with 3/2 interleaved comparators and high-order PVT stabilized HBT bandgap reference," *2019 IEEE International Symposium on Circuits and Systems (ISCAS)*, Sapporo, 2019, pp. 1–4.
[6] E. Temporiti, C. Weltin-Wu, D. Baldi, M. Cusmai, and F. Svelto, "A 3.5 GHz wideband DPLL with fractional spur suppression through TDC dithering and feedforward compensation," *IEEE J Solid-State Circuits*, vol. 45, no. 12, pp. 2723–2736, December 2010.
[7] E. Temporiti, C. Weltin-Wu, D. Baldi, M. Cusmai and F. Svelto, "An ultra-wideband low-power ADPLL chirp synthesizer with adaptive loop bandwidth in 65 nm CMOS," *2016 IEEE Radio Frequency Integrated Circuits Symposium (RFIC)*, San Francisco, CA, 2016, pp. 35–38.
[8] G. Marzin, S. Levantino, C. Samori, and A. Lacaita, "A 20Mb/s phase modulator based on a 3.6GHz digital PLL with −36dB EVM at 5mW power," *2012 IEEE J Solid-State Circuits*, San Francisco, CA, 2012, pp. 342–344.
[9] T. Chi, H. Wang, M. Huang, F. F. Dai, and H. Wang, "A bidirectional lens-free digital-bits-in/-out 0.57mm^2 terahertz nano-radio in CMOS with 49.3mW peak power consumption supporting 50cm Internet-of-Things communication," *2017 IEEE Custom Integrated Circuits Conference (CICC)*, Austin, TX, 2017, pp. 1–4.
[10] H. Wang, Z. Su, H. Zhao, Y. Wang, and F. Dai, "A 3.8 mW sub-sampling direct RF-to-digital converter for polar receiver achieving 1.94 Gb/s data rate with 1024-APSK modulation," *Symposium on VLSI Circuits*, Kyoto, 2019, pp. 1–2.
[11] H. Wang, "Integrated circuit for time domain mixed signal processing," PhD Dissertation, Auburn University, 2018.
[12] M. A. Thompson, M. W. Werner, R. R. Egan, P. D. Coan and P. E. Robl, "Free running time to digital converter with 1 nanosecond resolution," *IEEE Trans Nucl Sci*, vol. 35, no. 1, pp. 184–186, February 1988.
[13] P. Dudek, S. Szczepanski and J. V. Hatfield, "A high-resolution CMOS time-to-digital converter utilizing a Vernier delay line," *IEEE J Solid-State Circuits*, vol. 35, no. 2, pp. 240–247, February 2000.
[14] M. Lee and A. A. Abidi, "A 9b, 1.25ps resolution coarse–fine TDC in 90nm CMOS that amplifies a time residue," *IEEE J Solid-State Circuits*, vol. 43, no. 4, pp. 769–777, 2008.
[15] Z. Xu, M. Miyahara, K. Okada and A. Matsuzawa, "A 3.6 GHz low-noise fractional-N digital PLL using SAR-ADC-based TDC," *IEEE J Solid-State Circuits*, vol. 51, no. 10, pp. 2345–2356, October 2016.

[16] A. Sai, S. Kondo, T. T. Ta, H. Okuni, M. Furuta and T. Itakura, "19.7 A 65nm CMOS ADPLL with 360μW 1.6ps-INL SS-ADC-based period-detection-free TDC," *2016 IEEE Int. Solid-State Circuits Conf. (ISSCC)*, San Francisco, CA, 2016, pp. 336–337.

[17] T. Rahkonen, J. Kostamovaara, and S. Saynajakangas, "Time interval measurements using integrated tapped CMOS delay lines," *Proceedings of the 32nd Midwest Symposium on Circuits and Systems*, Champaign, IL, 1989, pp. 201–205, vol.1.

[18] J. Yu, F. F. Dai, and R. C. Jaeger, "A 12-bit Vernier ring time-to-digital converter in 0.13μm CMOS technology", *IEEE J Solid-State Circuits*, vol. 45, no. 4, pp. 830–842, 2010.

[19] M. Z. Straayer and M. H. Perrott, "A multi-path gated ring oscillator TDC with first-order noise shaping," *IEEE J Solid-State Circuits*, vol. 44, no. 4, pp. 1089–1098, 2009.

[20] T. Konishi, K. Okuno, S. Izumi, M. Yoshimoto, and H. Kawaguchi, "A 61-dB SNDR 700 μm^2 second-order all-digital TDC with low-jitter frequency shift oscillators and dynamic flipflops," *2012 Symposium on VLSI Circuits (VLSI)*, Honolulu, HI, 2012, pp. 190–191.

[21] D. Liao, H. Wang, F. F. Dai, Y. Xu, R. Berenguer, and S. M. Hermoso, "An 802.11a/b/g/n digital fractional-N PLL with automatic TDC linearity calibration for spur cancellation," *IEEE J Solid-State Circuits*, vol. 52, no. 5, pp. 1210–1220, May 2017.

[22] T. Tokairin, M. Okada, M. Kitsunezuka, T. Maeda, and M. Fukaishi, "A 2.1-to-2.8-GHz low-phase-noise all-digital frequency synthesizer with a time-windowed time-to-digital converter," *IEEE J Solid-State Circuits*, vol. 45, no. 12, pp. 2582–2590, December 2010.

[23] H. Wang and F. F. Dai, "A 14-bit, 1-ps resolution, two-step ring and 2D Vernier TDC in 130nm CMOS technology," *ESSCIRC 2017 - 43rd IEEE European Solid State Circuits Conference*, Leuven, 2017, pp. 143–146.

[24] A. Elkholy, S. Saxena, R. K. Nandwana, A. Elshazly and P. K. Hanumolu, "A 2.0–5.5 GHz wide bandwidth ring-based digital fractional-N PLL with extended range multi-modulus divider," *IEEE J Solid-State Circuits*, vol. 51, no. 8, pp. 1771–1784, August2016.

[25] H. Wang, F. F. Dai, and H. Wang, "A 330μW 1.25ps 400fs-INL Vernier time-to-digital converter with 2D reconfigurable spiral arbiter array and 2nd-order ΔΣ linearization," *2017 IEEE Custom Integr. Circuits Conf. (CICC)*, Austin, TX, 2017, pp. 1–4.

[26] L. Vercesi, A. Liscidini, and R. Castello, "Two-dimensions Vernier time-to-digital converter," *IEEE J Solid-State Circuits*, vol. 45, no. 8, pp. 1504–1512, 2010.

[27] J. W. M. Rogers, F. F. Dai, M. S. Cavin, and D. G. Rahn, "A multiband ΔΣ fractional-N frequency synthesizer for a MIMO WLAN transceiver RFIC," *in IEEE Journal of Solid-State Circuits*, vol. 40, no. 3, pp. 678–689, March 2005.

[28] D. Liao, H. Wang, F. F. Dai, Y. Xu, and R. Berenguer, "An 802.11 a/b/g/n digital fractional-N PLL with automatic TDC linearity calibration for spur

cancellation," *2016 IEEE Radio Freq. Integr. Circuits Symp. (RFIC)*, San Francisco, CA, 2016, pp. 134–137.

[29] Y. J. Chen, K. H. Chang, and C. C. Hsieh, "A 2.02–5.16 fJ/conversion step 10-bit hybrid coarse-fine SAR ADC with time-domain quantizer in 90 nm CMOS," *IEEE J Solid-State Circuits (JSSC)*, vol. 51, no. 2, pp. 357–364, 2016.

[30] T. Chi, H. Wang, M. Huang, F. Dai, and H. Wang, "A bidirectional lens-free digital-bits-in/-out 0.57mm^2 terahertz nano-radio in CMOS with 49.3mW peak power consumption supporting 50cm internet-of-things communication," *2018 IEEE Custom Integr. Circuits Conf. (CICC)*, San Diego, CA, 2018, pp. 1–4.

[31] A. Tang and M. F. Chang, "183GHz 13.5mW/pixel CMOS regenerative receiver for mm-wave imaging applications," *2011 IEEE Int. Solid-State Circuits Conf.*, San Francisco, CA, 2011, pp. 296–298.

[32] W. Yu, K. S. Kim, and S. H. Cho, "A 0.22 ps_{rms} integrated noise 15 MHz bandwidth fourth-order ΔΣ time-to-digital converter using time-domain error-feedback filter," *IEEE J Solid-State Circuits*, vol. 50, no. 5, pp. 1251–1262, 2015.

[33] S. Kim, W. Kim, M. Song, J. Kim, T. Kim, and H. Park, "15.5 A 0.6V 1.17ps PVT-tolerant and synthesizable time-to-digital converter using stochastic phase interpolation with 16× spatial redundancy in 14nm FinFET technology," *IEEE Int. Solid-State Circuits Conf. (ISSCC), Digest of Technical Papers*, San Francisco, CA, 2015, pp. 1–3.

[34] H. Wang, F. F. Dai, and H. Wang, "A reconfigurable Vernier time-to-digital converter with 2-D spiral comparator array and second-order ΔΣ linearization," *IEEE J Solid-State Circuits*, vol. 53, no. 3, pp. 738–749, 2018.

Chapter 6

Bang-bang digital PLLs for wireless systems

Salvatore Levantino[1] *and Carlo Samori*[1]

The phase-locked loop (PLL) is the ubiquitous block employed for the frequency synthesis in a wireless apparatus; its performance is often among the most difficult to achieve in these communication systems that feature limited power budgets and, at the same time, have tight noise specifications. This is partly due to the fundamental limitations in the voltage-controlled oscillator (VCO). The latter is, in fact, always implemented as an active element connected in positive feedback to balance the losses of an LC tank, whose quality factor ultimately sets the tradeoff between the oscillator (thermal) phase noise and its power dissipation.

Other figures are, however, also important in a frequency synthesiser: the in-band noise, whose contribution to the integrated jitter is usually at par with the contribution of the VCO phase noise, the level of the spurious tones, and the PLL bandwidth for agile switching, just to mention the key ones. These figures depend on both the PLL architecture and the PLL's other than VCO building blocks. The power dissipation of these circuits can be of the same order, or even larger, than the VCO one.

Most of the PLLs are still today implemented with the analog charge-pump architecture, a well-established design that guarantees very good performance in standard complementary-metal-oxide-semiconductor (CMOS) and bipolar technology. The situation is different for an aggressively scaled CMOS process (e.g. below 65 nm). The first problem is related to the use of an analog loop filter, whose capacitances essentially do not scale, leading to a significant area occupation. Then, the performance of the charge pump, for instance, its linearity, are affected by the reduction of the supply voltage, worsening both the spurs performance and the in-band noise. Even if these issues can be circumvented, for example, by exploiting dual-gate oxide transistors and a second higher voltage supply for the charge pump, a simpler approach is to move part of the analog processing to the digital domain, which could benefit from scaling.

The idea of the so-called digital PLL (DPLL) is to replace the analog loop filter with a digital one. To this purpose, an analog-to-digital conversion (ADC) is

[1]Dipartimento di Elettronica, Informazione e Bioingegneria (DEIB), Politecnico di Milano, Milan, Italy

needed to digitise the information of the phase error, an operation performed by a time-to-digital converter (TDC). The tuning of the oscillator, on the other hand, is controlled by the digital output of the filter, realising in this fashion a digitally controlled oscillator (DCO). This class of PLLs, to be employed as high spectral purity frequency synthesisers for radio-frequency (RF) communications, has appeared in the first decade of this century (even if digital PLLs were already in use from the 1960s, for different applications at lower frequencies [1,2]) and from an interesting research topic it has shortly turned into an important building block for future wireless transceivers.

As a matter of fact, it has quite soon become evident that, in addition to the reduced filter area, this solution presents other advantages. In particular, the digitally intensive approach allows for implementing calibration techniques, running in the background, that can improve several performances, e.g. spurious tones, and also enable fast and linear frequency modulations, for both communications and radar applications [3–16].

Of course, the implementation of such a circuit is anything but an easy task, being it essentially a mixed-signal circuit embedding both an ADC conversion (i.e. the TDC) and a digital-to-analog (DAC) conversion (i.e. the DCO). No wonder that the TDC design quite soon has appeared as one of the key problems: its non-linearity produces spurious tones, while its noise, both thermal and quantisation one, appears at the output multiplied by the division factor; moreover, TDC power dissipation can become an important fraction of the total DPLL dissipation. After all, in any ADC both linearity and noise are in trade-off with power.

To avoid these issues, DPLLs embedding a single-bit or bang-bang (BB) TDC have been proposed, and this is the topic of this chapter. Such an approach was already applied in DPLLs for other applications, e.g. clock and data recovery in wireline communications [5], but its use in wireless frequency synthesisers has required to solve a number of problems related to the peculiar nature of the BB-TDC, which is a phase detector constantly overloaded with no linear range [17]. In particular, how to get a clean wireless-class spectrum? how to synthesise a fractional-N channel? how to guarantee a fast lock for large phase errors?, and how to control the PLL's bandwidth? These problems are discussed in this chapter.

6.1 Analysis of digital PLLs

6.1.1 Standard TDC

We will build our DPLL starting from the popular charge-pump-based PLL. In the latter, the error signal detected by a phase-frequency detector (PFD) is converted to a charge signal injected into an analog loop filter. The filter then drives the tuning voltage of a VCO. In a DPLL, the filter is replaced with a digital one. Hence, a TDC, in place of the PFD and the charge pump, is employed to convert the time error into a digital word $e[k]$ driving the filter. The output of the filter is then interfaced with a DCO, which replaces the VCO, and it is used to digitally control the oscillation frequency.

In such a DPLL, whose block diagram is shown in Figure 6.1(a), we can identify two sources of quantisation. The first one is associated with the TDC whose ideal mid-tread conversion characteristic is shown in the figure, where Δt and e are the TDC input and output, respectively, and t_{res} its time resolution. We are assuming that the output of the TDC, $e[k]$, varies in steps of one. The second source of error arises from the finite resolution of the DCO and it is given by the truncation between the digital loop filter and the DCO. The ideal characteristic of the DCO, where f_{dco} is the DCO frequency and tw its input tuning word, is also depicted in the figure.

To properly analyse and design DPLLs, it is useful to derive a model similar to what it is conventionally done for analog PLLs. The equivalent model in Figure 6.1(b) is based on the time occurrences, $t_r[k]$ and $t_v[k]$, of the zero crossings of the reference and the DCO signals. The TDC is modelled as a block detecting the time error $\Delta t = t_r - t_v$, multiplied by a gain $K_{tdc} = 1/t_{res}$ (i.e. the slope of the ideal TDC characteristic), and passed through a staircase function $Q(x)$ with unitary average gain. The same $Q(x)$ function is used at the input of the DCO to model the

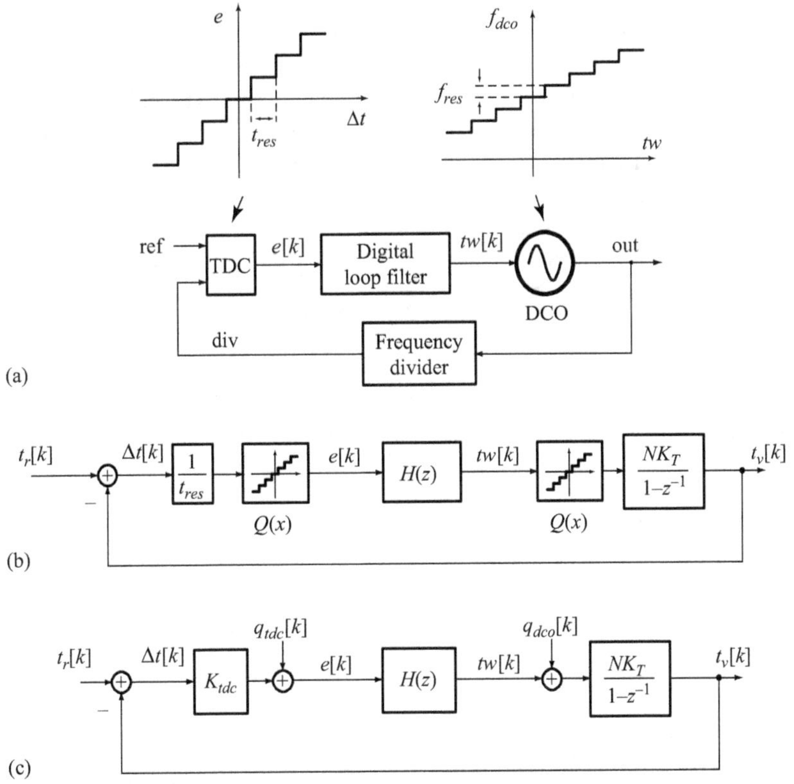

Figure 6.1 Digital PLL with mid-tread TDC: (a) block schematic; (b) equivalent nonlinear model; and (c) linear model

numerical truncation. If we denote f_{res} as the DCO frequency resolution (in hertz), the sensitivity of the DCO period, i.e. the variation of the period itself due to an increase of one least significant bit (LSB) in the DCO tuning word (measured in seconds/bit), is $K_T \approx T_{v0}^2 \cdot f_{res}$ (with T_{v0} being the DCO average period).

The presence of the two $Q(x)$ blocks makes the loop in Figure 6.1(b) a non-linear system. However, if the quantisation error introduced by those two quantisers, $q_{tdc}[k]$ and $q_{dco}[k]$, are random sequences independent from the quantiser inputs, the model can be simplified into the linear system depicted in Figure 6.1(c). The latter can be analysed using the theory of feedback systems.

To get a type-II PLL, we have to add a second integrator in the loop and a zero to stabilise the system. Thus, the loop filter will be a proportional-integral filter with transfer function, $H(z) = \beta + \alpha/(1 - z^{-1})$, where β is the proportional gain and α the integral one.

The loop gain of the model in Figure 6.1(b) can be written as:

$$G_{loop}(z) = \frac{N\beta K_T}{t_{res}} \cdot \frac{1 + \dfrac{\alpha}{\beta} - z^{-1}}{(1 - z^{-1})^2}, \tag{6.1}$$

Though a discrete-time system is properly described in z domain, it may be convenient to further simplify moving from the discrete- to the continuous-time domain. This is possibile since the PLL loop bandwidth will be typically much narrower than the reference frequency and since we are typically interested in the loop behaviour at frequency lower than the reference frequency. Thus, relying on the approximation $z^{-1} = e^{-j\omega T_r} \approx 1 - j\omega T_r$ (where the reference period, T_r, is much shorter than the loop time constants), we obtain the following simplified expression for the loop gain:

$$G_{loop}(\omega) \approx \frac{N\beta K_T}{t_{res}} \cdot \frac{\dfrac{\alpha}{\beta} + j\omega T_r}{(j\omega T_r)^2}, \tag{6.2}$$

which can be further simplified as:

$$G_{loop}(\omega) \approx \frac{\omega_u}{j\omega} \cdot \left(1 + \frac{\dfrac{\alpha}{\beta}}{j\omega T_r}\right), \tag{6.3}$$

where

$$\omega_u \approx \frac{N\beta K_T}{t_{res} T_r} \tag{6.4}$$

The latter ω_u is approximately equal to the crossover angular frequency of the loop gain, assuming that it is much larger than the angular frequency of the zero $(\alpha/\beta)(1/T_r) \ll \omega_u$.

Moreover, in a feedback loop, the crossover frequency is also an approximation of the closed-loop bandwidth. Therefore, a straightforward conclusion can be drawn from (6.4): the bandwidth of a DPLL is directly proportional to the gain βK_T and inversely proportional to t_{res}. This means that to widen the DPLL bandwidth, βK_T should be increased and/or t_{res} should be decreased. Unfortunately, βK_T could be increased only at the expense of a higher DCO resolution, which would increase the truncation error $q_{dco}[k]$. This is obvious because if we increase the gain between the TDC and the DCO, we are equivalently making the DCO coarser. The increased error would translate into extra phase noise and spurious tones in the output spectrum. So, the only way to widen the bandwidth in this class of PLLs is to reduce t_{res}, that is to improve TDC resolution.

The theoretical results can be verified relying on numerical simulations of the nonlinear system in Figure 6.1(b). The DPLL can be designed, plugging practical numbers into (6.4). Let us assume a DCO operating around 3.6 GHz with a sensitivity K_T of 0.9 fs/bit, that is equivalent to a DCO frequency resolution of about 12 kHz (realistic in a 65-nm CMOS node, if we assume to switch on and off MOS varactors). For the TDC, a resolution t_{res} of 20 ps is feasible in the same process node, using a simple flash-type TDC (i.e. a delay line tapped by a number of flip-flops). The division factor is $N = 72$, if the reference signal comes from a crystal oscillator at 50 MHz. From (4), the closed-loop bandwidth can be set to $f_u = \omega_u/2\pi$ of about 100 kHz by setting $\beta = 4$. The zero of the loop gain is instead located at about 10 kHz (about one decade earlier) by taking $\alpha = 5 \cdot 10^{-3}$.

The resulting spectrum, simulating the nonlinear system with the set of parameters just derived, does not has the shape expected but presents a large spurious tone at about 40-kHz offset from the carrier, which is 35 dB below the carrier. The explanation of the presence of this tone is that the error $q_{tdc}[k]$, resulting from TDC quantisation, instead of being random and with energy spread in the spectrum is a periodic sequence which produces a concentrated line in the spectrum. The presence of periodic patterns, also known as *limit cycles*, is typical of quantised feedback systems.

In this case, however, the existence of electronic noise comes to help. In fact, if we account for the random thermal noise contributed by the DPLL sub-blocks, and if this noise is sufficiently powerful, this limit cycle gets dithered. Superimposing a white Gaussian phase noise (at the reference input), whose standard deviation is equal to the standard deviation of the TDC quantisation error, $t_{res}/\sqrt{12}$, the simulated spectrum given in Figure 6.2 has no spurious tones, and shows a closed-loop bandwidth of approximately 100 kHz, as targeted in our design. This confirms the validity of our analysis.

Equation (6.4) revealed that the larger the DPLL bandwidth is, the finer the TDC resolution should be. To assess this property, TDC resolution t_{res} has been reduced from 20 to 2 ps, and the amount of random phase noise added to the reference decreased by 20 dB (to maintain the balance between limit cycle and random-noise power). The new spectrum, also plotted in Figure 6.2, demonstrates that the loop bandwidth is now increased from 100 kHz to 1 MHz, as expected.

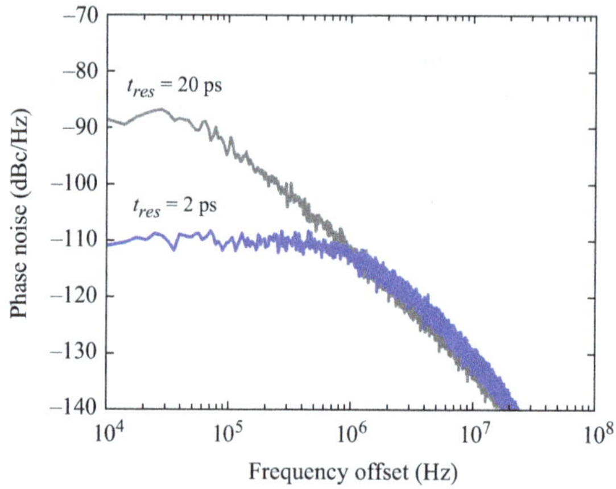

Figure 6.2 Simulated phase spectrum of the DPLL with mid-tread TDC in Figure 6.1(b) at different TDC resolutions

6.1.2 Moving to BB operation

One important conclusion of the analysis in the previous section is that to widen the bandwidth of a DPLL is necessary to refine TDC resolution by reducing t_{res}. This implies that, for the same TDC input range, the number of TDC bits has to increase and, in general, it becomes more difficult to guarantee the required TDC linearity. As a matter of fact, it is easy to verify that the TDC integral non-linearity (INL) is a source of spurious tone in the output spectrum. Of course, resolution and linearity are traded off with power dissipation, which makes the TDC a bottleneck in the design of a DPLL for wireless system, where the dissipation is a key requirement.

The most efficient way to deal with this issue is to replace the multi-bit, high-resolution TDC with a single-bit one. Such a circuit is a simple threshold element, equivalent to a comparator in the time domain, that detects which one of the two inputs occurs before the other one, i.e. it is a lead-lag phase detector. It is referred to as a *bang-bang phase detector* (BBPD) in the field of wireline circuits, and its output-vs-input characteristic is a $sign(\cdot)$ function centred at zero time error.

In what follows, instead of studying a PLL embedding a simple BBPD (featuring a two-level output), we can consider a TDC with a mid-rise characteristic, see Figure 6.3(a), and a very coarse time resolution t_{res}, it is evident that if t_{res} becomes very large the situation should not be very different with respect to the standard BBPD. It will become clear very soon how much large must become t_{res} for the mid-rise TDC to behave as a lead-lag phase detector; in the meantime, we can say that if the TDC resolution is coarse enough, its output, in PLL lock state, will consists in a sequence of ± 1, as in the case of a BBPD.

The linear PLL model employed in the previous section cannot be plainly applied to study the behaviour of the circuit in Figure 6.3(a), since the TDC equivalent model is a hard limiter. Instead, a non-linear analysis has been carried out in [18,19], where it was shown that in first- and second-order loops, the time

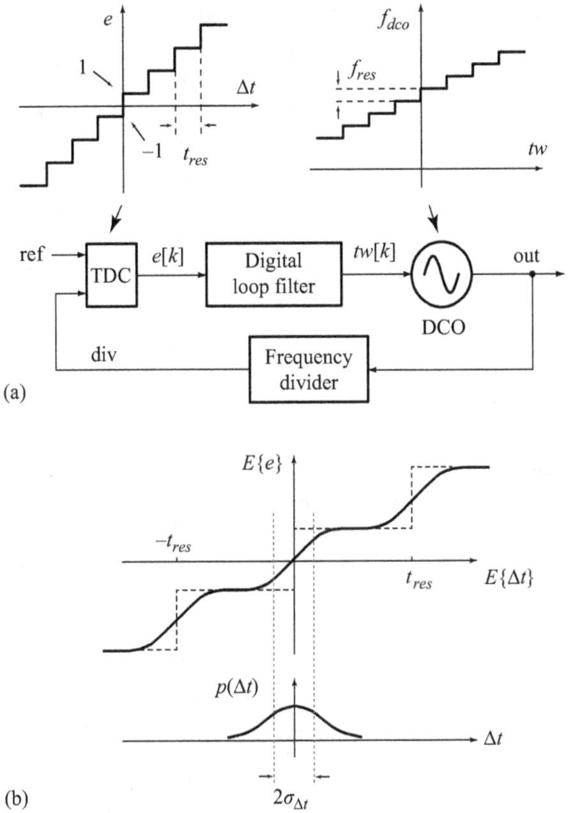

Figure 6.3 DPLLs with mid-rise TDC quantisation: (a) block schematic and (b) averaged detector input/output characteristic in random-noise regime

error Δt, between the input reference and the feedback signal, at steady state features a uniform distribution around the threshold. The peak-to-peak deviation of Δt depends on the time delay induced by the DCO frequency quantisation (often called frequency of granularity) and results $\Delta t_{pp} = 2(1+D)N\beta K_T$, where D is the number of clock cycles of loop-filter latency. Assuming Δt_{pp} to be much lower than $2t_{res}$, this result is also valid in the more general case of a mid-rise quantiser (note that this is a first condition about the width of the bin t_{res}). According to the classical theory about the quantisation noise, if the distribution of Δt is uniform, the rms value (i.e. the jitter at the detector input) becomes:

$$\sigma_{\Delta t} = \frac{\Delta t_{pp}}{\sqrt{12}} = (1+D) \cdot \frac{N\beta K_T}{\sqrt{3}}, \tag{6.5}$$

The key point is that, in this condition, the jitter is no longer limited by TDC time resolution, t_{res}, but just by DCO time quantisation (granularity) βK_T. The Δt changes at each clock cycle; in other words, it is a discrete-time signal. Even if its probability distribution is uniform, this signal follows a periodic or quasiperiodic pattern that gives rise to a limit cycle (this behaviour is often called a *limit-cycle regime* [20]). In addition to the theoretical interest for this phenomenon, this regime should be avoided since it always produces large spurious tones in the output spectrum. This regime, that would make the BBPD useless for many practical applications, can be avoided, thanks to the presence of random noise, usually both thermal and flicker noise, produced by the various circuit's components. As a matter of fact, the averaged output of a BBPD depends linearly on its averaged input values (over a certain range), if the BBPD input is dominated by random noise.

Of course, the output of the BBPD can be only ± 1, but its average can be a continuous value. In other words, assume that an input time shift Δt is enforced at BBPD input with no noise, the output would be a deterministic value either $+1$ or -1, let us imagine we are in the former situation as an example. The random noise adds to Δt a random time jitter that statistically could change the value of the input time shift, such that the output can also become -1. If the same measurement is repeated several times, with the same systematic Δt, it is intuitive that the BBPD average output will depend on the ratio between the Δt and the jitter variance $\sigma_{\Delta t}$. The same condition holds true for the multi-bit mid-rise TDC, provided that the TDC LSB is much wider than the jitter variance $\sigma_{\Delta t}$, this TDC characteristic is sketched in Figure 6.3(b). Note that we have now also answered to the previous question about how much large should t_{res} be, in order to have a BBPD-like behaviour.

When the PLL is locked, the time average of the BBPD is extracted by the loop filter. This is due to the fact that the closed-loop bandwidth, in any PLL either digital or analog, is well below the reference frequency, which means that Δt does not change appreciably but in several reference cycles. The key benefit of the linear behaviour of the BBPD is that it allows exploiting the advantages of a single-bit TDC, with no limit cycle, thus featuring a wireless-class output spectrum. This operating condition can be called *random-noise regime* [20].

Since we have shown that a coarse mid-rise multilevel TDC is practically equivalent to a single-bit TDC, one may ask why not just considering the latter case. One reason is that a multi-bit TDC will be not overloaded in case of a large input time error, such as the one that can occur in a PLL either at start-up or after a large disturbance. The TDC overload would produce a long sequence of $+1$ (or -1) slowing the recovery of the lock, similarly to the slew-rate mechanism in an op-amp. Avoiding it substantially speeds up the lock transient. Furthermore, since the extra levels of the TDC are not exploited during normal operation of the DPLL, the mid-rise TDC can dissipate about the same power of a single-bit TDC and, for the same reason, the INL is not an issue. An advanced exploitation of this concept is presented in a following section.

Going back to Figure 6.3(b), a rigorous mathematical derivation of the link existing between the expectations of the time shift, $E\{\Delta t\}$, and the one of the detector output, $E\{e\}$, can be obtained by assuming that the input has a probability density function $p(\Delta t)$, starting from:

$$E\{e\} = \left(2\int_{-E\{\Delta t\}}^{+\infty} p(x)dx \right) - 1. \tag{6.6}$$

For the standard case of Gaussian $p(\Delta t)$, the one represented in Figure 6.3(b), the characteristic is approximately linear around $E\{\Delta t\} = 0$, and the slope of the curve, obtained differentiating (6), is $2 \cdot p(0) = \eta/\sigma_{\Delta t}$, where $\eta = \sqrt{2/\pi}$. As a consequence, the larger the absolute jitter at BBPD input is, the lower the detector gain is.

The DPLL in this case can be analysed with the model in Figure 6.1(c), where now $K_{tdc} = \eta/\sigma_{\Delta t}$. The loop gain expression is unchanged with respect to (6.3), while the crossover angular frequency becomes:

$$\omega_u \approx \frac{\eta N \beta K_T}{\sigma_{\Delta t} T_r}. \tag{6.7}$$

Note the dependence of the bandwidth on the rms jitter. We can also evaluate the latter figure, or more precisely the jitter variance $\sigma^2_{\Delta t}$, assuming that $S_{t_r}(f)$ and $S_{t_v}(f)$ are the two-sided power spectral densities (PSDs) of the (time) noise affecting the reference and DCO timestamps, it is:

$$\sigma^2_{\Delta t} = \int_0^{1/T_r} \frac{S_{t_r}(f) + S_{t_v}(f)}{|1 + G_{loop}(f)|^2} df, \tag{6.8}$$

Equation (6.8) has to be solved in the unknown $\sigma_{\Delta t}$, because the loop gain $G_{loop}(f)$, as just pointed out, also depends on $\sigma_{\Delta t}$.

Let us consider the typical case of a white phase noise for the reference signal (with absolute jitter σ_{t_r}) and a random-walk phase noise for the DCO (with cycle jitter σ_{T_v}). Hence, $S_{t_r}(f) = T_r \sigma^2_{t_r}$ and $S_{t_v}(f) = \sigma^2_{T_v}/(2\pi^2 f^2 T_{v0})$. What (6.8) highlights is that the two noise sources are transferred to the detector input through a second-order high-pass transfer function, $1/|1 + G_{loop}(f)|^2$. So, if we approximate the loop gain $G_{loop}(f)$ as $f_u/(jf)$ (with f_u being the crossover frequency $\omega_u/2\pi$) and assume $f_u T_r \ll 1$, we get:*

$$\sigma^2_{\Delta t} \approx T_r \sigma^2_{t_r} \left[\frac{1}{T_r} - f_u \arctan\left(\frac{1}{f_u T_r}\right)\right] + \frac{\sigma^2_{T_v} \arctan\left(\frac{1}{f_u T_r}\right)}{2\pi^2 f_u T_{v0}} \tag{6.11}$$

$$\sigma^2_{\Delta t} \approx \sigma^2_{t_r}\left[1 - \frac{\pi}{2} f_u T_r\right] + \sigma^2_{T_v} \frac{1}{4\pi f_u T_{v0}} \tag{6.12}$$

$$\sigma^2_{\Delta t} \approx \sigma^2_{t_r} + \sigma^2_{T_v} \cdot \frac{N}{4\pi f_u T_r} \tag{6.13}$$

*Note that

$$\int \frac{a^2 x^2}{1 + a^2 x^2} dx = x - \frac{1}{a}\arctan(ax) \tag{6.9}$$

and

$$\int \frac{1}{1 + a^2 x^2} dx = \frac{1}{a}\arctan(ax). \tag{6.10}$$

To avoid misunderstanding, it is worthy to remind that the one above is the expression of the variance of the jitter at the input of the single-bit TDC, which must be used in (6.7), as a function of the reference and DCO jitters. Note that the absolute jitter at the PLL output is not $\sigma_{\Delta t}$, a fact that can be simply understood by recalling that the transfer functions of both the reference and DCO noise to the TDC input is different from the transfer functions to the DCO output.

As an example, let us consider the design of a 3.6-GHz DPLL starting from a reference of 50 MHz (i.e. $T_r = 20$ ns and $N = 72$), featuring a bandwidth $f_u = 1$ MHz and, what is more interesting, using a coarse mid-rise TDC with time resolution $t_{res} = 20$ ps. We are also assuming $\beta = 4$, $K_T = 0.9$ fs/bit; therefore, to get the desired bandwidth, the total jitter at BBPD input (including both random noise and limit cycle contribution) must be $\sigma_{\Delta t} = 1.65$ ps.

The limit-cycle component of jitter from (6.5) is only 244 fs, assuming a latency $D = 1$; hence, an extra reference phase noise has been added to reach the desired figure (for the sake of simplicity, no DCO noise was added). Figure 6.4 shows the simulation results, and the key point is the comparison of the spectrum obtained with the same DPLL using a TDC with the same time resolution but featuring a mid-tread characteristic. While in the former case the TDC gain is much larger, since it is inversely proportional to the jitter ($\sigma_{\Delta t} = 1.65$ ps, in our case), in the latter it is inversely proportional to $t_{res} = 20$ ps. In other words, the DPLL bandwidth (and its in-band noise) scales proportionally to the gain of the TDC and, in general, it does not necessarily depends on its resolution. This is a significant advantage, since the TDC is one of the most power hungry blocks, and achieving a fine resolution typically has a two-fold impact on the dissipation: both a finer resolution costs power and the linearity issues are exacerbated (hence to fix those issues, more power is required).

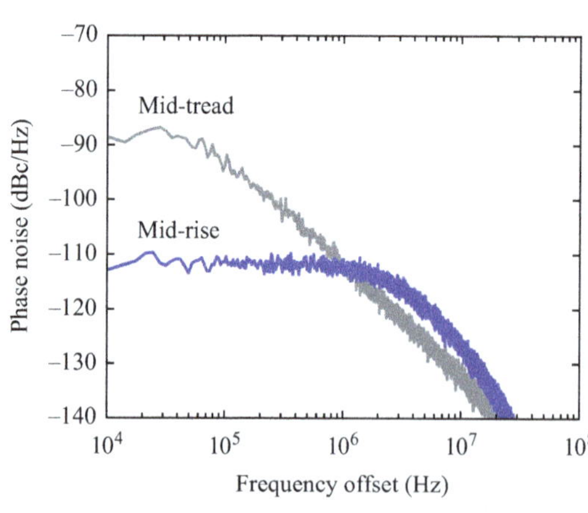

Figure 6.4 Simulated phase spectrum of the conventional DPLL (mid-tread) and the DPLL in Figure 6.3(a) working in random-noise regime (mid-rise). In both cases, TDC resolution is 20 ps

6.2 Fractional-N frequency synthesis

Our discussion has so far concerned integer-N frequency synthesis. Achieving BB operation in fractional-N synthesis is not straightforward. In fact, to synthesise a fractional-N frequency, a $\Delta\Sigma$ modulator is employed to dither the frequency-divider modulus control. This operation introduces a phase modulation into the frequency-divided signal, whose amplitude is proportional to the accumulated quantisation error $s[k]$ of the $\Delta\Sigma$. The sequence $s[k]$ is converted by the TDC and, in wideband PLLs, has to be subtracted from TDC output [21]. The block schematic of fractional-N DPLL is sketched in Figure 6.5, where the automatic calibration of the cancellation path relies on least-mean square (LMS) adaptive filtering: the quantisation noise $q[k]$ is first accumulated (to account for the frequency to phase intrinsic integration) producing $s[k]$, and then this signal is scaled by the gain $g[k]$ to cancel the quantisation noise at the TDC output. This gain $g[k]$ is the one calibrated by the LMS that correlates the loop filter input with $s[k]$, until the contribution of the latter is removed. A similar technique has been applied in analog PLLs, but the cancellation is intrinsically inaccurate because of the different shapes of the cancellation signal with respect to one of the injected quantisation noise. In DPLLs, the cancellation is only limited by the finite time resolution and nonlinearity of the TDC.

If, due to imperfect cancellation, a residual quantisation error leaks into the digital filter, it causes spurs and raise the output noise level. There are at least two sources of imperfect cancellation: (i) the limited resolution of the TDC and (ii) its integral nonlinearity. The simulated spectra for both a 20- and 2-ps resolution TDC

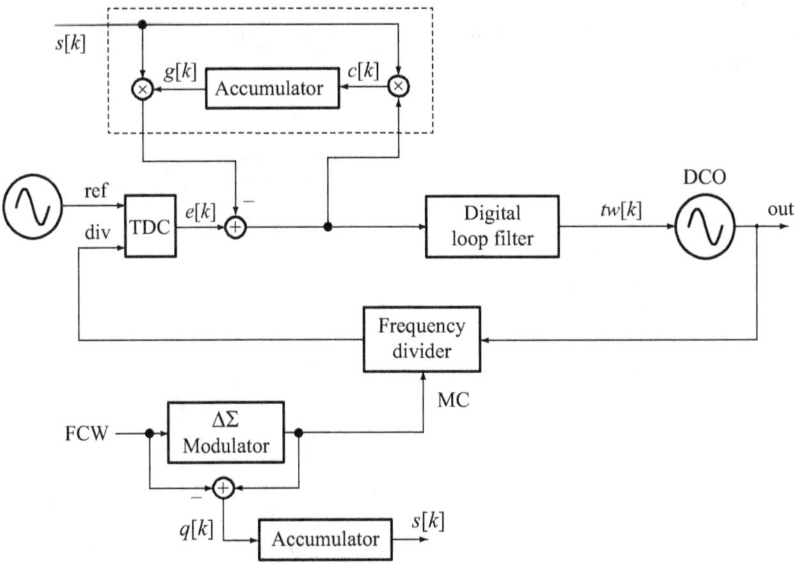

Figure 6.5 Fractional-N DPLL with a multi-bit TDC [21]

(with mid-tread characteristics) are plotted in Figure 6.6, where the level of the most powerful spur is reduced from −45 to −65 dBc. On the other hand, the TDC has to cover at least a time range larger than one or more DCO periods (depending on the order of the ΔΣ modulator used to dither the divider modulus). In practice, even up to a few nanoseconds, in the GHz range. This result highlights that the spur level scales down as TDC resolution and that a TDC with more than 12 equivalent bits may be required to achieve the typical spur level needed in wireless applications. At the same time, the TDC conversion characteristic has to be linear, that is with an integral nonlinearity (INL) better than one LSB. A TDC with such a dynamic range and linearity is typically the most complex and power hungry block of a DPLL. Although the phase-domain all-digital PLL architecture in [3] has no explicit ΔΣ quantisation noise injected into the loop, it shares the same trade-off between noise/spur performance and TDC's required dynamic range and linearity [6,7].

The use of a BBPD clearly relaxes most of these issues but, on the other hand, its use in fractional-*N* DPLLs makes it necessary the adoption of an additional building block, to achieve an output spectrum suitable for wireless applications. As a matter of fact, the quantisation noise of the ΔΣ modulator that interpolates the frequency control word (FCW), is translated into a time delay at the input of phase detector. The minimum quantisation step is one DCO period, but typically ΔΣ modulators featuring order higher than one are employed, to increase the order of noise shaping. As a consequence, the resulting time shift is equal to several periods T_{v0} and can be as large as few nanoseconds, thus saturating the single-bit phase detector. That does not mean that the DPLL is unlocked but that the output frequency could be periodically modulated, again leading to limit cycles. Note that, to solve this issue, it is necessary to somewhat cancel the quantisation induced by the ΔΣ, and it is however not possible to adopt the solution presented in Figure 6.5, given that now the TDC features a single bit.

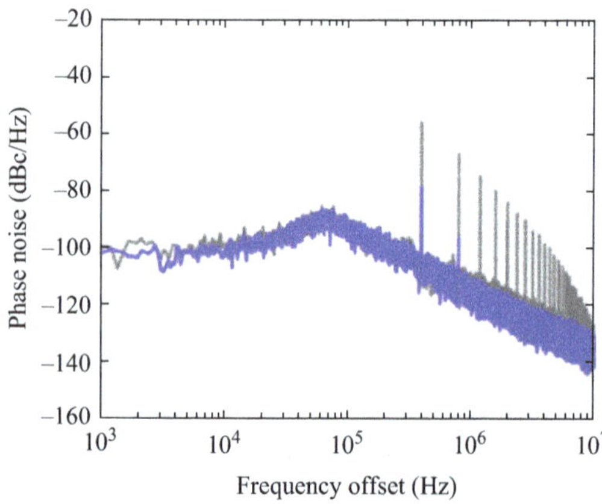

Figure 6.6 Simulated phase spectrum of the fractional-N DPLL in Figure 6.5

The additional building block, above mentioned, that is instead able to solve this problem is the digital-to-time converter (DTC). This element is essentially a delay cell realised by one or more buffers (as an example by a CMOS inverter chain or by a differential pair), whose time delay is digitally controlled by switching the capacitive load of the cell. This element is embedded in the feedback path, see Figure 6.7, and by regulating its delay it is possible to cancel the ΔΣ quantisation noise, [17]. To this purpose, the quantisation noise is first accumulated (to account for the link between frequency error and phase or time error), then the accumulator output, a digital signal proportional to the error to correct, drives the DTC. Of course, the DTC output is an analog variable, a time delay, therefore it is necessary to calibrate the DTC characteristic, whose input is the digital word and the output the desired time delay. The calibration is carried out by an LMS loop that correlates the single-bit error at BBPD output and the accumulated quantisation noise, Figure 6.7, thus implementing a sign-LMS algorithm. The time resolution achieved by today's DTCs is in the order of few hundreds of femtoseconds, which is typically well below the variance of the BBPD time jitter, in this fashion the random-noise regime can be recovered even when a fractional channel is synthesised.

This description immediately prompts the question about the real advantage of the use of a BBPD. After all, an analog-to-digital converter, the TDC, has been drastically simplified, but at the price of having employed a digital-to-analog converter, i.e. the DTC which also introduces both noise and non-linearity. The former impairment is completely equivalent to the input reference noise, and thus it may produce in-band noise in the PLL output spectrum. It is possible to demonstrate [22] that for a given power dissipation, the jitter introduced by the DTC is

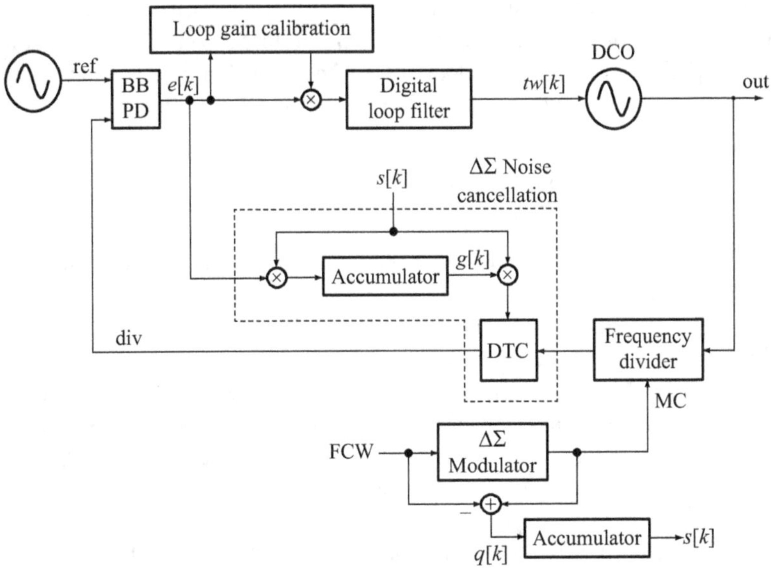

Figure 6.7 Fractional-N BB DPLL

proportional to the maximum time delay it can generate (the DTC output range). The latter, the non-linear DTC characteristic, is due to both the mismatch between capacitors and the systematic non-linear link between capacitive load and delay and usually produces output spurious tones.

Despite these limitations, the use of the DTC together with a BBPD is still advantageous with respect to an equivalent multi-bit TDC, in terms of area occupation and power dissipation. As a matter of fact, while the DTC noise is in the order of the reference one, which is often limited by an input integrated buffer, the DTC non-linearity can be significantly reduced by adopting digital predistortion algorithms, running in background, as the one in [23], which substantially decrease the level of fractional spurs, without adding significant power dissipation, when scaled CMOS technologies are employed.

6.3 Fast lock in BB PLLs

The single-bit TDC strongly reduces power dissipation, but we have just noticed that it can be seen as a block that always operates in the overload condition. As discussed earlier, this is not an issue when the PLL is locked, even for a fractional channel, since the TDC output is dithered by random noise. Things change when a large time error occurs at the TDC input, for instance, at the DPLL start-up, for a frequency step at channel switching, or when the synthesiser is affected by a large disturbance. In this case, a long sequence of 1's (or -1's) will appear at the TDC output, severely increasing the lock transient of the DPLL that could even fail to lock. A similar problem occurs in the so-called sub-sampling PLL.

Note that a trade-off exists between the DPLL settling time and the minimum output jitter, both important figures in modern wireless systems. Starting from the analysis in [24], it is possible to demonstrate that, for a large frequency step, the ratio between the lock time and the reference period is close to:

$$\frac{t_{lock}}{T_r} \approx \frac{\beta}{2\alpha} \cdot \left(\frac{T_{v0}}{N\beta K_T} \right)^2 \tag{6.14}$$

The previous expression highlights the intuitive fact that a larger βK_T product in the forward path speeds up lock, even when the loop behaves non-linearly because of the saturation of the TDC.

Unfortunately, to operate the DPLL in a random-noise regime, the time jitter at the TDC input, $\sigma_{\Delta t}$, has to be larger than or equal to $(N\beta K_T)/\sqrt{3}$. Thus, a low jitter implies a long transient time, or vice versa. As an example, given the typical values of α and β to get an rms jitter $\sigma_{\Delta t} \approx 100$ fs in an oscillator operating in the 10 GHz range, $T_{v0} \approx 100$ ps, implies that the transient would last about 10^6 reference cycles!

To partially relax this issue, practically any DPLL, and most of the standard analog PLLs, embed an extra servo loop (often referred to as frequency aid, coarse frequency loop or similar) that brings the output frequency close to the desired value. Usually, both the reference and the divided signal are fed to a simple circuit built around a counter, whose output is zero if the frequencies are identical. Otherwise, its integrated output is used to control the coarser of the DCO tuning banks. This is essentially just a frequency loop that does not enforce a zero phase between the reference and the divided signal, and the achieved reduction of lock time is often not enough.

An improved scheme has been presented in [25], and it is exemplified in the following. It essentially exploits the property discussed above: a mid-rise TDC with a coarse time resolution, i.e. larger than $\sigma_{\Delta t}$, behaves as a BB TDC. However, for a large time error, other TDC levels can be exploited, $\pm 3, \pm 5, \ldots$, so no overload occurs and the lock time can be reduced. This solution can be used in parallel with the standard frequency-aid circuit that will take care of very large frequency steps (start-up and channel switching).

The practical implementation of this idea is sketched in Figure 6.8, the main loop uses a "real" single-bit TDC, while a servo loop embedding a multilevel

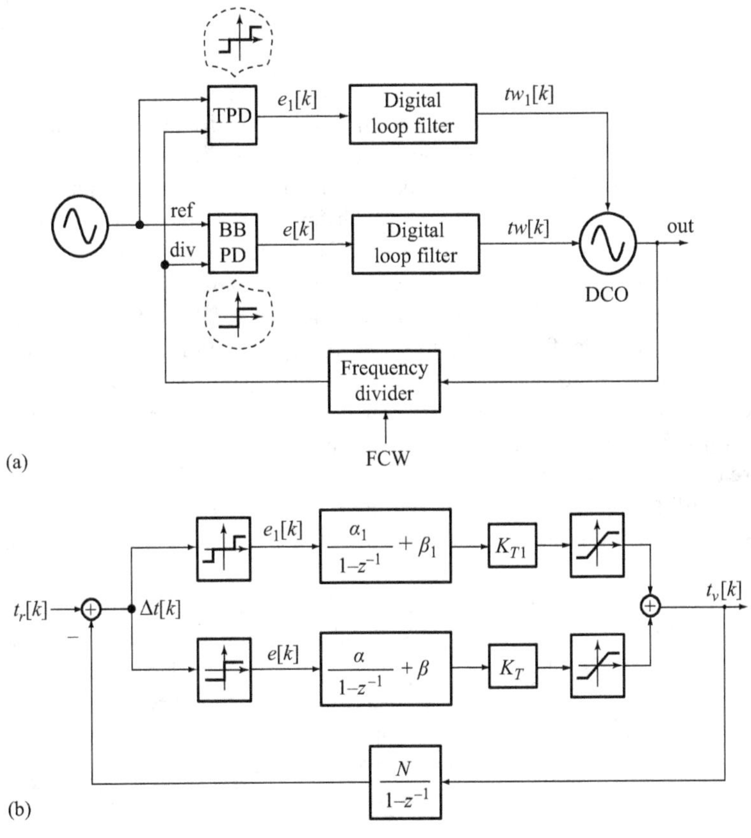

Figure 6.8 DPLL with frequency-aid principle: (a) block diagram and (b) equivalent model

coarse mid-tread TDC (which can be seen as a TDC with a dead zone) is placed in parallel to the first one. This element can be also referred to as a *ternary* phase detector (TPD). The reason for this choice is that the parameters of the additional loop can be optimised independently from the main one. The area overhead is negligible while, clearly, the additional loop is idle, consuming zero power, when the DPLL is locked.

Even with additional degrees of freedom, this design is not an easy task, as briefly exemplified in the following. First, considering the lock-range vs. jitter trade-off, it is natural to make the product between the proportional gain and the DCO gain in the additional path, say $\beta_1 K_{T1}$, larger with respect to the same product in the main loop, that is $\beta_1 K_{T1} > \beta K_T$. In this fashion, the DPLL can be optimised in terms of jitter, while the lock time becomes the task of the servo loop.

Unfortunately, this product cannot be increased too much, given the physical limitations of the capacitors' banks. As an example, the K_T in the main loop can be in the order of 1 fs/b, to limit the DCO quantisation noise. The bank driven by this gain usually features 32–64 capacitors, and a larger number of connections (often differential) would increase too much the layout difficulties and the parasitics. For this reason, the gain K_{T1} cannot exceed K_T by more than one order of magnitude; otherwise, no overlap between the banks would be guaranteed.

This relatively limited value of K_{T1} forces to choose a large α_1 value to speed up the transient; therefore, an even larger $\beta_1 > \alpha_1$ is required to ensure the stability of the system. On the other hand, the increase of β_1 is limited by the number of capacitors driven in the frequency loop. From these considerations, it appears evident that, even if the basic idea is simple, its practical implementation suffers for several constraints that depend on the particular design specifications.

The latter one on β_1, for instance, can be overcome by implementing the proportional path in a different fashion, see Figure 6.9. The output of the TDC is scaled by a suitable coefficient λ_1 and then added to the divider modulus. Given the intrinsic integration between the FCW signal and the phase error (or time shift) circulating in the loop, the behavior is equivalent to the one in Figure 6.8. The multiplicative coefficient of the integrator is $\lambda_1 T_{v0}$, given that any unitary increase of the FCW augments by one T_{v0} the time shift of the divider output. The net effect of this path is identical to have a proportional gain β_1, such that $N\beta_1 K_{T1} = \lambda_1 T_{v0}$, without the limitations of the β_1 related to the DCO capacitor bank size.

Clearly, the dead zone of the mid-tread TDC must be properly sized, in particular it should be larger than the peak-to-peak amplitude of the time jump induced by this path, i.e. $2\lambda_1 T_{v0}$, otherwise, the loop will give rise to a limit cycle, producing a sequence of $+1, -1, +1 \cdots$ at the mid-tread TDC output, see [25].

Further degrees of freedom, for an improved design, can be achieved by adding more than one servo loop embedding a dead-zone TDC. Clearly, the width of the dead zone of the second servo loop has to be increased (i.e. the mid-tread TDC is coarser, etc.).

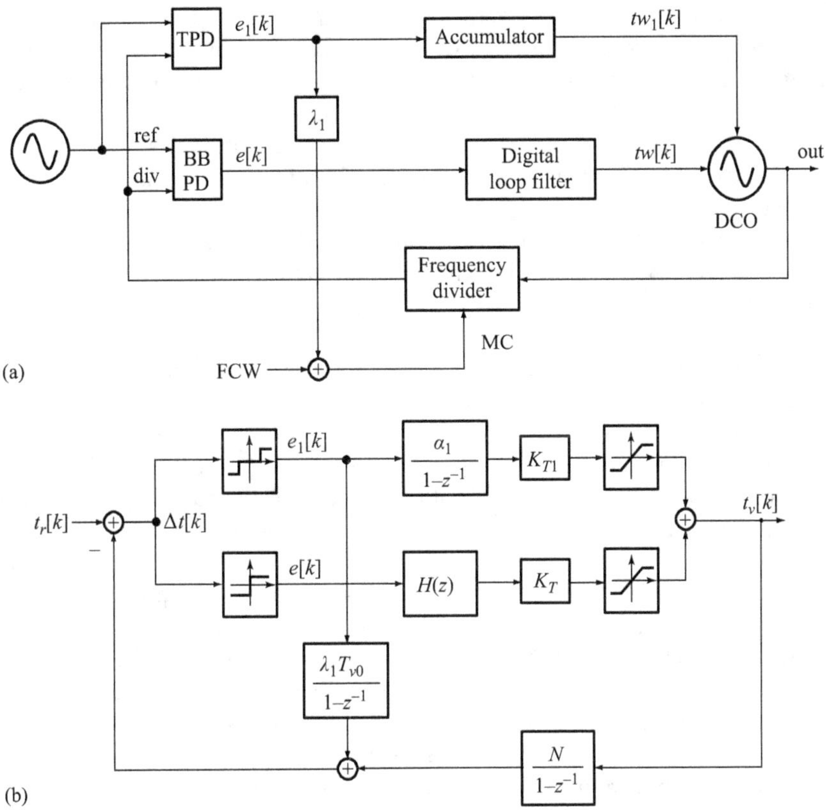

Figure 6.9 DPLL with frequency aid

6.4 Automatic bandwidth regulation

The bandwidth control is important to filter the oscillator flicker noise, to guarantee agile switching and to operate the PLL as a direct frequency modulator, for instance when the pre-emphasis technique is adopted. The dependence of the (linearised) TDC gain on the rms time jitter at the input of the TDC itself has been considered a serious limitation to the practical adoption of this solution. The reason is that the TDC gain enters in the expression of the loop gain, which in turn impacts on the bandwidth therefore on the noise transfer of the various sources in the loop, thus ultimately on the rms jitter at the TDC input.

Although this problem appears peculiarly related to this specific architecture, it is worthy to recall that the accurate control of the PLL bandwidth is also difficult in both digital PLLs adopting a multi-bit TDC and in analog charge-pump PLLs as well. In both cases, in fact, the bandwidth depends on the absolute value of analog parameters that cannot be controlled with a great accuracy over PVT variations. In analog PLLs, for example, the bandwidth is proportional to the charge-pump current and to the value of the loop filter resistance (the proportional gain). In a digital PLL, featuring either a multi- or single-bit TDC, the bandwidth is proportional to the product of the TDC gain K_{tdc} and the DCO gain K_T. Remember that for the multi-bit case it is $K_{tdc} = 1/t_{res}$, and the TDC time resolution t_{res} is an analog parameter as well. Therefore, the bandwidth control is an issue that is not particularly exacerbated in BB DPLLs with respect to other architectures. Indeed, the digital architecture enables an effective way to regulate the bandwidth over PVT spreads.

An exhaustive description of the many techniques proposed in the literature will require too much space. To simplify the matter, we can say that two mutually exclusive approaches exist for bandwidth regulation, in the first one, the bandwidth is kept constant, whatever is the input and DCO noise; in the second case, the regulation aims optimising the output jitter. The latter technique is usually adopted in PLLs for wireline circuits often embedding a ring oscillator (even if some exceptions exist) or in general showing poor jitter performance, some notable examples are [26–31]. It however appears difficult to apply this approach when the jitter to minimise is already very low.

We will instead focus on the former approach, given that in a wireless system it is desirable to have a constant-bandwidth PLL, for the reasons mentioned above; moreover, the jitter level required for today's application is quite low, even below 100 fs. Several techniques have been discussed to make the bandwidth insensitive to analog parameters. Some of them operate in foreground, which means that the PLL needs a calibration time in which it is not operating. One example is the work [32] in which two phase steps, featuring opposite polarity, are injected in the PLL, whose transfer function is calibrated. Other approaches are based on a fully analog design, not really suitable for scaled CMOS technologies.

Among the techniques that instead exploit the digital PLL architecture (although not the TDC-based one discussed so far), the one proposed by [33] operates in background, compensating the spread of the DCO gain. The latter is estimated by evaluating both the oscillator tuning word and the PLL FCW. A look-up-table is needed to reduce the approximation errors.

It is also worthy to mention for the interested readers the work in [34] that exploits an external modulation signal, which is however not available in every applications, and the work [35], where the output phase noise peak is probed in order to regulate the bandwidth, an approach that features a limited accuracy in low-phase-noise designs.

The idea discussed in this chapter is a simplified version of the one presented in [36]. We can start from (6.4), which gives approximately the closed-loop bandwidth of a PLL. It is worthy to rewrite it by lumping all the analog parameters (together with the division factor N) in a factor $G = NK_TK_{TDC}$. Of course, $K_{TDC} = 1/t_{res}$ is the TDC gain. Note that the loop gain in (6.3) is directly proportional to the factor G, while the bandwidth expression becomes $\omega_u \approx \beta G/T_r$.

The key idea is to desensitise the loop gain, thus the above bandwidth expression, from the analog factor G. This can be done by first producing a digital estimate of the parameter G and then by multiplying the output of the TDC by the inverse of this digital word. In this way, the loop gain will depend only on the digital parameters of the loop filter, α and β. Clearly, the estimate of G should account not only for the process spreads, but it has to track also the fluctuations of G produced by the temperature and voltage variations. Again, a background LMS circuit will provide an estimate with the characteristics required.

To see how this technique can be implemented, note from the simplified DPLL block scheme in Figure 6.10(a) that *considering the loop open* a signal $q[k]$ injected in the DCO tuning port goes across the DCO-divider-TDC cascade and appears at the TDC output multiplied by the transfer function $G/(1 - z^{-1})$, proportional to the factor G to estimate. It is for these reasons that N, which is not an analog parameter, is included in the definition of G. The integration of course is due to the DCO frequency-to-phase transfer function, an additional loop delay z^{-1} is usually present, but it is not important for this discussion and it has been neglected.

Figure 6.10(b) shows the operating principle of the LMS. The path from DCO input to TDC output, assuming the loop open, is sketched in parallel with a calibration path including a digital integrator and the LMS correction. While $f[k]$ is the standard signal circulating in the loop $q[k]$ is a digital training sequence going through both the paths. The output of the digital integrator $a[k]$ is multiplied by the

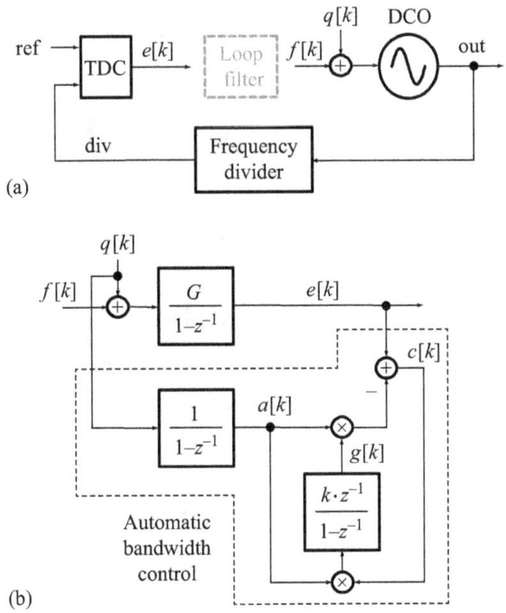

Figure 6.10 Principle of DPLL with automatic bandwidth control: (a) open-loop DPLL and (b) DPLL equivalent system and block diagram of the automatic bandwidth control

sequence $g[k]$ and the LMS loop, as usual, forces to zero the average difference between the product $a[k] \cdot g[k]$ and $e[k]$ (the TDC output). In steady-state conditions, $g[k]$ will be a constant (plus some noise) signal equal to the required parameter G, let us say $\hat{g}$, that can be easily inverted in the digital domain.

Even if a pseudorandom training sequence $q[k]$ can be generated and injected in the loop, this will increase the DPLL noise level. It is instead possible to exploit the quantisation noise of the $\Delta\Sigma$ modulator that interpolates the DCO tuning word. This is, as a matter of fact, a pseudorandom digital sequence, high-pass shaped, that is injected in the loop, and can be easily obtained as the difference between the output and the input of the $\Delta\Sigma$ itself. The resulting block scheme of the DPLL embedding this technique is sketched in Figure 6.11. Note that the multiplication at the TDC output is quite simple, being just a change of sign (either $1/(\hat{g})$ or $-1/(\hat{g})$). The closed-loop bandwidth is now equal to:

$$\omega_u \approx \frac{\beta}{T_r} \cdot \frac{G}{\hat{g}} \approx \frac{\beta}{T_r} \tag{6.15}$$

Even if ω_u is just an approximation of the closed-loop bandwidth, the loop gain is proportional to G without approximations. Thus, this technique will accurately desensitise the loop gain and, in turn, the value of the bandwidth from any analog parameter. An exact formula for the bandwidth is also evaluable, that is very complex and will not add any insight to this discussion, since it deviates from few percent with respect to the one above.

In a practical PLL, the automatic bandwidth-control algorithm is slightly modified, both to account for the closed-loop topology and for the presence of additional analog delays in the loop. For instance, the additional delay can be induced by the frequency divider or, if present, by an analog RC filter at the DCO

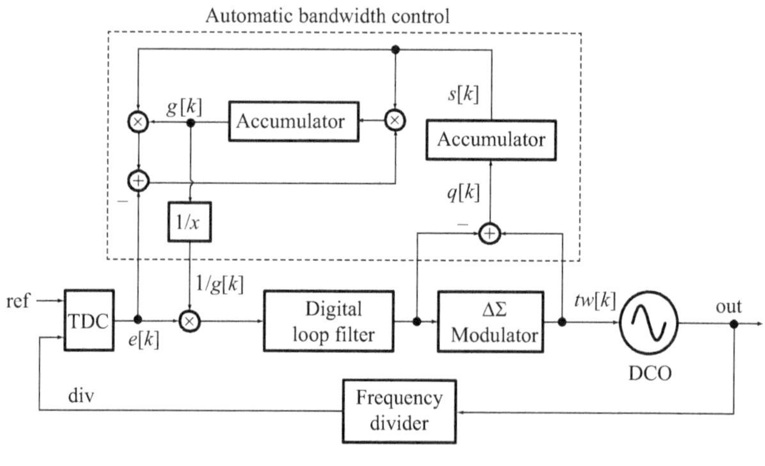

Figure 6.11 Diagram of DPLL with automatic bandwidth control

tuning port [36]. In this case, it is necessary to increase the number of taps (the coefficients) to estimate and then to combine them in an appropriate way. Usually, two gains may be enough to achieve the desired performance, see [36].

6.5 Practical implementations and measurements

Figure 6.12 shows a practical implementation of a fractional-N DPLL that embeds the solutions discussed in this chapter [25]. The circuit is based on a BBPD for low-power consumption, DTC and automatic gain regulation for fractional-N $\Delta\Sigma$ noise cancellation, and a frequency-aid loop for lock acquisition. For the sake of simplicity, some building blocks are simplified: the actual DTC digital control features multiple coefficients to reduce the DTC non-linearity, while the automatic bandwidth control is also present, but not included in the figure, that instead is focused on the frequency-aid system.

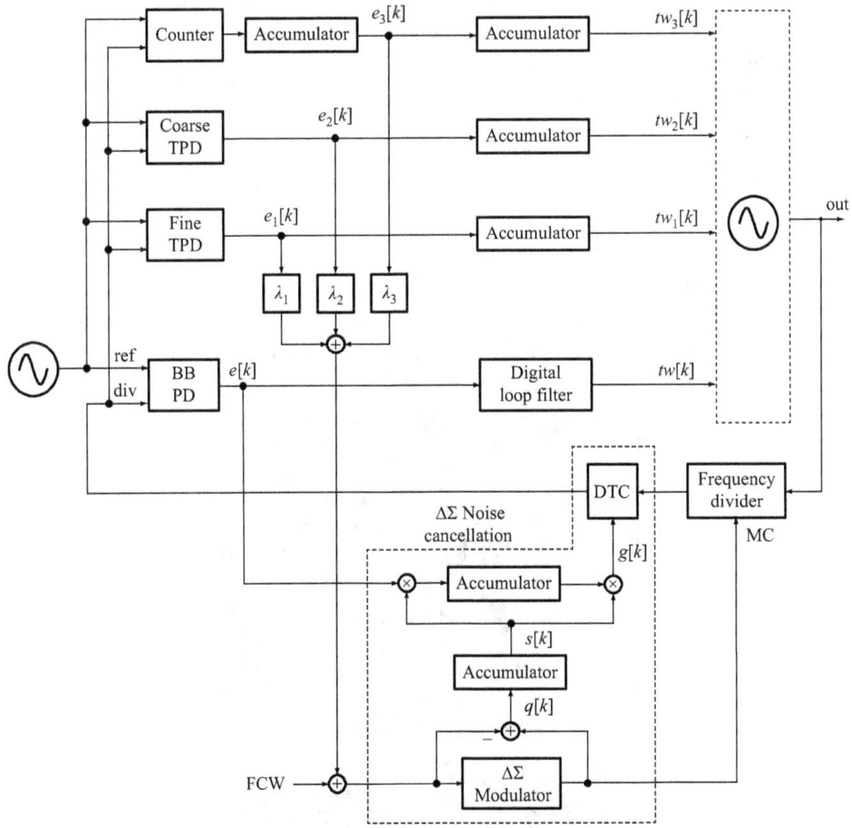

Figure 6.12 Practical DPLL implementation with frequency aid (three extra loops) and cancellation of ΔΣ quantisation errors

From this standpoint, two extra coarse DCO banks are used with respect to the basic scheme of Figure 6.9(a), so that, for a large frequency step, all the three loops of the frequency-aid circuit operate concurrently. However, since the coarser loop has a much larger gain, it practically sets the transient behavior of the PLL. When the time error is within the coarser dead zone, the coarser loop controlling the second tuning bank becomes idle, and the first tuning bank takes the control of the transient, until the phase error enters the dead zone of fine TPD.

The circuit shown has been fabricated in a 65-nm CMOS technology and occupies an active area of 0.61 mm^2. Figure 6.13 shows the die picture; the two small inductors do not belong to the main tank but are connected at the supply and at the ground, according to the well-known "tail filter" technique, resonating at twice the oscillation frequency.

The output signal can swing between 3.7 and 4.1 GHz, starting from a 52-MHz reference frequency, the resolution is below 200 Hz. The performance table and the phase noise spectrum measured in fractional-N mode are reported in Figure 6.14. The rms jitter integrated in the 1 kHz to 30 MHz range is about 182.5 fs, and a spot phase noise of −151 dBc/Hz at 20-MHz offset from the carrier, which makes it compliant with a GSM transceiver. The bandwidth was set at 150 MHz by means of the automatic bandwidth control.

This measurement shows that, even using a single-bit TDC, i.e. a BBPD, it is possible to get wireless-class phase noise, with an accurate control of the bandwidth, thanks to the digital techniques just discussed. Moreover, the circuit dissipates only 5.8 mW, thanks to the use of the BBPD, which is intrinsically low power.

The latter issue, the fast lock recovery with a BBPD, is also solved. As an example, for a frequency step as large as 346 MHz, if the frequency-aid circuit is disabled (just the counter-based loop is active besides the BB loop), the PLL cannot even reach the lock, as is evident from Figure 6.15(a). On the other hand, when the

Figure 6.13 Die photo

Reference (MHz)	52
Output (GHz)	3.7-4.1
Tuning range (%)	10
BW (MHz)	0.150
Settling time (μs)	5.6
Frequency hop (MHz)	364
Equivalent PN* (dBc/Hz)	–163.3
RMS jitter (fs)	183
Power (mW)	5.28
FoM (dB)	–247.5
Fractional spur (dBc)	–50
Spur frequency (MHz)	0.5
Area (mm^2)	0.61
Process (nm)	65

* at 900-MHz carrier and 20-MHz offset

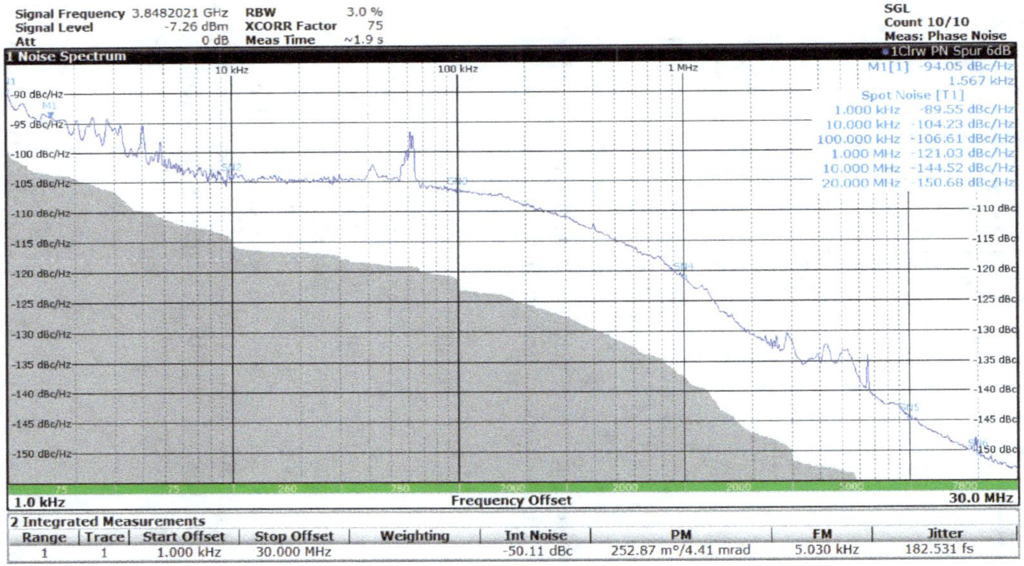

Figure 6.14 Performance table and measured phase noise spectrum in fractional-N mode

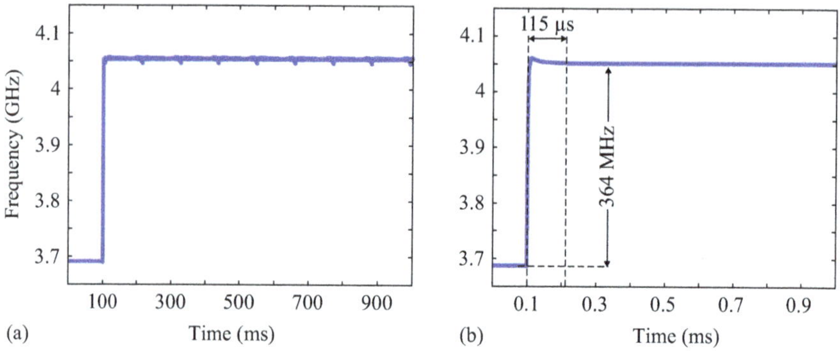

Figure 6.15 Settling behaviour for a 346-MHz frequency step, with the frequency-aid disabled (a) and enabled (b)

frequency aid is enabled, the circuits settles in 5.6 µs within 10 MHz from its final value, and in 115 µs within 364 kHz from its final value [see Figure 6.15(b)].

6.6 Conclusions

BB digital PLLs, thanks to the use of a single-bit TDC, i.e. a BBPD, dissipate much less power than their counterparts employing a multi-bit TDC. The main issues related to the adoption of such a hard non-linearity in the loop, i.e. how to avoid limit cycles in fractional-mode so as to obtain a wireless-class output spectrum, how to control the bandwidth, and how to guarantee a fast lock, have been discussed. We have shown that digitally intensive architectures enable powerful calibration circuits, mostly based on LMS techniques, to solve these issues, even in the presence of PVT variations.

References

[1] Lindsey WC and Chie CM. A Survey of Digital Phase-Locked Loops. Proc IEEE. 1981; 69(4):410–431.

[2] Dunning J, Garcia G, Lundberg J, *et al.* An All-Digital Phase-Locked Loop with 50-Cycle Lock Time Suitable for High-Performance Microprocessors. IEEE J Solid-State Circuits. 1995; 30(4):412–422.

[3] Staszewski RB, Muhammad K, Leipold D, *et al.* All-Digital TX Frequency Synthesizer and Discrete-Time Receiver for Bluetooth Radio in 130-nm CMOS. IEEE J Solid-State Circuits. 2004;39(12):2278–2291.

[4] Kim J, Yoon H, Lim Y, *et al.* A 76 fs_{rms} Jitter and –40 dBc Integrated-Phase-Noise 28-to-31 GHz Frequency Synthesizer Based on Digital Sub-Sampling PLL Using Optimally Spaced Voltage Comparators and Background Loop-Gain Optimization. In: 2019 IEEE International Solid-State Circuits Conference (ISSCC). IEEE; 2019. p. 258–260.

[5] Hsu TY, Shieh BJ, and Lee CY. An All-Digital Phase-Locked Loop (ADPLL)-Based Clock Recovery Circuit. IEEE J Solid-State Circuits. 1999;34(8):1063–1073.

[6] Temporiti E, Weltin-Wu C, Baldi D, *et al.* A 3.5 GHz Wideband ADPLL with Fractional Spur Suppression through TDC Dithering and Feedforward Compensation. IEEE J Solid-State Circuits. 2010;45(12):2723–2736.

[7] Zanuso M, Levantino S, Samori C, *et al.* A Wideband 3.6 GHz Digital ΔΣ Fractional-N PLL with Phase Interpolation Divider and Digital Spur Cancellation. IEEE J Solid-State Circuits. 2011;46(3):627–638.

[8] Rylyakov A, Tierno J, Ainspan H, *et al.* Bang-Bang Digital PLLs at 11 and 20 GHz with Sub-200 fs Integrated Jitter for High Speed Serial Communication Applications. In: IEEE ISSCC Dig. Tech. Papers. IEEE; 2009. p. 94–96.

[9] Nonis R, Grollitsch W, Santa T, *et al.* A 2.4 psrms-Jitter Digital PLL with Multi-Output Bang-Bang Phase Detector and Phase-Interpolator-Based Fractional-N Divider. In: IEEE ISSCC Dig. Tech. Papers. IEEE; 2013. p. 356–357.

[10] Talegaonkar M, Anand T, Elkholy A, *et al.* A 4.4–5.4 GHz Digital Fractional-N PLL using Frequency-to-Digital Converter. In: IEEE Symposium on VLSI Circuits. IEEE; 2014. p. 1–2.

[11] Elkholy A, Anand T, Choi WS, *et al.* A 3.7 mW 3 MHz Bandwidth 4.5 GHz Digital Fractional-N PLL with 106 dBc/Hz In-band Noise Using Time Amplifier based TDC. In: IEEE Symposium on VLSI Circuits. IEEE; 2014. p. 1–2.

[12] Weltin-Wu C, Zhao G, and Galton I. A Highly-digital Frequency Synthesizer using Ring-Oscillator Frequency-to-Digital Conversion and Noise Cancellation. In: IEEE ISSCC Dig. Tech. Papers. IEEE; 2015. p. 1–3.

[13] Gao X, Burg O, Wang H, *et al.* A 2.7-to-4.3 GHz, 0.16 psrms-jitter, −246.8 dB-FOM, Digital Fractional-N Sampling PLL in 28 nm CMOS. In: IEEE ISSCC Dig. Tech. Papers. IEEE; 2016. p. 174–175.

[14] Markulic N, Raczkowski K, Martens E, *et al.* A Self-Calibrated 10 Mb/s Phase Modulator with −37.4dB EVM Based on a 10.1-to-12.4 GHz, −246.6 dB-FOM, Fractional-N Subsampling PLL. In: *IEEE ISSCC Dig. Tech. Papers*; 2016. p. 176–177.

[15] Siriburanon T, Kondo S, Kimura K, *et al.* A 2.2 GHz 242 dB-FOM 4.2 mW ADC-PLL using Digital Sub-Sampling Architecture. In: IEEE ISSCC Dig. Tech. Papers. IEEE; 2016. p. 440–441.

[16] Seong T, Lee Y, Yoo S, *et al.* A −242 dB FOM and −75 dBc-Reference-Spur ring-DCO-based All-digital PLL Using a Fast Phase-error Correction Technique and a Low-power Optimal-threshold TDC. In: 2018 IEEE International Solid-State Circuits Conference (ISSCC). IEEE; 2018. p. 396–398.

[17] Tasca D, Zanuso M, Marzin G, *et al.* A 2.9-to-4.0 GHz Fractional-N Digital PLL with Bang-Bang Phase Detector and 560 fsrms Integrated Jitter at 4.5 mW Power. *IEEE J Solid-State Circuits*. 2011;46(12):2745–2758.

[18] Da Dalt N. Markov Chains-Based Derivation of the Phase Detector Gain in Bang-Bang PLL. IEEE Trans Circuits Systems II. 2006;53(11):1195–1199.

[19] Da Dalt N. Linearized Analysis of a Digital Bang-Bang PLL and its Validity Limits Applied to Jitter Transfer and Jitter Generation. IEEE Trans Circuits Systems I. 2008;55(11):3663–3675.

[20] Zanuso M, Tasca D, Levantino S, *et al.* Noise Analysis and Minimization in Bang-Bang Digital PLLs. IEEE Trans Circuits Systems II. 2009;56(11): 835–839.

[21] Hsu CM, Straayer MZ, and Perrott MH. A Low-Noise Wide-BW 3.6-GHz Digital $\Delta\sum$ Fractional-N Frequency Synthesizer with a Noise-Shaping Time-to-Digital Converter and Quantization Noise Cancellation. IEEE J Solid-State Circuits. 2008;43(12):2776–2786.

[22] Santiccioli A, Samori C, Lacaita AL, *et al.* Power-jitter Trade-off Analysis in Digital-to-Time Converters. Electron Lett. 2017;53(5):306–308.

[23] Levantino S, Marzin G, and Samori C. An Adaptive Pre-Distortion Technique to Mitigate the DTC Non-Linearity in Digital PLLs. IEEE J Solid-State Circuits. 2014;49(8):1762–1772.

[24] Da Dalt N. Theory and implementation of digital band-bang frequency synthesizers for high speed serial data communications. Rheinisch-Westfälischen Technischen Hochschule Aachen. Aachen, Germany; 2007. [Online] Available http://citeseerx.ist.psu.edu.

[25] Bertulessi L, Grimaldi L, Cherniak D, *et al.* A Low-Phase-Noise Digital Bang-Bang PLL with Fast Lock Over a Wide Lock Range. In: Digest of International Solid-State Circuits Conference (ISSCC); 2018. p. 252–253.

[26] Kuan T and Liu S. A Digital Bang-Bang Phase-locked Loop with Automatic Loop Gain Control and Loop Latency Reduction. In: 2015 Symposium on VLSI Circuits (VLSI Circuits); 2015. p. C138–C139.

[27] Kuan TK and Liu SI. A Loop Gain Optimization Technique for Integer-N TDC-Based Phase-Locked Loops. IEEE Trans Circuits Systems I 2015; 62(7):1873–1882.

[28] Kuan T and Liu S. A Bang Bang Phase-Locked Loop Using Automatic Loop Gain Control and Loop Latency Reduction Techniques. IEEE J Solid-State Circuits. 2016;51(4):821–831.

[29] Kwon S, Lee J, Lee J, *et al.* An Automatic Loop Gain Control Algorithm for Bang-Bang CDRs. IEEE Trans Circuits Systems I. 2015;62(12): 2817–2828.

[30] Jang S, Kim S, Chu S, *et al.* An Optimum Loop Gain Tracking All-Digital PLL Using Autocorrelation of Bang–Bang Phase-Frequency Detection. IEEE Trans Circuits SystEems II. 2015;62(9):836–840.

[31] Kim D, Song H, Kim T, *et al.* A 0.3–1.4 GHz All-Digital Fractional-N PLL With Adaptive Loop Gain Controller. *IEEE J Solid-State Circuits*. 2010 Nov;45(11):2300–2311.

[32] Ferriss M, Plouchart J, Natarajan A, *et al.* An Integral Path Self-Calibration Scheme for a Dual-Loop PLL. IEEE J Solid-State Circuits. 2013;48(4): 996–1008.

[33] Lin J and Yang C. A Fast-Locking All-Digital Phase-Locked Loop With Dynamic Loop Bandwidth Adjustment. IEEE Trans Circuits Systems I. 2015;62(10):2411–2422.
[34] McMahill D and Sodini CG. A 2.5-Mb/s GFSK 5.0-Mb/s 4-FSK Automatically Calibrated Frequency Synthesizer. IEEE J Solid-State Circuits. 2002;37(1):18–26.
[35] Joshi A and Midha G. Bandwidth Compensation Technique for Digital PLL. IEEE Trans Circuits Systems II. 2016;63(11):1044–1048.
[36] Mercandelli M, Grimaldi L, Bertulessi L, *et al.* A Background Calibration Technique to Control the Bandwidth of Digital PLLs. IEEE J Solid-State Circuits. 2018;53(11):3243–3255.

Chapter 7
Hybrid PLLs

Mark Ferriss[1] and Daniel Friedman[1]

Modern phase-locked loops (PLLs) can generally be categorized as either digital or analog; in this chapter, we will describe hybrid PLLs, in which we attempt to combine the best features of both. With this goal in mind, let us first discuss the two approaches with a view to gaining an understanding of the benefits and limitations of each approach.

7.1 Introduction: analog and digital PLL considerations

First, consider the classic analog PLL, as depicted on the left side of Figure 7.1. The PLL achieves frequency lock by adjusting the varactor voltage and voltage-controlled oscillator (VCO) frequency such that the frequency and phase of the feedback clock and reference clock match. While this approach may be attractive for many applications, it does bring with it several practical problems. First, the PLL bandwidth is proportional to the varactor gain, and the varactors found in modern complementary-metal-oxide-semiconductor (CMOS) processes are extremely non-linear; the small signal gain can easily vary by a factor of 3 across the control voltage range of the varactor. Furthermore, the capacitor itself may be non-linear, leading to an additional source of gain variation. An additional problem arises when we consider the charge pump. In particular, several hundred millivolts must be reserved between the maximum and minimum supported voltages and the supply rails to prevent the charge pump's current sources from entering their triode regions. In a sub-1 V design as is typical for advanced CMOS implementations, however, this headroom requirement leaves varactors with a significantly limited voltage tuning range. To compensate for the lack of voltage range, the VCO gain must be increased, leading to an extreme sensitivity to noise.

One approach to addressing these key challenges associated with the classic analog PLL architecture is the dual-path analog PLL approach, as depicted on the right side of Figure 7.1, an example of which can be found in [1]. In this architecture, the control loop is split into two independent paths, namely, the

[1]IBM Thomas J. Watson Research Center, Communication Circuits and Systems Department, Yorktown Heights, NY, USA

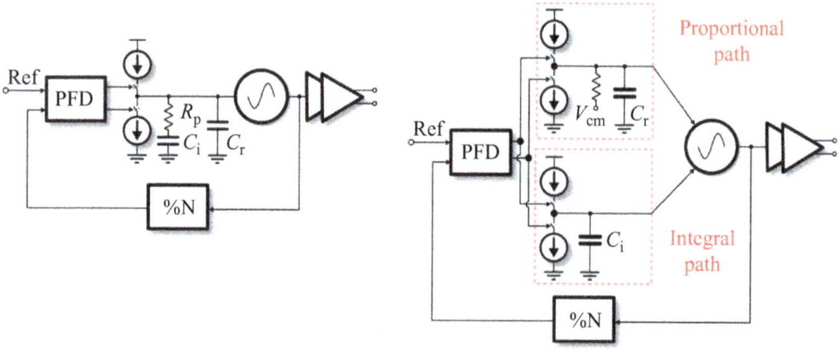

Figure 7.1 Classic analog PLL (left) and dual-path analog PLL (right)

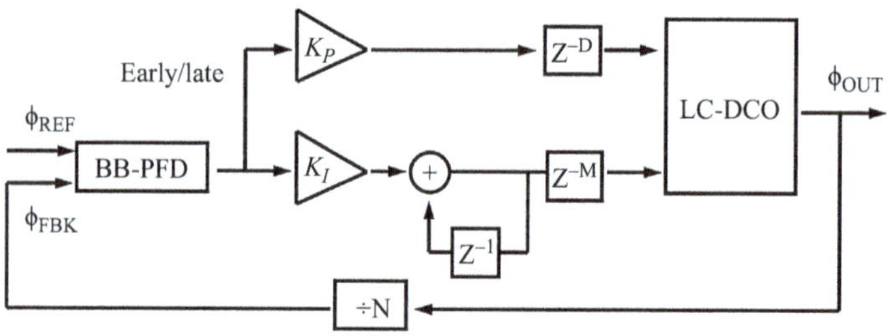

Figure 7.2 Digital PLL block diagram

proportional path and the integral path. When the PLL is in a locked condition or a near-locked condition, the voltage of the proportional path signal is centered around its common mode point. Once the PLL locked, the proportional path DC voltage barely moves, only making small excursions to correct for noise. Consequently, the proportional path suffers significantly less from key problems associated with the classic analog approach outlined above. Specifically, the proportional path's varactor is only exercised around one point on its transfer curve, so gain variation due to varactor non-linearity disappears. Similarly, even if the ripple capacitor is implemented with a high-density non-linear capacitor, this does not result in variation in the frequency of the resulting higher order pole. Unfortunately, the problems highlighted for the classic analog PLL approach still occur in the integral path. Problems like the foregoing, as well as other issues such as the physical size of the loop filter capacitor if implemented as an integrated element, often lead to the conclusion that a digital PLL is the answer.

In a digital PLL, the PLL's phase error is digitized and the PLL's loop filter is implemented in the digital domain (Figure 7.2). Digital PLLs offer substantial advantages in terms of area and programmability as compared to their analog counterparts and can be applied in a wide range of application areas and offer a ripe area for innovation [2–6]. As an example, this approach makes the transfer characteristics of the loop filter independent of component variation. These benefits, however, come at the price of potentially degraded spurious noise performance, as well as complex non-linear loop dynamics. Several major challenges must be addressed if the full potential of the digital PLL is to be realized. Bang-bang PLLs (BB-PLLs) have been demonstrated to be capable of good performance once they achieve phase lock [7], but the hard non-linearity introduced by the 1-bit bang-bang phase/frequency detector (BB-PFD) severely impacts locking dynamics and noise event response of BB-PLL [8], resulting in relatively slow frequency and phase acquisition times. The slow phase acquisition time, in conjunction with the narrow linear range of the BB-PFD, significantly affects the robustness of the BB-PLL to external noise events, especially as compared to that of an analog PLL; furthermore, the locking dynamics can exhibit significant ringing. The use of a time-to-digital converter (TDC)-based approach partially alleviates the limitations of BB-PLLs [9,10], but an important characteristic of a TDC is that the resolution of an inverter-based delay line implementation is limited to multiple picoseconds, whereas the phase errors seen in a high-performance integer-*N* PLL are typically at or below the picosecond level. While improved-resolution approaches have been explored, significant challenges remain in the TDC design. Finally, from an implementation perspective, digital PLL designs also place significant demands on VCO control and the management of the propagation delay through the control loop.

Returning to a consideration of the dual-path analog architecture, however, we note that there are several key differences between the proportional and integral paths, and further that the properties of the integral path in this context present attractive architectural opportunities. First, the gain of the integral path is much lower than that of the proportional path and consequently is less sensitive to propagation delays. Furthermore, this lower gain provides an opportunity to perform more filtering, or averaging, of the output from the phase detector, implying that even an imperfect phase detector might be suitable in the dual-path context. These considerations suggest the possibility of implementing just the integral path in the digital domain, while retaining the analog proportional path as in a dual-path analog PLL.

In this chapter, we describe a PLL architecture in which the control loop is implemented using a hybrid of digital and analog PLL techniques, shown in canonical form in Figure 7.3, examples of which can be found in [11,12]. This architecture maintains the key benefits of the all-digital approach (compact design and CMOS-friendly implementation), while at the same time preserving many of the linear phase response characteristics of conventional analog PLLs. In its basic form, the hybrid PLL architecture combines a linear analog proportional path control and bang-bang digital integral path control. The linearity of the proportional path control resolves the noise sensitivity issues of BB-PLLs outlined above, while the digital implementation of the integral path keeps the PLL small, programmable, and scalable. The proposed architecture is free of the two main challenges facing an analog PLL implementation in a CMOS process optimized for digital applications: there is no need for a large, low-leakage capacitor to perform the integrating function of the control loop, and no need for a wide dynamic range, low-noise charge pump.

Under the hybrid PLL approach, the proportional path design of the PLL can be greatly simplified as compared to that of a conventional charge-pump PLL.

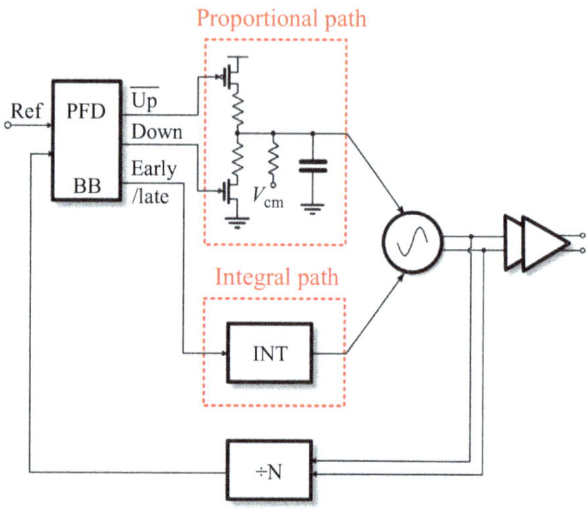

Figure 7.3 Hybrid PLL block diagram

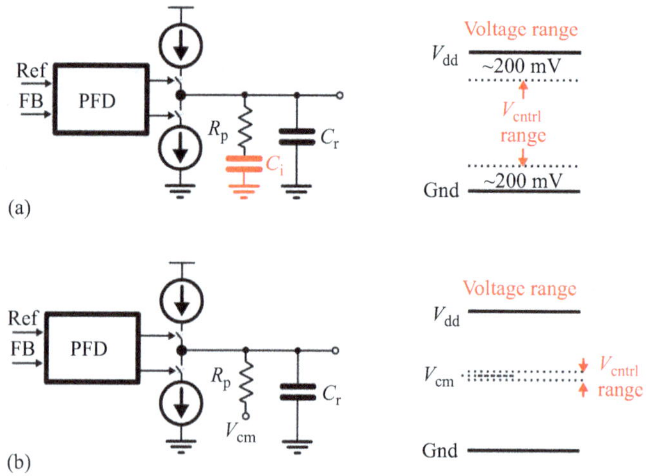

Figure 7.4 Proportional path comparison between analog and hybrid PLL approaches

In the proposed architecture, the proportional control voltage always operates around its common mode position when the loop is in phase lock simply because the integrating function is removed from the analog control path. This comparison is illustrated in Figure 7.4. The top portion of the figure corresponds to the proportional path in a classic analog PLL design; note that the charge pump must maintain performance over the entire common mode voltage excursion range required by the integral path's establishment of VCO operating frequency. In the hybrid PLL approach, by contrast, the proportional path common mode can be maintained in a very narrow range and can be nominally set to an easy-to-generate voltage such as half of the supply voltage. In this context, the charge pump's high impedance current sources used in the classic analog approach can be replaced by structures as simple as switched resistors. Figure 7.5 illustrates the concept; notice that the removal of the integrating cap removes the need for the troublesome common mode feedback loop.

Finally, the phase detector must be capable of producing two sets of outputs; the classic UP/DOWN pulses for the proportional path and a bang-bang output for the integral path. Fortunately, this is straightforward, Figure 7.6 illustrates how a

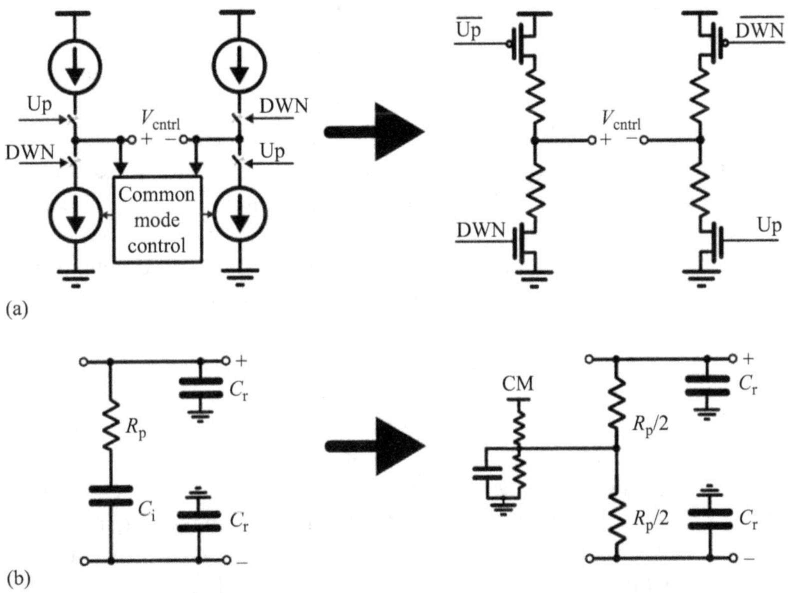

Figure 7.5 Charge pump and loop filter transformation

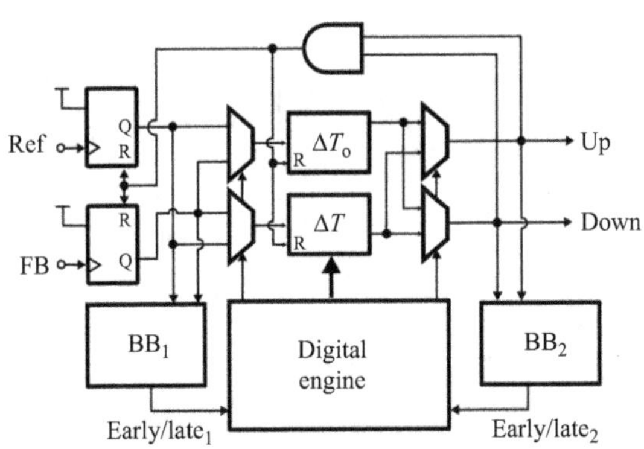

Figure 7.6 The phase detector

bang-bang can be added to the programmable delay generators. The purpose of the programmable delay generators will be explained in Section 7.3.

7.2 Hybrid VCO control with continuous tuning

In this section, we will discuss how a hybrid collection of analog and digital techniques can be used to solve a common problem; how do we design an oscillator that can be tuned over a wide frequency range without glitches, and without degrading the performance.

In both analog and digital PLLs, the VCO tuning range is typically divided into discrete bands, as illustrated on the left side of Figure 7.7. It is possible in principle to use a single large varactor to achieve the required tuning range, but this approach incurs significant penalties. First, note that in a low-voltage CMOS process, the voltage range of the control node is typically less than 1 V. If the entire VCO range is to be accessed from a single control signal in this context, the voltage-to-frequency gain of that node must be extremely high, resulting in a significant noise sensitivity on the control node. Second, large analog varactors are a well-known means by which amplitude modulation (AM) noise is converted to phase modulation (PM) noise inside the VCO. Consequently, the preferred approach is to use banks of digitally switched caps to achieve most of the target tuning range and to use an analog varactor for fine-tuning in the PLL itself. This traditional banding strategy leads to several practical problems. First, the entire VCO frequency range cannot be traversed while maintaining glitch-free phase lock, as changing bands involves simultaneously switching in and out large capacitors. Further, the per-band tuning range must be wide enough to guarantee that the mismatch between digitally switched capacitor banks does not result in holes in the frequency transfer curve of the VCO. Finally, sufficient tuning range to accommodate temperature- and supply variation-induced drift must be present within each band, placing stringent demands on adjacent band overlap and on band selection algorithms. Our

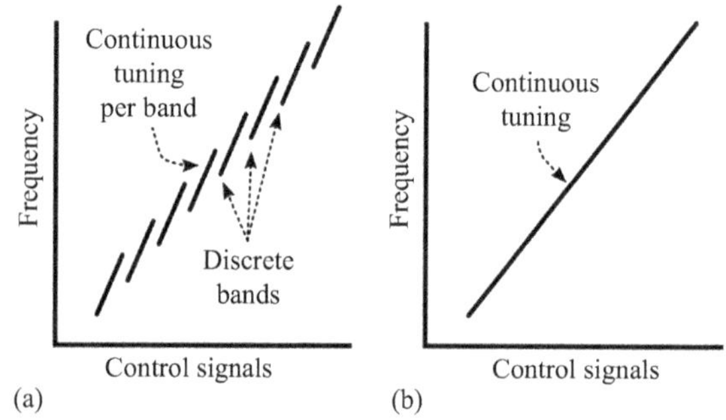

Figure 7.7 The left shows typical VCO characteristics, and on the right is our objective

goal in this section is to describe a system that achieves coarse band elimination, or access to the entire tuning range of the VCO without the use of high-gain varactors and without the glitches associated with switching between bands.

To understand how this objective can be achieved, we first briefly review a typical tuning scheme for a digital PLL. The digital PLL generates a control word based on detected phase errors; the objective of the frequency control scheme is to translate changes of the digital control word into changes in capacitance in the VCO. A popular way to implement this functionality, shown in Figure 7.8, is to use a simple oversampled sigma-delta and RC filter to generate an analog control signal, and to apply the resulting signal to a varactor. To prevent sigma-delta noise from dominating the noise of the VCO, the gain of the varactor must be kept small. However, low gain in a low-voltage process unfortunately corresponds to achieving a narrow tuning range. So, how do we extend the tuning range without increasing the varactor gain? One possible approach is to use a bank of varactors, in which the analog control signal is applied to one of the varactors and where the remaining varactors are switched to their maximum or minimum capacitance state. Then, when the analog control signal reaches its full scale value, we can simply pass that control signal to the next varactor in line, as illustrated in Figure 7.9. In this way, we can fold the transfer functions of the individual elements together to make one aggregate band, as demonstrated in [13]. This approach works because of the digital nature of the integral path; we know which integral path control levels correspond to the transitions between banks, so we can coordinate the switching of analog multiplexers with these transition points.

While this basic approach is conceptually sound, there are some practical issues that must be addressed if a truly wide band version of this architecture is to be implemented. The first of these issues is exactly how we should implement the tuning capacitor. There are two broad categories of tuning capacitors we could use: (1) continuously tunable analog varactors or (2) high-Q linear capacitors in series with switches. The capacitor/switch approach is preferred for most of the tuning

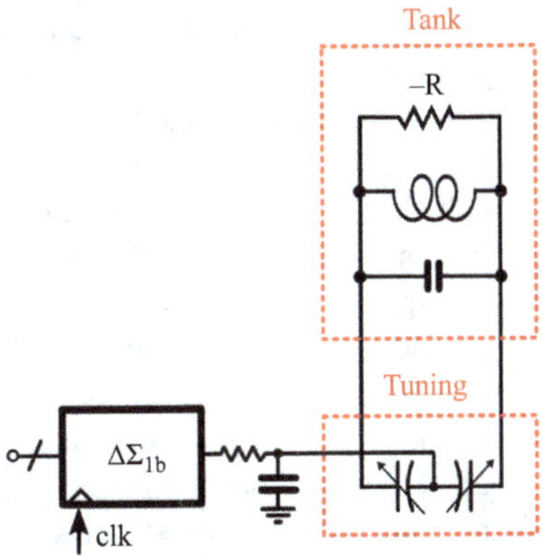

Figure 7.8 Digital tuning with a simple varactor

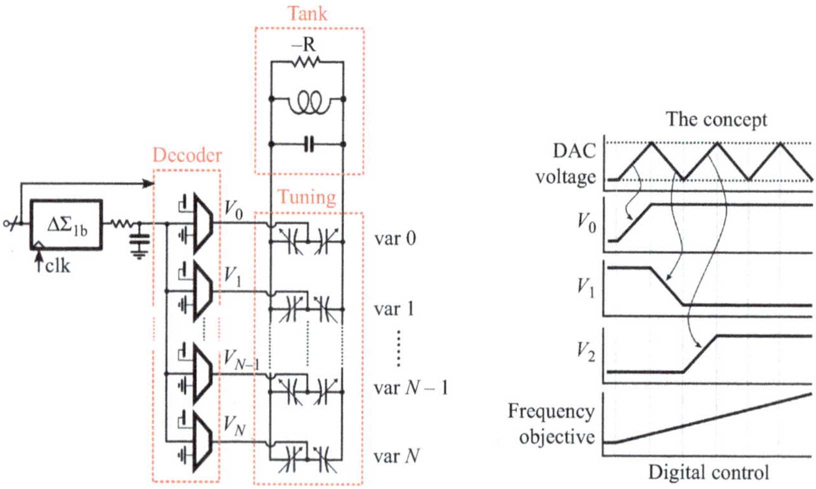

Figure 7.9 Architecture of varactor folding approach

range, but the varactor scheme as described in Figure 7.9 relies on continuously tuned capacitors.

In Figure 7.10, we compare these two approaches in order to gain some insight. Note that the figure is intended to illustrate basic characteristics of the two approaches and that detailed results will vary with, e.g., technology and supply. For this analysis, we first size the capacitors used in each scheme such that they provide matching *C*-on to *C*-off ratios. Next, the voltage-to-capacitance gain and quality factor for each approach are extracted from a periodic steady-state simulation, where the *x*-axis corresponds to the control signals being swept. Notice that while both varactor approaches are non-linear, the capacitor/switch version is dramatically more so; the gain is very high close to the *Vt* of the switch and saturates quickly. When the elements are in their highest capacitance state, the *Q* of the capacitor/switch is better than that of the varactor. In the minimum capacitance state, the varactor *Q* is better than that of the capacitor/switch combination, but this advantage for the varactor approach is less important because in this configuration the capacitors represent a negligible fraction of the total capacitance of the system. Crucially, the *Q* of the capacitor/switch is degraded when the impedance of the switch is close to that of the capacitor. This *Q* degradation occurs before the net varactor impedance is high; consequently, it can have a significant impact on the *Q* of the overall system. This analysis tells us why the capacitor/switch approach works well for hard switching as when used to implement coarse tuning elements but works poorly when used for continuously tunable capacitance. Note further that, because there is little residual gain at the extremes, the gain saturation properties of the capacitor/switch approach are attractive, making this approach very well suited for use with large banks of capacitors.

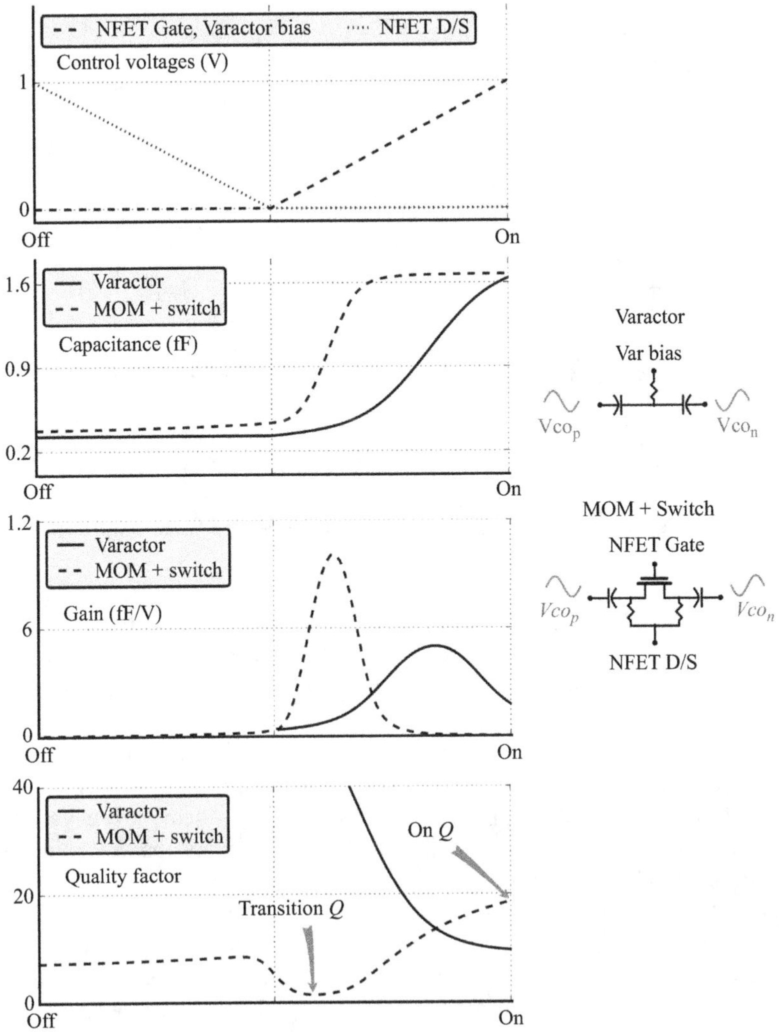

Figure 7.10 Varactor characteristics

As described above, while it has very attractive properties in terms of its on Q and gain flatness in saturation, the capacitor/switch combination appears to be a poor choice for continuously tunable structures both due to its degradation of Q during its transition between capacitance regimes and its very strong non-linearity. If these limitations can be mitigated, the capacitor/switch combination would, therefore, be an excellent choice as our base capacitive tuning element. First, consider the Q degradation. If one element in a bank of switch/cap elements is used as a continuously tunable capacitor, then how big a degradation to Q will occur?

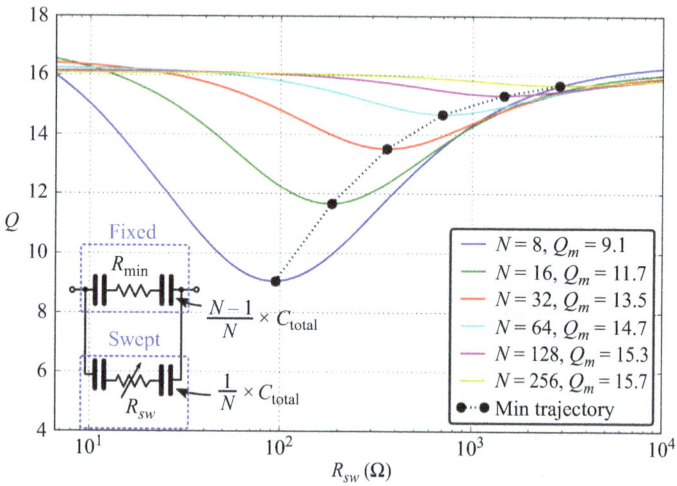

Figure 7.11 Q degradation versus number of elements

The answer depends on what fraction of the total capacitance suffers from the Q degradation, as illustrated in Figure 7.11. As can be seen in the figure, if the capacitance of the continuously tunable element is on the order of less than 1% of the total capacitance, the reduction in Q becomes negligible. With this constraint, we can use the capacitor/switch combination as a continuously tunable element provided that the gain is small. By itself this would result in a very small tuning range, but we can pair this approach with the folding scheme we have already discussed to make one continuous tuning band, enabling coarse band elimination in the resulting structure.

Next, how can we mitigate the strong non-linearity of the capacitor/switch combination? This issue can be addressed by using multiple bands in parallel, with each band offset with respect to each other. Using this approach, even while each band by itself is highly non-linear, the aggregate will be substantially linearized. Figure 7.12 shows an example of how the bands can be implemented, while Figure 7.13 shows how offsets can be added between the controls for each of the bands. Reading from left to right, the second band will be offset by 1/4 of the full scale, the third band by 1/2, and the last band will be offset by 3/4. Finally, in Figure 7.14, we see how the composite gain of multiple non-linear elements in parallel can yield an overall well-linearized response.

A practical implementation of this approach is illustrated in Figure 7.15, and a representative physical design of an implementation of this approach is shown in Figure 7.16. An implementation example can be found in [14]. In this case, the desired offset was implemented by selecting devices with offset threshold voltages with respect to each other. This approach simplifies the logic implementation at the expense of sensitivity to variations in Vt. This implementation also uses a mixture of NMOS and PMOS elements, for reasons that are beyond the scope of the present

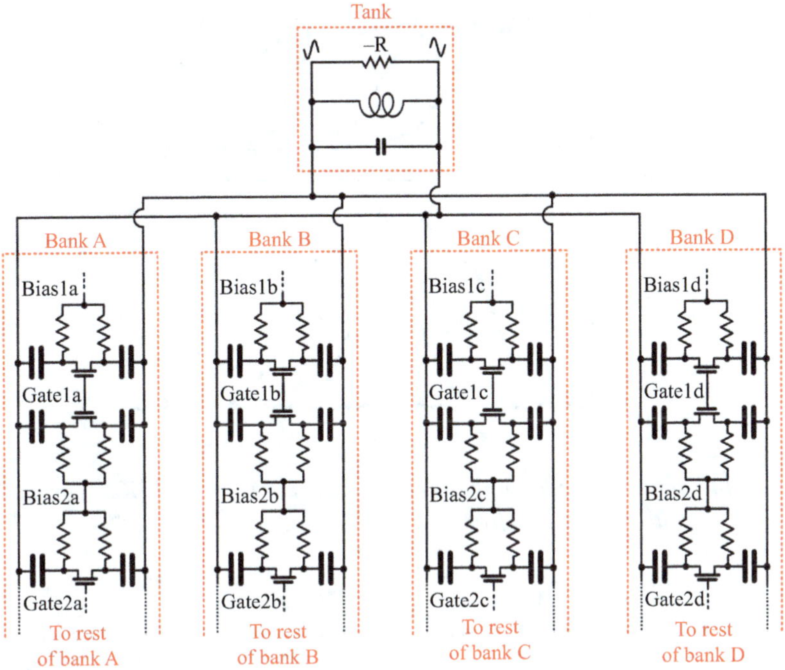

Figure 7.12 Multiple bands in parallel

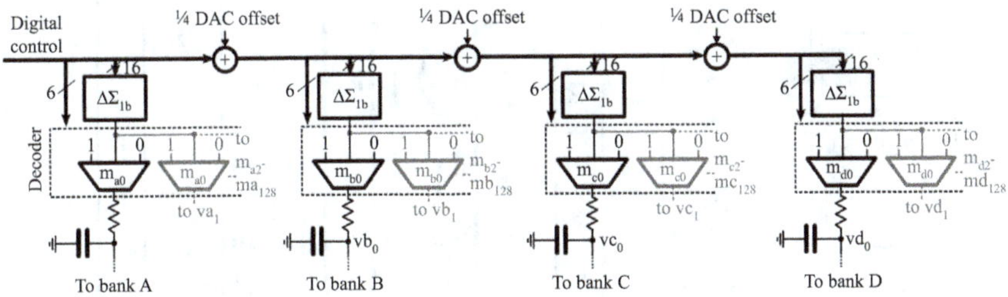

Figure 7.13 Achieving offset between the controls

discussion. Figure 7.17 shows a simplified view of a PLL architecture that incorporates the varactor folding approach using the capacitor/switch to support both hard switched and continuously tunable capacitance. In this design, we have added extra infrastructure to allow use of the PLL as an FMCW chirp generator. Our objective here is to take advantage of the extra tuning range enabled by the folding scheme implemented in concert with the capacitor/switch elements to generate chirps of magnitudes that would not be possible in a conventional design. In Figure 7.18, the measured tuning curve of the folding VCO with coarse band elimination is shown. In this measurement, the PLL loop is disabled, and the

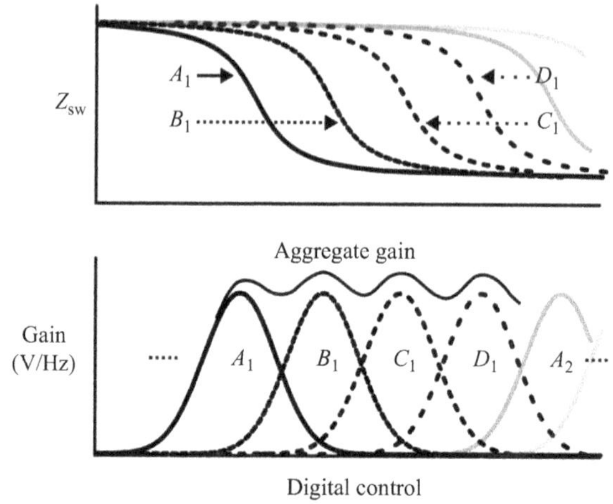

Figure 7.14 Impact on gain of the parallel elements © 2016 IEEE. Reprinted, with permission, from [16]

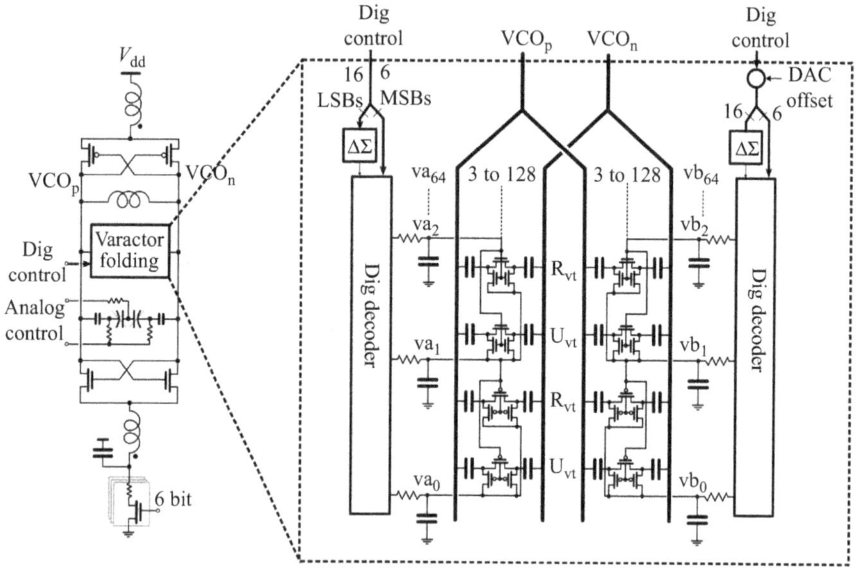

Figure 7.15 A practical implementation. © 2016 IEEE. Reprinted, with permission, from [14]

integral path control signal is programmed using the test chip's serial interface. The bottom portion of the plot shows the linearity of the gain. When we disable the offsets between bands, the transfer function is indeed extremely non-linear. With the offsets enabled, some degree of non-linearity remains, but the variation in gain

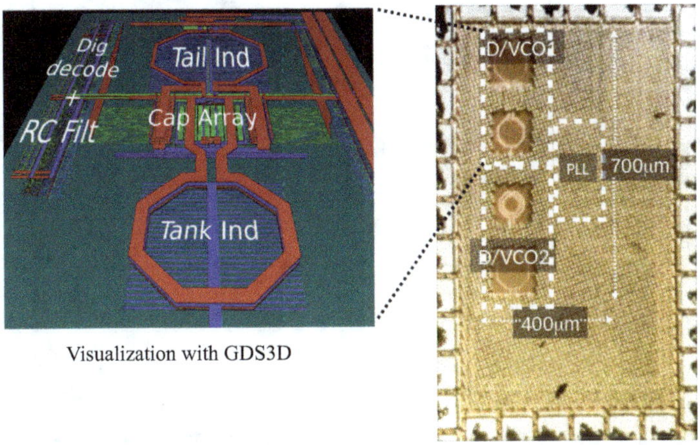

Figure 7.16 Die photo. © 2016 IEEE. Reprinted, with permission, from [14]

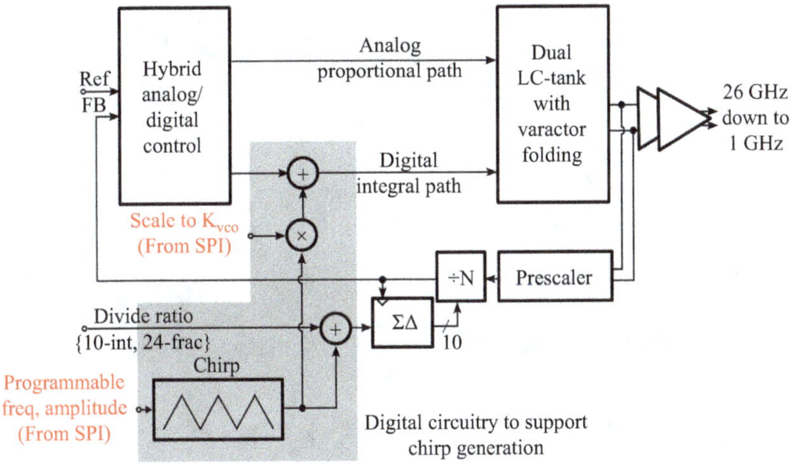

Figure 7.17 Architecture. © 2016 IEEE. Reprinted, with permission, from [14]

has been dramatically reduced. Figure 7.19 shows the measured phase noise for a hybrid PLL design realized using the folding scheme. The phase noise performance is similar to what would be expected without the folding scheme. Note that for high-frequency VCOs, the Q of the varactor capacitance dominates the total Q of the tank; consequently, designing high frequency, wide tuning range VCOs in low-voltage processes represents the most challenging design space from a phase noise performance perspective. Finally, Figure 7.20 shows a measured frequency chirp executed with this design. The folding scheme allows generation of chirps that span more than 8 GHz of tuning range, taking advantage of the full frequency range of the VCO.

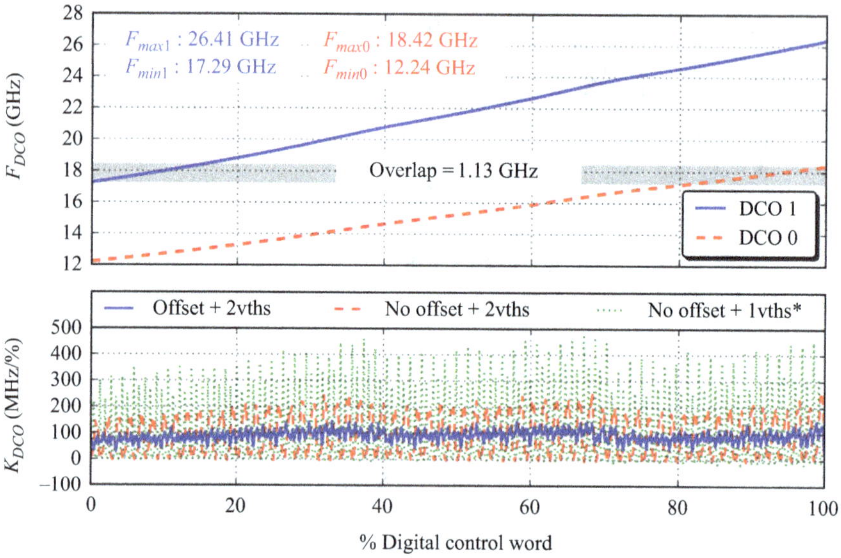

Figure 7.18 Measured tuning curve. © 2016 IEEE. Reprinted, with permission, from [14]

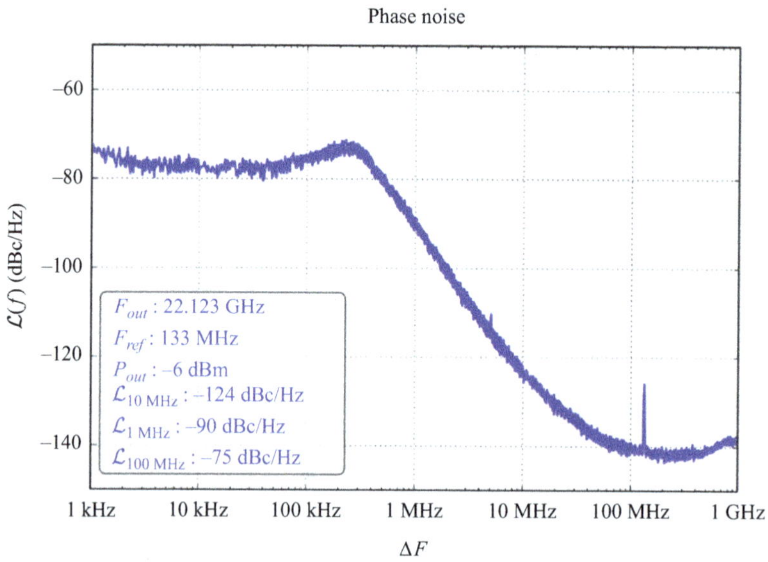

Figure 7.19 Phase noise. © 2016 IEEE. Reprinted, with permission, from [14]

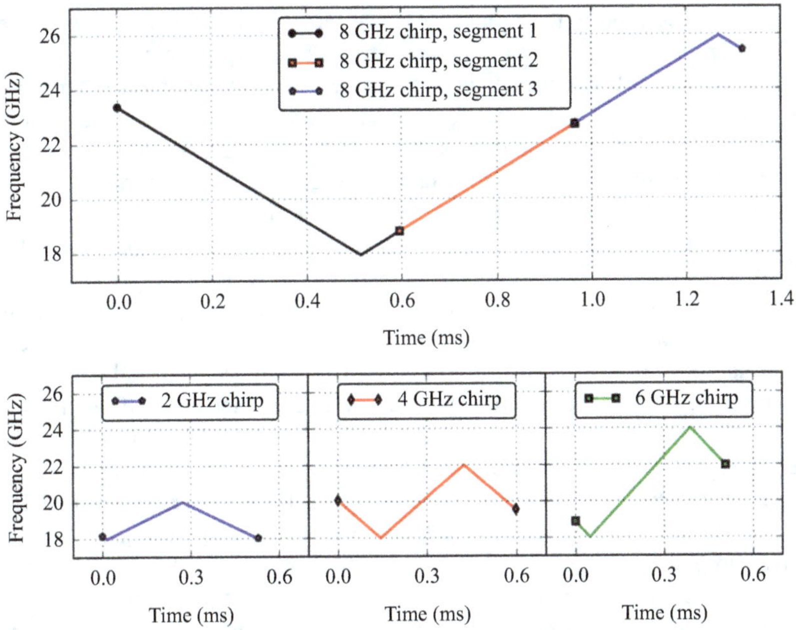

Figure 7.20 Measured frequency chirps. © 2016. IEEE. Reprinted, with permission, from [14]

7.3 Frac-*N* noise cancellation in a digital-friendly technology

The hybrid PLL architecture is well suited not only to integer-*N* PLL design but also to fractional-*N* designs. In this section, we will describe an approach to a fractional-*N* noise cancellation that can be implemented in a low-voltage CMOS, digital-friendly technology. In a fractional-*N* PLL, the feedback divider's division ratio is modulated from reference cycle to reference cycle such that the average division ratio is fractional even if the counter-based divider's instantaneous division ratio is restricted to integers. The noise associated with switching is shaped to high frequencies using a sigma-delta modulator. If the bandwidth of the PLL is sufficiently narrow and the reference frequency is sufficiently high, the PLL will filter out this noise before it contaminates the phase noise of the PLL. However, if the PLL bandwidth is pushed to a few megahertz, then frac-*N* noise suppression by filtering alone becomes impractical. A solution to this problem can be achieved by first recognizing that the noise for the sigma-delta is not random; instead, this noise is driven by the behavior of the digital sigma-delta modulator, and hence is fully deterministic. The instantaneous value of this noise can, therefore, be calculated, and a correcting signal can be added to the PLL control loop to compensate for this deterministic noise. Many techniques to implement such a scheme have been reported in the literature, e.g. in [15], often requiring elements such as high-precision current digital-to-analog converters that inject cancellation currents into the loop filter to compensate the error from the charge pump.

To elucidate, let us consider how to implement a cancellation scheme using digital standard cells, which simplifies the canceller's implementation in a digital-friendly process. First, consider how the frac-N noise impacts the time-domain inputs to the phase detector, as illustrated in Figure 7.21. Notice that the frac-N noise causes the feedback clock (FB) to be sometimes early and sometimes late with respect to the reference clock. This deterministic noise source results in modulation to the width of the up and down (UP/DWN) pulses. The additional duration of each pulse is deterministic; in the digital domain, we know the magnitude of the error, and we know for each reference period which of the two edges will be delayed. We can take advantage of this knowledge by adding a set of programmable digital-to-time converters to the phase detector itself, as illustrated in Figure 7.22. The objective of these elements is to simply delay the rising edge of the early signal by the duration of the digitally computed deterministic instantaneous error. Because the reset is still asynchronous, the falling edges of the UP and DWN pulses still occur simultaneously, irrespective of delay settings.

The simplicity of this approach hides some subtle issues that must be addressed. Notice that we have added two delay generators, one programmable and one fixed, and a set of multiplexers to switch the programmable delay generator between the two paths. For any set of rising edges, we need only delay one of the two paths; the fixed delay generator is used on the non-delayed path to mimic the minimum propagation delay of the programmable delay generator. It may seem simpler to have independent delay generators for each of the two paths and to remove the need for the multiplexers, but it turns out that such an approach would complicate the implementation of the control scheme for calculating the cancellation gain.

In particular, note that the sigma-delta noise has zero mean, so the net cancellation signal will also have zero mean, i.e. we will be cancelling the two paths by the same average amount. Consequently, if the cancellation gain is too large or small leading to under- or over-correction, the mean of the resulting phase error

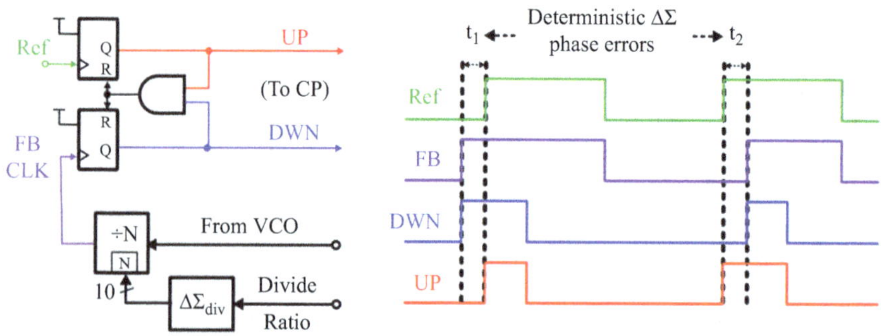

Figure 7.21 Time-domain phase detector inputs

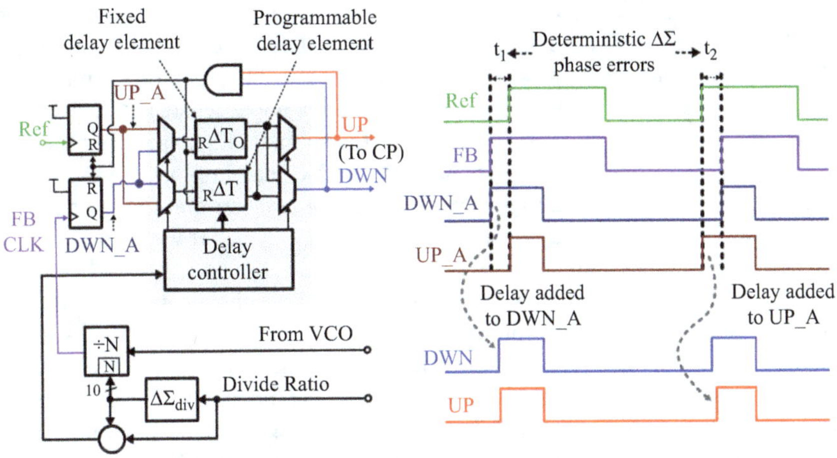

Figure 7.22 Time-domain phase detector inputs with programmable delay generators. © 2015 IEEE. Reprinted, with permission, from [16]

will still be zero. Small changes to the cancellation path gain will thus add only high-frequency phase errors to the PLL. If, in contrast, we were to implement two cancellation systems, one for each path, the result would be two control systems, one for each path. Crucially, the cancellation signals for the two independent paths would not have a zero mean error, an undesirable characteristic. To illustrate this point, consider an error sequence with the following repeating pattern: $\{+1.5, -0.75, -0.75, \ldots\}$. Notice that this pattern has zero mean; if the pattern is scaled by an arbitrary gain, it will still have a zero mean. However, if we split this into two separate sequences based on the polarity of the error signal, the result is the two sequences $\{+1.5, 0, 0, \ldots\}$ and $\{0, -0.75, -0.75, \ldots\}$. Notice that these sequences, considered independently, do not have zero mean; changes to the gain for one of these patterns changes its average. Thus, changes to the gain of one of the paths in the two-control-system approach will introduce a phase change in the PLL. The PLL will respond to correct this error, the cancellation path loop will, in turn, respond and affect overall PLL operation. In contrast, by using only one delay element and a set of multiplexers, changes to the gain of the delay generator add zero average phases to the PLL, which enables the implementation of the cancellation path control scheme to be orthogonal to PLL operation.

Next, we consider why it is desirable to implement the cancellation in the time domain. There are two main reasons why this approach is beneficial. First, in a deep submicron process, implementing programmable delay generators is more straightforward than implementing precise currents. By cancelling the noise in the time domain, the cancellation can be implemented with exclusively digital building blocks, even in the context of an analog control path; note that this approach does not require any quantization of the VCO/reference clock-induced phase error. The second reason this approach is beneficial is more subtle; it is related to the ease with which we can detect errors in the gain of the noise cancellation scheme. Consider the case where cancellation is implemented by injecting noise pulses into the loop filter. In this scenario, we can attempt to detect if the loop filter voltage has a negative or positive correlation with the sigma-delta noise, where zero correlation corresponds to perfect cancellation. However, the sigma-delta noise is not the only noise source in our system; the phase noise of the VCO and of the reference clock also impacts the result. Assuming the phase noise components are truly random, we must rely on long-term averages over many reference cycles if we are to successfully detect and compensate for gain errors.

The same limitations apply to the time-domain implementation, with one crucial difference: the detection comparator (a bang-bang detector, seen in Figure 7.6) is placed in *front* of the loop filter, where the sigma-delta noise is easier to distinguish from the phase noise of the other components. For an intuitive understanding of why the sigma-delta noise must be easier to distinguish in this case, consider the scenario in which the sigma-delta noise contributes the same net disturbance to the loop filter as the phase noise components. The phase noise components generally have their noise concentrated at low frequencies, whereas the sigma-delta noise is shaped to high frequencies. The loop filter itself is a low-pass filter, consequently, for the two types of noise to have the same net contribution after the loop filter, the sigma-delta noise must be significantly larger before the loop filter.

The digital state machine, an overview of which is shown in Figure 7.23, can be used to address several practical problems that occur in the delay generator, an overview of which is shown in Figure 7.24. The delay generator is composed of a

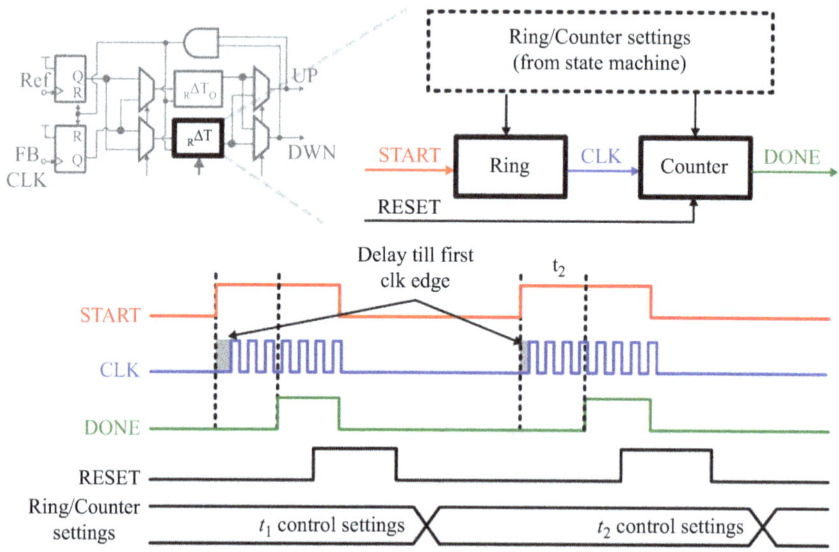

Figure 7.23 Architecture of delay generation

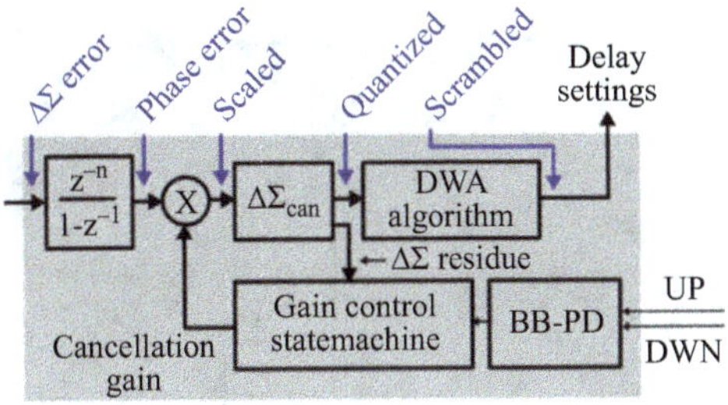

Figure 7.24 Overview of cancellation state machine

simple ring oscillator. The ring oscillator is held in reset before the arrival of the incoming edge. The rising edge of this incoming signal enables the ring. One of the nodes of the ring is used to clock a downstream decrementing counter; when the decrement reaches zero, the output goes to zero, thus completing the synthesis of the delay. Any of the nodes in the ring could be used to generate the clock; it is the selection of which ring node is used as the clock that determines the LSB size. The initial state of the ring, along with the initial state of the counter, together determines the magnitude of the generated delay. Thus, the delay generator is implemented with logic standard cells, with an LSB step size close to 10 ps easily achievable in a modern deep sub-micrometer process.

The magnitude of the LSB varies with PVT, but the digital representation of the delay can be calibrated to the actual differential delay created by the delay elements. To this end, the PLL includes an integrated automatic compensation loop, driven by a monitoring BB-PD placed at the output of the delay elements. Errors in cancellation path gain are detected by correlating the sign of the time delay with the early/late output of the monitor BB-PD, then adjusting the gain to drive the correlation to zero. What about the mismatch between the elements of the ring? To address this issue, the state machine also implements a simple dynamic element matching scheme that scrambles the delay generators between edges, such that the mismatch error itself is mapped to high-frequency error. A standard barrel shifting algorithm is used to implement the dynamic element matching aspect of the design.

An implementation example can be found in [16]. Figure 7.25 shows the impact of the scheme, when disabled, enabled, and finally with the control loop determining the gain disabled. As can be seen in the figure, when the phase-noise compensation system is enabled, the phase noise is dramatically reduced.

Finally, in Figures 7.26 (concept) and 7.27 (measurement), an overview of a fast frequency switching mechanism is shown. The concept, which takes advantage of the highly digital nature of the frequency control information within the hybrid PLL architecture, is simple enough; when switching the PLL between different frequencies, we store the state-machine's integral path control word such that when the PLL jumps back to that frequency, the measured control word can be loaded directly instead of re-found through the standard PLL lock process. Again, this approach works because of the digital nature of the integral path, despite the fact

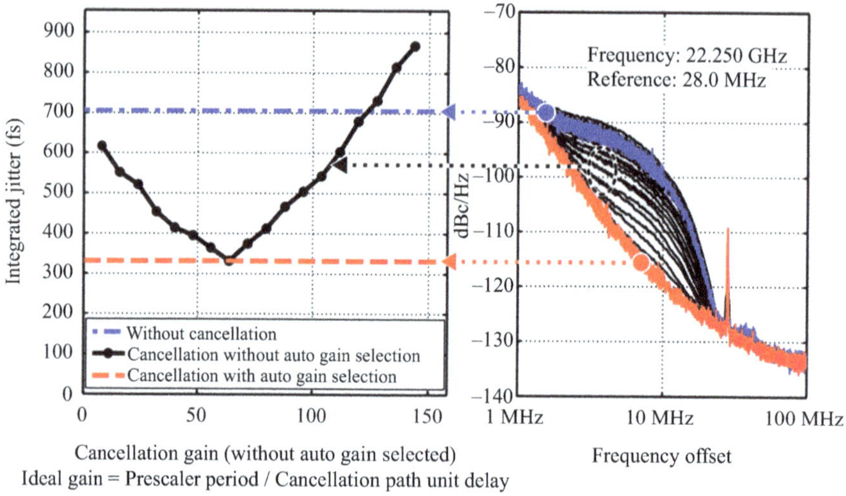

Figure 7.25 Phase noise plot of cancellation scheme. © 2015 IEEE. Reprinted, with permission, from [16]

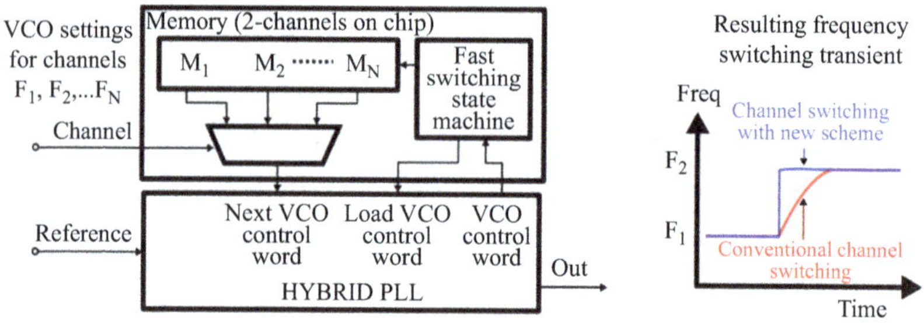

Figure 7.26 Fast reply concept

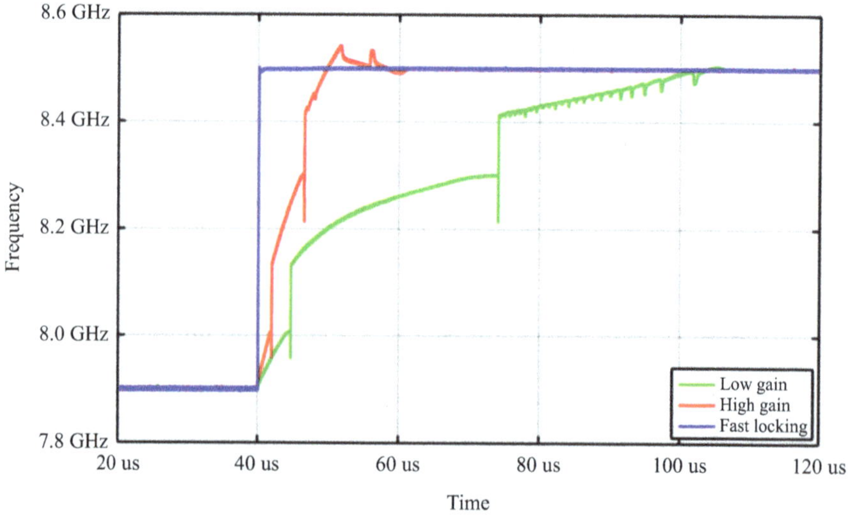

Figure 7.27 Fast replay results

that the PLL retains its analog characteristics. Thus, despite the partial analog nature of the PLL, we can still implement schemes more commonly associated with digital PLLs.

7.4 Conclusion

In this chapter, we have presented a collection of PLL techniques that have been used to address common PLL problems. In part by avoiding the traditional digital/analog dichotomy, these techniques enable a best-of-both-worlds approach to PLL design. In our view, state of the art of PLL design is advanced when solutions are found based on the strengths and weaknesses of the underlying technology and the characteristics of the particular application. Thus, the optimum solution often is a hybrid analog/digital implementation. Furthermore, even with a true analog proportional path, the digital nature of the integral path keeps open the possibility of implementing more advanced control and optimization systems, an example of which can be found in [17].

References

[1] M. Ferriss, J.-O. Plouchart, A. Natarajan *et al.*, "An integral path self-calibration scheme for a dual-loop PLL," *IEEE Journal of Solid-State Circuits*, vol. 48, no. 4, pp. 996–1008, 2013.

[2] J. A. Tierno, A. V. Rylyakov, and D. J. Friedman, "A wide power supply range, wide tuning range, all static CMOS all digital PLL in 65 nm SOI," *IEEE Journal of Solid-State Circuits*, vol. 43, no. 1, pp. 42–51, 2008.

[3] A. Rylyakov, J. Tierno, H. Ainspan *et al.*, "Bang-bang digital PLLs at 11 and 20GHz with sub-200fs integrated jitter for high-speed serial communication applications," *ISSCC Dig. Tech. Papers*, San Francisco, CA, Feb. 2009, pp. 94–95.

[4] J. Zhuang, Q. Du, and T. Kwasniewski, "A 4 GHz low complexity ADPLL-based frequency synthesizer in 90 nm CMOS," *CICC Dig. Tech. Papers*, San Jose, CA, 2007, pp. 543–546.

[5] D. Tasca, M. Zanuso, G. Marzin, S. Levantino, C. Samori, and A. L. Lacaita, "A 2.9–4.0-GHz fractional-N digital PLL with bang-bang phase detector and 560-integrated jitter at 4.5-mW power," *IEEE Journal of Solid-State Circuits*, vol. 46, pp. 2745–2758, 2011.

[6] Z. Ru, P. Geraedts, E. Klumperink, X. He, and B. Nauta, "A 12 GHz 210 fs 6 mW digital PLL with sub-sampling binary phase detector and voltage-time modulated DCO," *Symp. VLSI Circuits*, Kyoto, Japan, pp. C194–195, Jun. 2013.

[7] J. Tierno, A. Rylyakov, D. Friedman *et al.*, "A DPLL-based per core variable frequency clock generator for an eight-core POWER7^{TM} microprocessor," *2010 Symposium on VLSI Circuits*, Honolulu, HI, 2010, pp. 85–86.

[8] N. Da Dalt, "Markov chains based derivation of the phase detector gain in bang-bang PLLs," *IEEE Transactions on Circuits Systems* II, vol. 53, no. 11, pp. 1195–1199, 2006.

[9] M. Lee and A. Abidi, "A 9 b, 1.25 ps resolution coarse-fine time-to-digital converter in 90 nm CMOS that amplifies a time residue," *IEEE Journal of Solid-State Circuits*, vol. 43, no. 4, pp. 769–777, 2008.

[10] V. Kratyuk, P. Hanumolu, K. Ok, U.-K. Moon, and K. Mayaram, "A digital PLL with a stochastic time-to-digital converter," *IEEE Transactions on Circuits Systems I*, vol. 56, no. 8, pp. 1612–1621, 2009.

[11] W. Yin, R. Inti, A. Elshazly, B. Young, and P. Hanumolu, "A 0.7-to-3.5 GHz 0.6-to-2.8 mW highly digital phase-locked loop with bandwidth tracking," *IEEE Journal of Solid-State Circuits*, vol. 46, no. 8, pp. 1870–1880, 2011.

[12] M. Ferriss, A. Rylyakov, J. A. Tierno, H. Ainspan, and D. J. Friedman, "A 28 GHz hybrid PLL in 32 nm SOI CMOS," *IEEE Journal of Solid-State Circuits*, vol. 49, no. 4, pp. 1027–1035, 2014.

[13] P. Wang, J. C. Zhan, H. Chang, and H. S. Chang, "A digital intensive fractional-N PLL and all-digital self-calibration schemes," *IEEE Journal of Solid-State Circuits*, vol. 44, no. 8, pp. 2182–2192, 2009.

[14] M. Ferriss, B. Sadhu, A. Rylyakov, H. Ainspan, and D. Friedman, "10.8 A 12-to-26GHz fractional-N PLL with dual continuous tuning LC-D/VCOs," *2016 IEEE International Solid-State Circuits Conference (ISSCC)*, San Francisco, CA, 2016, pp. 196–198.

[15] A. Swaminathan, K. J. Wang, and I. Galton, "A wide-bandwidth 2.4 GHz ISM band fractional-N PLL with adaptive phase noise cancellation," *IEEE Journal of Solid-State Circuits*, vol. 42, no. 12, pp. 2639–2650, 2007.

[16] M. Ferriss, B. Sadhu, A. Rylyakov, H. Ainspan, and D. Friedman, "10.9 A 13.1-to-28GHz fractional-N PLL in 32nm SOI CMOS with a ΔΣ noise-cancellation scheme," *2015 IEEE Int. Solid-State Circuits Conf. (ISSCC) Dig. Tech. Pap.*, San Francisco, CA, 2015, pp. 1–3.

[17] M. Ferriss, B. Sadhu, and D. Friedman, "A gradient descent bias optimizer for oscillator phase noise reduction demonstrated in 45 nm and 32 nm SOI CMOS," *2018 IEEE Radio Frequency Integrated Circuits Symposium (RFIC)*, Philadelphia, PA, 2018, pp. 344–347.

Chapter 8
Spur mitigation techniques for DPLL architecture

Cheng-Ru Ho[1] *and Mike Shuo-Wei Chen*[2]

8.1 Introduction

Phase-locked loops (PLLs) are widely deployed in most electronic systems to generate a desired clock frequency, to perform clock data recovery (CDR), and to achieve frequency or phase modulation depending on the intended applications. Recently, the trend of system on a chip (SoC) has resulted in the need for a low-cost, robust, and highly reconfigurable PLL architecture [1–15]. This large-scale integration for advanced application requires a deeply scaled integrated-circuit technology with a low supply voltage, but it inevitably results in a design bottleneck for analog PLLs, such as voltage headroom or reliability issues. Hence, interest in digital PLL (DPLL) architecture has increased, as its digital nature offers the advantage of technology scaling in a way that analog PLLs do not. In addition, DPLL architecture is more flexible and reduces area and power consumption, as its compact digital logics replace bulky and passive analog components.

The conventional analog PLL [Figure 8.1(a)] consists of an analog phase frequency detector (PFD), a charge-pump (CP), a passive loop filter, and a voltage-controlled oscillator (VCO). To maintain PLL locking, the analog PFD detects the phase error between the reference and the VCO clock; it then outputs a sequence of pulses with a width that is proportional to the difference of the two phases. To avoid the direct modulation of these pulses to the VCO, the CP and the low-pass loop filter are used to suppress the high-frequency component at the PFD output and to generate an averaged value at the oscillator's voltage control node in order to adjust the oscillation frequency. Finally, the VCO output is fed to the PFD via a feedback divider. In the steady state, the frequency of the PLL is locked to the multiplication of the reference clock frequency (F_{REF}) and feedback division ratio (N) for both analog and digital architectures. To illustrate what makes the DPLL architecture special, a simplified block diagram of DPLL is depicted in Figure 8.1(b). The implementation complexity and the design tradeoffs between the two

[1]Inphi Corporation, Santa Clara, CA, USA
[2]Ming Hsieh Department of Electrophysics, University of Southern California, Los Angeles, CA, USA

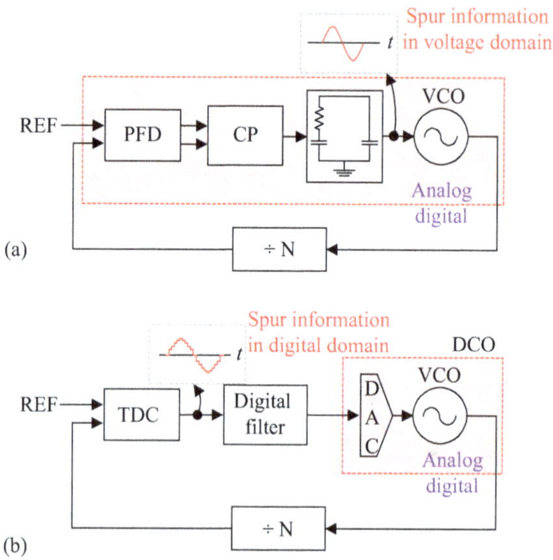

Figure 8.1 Block diagrams of (a) analog phase-locked loop and (b) digital phase-locked loop

architectures are evident. In the DPLL architecture, a time-to-digital converter (TDC) is used to quantize the output phase of the oscillator into digital codes. To process the digital information, the original PFD and the loop filter can be replaced with a digital loop filter (DLF), which consists of a proportional path to correct instantaneous phase errors and an integral path to set averaged oscillation frequency. By eliminating the analog loop filter, this reduces not only the large area consumption of the passive loop filter but also typical analog circuit non-idealities such as mismatch, finite output impedance, and current leakage from the PFD and CP circuitry. Finally, to bridge the signal between analog and digital domains, the digital-to-analog converter (DAC) converts digitized information back to the voltage domain to control the frequency of the oscillator, forming a digitally controlled oscillator (DCO) structure [16].

Although DPLL architecture itself already alleviates many analog design constraints, it has certain mechanisms that generate spurious tones that are similar to those of analog PLL architecture. The key difference is that the spurious tone information can be more easily extracted in the case of DPLL architecture. For example, when a single-tone spur is detected at the PLL output, a sinusoidal phase error will appear at the input of the VCO in the analog PLL case or at the input of the DLF in the digital PLL case, as depicted in Figure 8.1. Because the phase error in a DPLL contains the *digitized* version of spur information, this digital-intensive topology allows for the use of digital signal processing (DSP) techniques, which are challenging for an analog CP PLL due to the need for additional analog-to-digital and digital-to-analog conversions. Thanks to the aforementioned advantage, the aim of this chapter is to explore DSP opportunities in order to achieve an ultra-low spurious performance in the DPLL topology.

The structure of the chapter is organized as follows. In Section 8.2, the impact of the PLL spurs to the system is overviewed. The generation of DPLL spurs is discussed in Section 8.3. Next, the potential and the rationale of the DSP technique are elaborated upon in Section 8.4. To illustrate the efficacy of the DSP technique for spur mitigation, two DPLL design examples in 65-nm CMOS will be studied in Section 8.5. Finally, Section 8.6 concludes.

8.2 The impact of spurious tones

Before beginning a detailed description of where the spur comes from, it is important to understand the impact of spurs on the whole SoC environment. From the system-level perspective, it is critical that the system can meet a general specification that considers all types of jittery effects from a variety of building blocks. Hence, it is difficult to decouple the PLL spurious effect without knowing more about the macroarchitectural design. In other words, if the spur budget is consumed, then the targeted PLL spurious performance should be adjusted accordingly. The impact of the spurious tone on the wireless and wireline systems should be examined to illustrate how to define the specification of the PLL spur.

8.2.1 Specification of spurious performances in a wireless system

An example of local oscillator (LO) generation in a conventional wireless application is shown in Figure 8.2. Ideally, a pure sinusoid should be generated, but unfortunately, other spectral impurities are synthesized in a real PLL. If the output spectrum of a real PLL is inspected, one or multiple spikes (i.e., spurs) will most likely be observed above the phase noise floor near the desired tone—in this case, the LO frequency. If this real PLL is applied to a wireless communication system, it will introduce unwanted reciprocal mixing between the spurs and the incoming signal, degrading the overall performance. As depicted in Figure 8.2(a), on the transmitter side, after the mixer, the input signal is up-converted and generates additional spurious tones, which may violate the emission mask issued by the Federal Communication Commission or other regulations. On the receiver side [Figure 8.2(b)], the incoming signal may include blockers, and if the frequency of the blockers coincides with that of the PLL spurs, the blockers will be down-converted to the baseband. Consequently, the signal-to-noise ratio (SNR) performance deteriorates.

In wireless communication, the PLL spur specification can be determined by evaluating the worst-case spur level normalized to the carrier power across different frequency offsets to ensure that the resulting system degradation is within the tolerance margin. By examining the transmitter's emission mask and the receiver's selectivity of the blockers, the maximum allowable spurious tone can be back-calculated. Assuming that the receiver requires a minimum SNR of 15 dB, Figure 8.3 shows the transceiver's

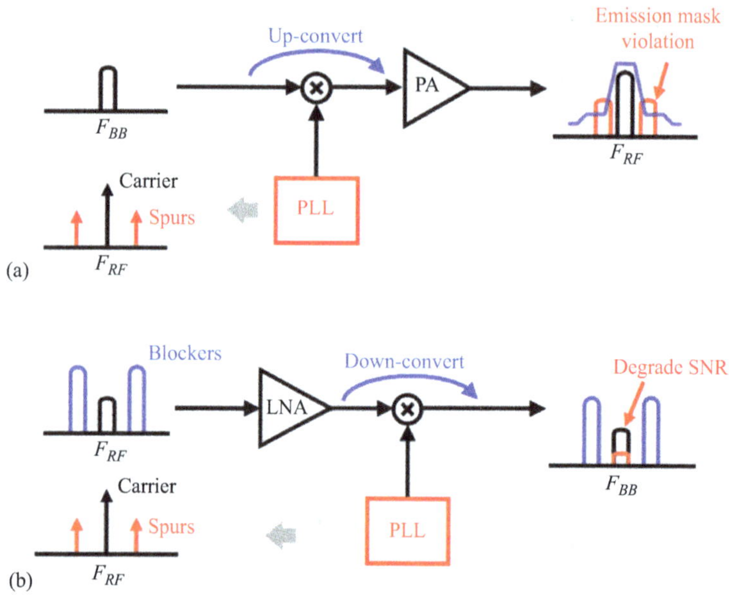

Figure 8.2 Illustration of spur-induced reciprocal mixing in a wireless transceiver

targeted PLL spurious masks of GSM 900 Class-4 Small Mobile Station standard [Figure 8.3(a)] and that of Bluetooth Class-3 standard [(Figure 8.3(b)].

8.2.2 Specification of spurious performances in a wireline system

Note that the spur-induced mixing issue is just one of many unwanted outcomes. Another example is that the spur will increase the deterministic jitter (DJ), which is problematic in the wireline application. Due to increasing bit rates, the time period of the data [defined as a unit interval (UI)] has reduced dramatically, leaving a stringent margin for the signal integrity between the transmitted data and the recovered clock. Consequently, when the effect of the jitter kicks in, it results in the closure of eye diagrams and increases the bit error rate (BER). The impacts of random jitter (RJ) and DJ on the eye diagram are depicted in Figure 8.4.

In the high-speed link, jitter tolerance defines the margin for all types of jittery sources. Under these circumstances, the receiver will operate properly with a decent BER figure (e.g., BER $< 10^{-12}$ or lower). To illustrate this, the jitter tolerance mask of the receiver shown in Figure 8.5 defines the minimum allowable phase displacement, including RJ and DJ, between the transmitted data and the recovered clock in the 10G Ethernet with 802.3 ae standard [Figure 8.5(a)] and synchronous optical network transport with OC-768 standard [Figure 8.5(b)]. In general, the high-frequency jitter cannot be tracked using the CDR loop; hence, any jittery effect out of the CDR bandwidth, normally in the order of a few megahertz,

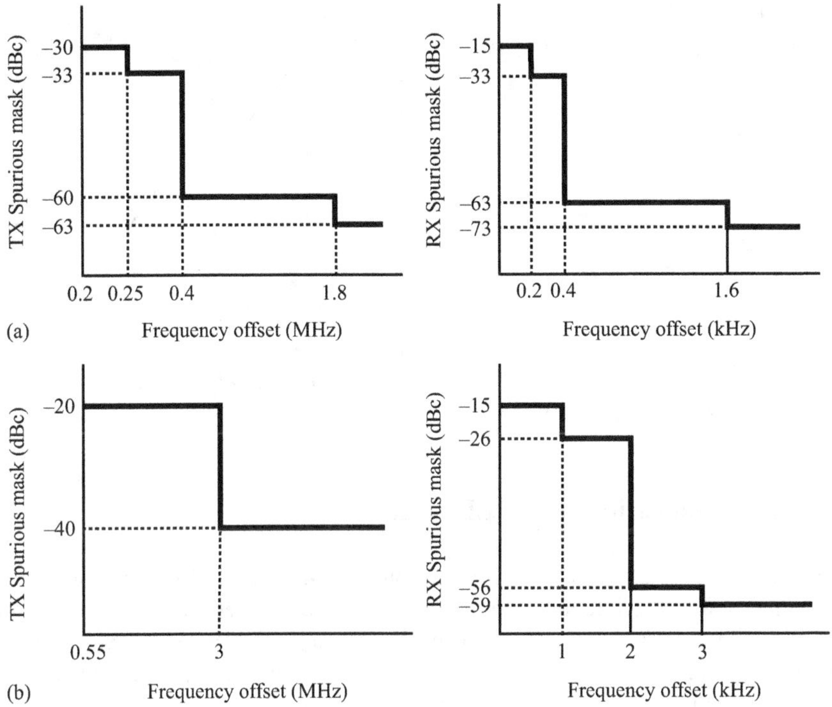

Figure 8.3 The PLL spurious mask with (a) GSM 900 Class-4 Small MS standard and (b) Bluetooth Class-3 standard

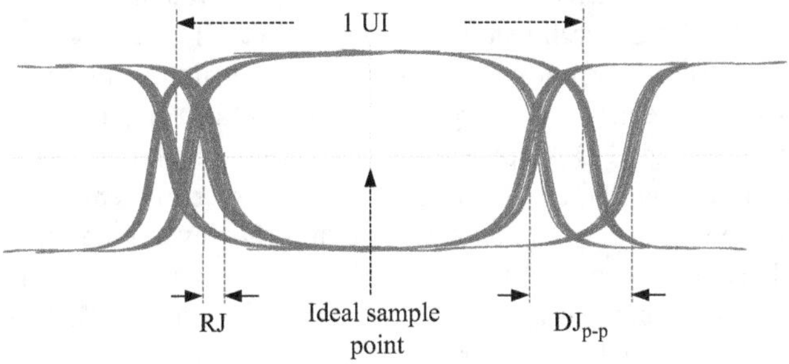

Figure 8.4 The impact of RJ and DJ to the eye diagram

will close the eye and degrade the BER. To meet the requirement of Figure 8.5, the jitter tolerance of the system should be higher than 0.1 and 0.15 $UI_{p\text{-}p}$ for 802.3 ae and OC-768, respectively; the minimum DJ requirement can then be derived from this point. In reality, the requirement for the PLL spurious tone is expected to be much more stringent because other blocks may further contribute DJ and RJ to consume the margin.

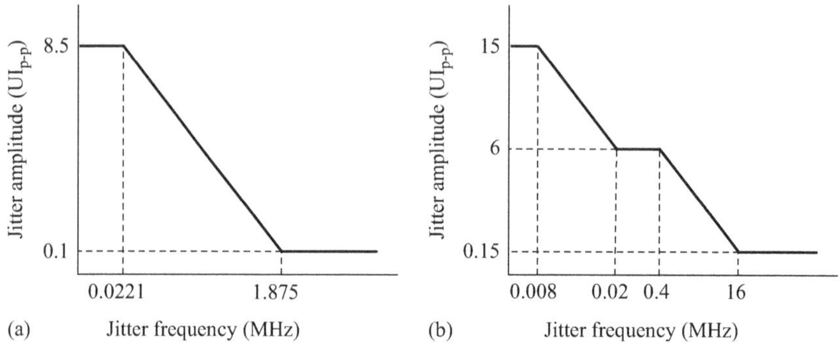

Figure 8.5 The PLL jitter tolerance mask from (a) 802.3ae standard and (b) OC-768 standard

8.3 The generation of DPLL spurs

There are two mechanisms that generate spurious tones in DPLL. First, spurs can be created due to the internal operation of the PLL, referred to as *internal spurs*; these include the fractional spur and the reference spur. Second, when the error source originates outside of the PLL, it is referred to as an *external spur*, mainly because of the unwanted coupling from peripheral blocks.

8.3.1 Internal spurs due to fractional-N operation

Fractional spurs originate internally from the fractional-*N* operation of PLLs [17]; the mechanism of spur generation can be examined through a general frequency accumulation-type DPLL architecture (see Figure 8.6). Rather than using a multi-modulus divider [18], the fractional-*N* operation in this architecture is derived from the phase error between the accumulation path of the frequency control word (FCW) and the quantized DCO phase after the TDC, which consists of an integer counter and a fractional TDC. This allows for a simpler implementation and prevents the dither noise of the delta-sigma modulator; however, the finite word length mismatches between the accumulated FCW path and the TDC quantization step inevitably introduces a periodic sawtooth error pattern at the DLF input, which modulates the DCO to generate spurs.

Interestingly, the periodicity of the sawtooth waveform is related to the fractional part of the FCW value. If the fractional part of the FCW is assumed to be *df*, the sawtooth waveform will repeat depending on the smallest bit "1" (SO) of *df*, referred to as SO(*df*) [2]. For example, if *df* is set to $1/2^7 + 1/2^{12}$, the periodicity is decided by $\mathrm{SO}(df) = 1/2^{12}$. Hence, the periodicity, defined as D, of the fractional spur pattern can be deterministically calculated according to (8.1):

$$D = \frac{F_{REF}}{F_{SPUR}} = \frac{F_{REF}}{F_{REF} \cdot \mathrm{SO}(df)} = \frac{1}{\mathrm{SO}(df)} \tag{8.1}$$

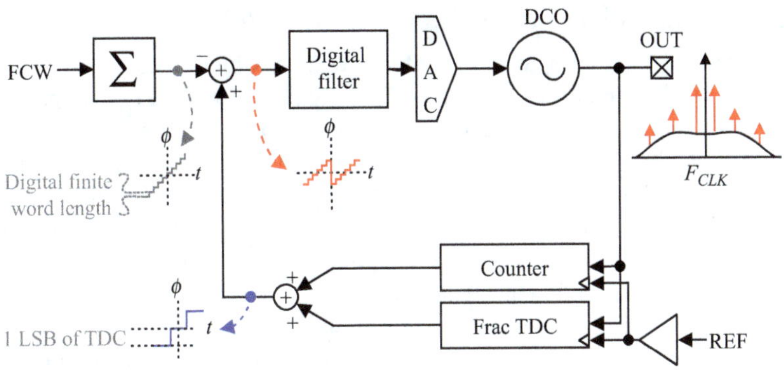

Figure 8.6 The block diagram of the frequency accumulation-type DPLL architecture

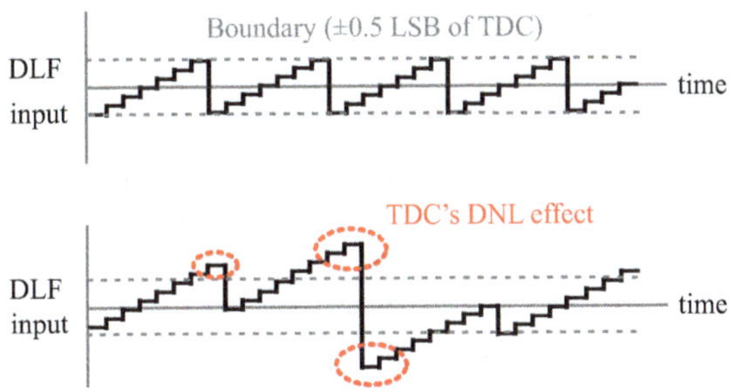

Figure 8.7 Waveforms of a fractional spur pattern shown at the DLF input with and without TDC's DNL effect

where F_{REF} and F_{SPUR} denote the reference clock frequency and the fundamental frequency of the fractional spur, respectively. Note that because the fractional spur period automatically tracks the reference clock, its periodicity is known *a priori*. The mismatch between the TDC quantization levels [i.e., the differential non-linearity (DNL) effect] can change the shape of the sawtooth waveform, as shown at the DLF input depicted in Figure 8.7. The unequal quantization step results in an asymmetric amplitude near the center DC level and across the boundary of ±0.5 least significant bit (LSB) of TDC; however, the waveform still repeats every 1/SO(*df*) samples.

For a multi-modulus-type DPLL architecture [3], the generation of the fractional spur is similar. The finite quantization step and the non-linearity of the TDC will degrade the in-band noise and spur by folding the shaped quantization noise of the delta-sigma modulator from the multi-modulus divider. Note that no matter which DPLL architecture is used, it is difficult to reject the in-band fractional spurs, as the loop filter cannot suppress them [19].

8.3.2 Internal spurs due to integer-N operation

Another type of internal spur is the reference spur, which is essentially a phase disturbance that occurs at the reference clock frequency when PLL is operated in the integer-*N* mode. In the conventional charge-pump PLL, the reference spur is a severe issue, and its generation can be ascribed to the analog circuit non-idealities, including PFD delay, CP switching delay, CP up and down current mismatch, and CP current leakage [18]. For example, if there is a timing skew effect between the up (I_{UP}) and down (I_{DN}) CP current, a periodic perturbation appears at the input of VCO every $1/F_{REF}$ second, which continuously modulates the VCO to generate the reference spur. Other non-idealities have a similar effect.

Fortuitously, in comparison with an analog CP PLL, the reference spur issue in DPLL architecture is relatively relaxed because all the circuit non-idealities can be waived after replacing the PFD, CP, and passive loop filter with digital logics; however, the digital core of the DPLL is clocked at the reference clock frequency, and its switching noise can still crosstalk to the DCO electrically or magnetically to cause a deviation of the DCO phase, which is the main source of the reference spur in the DPLL architecture.

8.3.3 External spurs due to interferences

There are several aggressors in a modern SoC platform, such as power amplifier (PA) output, digital processors, clock buffers from other domains, or simply other PLLs [20]. Depending on the aggressors, the interference can occur in different waveforms. As shown in Figure 8.8, when other clock domains or other PLLs interfere with the victim PLL, the aggressor signal will cause a sinusoidal disturbance to generate multi-tone spurs. In this case, the interference usually occurs

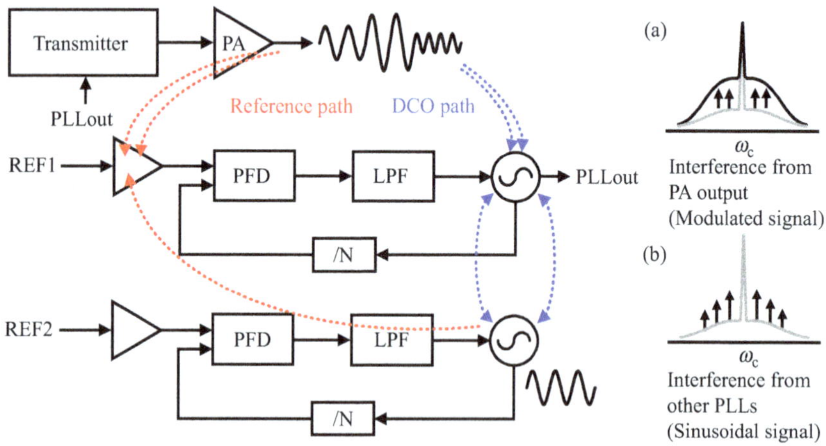

Figure 8.8 Output spectrum of the victim PLL interfered by (a) the transmitter output as a modulated disturbance or by (b) nearby PLLs as a sinusoidal disturbance

in a periodic waveform, such as a sinusoidal or multi-tone pattern. Another example is when an aggressor signal that originated from the PA output in a transmitter couples to the victim PLL—in this case, the interference will not be a periodic waveform any longer but a modulated signal, elevating the noise floor. Hence, in terms of external spurs, performance degradation is not only limited to the generation of spurious tones but also to the degradation of the phase noise profile.

When the interference couples through the reference path, the aggressor signal electrically disturbs the reference clock. Usually, the reference clock is generated from a high-quality-factor crystal oscillator off-chip, which is resilient to aggressors. Therefore, this discussion mainly focuses on the contamination of the on-chip reference clock buffer. When the interference modulates the power supply or the substrates of the clock buffer, the output phase of the clock buffer becomes perturbed. Another prominent mechanism for the interference to induce spurs is through the oscillator of a PLL (i.e., VCO or DCO). In this case, the coupling path of the oscillator can differ from that of the reference clock buffer. The interference can be electrically coupled, such as interference through the DCO power supply, substrates, and current biases. Alternatively, the interference can be magnetically coupled, including interfering through on-chip inductors or associated bond wires; this phenomenon [4–7] is now extensively observed in radio-frequency, millimeter-wave, and high-data-rate wireline applications.

8.4 DSP-assisted spur mitigation techniques leveraging the adaptive filter algorithm

This section introduces the emerging DSP-assisted techniques (i.e., the adaptive spur mitigation) that leverage the digital-intensive core of DPLL to perform functionality beyond the capability of analog PLL. Furthermore, it is expected that this technique will provide a new design direction for the PLL designer and eventually popularize the adoption of DPLL architecture by all electronic systems.

The concept of the adaptive spur mitigation scheme in the DPLL architecture is depicted in Figure 8.9. Assuming a single-tone sinusoidal disturbance coupling to the supply of the reference clock buffer, given the digital nature of DPLL operation, the interference-induced spur will appear as a phase disturbance inside the loop filter with accessibility in the digital form. The direct interference mitigation can be realized if there is a DSP module that is capable of analyzing this disturbance and generating the compensation signal. Note that the adaptive spur mitigation scheme can be extended toward other spur scenarios, including fractional spurs and multi-tone/modulated interference via different coupling paths. In Section 8.5, two design examples are studied to reveal how to use this technique to mitigate fractional spurs and multi-tone interference.

8.4.1 Adaptive spur mitigation algorithm

Following up on Figure 8.9, the DSP module can be realized as a gradient-based adaptive filter algorithm [8,9], illustrated in Figure 8.10 as a representative example.

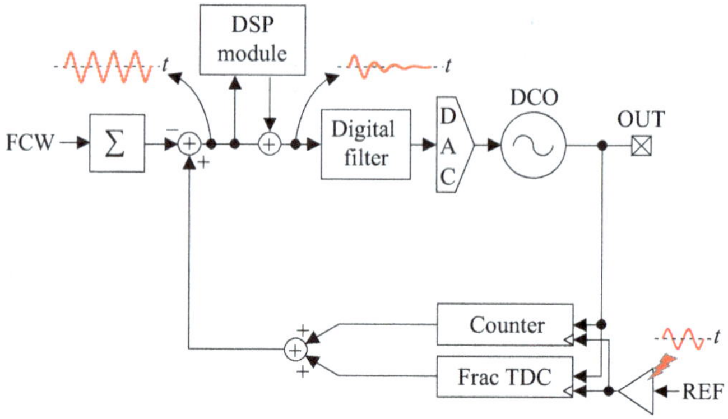

Figure 8.9 The concept of adaptive spur mitigation technique in the DPLL architecture

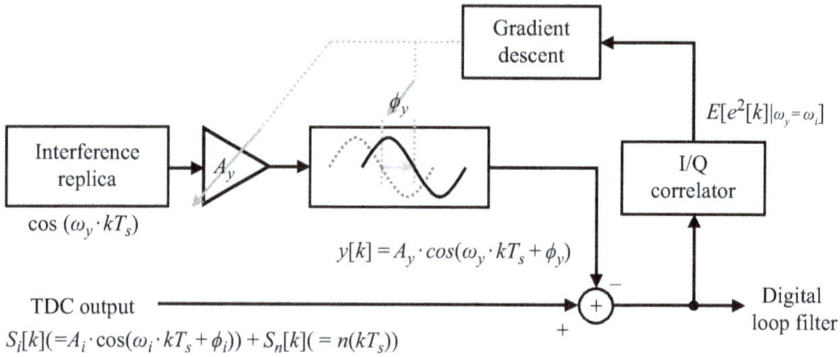

Figure 8.10 The rationale of adaptive spur mitigation via a gradient-based adaptive filter algorithm

In a steady state without mitigation, the input of the mitigation loop at the TDC output contains the interference part (referred as $S_i[k]$) coupled from the reference clock buffer and the remaining phase information (referred as $S_n[k]$), such as DCO phase noise. The key objective of the mitigation loop is to generate a compensation signal $y[k]$ that matches $S_i[k]$ and to subtract it prior to the DLF. The following discussions use the variables A_i, ϕ_i, and ω_i to denote the amplitude, phase, and frequency of the interference $S_i[k]$, respectively. Similarly, A_y, ϕ_y, and ω_y represent the amplitude, phase, and frequency of the generated compensation signal $y[k]$, respectively. For now, it is assumed that the difference between ω_y and ω_i is negligible—nevertheless, the frequency mismatch is not an issue in practice for the adaptive spur mitigation algorithm. Theoretically, if the estimated $y[k]$ is perfectly matched with $S_i[k]$, no interference disturbance will be passed on to the DCO, resulting in perfect spur cancellation. However, this cannot be achieved in practice; instead, the following cost function should be minimized:

$$\min_{A_y,\theta_y} E\left[|e[k]|^2\right] = \min_{A_y,\theta_y} E\left[|S_i[k] - y[k]|^2\right] \tag{8.2}$$

According to (8.2), the goal is to minimize the mean square value of the residue error energy, defined as $|e[k]|^2$, which is the difference between the real and estimated interference. Since the mitigation loop should operate in the background during normal operation, the adaptive filter algorithm continuously monitors the residue error energy in order to minimize the mean square error in real time.

The detail of the adaptive filter algorithm, realized with the 2-D gradient descent search, is illustrated in Figure 8.11(a). To reach the minimum residual error energy, the optimal values of amplitude and phase can be approached independently. Note that a good initial guess can reduce the convergence time, but it does not affect the final convergence value. The updating function of amplitude and phase in the conventional gradient steepest descent is expressed in the following equations:

$$A_y[k+1] = A_y[k] + A_{step} \cdot e[k] \cdot \frac{\partial y[k]}{\partial A_y[k]} \tag{8.3}$$

$$\phi_y[k+1] = \phi_y[k] + \phi_{step} \cdot e[k] \cdot \frac{\partial y[k]}{\partial \phi_y[k]} \tag{8.4}$$

where the updated equation is a function of residue error, step size, and compensation signal. The spur mitigation scheme can utilize the coordinate descent algorithm [21] to reduce the hardware cost of derivative operation $\partial y[k]/\partial A_y[k]$ and $\partial y[k]/\partial \phi_y[k]$; it is greatly simplified from the gradient descent search algorithm. Instead of updating the descent direction according to the true gradient vector, the

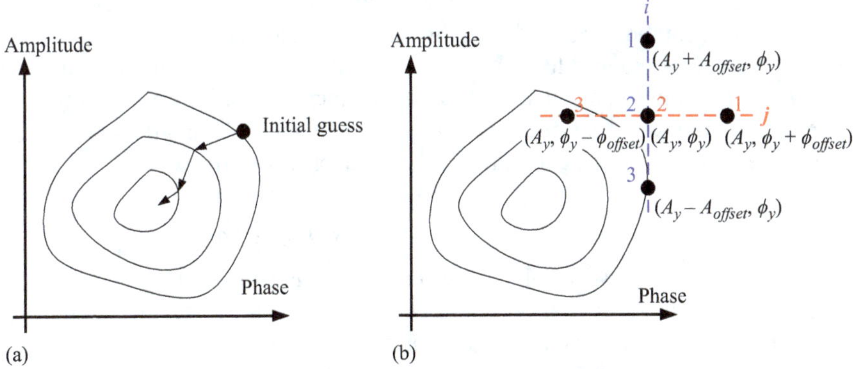

Figure 8.11 The loci of the updates using (a) the gradient descent algorithm and (b) the coordinate descent algorithm via five computation channels with skewed amplitude and phase

coordinate descent algorithm selects one of the two perpendicular descent directions with a fixed step (i.e., either amplitude or phase at a time):

$$A_y[k+1] = A_y[k] + A_{step} \cdot \left\{2 - \operatorname{argmin}\left(E\left[e_i^2(A_y)\right]\right)\Big|_{i\in\{1,2,3\}}\right\} \tag{8.5}$$

$$\phi_y[k+1] = \phi_y[k] + \phi_{step} \cdot \left\{2 - \operatorname{argmin}\left(E\left[e_j^2(\phi_y)\right]\right)\Big|_{j\in\{1,2,3\}}\right\} \tag{8.6}$$

To determine the descent direction, five parallel computation channels can be implemented to calculate the residue error energies corresponding to different amplitudes and phases of the compensation signal. For illustrative purposes, the five computation channels are mapped onto a 2-D coordinate system, as depicted in Figure 8.11(b), where amplitude is plotted in the Y-axis and phase in the X-axis. To simplify the mathematical expression, three channels with compensation signal amplitudes of $A_y \cdot (1 + A_{offset})$, A_y, and $A_y \cdot (1 - A_{offset})$ are grouped in the same basis with channel index numbers $i = 1$, 2, and 3. These three channels compute the corresponding residue error energy values: $E\left[e_1^2(A_y)\right]$, $E\left[e_2^2(A_y)\right]$, and $E\left[e_3^2(A_y)\right]$, respectively. Similarly, channels with compensation signal phases of $(\phi_y + \phi_{offset})$, (ϕ_y), and $(\phi_y - \phi_{offset})$ are assigned to another basis with channel index numbers $j = 1$, 2, and 3; their corresponding residue error energy values are $E\left[e_1^2(\phi_y)\right]$, $E\left[e_2^2(\phi_y)\right]$, and $E\left[e_3^2(\phi_y)\right]$. As Figure 8.11(b) shows, the center channel with index numbers of $i = 2$ and $j = 2$ should yield the lowest residue error energy after reaching the steady state of the coordinate descent algorithm; therefore, only the compensation signal from this center channel is injected inside the DPLL loop. The iterative algorithm is based on updating either the amplitude or phase [i.e., (8.5) and (8.6)].

The descent path of the update that yields the least residue error energy will be selected. In other words, if the residue error energy due to a change of amplitude is less than the residue error energy due to a change of phase, the update based on (8.5) will be chosen instead of (8.6) and vice versa. Note that the demonstrated coordinate descent algorithm utilizes five parallel computation channels, but this mode of implementation does not limit this spur mitigation technique. For instance, the hardware can be reduced to a single channel with time-division multiplexing. In other words, the single channel sequentially computes the residue error energy with different amplitudes and phases. Ultimately, the mitigation loop still achieves the same cancellation accuracy but with a longer convergence time.

8.4.2 I/Q correlator: minimization of residue error energy

Since the coordinate descent algorithm aims to minimize certain spurious tone energies, the residue error energy should be evaluated at the target spur frequency to avoid potential disturbance from other interferences or random noise. Inaccurate residue error energy estimation due to other interference/noise disturbance can lead to an erroneous descent direction, and hence, convergence instability. A matched filter and averaging over N_{win} cycles are thus performed via I/Q correlators against sinusoids at the estimated spurious frequency, ω_i, as shown in

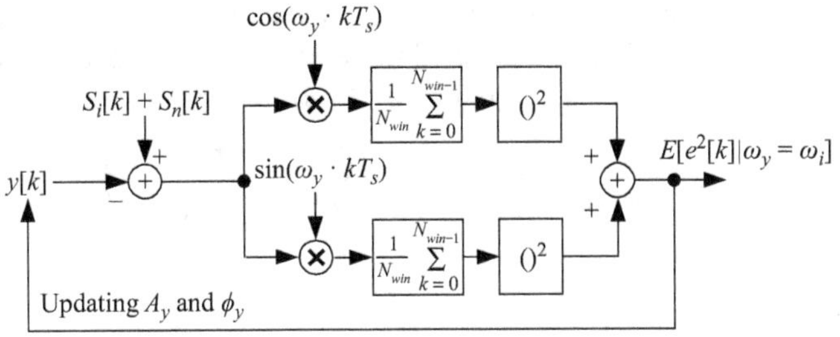

Figure 8.12 The block diagram of the residue error energy computation

Figure 8.12. The mathematical expression of residue error energy at ω_i can be derived using the following:

$$E[e^2[k]|_{\omega_y=\omega_i}] = \left\{ \frac{1}{N_{win}} \sum_{k=0}^{N_{win}-1} [(S_i[k] + S_n[k] - y[k]) \cdot \cos(\omega_i \cdot kT_s)] \right\}^2 + \left\{ \frac{1}{N_{win}} \sum_{k=0}^{N_{win}-1} [(S_i[k] + S_n[k] - y[k]) \cdot \sin(\omega_i \cdot kT_s)] \right\}^2 \tag{8.7}$$

Here, $S_n[k]$ is substituted with discrete-time samples $n(kT_s)$; likewise, $S_i[k]$ is substituted with $A_i \cdot \cos(\omega_i \cdot kT_s + \phi_i)$ and $y[k]$ is substituted with $A_y \cdot \cos(\omega_y \cdot kT_s + \phi_y)$. Equation (8.7) becomes:

$$E[e^2[k]|_{\omega_y=\omega_i}] = \left\{ \left[\frac{A_i}{2}\cos(\phi_i) - \frac{A_y}{2}\cos(\phi_y)\right] + \frac{1}{N_{win}} \sum_{k=0}^{N_{win}-1} \left[\frac{A_i}{2}\cos(2\omega_i \cdot kT_s + \phi_i) - \frac{A_y}{2}\cos(2\omega_i \cdot kT_s + \phi_y) + n(kT_s) \cdot \cos(\omega_i \cdot kT_s)\right\}^2 + \left\{ \left[-\frac{A_i}{2}\sin(\phi_i) + \frac{A_y}{2}\sin(\phi_y)\right] + \frac{1}{N_{win}} \sum_{k=0}^{N_{win}-1} \left[\frac{A_i}{2}\sin(2\omega_i \cdot kT_s + \phi_i) - \frac{A_y}{2}\sin(2\omega_i \cdot kT_s + \phi_y) + n(kT_s) \cdot \sin(\omega_i \cdot kT_s)\right\}^2 \tag{8.8}$$

According to (8.8), if the average window duration (N_{win}) is an integer multiple of the period $1/\omega_i$, the frequency-dependent terms, such as $(A_i/2) \cdot \cos(2\omega_i \cdot kT_s + \phi_i)$ and $n(kT_s) \cdot \cos(\omega_i \cdot kT_s)$, will be completely nullified. In contrast, if N_{win} is not exactly an integer multiple of the period $1/\omega_i$, the frequency-dependent terms will be nonzero but bounded, as they are periodic and zero-mean sinusoids. Moreover, the noise terms, $n(kT_s)$, are correlated against the spur frequency, which also helps to reduce the variation. As for the other frequency-independent terms containing pure spur phase and amplitude information, such as $(A_i/2) \cdot \cos(\phi_i)$ and $(A_y/2) \cdot \cos(\phi_y)$, they are independent of N_{win}, unlike the frequency-dependent terms that are scaled down by $1/N_{win}$. As a result, the frequency-dependent terms become less significant as

N_{win} increases. In the case that N_{win} is sufficiently large, the frequency-dependent terms in (8.8) become negligible compared to the frequency-independent terms after matched filtering and averaging and thus can be further approximated as:

$$\begin{aligned}&\left[\frac{A_i}{2}\cos(\phi_i)-\frac{A_y}{2}\cos(\phi_y)\right]^2+\left[-\frac{A_i}{2}\sin(\phi_i)+\frac{A_y}{2}\sin(\phi_y)\right]^2\\&=\frac{1}{4}(A_i-A_y)^2\ \text{if}\ |\phi_i-\phi_y|=0\\&=\frac{A_i^2}{4}\left[2-2\cos(\phi_i-\phi_y)\right]\text{if}|A_i-A_y|=0\end{aligned}\tag{8.9}$$

Equation (8.9) illustrates that there exists only one global minimum point over $|A_i - A_y|$ and $|\phi_i - \phi_y|$ when (8.9) is drawn on the 3-D surface plot for every 2π period of $|\phi_i - \phi_y|$. Another perspective is to examine (8.9) from the plane of $|A_i - A_y| = 0$ or $|\phi_i - \phi_y| = 0$. The residue error energy always monotonically increases around the global minimum point. Therefore, in steady state, the minimum error is achieved when A_y is equal to A_i and ϕ_y is equal to ϕ_i, which suggests that the compensation signal matches the real one to be mitigated.

8.4.3 Convergence of the spur mitigation scheme

Several time-domain numerical simulations are performed and shown in this subsection to verify the adaptive algorithm. Due to the implementation based on the cost function equation (8.1), the adaptive filter algorithm mainly focuses on mitigating the interference coupled to the supply of the reference clock, TDC, and divider. Hence, the phase disturbance is modeled to originate from the reference path.

Without enabling the mitigation loop, the TDC output shows a periodic phase disturbance due to the coupled interference. When enabling the cancellation loop, the phase disturbance gradually diminishes as the amplitude and phase of the replica interference approaches that of the actual interference, as shown in Figure 8.13(a). As a result, the spurious tone energy at the PLL output is reduced. Note that if the interference is absent but the mitigation loop is enabled, the adaptive algorithm will converge toward zero amplitude in the steady state. Therefore, it does not disturb the normal PLL loop operation if A_{step} in (8.5) is set to be sufficiently small, as shown in Figure 8.13(b). In addition, supplementary digital logic can be implemented to stop the adaptation of the mitigation loop in order to reduce power consumption once the interference is removed. In other words, the mitigation loop will not cause unwanted phase noise degradation, and it is effectively activated only when the interference is present. Figure 8.14(a) and (b) shows the numerically simulated tradeoffs between the convergence speed and the final convergence accuracy for several representative values of A_{step} and ϕ_{step}. One can observe that the convergence speed is inversely proportional to the step size. The impact of A_{offset} and ϕ_{offset} is shown in Figure 8.15(a) and (b), respectively. Since the offset value does not affect the convergence speed, only the steady-state spurious levels post-cancellation are plotted. Better convergence accuracy can be

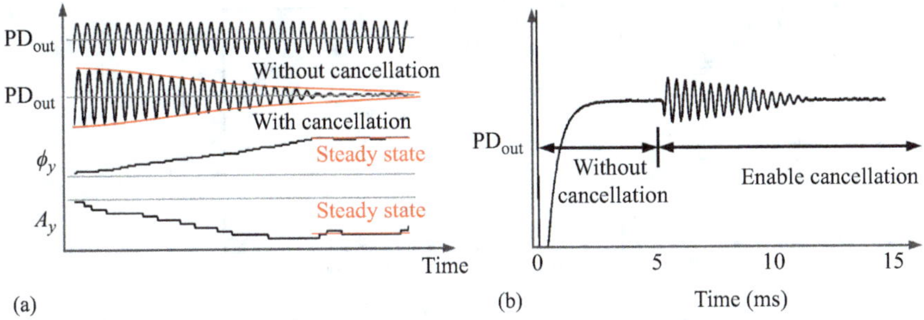

Figure 8.13 *(a) Numerical simulation of TDC output without (top) and with (bottom) enabling the adaptive spur mitigation scheme, showing the convergence of estimated interference parameters. (b) Numerical simulation of TDC output before and after enabling cancellation in the absence of interference*

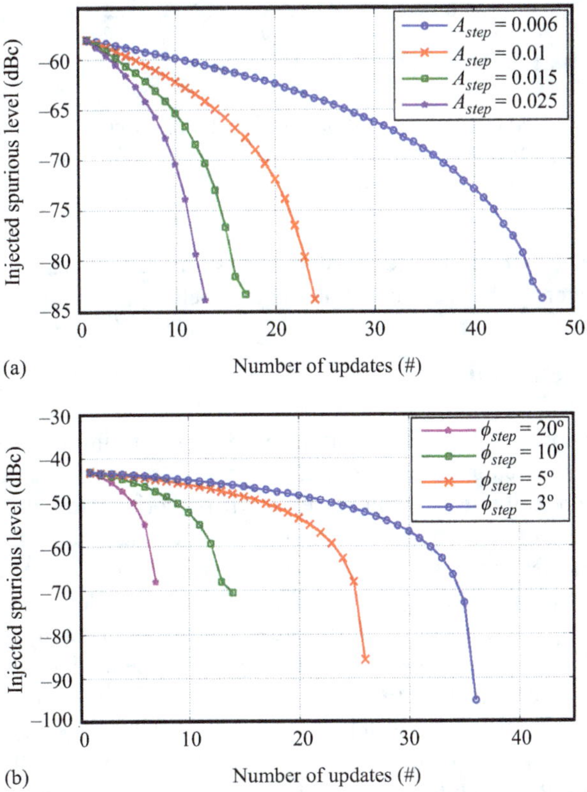

Figure 8.14 *(a) Spur level versus number of updated cycles with different A_{step} (normalized to the full swing of DDS) and (b) spur level versus number of updated cycles with different ϕ_{step}*

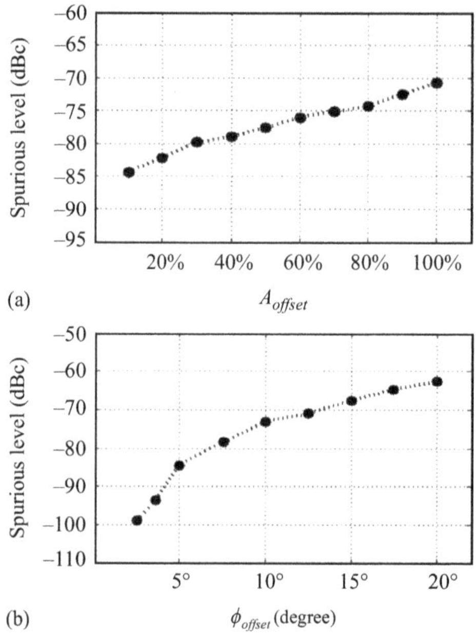

Figure 8.15 (a) The impact of amplitude offset A_{offset} (normalized to the full swing of DDS) on the cancellation accuracy and (b) the impact of phase offset ϕ_{offset} on the cancellation accuracy

observed when offset values are smaller; this is due to the smaller difference of residue error energies between computation channels.

8.5 Design example

While a 2-D coordinate descent algorithm was chosen to illustrate the spur mitigation concept in Section 8.4, it is possible to extend it using other adaptive filter algorithms, such as the least mean square (LMS) algorithm for simpler hardware implementation. For LMS, extra attention should be paid to selecting the appropriate updating step size, u_{step}, to guarantee loop stability [22]. Since the LMS algorithm approximates the expectation function of the random process, the converged states may not reach the optimal values and can fluctuate in the steady state. For instance, if u_{step} is not chosen properly, there will be a wider variation of the state variable, which will lead to a raised phase noise floor or compromised spur cancellation. Hence, the designer might need to thoroughly investigate the performance tradeoffs when choosing an applicable adaptive filter algorithm. To understand how the DPLL architecture can benefit from this adaptive spur mitigation technique, two prototypes leveraging *direct spur mitigation* and *dither-based spur mitigation* techniques for external and internal spur cancellation, respectively, are studied in this section.

8.5.1 DPLL with a direct spur mitigation technique

To avoid being confined by the source of the coupling path and mechanism, various previous works [2,4–11] have addressed the spur issue by applying a direct spur cancellation technique. The key purpose of this approach is to minimize unwanted phase errors within the digital embodiment, which will no longer propagate to the DCO output to generate spurs in the spectrum. To clarify, a silicon prototype [2] proposing a fractional-*N* DPLL with a feedforward multi-tone spur cancellation loop in 65-nm CMOS technology is introduced. The proposed scheme is capable of suppressing both internal spurs (i.e., fractional-*N* spur) and externally coupled spurs from input paths. It can be further extended for use in multi-stage operations to nullify multiple spur sources. Both theoretical analysis and experimental results are provided to explore the design tradeoffs of the technique.

8.5.1.1 Overall DPLL architecture

The block diagram of the fractional-*N* DPLL with the feedforward spur cancellation loop is illustrated in Figure 8.16. An LC-based DCO is used for better out-of-band phase noise. The phase of DCO is quantized by an injection-locked (IL) TDC [9] that generates 28 fine quantization levels within one DCO period, providing ~10 ps time resolution. This fine phase code is then combined with the 8-bit integer counter output to generate the complete TDC information. The IL TDC is mainly composed of a seven-stage ring oscillator (RO) that is IL by the DCO. Because it formulates the first-order PLL, the in-band phase noise of the IL RO is dominated by the DCO, as its

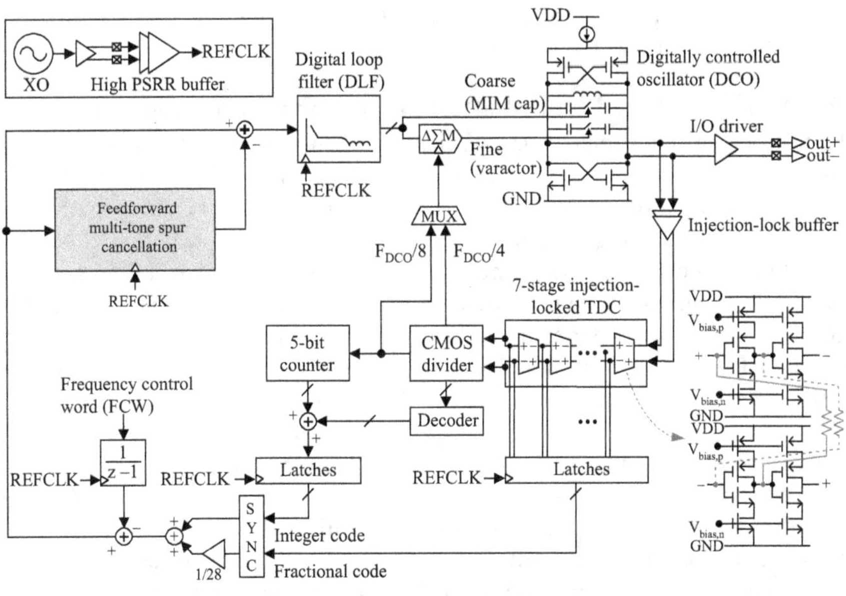

Figure 8.16 The block diagram of the fractional-N DPLL architecture with feedforward multi-tone spur cancellation loop

phase noise is low-passed within the IL bandwidth, whereas the RO phase noise still dominates beyond the IL bandwidth. To ensure proper operation of the DPLL and to minimize the RO's phase noise contribution, the IL bandwidth is designed to be much greater than the DPLL bandwidth. In this prototype, the designed IL bandwidth can be programmed up to 150 MHz and is tunable within 30%. The feedforward multi-tone spur cancellation loop is inserted at the output of the FCW accumulator such that the spurious signals from either external or internal sources can be removed prior to modulating the DCO. Alternatively, the spur compensation can be inserted right after the DLF; however, this requires duplicating the DLF response in the feedforward path and hence was not used in the prototype.

The finite word length of the entire digital logic is carefully chosen and simulated to ensure that the expected performance matches with the behavioral simulation result in MATLAB, using floating-point representation. For instance, if the finite word length in the feedforward cancellation path is not sufficient, it will limit the final cancellation accuracy, as the spur replica cannot precisely duplicate the pattern of the original spur. As discussed in (8.1), to compensate for the TDC's DNL effect, the cancellation loop should provide a fine enough resolution that is related to the least TDC quantization step. In contrast, the maximum bound of the cancellation loop is determined by the largest magnitude of the coupled external spurs or by the TDC's DNL. Based on these design considerations, the feedforward loop is chosen to maintain a 14-bit dynamic range with proper signal truncations, avoiding excessive digital logic. The DLF is composed of proportional gain and integral paths with programmable type I and type II responses for prototyping purposes. Following the loop filter, the delta-sigma DAC interface is used to modulate the varactor bank of LC-based DCO and provide fine frequency resolution on the order of 100 Hz.

8.5.1.2 Feedforward multi-tone spur cancellation scheme

The feedforward multi-tone spur cancellation scheme is based on the observation that either externally coupled or internally generated spurs should always present some kind of a periodic pattern at the phase detector output (TDC output, in the case of a DPLL). To illustrate the basics of the proposed cancellation concept, a sawtooth-like spur pattern with a periodicity of D is examined as a representative example in Figure 8.17, which would generate a series of spurious tones at frequencies, F_{ref}/D, and integer harmonics. As shown in Figure 8.18, if the spur pattern ($x[n]$) is intentionally delayed by precisely D delays, this delayed spur replica ($y[n]$) should coherently match with the original pattern. By subtracting the original spur pattern from this replica, the spur pattern can be completely removed before modulating the DCO; hence, the spur harmonics are canceled. From the time-domain perspective, since harmonics repeat at the exact same point, the delayed spur replica naturally aligns all the harmonic waveforms and potentially allows complete multi-tone cancellation. From the frequency-domain perspective, the signal components at DC and integer multiples of F_{ref}/D are essentially notched out.

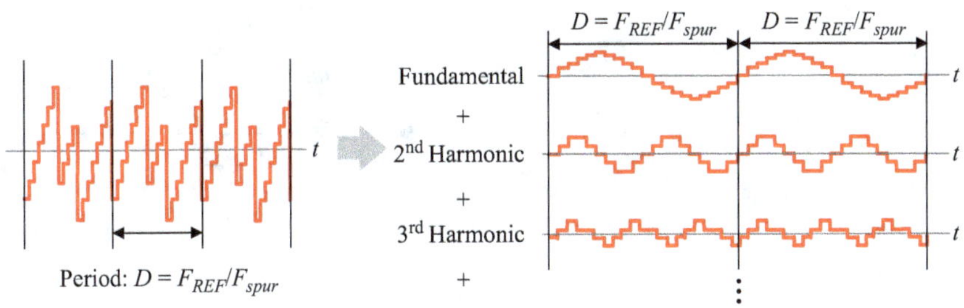

Figure 8.17 The property of multi-tone spurs and its harmonic decomposition

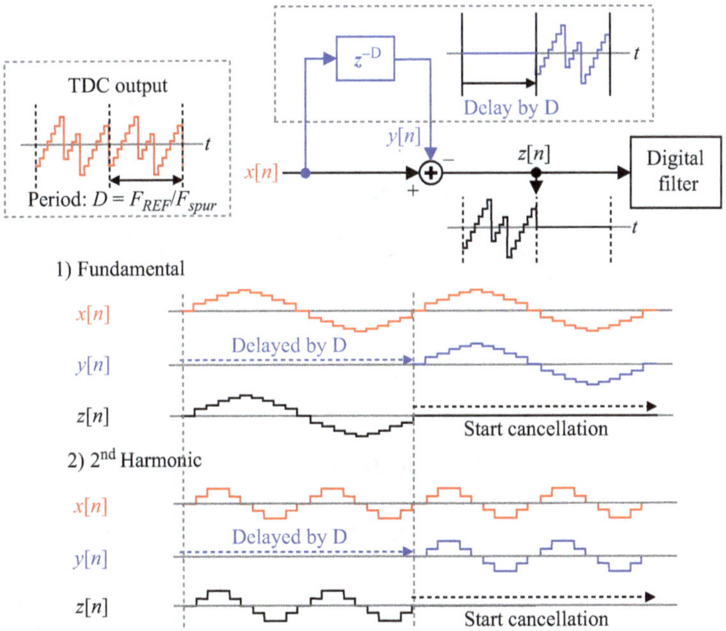

Figure 8.18 Basic concept of feedforward multi-tone spur cancellation in time-domain view

8.5.1.3 Derivation of the cancellation algorithm

While the concept of the feedforward cancellation scheme may work in the open-loop scenario, it has not considered the closed-loop operation of the PLL as well as the noise component at the TDC output; therefore, the cancellation algorithm requires further modifications to be feasibly embedded inside the PLL loop. As illustrated in Figure 8.19, the *z*-domain transfer function of the DPLL loop [23] is examined to understand the design. Assuming that a second-order type II loop response is combined with a D-cycle delay and subtraction—in other words, a

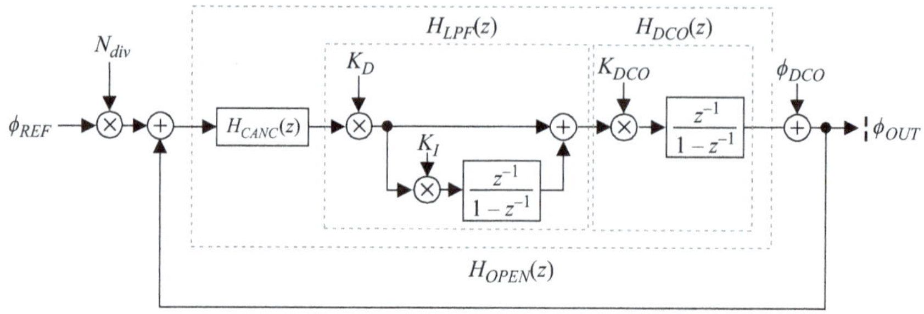

Figure 8.19 z-domain model of the DPLL with feedforward cancellation path $H_{CANC}(z)$

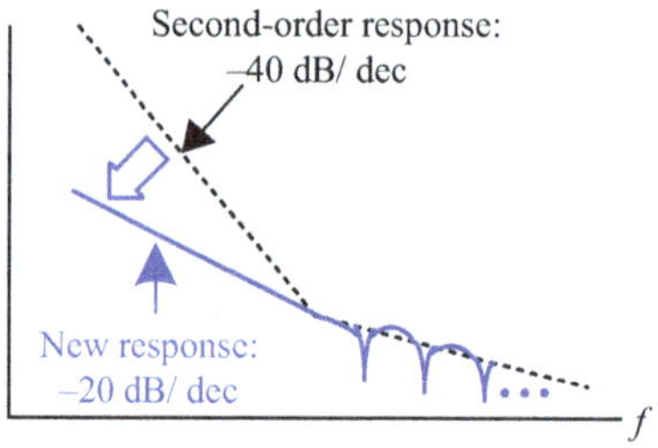

Figure 8.20 The open-loop response of DPLL with $H_{CANC}(z) = 1 - z^{-D}$

comb filter, or $H_{CANC}(z) = 1 - z^{-D}$—the DPLL open-loop response $H_{OPEN}(z)$ is expressed as:

$$\begin{aligned} H_{OPEN}(z) &= H_{CANC}(z) \cdot H_{LPF}(z) \cdot H_{DCO}(z) \\ &= (1 - z^{-D}) \cdot \left(\frac{z^{-1}}{1 - z^{-1}}\right) \cdot \left(1 + \frac{z^{-1}}{1 - z^{-1}} \cdot K_I\right) \cdot K_I \cdot K_D \end{aligned} \tag{8.10}$$

where K_{DCO}, K_D, and $K_D \cdot K_I$ represent the DCO gain, proportional gain, and integral gain of the loop filter, respectively. Note that the introduction of $H_{CANC}(z)$ changes the frequency response of the DPLL open loop. In the low-frequency region, one of the two poles introduced by digital LPF and DCO is canceled; hence, the overall roll-off becomes −20 dB/dec rather than −40 dB/dec (see Figure 8.20). This undesirably changes the DPLL response to the first-order loop; it can be observed by plotting the frequency response of the DCO noise transfer function.

To solve the degeneration of loop response as indicated from (8.10), a high-pass filter with the response of $(1 - z^{-1})/(1 - p \cdot z^{-1})$ is inserted into the D-cycle delay path [see Figure 8.21(a)]; it provides a pole/zero pair [see Figure 8.21(b)]. As a result, the transfer function of $H_{OPEN}(z)$ to the output is modified as:

$$H_{OPEN}(z) = \left(1 - \left(\frac{1 - z^{-1}}{1 - p \cdot z^{-1}}\right) \cdot z^{-D}\right) \cdot \left(\frac{z^{-1}}{1 - z^{-1}}\right) \cdot \left(1 + \frac{z^{-1}}{1 - z^{-1}} \cdot K_I\right) \cdot K_I \cdot K_D \tag{8.11}$$

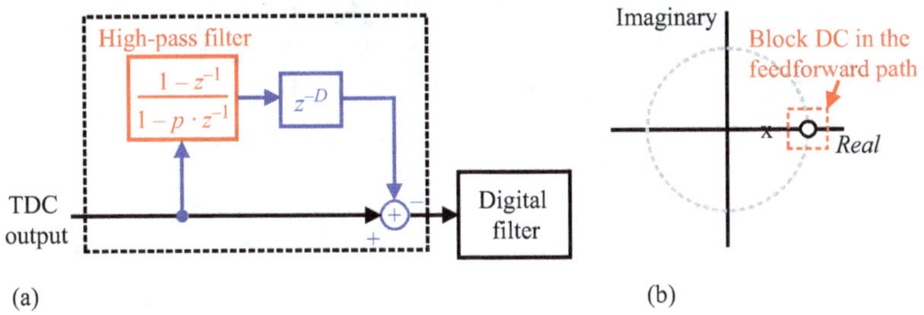

Figure 8.21 (a) Block diagram of feedforward cancellation path with high-pass filter and (b) zero-pole diagram of the high-pass filter

If the low-frequency region of the transfer function in (8.11) is examined, the overall roll-off can still provide −40 dB/dec. The zero-pole diagram of the open-loop transfer function given the original second-order type-II DPLL is examined to gain an intuitive understanding—there are two poles at DC and one zero within the unit circle, as shown in Figure 8.22(a). The $(1 - z^{-D})$ block, where $D = 4$ is just an example; four zeros will be added at 0°, 90°, 180°, and 270° on the unit circle; another four poles will be added at the origin [see Figure 8.22(b)]. When the zeros and poles from Figure 8.22(a) and (b) are combined, the zero at DC (0°) cancels out one of the poles [see Figure 8.22(c)]. As shown in Figure 8.22(d), by inserting the high-pass filter to block the DC in the D-cycle delay path, the zero-pole diagram indicates that the direct zero-pole cancellation at DC is avoided, which helps preserve the second-order characteristic of the DPLL response in the low-frequency region. Note that the value of p determines the frequency range of the slope correction.

Another design consideration is that the noise component at the input of the D-cycle delay path should be rejected in an ideal scenario—in other words, only the spur information should be extracted at the output of the delay path. In the case of simple delay and subtraction, some part of the frequency spectrum will experience noise enhancement, as evident from its frequency response (Figure 8.23). To resolve this issue, the SNR between the spur and the noise component in the D-cycle delay path should be increased. Hence, a simple averaging technique [24] can be used to provide the processing gain and effectively suppress the noise feedthroughs. The transfer function of the averaging operation in the z-domain is expressed as:

$$H_{AVG}(z) = \frac{1}{N_{win}} \sum_{i=0}^{N_{win}-1} z^{-i \cdot D} \tag{8.12}$$

Its implementation is illustrated in Figure 8.24(a). Note that increasing the averaging cycle (N_{win}) proportionally improves the SNR. Essentially, the D-cycle delay path that includes the averaging operation formulates a bandpass filter response. If the passband area is integrated and normalized to that of an all-pass

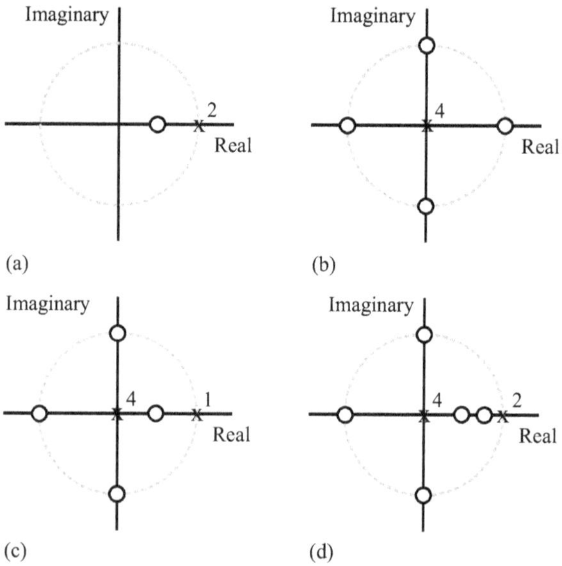

Figure 8.22 Zero-pole diagram of (a) open-loop response of second-order type II DPLL, (b) $H_{CANC}(z) = 1 - z^{-4}$, (c) $H_{OPEN}(z)$ with $H_{CANC}(z) = 1 - z^{-4}$, and (d) $H_{OPEN}(z)$ with $H_{CANC}(z) = 1 - ((1 - z^{-1})/(1 - p \cdot z^{-1})) \cdot z^{-4}$

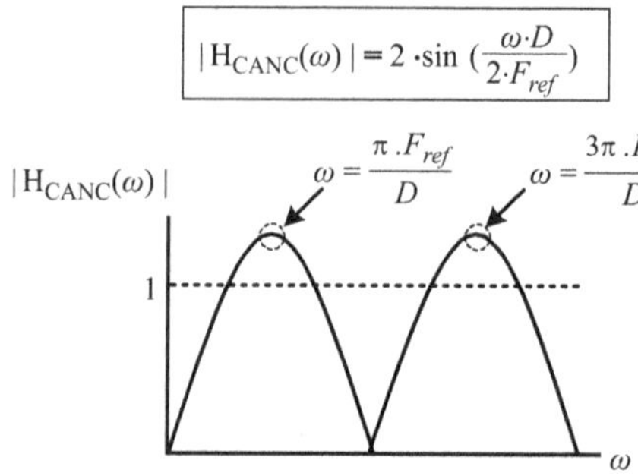

Figure 8.23 The magnitude response of $H_{CANC}(z) = 1 - z^{-D}$

filter (given the same frequency band), the ratio is approximately $1/N_{win}$. This suggests that roughly $100(1 - 1/N_{win})\%$ of the noise energy is rejected via the D-cycle delay path through the averaging operation. From another perspective, the averaging operation effectively sharpens the filter response of $H_{CANC}(z)$ around the notch frequencies and gains response as the passbands become flat (i.e., unity). Figure 8.24(b) shows the effect of frequency responses with different numbers of

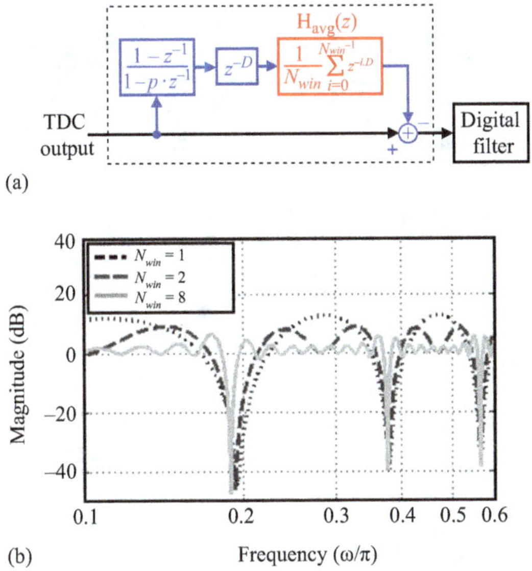

Figure 8.24 *(a) The block diagram of feedforward cancellation path with high-pass filter and signal averaging and (b) its frequency response with different N_{win}*

averaging cycles. To conclude, the complete transfer function of the open-loop $H_{OPEN}(z)$ response of the final implementation can be expressed as:

$$H_{OPEN}(z) = \left(1 - \left(\frac{1-z^{-1}}{1-p\cdot z^{-1}}\right)\cdot z^{-D}\right)\cdot\left(\frac{z^{-D}+z^{-2\cdot D}+\cdots+z^{-D\cdot N_{win}}}{N_{win}}\right)$$
$$\left(\frac{z^{-1}}{1-z^{-1}}\right)\cdot\left(1+\frac{z^{-1}}{1-z^{-1}}\cdot K_I\right)\cdot K_I\cdot K_D \tag{8.13}$$

The lab measurements have proven the effectiveness of the proposed signal averaging, where the phase noise profile is given with different averaging cycles.

8.5.1.4 Integer and fractional delay

In Section 8.5.1.3, the spur frequency is presumably an integer multiple of the reference clock frequency. Therefore, the delay D is an integer number, which can be simply achieved via a chain of flip-flops, assuming that they are clocked by the reference clock. This is always valid in the case of fractional-N spur. However, in the case of an externally coupled spur, the aggressor frequency may not be the integer multiple of the reference clock. In other words, delay D can be a non-integer number. If an integer delay chain is still applied, the delayed spur pattern will not be aligned with the original one, which will lead to an incomplete cancellation (see Figure 8.25). In this example, D is assumed to be 4.5, and the time-domain waveform suggests that a fractional delay of 4.5 is necessary to perfectly

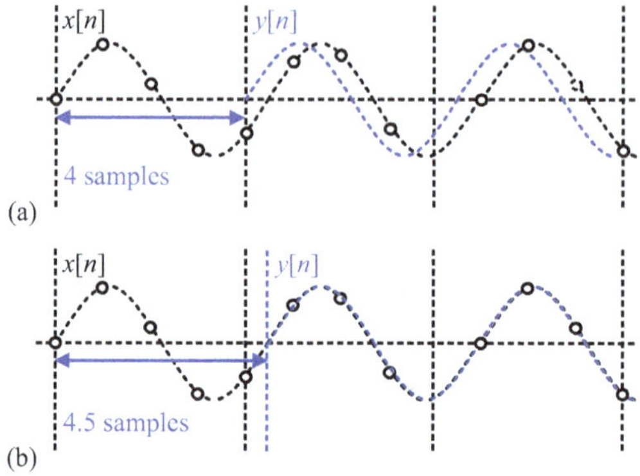

Figure 8.25 Illustration of phase alignment between the actual spur and its replica with (a) 4-cycle delay and (b) 4.5-cycle delay

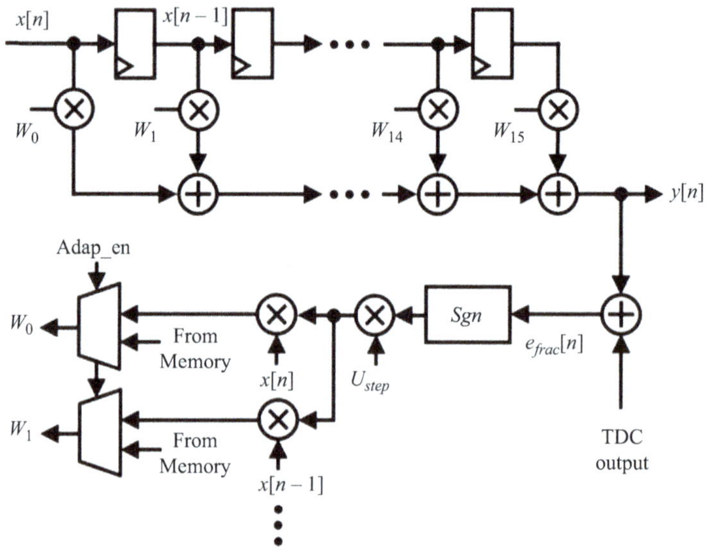

Figure 8.26 Block diagram of the Lagrange fractional delay filter with adaptability

align the delayed spur phase. To accommodate such a scenario for prototyping purposes, a Lagrange interpolation fractional delay (FD) filter [25] with some adaptive capability is used, as shown in Figure 8.26. By setting proper filter coefficients, it can provide the fractional part of the delay, while the integer part is still achieved via an integer number of flip-flops. The effectiveness of the proposed signal averaging has been proven in lab measurements, where the phase noise profile is given with different averaging cycles.

8.5.1.5 Complete feedforward loop and adaptability

The complete block diagram of the proposed feedforward multi-tone spur cancellation scheme is shown in Figure 8.27. It consists of four key building blocks: high-pass filter, integer delay chain, FD filter, and averaging block. So far, the spur frequency is assumed to be known *a priori*, which is almost always true in practical application, as the internal fractional-N spur frequency can be deterministically calculated, and the external aggressors are typically known during the system-level frequency planning. To accommodate the rare case that external spur frequency is not known *a priori*, two adaptive loops can be utilized to determine the spur frequency in sequence [22,26]. The first adaptive loop is applied to the integer delay chain to find the best integer delay setting. This estimation process is completed by observing the error between the output of the integer delay chain and the input of the feedforward cancellation path ($e_{int}[n]$, annotated in Figure 8.27) by sweeping the delay setting of the integer delay cell to determine the value that leads to the least error energy. Once the integer delay is set, the second adaptive loop is enabled to determine the best FD setting by enabling the *Adap_en* signal. Essentially, the FD filter coefficient is adjustable via an adaptive feedback loop; this allows the automatic estimation of the delay setting in the cancellation loop. The weight is adjusted by observing the error $e_{frac}[n]$ at the output of the Lagrange FD filter. The adaptation process is done via a sign LMS algorithm for lower implementation complexity at the cost of slower convergence speed. With the implemented Lagrange FD filter, the achievable fractional delay is always a positive value. As a result, some minor implementation detail between integer and fractional delay should be considered. For example, if D is 4.9, the integer delay achieving the least error energy would be 5 rather than 4. To prevent negative fractional delay from the FD filter, the programmed integer delay should be floored to the closest integer (4 in this case) such that the FD filter will always provide a positive fractional delay.

8.5.1.6 Implementation highlight: integer and fractional averaging scheme

As discussed in Section 8.5.1.4, the spur frequency may not be an integer multiple of the reference clock, which necessitates both a fractional delay and averaging in

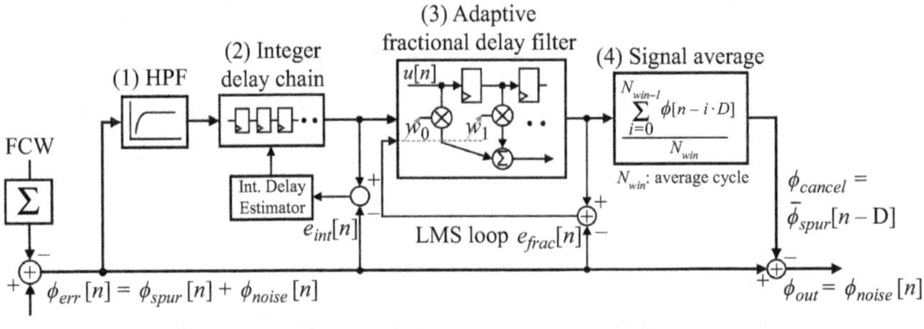

Figure 8.27 Complete block diagram of the feedforward multi-tone spur cancellation loop

the spur cancellation loop. Therefore, a signal-averaging scheme can be used to support both integer and fractional averaging. In principle, the incoming samples should first be down-sampled by the fundamental spur frequency F_{ref}/D. If D is the integer, the down-sampler is essentially a flip-flop that is clocked at every D cycle, as shown in Figure 8.28(a); it is referred to as the integer-averaging mode. In this case, the following infinite impulse response (IIR)-based accumulator can constructively sum the subsampled spur. Because the averaging block is instantiated at every internal node of the delay chain, a finite state machine is used to serialize the outputs of all the IIR-based accumulators in the correct sequence to reconstruct the spurious pattern with the right phase, as illustrated in Figure 8.28(b).

If D is a non-integer value, the integer-averaging mode no longer constructively aligns the summation phase properly (see Figure 8.29). In this example, D is 10.25, and the integer down-sampler is set to the nearest integer (i.e., 10)—the averaged waveform is noticeably distorted because the phase error is accumulated over time. As a result, fractional averaging should be used, where the key component is the fractional down-sampler. In the proposed implementation, the down-sampler is decimated between two integer numbers: M and $M+1$ for the P and Q cycles, respectively. The averaged down-sampling rate is essentially interpolated between M and $M+1$ and should approximate D as:

$$D = \frac{[P \cdot M + Q \cdot (M+1)]}{P+Q} \tag{8.14}$$

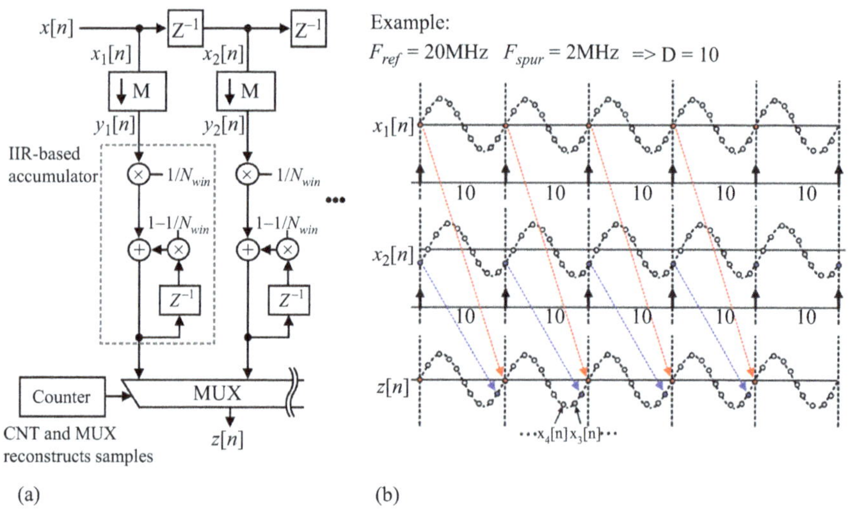

Figure 8.28 (a) Averaging scheme in integer mode and (b) representative examples of integer spur period with averaging in integer mode and the reconstruction of spur pattern at z[n]

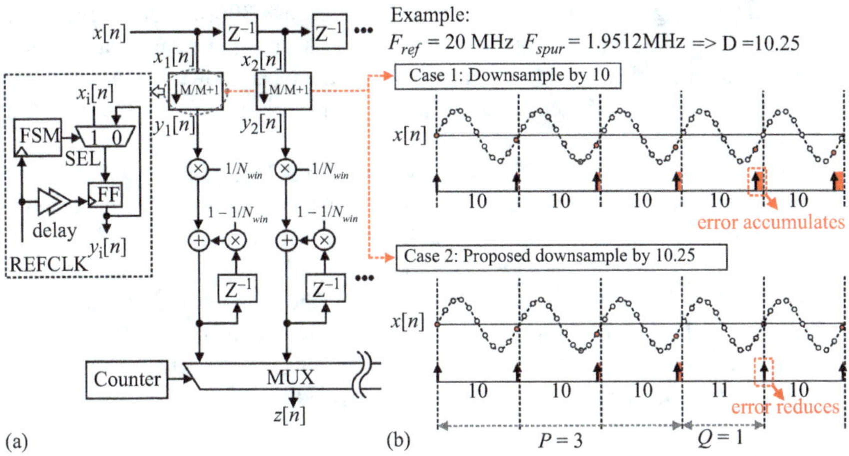

Figure 8.29 (a) Averaging scheme in fractional mode and (b) representative examples of fractional spur period with averaging in integer mode and fractional mode

Additionally, the z-domain transfer function of the fractional averaging is modified as:

$$H_{AVG}(z) = \frac{1}{N_{win}} \sum_{i=0}^{N_{win}-1} \left\{ \left(\frac{P}{P+Q}\right) \cdot z^{-i \cdot M} + \left(\frac{Q}{P+Q}\right) \cdot z^{-i \cdot (M+1)} \right\} \tag{8.15}$$

In this prototype, the value of $(P+Q)$ can be tuned between 1 and 16, and it is capable of interpolating 6.25% of one sample delay. To simplify the circuit implementation, the switching pattern is fixed (i.e., P cycles first and then Q cycles). Due to the incomplete cancellation, as shown in Figure 8.29(b), the accumulated phase error pattern is still periodic if the switching pattern is not randomized, regenerating the spurious tones by itself.

8.5.1.7 Modular extension: cascaded cancellation loops

So far in this design example, spurs in a harmonic relation have been considered, referred to as one series of spurs. However, it is possible that the spurs are in a non-harmonic relation (i.e., a mixture of a multiple series of spurs). For instance, if there are two series of spurs with the spur periods of D_1 and D_2, respectively, the combined spur period can be expressed and bounded by:

$$\min(D_1, D_2) \le D = LCM(D_1, D_2) \le \prod_{n=1}^{2} D_n, \tag{8.16}$$

where LCM represents the least common multiple operator. According to (8.16), the worst case occurs when D_1 and D_2 are co-prime with each other. In this case, the spur cancellation loop should be programmed with a delay of $D_1 \cdot D_2$, which may require a long delay chain, as shown in Figure 8.30(a). Alternatively, the proposed cancellation scheme can be extended by cascading two cancellation loops with their delay individually set to D_1 and D_2, respectively, as shown in Figure 8.30(b). In this case, the implemented delay chain will reduce to $D_1 + D_2$, which is significantly smaller than $D_1 \cdot D_2$.

In the prototype, the maximum delay stage of 128 for both D_1 and D_2 is implemented, which suggests that the lowest spur frequency to be canceled is ~234 kHz, given a reference clock frequency of 30 MHz. As indicated in (8.1), if the reference clock frequency is lowered, the cancellation loop can proportionately address even lower spur frequencies. Alternatively, the number of delay stages can be increased at the cost of additional hardware. Therefore, it is a tradeoff between hardware and reference clock frequency to set the minimum correctable spur frequency bound, depending on the need of the intended application.

8.5.1.8 Measurement results

The silicon prototype was implemented in a 65-nm CMOS process with an active area of 0.77 mm^2 (see Figure 8.31). The analog part of the digital PLL consumes 17.2 mA, whereas the output buffers consume 4.5 mA from the 1 V supply. The digital part dissipates 1.7 mA in normal DPLL operations, and each of the spur cancellation loops consumes an additional ~2 mA.

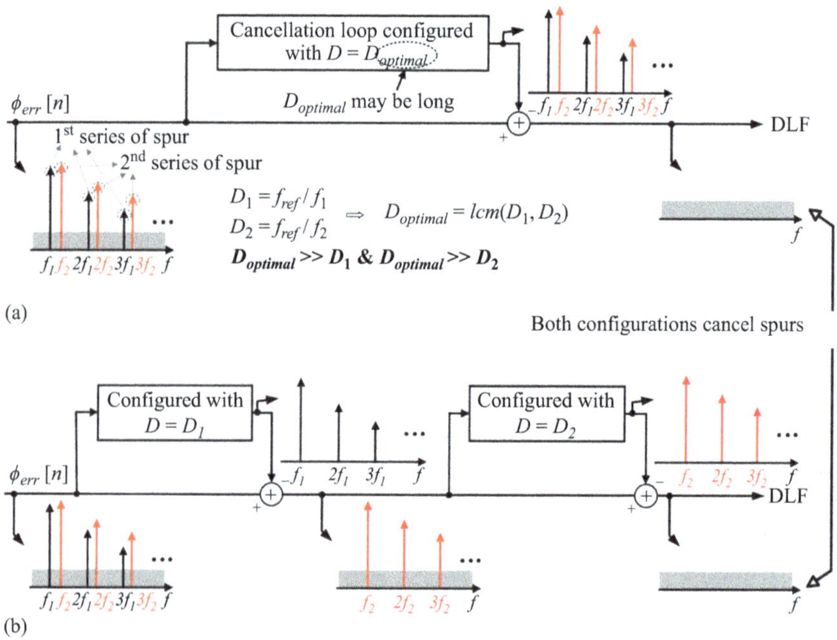

Figure 8.30 Illustration of feedforward multi-tone spur cancellation loop with (a) single-stage configuration and (b) two-stage configuration for two series of spurs

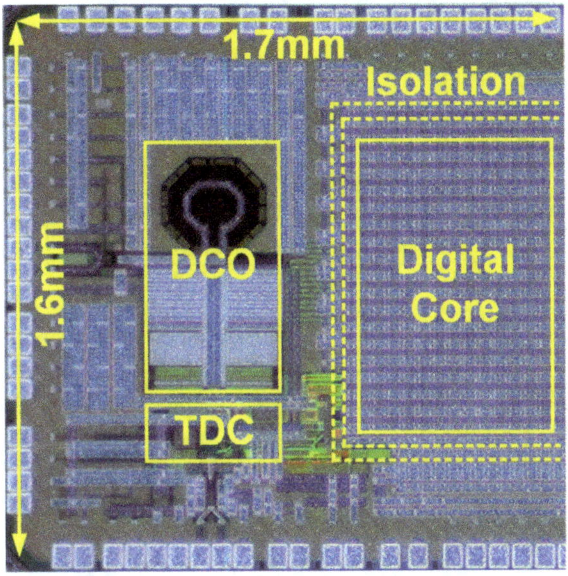

Figure 8.31 Die micrograph of Ref. [2]

The effectiveness of performing signal averaging in the cancellation loop is shown in Figure 8.32. In this case, an external sawtooth disturbance is also injected at 234-kHz frequency, offset from a carrier frequency of 3.57 GHz; this is within the DPLL bandwidth. With a smaller N_{win} value, such as 2, severe noise degradation is observed in the phase noise profile not only within the DPLL bandwidth but also outside the band. When the N_{win} value increases, the noise degradation is mitigated, as shown in Figure 8.32(b). The measured phase noise profiles confirm the effective use of the averaging block to avoid elevation of the phase noise floor.

The spectrum snapshot of the fractional-N spur before and after enabling the spur cancellation scheme is shown in Figure 8.33(a). In Figure 8.33(b), fractional-N spurs are measured across different fractional code settings, showing that the worst-case spur ranges from −73.66 to −117 dBc, indicating a 20–50 dB improvement. To demonstrate the modular extension capability of the proposed cancellation scheme, the external spur reduction is measured under single-stage and cascaded two-stage configurations. For a single-stage cancellation loop, a sawtooth disturbance is injected intentionally via the input buffer supply. As can be seen in Figure 8.34(a) and (b), the spurious tone is reduced by 15–35 dB on both its fundamental and second harmonic. The higher-order harmonics are not shown, as they are buried underneath the phase noise floor after cancellation. For cascaded two-stage cancellation, two independent series of triangular spurious tone located at 312- and 495-kHz offset from the carrier frequency are combined off-chip and then injected. After cancellation, both series of spurs are nullified under the noise floor (see Figure 8.35).

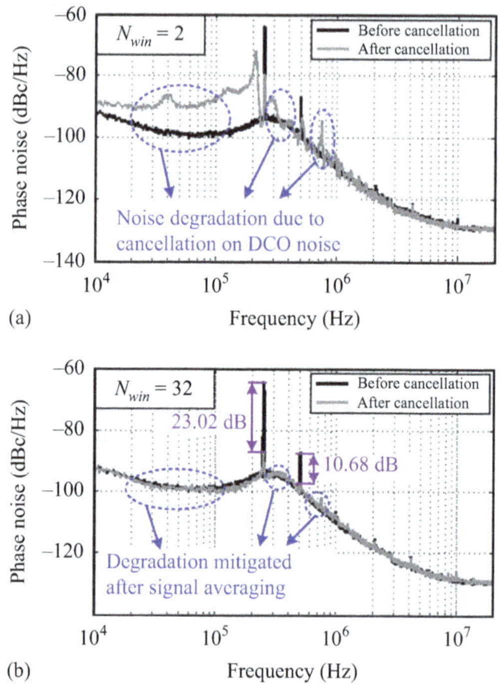

Figure 8.32 Measured DPLL phase noise profile given average cycle (N_{win}) of (a) 2 and (b) 32

8.5.2 DPLL with a dither-based spur mitigation technique

If the frequency accumulation-type DPLL and the conventional sigma-delta-modulated DPLL are compared, the latter has a better fractional spurious performance due to the intrinsic randomization of the delta-sigma modulator. A similar concept is used for the frequency accumulation-type DPLL to mitigate fractional spurs by incorporating reference-path dithering [12–15]. The reference clock dithered via a digital-to-time converter (DTC) will randomly trigger the TDC quantization step to scramble the TDC non-linearities, hence reducing the spur magnitude, as shown in Figure 8.36. Nevertheless, a large dither noise at the TDC output will significantly degrade the PLL phase noise profile, necessitating the subsequent dither noise cancellation scheme. Similar to the design example shown in Section 8.5.1, the prototype in [12] proposes a dither-based spur mitigation scheme to improve the near-carrier fractional spurs by using an adaptive filter algorithm. However, instead of directly mitigating the spurious tones, this technique uses dithering to spread out the spurious energy and then mitigate the dither noise.

8.5.2.1 Overall DPLL architecture

The block diagram of the DPLL architecture featuring the background dither noise cancellation scheme [12] is given in Figure 8.37. Input reference path dithering is implemented via a tunable delay buffer chain controlled by a Hadamard code

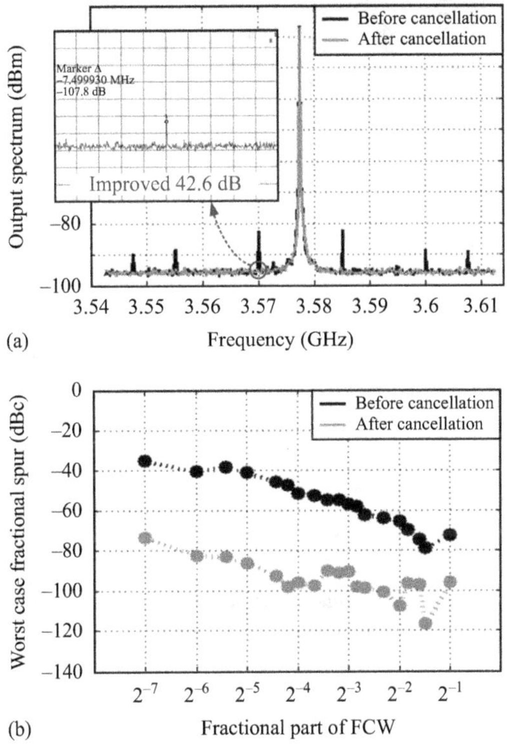

Figure 8.33 *(a) Spectrum snapshot of fractional spurs at FCW = 119.25 and (b) measured worst-case fractional spur level across different FCW settings*

generator to delay the reference clock phase before comparison with the LC-DCO phase. An LMS-based noise cancellation loop is inserted before the DLF to cancel the induced dithering noise. The DLF output controls the oversampled delta-sigma modulator, toggling a bank of varactors in the LC-DCO; this helps to reduce in-band frequency quantization noise. Then, the IL TDC tracks DCO frequency and phase variation over process–voltage–temperature (PVT) variations and provides 28 levels of fine phase quantization within one DCO period. Finally, the accumulated FCW is subtracted at the TDC output for fractional-*N* operation. The details of the DLF, IL TDC, and DCO are included in Section 8.5.1.1.

8.5.2.2 Dither noise cancellation scheme

As the TDC non-linearities appear in a sawtooth pattern, they will modulate the DCO to generate harmonic tones with a fundamental frequency of F_{SPUR}, which is defined in (8.1). The TDC non-linearities can be modeled as the superposition of Fourier components:

$$\phi_{frac}[k] = \sum A_n \cdot \cos\left(2\pi \cdot nF_{spur} \cdot kT_{REF}\right) \tag{8.17}$$

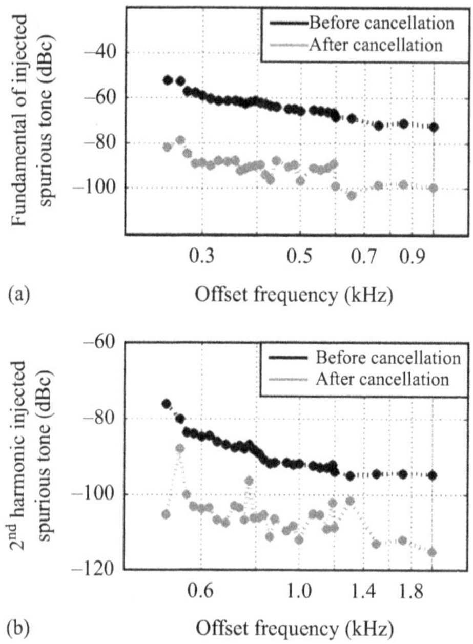

Figure 8.34 Measured (a) fundamental and (b) second harmonics of external spur reduction in single-stage cancellation

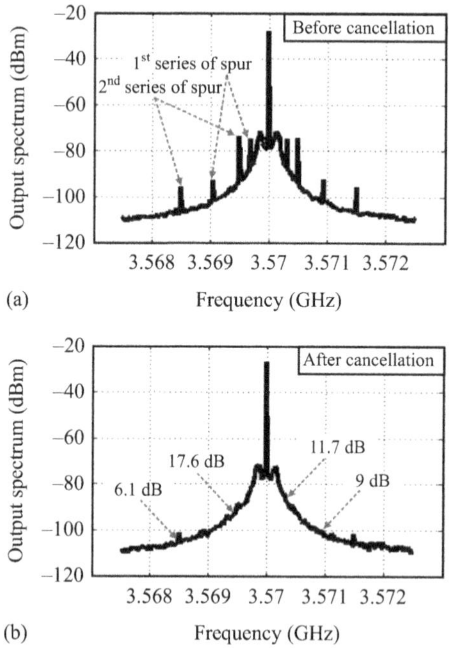

Figure 8.35 Measured external spur reduction in cascaded two-stage cancellation configurations (a) without and (b) with activating the spur cancellation scheme

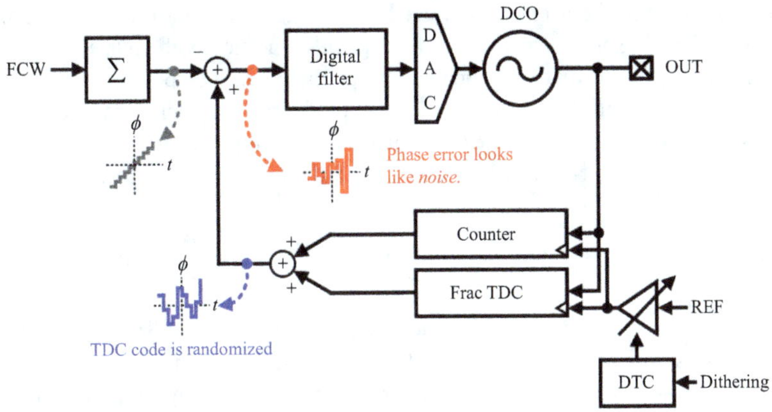

Figure 8.36 DPLL architecture with reference-path dithering to mitigate spurious tones

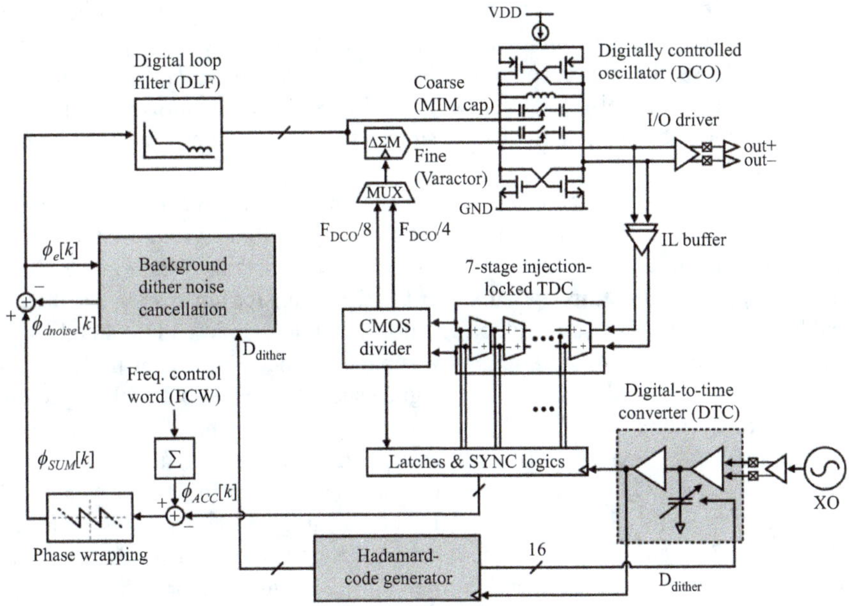

Figure 8.37 The block diagram of the fractional-N DPLL architecture with background dither noise cancellation loop

where A_n represents the magnitude of the nth harmonic fractional spur normalized to 1 LSB of the TDC's quantization step. Although this TDC's non-linearities are ideally confined within 1 LSB of the TDC's quantization level, if the TDC's DNL effect is considered, each fractional spur's magnitude can be approximated to have additional gain fluctuation on the amplitude variable A_n in (8.17).

To minimize the spur energy deduced in (8.17), one can introduce an additional term inside the sinusoidal expression; consequently, the sawtooth spur pattern cannot preserve its original format and concentrate its energy in the spectrum. If this additional term is time-varied and Gaussian-distributed, the spur energy can be spread as noise. As a result, the new phase error at the TDC output can be rewritten as the summation of the dither-induced noise (ϕ_n) and the randomized spurs:

$$\phi'_{frac}[k] = \phi_n + \sum A_n \cdot \cos\left(2\pi \cdot nF_{spur} \cdot kT_{REF} + \frac{T_{Dither}[k]}{T_{DCO}}\right) \quad (8.18)$$

where $T_{Dither}[k]$ denotes a dither-induced timing shift at the kth sample, which can be modeled as an injected dither code $D_{Dither}[k]$ at the kth sample multiplying the DTC gain, G_{DTC}. Note that this timing shift should be normalized to one DCO period (T_{DCO}) and fundamental spur frequency. Fortuitously, the induced phase shift term ($T_{Dither}[k]/T_{\mathrm{DCO}}$) phase modulates the sawtooth pattern in (8.18). Based on the numerical simulation, if the phase shift term is random for every sample, it helps to whiten spurious energy. Note that the existence of dither-induced noise (ϕ_n) will significantly raise the noise floor, necessitating the followup dither noise cancellation scheme. Assuming the dither-induced noise can be removed completely in (8.18), the entire spurious tone energy spreads out in the spectrum, appearing as random noise confined within 1 LSB of TDC quantization. In other words, the fundamental noise floor limit should still be determined by the designed TDC resolution.

The adaptive spur mitigation technique shown in Section 8.5.1 can be utilized here. Figure 8.38 contains the rationale of the proposed background dither noise cancellation scheme. The injected dither code D_{dither} is first converted in the phase domain as ϕ_{Dither} and multiplied with a certain DTC transfer function $H_{DTC}(z)$ via a tunable delay buffer. The same dithering information ϕ_{Dither} is fed into an adaptive filter, and its output is added after FCW subtraction node ϕ_{sum}. For perfect cancellation, the adaptive filter should be designed such that the output of the injection and the compensation path should match precisely at ϕ_{sum}. Therefore, the transfer function from ϕ_{Dither} to ϕ_{sum} [denoted as $H_{Loop}(z)$], which is a function of DLF response, DTC, DCO, and TDC gain, is examined. According to the analytical derivation, it presents a high-pass transfer function, and the frequency response can vary over PVT because the gain values change. On activating the cancellation loop, the frequency response of the adaptive filter $H_{can}(z)$ gradually approaches $H_{Loop}(z)$. The adaptive filter uses LMS algorithms that minimize the mean square error at DLF input node ϕ_e. Thus, the convergence speed and the settling accuracy depend on the step size of the iterative loop. The background operation compensates for the frequency response variation due to PVT drift.

8.5.2.3 Implementation highlight: dither noise cancellation loop

Figure 8.39 contains the block diagram of the implemented cancellation loop. The dither signal is produced using a Hadamard code generator, creating a random

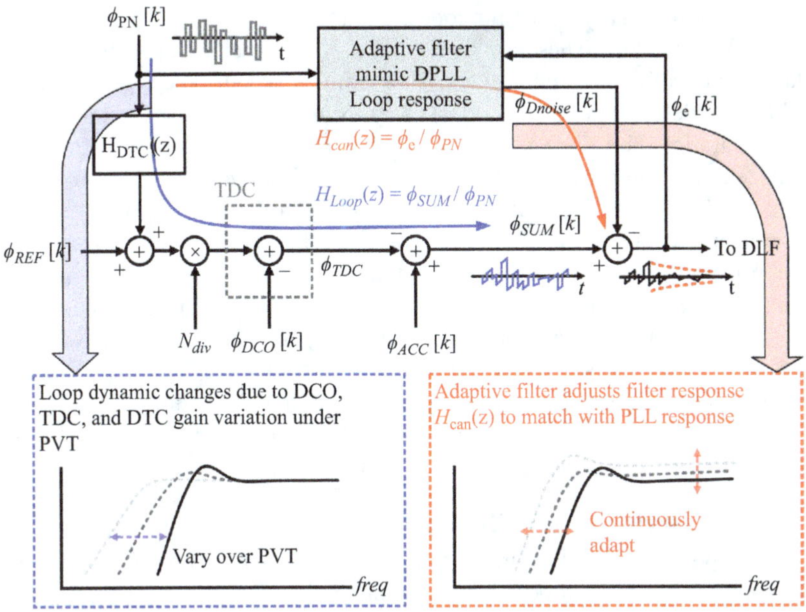

Figure 8.38 DPLL mathematical model with injected dither signal and the corresponding cancellation path in z-domain

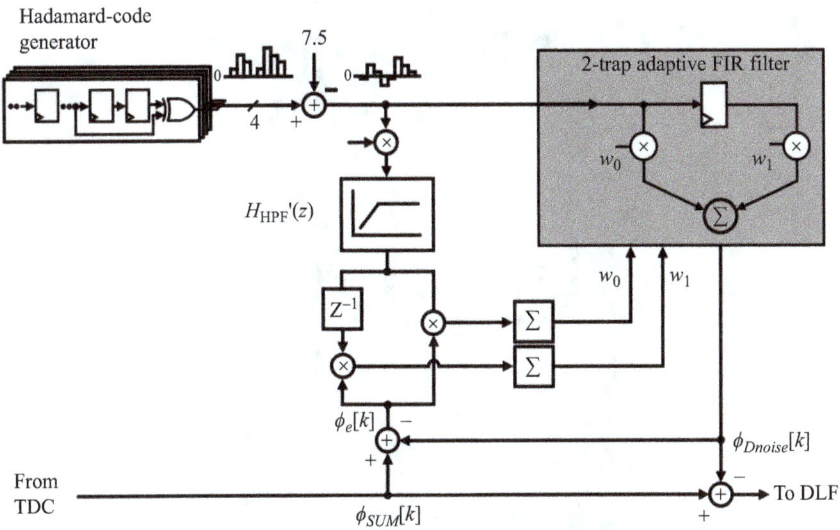

Figure 8.39 Complete implementation of the background dither noise cancellation with a two-tap configuration

sequence of 4-bit binary dithering codes, controlling a bank of 16 unitary metal-oxide-metal capacitors. These are converted to thermometer codes and retimed to the falling edge of the reference clock. This retiming scheme keeps glitches (due to switching the delay of the CML buffer) away from the rising edges of the reference clock. In parallel, the DC portion of D_{dither} is removed and fed into the LMS filter. The cancellation filter response $H_{can}(z)$ is composed of two parts: (1) a fixed high-pass filter $H_{HPF}'(z)$ to duplicate DPLL close-loop high-pass response, and (2) a two-tap adaptive FIR filter that adjusts the gain and phase to compensate for the variation from the analog portion of the dithering circuit. The DPLL loop filter is deterministic and depends only on the programmed DLF parameters (i.e., proportional and integral path gain for type II response), so it will not drift with PVT. Therefore, it is not necessary to adapt the high-pass filter coefficient of $H_{HPF}'(z)$ in real time. The coefficient of the two-tap FIR filter should be adjusted based on the product of phase error ϕ_e and the output of $H_{HPF}'(z)$, which leads to better convergence stability according to numerical simulations.

8.5.2.4 Measurement results

The prototype was fabricated in a 65-nm CMOS process with an active area of 0.334 mm^2; here, the proposed background noise cancellation loop occupied 0.017 mm^2, as shown in Figure 8.40. The analog core of the DPLL consumed 16.4 mW of energy from a 1-V supply. The digital part consumed 1.1 mW with the

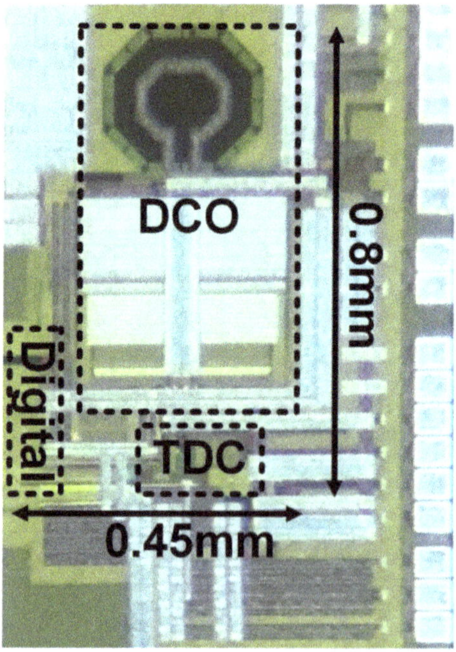

Figure 8.40 Die micrograph of Ref. [12]

normal DPLL operation, and the proposed dither and dither noise removal scheme dissipated an additional 0.6 mW.

To test the efficacy of the proposed dither and dither cancellation scheme, the phase noise profile of DPLL locking at an FCW of $120 + 2^{-10}$ was measured (see Figure 8.41). Specifically, Figure 8.41(a) has two curves that indicate the phase noise profile of the DPLL operating at an FCW of $120 + 2^{-10}$ without and with the dithering scheme, respectively. After cancelling the background dither noise, a suppressed noise profile was generated that demonstrated the worst-case spur and dither noise improvements [Figure 8.41(b])].

The spectrum figure regarding the worst-case fractional spur with an FCW of $120 + 2^{-14}$ (Figure 8.42) represents a near-carrier case. Before enabling the dither and dither removal scheme, the worst-case spurious level was −22 dBc at 1.83 kHz

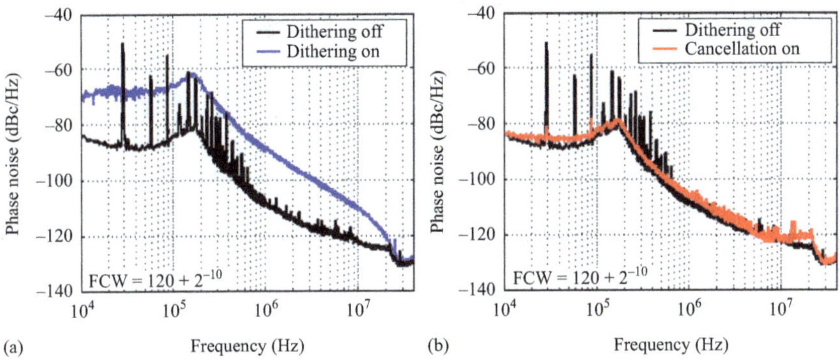

Figure 8.41 Measured phase noise profile with the dither and dither noise cancellation lop at an FCW of $120 + 2^{-10}$. The spur improvement of 30 dB and the noise improvement of 23 dB were noted

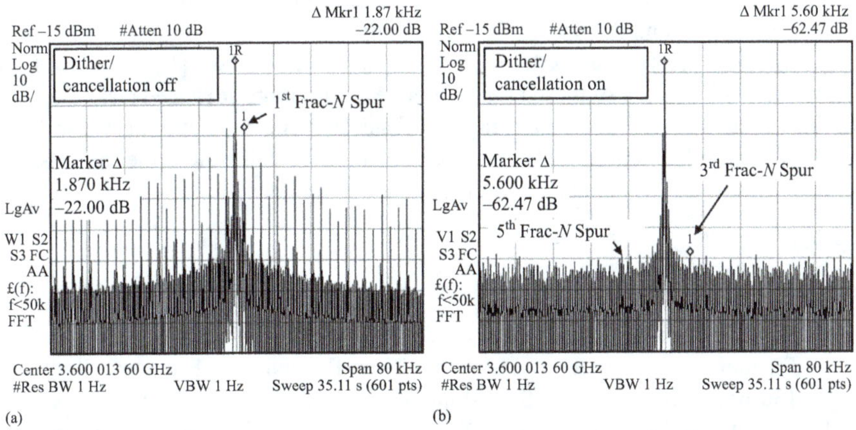

Figure 8.42 Measured worst-case fractional spur with and FCW of $120 + 2^{-14}$ (a) before and (b) after enabling the dither noise cancellation loop

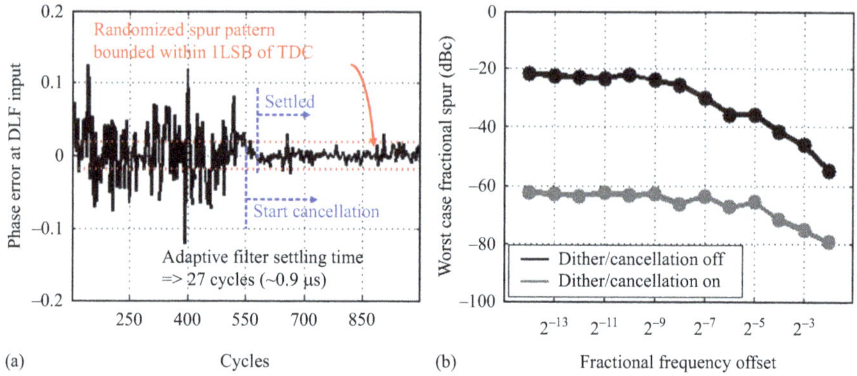

Figure 8.43 Measured settling behavior of a DPLL with an FCW of 120 + 2^{-14} and (b) the worst-case fractional spur with different FCW settings

offset. After enabling the technique, the level decreased to −62.47 dBc, whereas other harmonics were beneath the noise floor. During dither noise cancellation, the internal state before the DLF was captured in Figure 8.43(a)—the measurement shows that the adaptive filter loop tool took 27 cycles to settle (i.e., ~0.9 μs convergence time). Fractional-N spur levels were measured [Figure 8.43(b)] with different fractional frequency offsets ranging from 2^{-14} to 2^{-2}, demonstrating at least $\geq$ 20 dB improvements.

8.6 Conclusion

This chapter provides an overview of the generation of PLL spurs and their potential impact on wireless and wireline the system. In order to mitigate the spurs, the new DSP-assisted technique can be applied in the digital PLL platform to push the performance limit beyond the analog PLL capability. Two design examples with *direct spur mitigation* and *dither-based spur mitigation* techniques are introduced to demonstrate the efficacy of these DSP techniques. With more advanced technology to come and accommodate their advantage for faster digital computation, it is expected that DSP-assisted techniques, also summarized in [27], will enable the larger-scale adoption of this DPLL architecture in the design community.

References

[1] R.B. Staszewski, K. Muhammad, D. Leipold, *et al.* All-Digital TX Frequency Synthesizer and Discrete-Time Receiver for Bluetooth Radio in 130-nm CMOS. IEEE J. Solid-State Circuits. 2004; 39:2278–91.

[2] C.-R. Ho and M.S.W. Chen. A Digital PLL with Feedforward Multi-Tone Spur Cancellation Loop Achieving <−73dBc Fractional Spur and <−110dBc Reference Spur in 65nm CMOS. IEEE J. Solid-State Circuits. 2016; 51:3216–30.

[3] C.M. Hsu, M.Z. Straayer and M.H. Perrott, A Low-Noise Wide-BW 3.6-GHz Digital Fractional-N Frequency Synthesizer with a Noise-Shaping Time-to-Digital Converter and Quantization Noise Cancellation. IEEE J. Solid-State Circuits. 2008; 43:2776–86.

[4] A. Mirzaei and H. Darabi. Pulling Mitigation in Wireless Transmitters. IEEE J. Solid-State Circuits. 2014; 49:1958–70.

[5] G.L. Puma and C. Carbonne. Mitigation of Oscillator Pulling in SoCs. IEEE J. Solid-State Circuits. 2016; 51:348–56.

[6] R. Winoto, A. Olyaei, M. Hajirostam, *et al.* A 2 × 2 WLAN and Bluetooth Combo SoC in 28 nm CMOS with On-Chip WLAN Digital Power Amplifier, Integrated 2G/BT SP3T Switch and BT Pulling Cancelation. ISSCC Dig. Tech. Papers. 2016 Feb; San Francisco, CA, USA. p. 170–1.

[7] C.-R. Ho and M.S.W. Chen. A Digital Frequency Synthesizer with Dither-Assisted Pulling Mitigation for Simultaneous DCO and Reference Path Coupling. ISSCC Dig. Tech. Papers. 2018 Feb; San Francisco, CA, USA. p. 254–5.

[8] C.-R. Ho and M.S.W. Chen. A Fractional-N DPLL with Calibration-Free Multi-Phase Injection-Locked TDC and Adaptive Single-Tone Spur Cancellation Scheme. IEEE Trans. Circuits Syst. I: Reg. Papers. 2016; 63:1111–22.

[9] C.-R. Ho and M.S.W. Chen. Interference-induced DCO Spur Mitigation for Digital Phase Locked Loop in 65-nm CMOS. Proc. ESSCIRC. 2016 Sep; Lausanne, Switzerland. p. 213–6.

[10] F. Opteynde. A 40 nm CMOS All-Digital Fractional-N Synthesizer without Requiring Calibration. ISSCC Dig. Tech. Papers. 2012 Feb; San Francisco, CA, USA. p. 346–7.

[11] H. Kim, J. Sang, H. Kim, *et al.* A 5 GHz −95 dBc-Reference-Spur 9.5 mW Digital Fractional-N PLL Using Reference-Multiplied Time-to-Digital Converter and Reference-Spur Cancellation in 65 nm CMOS. ISSCC Dig. Tech. Papers. 2015 Feb; San Francisco, CA, USA. p. 258–9.

[12] C.-R. Ho and M.S.W. Chen. A Fractional-N Digital PLL with Background-Dither-Noise Cancellation Loop Achieving <−62.5 dBc Worst-Case Near-Carrier Fractional Spur in 65 nm CMOS. ISSCC Dig. Tech. Papers. 2018 Feb; San Francisco, CA, USA. p. 394–5.

[13] E. Temporiti, C.W.-Wu, D. Baldi, R. Tonietto and F. Svelto. A 3.5 GHz Wideband ADPLL with Fractional Spur Suppression Through TDC Dithering and Feedforward Compensation. IEEE J. Solid State Circuits. 2010; 45:2723–36.

[14] R.B. Staszewski, K. Waheed, F. Dulger and O. Eliezer. Spur-Free Multi-rate All-Digital PLL for Mobile Phones in 65 nm CMOS. IEEE J. Solid-State Circuits. 2011; 46:2904–19.

[15] Y. He, Y.H. Liu, T. Kuramochi, *et al.* A 673 μW 1.8-to-2.5 GHz Dividerless Fractional-N Digital PLL with an Inherent Frequency-Capture Capability and a Phase-Dithering Spur Mitigation for IoT Applications. ISSCC Dig. Tech. Papers. 2017 Feb; San Francisco, CA, USA. p. 420–1.

[16] R.B. Staszewski, C.-M. Hung, N. Barton, M.-C. Lee, and D. Leipold. A Digitally Controlled Oscillator in a 90 nm Digital CMOS Process for Mobile Phones. IEEE J. Solid-State Circuits. 2005; 40:2203–11.
[17] S.D. Vamvakos, R.B. Staszewski, M. Sheba and K. Waheed. Noise Analysis of Time-to-Digital Converter in All-Digital PLLs. Design, Applications, Integration and Software, 2006 IEEE Dallas/CAS Workshop on. 2006 Oct; Dallas, TX, USA. p. 87–90.
[18] B. Razavi, *RF Microelectronics*, 2nd ed. Upper Saddle River (NJ): Prentice Hall; 2011.
[19] R.B. Staszewski and P.T. Balsara, All-Digital Frequency Synthesizer in Deep-Submicron CMOS, 1st ed. New York: Wiley-Interscience; 2006.
[20] O.E. Eliezer, R.B. Staszewski, I. Bashir, S. Bhatara, and P.T. Balsara. A Phase Domain Approach for Mitigation of Self-Interference in Wireless Transceivers. IEEE J. Solid-State Circuits. 2009; 44:1436–53.
[21] Y. Nesterov. Efficiency of coordinate descent methods on huge-scale optimization problems. SIAM J. Optimiz. 2012; 22:341–62.
[22] Simon Haykin. Adaptive Filter Theory. 4th ed. Upper saddle River: Prentice Hall; 1996.
[23] S. Mendel, C. Vogel and N. Da Dalt. Signal and Timing Analysis of a Phase-domain All-digital Phase-locked Loop with Reference Retiming Mechanism. Proc. Mixed Design Integr. Circuits & Systems. 2009 June; Lodz, Poland. p. 681–7.
[24] A. Gaita, Y.-D. Lee and W.-Y. Chung. ECG Signal De-Noising with Signal Averaging and Filtering Algorithm. Proc. 3rd Int. Conf. Converg. Hybrid Inf. Technol. (ICCIT). 2008 Nov; Busan, South Korea. p. 409–15.
[25] P. Depalle and S. Tassart. Fractional Delay Lines Using Lagrange Interpolators. International Computer Music Conference. 1996 Sep; Hong Kong, China. p. 341–3.
[26] B. Widrow, J.R. Glover, J.M. McCool, *et al.* Adaptive Noise Cancelling: Principles and Applications. Proc. IEEE. 1975; 63:1692–716.
[27] C.-R. Ho and M. Chen. Smoothing the Way for Digital Phase-Locked Loops: Clock Generation in the Future with Digital Signal Processing for Mitigating Spur and Interference. IEEE Microwave Magazine. 2019; 20:80–97.

Chapter 9

Fully synthesized digital PLL

Bangan Liu[1], Wei Deng[1], and Kenichi Okada[1]

As CMOS technology scaling advances, traditional analog circuits have moved toward digital-intensive or all-digital designs during the past few years. While taking advantage of digital circuits in scaled complementary metal-oxide-semiconductor (CMOS) technology, digital-intensive or all-digital designs still cannot be absorbed in advanced design automation used by digital very-large-scale integration circuits. This chapter investigates a design methodology for analog circuits that is referred to as synthesizable analog design. In this design methodology, all circuit building blocks are implemented in all-digital architecture and expressed in a hardware description language (HDL), which can then be synthesized from commercial standard-cell libraries and automatically perform place and route (P&R) using electronic design automation tools.

Compared to conventional pure analog or digital-intensive analog implementations, synthesizable analog implementations significantly shorten the design time and cost. In addition, synthesizable architectures enhance the portability and scalability of analog circuits for various applications and different design nodes because the functionality of cells remains unaltered between different process nodes or between different process technologies. Moreover, design rules for deep submicron technology nodes become much more complex and only allow restricted forms of circuit layout [1], which degrades the productivity of conventional analog design, whereas synthesizable analog design has the potential to circumvent such restrictions on circuit layout. Finally, fully synthesizable analog designs can be better integrated with digital systems. Functions such as calibration logic, built-in-self-test, and boundary scan can be readily integrated in order to monitor the internal circuit state, which is very helpful in circuit design and de-bugging. Therefore, the fully synthesizable design methodology has been applied to various applications, including all-digital transmitters [2,3], all-digital phase-locked loops (PLLs) [4–7], time-to-digital converters (TDCs) [8], and analog-to-digital converters [9–12].

While a number of fully synthesizable PLLs have been reported recently [4–7], there is still a considerable room for improvement in terms of power consumption and chip area. Moreover, a considerable amount of custom-designed cells are

[1]Department of Electrical and Electronic Engineering, Tokyo Institute of Technology, Tokyo, Japan

required in previously synthesized PLLs, which introduce additional design and P&R procedures [5,6,13]. These additional design procedures lead to a performance degradation in terms of portability, integration, and scalability.

In order to address the abovementioned issues, this chapter presents two fully synthesizable PLLs, which do not use any custom design but are purely designed using a foundry-provided standard-cell library by a standard digital design flow. In order to mitigate the mismatches and variations in fully synthesizable designs, a holistic optimization of system design, circuit topology, and implementation flow is adopted.

The remainder of this chapter is organized as follows. Section 9.1 explains the digital compatible design process of synthesizable PLLs with illustrative design examples. Section 9.2 describes the overall architecture of the PLL and relevant considerations. Section 9.3 presents a fully synthesizable all-digital PLL with an interpolative phase-coupled oscillator, a current-output digital-to-analog converter (DAC), and a fine-resolution digital varactor with a gated edge-injection technique. Section 9.4 presents a synthesizable injection-locked PLL (IL-PLL) with self-clocked nonoverlap update and a slope-balanced subsampling bang-bang phase detector (SS-BBPB). Finally, Section 9.5 concludes.

9.1 Design process

9.1.1 Digital-compatible design flow

Figure 9.1 shows the design flow of a fully synthesizable PLL. Analog circuits, such as digitally controlled oscillators (DCOs), and phase detectors are designed with standard cells and are described in gate-level HDL. The digital control logic, such as loop filters and calibration blocks, are described in register-transfer level (RTL) HDL and are mapped to a gate-level netlist using digital synthesis tools. The netlists from the two parts are combined and converted to layout using automated layout synthesis. When migration to a new process is needed, only the analog circuits need to be redesigned by replacing standard cells with appropriate timing

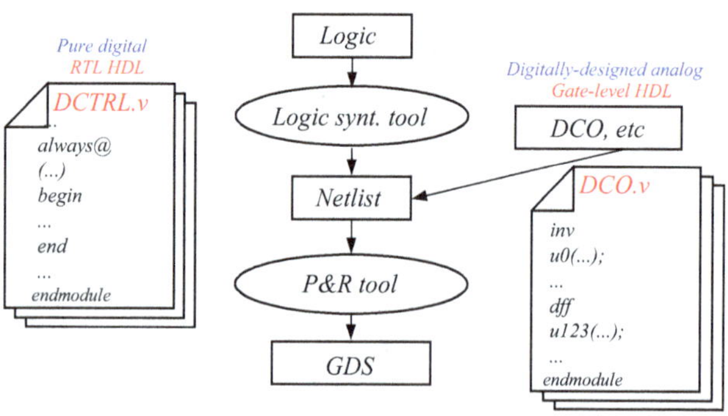

Figure 9.1 Design process of a fully synthesizable phase-locked loop (PLL)

Table 9.1 Design times of the synthesizable phase-locked loop and the conventional analog phase-locked loop

	Analog PLL	**Synthesizable PLL**
System-level design	4 h	1 h
Block design[a]	1 month	1 week
Layout design[a]	1 month	1 h

[a]For a new process.

characteristics, whereas the digital logics can be readily resynthesized. Thus, the design migration is significantly accelerated. In order to mitigate the mismatches and variations caused by automatic synthesis, techniques such as relative placement can be used. With such design constraints, reasonably good performance can be expected. Moreover, these design constraints are process independent and can be readily used across different processes.

One interesting specification that merits examination is the total time used for this design. Here the design time of both a conventional analog PLL and the proposed synthesizable PLL is summarized in Table 9.1. Compared to a conventional analog PLL, the synthesizable PLL enables faster circuit design for a new process, since only resizing of standard cells is required. In addition, the layout design is greatly accelerated, as automatic P&R can be used to eliminate the burden of re-learning complex design rule check rules, especially for advanced technologies.

9.1.2 Hardware description language-based circuit design

As a demonstration, the implementation of a BBPD-based PLL is designed, and example codes of the loop filter and DCO are shown. Code listing 9.1 shows a digital loop filter (DLF) for a PLL with a BBPD. The DLF accepts the binary output from the BBPD, and the output is used to control the DCO. Due to the digital implementation, the DLF has a compact area and can be easily reconfigured to realize different transfer functions. The DLF used in this design has two paths: the proportional path and the integral path. The outputs of the two paths are combined in a digital domain. The RTL Verilog code implementation of the DLF is shown in code Listing 9.1.

Listing 9.1 Verilog code of digital loop filter

```
//DLF.V
module DLF (CLK, PD, Dout, RSTB);
    parameter beta = 6;
    parameter alpha = 4;
    input   CLK, PD, RSTB;
    output   [4:0]   Dout;
```

```
    reg  signed [30:0]  integ;
    wire    signed [30:0]  Dout_temp, i_path, p_path;
//
    assign  i_path = integ <<< alpha;
    assign  p_path = { {10{PD} } } <<< beta;
    assign  Dout_temp = i_path + p_path + 31?00100000;
    assign  Dout = Dout_temp[20:16];
//
    always  @( posedge CLK ) begin
    if (!RSTB) begin
       integ < = 0;
       end
       else begin
          if (DIN = = 0) begin
             integ < = integ - { { 20{1'b0} }, {10?3FF} };
          end
          else begin
             integ < = integ + { {201'b0} }, {10?3FF} };
             end
          end
       end
endmodule
```

As one of the most important blocks of a PLL, the DCO determines the out-of-band phase noise and directs the PLL design. Code listing 9.2 shows a multistage inverter-based ring oscillator with not-and (NAND)2-based digital varactors. Due to this standard-cell-only implementation, the design can be easily ported to other processes. The DCO is described in gate-level Verilog, and the netlist is shown in code Listing 9.2. In order to reduce the mismatches and variations in automatic P&R, design constraints such as relative placement can be applied, as shown in code Listing 9.3. Note that such design constraints are independent of technology niceties and can be readily reused across different processes.

Listing 9.2 Verilog code of the digitally controlled oscillator

```
// DCO_BBPLL.V
module DCO_BBPLL (DCO_out, Din_dec);
     input [30:0] Din_dec;
     output DCO_out;
   //
     NAND2X6BA10TR I47 ( .VDD(VDD), .VSS(VSS),
          .B(Din_dec[18]), .Y(net596), .A(FINE3));
     NAND2X6BA10TR I46 ( .VDD(VDD), .VSS(VSS),
          .B(Din_dec[19]), .Y(net597), .A(FINE4));
. . .
INVX3BA10TR I1 ( .VDD(VDD), .VSS(VSS), .Y(FINE2), .A(FINE1));
INVX3BA10TR I0 ( .VDD(VDD), .VSS(VSS), .Y(FINE1), .A(DCO_IN));
endmodule
```

Listing 9.3 Digitally controlled oscillator relative placement commands

```
create_rp_group    DCO_RP -column M -row N
add_to_rp_group    DCO_BBPLL::DCO_RP -column C1 -row R1
         -leaf DCO_BBPLL/I0
add_to_rp_group    DCO_BBPLL::DCO_RP -column C2 -row R2
         -leaf DCO_BBPLL/I1
. . .
add_to_rp_group    DCO_BBPLL::DCO_RP -column C3 -row R3
         -leaf DCO_BBPLL/I46
add_to_rp_group    DCO_BBPLL::DCO_RP -column C4 -row R4
         -leaf DCO_BBPLL/I47
```

9.2 PLL architecture

9.2.1 Proposed PLL architecture

All-digital PLLs, as shown in Figure 9.2, generally have TDC-based architecture [14]. Although some all-digital PLLs [15] can be fully described in HDL, the layout design, especially the design of DCOs and TDCs, must still be performed by conventional custom layout drawing. This is because these critical building blocks are prone to layout uncertainty and impairment resulting from automatic P&R, which leads to linearity degradation causing poor in-band and out-of-band phase noise in all-digital PLLs. In order to overcome this problem, a plugin unit-cell technique was proposed in a previous study [6], but the additional procedures required in unit-cell design P&R negates any advantages promised by this method.

On the other hand, ILPLLs feature a feed-forward reference injection path to correct the oscillator error, as shown in Figure 9.3. Thus, problems with TDC, such as linearity and power-resolution trade-off, can be avoided. Unlike conventional all-digital PLLs that adopt TDC for phase locking, feed-forward reference injection could provide much wider noise filtering bandwidth. For a conventional type-II TDC-based PLL, the PLL suppresses the phase noise of a free-running oscillator only within its loop bandwidth, which is typically less than 0.1 times $f_{\rm fref}$ due to the loop-stability constraint. The limited loop bandwidth results in a poor in-band phase noise. On the other hand, periodic phase alignment by injecting a clean reference into the oscillator makes use of the stable reference clock over a much wider bandwidth, thereby improving the suppression of free-running oscillator phase noise to obtain low in-band phase noise.

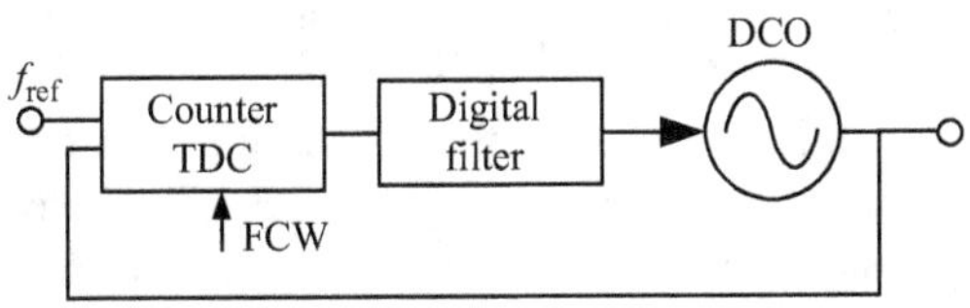

Figure 9.2 Simplified diagram of a conventional TDC-based all-digital PLL

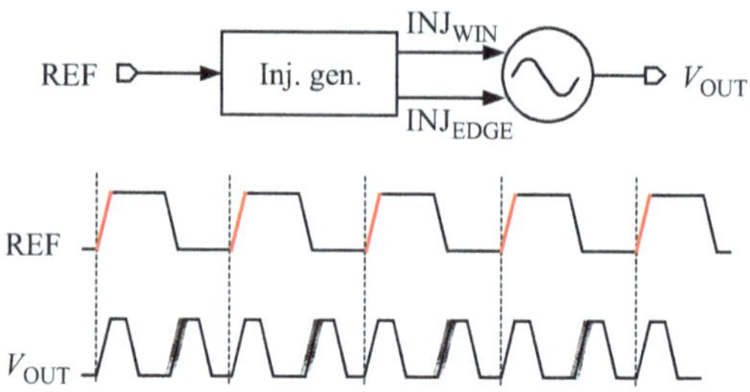

Figure 9.3 Conceptual operation of a generic injection-locked PLL (ILPLL)

A detailed phase noise analysis of this kind of edge IL oscillator was first performed in [16]. Further analysis [17] revealed a 3-dB increase in the high-frequency phase noise of the IL oscillator based on continuous time-domain approximation. In contrast, a different phase noise analysis method using a discrete time-domain analysis in which the phase noise expression is valid for any multiplication ratio N is provided in [18]. Note that the single side band (SSB) phase noise power spectrum density (PSD) $\mathcal{L}(f)_{\text{free-running}}$ of a free-running oscillator at frequency offset f is expressed as $\sigma_p^2 N f_{\text{ref}}/(8\pi^2 f^2)$. The SSB phase noise PSD $\mathcal{L}(f)_{\text{injection}}$ of an edge IL oscillator is given as:

$$\mathcal{L}(f)_{\text{injection}} = \mathcal{L}(f)_{\text{free-running}} \cdot \frac{2\pi^2(N-1)(2N-1)}{3f_{\text{ref}}^2 N^2} \cdot \frac{f^2}{1+(f/f_{\text{inj}})^2} \tag{9.1}$$

and the IL bandwidth $\mathcal{L}(f)_{\text{inj}}$ is:

$$f_{\text{inj}} = f_{\text{ref}} \cdot \frac{1}{\pi} \cdot \sqrt{\frac{3N}{2(N-1)}} \tag{9.2}$$

where f_{ref} is the injection reference, N is the multiplication factor, and σ_p^2 is the variance of Gaussian random variables of phase errors.

At high frequency offset, (9.1) can be approximated as:

$$\mathcal{L}(f)_{\text{injection}} = \mathcal{L}(f)_{\text{free-running}} \cdot \frac{2N-1}{N} \approx \mathcal{L}(f)_{\text{free-running}} \cdot 2 \tag{9.3}$$

Equations (9.2) and (9.3) reveal several things. First, at a high value of N, f_{inj} is close to 0.4 times f_{ref}, which is much wider than PLL bandwidth. In addition, the phase noise of an edge IL oscillator is approximately 3 dB higher than that of a free-running oscillator at high offset frequency.

It is necessary to point out that output phase noise due to the reference is increased by 20 log(N). Fortunately, the phase noise of reference clocks is usually much better than the intrinsic oscillator noise over the entire frequency range of interest, which will not affect the overall performance. Details are shown in Figure 9.4.

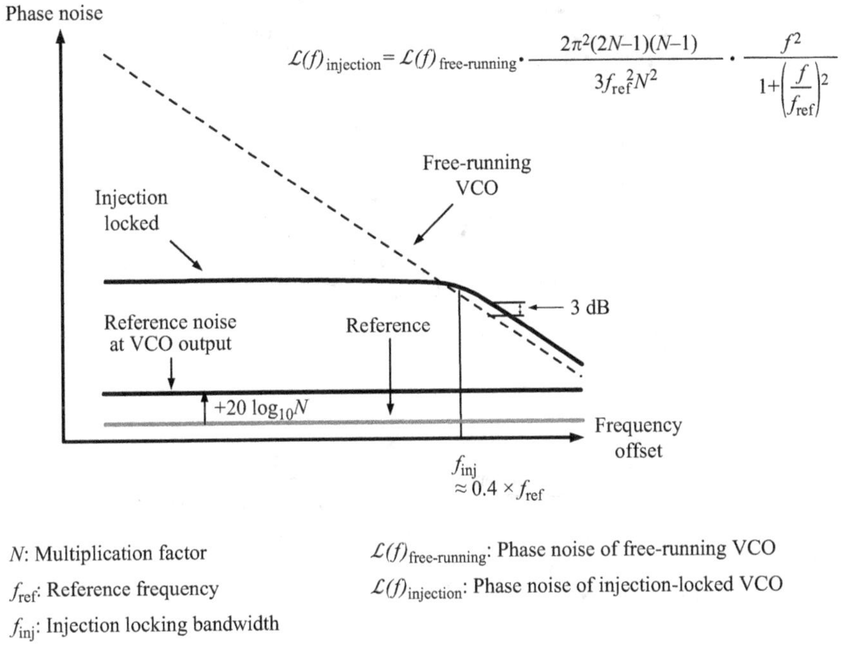

Figure 9.4 Concept diagram of phase noise characteristics

9.3 Design example I: injection-locked PLL with an interpolative phase-coupled oscillator

Figure 9.5 shows the simplified architecture of the proposed PLL. An interpolative phase-coupled oscillator incorporated in the design helps to minimize the output phase imbalance from automatic P&R, and a current-output DAC reduces the power consumption and chip area. A standard-cell-based varactor ensures fine resolution, and a gated edge-injection technique avoids problems related to injection-pulse width.

As for the injection method, conventional pulse injection [19] imposes severe timing requirements on the injection-pulse width, whereas the proposed edge injection does not have strict demands on injection-window width, which considerably simplifies timing design. Regarding the frequency locking, a dual-loop PLL architecture [20,21] is used and improved in this design in order to provide continuous tracking of voltage and temperature variations and avoid timing problems in the conventional ILPLL, as shown in Figure 9.6(b). A detailed block diagram of the proposed PLL with a DAC and an interpolative phase-coupled oscillator is given in Figure 9.7. All circuits that make up the PLL, including the DAC and DCO, are implemented using digital standard cells in the automated design procedure. The oscillation frequencies of the two oscillators are digitized by two 14-bit counters. A digital signed adder/subtractor with a dead zone compares

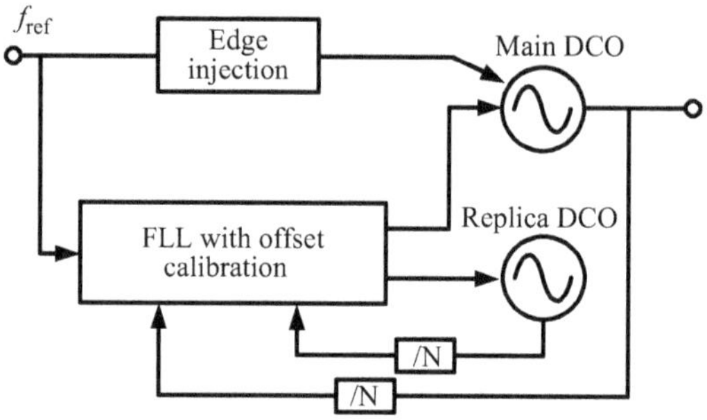

Figure 9.5 Simplified diagram of the proposed fully synthesized PLL

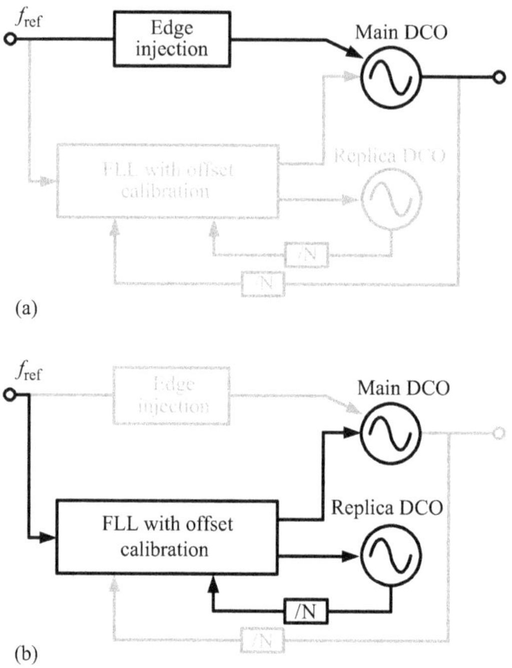

Figure 9.6 (a) Phase-locking and (b) frequency locking of the fully synthesized dual-loop ILPLL

the digitized frequency with a predefined frequency control word and generates a frequency difference, which is then fed to a DLF consisting of a proportional path and an integral path. The designed loop filter bandwidth is around 800 kHz. The filter output determines whether to increase, decrease, or hold the DCO oscillation frequency. Considering that the oscillator oscillates from 0.39 to 1.41 GHz, a 14-bit frequency control scheme, which is composed of 4-bit coarse tuning, 6-bit medium

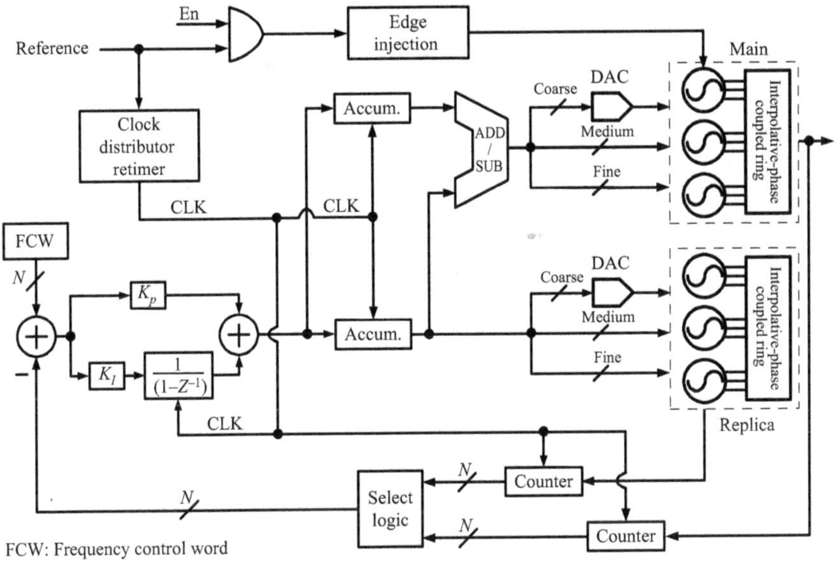

Figure 9.7 Block diagram of the proposed fully synthesizable PLL with a DAC and an interpolative phase-coupled oscillator

tuning, and 4-bit fine-tuning, yields an oscillator least significant bit resolution of 100 kHz. Such a fine resolution helps to achieve a low spur level [21].

9.3.1 Interpolative phase-coupled oscillator

As shown in Figure 9.8, there are various standard-cell-based DCO topologies, each of which entails different pros and cons. Conventional standard-cell-based DCO, as shown in Figure 9.8(a), adopts a digital varactor [22] for frequency tuning. While benefiting from high linearity, this method suffers from large power consumption and limited resolution. A path-selection method in Figure 9.8(b) is proposed as the coarse tuning in synthesized oscillators [23] in order to reduce power consumption. However, this method suffers from poor linearity. In order to maintain the control code resolution and frequency linearity, the oscillator is designed to operate with coarse-tuning circuitry controlled by a DAC and medium-/fine-tuning circuitry controlled by digital varactors, as shown in Figure 9.8(c). The circuit will be revisited later for implementation details.

Recall that the PLL calibration system operates in three phases [21]. The first two phases are used upon startup for compensating the frequency offset between the main and replica oscillators due to process variations at a certain temperature. The third calibration phase is continuously operating in order to track voltage and temperature variations. However, a frequency mismatch between the main and replica oscillators might also occur, because the frequency of the two automatic P&R oscillators does not change equally with respect to the temperature effect. Such a frequency mismatch is not compensated in the third calibration phase [21],

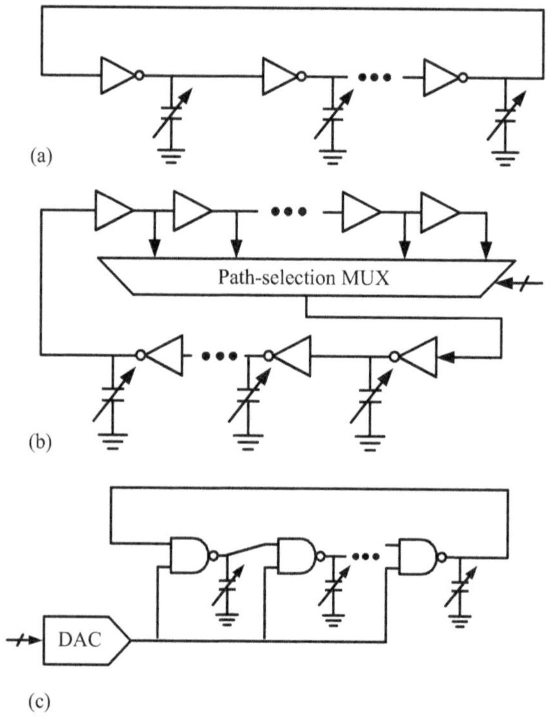

Figure 9.8 Simplified diagram of synthesizable digitally controlled oscillators (DCOs) using (a) varactor, (b) varactor and path selection, and (c) varactor and DAC for frequency tuning

and, thus, there will be an increase in the phase error of the main oscillator under injection locking. In order to relax this problem, an interpolative phase-coupled oscillator is developed, which has its roots in the concept presented in a previous study [24]. Figure 9.9 illustrates the proposed oscillator architecture that is constructed by three three-stage oscillators used in both main and replica VCOs. Figures 9.10(a), (b), and (c) show a block diagram of the separated DCOs. Figure 9.10(d) shows a block diagram of the interpolative phase-coupled ring. The phase interpolator is constructed using two inverters. If the oscillator does not have an auxiliary ring, the phases of the three oscillators are independent of each other. The auxiliary ring couples every main oscillator to ensure an accurate oscillation phase. As a result of its internal feedback and feed-forward control by the coupled interpolators, the phase difference within the ring and between all adjacent rings will remain fixed with time, leading to balanced output phases.

Simulations confirm the effectiveness of the proposed interpolative phase-coupled oscillator. Assuming the simulation conditions of a reference injection from a 200-MHz source, a multiplication ratio of 10, and 0.5% mismatch in free-running frequency, as shown in Figure 9.11, the maximum phase error is improved

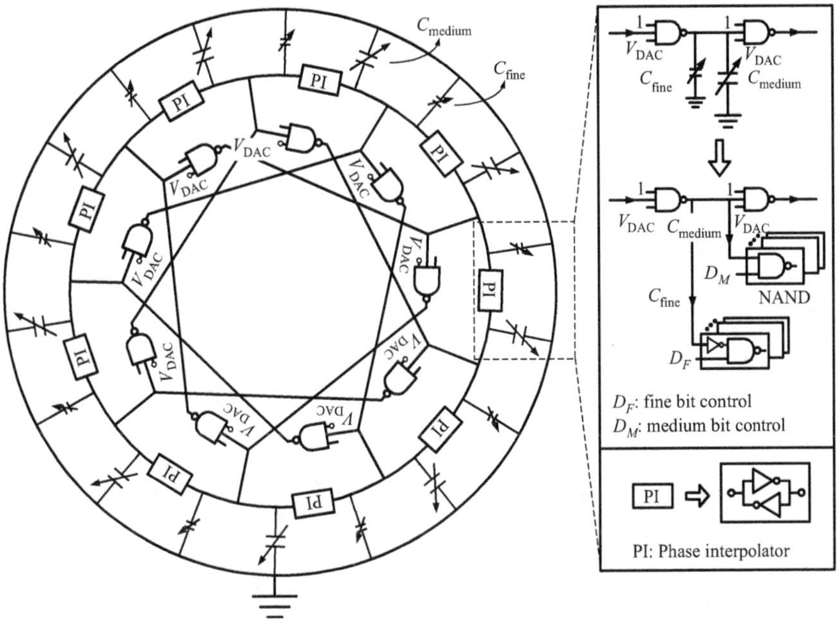

Figure 9.9 Block diagram of the interpolative phase-coupled oscillator

from 6% in the case of a conventional three-stage ring oscillator to 2.8% in the case of the proposed interpolative phase-coupled oscillator, where the phase error is defined as the deviation in percentage from its mean value.

9.3.2 Current-output DAC converter

Basically, a DAC can be used to control the coarse-tuning part of the oscillator. If the control code of the DAC is changed, then the oscillator frequency can be changed accordingly. DAC control has the added advantage of balancing the loading at each node. Figure 9.12(a) shows a simple inverter that can be considered as a 1-bit voltage-output DAC. If D_0 is 0, then the output voltage V_{out} becomes V_{DD}. Similarly, this concept can be extended to a two-bit voltage-output DAC by voltage interpolation. If D_0D_1 is 11, then V_{out} becomes 0. If D_0D_1 is 10 or 01, then V_{out} becomes $V_{DD}/2$, and if D_0D_1 is 00, then V_{out} becomes V_{DD}. By further extending this concept, a 4-bit DAC with binary coding can be designed, as shown in Figure 9.12(c). Thus, an ideal voltage-linear DAC can be obtained by voltage interpolation. However, considering that the interpolative phase-coupled oscillator is based on current-starving topology, a current-linear DAC rather than a voltage-linear DAC is desired.

In this design, an equivalent current-output DAC constructed entirely from standard CMOS NAND gates is proposed as the coarse-tuning circuitry. The proposed implementation contributes to the reduction in the overall power consumption and chip area without sacrificing phase balance. Tristate buffers/inverters

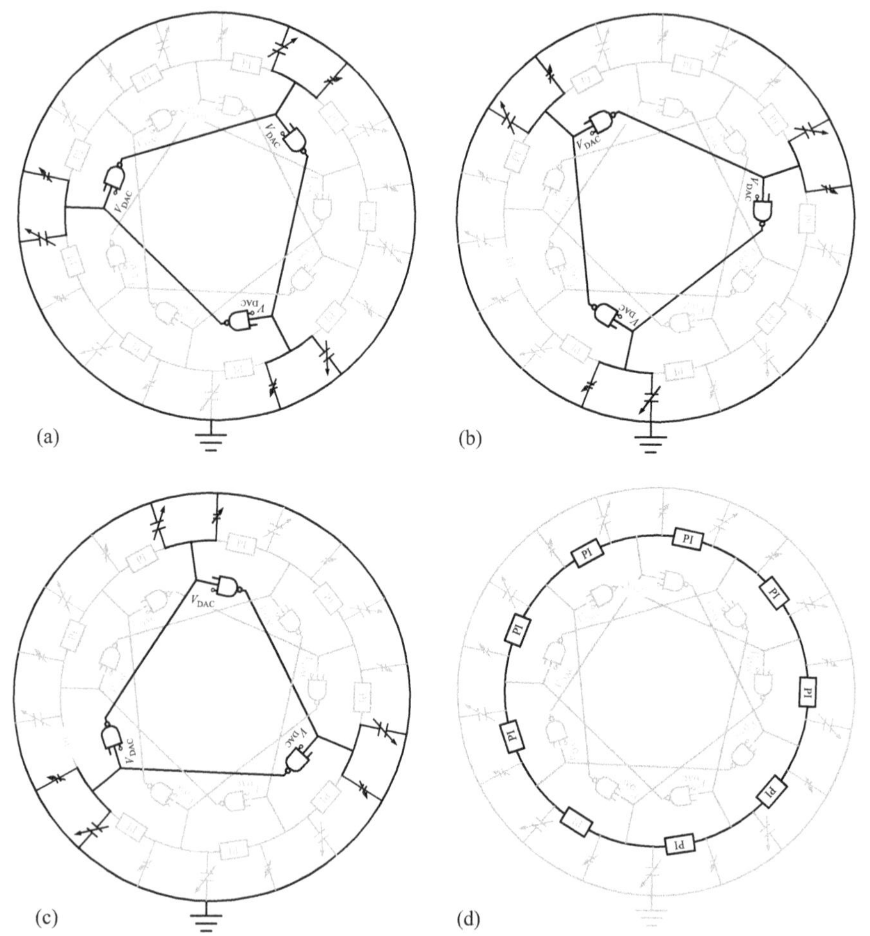

Figure 9.10 Block diagram of (a), (b), (c) DCO built by NAND gates and digital varactors, and (d) phase-coupled ring

[4–6] are not used because they are sometimes not included in standard-cell libraries. For ease of understanding, Figure 9.13 shows a conceptual diagram of a 4-bit current-output DAC constructed from five two-input NAND gates. The proposed DAC consists of a PMOS-current-source array and an NMOS current mirror. The PMOS-current-source array is built by driving one of the two NAND gate inputs by a digital control word, and the other NAND gate input is driven by D4 and by connecting the outputs of four NAND gates. The NMOS current mirror is built by connecting one input of an NAND gate to its output and the other input to logic HIGH. As shown in Figure 9.14, the proposed current-output DAC is combined with a current-starved ring oscillator using an NAND logic gate as its delay cell to form a DCO.

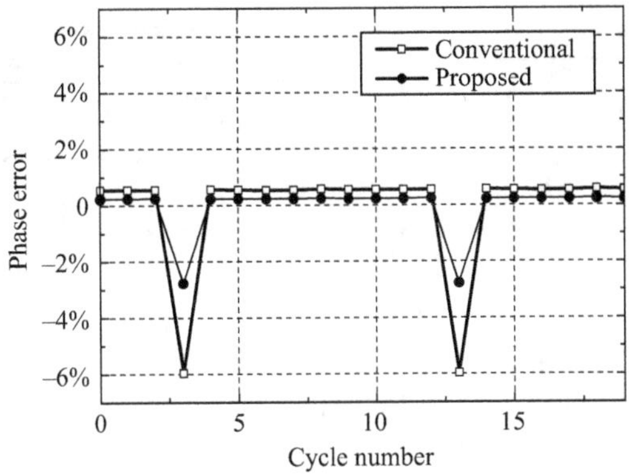

Figure 9.11 Simulated phase error in the case of the conventional and proposed oscillator

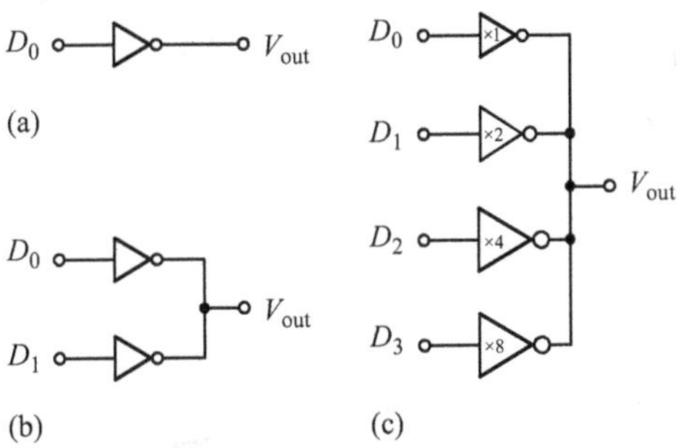

Figure 9.12 Conceptual diagrams of (a) one-bit, (b) two-bit, and (c) four-bit binary coding voltage-linear DACs

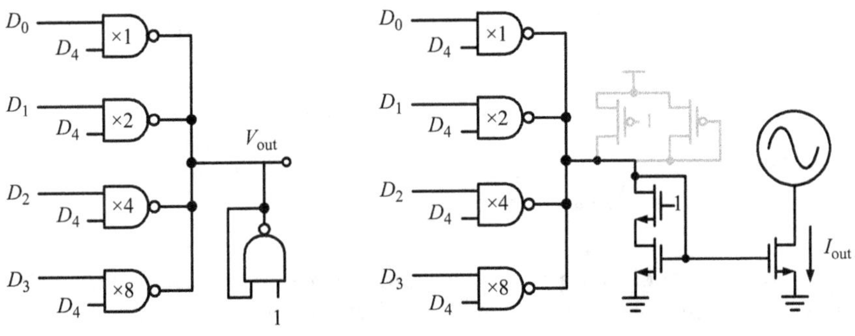

Figure 9.13 Conceptual diagram of the proposed current-output DAC

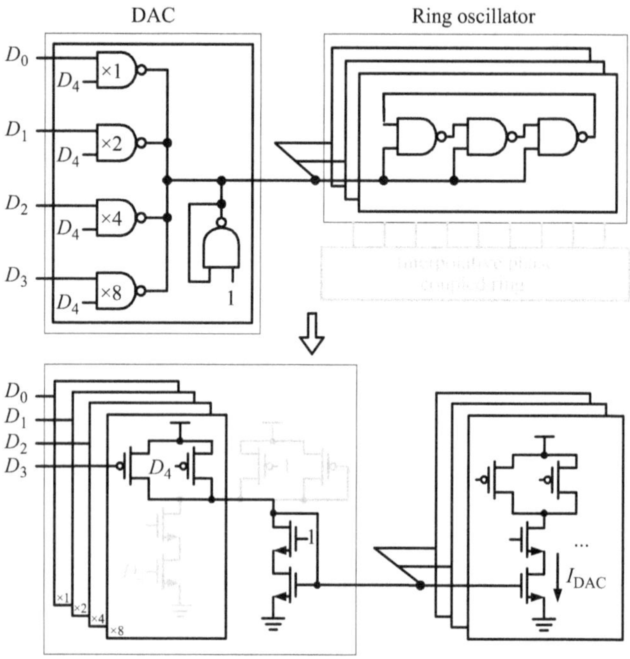

Figure 9.14 Block diagram of the proposed synthesizable DAC with a current-linear output

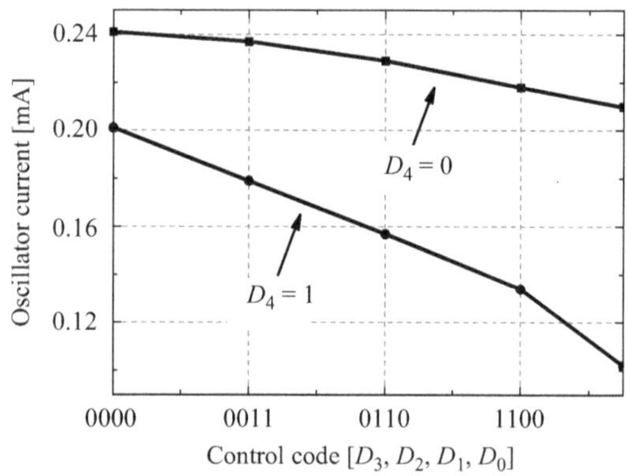

Figure 9.15 Simulated current consumption of the oscillator using the current-output DAC

Several simulations are performed in order to verify the effectiveness of the proposed current-output DAC. Figure 9.15 shows the simulated current consumption of the current-starved oscillator using the proposed current-output DAC. Figure 9.16 shows the simulated current consumption and frequency of the

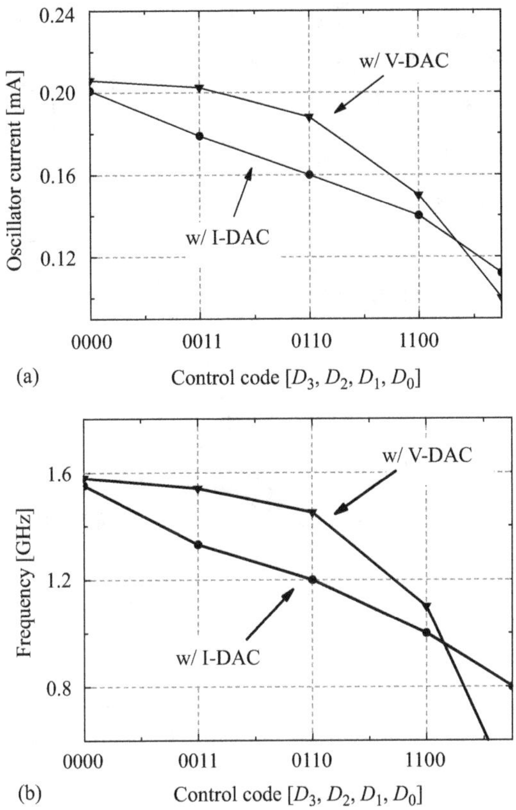

Figure 9.16 Comparison of (a) current consumption (b) frequency of the oscillator using the current-output DAC and voltage-output DAC when D4 connects to logic high

oscillator using current-output and voltage-output DACs when D4 is connected to logic HIGH. These two graphs show that the proposed current-output DAC is a better option in terms of oscillator frequency linearity. The simulated DAC power consumption is less than 60 μW.

9.3.3 Standard-cell-based varactor

In order to achieve better resolution for reducing spurs, medium- and fine-tuning circuitry using a standard-cell-based varactor are required. Conventionally, a standard-cell-based varactor [22] is constructed using an NAND or not-or (NOR) gate. In medium-tuning circuitry, a digitally controlled varactor using an NAND logic gate is adopted in order to introduce a capacitance difference between the two delay stages. As shown in Figure 9.15(a), the marked NMOS transistor produces this capacitance difference using different values of D_m. In general, finer resolution can be achieved by reducing the size of the marked NMOS transistor or increasing

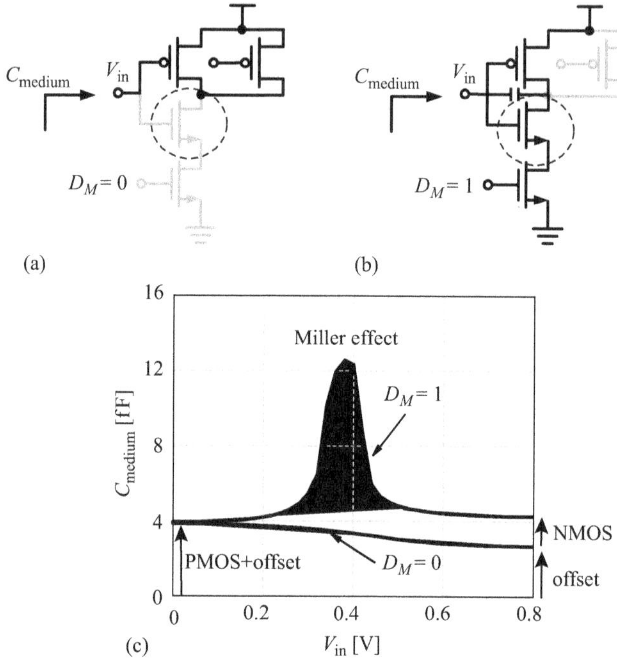

Figure 9.17 (a) Equivalent circuit of the medium-tuning capacitor and (b) simulated medium-tuning capacitance over V_{in}

the size of the logic gate in the previous stage. However, the transistor size is limited in standard-cell-based design. Therefore, an improved standard-cell-based varactor is developed for the fine-tuning circuitry.

Figure 9.17(b) shows the simulated medium capacitance C_{medium} with respect to V_{in} at different D_m. Interestingly, but not surprisingly, the equivalent input capacitance C_{medium} increases due to the Miller effect, which happens at around $V_{in} = 0.4$ V for $D_m = 1$. In order to investigate the possibility of achieving finer resolution using the Miller effect, a test bench is adopted, as shown in Figure 9.18(a). The simulation results in Figure 9.18(b) indicate that the Miller effect sensitivity can be varied by varying V_{out}, which cannot be controlled directly. However, if a varactor is placed at the output node, as shown in Figure 9.19, the capacitance difference will cause a transient variation of V_{out}, which in turn, changes the Miller effect sensitivity.

Based on this concept, the fine-tuning circuitry is realized by another type of digitally controlled varactor using inverters and NAND gates. The digitally controlled NAND gate introduces a capacitance difference at the inverter output node, which alters the rising and falling slopes, thereby changing the Miller effect sensitivity. Thus, the effective capacitance observed from the ring oscillator is also changed. This capacitance difference is used for the fine-tuning stage.

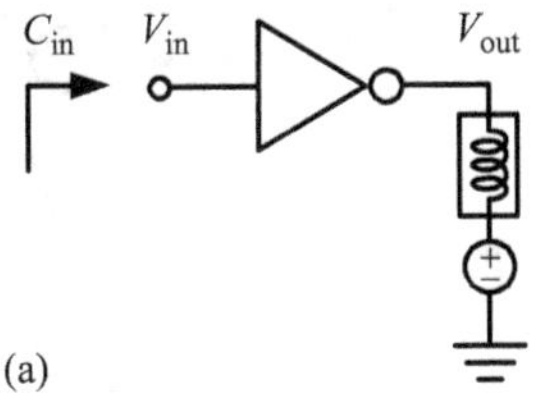

(a)

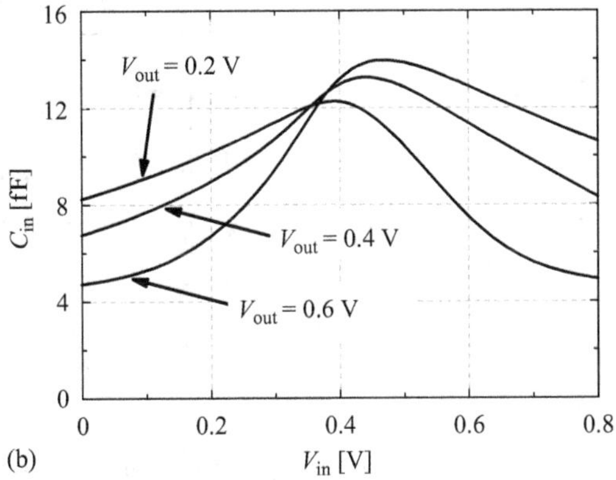

(b)

Figure 9.18 *(a) Test bench for the Miller effect sensitivity and (b) simulated input capacitance with respect to V_{in} at various V_{out}*

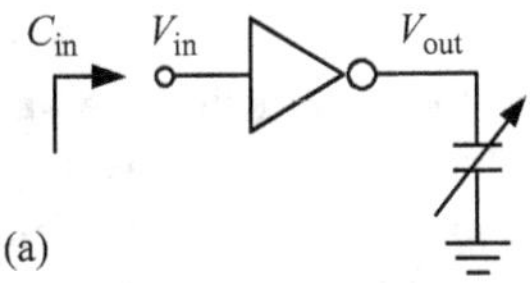

(a)

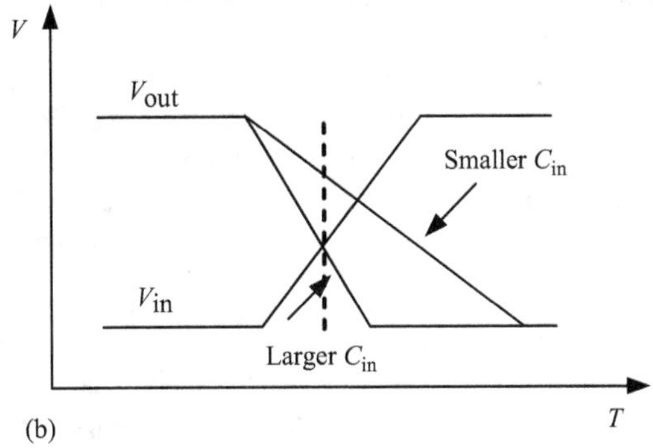

(b)

Figure 9.19 *(a) Equivalent circuit of the fine-tuning capacitor and (b) conceptual diagram of the time-domain waveform of V_{out} and V_{in}, indicating that a transient variation in V_{out} can make a fine capacitance difference in C_{in}*

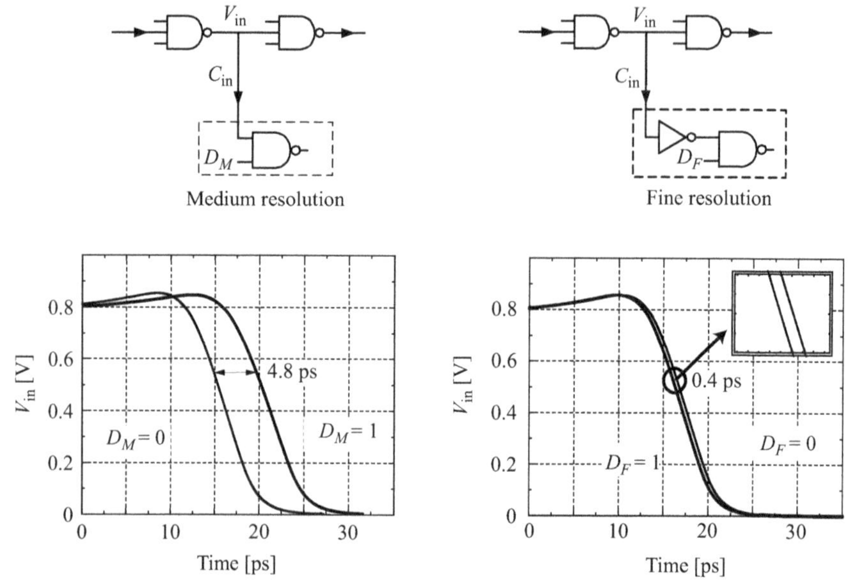

Figure 9.20 Simulated input capacitances with respect to input voltage V_{in} and the resulting delay resolutions for the medium-tuning and the proposed fine-tuning varactors

Figure 9.20 shows the input capacitances with respect to the input voltage, V_{in}, and the resulting delay resolutions for the medium-tuning and the proposed fine-tuning varactors. A digitally controlled varactor using NAND logic gates is adopted in the medium-tuning circuitry, providing a 4.8 ps delay resolution. As for the fine-tuning circuitry, the simulation results show that a 0.4-ps delay resolution is achieved. The DCO nonmonotonic response would introduce the unlock phenomenon in the frequency-locked loop and increase the spur level. Although the monotonic characteristic of the medium-tuning circuitry was studied in a previous study [22], the monotonic response of the proposed fine-tuning circuitry is verified by the post-layout simulation. Figure 9.21 shows that the simulated differential nonlinearity ranges from −0.6 to 1.5 least significant bits and the DCO frequency varies from 1.234 to 1.237 GHz, which proves the monotonic response of the synthesized DCO.

9.3.4 Gated edge injection

Injecting a stream of narrow pulses into the oscillator is widely used to improve its jitter performance by resetting the oscillator at every reference cycle [19]. However, the injection-pulse width requires additional calibration in order to guarantee its robust operation over environmental variations, because an excessively narrow pulse width causes a failure to lock, or an overdesigned pulse width causes a strong periodical disturbance to the oscillator phase and amplitude, degrading its deterministic jitter, as shown in Figure 9.22. The conventional method uses a variable setting on the

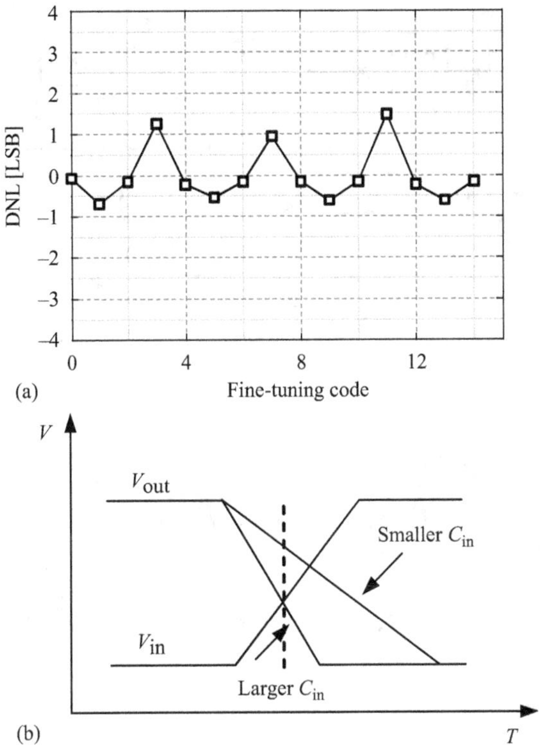

Figure 9.21 *Post-layout simulated (a) differential nonlinearity (DNL) and (b) DCO frequency with respect to fine-tuning code*

pulse generator to manually adjust the injection-pulse width, which inevitably lacks robust operation over process–voltage–temperature (PVT) variations. Performing background calibration on the pulse-injection technique requires severe timing design for injection-pulse width, which is not feasible for synthesizable analog circuits.

As shown in Figure 9.23, a gated edge-injection technique is proposed in order to avoid the injection-pulse width issue [25]. Assuming that the oscillator frequency is N times higher than the reference frequency, an inverted injection-window signal forces the oscillator to cease oscillation at the rising edge of the injection-window signal. Then an injection signal with clean edges is forwarded to the oscillator, which replaces the noisy edge, V_y, thus resetting the jitter accumulated for several cycles. Finally, the oscillator resumes normal operation after the falling edge of the injection-window signal, with its phase aligned to the phase of the injection signal. Compared to the conventional pulse injection requiring severe timing design of the injection-pulse width, the proposed gated edge injection does not have strict demands on its injection-window width, which virtually eliminates the timing design problems. The entire phase replacement and alignment can be performed in one reference cycle.

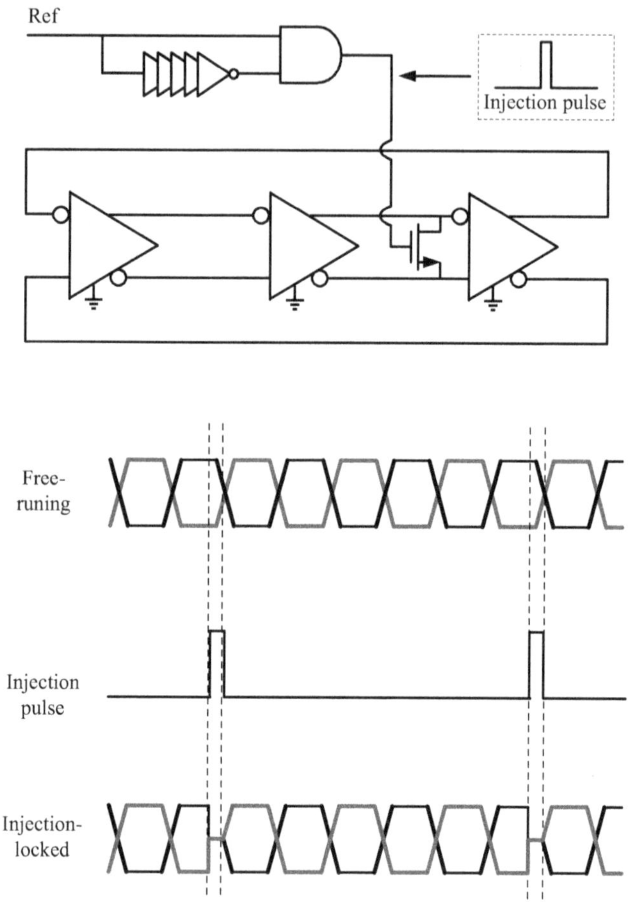

Figure 9.22 Block diagram and locking transient of the conventional pulse injection locking

9.3.5 Measurement results

The proposed fully synthesizable PLL is fabricated in a 65-nm digital CMOS technology. Figure 9.24 shows a micrograph of the PLL, which occupies a layout area of only 110 μm × 60 μm. The phase noise is evaluated using a signal source analyzer (Agilent E5052B), and the spectrum is measured using a spectrum analyzer (Agilent E4407B). The measured frequency tuning range of the PLL is 0.39–1.41 GHz. At a 0.9-GHz output, the power consumption is 780 μW, excluding the output buffers from a 0.8-V power supply. Figure 9.25 shows the measured phase noise at 0.9 GHz output using a 150-MHz reference clock. The phase noise in free-running mode is measured in the third calibration phase with reference injection disabled. Considering (9.1) and assuming a 150-MHz reference clock, a multiplication factor of 6, and an output frequency of 900 MHz, the theoretical phase

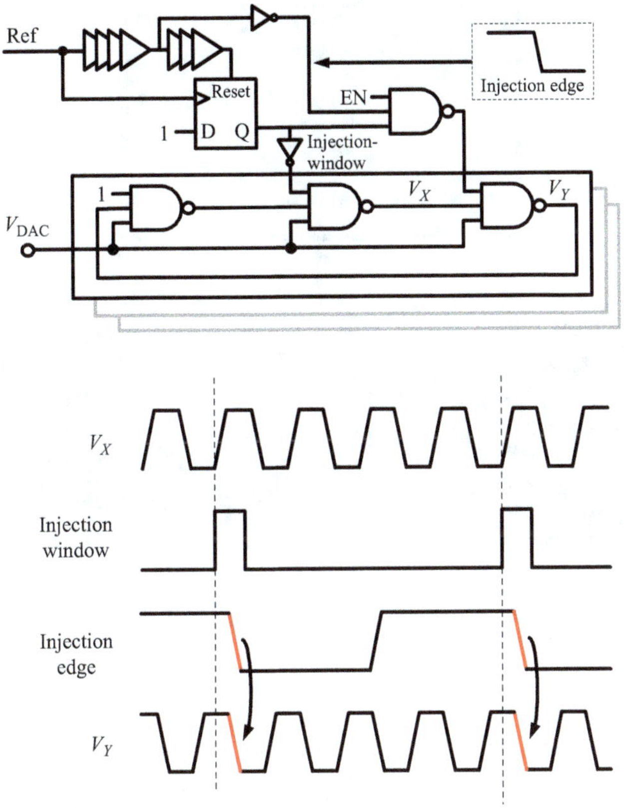

Figure 9.23 Block diagram and locking transient of the proposed gated edge injection locking

noise is obtained. Figure 9.25 shows that approximately a 3-dB difference is observed between the measured and theoretical calculated results. The phase noise maps to a 1.7-ps jitter when integrated from 10 kHz to 40 MHz. All of the abovementioned measurements were performed at room temperature.

In order to assist in the evaluation of spur level in edge injection, an ILPLL using a pulse-injection method is chosen for comparison. The pulse width is tunable in order to achieve optimum performance for fair comparison. As shown in Figure 9.26, both injection methods have a similar first reference spur, which is approximately −41 dBc below the carrier. As for the second reference spur at $2 \times f_{\mathrm{ref}}$, the edge-injection PLL has a −51-dBc spur level, while the pulse-injection PLL achieves a −42-dBc spur level. The reason for this is that the edge-injection method introduces a weaker periodical disturbance by replacing the oscillator noise edge.

As mentioned earlier, the fully synthesized PLL achieves 1.7 ps root-mean-square (RMS) jitter while consuming 780 μW of power, resulting in a figure of merit (FoM) of −236.5 dB, where FoM is defined as $10\log\left[(\sigma_t/1\mathrm{s})^2 \cdot (P_{\mathrm{DC}}/1\mathrm{mW})\right]$, σ_t is

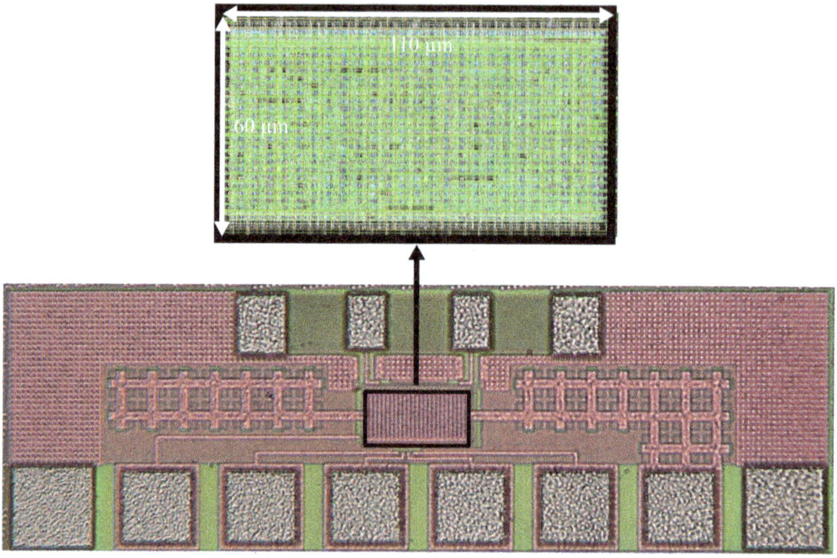

Figure 9.24 Die microphotograph

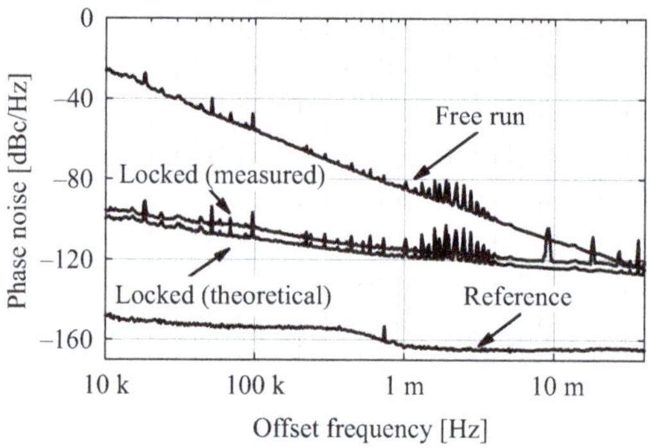

Figure 9.25 Measured phase noise characteristic at a carrier frequency of 0.9 GHz

the integrated jitter, and P_{DC} is the DC power consumption. The hierarchically synthesized PLL achieves 2.32 ps RMS jitter for a power consumption of 640 μW, resulting in an FoM of −234.6 dB at the same output frequency. The difference in power consumption and jitter can be explained as follows. The oscillator in the fully synthesized PLL suffers from more parasitic capacitances from automatic P&R, which leads to a lower oscillation frequency compared to that in the hierarchical synthesized PLL for the same power budget.

Thus, a larger current is necessary for the oscillator in the fully synthesized PLL in order to obtain the same oscillation frequency, which contributes to

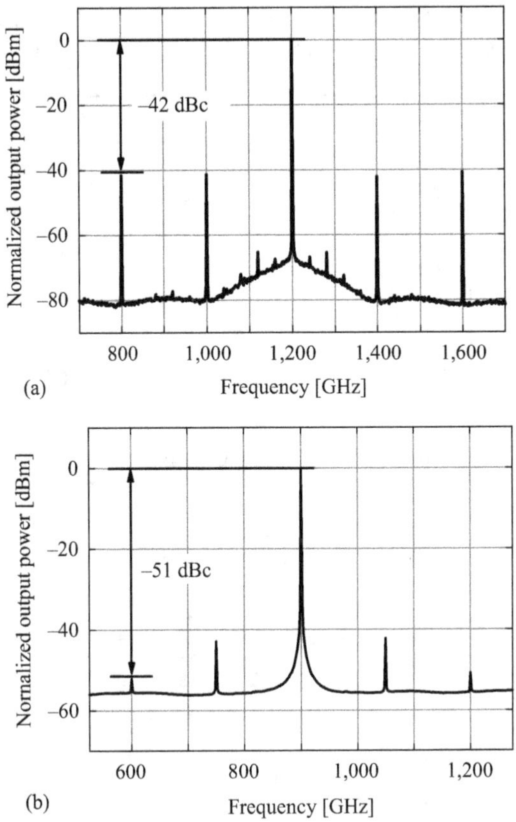

Figure 9.26 Measured second spur level of (a) conventional pulse injection and (b) edge injection at a multiplication factor of 6

achieving lower phase noise. Although some differences are observed in power consumption and jitter due to parasitic capacitances from automatic P&R, both fully synthesized and hierarchically synthesized PLLs achieve similar performances in terms of FoM, which demonstrates that fully synthesized design does not degrade the overall performance. On the other hand, the output jitter performance might vary from one layout synthesis procedure to another due to variation in parasitic capacitances, as mentioned earlier. DCO phase noise can be simulated. Then the jitter of the PLL can be calculated using (9.1), (9.2), and (9.3). Basically, the overall jitter performance is determined primarily by the DCO phase noise, which can be improved by adding more capacitances and increasing the driving current.

Table 9.2 compares the results for the synthesizable PLLs of previous studies [4–6] and that proposed herein. The proposed synthesizable PLL achieves the best performance in terms of power, jitter, and area, as compared to previously proposed synthesizable PLLs, even those implemented using advanced technology.

Table 9.2 Performances of the proposed fully synthesizable phase-locked loop and state-of-the-art synthesized phase-locked loops

	This work	**[4]**	**[6]**	**[5]**
Frequency [GHz]	0.39–1.41	1.5–2.7	0.25–1.65	0.4–0.46
Ref. [MHz]	40–350	10	25	40.3
Power [mW]	0.78@900 MHz	13.7@ 2.5 GHz	3.1@250 MHz	2.1@403 MHz
Area [mm^2]	**0.0066**	0.042	0.032	0.1
Normalized area	1	6.36	4.84	15.15
Integ. jitter [ps]	1.7[a]	N.A.	30	N.A.
Jitter RMS [ps]	2.8	3.2	N.A.	13.3
FOM [dB]	**−236.5**	−218.6[b]	−205.5	−214[b]
CMOS Tech.	65 nm	65 nm	28 nm	65 nm
w/custom cells?	No	No	Yes	Yes
Topology	Injection locking	TDC based	TDC based	TDC based

[a]With a 150-MHz reference clock.
[b]FOM is calculated based on the RMS jitter.
The bold values here highlight the smallest area and the best FOM

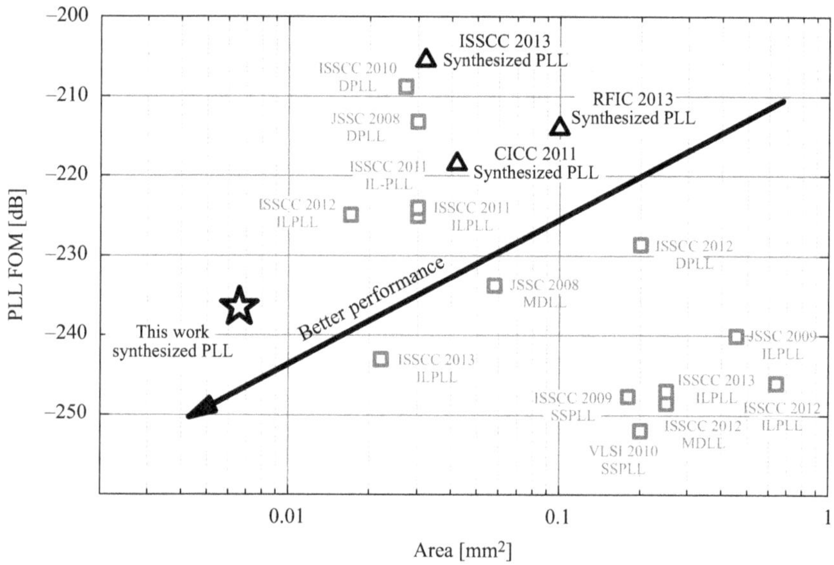

Figure 9.27 Performance comparison of the proposed synthesizable PLL and various state-of-the-art synthesizable PLLs

Figure 9.27 shows the performance-area trade-off for the proposed fully synthesized PLL and other synthesized PLLs. The proposed fully synthesized PLL achieves the best performance-area trade-off in terms of FoM and area when compared to other approaches.

9.4 Design example II: injection-locked PLL with bang-bang phase detector-based frequency and phase calibration

The synthesizable PLL in Section 9.3 used a replica oscillator to track the frequency mismatch. However, the replica oscillator incurs additional power consumption. In addition, the main oscillator and the replica oscillator inevitably have mismatches due to both device statistical variations and automatic P&R. Moreover, the different working conditions of the two oscillators also introduce a frequency mismatch, as explained in Section 9.3.1. Even with interpolative DCO, a nonzero frequency mismatch persists in the oscillators. Also, the phase relationship between the injection signal edge and the DCO edge is unsynchronized with only the frequency-locked loop.

Therefore, in this design, a frequency-/phase-synchronized ILPLL is proposed. A novel DCO architecture is proposed to realize low-power operation. By using a fully symmetrical multiplexer (MUX) and an SS-BBPD, the proposed DCO can realize simultaneous injection and phase comparison without using an additional replica oscillator, thus significantly reducing power consumption. In order to minimize the reference spur, a self-clocked nonoverlap update scheme is proposed in order to reduce the reference spur caused by digital logic clocking. In addition, a slope equalization technique is proposed for the SS-BBPD, which minimizes the slope-induced delay offset.

9.4.1 PLL architecture and noise analysis

The architecture of the proposed fully synthesizable PLL is shown in Figure 9.28(a). Three paths exist in the PLLs, the injection-lock path, the frequency-lock path, and phase-lock path. The injection-lock path has wide bandwidth and could effectively suppress the high phase noise of ring oscillators. However, ILPLL is sensitive to frequency and phase offset, and thus proper frequency and phase lock are required to mitigate PVT variations [26–28]. In this PLL design, both the frequency-lock path and phase-lock path are implemented in a digital domain, making them area compact and synthesizable. The counter-based frequency-lock path controls the DCO coarse and medium stages. In addition, the phase-lock path uses an SS-BBPD and a DLF to control the DCO fine stage. In order to further enhance the effective resolution, a delta–sigma modulator, which is clocked by a divided DCO output, is used to quantize the fractional bits.

Unlike the FLL-based frequency calibrated ILPLL described in Section 9.3, the SS-BBPD phase-lock path and the injection path work simultaneously. However, the total output phase noise is jointly determined by the injection path and the phase-lock path in order to quantize the noise contribution of each noise source. The linearized noise model of proposed PLL is shown in Figure 9.28(b). The input reference is 100 MHz, and the frequency multiplication ratio (FMR) N is 10. The input reference phase is denoted as $\Theta_{n,\mathrm{REF}}$. For the reference injection path, a model similar to that in [29] is adopted with the phase realignment factor β set to 1.

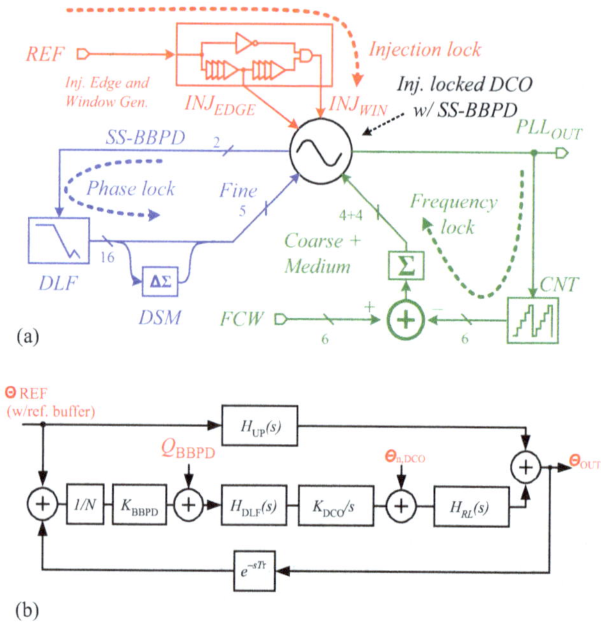

Figure 9.28 *(a) System architecture and (b) linearized noise model of the proposed fully synthesizable ILPLL*

This model is described by:

$$H_{\mathrm{UP}}(s) = N * e^{-sT_r/2} * \mathrm{sinc}\,(\omega T_r/2) \tag{9.4}$$

$$H_{\mathrm{RL}}(s) = 1 - e^{-sT_r/2} * \mathrm{sinc}(\omega T_r/2) \tag{9.5}$$

where H_{UP} represents the up-conversion of the reference noise to the PLL output, and H_{RL} is the effect of phase resetting by injection. The symmetrical SS-BBPD and MUX operation is modeled by the time delay e^{-sT_r}. The SS-BBPD is modeled as a linear gain plus output quantization noise [30,31], as:

$$K_{\mathrm{BBPD}} = \sqrt{\frac{2}{\pi}} \times \frac{1}{2\pi f_{\mathrm{ref}} \sigma_t} \tag{9.6}$$

$$Q_{\mathrm{BBPD}}(f) = \left(1 - \frac{2}{\pi}\right)\frac{2}{f_{\mathrm{ref}}} \times \left(\mathrm{sinc}\left(\frac{f}{f_{\mathrm{ref}}}\right)\right)^2 \tag{9.7}$$

where σ_t is the time-domain RMS jitter at SS-BBPD input. In this noise model, σ_t is approximated by PLL output jitter. The DLF with 1 MHz bandwidth is modeled as its continuous-time equivalent $H_{\mathrm{DLF}}(s) = K_P + K_I/s$, and the DCO is modeled as a

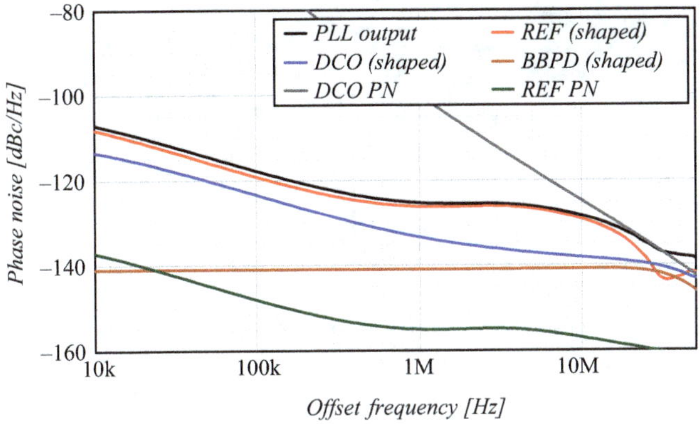

Figure 9.29 Total output noise and noise contribution from each noise source of the proposed ILPLL based on the linear noise model

continuous-time integrator K_{DCO}/s. The open-loop gain $T(s)$ and noise transfer functions are as follows:

$$\mathrm{T\&}(s) = e^{-sT_r} * \frac{K_{\mathrm{BBPD}}K_{\mathrm{DCO}}}{Ns} H_{\mathrm{DLF}}(s)H_{\mathrm{RL}}(s) \tag{9.8}$$

$$\mathrm{NTF}_{\mathrm{REF}}(s) = H_{\mathrm{UP}}(s) + \frac{T(s) \times e^{sT_r}}{1 + T(s)} \tag{9.9}$$

$$\mathrm{NTF}_{\mathrm{BBPD}}(s) = \frac{H_{\mathrm{DLF}}(s)H_{\mathrm{RL}}(s)K_{\mathrm{DCO}}/s}{1 + T(s)} \tag{9.10}$$

$$\mathrm{NTF}_{\mathrm{DCO}}(s) = \frac{H_{\mathrm{RL}}(s)}{1 + T(s)}$$

Figure 9.29 shows the total output noise of the PLL and the noise contributions. Both DCO and reference phase noise are related to the measurement results. The modeled DCO phase noise is −100 dBc/Hz at a frequency offset of 1 MHz, with a flicker noise corner of 3 MHz. The integrated reference jitter from 10 kHz to 10 MHz is 0.25 ps, including the jitter of the reference buffers.

9.4.2 Highly linear DCO with self-clocked nonoverlap update

In order to satisfy the power, phase noise, and tuning range requirements, a multistage ring DCO is proposed, as shown in Figure 9.30. The implementation of a multistage IL ring DCO is shown in Figure 9.30. The DCO coarse stage features a delay cell based on path selection, whereas the medium and fine stages use NAND3-based digital varactors to tune the delay. The proposed varactor applies

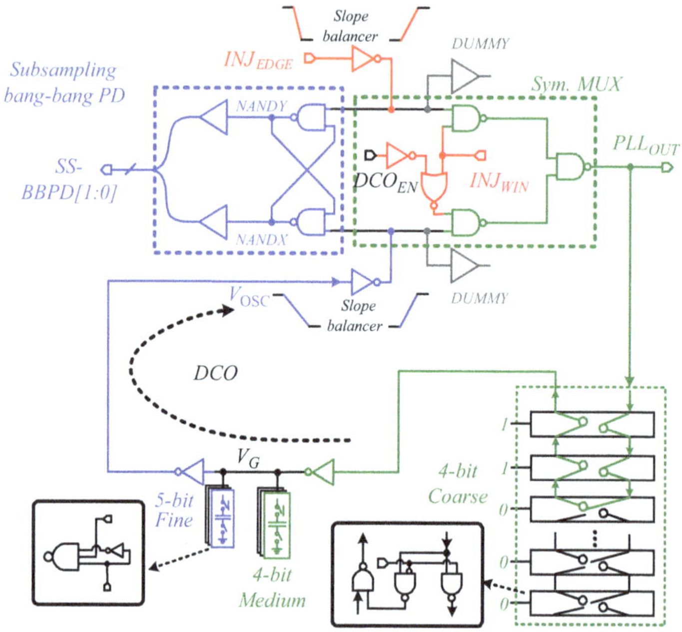

Figure 9.30 Implementation of the proposed injection-locked DCO with a subsampling bang-bang phase detector (SS-BBPD)

two complementary control signals to control the lowest and highest NMOSs in the NAND3 cell.

In order to reduce the digital clocking-induced spur, a novel self-clocked nonoverlap update scheme is proposed, as shown in Figure 9.31. The capacitance of a digital varactor is a function of both load voltage and the digital control code. Thus, the control code update time has a significant effect on capacitance discontinuity. In Figure 9.32, two fine-stage varactors have the same NAND3 implementation and control code but with different update timings. In the conventional update, the control code is synchronized to a global signal, such as PLL_{OUT}. Since the delay between PLL_{OUT} and V_G depends on the DCO coarse stage delay and PVT conditions, there is a possibility that the control code updates at nonzero V_G. This will induce a capacitance discontinuity and degrade the reference spur, as shown in Figure 9.33. In contrast, in the proposed self-clocked nonoverlap update scheme, the fine-stage control code is locally synchronized to its output V_G. Thus, the control code is guaranteed to update when V_G is zero. In this way, capacitance discontinuity is eliminated, as shown in Figure 9.33. As an additional bonus, the proposed update scheme eliminates short circuit currents in the NAND3 varactor, which enables a power reduction of 30% as compared to the conventional NAND2 varactor while having the same resolution.

The timing relationship among REF, V_{OSC}, and V_G is shown in Figure 9.31. With falling edge injection, the falling edges of REF and V_{OSC} are aligned, and the

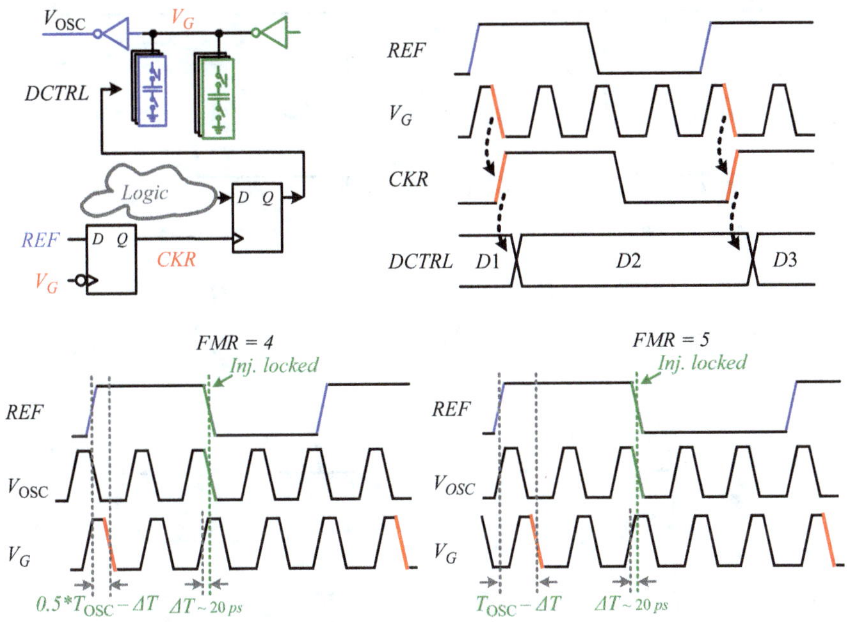

Figure 9.31 *Circuit implementation of the proposed self-clocked nonoverlap update scheme*

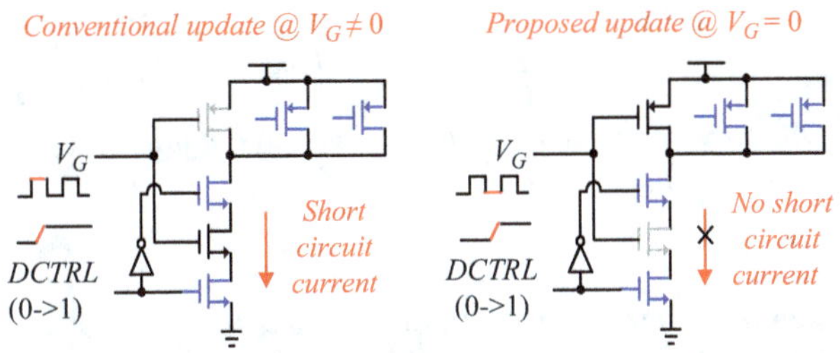

Figure 9.32 *Time-domain operation of the proposed self-clocked nonoverlap update scheme*

rising edge of V_G is $\Delta\tau$ ahead, where $\Delta\tau$ is one inverter delay (approximately 20 ps). The timing margin is $0.5 \cdot T_{OSC} - \Delta\tau$ and $T_{OSC} - \Delta\tau$ for even and odd FMRs, respectively, where T_{OSC} is the DCO period. The worst-case additional delay is less than one DCO period, which has a negligible effect on stability. Moreover, in order to mitigate layout-induced mismatches and variations, relative placement design constraints are applied to standard digital P&R, and a reasonably good linearity was achieved, as shown in Figure 9.34.

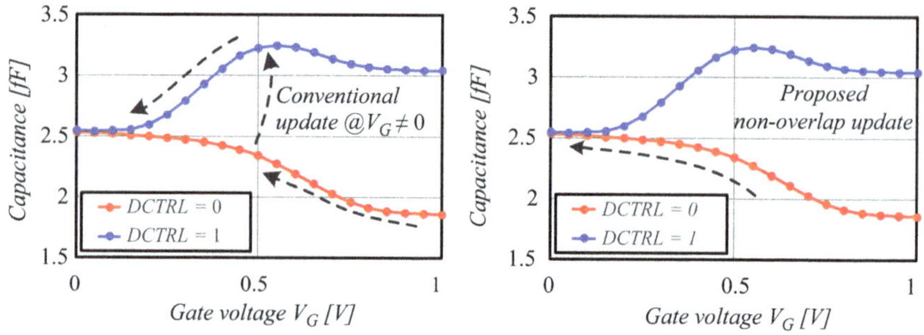

Figure 9.33 Varactor capacitance characteristic of the proposed self-clocked nonoverlap update scheme

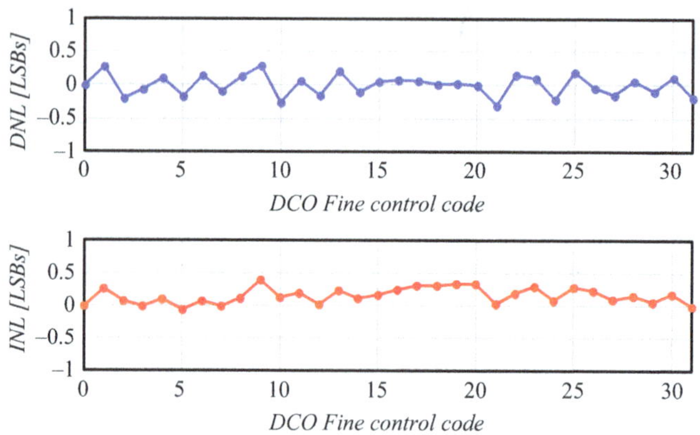

Figure 9.34 Digitally controlled oscillator fine-stage nonlinearity obtained from post-layout simulation

9.4.3 Fully symmetrical low-offset MUX and SS-BBPD design

Since the ILPLL is sensitive to delay mismatch between the injection path and the phase detection path [28,32], single-ended fully symmetrical MUX and SS-BBPD are integrated into the DCO in order to simplify routing and reduce the mismatch. In order to minimize path mismatch between the MUX and the SSPD, the two parts are designed jointly and are optimized for low delay offset. As shown in Figure 9.30, both the SS-BBPD and the MUX are implemented with NAND2 standard cells so as to ensure symmetry. In order to minimize the slope-induced delay offset in the SS-BBPD, a pair of slope balancers are placed on the V_{OSC} and the INJ_{EDGE} path. In addition, two dummy loads are added in order to reduce the percentage of capacitance variation with different INJ_{WIN} conditions. The layout mismatches and variations are minimized with compact area and

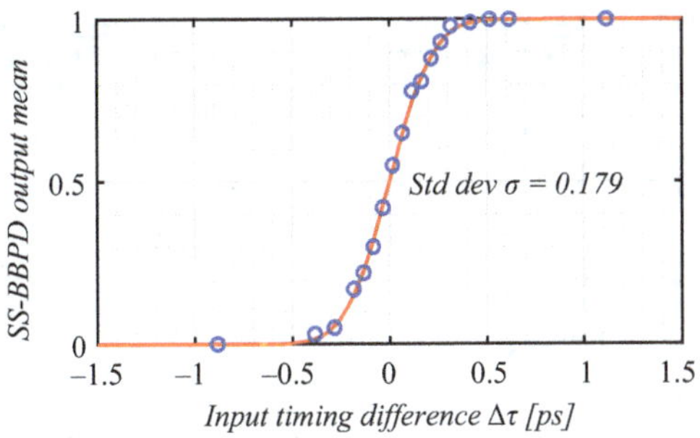

Figure 9.35 Subsampling bang-bang phase detector input offset obtained from Monte-Carlo post-layout simulation

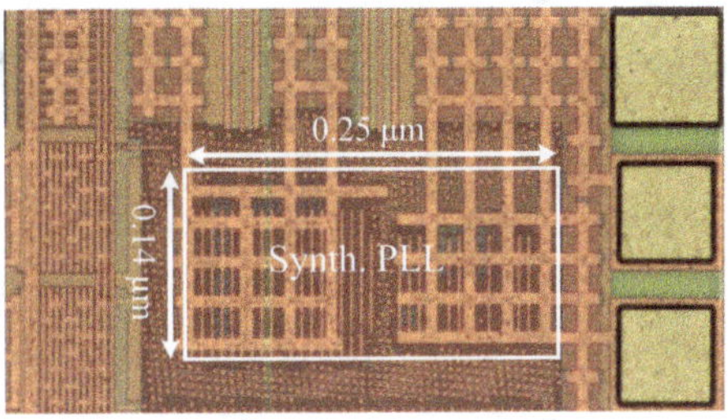

Figure 9.36 Die micrograph of the fully synthesizable ILPLL

constraint-directed layout synthesis. In order to further reduce slew-rate mismatch, background calibration can be implemented [33].

The intrinsic delay of the proposed MUX is only three inverter delays, which is beneficial in order to achieve a high oscillation frequency of 1.8 GHz in this design. The delay offset of the proposed SS-BBPD in the presence of device mismatch is obtained through Monte-Carlo simulation. Two inputs V_{OSC} and INJ_{EDGE} with timing differences were swept, and the occurrence of output 10/01 was recorded in order to obtain the mean value. Figure 9.35 shows the output mean value under each timing setting. An RMS phase offset of 0.18 ps was obtained by fitting the result to a Gaussian error function, which translates to an offset-induced reference spur of <−74 dBc.

9.4.4 Measurement results

The entire PLL is synthesized using commercial digital design tools with a non-modified standard-cell library. The core area of the PLL is 0.035 mm^2, which is fabricated in a 65-nm CMOS process. Figure 9.36 shows the die photograph. The phase noise is measured by a signal source analyzer (Keysight E5052B), and the spectrum is measured by a spectrum analyzer (Anritsu MS2830A). Figures 9.37 and 9.38 show the measured phase noise and spectrum in integer-N mode with a 100-MHz reference clock. The integrated jitter is 0.4 ps jitter from 10 kHz to 10 MHz. The DCO achieves −98 dBc/Hz at a frequency offset of 1 MHz and a

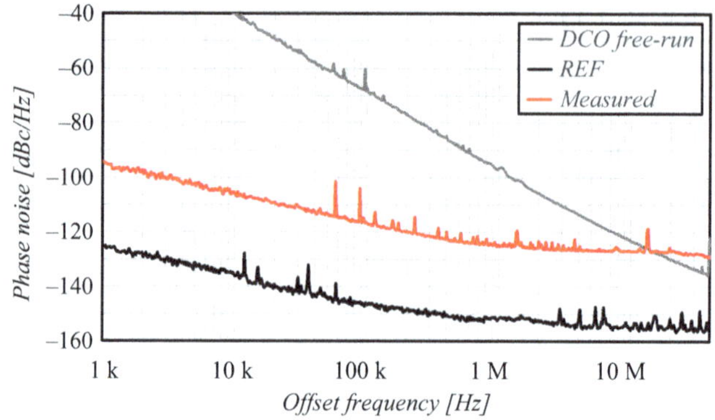

Figure 9.37 Measured PLL output phase noise at an output frequency of 1.0 GHz with an input reference of 100 MHz

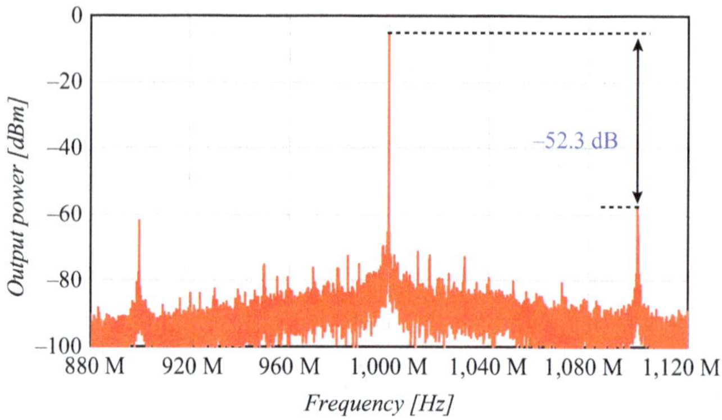

Figure 9.38 Measured PLL output spectrum at an output frequency of 1 GHz with an input reference of 100 MHz

power consumption of 1.0 mW. Note that the spurs around 10 kHz, 100 kHz, and 10 MHz are caused by nonideal grounding of measurement equipment. The reference spur at a frequency offset of 100 MHz is around −52 dBc, which is 7 dB better than that in a previous study [34]. In order to investigate the higher than expected reference spur, the DCO supply was changed by ±50 mV. The reference spur was barely changed, indicating that the phase offset between MUX and the SS-BBPD is not the limiting factor. On the other hand, the reference spur could be improved by 2 dB by reducing the input reference power from 12 to 7 dBm, albeit with much degraded phase noise. Thus, the limiting factor could be parasitic coupling through input pads, probe pins, or supply.

The PLL performance is summarized and compared in Table 9.3. The −52 dBc reference spur is the best among the synthesizable PLLs but is worse than that reported in a previous study [33]. In addition, the proposed PLL achieves the

Table 9.3 Performance summary and comparison

	This work	**[34]**	**[13]**	**[32]**	**[33]**
Arch.	ILPLL	ILPLL	TDC PLL	MDLL	ILCM
Synth.?	**Yes**	Yes	Yes	No	No
Marco?	**No**	No	Yes	–	–
Type	Int-N	Frac-N	Int-N	Int-N	Int-N
Tech. [nm]	65	65	14	28	65
Area [mm^2]	0.035	0.012	0.009	0.0056	0.055
Power [mW]	1.2	2.2[a]	9.7	1.45	11.0
Ref. [MHz]	100	100	50	200	100
FMR	10	10	100	15	24
Int. jitter [ps]	0.4	0.3[a]	4.71	0.292	0.14
Int. BW [MHz]	0.01–10	0.01–10	–	0.01–40	0.01–30
Ref. spur [dBc]	**–52**	–45	–40	–44	–72
FoM[b] [dB]	–247.2	–246.7[a]	–216.8	–249.1	–246.7

[a]Integer-N mode w/reference doubler.
[b]FoM = $10 \log[(\sigma_t/1\text{s})^2 (P_{DC}/1 \text{ mW})]$.
The bold text and values here are used to indicate the feature/advantage of this work. The usage is intentional.

highest jitter-power FoM among the synthesizable PLLs, even without using a reference doubler.

9.5 Summary and conclusion

This chapter details the motivation for adopting fully synthesizable design. The design considerations and trade-offs of ILPLLs are presented along with a detailed noise analysis.

As one of the most popular research areas, two synthesizable PLLs are presented as illustrative examples in order to demonstrate the potential of fully synthesizable design. In the first example, the PLL used an interpolative phase-coupled oscillator, a current-output DAC, and a fine-resolution digital varactor. Implemented in a 65-nm digital CMOS process, the PLL occupies a layout area of only 110 μm × 60 μm, which is the smallest among the PLLs reported thus far (to the best of our knowledge). The measurement results show that a 1.7-ps RMS jitter was achieved at an output frequency of 900 MHz while consuming 780 μW DC power.

In the second example, a single-phase multistage oscillator with a fully symmetrical phase detector and MUX was proposed. In order to minimize the reference spur, a self-clocked nonoverlap update scheme was proposed in order to reduce the reference spur caused by digital logic clocking. Moreover, a slope-balanced fully symmetrical MUX and an SS-BBPD were proposed. With constraint-directed automatic layout synthesis, the delay offset is greatly reduced. Implemented in a 65-nm CMOS process, the PLL achieved 0.4 ps integrated jitter at an output frequency of 1 GHz with a −52-dBc reference spur. The power consumption was 1.2 mW, corresponding to an FoM of −247.2 dB.

Thanks to the holistic design optimization of the system architecture, circuit topology, and layout physical implementation, both PLLs achieved the state-of-the-art performance in terms of jitter and reference spur, validating the feasibility of using fully synthesizable design in high-performance applications.

References

[1] Calhoun BH, Yu C, Xin L, *et al.* Digital Circuit Design Challenges and Opportunities in the Era of Nanoscale CMOS. Proc IEEE. 2008;96: 343–365.
[2] Park Y, and Wentzloff DD. An All-Digital 12 pJ/pulse IR-UWB Transmitter Synthesized from a Standard Cell Library. IEEE J Solid-State Circuits. 2011;46:1147–1157.
[3] Li Y, Dhwaj K, Wong C, *et al.* A Novel Fully Synthesizable All-Digital RF Transmitter for IoT Applications. IEEE Trans Comput Aided Des Integr Circuits Syst. 2018;37:146–158.
[4] Park Y, and Wentzloff DD. An All-Digital PLL Synthesized from a Digital Standard Cell Library in 65nm CMOS. In: Proc. IEEE CICC; 2011. p. 1–4.
[5] Faisal M, and Wentzloff DD. An Automatically Placed-and-Routed ADPLL for the Medradio Band Using PWM to Enhance DCO Resolution. In: IEEE RFIC Symp. Dig. Papers; 2013. p. 115–118.
[6] Kim W, Park J, Park H, *et al.* Layout Synthesis and Loop Parameter Optimization of a Low-Jitter All-Digital Pixel Clock Generator. IEEE J Solid-State Circuits. 2014;49:657–672.
[7] Su R, Lanzisera S, and Pister KSJ. A 2.6 ps Rms-Period-Jitter 900 MHz All-Digital Fractional-N PLL Built with Standard Cells. In: Proc. ESSCIRC; 2011. p. 455–458.
[8] Park Y, and Wentzloff DD. A Cyclic Vernier TDC for ADPLLs Synthesized from a Standard Cell Library. IEEE Trans Circuits Syst I, Reg Papers. 2011;58:1511–1517.
[9] Weaver S, Hershberg B, and Un-Ku M. Digitally Synthesized Stochastic Flash ADC Using Only Standard Digital Cells. IEEE Trans Circuits Syst I, Reg Papers. 2014;61:84–91.
[10] Seo M, Roh Y, Chang D, *et al.* A Reusable Code-Based SAR ADC Design With CDAC Compiler and Synthesizable Analog Building Blocks. IEEE Trans Circuits Syst II, Exp Briefs. 2018;65:1904–1908.
[11] Chang E, Han J, Bae W, *et al.* BAG2: A Process-portable Framework for Generator-based AMS Circuit Design. In: Proc. IEEE CICC; 2018. p. 1–8.
[12] Wulff C, and Ytterdal T. A Compiled 9-bit 20-MS/s 3.5-fJ/conv. Step SAR ADC in 28-nm FDSOI for Bluetooth Low Energy Receivers. IEEE J Solid-State Circuits. 2017;52:1915–1926.
[13] Moore DM, Xanthopoulos T, Meninger S, and Wentzloff DD. A 0.009 mm^2 Wide-Tuning Range Automatically Placed-and-Routed ADPLL in 14-nm FinFET CMOS. IEEE Solid-State Circuits Letters. 2018 Mar;1(3):74–77.

[14] Staszewski RB, Wallberg JL, Rezeq S, *et al.* All-Digital PLL and Transmitter for Mobile Phones. IEEE J Solid-State Circuits. 2005;40:2469–2482.

[15] Wu CT, Shen WC, Wang W, *et al.* A Two-Cycle Lock-In Time ADPLL Design Based on a Frequency Estimation Algorithm. IEEE Trans Circuits Syst II, Exp Briefs. 2010;57:430–434.

[16] Sheng Y, Jansson L, and Galton I. A Multiple-Crystal Interface PLL with VCO Realignment to Reduce Phase Noise. IEEE J Solid-State Circuits. 2002;37:1795–1803.

[17] Gierkink SLJ. Low-Spur, Low-Phase-Noise Clock Multiplier Based on a Combination of PLL and Recirculating DLL With Dual-Pulse Ring Oscillator and Self-Correcting Charge Pump. IEEE J Solid-State Circuits. 2008;43:2967–2976.

[18] Dalt ND. An Analysis of Phase Noise in Realigned VCOs. IEEE Trans Circuits Syst II, Exp Briefs. 2014;61:143–147.

[19] Helal BM, Chun-Ming H, Johnson K, *et al.* A Low Jitter Programmable Clock Multiplier Based on a Pulse Injection-Locked Oscillator With a Highly-Digital Tuning Loop. IEEE J Solid-State Circuits. 2009;44:1391–1400.

[20] Deng W, Musa A, Siriburanon T, *et al.* A 0.022 mm^2 970 μW Dual-Loop Injection-Locked PLL with 243 dB FOM Using Synthesizable All-Digital PVT Calibration Circuits. In: IEEE ISSCC Dig. Tech. Papers; 2013. p. 248–249.

[21] Musa A, Deng W, Siriburanon T, *et al.* A Compact, Low-Power and Low-Jitter Dual-Loop Injection Locked PLL Using All-Digital PVT Calibration. IEEE J Solid-State Circuits. 2014;49:50–60.

[22] Chen PL, Chung CC, and Lee CY. A Portable Digitally Controlled Oscillator Using Novel Varactors. IEEE Trans Circuits Syst II, Exp Briefs. 2005;52: 233–237.

[23] Sheng D, Chung CC, and Lee CY. An Ultra-Low-Power and Portable Digitally Controlled Oscillator for SoC Applications. IEEE Trans Circuits Syst II, Exp Briefs. 2007;54:954–958.

[24] Matsumoto A, Sakiyama S, Tokunaga Y, *et al.* A Design Method and Developments of a Low-Power and High-Resolution Multiphase Generation System. IEEE J Solid-State Circuits. 2008;43:831–843.

[25] Park D, and Cho S. A 14.2 mW 2.55-to-3 GHz Cascaded PLL With Reference Injection and 800 MHz Delta-Sigma Modulator in 0.13 μm CMOS. IEEE J Solid-State Circuits. 2012;47:2989–2998.

[26] Levantino S, Marucci G, Marzin G, Fenaroli A, Samori C, and Lacaita AL. A 1.7 GHz Fractional-N Frequency Synthesizer Based on a Multiplying Delay-Locked Loop. IEEE J Solid-State Circuits. 2015;50:2678–2691.

[27] Elkholy A, Elmallah A, Ahmed MG, and Hanumolu PK A 6.75–8.25-GHz -250-dB FoM Rapid ON/OFF Fractional-N Injection-Locked Clock Multiplier. IEEE J Solid-State Circuits. 2018;53:1818–1829.

[28] Cho S, Kim S, Choo M, *et al.* A 5-GHz Subharmonically Injection-Locked All-Digital PLL with Complementary Switched Injection. In: Proc. ESSCIRC; 2015. p. 384–387.

[29] Sheng Y, Jansson L, and Galton I A Multiple-crystal Interface PLL with VCO Realignment to Reduce Phase Noise. IEEE J Solid-State Circuits. 2002;37:1795–1803.

[30] Dalt ND. Linearized Analysis of a Digital Bang-Bang PLL and its Validity Limits Applied to Jitter Transfer and Jitter Generation. IEEE Trans Circuits Syst I, Reg Papers. 2008;55:3663–3675.

[31] Xu H, and Abidi AA Design Methodology for Phase-Locked Loops Using Binary (Bang-Bang) Phase Detectors. IEEE Trans Circuits Syst I, Reg Papers. 2017;64:1637–1650.

[32] Yang S, Yin J, Mak P, and Martins RP. A 0.0056 mm^2 All-Digital MDLL Using Edge Re-extraction, Dual-Ring VCOs and a 0.3 mW Block-Sharing Frequency Tracking Loop Achieving 292fs_{rms} Jitter and −249 dB FOM. In: IEEE ISSCC Dig. Tech. Papers; 2018. p. 118–120.

[33] Yoo S, Choi S, Lee Y, Seong T, Lim Y, and Choi J A 140fs_{rms}-Jitter and −72dBc-Reference-Spur Ring-VCO Based Injection-Locked Clock Multiplier Using a Background Triple-Point Frequency/Phase/Slope Calibrator. In: IEEE ISSCC Dig. Tech. Papers; 2019. p. 490–492.

[34] Liu B, Ngo HC, Nakata K, *et al.* A 1.2 ps-Jitter Fully-Synthesizable Fully-Calibrated Fractional-N Injection-Locked PLL Using True Arbitrary Nonlinearity Calibration Technique. In: Proc. IEEE CICC; 2018. p. 1–4.

Chapter 10
Ultra-low-power ADPLL

Hanli Liu[1] *and Kenichi Okada*[1]

10.1 Introduction

Internet of things (IoT) covers huge application scenarios. Machines such as vehicles can be connected over the air and "talk" to each other to ensure safe driving. Devices on or inside a human's body can help monitor both the mental and physical state of a person and share the data with doctors or family members. Sensors can be placed inside the house to help manage our daily life, for example, ambient light sensors and proximity sensors can automatically turn on/off the light or change the level of the light automatically; humidity sensors, temperature sensors, and air quality sensors will control the air conditioner and the air purifier to regulate the room temperature and clean the harmful particles inside the house. By those interconnected "smart" devices in the house, the house itself can become intelligent and more comfortable for people who live in. Moreover, the city infrastructures will also be connected and form a huge network, for example, devices with the beacon mode can transform into a very accurate navigation network, which can help people doing indoor navigation, etc. After researching and developing for many years, lots of wireless standards are carried out and developed to enable the IoT era. Different standards such as Bluetooth, Wi-Fi, ZigBee, WirelessHart, DigMesh, NFC, ANT, EnOcean, RFID, NB-IoT, etc. are used for different scenarios in a complex real-world radio environment. The standard-based wireless systems are required to consume as little energy as possible, targeting one main purpose: extending the battery life to reduce the cost. Since billions of devices will be driven by small-size batteries in the future, constantly replacing the battery will significantly increase the cost and potentially harm the nature.

The integrated wireless transceiver (TRX) is the key to realize the wireless communication. Since the TRX consumes a significant amount of power in the wireless system, the ultra-low-power (ULP) operation is the most important specification for the TRX utilized for IoT applications. The radio-frequency phase-lock loop (RF-PLL), one of the most important elements in TRX, consumes a significant amount of power due to the phase noise and spurious requirements [1].

[1]Department of Physical Electronics, Tokyo Institute of Technology

Hence, a reduction in the PLL's power will greatly lower the power consumption of the TRX. All-digital PLLs (ADPLLs), which take advantage of the scaling of complementary metal oxide semiconductor (CMOS) technology, become more promising than its analog counterpart in advanced CMOS process [2–8]. It shrinks the required chip size while providing easily accessed analog/digital inputs/outputs (IOs). Those IOs can be used for digital-intensive calibrations and modulations. While the benefits are very obvious for the ADPLL, still there are barriers to realize the ULP operation of the ADPLL in a fractional-N mode. Two key performance indicators of the fractional-N PLL used for wireless communications are generally evaluated, i.e., the phase noise and the fractional spurs. To realize a good phase noise and low fractional spurs, significant amount of power is required, which is in conflict with the concept of ULP operation.

10.2 Requirements and architecture considerations for low-power ADPLL

10.2.1 Phase noise and spur considerations for IoT applications

Among those IoT standards, Bluetooth technology is one of the most popular wireless standards in short- and middle-range communications. The biggest advantage of Bluetooth standard is that it has been integrated in every smart phone nowadays to replace the traditional wire connections. The Bluetooth standard defines two modes for two different purposes. The classic mode (Bluetooth classic) targets at a high data rate, and the low-energy mode [Bluetooth low energy (BLE)] focus at ULP operation. The BLE is more popular in terms of the IoT applications because of the extended battery life enabled by its low-energy operation. The moderate data rate can also meet most of the communication requirements. In this section, our analysis of the PLL requirements of the phase noise and spurs will base on the BLE standard.

Table 10.1 lists some of the most important requirements for the BLE receiver (RX) design [9]. To successfully communicate with other IoT units over the air, the RX should also have a strong anti-interference performance (blocker immunity). The blocker performances are of two kinds: the in-band blocker tolerance/the

Table 10.1 Receiver specifications for BLE

Sensitivity	<-70 dBm	
Bit error rate (BER)	<0.1%	
Adjacent channel rejection	0 MHz	−21 dB
	1 MHz	−15 dB
	2 MHz	17 dB
	$\geq$ 3 MHz	27 dB
Blocker power	30–2,000 MHz	−30 dBm
	2,000–2,400 MHz	−35 dBm
	2,500–3,000 MHz	−35 dBm

adjacent channel rejection (ACR) and the out-band blocker tolerance. The blocker signals are specified as ($S_{blocker}$) with different offset frequency from the frequency of S_{desire}. S_{desire} is specified as 3 dB higher than the specified sensitivity level of −70 dBm, which is −67 dBm. S_{desire} and $S_{blocker}$ are both specified as 1 Mbps GFSK signals with modulation index of 0.5 and Bluetooth of 0.5. In order to give the phase noise and spur requirements for the PLL based on blocker performances, it is necessary to answer the following question: how do the blockers affect the RX performance?

Figure 10.1 shows the frequency translation process inside the radio-frequency front end (RF-FE). A mixer acts as a multiplier that moves the high-frequency signals to a lower frequency by using a local oscillator (LO) signal. As shown in Figure 10.2 (a), if an ideal LO is used, the wanted signal S_{desire} will be down-converted to an intermediate frequency (IF) signal at (f_{IF}), while the unwanted signal such as a blocker $S_{blocker}$ is down-converted to the frequency of $f_{blocker} - f_{LO} + f_{IF}$. However, the LO such as an oscillator or a PLL will produce extra energy around the idea LO frequency, i.e., phase noise. As shown in Figure 10.2(b), the unwanted signal will also be down-converted to f_{IF} and the component $S_{mx,noise}$ cannot be distinguished after conversion. This process is well known as reciprocal mixing [10]. Because of the nonideality from the LO, the phase noise level and the spur energy from the LO should be well constrained at blocker frequencies. To satisfy the ACR performance, the phase noise (PN) is calculated by using the following equation:

$$PN = S_{desire} - S_{blocker} - SNR_{demod} - 10\log_{10} BW \tag{10.1}$$

SNR_{demod} is the required minimum signal to noise ratio (SNR) for the baseband demodulator to properly demodulate the data with a specified bit error rate. For BLE standard, 15 dB can be assumed for SNR_{demod} with a sufficient margin.

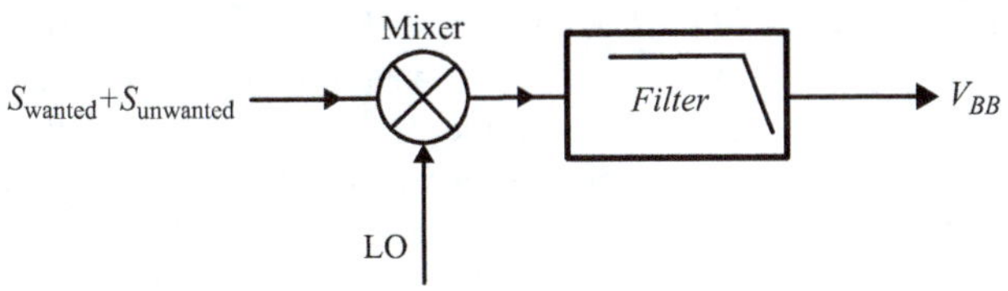

Figure 10.1 S_{desire} *and* $S_{blockers}$ *are simultaneously down converted by RX mixers*

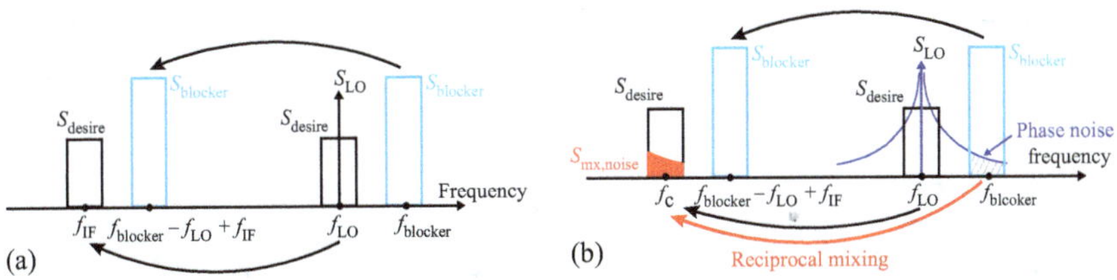

Figure 10.2 *(a) Ideal down-conversion (b) down-conversion with LO phase noise*

BW is the signal bandwidth and it is 1 MHz for BLE. Based on (10.1), the phase noise at ≥ 3 MHz offset should be lower than -102 dBc/Hz. The spurs from the LO, such as the fractional spur and the reference spur, will also cause reciprocal mixing, which leads to potential degradation of the desired signal on SNR. The maximum spur (*Spr*) level can be calculated by using the following equation:

$$\text{Spr} = S_{\text{desire}} - S_{\text{blocker}} - \text{SNR}_{\text{demod}} \tag{10.2}$$

And the calculated spur level $\geq$ 3MHz is about -42 dBc. Note that those phase noise and spur requirements are calculated by only considering the effect of the phase noise and spurs, any other circuit nonlinearity in RX will also degrade the blocker performances. Hence, more margins should be left to the phase noise and spur requirements based on (10.1) and (10.2). In low-power RX design, the phase noise and spurs of the PLL are always limited by the tightened power budget. Thus, any improvement in the phase noise and spurs will significantly improve the blocker performances.

10.2.2 Low-power fractional ADPLL architectures

Analog charge-pump PLL (CPPLL) has been the most preferred ADPLL architecture to implement fractional-N frequency synthesizing [11]. CPPLL can realize the excellent jitter and spurious performance with less power consumption, i.e., high figure-of-merit (FOM). However, an analog PLL filter requires a large area and is also very difficult to reconfigure. Furthermore, for advanced CMOS technology, digital circuits are more preferred to replace the analog counterparts such as filters, calibrations, etc. To overcome these shortcomings, digital fractional-N PLL [2–8] is investigated and studied, which requires no large capacitors. Thanks to their intensive digital implementation, loop characteristics are much easier to reconfigure. Also, the digital PLL is much easier to shift from one process to another.

A famous fractional-N ADPLL architecture [12,13] is shown in Figure 10.3(a). This architecture is very similar to a fractional-N CPPLL using a delta–sigma modulator (DSM) and a multi-modulus divider (MMDIV) at the feedback path. The MMDIV performs as a frequency divider and a phase modulator. The frequency control word (FCW) is to feed the DSM oversampled by a clock at a much higher frequency than that of the FCW. The long bit-width FCW will be truncated into a short bit-width by DSM to control the division ratio of the MMDIV. If the

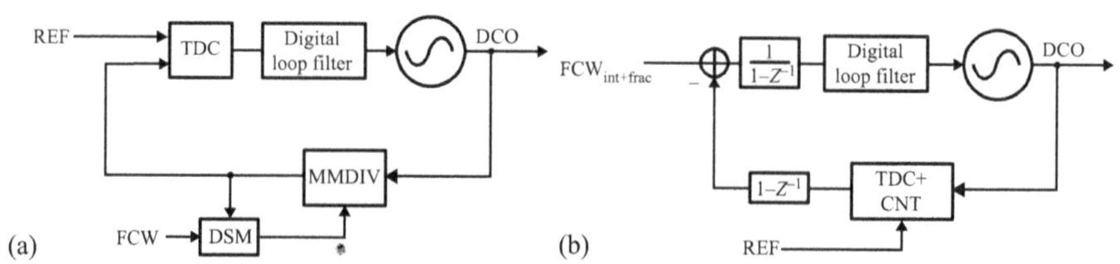

Figure 10.3 (a) Divider-based fractional-N ADPLL and (b) phase-domain fractional-N ADPLL

frequency and the phase of the PLL are locked, the average frequency and phase generated from MMDIV should equal to the reference frequency and phase. The DSM and MMDIV operation will generate a peak phase error of one digital-controlled oscillator (DCO) period at MMDIV output when a first-order DSM is applied. If the phase from MMDIV is not equal to the reference phase, the time-to-digital converter (TDC) will produce a corresponded phase error. The digital loop filter (DLF) will filter the phase error and control the DCO accordingly. The TDC is similar to a phase frequency detector (PFD) and charge pump in CPPLL. The MMDIV automatically translate the frequency of DCO to a reference rate, hence the frequency error is also detected at TDC. This architecture is known as the divider-based ADPLL.

Figure 10.3(b) shows another type of the fractional-N ADPLL implementation [2]. It uses a high-speed counter (CNT) and a TDC to count the phase values from the oscillator. A reference clock is used to perform a clean timer to read out the counted phases. The integer counter, which works at oscillator frequency, will count every oscillator period. Its output can be represented as an integer phase. However, if only the counter is implemented, then the residue phase information within one oscillator period cannot be extracted due to the limited resolution of the integer counter. Hence, a fractional counter with much finer resolution is required to cover one oscillator period to assist the integer counter operation. The fractional counter is also named as TDC. The integer phase and fractional phase of the oscillator within one reference period are quantized and synchronized by reference clock. Because an FCW is more commonly used than the phase control word (PCW) in frequency synthesizer applications, the integer phase and fractional phase will be differentiated into the integer frequency and fractional frequency information by a digital differentiator. Then it compares with the desired FCW at PLL input. The error information is the frequency error, and the phase error is derived after the integration operation inside the DLF. A DCO is implemented and the oscillation frequency is controlled by digital codes. This architecture is well known for its pure phase-domain operation, and the CNT and TDC have to work together to derive the phase and frequency information from the DCO output.

Several differences are involved in designing these two architectures. The phase-domain ADPLL requires an absolute synchronized operation of the counter and the TDC to count the DCO phase within one reference period. And the TDC range should be exactly one DCO period. Any mismatch in the TDC range and DCO period will cause significant fractional spurs in fractional mode. As for the divider-based ADPLL, the TDC range should be over one DCO period if a first-order DSM is applied. A higher DSM order will result in a much wider TDC range. As a phase error converter, the linearity is very important for achieving lower fractional spurs. It causes high power consumptions for TDC design. Fundamentally, this high phase error is caused by the MMDIV, which generates a peak phase error of one DCO period for a first-order DSM. So if we can minimize this error, the phase error at the TDC input could be reduced. A digital-to-time converter (DTC) is proposed to produce the fractional phase that reduces the quantization noise from the MMDIV operation, as shown in Figure 10.4(a). As a

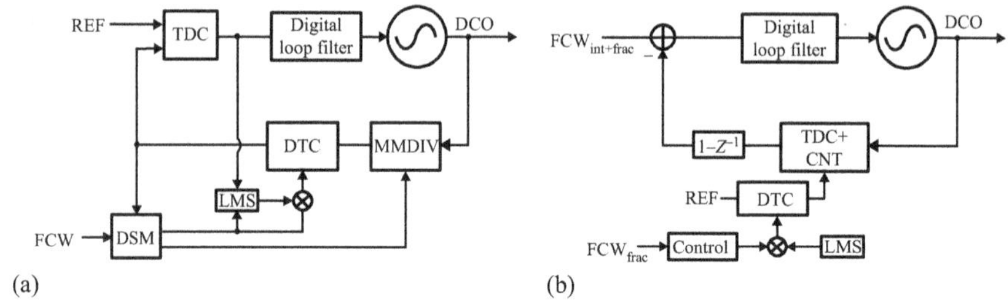

Figure 10.4 DTC-based fractional-N ADPLL implementation (a) divider based (b) phase domain

result, the DTC can reduce the quantization error to only several DTC resolution. Hence, a narrow-range TDC can be implemented. This greatly cut down the TDC power consumption. At an extreme case, even 1 bit TDC bang-bang phase detector (BBPD) can be used for fractional operation [5]. However, as discussed in [4], a BBPD suffers from the very limited quantizing range and an ill-defined gain. The limited quantizing range causes the slow convergence speed of the PLL by the well-known cycle-slip effect and also causes slow settling speed of the DTC-gain calibration. A multi-level TDC is desired for achieving faster converge speed and a well-defined phase detector (PD) gain, which is more preferable for wireless applications [4]. The same methods can be applied to the phase-domain fractional-N ADPLL in Figure 10.3(b), as shown in Figure 10.4(b). A DTC is used to produce a fractional reference phase by delaying the reference input, and it can also greatly reduce the TDC range. After phase is locked, the counter can be turned off to further save the power consumption.

10.2.3 DTC for low-power ADPLL

As discussed above, to reduce the dynamic range of the TDC, DTC is used to assist the operation of the TDC with narrower dynamic range. It improves the TDC resolution and linearity while achieving high-energy efficiency on phase quantization process. Similar to TDC, the power consumption, linearity, resolution, and noise are also important for DTC design.

In general, DTC is a circuit that generates delays instead of quantizing phase errors. It can be designed in a much more efficient way compared to TDC. To get a delayed rising edge, a slope with a slew rate (SR) of SR_{Ramp} and a comparator with a threshold of V_{TH} are required as shown in Figure 10.5. The delay can be written as:

$$t_{delay} = \frac{V_{TH} - V_{Ref,p}}{SR_{Ramp}} \tag{10.3}$$

From (10.3), the delay can be varied by changing the comparator V_{TH}, the starting voltage of the slope $V_{Ref,p}$, or the SR of the slope SR_{Ramp}. Each of which has different design trade-offs, and we will discuss it later.

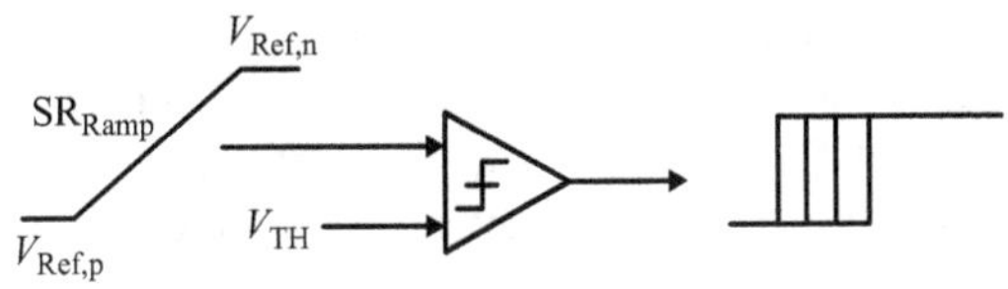

Figure 10.5 Concept of DTC

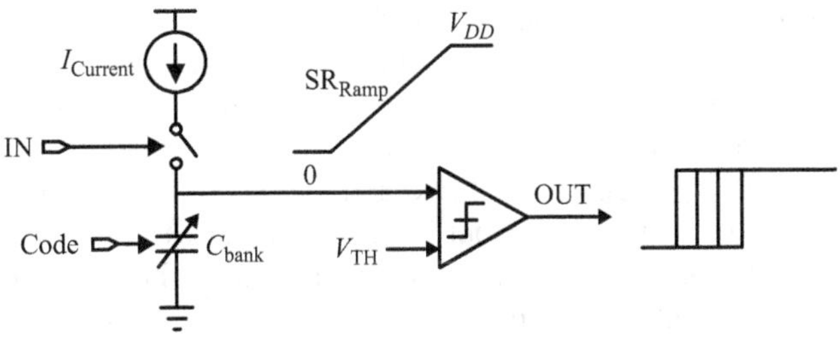

Figure 10.6 Variable-slope DTC

10.2.3.1 Variable SR (variable-slope) DTC

As for the most commonly seen DTC in [14] shown in Figure 10.6, it utilized the so-called variable-slope method to acquire different delays. A current source is used to generate the ramp signal triggered by input. A digital controlled capacitor bank is used to vary the SR of:

$$\text{SR}_{\text{Ramp}} = \frac{I_{\text{current}}}{C_{\text{bank}}} \tag{10.4}$$

We can derive the achievable delay range DR_{DTC} as:

$$\text{DR}_{\text{DTC}} = \frac{V_{\text{TH}}}{\frac{I_{\text{current}}}{C_{\text{bank}}}} = V_{\text{TH}} \frac{C_{\text{bank}}}{I_{\text{current}}} \tag{10.5}$$

If we assume the current is constant during charging and discharging the C_{bank}, we can write the DTC power consumption as:

$$P_{\text{DTC}} = V_{DD} I_{\text{current}} f \frac{V_{DD}}{\frac{I_{\text{current}}}{C_{\text{bank}}}} = V_{DD}^2 C_{\text{bank}} f \tag{10.6}$$

f is the operation frequency of the DTC. From the above equation, double the delay range while keeping the same power consumption, C_{bank} cannot be changed as shown in (10.6). Hence, I_{current} should be halved as shown in (10.5). The variance in the timing jitter of the DTC can be analyzed as shown in [15]. We can write the jitter variance equation of the DTC as:

$$\delta_{\text{DTC}}^2 = 4\,kT \frac{\gamma_{\text{current}} \text{DR}_{\text{DTC}}}{I_{\text{current}}(V_{\text{DD}} - V_{\text{TH}})} + \frac{kTC_{\text{bank}}}{I_{\text{current}}^2} = \frac{kTC_{\text{bank}}}{I_{\text{current}}^2} \left(1 + \frac{4\gamma_{\text{current}}}{\frac{V_{\text{DD}}}{V_{\text{TH}}} - 1} \right) \tag{10.7}$$

where k is the Boltzmann constant and T is the temperature. Now it has to be found if the current is reduced by half, will the jitter double. This is the well-known delay range and jitter trade-off for delay elements. To reduce the jitter from DTC, the current and load capacitor banks should increase simultaneously while leaving their ratio the same.

10.2.3.2 Variable starting voltage (constant-slope) DTC

The variable-slope architecture shows a simplicity in design while bringing a major problem. As shown in Figure 10.7(a), a comparator output reversed edge from its input at the threshold voltage. Internally, the comparator is an open-loop amplifier that has a very high gain around its reference voltage. Any small voltage difference between the IN and V_{TH} will be amplified and the output will be saturated. If the gain is infinite, no matter what the input SR is, the output will instantaneously drop to zero at the first input and produce its corresponding rising edge at the second stage. However, practically, the gain cannot be infinite. The finite gain will produce a finite SR at the output of the first stage if the input SR varies. As a result, the variable delay is dependent on the input SR. This is a code-dependent quantity, which is very difficult to remove. It is the major nonlinear source for the variable-slope DTC.

However, as shown in Figure 10.7(b), if we vary the starting voltage while using the same ramp for every comparison, the nonlinearity due to the comparator imperfection can be fundamentally removed [16]. This is the so-called constant-slope DTC. The circuit implementation is as shown in Figure 10.8. The starting voltages can be changed by using a digital-to-voltage converter (DAC). This method achieves excellent linearity compared with conventional variable-delay DTCs [16].

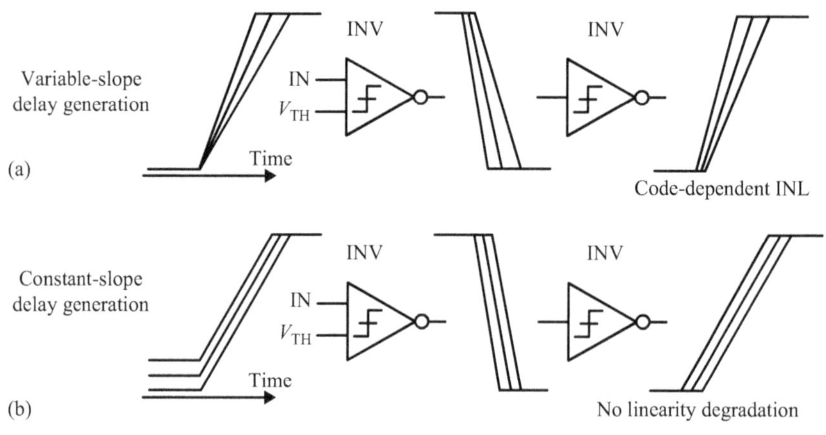

Figure 10.7 Variable-slope delay generation vs. constant-slope delay generation

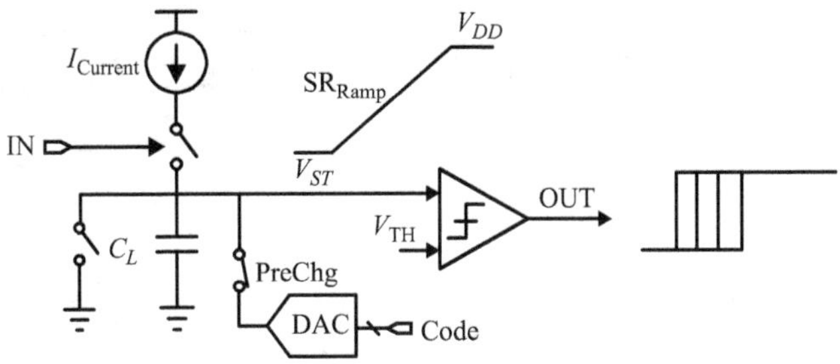

Figure 10.8 Constant-slope DTC

10.2.3.3 Variable threshold (ramp division) DTC

From (10.3), the last thing we can do is to vary the threshold of the comparator. The slope still has a constant SR during the comparison, which mitigates the finite gain-induced nonlinearity. One of the comparator input can be connected to a DAC, which generated code-dependent voltage. The V_{TH} decides the threshold of the comparator. Another side is connected to the constant ramp. When the ramp reaches the V_{TH}, a rising edge will be produced. However, practically, the common-mode voltage of the comparator will also vary according to the variable V_{TH}. The common mode voltage will greatly affect the intrinsic delay of the comparator. This intrinsic delay could vary more than several ps that greatly degrade the linearity. It may show a worse integral nonlinearity (INL) than the variable-slope DTC.

10.3 Proposed ULP ADPLL architecture

10.3.1 System overview

To realize low-power, low-jitter, and low-spur performances simultaneously in a fractional-N ADPLL, the selection of the architecture and the specifications of each building block will be crucial in system-level design. Figure 10.4(b) shows one of the way to realize the sub-mW fractional-N ADPLL [6,7], it basically mimics the operation of the analog sub-sampling operation [17,18]. In order to save the power consumption, CNT is completely shut down after phase lock [7], which equivalently means the frequency locked loop is turned off in the analog sub-sampling PLL. This will potentially lead the PLL to lock a wrong frequency if high frequency and phase disturbance are presented during its operation. According to [6,7] another issue is that the poor resolution of the TDC causes degradation of in-band phase noise and also the jitter performance of the ADPLL. Furthermore, the DTC linearity limits the spur performance, which needs the special technique to deal with [7] and it consumes power as well as sacrifice jitter performance.

A detailed architecture-level block diagram of the proposed ADPLL is shown in Figure 10.9. To improve the in-band phase noise, a 4b 2 ps-resolution time amplifier (TA) TDC [4] and a reference doubler are implemented. The duty cycle issue of the doubler is calibrated by using the method proposed in [11]. A proposed

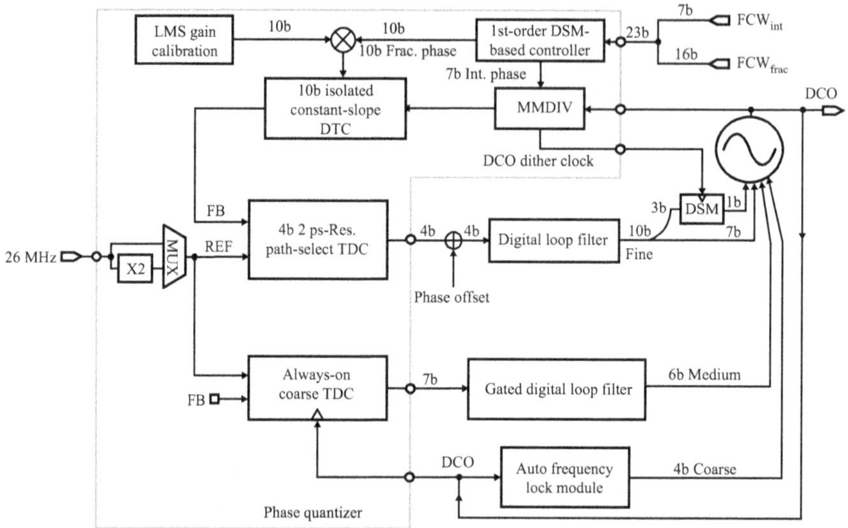

Figure 10.9 Detailed block diagram of the proposed ULP fractional-N ADPLL

10b isolated constant-slope DTC supports the narrow-range TDC operation. Theoretically, with the assistance of the fine-resolution DTC, the range of the TDC can be reduced to 1 bit. However, in practice, 1-bit TDC or BBPD will cause other issues. One issue is that it significantly degrades the locking speed of the ADPLL. Another important issue is that it degrades the convergence speed of the DTC-gain calibration as discussed in [4]. An MMDIV is used to perform phase accumulation and frequency division simultaneously. The frequency/phase of DCO will be monitored by MMDIV, hence any phase/frequency change will be reflected at TDC input. Both MMDIV and DTC are controlled by the DSM-based controller [5].

Figure 10.10 shows the implementations and the mathematic models of the DSM-based fractional controller, MMDIV, and DTC. In Figure 10.10(a), $\mathrm{FCW}_{\mathrm{int+frac}}$ is separated into 7b most significant bit (MSB) ($\mathrm{FCW}_{\mathrm{int}}$) and 16b LSB ($\mathrm{FCW}_{\mathrm{frac}}$). $\mathrm{FCW}_{\mathrm{frac}}$ is modulated by a first-order DSM that produces 1-bit output F_{DSM}, and F_{DSM} is added to $\mathrm{FCW}_{\mathrm{int}}$. As shown in Figure 10.10(b), the phase-domain model of MMDIV performs as an integrator. The $\mathrm{FCW}_{\mathrm{int}} + F_{\mathrm{DSM}}$ is accumulated by MMDIV into 7b phase control code $\mathrm{PCW}_{\mathrm{int}}$. Then $\mathrm{PCW}_{\mathrm{int}}$ is multiplied by 2π inside MMDIV, where 2π is normalized to one DCO period, which produces the integer phase $\Phi_{\mathrm{int}} = 2\pi\mathrm{PCW}_{\mathrm{int}}$. The fractional frequency error (DSM quantization error) Q_{frac} is extracted by subtracting F_{DSM} from $\mathrm{FCW}_{\mathrm{frac}}$ as shown in Figure 10.10(a), then it is accumulated into fractional phase error $\mathrm{PCW}_{\mathrm{frac}}$ by a digital accumulator. It is multiplied by 2π phase at DTC to acquire the fractional phase $\Phi_{\mathrm{frac}} = 2\pi\mathrm{PCW}_{\mathrm{frac}}$. Finally, Φ_{int} and Φ_{frac} sum at DTC and produce the desired fractional feedback phase of $\Phi_{\mathrm{FB}} = \Phi_{\mathrm{int}} + \Phi_{\mathrm{frac}} = 2\pi\mathrm{PCW}_{\mathrm{int+frac}}$. To realize fractional-N phase and frequency synthesis, the reference (REF) phase Φ_{REF}

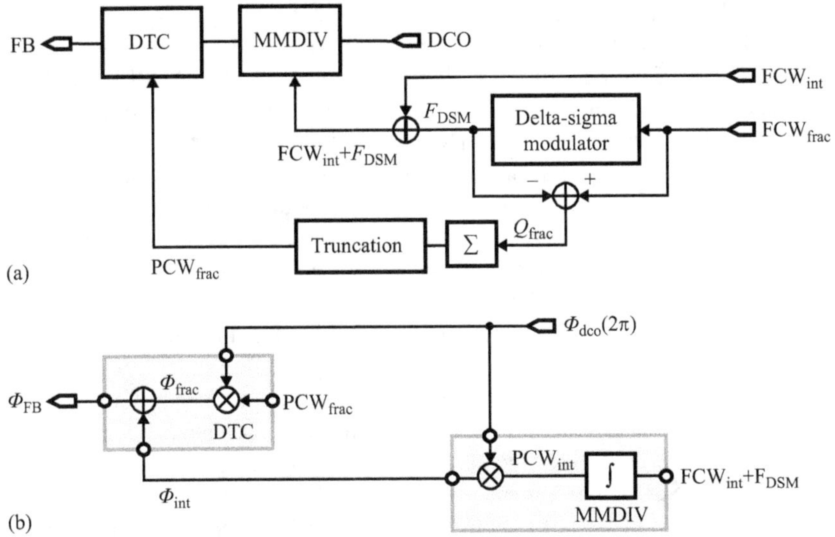

Figure 10.10 (a) DSM-based fractional controller and (b) phase models of MMDIV and DTC

should be exactly equal to $2\pi \mathrm{PCW}_{\mathrm{int+frac}}$. After phase locking, the frequency will be automatically locked to $\mathrm{FCW}_{\mathrm{int+frac}} \cdot f_{\mathrm{REF}}$.

As compared with the second-order DSM used in [5], a first-order DSM-based architecture can accept a much simpler digital implementation, which directly saves power consumption and area. Furthermore, the higher-order DSM-based controller will increase the required dynamic range of DTC and potentially increases jitter contribution. However, the higher-order DSM randomizes the DTC control code, which spreads the spur energy caused by DTC INL into a white spectrum that can be filtered by PLL bandwidth. In order to discuss the power, jitter, and spur trade-offs using the different orders of the DSM, we compared the performance differences between the first-order DSM and the second-order DSM. As the DSM order increases, the $\mathrm{PCW}_{\mathrm{int}}$ jump between each code will be higher, thereby increasing the required DTC delay range. The first-order DSM only requires a DTC with 2π range, while the second-order DSM doubles the required DTC range to 4π.

As discussed in Section 10.2.3, doubling the delay range $\mathrm{DR}_{\mathrm{DTC}}$ will halve I_{current}. Equation (10.6) shows that it will also double the jitter contribution if I_{current} is halved. This is a trade-off between the delay range and jitter of the delay element such as a DTC. For the low-power design, the jitter from each component should be optimized by considering the power budget. For a system-level estimation and a rough transistor-level simulation, a worst-case rms jitter of 0.7 ps is expected at DTC output for the first-order DSM, while the rms jitter will become 1.4 ps using the second-order DSM. All blocks shown in Figure 10.9 are modeled with pre-determined parameters, by changing only the DTC as well as the order of the DSM.

The DTC INL is modeled in a sinusoid shape with a lookup table. We sweep the relative INL of both DTCs and record the rms jitter and the worst spurs (in-band fractional spurs) of the ADPLL as shown in Figure 10.11. The ADPLL bandwidth is optimized to 600 kHz at a 52-MHz reference and the first in-band fractional spur is located at 200 kHz. At very small INL, the ADPLL demonstrates around 190 fs better rms jitter using the first order DSM. This is expected because the periodic jitter is not dominant in the output jitter, and the doubled DTC jitter contributes more to the output jitter in the second-order DSM case. However, when we increase the INL over 0.4%, the periodic jitter caused by spurs becomes dominant, and the output jitter in the first-order DSM case becomes worse than that in the second-order DSM case. For achieving better FOM performance, the rms jitter should be kept as low as possible, which motivates us in using the first-order DSM. The INL specification of DTC is from 0.05% to 0.4%, which contributes to a 2.8-dB improvement in FOM for a 1-mW operation of the ADPLL. The difference in FOM will increase if the DTC jitter increases, which is a common situation in a noisy system-on-chip (SoC) environment.

To estimate the phase noise, both phase-domain and time-domain methods are used. As shown in Figure 10.12, the four major noise contributors are TA noise, TDC quantization noise, DTC noise, and voltage-controlled oscillator (VCO) noise. The well-known reference noise is included in the TDC quantization noise. The red line shows the output phase noise of the fractional-N ADPLL. The black line shows the time-domain simulation results of the proposed fractional-N ADPLL, which operates at 2442 MHz using a 52-MHz reference clock. The fractional spurs are expected to be at around 2 MHz and its harmonics can be observed in the figure as well. The time-domain simulation matches with the phase-domain simulation, which proves to be a good estimation of the overall output phase noise. The overall power consumption is estimated as 1 mW at a 52-MHz reference, according to the post-layout simulation.

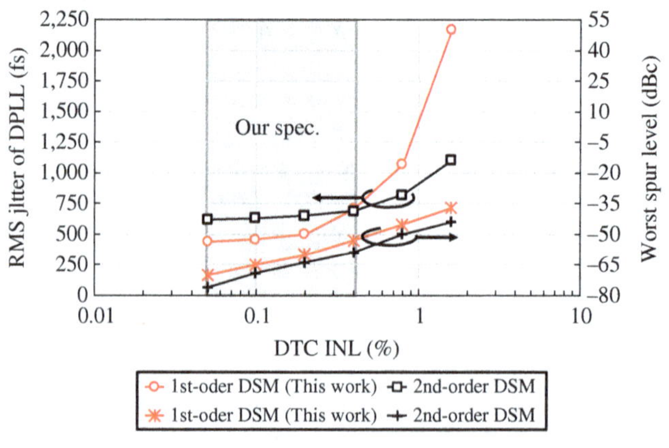

Figure 10.11 System-level analysis of the first-order and second-order DSM-based fractional controller for low-power ADPLL

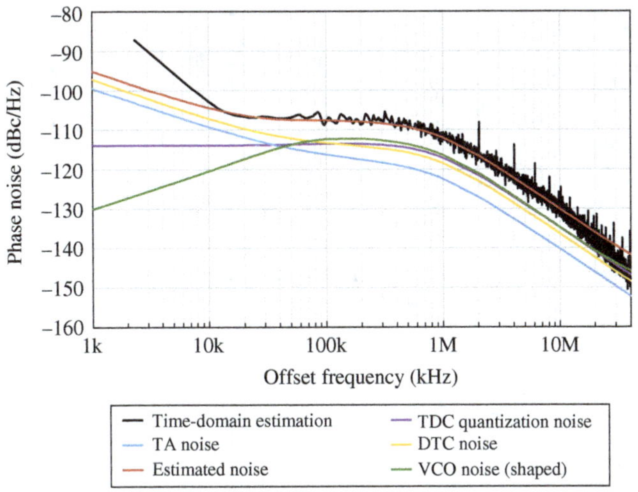

Figure 10.12 Phase noise estimation of the ADPLL in the phase-domain (without fractional spurs) and the time-domain phase noise simulation at 2442 MHz using 52-MHz reference (with fractional spurs)

10.3.2 Low-power highly linear DTC

10.3.2.1 Concept operation

Since the DTC linearity performance will greatly influence the jitter and spur performances of the proposed ADPLL using a first-order DSM, the DTC design becomes more challenging than TDC at a limited power budget. Constant-slope charging method is proposed in [16] to mitigate the inverter-induced nonlinearity. It demonstrates a fundamental improvement in the linearity of the delay generation over the conventional variable-slope method [5–7], which generates delays by using variable SR slopes at the input of the inverter-based comparator. Figure 10.13(a) shows the implementation of the conventional constant-slope DTC. In the concept of the original constant-slope DTC, as shown in Figure 10.13(b), the digital controlled delays are acquired by varying the starting voltage V_{ST} of the slopes generated by a fixed current source. A current-type digital-to-converter (I-DAC) is used to acquire V_{ST} by precharging the loading capacitor C_L before the input signal, triggering the current source. Since the charging slopes across the inverter $V_{TH,inv}$ shares the same SR, the inverter-induced nonlinearity will be mitigated.

Figure 10.14(a) shows the proposed 10b DTC utilizing constant-slope method. Instead of varying the V_{ST} of the constant slopes that will potentially distort the current source, a ramp division architecture is adopted instead, as shown in Figure 10.14(b), wherein the $V_{TH}(n)$ of the comparator is shifted. The comparator will output a corresponding edge at t_n, producing the delay of t_n–t_0. By always using the same ramp generated by a current source, the linearity degradation from

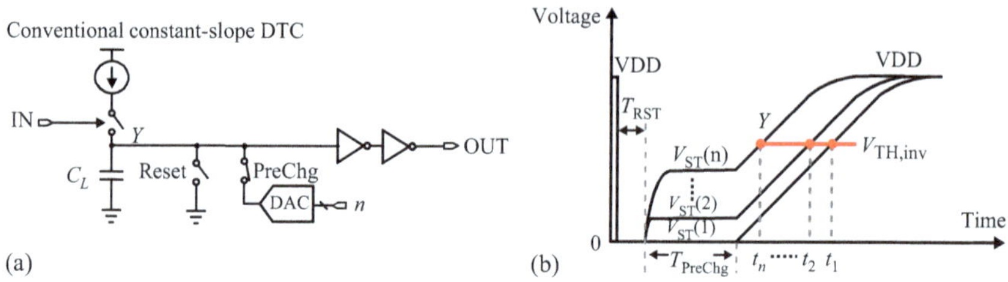

Figure 10.13 (a) Conventional constant-slope DTC (b) its operation

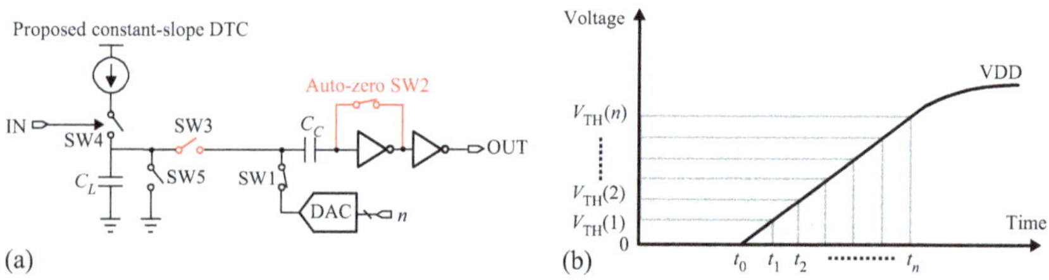

Figure 10.14 (a) Proposed isolated constant-slope DTC and (b) concept operation

the current source will be mitigated and higher voltage supply is not required. Furthermore, the ramp information can be used for nearly 600 mV/1 V = 60%, comparing with only 200 mV/1.2 V ≈ 17% in [16]. The V_{TH} (n) is shifted by isolating the precharge step from the comparing step by using a series capacitor C_C and a DAC. C_C is small enough not to degrade the precharge time on both sides of the C_C by DAC and Φ_3. The reduced precharging capacitance can minimize the DAC power consumption even in a high-speed operation. Another important issue raised from the comparator, which essentially just a simple inverter in most of the state-of-the-art DTCs. As shown in Figure 10.15, because of the off-chip inductance and resistance on the supply line from the power management IC, it causes ripples on the supply line. Those ripples will remain because of the limited area for decoupling capacitor and limited power budget for on-chip regulators in IoT applications and will strongly couple to the threshold of the inverter-based comparator. The linearity will be greatly influenced by the threshold variation and degrades the ADPLL jitter performance when the first-order DSM architecture is utilized. In the proposed architecture, this issue is solved by auto-zero switch Φ_3, which mitigates the inverter $V_{TH,inv}$ offset in every conversion, hence greatly improves the INL.

Figure 10.16(a) shows the detailed implementation of the proposed isolated constant-slope DTC. Figure 10.16(b) shows the detailed operation steps of the proposed DTC. In precharge step, the nodes A and B of C_C will discharge from the saturated voltage of previous slopes. Node A will discharge to the desired DAC voltage and node B will discharge to $V_{TH,inv}$. As we noticed, the precharge step is

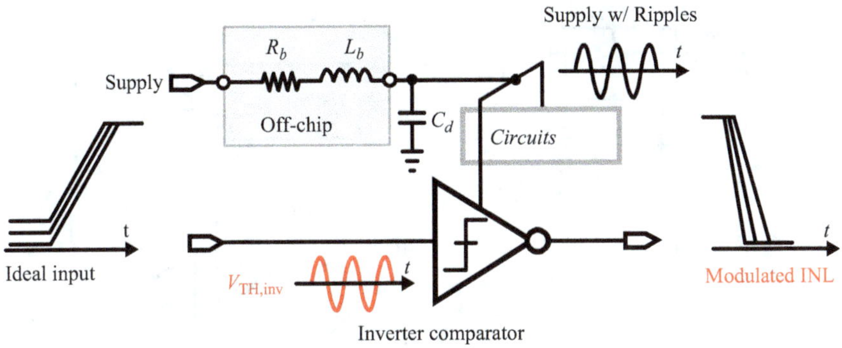

Figure 10.15 *V_{TH} offset caused INL, which was induced by noisy supply*

actually a discharge process that causes no extra power consumption from DAC. The precharge speed will be limited by the R-DAC and C_C. Because C_C is small, R-DAC can be chosen to be large to minimize the power consumption from DAC. In set step, the Φ_1 and Φ_4 are closed to short-node A to 0 V. Node B drops the same amount of voltage, resulting in a new $V_{ST} = V_{TH,inv} - V_{DAC}$ at inverter input. The conversion time of the set step is minimized for not degrading the operation speed of the proposed DTC but for minimizing the leakage charge from Φ_2. At the final step/compare step, Φ_1 is closed and Φ_5 is triggered by input rising edge. The current source starts to charge C_{load} to acquire a ramp at node A from 0 V to VDD, and node B will copy the ramp of node A while starting from V_{ST} generated by set step. The rising edge will reach the decision point of the inverter-based comparator and produces variable delays. If we consider a supply noise environment as shown in Figure 10.15, a $V_{n,Supply}$ is presented at the supply line. The digital controlled $V_{TH}(n)$ can be written as:

$$V_{TH}(n) = V_{TH,inv}(t_{N+1}) - \left(V_{TH,inv}(t_N) - V_{DAC}(n)\right) = \alpha(f)V_{n,Supply} + V_{DAC}(n) \tag{10.8}$$

where $\alpha(f)$ is a frequency-dependent factor with a value of less than one, and it also depends on the difference between t_{N+1} and t_N. Ideally, if $t_{N+1} = t_N$, $\alpha(f)$ will be 0. The variable delay for each control code will be:

$$t_n = V_{TH}(n)\frac{C_{bank}}{I_{current}} = \left(\alpha(f)V_{n,Supply} + V_{DAC}(n)\right)\frac{C_{bank}}{I_{current}} \tag{10.9}$$

Equation (10.9) shows that ideally the delay is no longer determined by the inverter threshold, thanks to the auto-zero function.

10.3.2.2 Nonlinear sources and circuit implementations

Since the linearity of the DTC affects both the rms jitter and fractional spurs of the proposed ADPLL, the linearity degradation from nonlinear sources should be minimized. The detailed DTC core implementation is shown in Figure 10.16(a). A

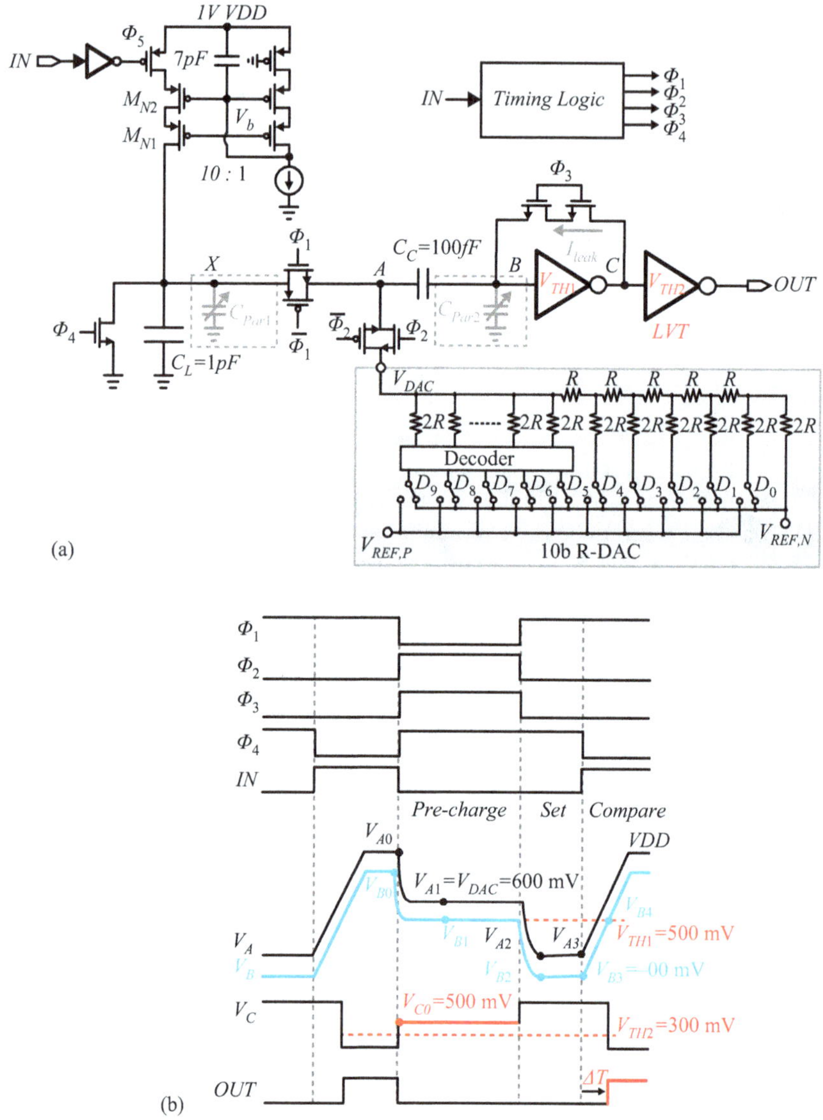

Figure 10.16 Detailed circuit of (a) isolated constant-slope DTC and (b) its timing chart

cascode current source is adopted to improve the current source linearity. As shown in Figure 10.14(b), the slope will be interpolated by $V_{\mathrm{TH}}(n)$. Any nonlinearity in the slope will be transferred to the DTC INL. Long-channel devices of M_{N1} and M_{N2} are chosen to minimize this error. Since the utilized slope information as shown in Figure 10.14(b) contributes to most of the nonlinearity of the proposed DTC, any improvement in the current source linearity will directly improve the DTC linearity.

Another major nonlinear source is from the junction capacitors C_{Par1} and C_{Par2} at node X and node B where all transistors connected to these nodes will contribute to C_{Par1} and C_{Par2}. $C_{\text{Para1}}(V_X)$ will be negligible if C_L is sufficiently higher and will not degrade the INL. While as for C_{Par2}, it acts as a voltage divider capacitor in series with C_C. In other words, the slope at node B in the compare step will not exactly follow the slope at node A. The waveform distortion at node B will degrade the INL of the DTC if the value of C_C is not properly sized. A high C_C is desired to minimize the effect from C_{Par2}. However, a high C_C will potentially increase the settling time and the power consumption of the DAC, which limits the maximum operation frequency. C_C is optimized to a value of 100 fF when considering sufficient margin to cover PVT variations of the above issues. Furthermore, node A always drops from $V_{\text{DAC}}(n)$ to 0 V, hence, C_{Par2} is a function of $V_{\text{DAC}}(n)$, i.e., C_{Par2} $(V_{\text{DAC}}(n))$. This dependency limits the maximum output range from DAC. In this design, an optimized range of 350 mV is chosen for the better DTC linearity when both the current source linearity and effects from C_{Par2} are considered.

One more major source of nonlinearity is from the leakage current I_{leak} of the auto-zero switch Φ_3 during set step. Φ_3 is opened to hold V_{ST} values before the slope arrives. However, due to the limited off-resistance r_{off} of the CMOS switch, the current will leak from node C to node B from the inverter's supply. It charges C_{Par2} and C_C simultaneously and causes nonlinear error voltages at node B. This error is highly dependent on $V_{\text{DAC}}(n)$ as well as the operation time of the set step and compare step $(t_{N+1} - t_N)$. To minimize this error, two switches are implemented in series to increase the effective r_{off} while the $t_{N+1} - t_N$ is minimized to 1/10 of the DTC period. The shortened $t_{N+1} - t_N$ also contributes to the supply noise suppression.

Last but not least is the nonlinearity from the 10b resistor-DAC (R-DAC). The binary code of 5b is designed for LSB to save the chip area, while 5b thermal code is designed for MSB to maintain a good linearity. The mismatch of the resistors and the nonideal reference voltage will cause the linearity degradation of the R-DAC. It will directly transfer to DTC INL. Other nonmajor nonlinear sources such as charge sharing of CMOS switches can be minimized by proper sizing of the transistors in simulations.

The detailed timing chart of the proposed DTC is shown in Figure 10.16(b). When the auto-zero switch is closed in the precharge step, node B will be set to the inverter V_{TH1}, which is around 500 mV. The voltage of the node C will shorted with node B and generate a 500mV at note B and C. If the second inverter V_{TH2} is also around 500 mV, the DTC output OUT will produce multiple zeros and ones due to the noise. To maintain a robust operation, an LVT inverter, whose $V_{\text{TH}} = 300$ mV, is placed at the output of the DTC.

10.3.2.3 Simulation results

The simulated INL of the proposed DTC are shown in Figure 10.17. In typical–typical corner with 1.0 V VDD and a temperature of 25°C, it achieves a 10b range of 560 ps with a 550 fs resolution. The peak INL is 0.05% (200 fs) at 52 MS/s with 140 μW power. Other corner and temperature conditions are also applied as shown in Figure 10.17(a). The worst case is observed when using a 0.9 V supply voltage. It is mainly because of the linearity degradation from the current source. The post-layout Monte-Carlo simulations are performed as shown in Figure 10.17(b). These simulations indicate a +450 fs peak INL. The linearity degradation shown in Monte-Carlo simulations is mainly caused by the linearity degradation from R-DAC. To evaluate the effect from the supply noise, the deterministic jitter variance δ^2 with and without auto-zero offset switch are shown in Figure 10.18. In the post-layout simulation, 20 mV_{PP} sine waves with different frequencies are applied to the supply of DTC core. The jitter variance as well as the corresponding suppression in dB with and without auto-zero function is recorded. From the simulation results, the suppression is higher if the supply noise frequency is lower. This matches with the analysis of the above (10.8). The smaller the noise frequency, the smaller will be the $V_{\mathrm{TH,inv}}(t_{N+1}) - V_{\mathrm{TH,inv}}(t_N)$. The simulated jitter of the DTC is around 560 fs. The current source contributes 82% of the jitter, while the other noise sources, such as the DAC noise, the switching noise and the inverter noise, contribute a total of around 18%.

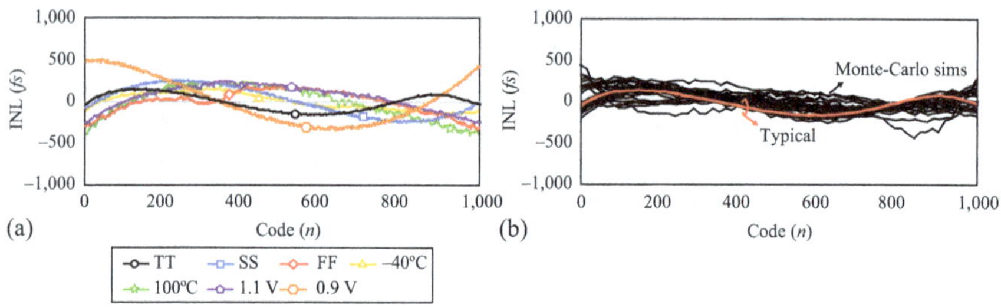

Figure 10.17 (a) Post-layout INL simulations of the proposed DTC with process, temperature, and voltage variations and (b) the Monte-Carlo simulations of the proposed DTC INL

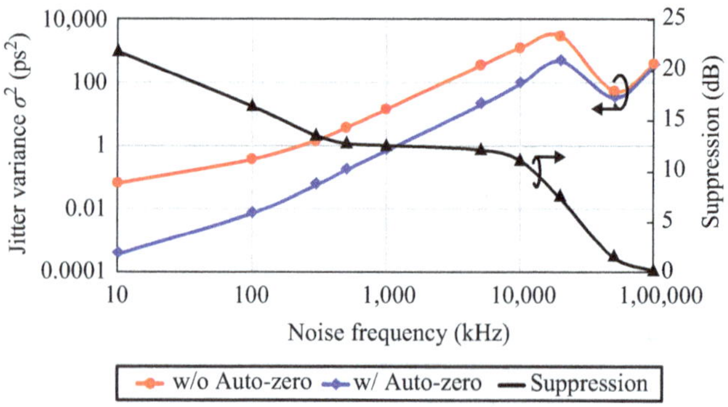

Figure 10.18 Simulated deterministic jitter power with and without auto-zero switch when noisy supply with different frequencies are presented, and the deterministic jitter power suppression with auto-zero switch

The proposed isolated constant-slope DTC can be used in many applications other than low-power fractional-N ADPLL. For example, it can be used for ultra-low-jitter low-spur fractional-N ADPLLs. It can also be applied to time-domain ADCs and digital low dropout regulators (LDOs) which require highly linear digital controlled delay units.

10.3.3 Narrow-range high-resolution TDC

A path-select TA-TDC is implemented as shown in Figure 10.19. A BBPD derives the sign of the phase error after TA, a path select logic is used to switch the up and down paths of TA output based on the results from BBPD. If T_{ON} leads T_{OP}, the up-down switch will be transparent for both signals. If T_{ON} lags T_{OP}, the up-down switch will switch the two signal paths to avoid the output of all zeros. The up-down switch function can make sure both lead and lag conditions between T_{ON} and T_{OP} are quantized by a 3b 16ps-resolution coarse TDC. At the output of the TDC, the quantizer output will be combined with 4b of the BBPD output. As compared with [3] which adopted two 3b quantizers for the same purpose, the path-select technique saves almost half of the power and area. Notice that in order to properly switch up and down signals, the BBPD should output selection signal before T_{OP} and T_{ON} arriving at the up-down switch. While in practice, the BBPD takes $\triangle t_{BBPD}$ to derive the selection signal. Two extra delays of $\triangle t_{SEL}$ are added before two inputs of the up-down switch. If $\triangle t_{SEL}$ is longer than $\triangle t_{BBPD}$, T_{OP} and T_{ON} can be properly switched. However, the path mismatch introduced by two extra delays bring a time error at 0 code and cause INL degradation. The TA can minimize the time error using its gain. In order to completely mitigate this issue, a constant offset code is added at the TDC output. After phase locking, the TDC will not use codes around 0 to avoid the potential linearity degradation. This constant offset is decided by the jitter of the ADPLL itself. A post-layout simulation of the path-select TDC shows a +50 fs/−220 fs peak INL with around 2.1 ps/LSB resolution.

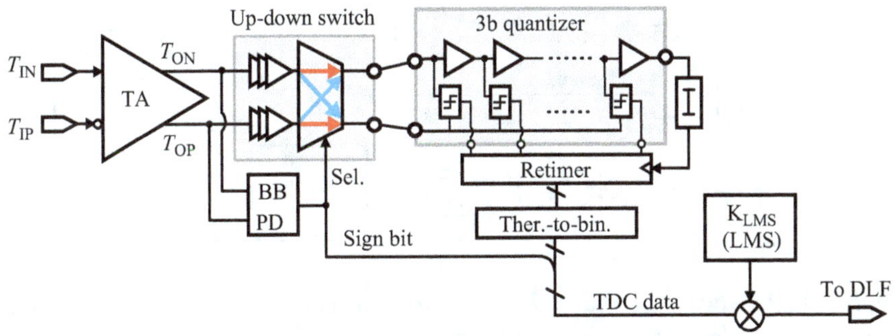

Figure 10.19 Path-select TDC

10.3.4 Coarse DPLL

For narrow-range TDC, the phase-locked time is generally very long when a high frequency error is presented. For ADPLL with a bang-bang PD [5], a 1 ms is reported for phase lock in case of a high frequency step. A lock time of around 40 μs is required for a TDC with 16 ps range even when no frequency error is presented [4]. In ADPLLs using the narrow-range TDC, the frequency locked loop is shut down for further saving the power consumption, which makes the loop easily suffer from high frequency and phase jump.

In our design, to lock the frequency of ADPLL, a 4b coarse bank of the DCO with a resolution of 50 MHz/LSB is controlled by autofrequency control function. The rest of the frequency error will be covered by a medium bank with a resolution of 1.25 MHz/LSB and a fine bank with a resolution of 7 kHz/LSB, thanks to the dither operation of DSM. However, even when the frequency is locked very close to the desired frequency, the high phase error may present at the narrow-range TDC. The narrow-range TDC will be saturated, and an open-loop gain will become zero. The saturated narrow-range TDC will degrade the converge speed of the main PLL. In our design, an always-on coarse-DPLL shown in Figure 10.20 works simultaneously with the main PLL. A dead-zone logic is inserted after the PFD which produces an enable signal to the counter running at DCO frequency. When the magnitude of the phase error is higher than the dead zone of 64 ps, the counter will be triggered by EN signal which has the same length of the phase error. The 4b narrow-range TDC will be saturated by the high phase error and the main PLL will be idle. When the phase error is sufficiently small and within narrow-range TDC quantization, the main loop will dominate the phase lock of the residue phase error. The EN signal is low after the phase lock and the counter stopped working. The coarse loop filter will be automatically disabled and the gating logic will minimize the digital power consumption from the loop filter. Since this loop will never

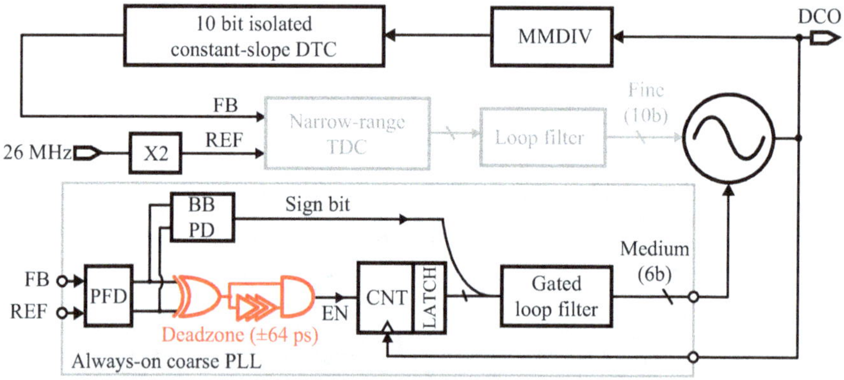

Figure 10.20 Always-on coarse PLL with a dead zone of ±64 ps which consumes almost zero power after phase lock

be turned off, the ADPLL will not suffer from the sudden high frequency and phase jump. The simulated power consumption of the coarse-DPLL is 5 μW after phase lock.

10.4 Measurement results

The proposed fractional-N ADPLL prototype was fabricated in a 65-nm CMOS process. The chip photograph of the fractional-N ADPLL is shown in Figure 10.21. The proposed 10b isolated constant-slope DTC and path-select TDC are also fabricated in 65 nm CMOS process as individual test circuits for INL measurement. The path-select TDC realized a 4b with 2.15 ps resolution at 52 MS/s. The peak INL is around 0.65 ps, thanks to the reduction of TDC range by the assist of DTC. The linear operation of the TA and TA gain also helps to reduce the linearity degradation due to the coarse quantizer. Sub-ps resolution DTC is not easy to measure due to the finite sampling frequency of the oscilloscope. A frequency-domain-based measurement method was introduced in [19]. The DTC achieves a 580 fs time resolution with a peak INL of 0.87 ps. It corresponds to an effective resolution of 9.4b in terms of linearity performance, and the DTC only consumes 140 μW at 52 MS/s. The proposed fractional-N ADPLL prototype was fabricated in a 65-nm CMOS process.

The fractional-N ADPLL is measured under one of the BLE channels at 2442 MHz as shown in Figure 10.22(a) and (b). To save the power consumption, reference doubler is bypassed by the MUX logic. The 26-MHz reference is directly used for DTC, TDC, and digital circuits. The phase noise plot is shown in Figure 10.22(b), and an integrated jitter from 10 kHz to 10 MHz of 1.00 ps is achieved. The measured fractional spurs are shown in Figure 10.23 by sweeping the FCW. A worst-case spur of −50 dBc is achieved without reference doubler. The

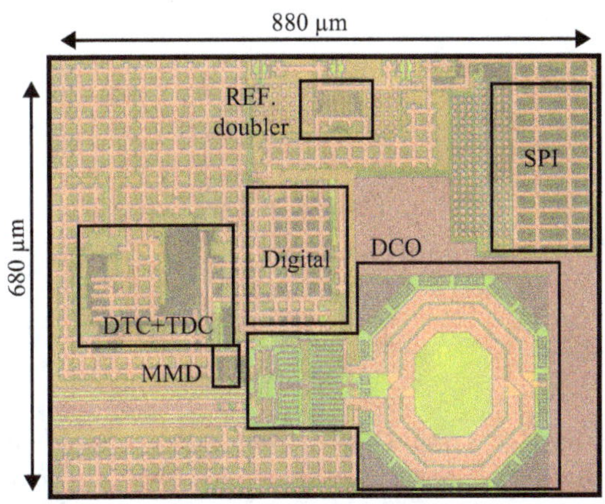

Figure 10.21 Chip photo

power consumption is extremely low for the achieved spur and jitter performances, which can be adopted to BLE applications. To boost the effective resolution of the TDC by increasing the sampling frequency, reference doubler is utilized. Figure 10.22(a) shows the phase noise plot at the same BLE channel of 2442 MHz. An integrated jitter of 535 fs is achieved. The integrated phase noise from 10 kHz to 10 MHz is −44 dBc, which is demanded for IEEE802.11b/g/n applications. A worst fractional spur of −56 dBc is measured as shown in Figure 10.23 with the reference doubler. When a small FCW of 47.000112 is used, the measured worst integrated jitter is 590 fs, as shown in Figure 10.22(d). And when we bypass the reference doubler, the measured phase noise is shown in Figure 10.22(c). However, for the target applications, such as BLE and Wi-Fi, such a small fractional number is not required. Hence, we can avoid using such small FCW to avoid jitter degradation.

The detailed power breakdown of each building block in the signal path of the proposed ADPLL is measured. The supply voltage for DTC and TDC are 1.0 V, so that a good linearity is maintained. A 0.8 V supply is assigned to the DCO and the

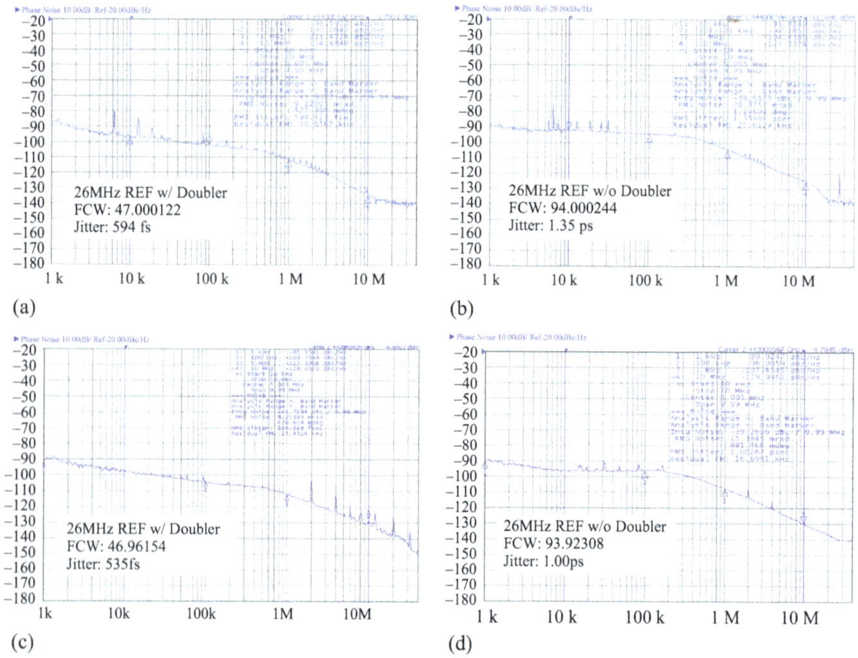

Figure 10.22 Measurement phase noise of the proposed fractional-N ADPLL with in-band fractional spurs and out-of-band fractional spurs: (a) and (c) the proposed ADPLL with reference doubler; (b) and (d) measurement result of the proposed ADPLL without reference doubler

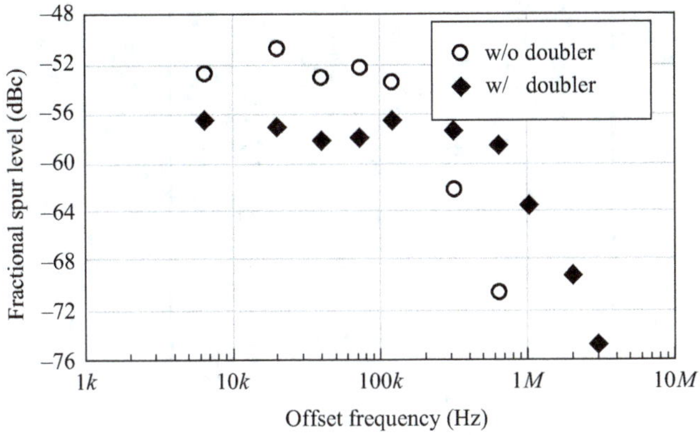

Figure 10.23 Measurement result of the fractional spurs vs. spur frequencies

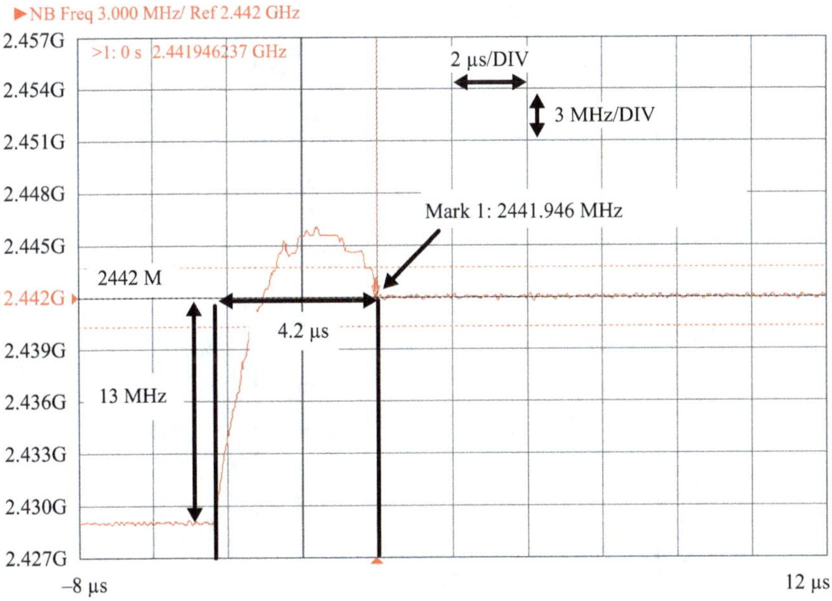

Figure 10.24 Measured lock transient from an initial frequency error of 13 MHz

digital parts, which include the DCO buffer, MMDIV, and the synthesized digital circuits, in order to maintain a low-power consumption. At a sampling rate of 26 MS/s, the DTC and TDC consume 98 and 80 μW power, respectively. The DCO bias is optimized for very low power operation of 285 μW. The digital parts consume 190 μW power. The total power is 0.65 mW for the jitter performance as shown in Figure 10.22(b). For a sampling rate of 52 MS/s, the DTC and TDC consumes 142 and 140 μW power, respectively. An additional power of 112 μW

Table 10.2 Comparison table of the state-of-the-art fractional-N ADPLLs

Reference	This work		[20]	[6]	[7]	[3]
Technology	65 nm		28 nm	40 nm	40 nm	65 nm
Architecture	Isolated constant-slope DTC + TDC		Full-range TDC	VS[a]-DTC +TDC	VS-DTC +TDC	VS-DTC +TDC
Ref. frequency	26 MHz w/doubler	26 MHz w/o doubler	40 MHz	32 MHz	N.A.	50 MHz
Frequency	2.0–2.8 GHz		2.05–2.55 GHz	2.1–2.7 GHz	1.8–2.5 GHz	4.4–5.2 GHz
Integrated jitter (Integrated BW)	0.53 ps (10k–10M)	1.00 ps (10k–10M)	0.86 ps (10k–10M)	1.71 ps (1k–100M)	1.98 ps (10k–10M)	0.49 ps (10k–20M)
In-band fractional spur	−56 dBc	−50 dBc	N.A.	−37 dBc	−56 dBc	−51.5 dBc
Power	0.98 mW	0.65 mW	1.6 mW	0.86 mW	0.67 mW	3.7 mW
FOM	−246 dB	−242 dB	−239.3 dB	−236 dB	−236 dB	−240.5 dB
Active area	0.23 mm^2		0.33 mm^2	0.2 mm^2	0.18 mm^2	0.22 mm^2

[a]Variable slope.

from the reference doubler is consumed to double 26 to 52 MHz with a 1.0 V supply. The DCO bias is optimized for achieving a better out-of-band phase noise. The digital parts consume 283 μW. The total consumed power is 0.98 mW for the jitter performance as shown in Figure 10.22(a).

Figure 10.24 shows the measured phase locking transient of the ADPLL. A 13-MHz frequency error is an input to the FCW of the ADPLL, which is over twice the entire frequency coverage of the DCO fine bank. With the help of the proposed coarse PLL, a measured lock-up time of 4.2 μs is achieved when the ADPLL locks 54 kHz away from the target frequency. The fast-phase converge speed can be adopted in frequency hopping applications such as BLE. The frequency hopping will cause the LMS calibration of the DTC gain to re-lock for the new synthesized frequency. The simulated re-lock time for the LMS calibration takes less than 15 μs to converge to a 0.3% gain error with 13 MHz frequency jump. However, even if the LMS does not converge to the final value, the PLL will still lock to the target frequency without any issue, while the fractional spur will be degraded during settling. The detailed performance comparison with the state-of-the-art fractional-N ADPLLs is shown in Table 10.2. The proposed ADPLL achieves a 10-dB better FOM than the conventional sub-mW ADPLLs [6,7].

10.5 Conclusion

To realize ULP fractional-N ADPLL with low jitter and low spurs, the first-order DSM-based fractional controller works in conjunction with a highly linear DTC. The rms jitter can be improved when compared to using higher-order DSM, and for this a DTC with high linearity is required. To realize a linear and high-energy-efficient DTC, an isolated constant-slope method is proposed. Thanks to the isolated operation of DTC, the proposed DTC can potentially work at a high sampling frequency with small power consumption while maintaining good linearity with high energy efficiency. Furthermore, the auto-zero offset switch mitigates part of the supply noise, which can improve the linearity in SoC environment. The proposed fractional-N ADPLL achieves good fractional spurs while maintaining a low jitter performance and low power, which proves the linearity and power efficiency of the DTC. The gain calibration of TA demonstrates a steady in-band phase noise of the ADPLL over the temperature variations. The measurement of lock time proves the effectiveness of the always-on coarse PLL in the feedback loop.

References

[1] H. Liu, Z. Sun, D. Tang *et al.*, “A DPLL-Centric Bluetooth Low-Energy Transceiver With a 2.3-mW Interference-Tolerant Hybrid-Loop Receiver in 65-nm CMOS,” IEEE Journal of Solid-State Circuits, vol. 53, no. 12, pp. 3672–3687, 2018.

[2] R. B. Staszewski, J. L. Wallberg, S. Rezeq *et al.*, "All-Digital PLL and Transmitter for Mobile Phones," IEEE Journal of Solid-State Circuits, vol. 40, no. 12, pp. 2469–2482, 2005.

[3] H. Liu, D. Tang, Z. Sun *et al.*, "A Sub-mW Fractional-N ADPLL With FOM of −246 dB for IoT Applications," IEEE Journal of Solid-State Circuits, vol. 53, no. 12, pp. 3540–3552, 2018.

[4] A. Elkholy, T. Anand, W. Choi, A. Elshazly and P. K. Hanumolu, "A 3.7 mW Low-Noise Wide-Bandwidth 4.5 GHz Digital Fractional-N PLL Using Time Amplifier-Based TDC," IEEE Journal of Solid-State Circuits, vol. 50, no. 4, pp. 867–881, 2015.

[5] D. Tasca M. Zanuso G. Marzin S. Levantino C. Samori and A. L. Lacaita *et al.*, "A 2.9–4.0 GHz Fractional-N Digital PLL With Bang-Bang Phase Detector and 560-fs_{rms} Integrated Jitter at 4.5-mW Power," IEEE Journal of Solid-State Circuits, vol. 46, no. 12, pp. 2745–2758, 2011.

[6] V. K. Chillara, Y. Liu, B. Wang *et al.*, "An 860_W2.1-to-2.7 GHz All-Digital PLL-Based Frequency Modulator with a DTC-Assisted Snapshot TDC for WPAN (Bluetooth Smart and ZigBee) Applications," Int. Solid-State Circuits Conf. (ISSCC) Dig. Tech. Papers, pp. 172–173, Feb. 2014.

[7] Y. He, Y. Liu, T. Kuramochi *et al.*, "A 673_W 1.8-to-2.5GHz Dividerless Fractional-N Digital PLL with an Inherent Frequency-Capture Capability and a Phase-Dithering Spur Mitigation for IoT Applications," Int. Solid-State Circuits Conf. (ISSCC) Dig. Tech. Paper, pp. 420–421, Feb. 2017.

[8] H. Liu, Z. Sun, H. Huang *et al.*, "A 265 μW Fractional-N Digital PLL with Seamless Automatic Switching Subsampling/Sampling Feedback Path and Duty-Cycled Frequency-Locked Loop in 65 nm CMOS," IEEE International Solid-State Circuits Conference (ISSCC), San Francisco, CA, USA, pp. 256–258, 2019.

[9] "Specification of the Bluetooth System v4.2," Bluetooth Special Interest Group.

[10] H. Darabi, Radio Frequency Integrated Circuits and Systems. Cambridge University Press, 2015.

[11] Y. L. Hsueh, L. Cho, C. Shen *et al.*, "A 0.29 mm2 Frequency Synthesizer in 40 nm CMOS with 0.19 ps rms Jitter and <−100 dBc Reference Spur for 802.11ac," Int. Solid-State Circuits Conf. (ISSCC) Dig. Tech. Papers, pp. 472–473, Feb. 2014.

[12] S. K. Lee, Y. Seo, H. Park *et al.*, "A 1 GHz ADPLL with a 1.25 ps Minimum-Resolution Sub-Exponent TDC in 0.18_ CMOS," Int. Solid-State Circuits Conf. (ISSCC) Dig. Tech. Paper, pp. 482–483, Feb. 2010.

[13] C. W. Yao, R. Ni, C. Lau *et al.*, "A 14-nm 0.14-psrms Fractional-N Digital PLL With a 0.2-ps Resolution ADC-Assisted Coarse/Fine-Conversion Chopping TDC and TDC Nonlinearity Calibration," IEEE Journal of Solid-State Circuits, vol. 52, no. 12, pp. 3446–3457, 2017.

[14] N. Markulic, K. Raczkowski, P. Wambacq, and J. Craninckx, "A 10-bit, 550-fs Step Digital-to-Time Converter in 28 nm CMOS," European Solid State Circuits Conference (ESSCIRC), pp. 79–82, Sep. 2014.

[15] A. A. Abidi, "Phase Noise and Jitter in CMOS Ring Oscillators," IEEE Journal of Solid-State Circuits, vol. 41, no. 8, pp. 1803–1816, 2006.
[16] J. Z. Ru, C. Palattella, P. Geraedts, E. Klumperink and B. Nauta, "A High-Linearity Digital-to-Time Converter Technique: Constant-Slope Charging," IEEE Journal of Solid-State Circuits, vol. 50, no. 6, pp. 1412–1423, 2015.
[17] X. Gao, E. A. M. Klumperink, M. Bohsali and B. Nauta, "A Low Noise Sub-Sampling PLL in Which Divider Noise is Eliminated and PD/CP Noise is Not Multiplied by N^2," IEEE Journal of Solid-State Circuits, vol. 44, no. 12, pp. 3253–3263, 2009.
[18] K. Raczkowski, N. Markulic, B. Hershberg, J. Van Driessche and J. Craninckx, "A 9.2–12.7 GHz Wideband Fractional-N Subsampling PLL in 28 nm CMOS with 280fs RMS Jitter," IEEE Journal of Solid-State Circuits, vol. 50, no. 5, pp. 1203–1213, 2015.
[19] C. Palattella, E. A. M. Klumperink, J. Z. Ru and B. Nauta, "A Sensitive Method to Measure the Integral Nonlinearity of a Digital-to-Time Converter Based on Phase Modulation," IEEE Transactions On Circuits and Systems II: Express Briefs, vol. 62, no. 8, pp. 741–745, 2015.
[20] N. Pourmousavian, F. Kuo, T. Siriburanon, M. Babaie and R. B. Staszewski, "A 0.5 V 1.6 mW 2.4 GHz Fractional-N All-Digital PLL for Bluetooth LE with PVT-insensitive TDC Using Switched-Capacitor Doubler in 28 nm CMOS," IEEE Symp. VLSI Circuits (VLSIC), pp. C178–C179, Jun. 2017.

Part III

Low-noise frequency generation and modulation

Chapter 11
Integrated LC oscillators

Jun Li[1] *and Yiwu Tang*[2]

The integrated inductor capacitor (LC) oscillator is one of the main oscillator topologies to generate low-noise local oscillator (LO) signals in wireless transceivers. Compared to inductor-less oscillators, such as the ring oscillator or the relaxation oscillator, the LC oscillator can achieve better phase noise and figure of merit (FOM) due to its high Q resonant tank. Figure 11.1 shows LC oscillators in modern radio frequency (RF) transceivers functioning as LO signal generators that drive the down conversion mixers in the receiver and the up conversion mixers in the transmitter.

11.1 LC oscillator design challenges in RF transceivers

A radio front-end must support new features such as software defined radio (SDR), carrier aggregation (CA), and multiple-in multiple-out (MIMO) links. Advanced cellular communication technologies, including 4G long-term evolution (LTE) and 5G new radio (NR), require superior phase noise performance for wideband communication. However, area and power consumption limit the number of oscillators in a single chip solution. Moreover, oscillators on the same chip can have coupling issues that generate many unwanted spurs along with the desired LO signal. These spurs degrade the phase lock loop (PLL)-integrated phase noise (IPN), resulting in worse signal to noise ratio (SNR) or error vector magnitude (EVM) performance. Therefore, modern radio transceivers require wideband and low-noise oscillators.

11.1.1 WiFi, Bluetooth low energy, and Internet of things

Nowadays, wireless connectivity standards, such as WiFi and Bluetooth low energy (BLE), are widely supported. 5G applications, such as narrowband Internet of things (NB-IoTs), have also grown more and more popular. All these standards require highly integrated oscillators for frequency translation of the modulated signal. Figure 11.2 shows the example frequency plans for WiFi, BLE, and IoT applications.

[1]Anokiwave, Inc., San Diego, CA, USA
[2]Qualcomm Technologies, Inc., San Diego, CA, USA

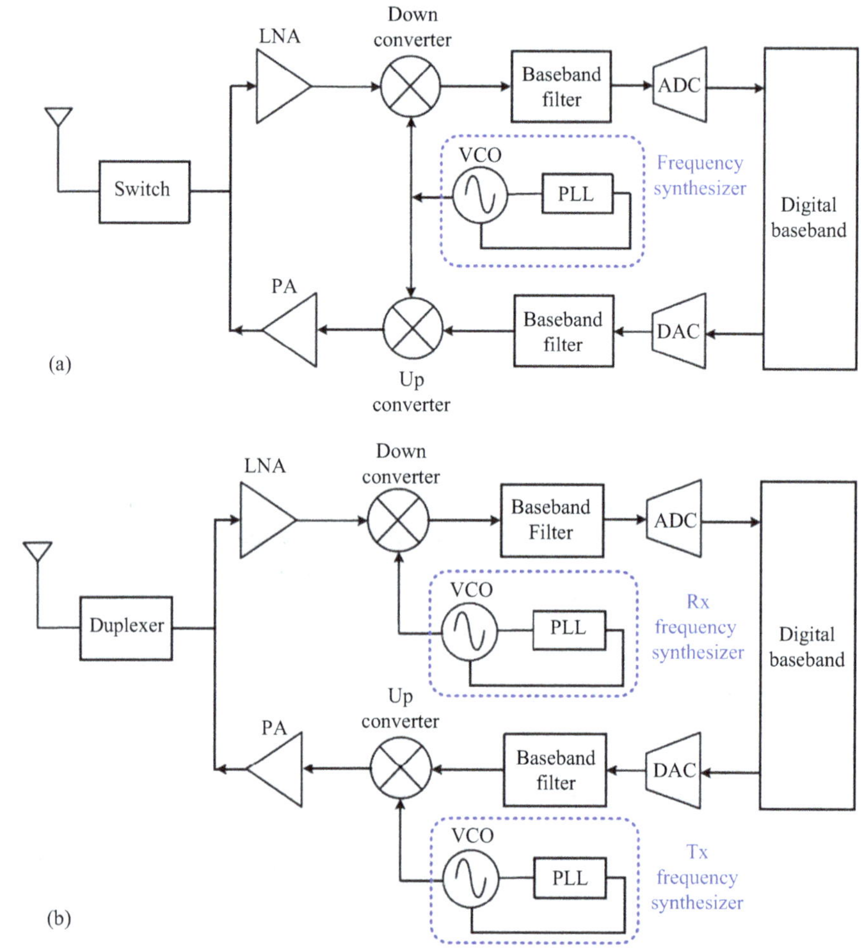

Figure 11.1 Integrated LC oscillators in wireless transceivers: (a) TDD (b) FDD

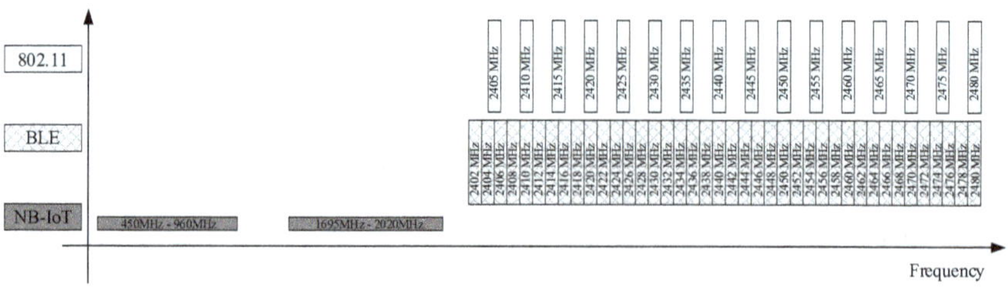

Figure 11.2 Frequency plan examples for WiFi, BLE, and IoT applications

Table 11.1 802.11 Standards and frequency bands

WiFi standards	Frequency bands	BW (MHz)	Modulation	PHY speeds (Mbps)
802.11 a	5 GHz	20	OFDM	6–54
802.11 b	2.4 GHz	22	CCK, DSSS	1–11
802.11 g	2.4 GHz	20	DSSS, OFDM	6–54
802.11 n	2.4/5 GHz	40	OFDM	600
802.11 ac	<6 GHz	160	OFDM	6,930
802.11 ad	60 GHz	2,160	SC, OFDM	6,760
802.11 af	<1 GHz	8	SC, OFDM	26.7
802.11 ah	<1 GHz	16	SC, OFDM	40
802.11 ax	2.4/5 GHz	160	OFDM	9607

Table 11.2 EVM and IPN specification for different modulation schemes

Modulation scheme	Code rate	EVM (dB)[1]	IPN (dBc)
QPSK	3/4	−13	−20
16QAM	3/4	−19	−25
64QAM	5/6	−28	−34
256QAM	5/6	−34	−40
1024QAM	NA	−46	−50

[1]Transmitter to receiver full link EVM budget example.

11.1.1.1 WiFi IEEE 802.11

WiFi IEEE 802.11 is used in many devices from smartphones, laptops, and tablets to remote sensors, actuators, televisions, and so on. Table 11.1 summarizes the frequency bands used by the WiFi 802.11 standard.

Increased data throughput and wider bandwidth require higher-order modulation schemes and MIMO techniques. For example, 802.11 ac achieves data rates up to 6.93 Gb/s with 160 MHz bandwidth, using 256 QAM modulation and MIMO configuration. Table 11.2 lists the typical phase noise specifications for different modulation schemes. The higher the modulation order, the lower is the required IPN. For 1024 QAM, the IPN specification is as low as −50 dBc, with an integration bandwidth from 7.5 kHz to 80 MHz.

11.1.1.2 Bluetooth low energy

BLE operates in the same spectrum range as the classic Bluetooth 2.4 GHz–2.4845 GHz Industrial, Scientific, and Medical (ISM) band, with different channel allocations. Instead of 79 channels of 1 MHz each as in Bluetooth, BLE has 40 channels of 2 MHz each. Within each channel, the Gaussian frequency shift keying (GFSK) modulation transmits data with 0.5 index and −70 dBm sensitivity, similar to the classic Bluetooth's basic rate scheme. The bit rate is 1 Mb/s, with an option to enable 2 Mb/s in Bluetooth. The maximum transmit power is 10 and

100 mW in BLE and Bluetooth, respectively. Table 11.3 shows the main characteristics of ZigBee, BLE, and classic Bluetooth. Note that BLE uses frequency hopping to counteract narrowband interference problems. Instead of a frequency-hopping spread spectrum (FHSS), BLE frequency-hopping technology is a direct-sequence spread spectrum (DSSS). To support frequency hopping, fast settling is imposed on the oscillator and phase locked loop designs.

BLE requires a 10^{-3} bit error rate (BER), implying an SNR higher than 6 dB based on an additive white Gaussian noise model. A better BER or worse channel condition dictates higher SNR performance. Therefore, the minimum specification for LO phase noise at 1 MHz offset would be <−110 dBc/Hz. Table 11.4 shows PLL phase noise examples from recent publications.

Table 11.3 Main characteristics of ZigBee, BLE, and classic Bluetooth

		ZigBee	**BLE**	**Classic Bluetooth**
Physical layer	RF band (MHz)	868/915/2,400	2,400	2,400
	Bit rate (kbps)	20/40/250	1,000	721/3,000/24,000
	Modulation	BPSK O-QPSK	GFSK	GFSK Π/4-DQPSK 8DPSK
	Spreading technique	DSSS	FHSS (2MHz)	FHSS (1MHz)
	Receiver sensitivity	−85 dBm (ISM) −92 dBm (900MHz)		<−70 dBm
	Transmit power (dBm)	−32 to 0	−20 to 10	20/4/0 (class 1/2/3)
Link layer	MAC	TDMA+CSMA/CA (beacon mode) CSMA/CA (beaconless mode)	TDMA	TDMA
	Message size (bytes)	127	8–47	358
	Error control	16-bit CRC	24-bit CRC	8-bit CRC 16-bit CRC + 2/3 FEC
	Latency (ms)	<5	<3	<100

Table 11.4 Typical phase noise performance for BLE

Reference	**[1]**	**[2]**	**[3]**	**[4]**
Technology (nm)	65	65	90	40
Frequency (GHz)	2.4	2.0	2.4	2.4
PLL Type	Analog Int-N	Digital Int-N	Analog Frac-N	Digital Frac-N
Supply (V)	0.68	0.78	1.2	1
Power (mW)	0.68	0.78	1.24	0.86
PN (dBc/Hz @1 MHz)	−110	−102	−113	−109

11.1.1.3 Narrowband IoTs

As part of 5G wireless communication standards, NB-IoT attracts a lot of attention. Also known as LTE Category NB1, it is a low-power wide area technology operating in half-duplex FDD (HD-FDD) mode. It supports many emerging applications such as industry fabrication 4.0, automotive, and smart homes. Table 11.5 lists the main bands assigned to the IoT standards.

The LO phase noise requirement at the offset frequency f_m can be calculated by (11.1):

$$\mathcal{L}(f_m) = C - S - I - 10\log(B) \tag{11.1}$$

where C is the desired signal in dBm, S is the selectivity in dB, I is the interference signal in dBm, and B is the signal bandwidth. The three requirements in the NB-IoT receiver specifications determine the PLL phase noise specifications: (1) the adjacent channel selectivity (ACS), (2) the in-band blocker, and (3) the out-of-band blocker. Table 11.6 shows the phase noise requirements derived from these requirements.

Table 11.7 summarizes the LO phase noise requirement of the transmitter. The far out phase noise at the duplex frequency is around −133 dBc/Hz determined by the SNR requirement and the transmitter (Tx) noise budget in the receiver (Rx) signal band.

Table 11.5 LO signal specifications for Internet of things' (IoTs) applications [5]

NB-IoT	Uplink band (MHz)	Downlink band (MHz)	Bandwidth (MHz)	Duplex mode
B1	1,920–1,980	2,110–2,170	60	HD-FDD
B2	1,850–1,910	1,930–1,990	60	HD-FDD
B3	1,710–1,785	1,805–1,880	75	HD-FDD
B4	1,710–1,755	2,110–2,155	40	HD-FDD
B5	824–849	869–894	25	HD-FDD
B8	880–915	925–960	25	HD-FDD
B11	1,427.9–1,447.9	1,475.9–1,495.9	20	HD-FDD
B12	699–716	729–746	17	HD-FDD
B13	777–787	746–756	10	HD-FDD
B14	788–798	758–768	10	HD-FDD
B17	704–716	734–746	12	HD-FDD
B18	815–830	860–875	15	HD-FDD
B19	830–845	875–890	15	HD-FDD
B20	832–862	791–821	30	HD-FDD
B25	1,850–1,915	1,930–1,995	65	HD-FDD
B26	814–849	859–894	35	HD-FDD
B28	703–748	758–803	45	HD-FDD
B31	452.5–457.5	462.5–467.5	5	HD-FDD
B66	1,710–1,780	2,110–2,200	70/90	HD-FDD
B70	1,695–1,710	1,995–2,020	25	HD-FDD
B71	663–698	617–783	65	HD-FDD
B72	451–456	461–466	5	HD-FDD
B73	450–455	461–465	5	HD-FDD
B74	1,427–1,470	1,475–1,518	43	HD-FDD
B85	698–716	728–746	20	HD-FDD

Table 11.6 Phase noise requirements due to blockers using (11.1)

Requirement	Blocker	Frequency offset (kHz)	Phase noise (dBc/Hz)
ACS1	GSM	110	−75.4
	E-UTRA	250	−83.6
ACS2	GSM	110	−73.2
	E-UTRA	250	−71.5
IBB1	E-UTRA	5,255	−96
IBB2	E-UTRA	10,250	−109
Out of band blocking	729 MHz–1 GHz	15,000	−110.8
	1,475.9 MHz–2,200 MHz	60,000	−124.8
	462.5 MHz–467.5 MHz	85,000	−139.8

Table 11.7 Transmitter LO phase noise requirements

Specifications	Band	Frequency offset (kHz)	Phase noise (dBc/Hz)
ACLR-GSM	N/E	110	−25.2
ACLR-UTRA	N/E	580	−26.8
Spurious emission	1	30,000	−133
	2	30,000	−133
	3	20,000	−133
	5	10,000	−110
		20,000	−133
	8	10,000	−133
	11	28,000	−133
	12	18,000	−133
	13	2,000	−96
		21,000	−133
	17	18,000	−133
	18	12,000	−123
		16,000	−133
	19	27,000	−133
	20	44,000	−133
	21	37,700	−133
	25	15,000	−133
	26	10,000	−133
	28	9,000	−117
		10,000	−115
		25,000	−133
	31	5,000	−133
		12,500	−134
	66	150,000	−133
	70	136,000	−133
Out-of-band emissions	N/E	0	−41.8
	N/E	100	−72.8
	N/E	150	−75.8
	N/E	300	−96.8
	N/E	500–1,700	−102.8
In-band emission	N/E	0	−30

Table 11.8 4G LTE and 5G NR bands

Bands	Uplink band (MHz)	Downlink band (MHz)	Divide Ratio	VCO1 (MHz)	VCO2 (MHz)
n71	663–698	617–652	12		7,956–8,376
B26	814–849	859–894	4	3,256–3,396	
B18	815–830	860–875	4	3,260–3,320	
UMTS850	824–849	869–894	4	3,296–3,396	
UMTS850	830–840	875–885	4	3,320–3,360	
UMTS900	880–915	925–960	4	3,520–3,660	
UMTS1800	1,710–1,785	1,805–1,880	2	3,420–3,570	
UMTS1700	1,750–1,785	1,845–1,880	2	3,500–3,570	
UMTS1900	1,850–1,910	1,930–1,995	2	3,700–3,820	
UMTS2600	2,500–2,570	2,620–2,690	2	5,000–5,140	
TD-LTE	2,496–2,690	2,496–2,690	2	4,992–5,380	
n78	3,300–3,800	3,300–3,800	2		6,600–7,600
n77	3,300–4,200	3,300–4,200	2		6,600–8,400
n79	4,400–5,000	4,400–5,000	2		8,800–10,000
		VCO tuning range		3,256–5,380 49.2%	6,600–10,000 42%

11.1.2 4G Long-term evolution and 5G NR

Current wireless communication networks widely deploy 4G LTE. 5G NR will be available in 2019. In 4G LTE and 5G NR, carrier aggregation supports gigabit per second (Gbps) data throughput. Hundreds of CA frequency band combo options entail 2–5 Rx downlink pipes and 2–3 Tx uplink pipes that require multiple PLLs on the same chip to provide multiple LO signals simultaneously. Each PLL needs to support as many bands as possible. Therefore, wideband oscillators are desired to minimize the number of voltage-controlled oscillators (VCOs) needed for each PLL and save die area. Table 11.8 provides a 4G LTE and 5G NR frequency plan example that use two VCOs to cover the frequency bands with various LO divide ratios.

Figure 11.3 shows a transceiver system on chip (SOC) example that supports 5G NR and 4G LTE CA features [6]. The chip integrates six Rx synthesizers and two Tx synthesizers for a total of eight frequency synthesizers on the same die, which are highlighted in Figure 11.3 chip block diagram and die photo. The proposed frequency plan requires multiple wide tuning range oscillators in order to reduce the number of VCOs and the die area. In the CA mode, multiple VCOs operate simultaneously. Chip-level floor plan maximizes VCO-to-VCO distance to mitigate coupling and spur issues.

11.2 LC oscillator principles of operation

11.2.1 Oscillator start-up condition, frequency tuning, and phase noise

11.2.1.1 Start-up condition

An oscillator can be modeled by the feedback system shown in Figure 11.4(a). When Barkhausen's criteria for oscillation given by (11.2) are met, the feedback

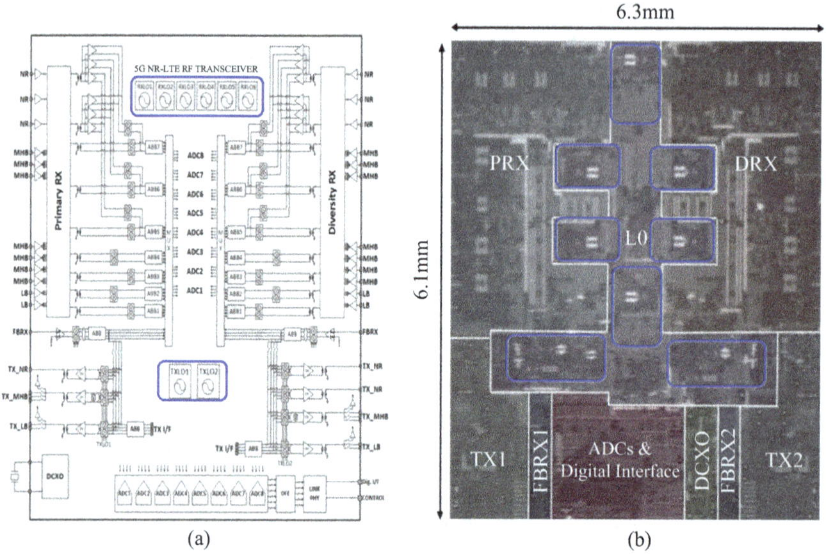

Figure 11.3 4G and 5G RF transceiver [6] (a) block diagram (b) die photo

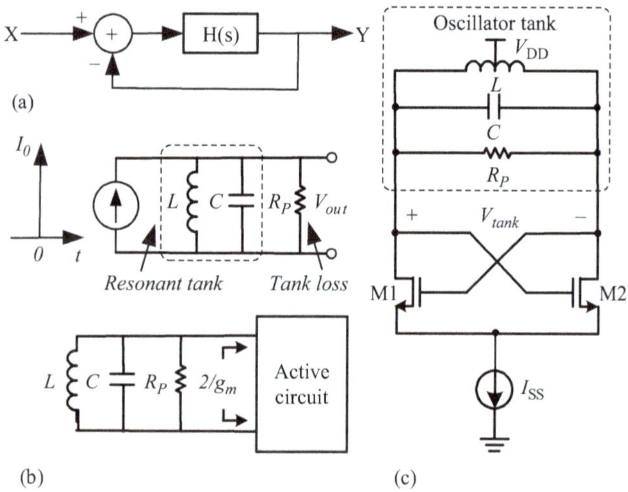

Figure 11.4 (a) Feedback system model, (b) small signal model of integrated LC oscillator, and (c) NMOS LC oscillator schematic

becomes positive and the loop becomes unstable and oscillates. The frequency ω_{osc} at which the criteria are met is the oscillation frequency:

$$|H(s = j\omega_{osc})| = 1 \tag{11.2a}$$

$$\angle H(s = j\omega_{osc}) = 180^{\circ} \tag{11.2b}$$

In other words, at the oscillation frequency, the loop is a positive feedback system with 360° phase shift around the loop. A classical LC oscillator implements the positive feedback with an n-channel metal oxide semiconductor (NMOS) cross-coupled differential pair as shown in Figure 11.4(c). The LC tank generates a 360° phase shift at its resonant frequency that determines the oscillation frequency.

From a one-port view, Figure 11.4(b) shows an LC oscillator small signal model. The cross-coupled NMOS pair generates negative impedance $-2/g_m$, where g_m is the transconductance of each NMOS transistor. To ensure oscillation, the magnitude of the negative impedance needs to be smaller than the tank impedance R_p expressed by (11.3):

$$R_p = Q\sqrt{L/C} \tag{11.3}$$

where Q is the LC tank quality factor, L is tank inductance, and C is tank capacitance. For robust start-up, the feedback loop small signal start-up gain, is usually designed to be greater than 2 over the process corners and temperature. In other words, oscillation start-up condition can be described as

$$2/g_m \leq R_p \tag{11.4}$$

In steady state, the negative impedance magnitude equals the tank impedance R_p expressed by (11.3), so that the active device compensates for tank loss.

The steady-state oscillation swing V_{tank} and oscillation frequency ω_{osc} can be calculated using (11.5a) and (11.5b), respectively:

$$V_{\text{tank}} = \frac{2}{\pi} I_{ss} R_p \tag{11.5a}$$

$$\omega_{\text{osc}} = \frac{1}{\sqrt{L(C + C_{\text{para}})}} \tag{11.5b}$$

where I_{SS} is the tail bias current and C_{para} is the tank parasitic capacitance.

11.2.1.2 Frequency tuning

Adding a voltage-controlled varactor to the oscillator tank supports tuning of the oscillation frequency. The inductor, the capacitor bank, and the varactor determine the overall effective tank Q. Figure 11.5 shows the tuning characteristic of an NMOS accumulation varactor. In general, the varactor capacitance and Q value can be expressed by (11.6) and (11.7), respectively. Design trade-offs need to be made between the tuning range and the tank quality factor. Minimizing the varactor channel length L gives better Q, while reducing the tuning range. For a wide fine-tuning frequency range, the varactor requires a large $C_{\max} - C_{\min}$ difference, leading to a high value of $W \cdot L$. In general, the minimum channel length will be used for better varactor Q and W is increased to meet the $C_{\max}/C_{\min}$ requirement:

$$C_{\max} = C_{OX} WL \tag{11.6a}$$

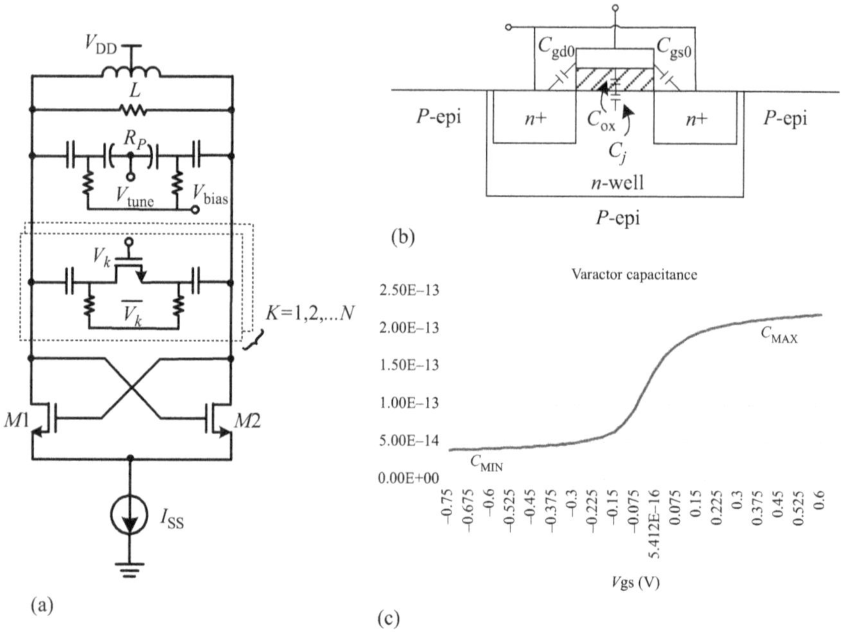

Figure 11.5 (a) NMOS LC VCO, (b) NMOS varactor, and (c) capacitance vs Vbias

$$C_{\min} \approx C_{\text{gdo}} W \tag{11.6b}$$

$$Q_{\text{var}} \approx \frac{12k_p(V_{GS} - |V_T|)}{\omega_0 C_{\text{ox}} L^2} \tag{11.7}$$

A discrete capacitor bank provides additional capacitor tuning range. As shown in Figure 11.5(a) LC oscillator schematic, the capacitor bank consists of a group of capacitors and switches that can be individually turned on or off to vary the total tank capacitance. Together with the tank inductor, the tank total maximum and minimum capacitance, including both the capacitor bank and the varactor, determines the final oscillator frequency range. Frequency tuning range (FTR) can be expressed by (11.8):

$$\text{FTR}(\%) = \frac{f_{\max} - f_{\min}}{(f_{\max} + f_{\min})/2} \times 100\% \tag{11.8}$$

11.2.1.3 Phase noise

Phase noise is an important wireless transceiver performance specification to meet SNR requirements. It is measured as the power per 1-Hz unit frequency at an offset frequency from the carrier. It is also normalized to the carrier power using a unit of dBc/Hz, where "c" stands for the carrier power normalization. The modulation scheme determines the close-in or in-band phase noise requirement. High-order modulation, such as 1024 QAM, dictates an SNR around 46 dB that leaves an in-band IPN budget of 50 dB. The out-of-band spot phase noise (SPN) folds into the signal band by the out-of-band jammers through reciprocal mixing. Therefore, the SPN specification is derived from the jammer offset frequency, jammer power level, and SNR requirement. An FDD transceiver's own Tx signal is usually the strongest jammer at the far out duplex frequency. For the Rx VCO, the phase noise at the duplex frequency degrades receiver SNR through reciprocal mixing of its Tx jammer. For the Tx VCO, the phase noise at this offset directly leaks into the Rx signal band.

Several main noise sources exist in the LC oscillator, including the thermal noise of the tank resistance R_p that models the tank loss, the thermal and flicker noise of the cross-coupled G_m cell as well as the tail current source. Based on the well-known Leeson's equation (11.9), the phase noise increases at higher oscillation frequency and lower offset frequency. To reduce the phase noise, the carrier power or oscillator tank swing needs to be increased and the tank Q maximized:

$$L(\Delta\omega) = 10 \cdot \log\left(\frac{FkT}{2P_{\text{sig}}}\frac{1}{Q^2}\left(\frac{\omega_0}{\Delta\omega}\right)^2\right) \tag{11.9}$$

Not included in Leeson's equation is the tail current noise, especially the flicker noise, that is up converted into the phase noise around the carrier frequency (Figure 11.6). Meanwhile, the noise around $2\omega_0$ will be down converted to the vicinity of the carrier frequency. The degeneration inductor provides a high common mode

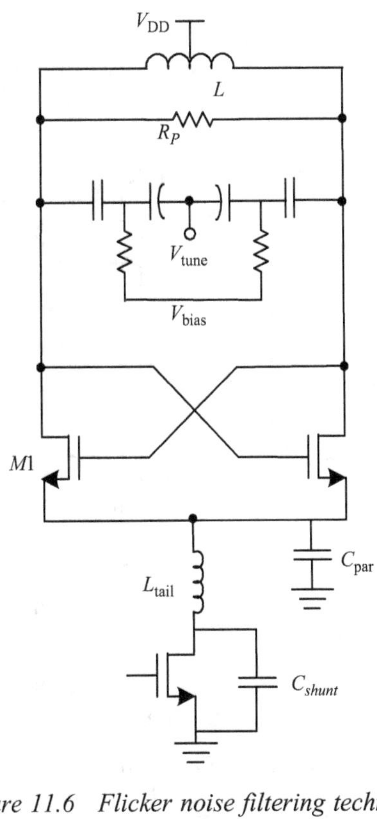

Figure 11.6 Flicker noise filtering technique

rejection ratio at $2\omega_0$ which removes the noise down conversion mechanism. Ideally, the degeneration inductor and its shunt capacitor at the drain node are chosen to resonate at $2\omega_0$. To vary the oscillation frequency in a wideband VCO, the degeneration tank uses a tunable shunt capacitor (C_{shunt}) to tune its resonant frequency as shown in Figure 11.6.

11.2.2 Topology comparison of NMOS VCO and CMOS VCO

Two types of widely used LC VCO topologies are NMOS VCO and complementary metal oxide semiconductor (CMOS) VCO. In the NMOS VCO shown in Figure 11.7, the negative G_m cell consists of an NMOS cross-coupled pair; while in the CMOS VCO, the negative G_m cell consists of both an NMOS pair and a PMOS pair.

The CMOS VCO tank voltage is limited to 0 ~ V_{DD} due to the PMOS pair, so the maximum swing is $V_{\text{DD}}/2$. For the NMOS VCO, the tank voltage can go above the supply voltage to $2V_{\text{DD}}$. In the current limited regime, where the tank swing is below the maximum, the CMOS VCO has a swing twice as much as that of the NMOS VCO for the same bias current, as illustrated in Figure 11.8. The center of the tank impedance R_p is a virtual ground. In the NMOS VCO, the current of each $R_p/2$ varies from 0 to I_{ss}. In the CMOS VCO, the current of R_p varies from $-I_{ss}$ to I_{ss}.

Because the CMOS VCO tank swing is twice that of the NMOS VCO for the same bias current in the current limited regime, its phase noise performance is superior to that of the NMOS VCO. However, its best achievable phase noise is worse than that of the NMOS VCO because its maximum tank voltage amplitude is limited to $V_{\text{DD}}/2$. Table 11.9 provides a comparison between the NMOS VCO and the CMOS VCO.

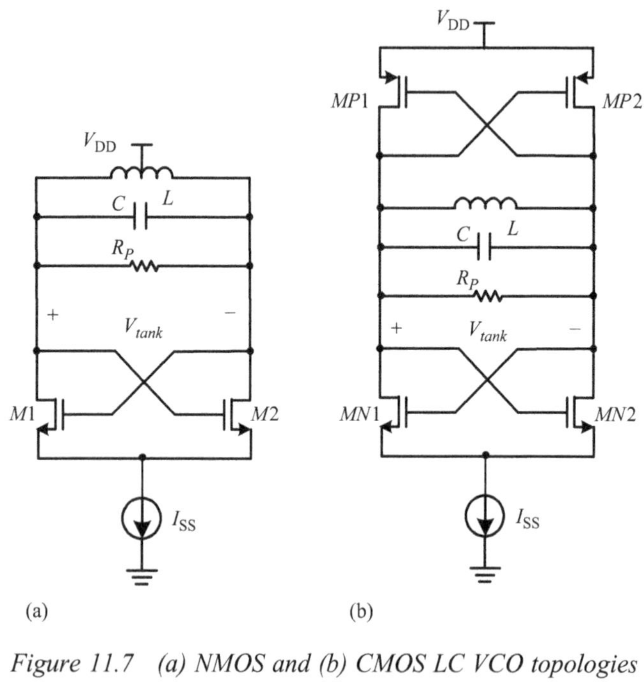

Figure 11.7 (a) NMOS and (b) CMOS LC VCO topologies

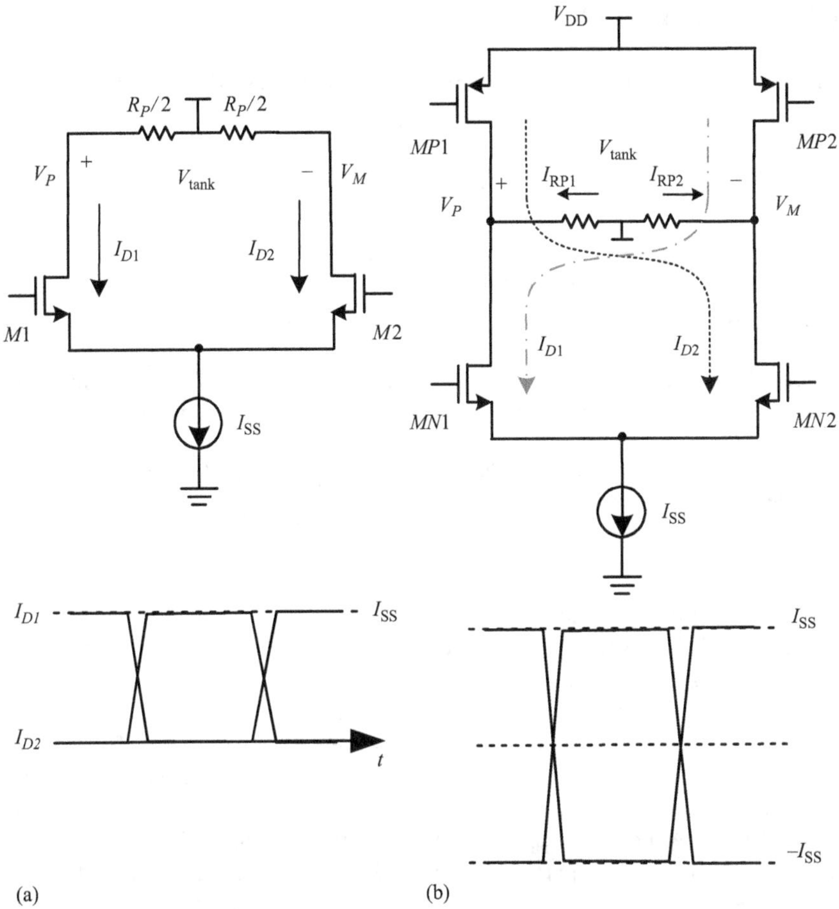

Figure 11.8 (a) NMOS VCO and (b) CMOS VCO current injection

Table 11.9 Topology comparison of NMOS and CMOS VCO

	NMOS VCO	CMOS VCO
G_m devices	NMOS pair	NMOS and PMOS pairs
Swing at same current	$\frac{2}{\pi} I_{SS} R_P$	$\frac{4}{\pi} I_{SS} R_P$
Maximum achievable swing	V_{DD}	$0.5 \cdot V_{DD}$
Parasitic capacitance from G_m	Low	High
PN at same current	High	Low
Best achievable PN at same VDD	Low	High
Start-up gain at same current	$g_{mN} R_P$	$(g_{mN} + g_{mP}) R_P$

11.3 Wideband LC oscillator

VCO performance is usually benchmarked using FOM specified by (11.10):

$$\mathrm{FOM} = 10\log\left[\left(\frac{f_0}{\Delta f}\right)^2 \frac{1}{L(\Delta f)P_{\mathrm{DC,mW}}}\right] \tag{11.10}$$

where $L(\Delta f)$ is the phase noise at a carrier frequency of f_0 and an offset frequency of Δf. $P_{\mathrm{DC,mW}}$ is the power consumption in mW.

As discussed in Section 11.1, wideband LC oscillators are highly desirable to save area and reduce cost of multiband wireless transceivers, in which the LO frequency to be generated spans across multiple octaves. To quantify tuning range performance, FOM_T, the FOM with tuning range, is expressed by (11.11):

$$\mathrm{FOM}_T = \mathrm{FOM} + 20\log\left(\frac{\mathrm{FTR}}{10}\right) = 10\log\left[\left(\frac{f_0}{\Delta f}\cdot\frac{\mathrm{FTR}}{10}\right)^2 \frac{1}{L(\Delta f)P_{\mathrm{DC,mW}}}\right] \tag{11.11}$$

where FTR is defined by (11.8).

Many published works [7–13] have demonstrated wide oscillator tuning range while achieving reasonable phase noise performance. To minimize phase noise degradation due to widened tuning range, the oscillator must maintain the tank Q or minimize the Q degradation. If the tuning range were achieved by simply increasing the coarse tune bank capacitance, it would reduce the tank inductance L, R_p, and tank Q due to finite switch resistance in the on state and large parasitic capacitance in the off state. Alternative approaches include the switchable inductor [7,8] that makes the tank inductance part of the tuning variable, the multi-resonator tank that alters the tank inductance by magnetic coupling [9–12], or the dual mode oscillator that exhibits different resonant frequencies in different oscillation modes [13].

11.3.1 Switchable inductor oscillator design

In addition to the conventional capacitive tuning, switchable inductor oscillators switch the tank inductor to two or more different values for frequency tuning. At each inductance value, the oscillator relies on the capacitor bank and the varactor for coarse and fine frequency tuning, respectively. The inductive tuning step is usually large due to the inductor size and should be covered by the capacitive tuning to support the overall wide contiguous tuning range. Usage of a switchable inductor can achieve an f_{max} to f_{min} ratio larger than 2. Moreover, with the help of LO dividers, a single oscillator can support LO frequency range from tens of MHz to tens of GHz, thus enabling low-cost SDR designs with a single oscillator.

Figure 11.9 shows an example of a switchable inductor-based dual-band LC oscillator [7]. In the low-frequency band, the inductor L_2 is shorted by turning on the switches M_5 and M_{10}, while its alternating current (AC) coupling capacitors

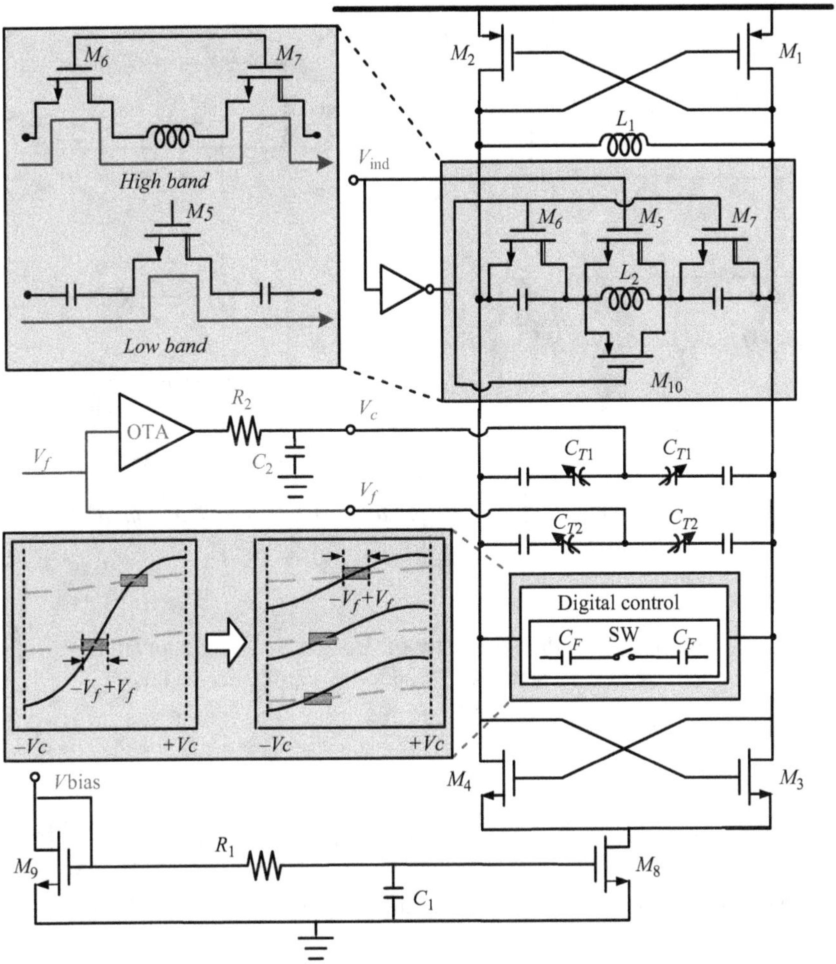

Figure 11.9 Switchable inductor oscillator

are enabled by turning off the switches M_6 and M_7. The effective tank inductor becomes L_1. In the high-frequency band, the inductor L_2 is enabled by turning off the switches M_5 and M_{10}, while its AC coupling capacitors are shorted by tuning on the switches M_6 and M_7. The effective tank inductor reduces to $L_1//L_2$ that increases the resonant frequency. The two-tank inductance values and 64-capacitor banks together achieve a wideband frequency tuning from 2.74 to 5.37 GHz. With LO dividers, 50 MHz–5 GHz ultra wideband LO is available either directly from the VCO buffer or from the divider output buffers.

Figure 11.10 shows the resonator models in the low-frequency band (a) and the high-frequency band (b), respectively, for the tank Q analysis. In the figure, C_1 is the AC coupling capacitors of the inductor L_2, $R_{on,sw1}$ is the combined resistance of

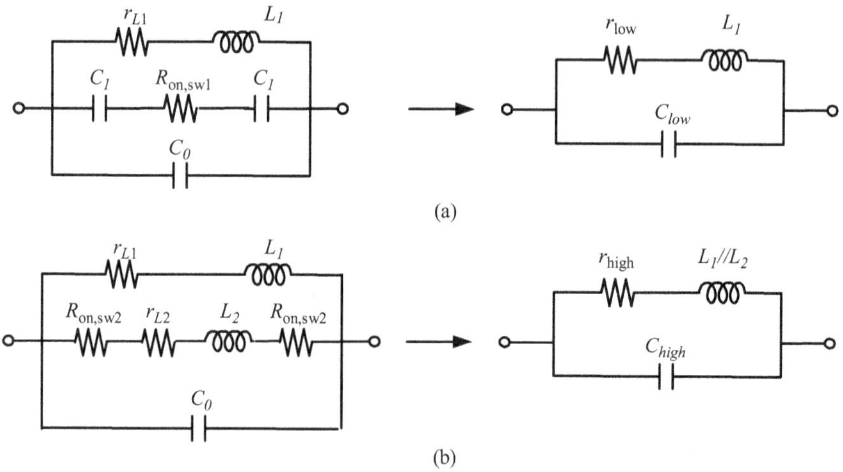

Figure 11.10 Switchable inductor tank models (a) low-frequency band (b) high-frequency band

M_5 and M_{10}, and $R_{on,sw2}$ is the resistance of M_6 and M_7. The equivalent low and high band inductor series resistances r_{low} and r_{high} can be derived as

$$r_{low} \cong r_{L_1} + R_{on,sw_1} \tag{11.12}$$

$$r_{high} \cong \frac{r_{L_1} + (r_{L_2} + R_{on,sw_2})(L_1/L_2)^2}{(1 + L_1/L_2)^2} \tag{11.13}$$

Assuming the capacitor load is C_{load}, the tank Q in the two bands is:

$$C_{load} = C_{load,min} + \Delta C \tag{11.14}$$

$$Q_{low} \cong \frac{1}{r_{low}} \sqrt{\frac{L_1}{C_{load} + C_1/2}} \tag{11.15}$$

$$Q_{high} \cong \frac{1}{r_{high}} \sqrt{\frac{L_1 L_2}{C_{load}(L_1 + L_2)}} \tag{11.16}$$

Where $C_{load,min}$ is the minimum load capacitance, Q_{high} and Q_{low} are the quality factor values of the VCO tank in the high band and the low band, respectively. Because r_{high} is usually larger than r_{low}, additional capacitor C_1 is inserted to balance the Q difference between the two bands in order to minimize the Q degradation in the high-frequency band. In order to achieve continuous tuning, the overlap between high band minimum frequency $\omega_{0,high\_min}$ and low band maximum frequency ω_{0,low_max} needs to be ensured. By using minimum capacitor load C_{min} and maximum capacitor load C_{max}, these two frequencies can be calculated. The low band and high band oscillation frequencies are

$$\omega_{0,\text{low}} \cong \sqrt{\frac{1}{L_1(C_{\text{load}} + C_1/2)}} \tag{11.17}$$

$$\omega_{0,high} \cong \sqrt{\frac{1}{(L_1//L_2)C_{\text{load}}}} \tag{11.18}$$

When $\omega_{0,\text{high\_min}} \leq \omega_{0,\text{low\_max}}$ for continuous tuning, the inductor ratio L_2/L_1 must meet:

$$\frac{L_2}{L_1} \geq \frac{C_{\text{min}} + C_1/2}{\Delta C - C_1/2} \tag{11.19}$$

If $r_{\text{high}} = r_{\text{low}}$, a condition of keeping Q constant can be expressed as

$$\frac{L_1}{L_2} \cong \frac{C_1}{2C_0} \tag{11.20}$$

With the switchable inductor, the dual-band VCO achieves an FOM of −182 dBc and FOM_T of −198.3 dBc at 4.6 GHz. Similar techniques applied to the SiGe process has also demonstrated larger than 85% tuning range [8].

11.3.2 Multi-resonator oscillator design

The multi-resonator oscillator is another wideband oscillator topology that supports multiple oscillation bands [9–12]. It finds application in SDR and millimeter wave frequency generation. It is a magnetic tuning technique, where magnetic coupling alters the tank inductance. Figure 11.11 shows a multi-resonator design example [9]. When the switch M_3 is off, there is no current induced in the inductor L_2, so the tank inductance is L_1. When the switch M_3 is on, the current in the inductor L_1 induces a current flowing through L_2 by magnetic coupling. By Lenz's Law, the induced current generates a magnetic field that partially cancels the one from the current in L_1, thus reducing the effective inductance, as calculated by (11.21):

$$L_{\text{eff}} = L_1 - \frac{M^2}{L_2} \tag{11.21}$$

M is the mutual inductance between L_1 and L_2. Similar to the switchable inductor, the effective Q of L_1 degrades due to the resistance loss from both L_2 and M_3 when M_3 is on. In addition, the effective inductance is reduced according to (11.21), so the effective Q is reduced even if L_2 and M_3 are lossless.

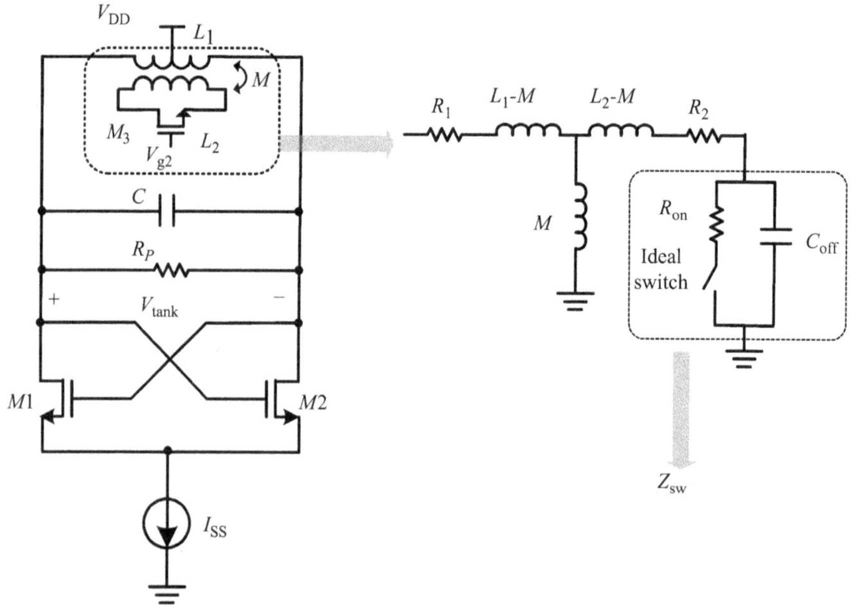

Figure 11.11 Multi-resonator oscillator

11.3.3 Dual-mode oscillator design

In addition to the switchable inductor and multi-resonator techniques, dual-mode oscillators achieve better phase noise by eliminating the lossy switches. Figure 11.12 shows a dual-mode VCO design example [13].

The *LC* resonator has two identical *LC* tanks coupled through inductor magnetic coupling M and capacitive coupling C_C. As a fourth-order system, it has two resonant modes with different resonant frequencies. In the even mode, the voltages across the two LC tanks, as well as the currents in the two coils, have the same amplitude and phase. As illustrated in Figure 11.13(a), each coupling capacitor C_C experiences zero voltage drop and zero current and thus can be removed. Because the currents in both coils are in phase, the effective inductance is $L + M$. Thus, the even-mode resonant frequency is

$$\omega_{\text{even}} = \frac{1}{\sqrt{(L+M)C}} \tag{11.22}$$

In the odd mode, the voltages across the two *LC* tanks and the currents in the two coils have the same amplitude but are 180° out of phase. As illustrated in Figure 11.13(b), each coupling capacitor C_C experiences a differential voltage and thus can be modeled as two $2C_C$ in series with a virtual ground in between. Hence, each *LC* tank sees a series of two $2C_C$, leading to an effective tank capacitance of

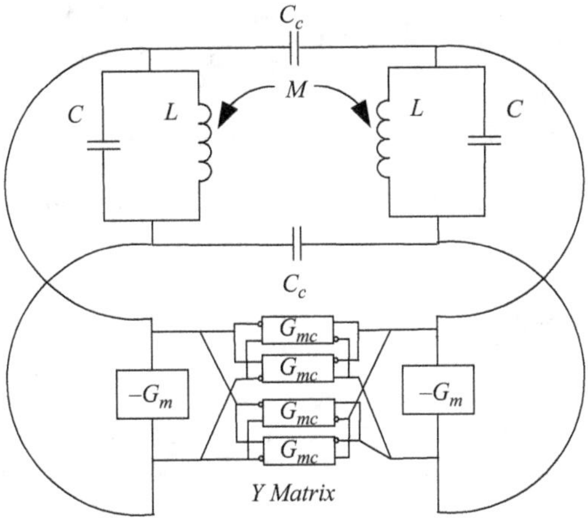

Figure 11.12 Dual-mode oscillator

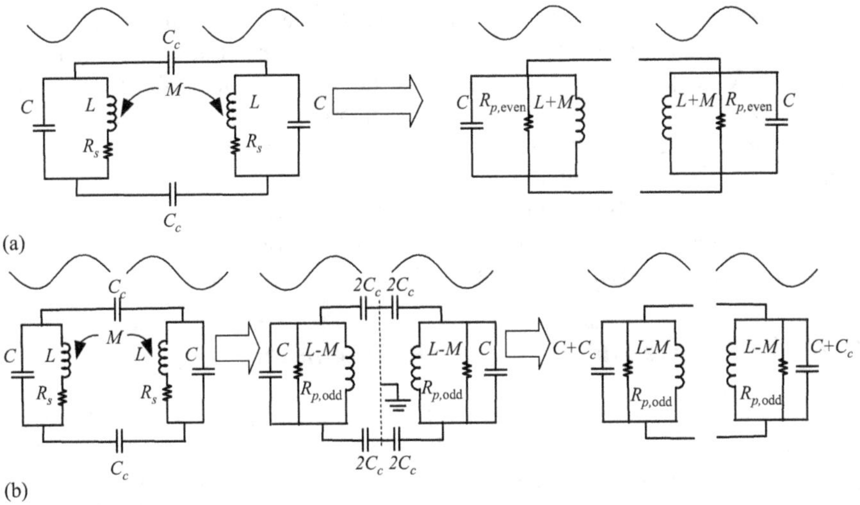

Figure 11.13 Dual-mode oscillator (a) even mode and (b) odd mode

$C + C_C$. Due to the out-of-phase currents in the two inductors, the effective inductance is $L - M$. As a result, the odd-mode resonant frequency is

$$\omega_{\text{odd}} = \frac{1}{\sqrt{(L - M)(C + C_C)}} \tag{11.23}$$

Assuming the series resistance of two coils, R_s dominates in the resonator's energy loss, the equivalent parallel resistance of the two resonator modes is

$$R_{p,\text{even}} = \frac{L + M}{C * R_s} \tag{11.24}$$

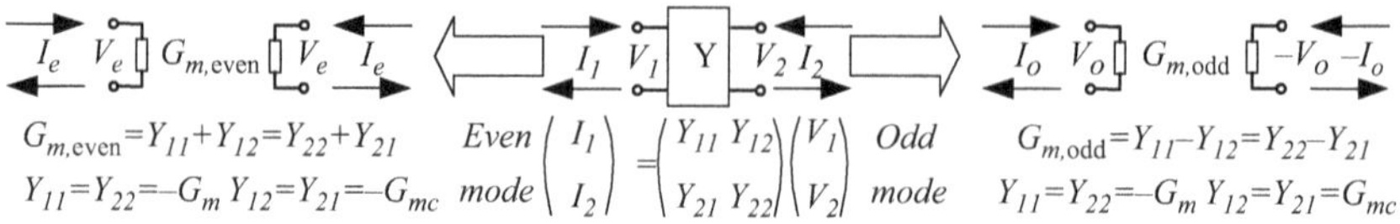

Figure 11.14 Y matrix in even and odd modes

$$R_{p,\text{odd}} = \frac{L - M}{(C + C_C) * R_s} \tag{11.25}$$

Note that negative M is preferred, which makes it possible to keep $R_{p,\text{even}} = R_{p,\text{odd}}$. The oscillation mode can be controlled by a switched transistor network. Figure 11.14 shows the even-mode and odd-mode Y matrix. In order to enhance one oscillation mode and damp the other mode, in addition to the two main G_m cells, two pairs of coupling G_m cells (G_{mc}) are included as shown in Figure 11.12. Their transconductance designed is larger than that of the main G_m cells. By properly connecting the two-port network between two G_m cells, Y_{12} can be set as G_{mc} or $-G_{mc}$. In the even mode, the top G_{mc} pair is off and the bottom pair is on. As a result, $-G_m - G_{mc}$ is negative with a large amplitude enhancing the even oscillation mode, while $-G_m + G_{mc}$ is positive damping the odd oscillation mode. Similarly in the odd mode, the top G_{mc} pair is on and the bottom pair is off, which enhances the odd-mode oscillation and damps the even-mode oscillation.

The prototype is fabricated in 65 nm CMOS and covers a frequency range from 2.48 to 5.62 GHz. FOM is between 189.4 and 193.7 dBc depending on the oscillation mode with a current consumption of 16.4–25.3 mA from 0.6 V.

11.4 Low-noise and low-power LC oscillator

Modern wireless communication standards such as 4G LTE and 5G NR extend the signal bandwidth to 100 MHz or more using advanced modulation schemes, such as 256 QAM or 1024 QAM. They require low EVM or high SNR for low BER that dictates low IPN over wide signal bandwidth for LO signals. Therefore, a low-noise oscillator is a key block for modern wireless communication systems. Low-noise LC oscillator topologies have been widely studied in publications. In this section, we will discuss coupled oscillators, as well as class C and class F oscillators, some of which are practical topologies adopted in wireless transceiver production chips.

11.4.1 Coupled oscillator

As CMOS technology feature size continues to shrink for high integration and low power, so does the V_{DD} supply voltage. Because the oscillator phase noise is the ratio between the noise power and carrier power, the oscillator swing and supply voltage limit phase noise performance. By applying two oscillators coupled in phase, we can enhance signal power by 6 dB, while uncorrelated noise power

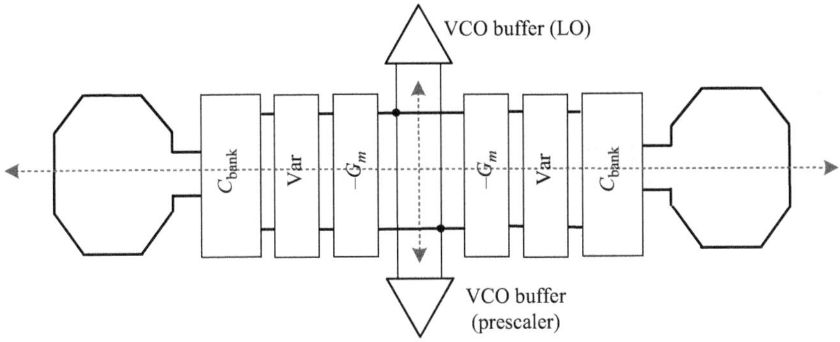

Figure 11.15 Coupled oscillator [14]

increases by 3 dB, resulting in a 3-dB phase noise improvement. Because the total power consumption is doubled, the FOM remains the same. Although the FOM is not improved, the best achievable phase noise is 3 dB better. Figure 11.15 shows the block diagram of a coupled oscillator example [14] with two sets of inductors, capacitor banks, varactors, and G_m cells, which is also called the dual-core oscillator. The coupled oscillator often finds its application in high-frequency VCOs or millimeter-wave VCOs, for 256 QAM or 1024 QAM requiring very low phase noise.

When two oscillators are coupled in phase, the currents flowing through the two inductors take opposite directions, one clockwise and the other counter-clockwise. Their magnetic fields cancel each other along the vertical symmetry line. If another VCO is placed along this line, the VCO-to-VCO coupling can be minimized. This is a valuable benefit in highly integrated transceivers where many VCOs may operate at the same time to support CAs.

To reduce tank loading and power consumption, there is usually only one VCO buffer that is directly connected to the tank. In the coupled oscillator, a practical yet important issue to consider is how to route the tank signal to the VCO buffers that drive the LO path and PLL N-divider path. If there were only one VCO buffer connected to the center of the oscillator from one side, the oscillator would no longer be symmetric with respect to the horizontal symmetry line. The asymmetric coupling causes tank differential voltage phase and amplitude error, resulting in inphase/quadrature (I/Q) mismatch and residual side band (RSB) performance degradation after the LO divider. To maintain the symmetry, two VCO buffers can be applied symmetrically to both sides, one for the LO path and the other for the N-divider side. The penalty is some extra power consumption.

11.4.2 Class C oscillator

Class C and Class F oscillators have demonstrated good phase noise and low power [15–17]. Figure 11.16 shows an example of a Class C oscillators.

In the conventional LC oscillator, the bias current is steered between the two transistors in the G_m cell, and each transistor conducts current during half of a

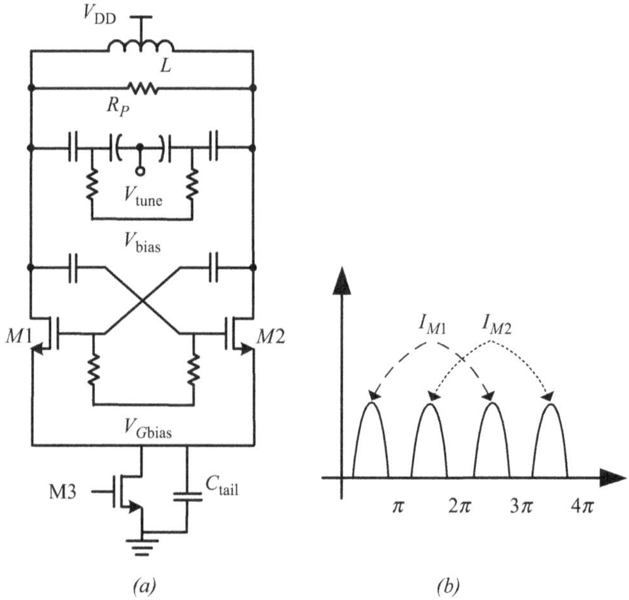

Figure 11.16 (a) Class C oscillator and (b) G_m device drain currents

period or at 180°. Following the power amplifier (PA) naming convention, it is therefore called a Class B oscillator. Because a Class C PA exhibits higher power efficiency than its Class B counterpart due to smaller transistor current conduction angle (<180°), the Class C oscillator was proposed [15]. As shown in Figure 11.16 (a), the gates and drains of the cross-coupled transistors M_1 and M_2 are AC coupled, so that their gates can be independently biased. If V_{Gbias} is set below V_{DD}, during part of each period, neither M_1 nor M_2 is conducting. A large tail capacitor C_{tail} is needed in the Class C oscillator to sink current when M_1 or M_2 current goes above I_{bias} or to source current otherwise. It also provides an AC ground at the source of the G_m cell. The source voltage therefore remains nearly constant, contrasting with the Class B oscillator, whose source voltage has a ripple twice that of the oscillator frequency.

In the Class B oscillator, where the drain current is nearly a square wave, the fundamental frequency component is $(2/\pi) \cdot I_{\text{bias}}$. Whereas in the Class C oscillator, it is approximately I_{bias}. Therefore, for the same phase noise, the theoretical power consumption reduction is 3.9 dB compared with the Class B oscillator. Or in other words, with the same current consumption, the theoretical phase noise performance improves by 3.9 dB. The advantage of the Class C oscillator therefore stems from its small current conduction angle that results in a higher conversion efficiency of the direct current (DC) current to the fundamental oscillation current.

The Class C oscillator has another advantage in that it prevents the G_m transistors from entering the triode region, thanks to gates biased below the drain voltage. As a result, it prevents the tank Q degradation and improves the phase noise performance. Because the gates of the G_m cells are biased at a lower voltage than its Class B counterpart to gain the Class C operation benefits, oscillator start-up becomes more difficult. Adaptive gate biasing using an amplitude feedback loop was proposed [17] to ensure robust start-up without a current penalty in the steady state.

11.4.3 Class F Oscillator

The Class C oscillator improves the power efficiency and phase noise performance by reducing the conduction angle of the G_m transistor drain current similar to the Class C PA technique. The Class F oscillator [18], on the other hand, focuses on shaping the tank voltage waveforms, mimicking the Class F PA that tunes the resonator impedance to odd harmonics and shorts the even harmonics. Therefore, the output waveform ideally is a square wave with a maximum flat region. In practice, it is feasible to introduce a third harmonic resonance to the tank in addition to the fundamental peak impedance. Higher-order resonance is not practical to implement. As shown in Figure 11.17, the transformer primary inductor and secondary inductors form a fourth-order system with two-capacitor banks C_1 and C_2. It exhibits two peak resonant frequencies ω_1 and ω_3 with the peak impedance of R_{p1} and R_{p3}, respectively. By choosing the proper inductance and capacitance, a relationship of $\omega_3 = 3\omega_1$ can be obtained, as shown in (11.26–11.28). To tune the oscillation frequency, both ω_1 and ω_3 need to be tuned to vary the fundamental oscillation frequency and maintain the ω_3 and ω_1 ratio:

$$\omega_1^2 = \frac{1}{L_pC_1 + L_sC_2} \tag{11.26}$$

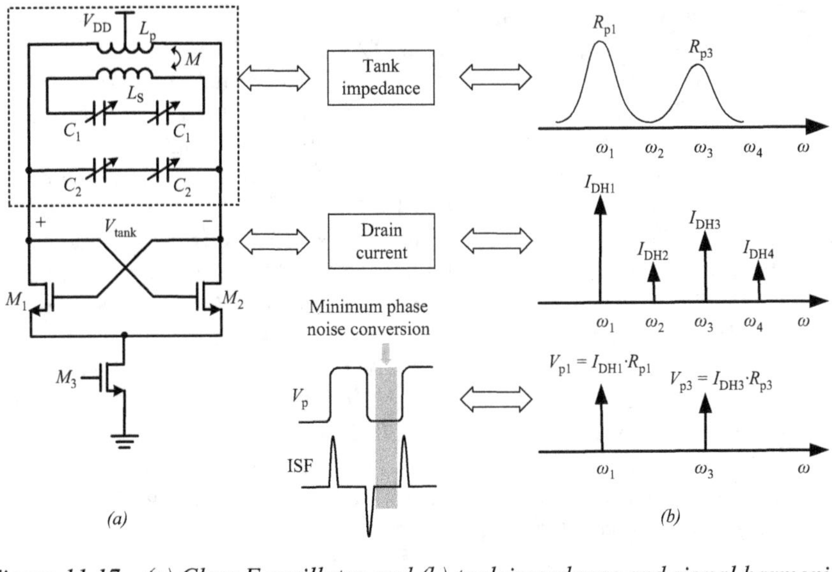

Figure 11.17 (a) Class F oscillator and (b) tank impedance and signal harmonics

$$\omega_3^2 = \omega_1^2 \frac{1 + X + \sqrt{1 + X^2 + X(4k^2 - 2)}}{1 + X - \sqrt{1 + X^2 + X(4k^2 - 2)}} \tag{11.27}$$

$$X = \frac{L_s C_2}{L_p C_1} \tag{11.28}$$

Due to the pseudo-square waveform, the impulse sensitivity function (ISF) becomes minimum during most of the period when the tank voltage is flat as shown in Figure 11.17. With low ISF, less drain current noise injected into the tank is converted to phase noise. When ω_1 and ω_3 deviate from the optimal 1:3 ratio, the ISF rms value is also worse than optimum, thus increasing drain current noise to phase noise conversion. In addition, it also causes more asymmetry in the voltage waveform rise and fall times, resulting in a nonzero ISF DC value that increases the G_m devices flicker noise up conversion and degrades the phase noise performance.

References

[1] A. Paidimarri, N. Ickes and A. P.Chandrakasan, "A 0.68V 0.68mW 2.4GHz PLL for ultra-low power RF systems," *2015 IEEE Radio Frequency Integrated Circuits Symposium (RFIC)*, pp. 397–400, 2015.

[2] J. Silver, K.Sankaragomathi and B.Otis, "An ultra-low-voltage all-digital PLL for energy harvesting applications," *Proc. IEEE Eur. Solid-State Circuits Conf.*, pp. 91–94, 2014.

[3] S.-A. Yu and P.Kinget, "A 0.65V 2.5GHz fractional-N frequency synthesizer in 90 nm CMOS," *IEEE Int. Solid-State Circuits Conf. (ISSCC) Dig. Tech. Papers*, pp. 304–604, 2007.

[4] V. Chillara, Y. H. Liu, B.Wang, *et al.*, "An 860_W 2.1-to-2.7GHz all-digital PLL-based frequency modulator with a DTC-assisted snapshot TDC for WPAN (Bluetooth smart and ZigBee) applications," *IEEE Int. Solid-State Circuits Conf. (ISSCC) Dig. Tech. Papers*, pp. 172–173, 2014.

[5] IEEE, "Standardization of NB-IOT completed." *3gpp.org*. 3GPP. June 22, 2016. p. 1. Retrieved August 19, 2016.

[6] J. Lee, S.Han, J. Lee *et al.*, "A sub-6GHz 5G new radio RF transceiver supporting EN-DC with 3.15Gb/s DL and 1.27Gb/s UL in 14nm FinFET CMOS," *2019 IEEE International Solid-State Circuits Conference – (ISSCC)*, San Francisco, CA, USA, 2019, pp. 354–356.

[7] Y. Sun, Jun Li, Zhuo Zhang *et al.*, "A 2.74–5.37GHz boosted-gain type-I PLL with <15% loop filter area," *2012 IEEE Radio Frequency Integrated Circuits Symposium, Montreal, QC*, 2012, pp. 181–184.

[8] B. Sadhu, J. Kim and R. Harjani, "A CMOS 3.3–8.4 GHz wide tuning range, low phase noise LC VCO," *2009 IEEE Custom Integrated Circuits Conference, Rome*, 2009, pp. 559–562.

[9] M. Demirkan, S. P. Bruss and R. R. Spencer, "Design of wide tuning-range CMOS VCOs using switched coupled-inductors," *IEEE Journal of Solid-State Circuits*, vol. 43, no. 5, pp. 1156–1163, 2008.

[10] S. Rong, and H. C. Luong, "Analysis and design of transformer-based dual-band VCO for software-defined radios," *IEEE Transactions on Circuits and Systems I: Regular Papers*, vol. 59, no. 3, pp. 449–462, 2012.

[11] J. Yin, and H. C. Luong, "A 57.5–90.1-GHz magnetically tuned multimode CMOS VCO," *IEEE Journal of Solid-State Circuits*, vol. 48, no. 8, pp. 1851–1861, 2013.

[12] Seong-Mo Yim and K.K. O, "Switched resonators and their applications in a dual-band monolithic CMOS LC-tuned VCO," *IEEE Transactions on Microwave Theory and Techniques*, vol. 54, no. 1, pp. 74–81, 2006.

[13] G. Li, Li Liu, Yiwu Tang, and E. Afshari, "A low-phase-noise wide-tuning-range oscillator based on resonant mode switching," *IEEE Journal of Solid-State Circuits*, vol. 47, no. 6, pp. 1295–1308, 2012.

[14] S. A. Ahmadi-Mehr, M. Tohidian and R. B. Staszewski, "Analysis and design of a multi-core oscillator for ultra-low phase noise," *TCAS-I*, 2016.

[15] A. Mazzanti, and P. Andreani, "Class-C harmonic CMOS VCOs, with a general result on phase noise," *IEEE J. Solid-State Circuits*, vol. 43, no. 12, pp. 2716–2729, 2008.

[16] L. Fanori and P. Andreani, "Highly efficient Class-C CMOS VCOs, including a comparison with Class-B VCOs," *IEEE J. Solid-State Circuits*, vol. 48, no. 7, pp. 1730–1740, 2013.

[17] W. Deng, K. Okada and A. Matsuzawa, "Class-C VCO with amplitude feedback loop for robust start-up and enhanced oscillation swing," *IEEE J. Solid-State Circuits*, vol. 48, no. 2, pp. 429–440, 2013.

[18] M. Babaie and R. B. Staszewski, "A Class-F CMOS oscillator," *IEEE J. Solid-State Circuits*, vol. 48, no. 12, pp. 3120–3133, 2013.

Chapter 12
Mm-wave and sub-THz CMOS VCOs

Xiaolong Liu[1] and Howard C. Luong[1]

Millimetre-wave (mm-Wave) and sub-terahertz (sub-THz) oscillators are the most critical elements of mm-Wave/sub-THz signal-generation systems as their performance, in terms of tuning range, phase noise and output power, significantly affects the system performance. With the continuous scaling of the complementary metal-oxide-semiconductor (CMOS) technologies, the realization of the fundamental CMOS voltage-controlled oscillators (VCOs) operating at mm-Wave/sub-THz frequency ranges has made understanding more feasible and attractive. However, it is still very challenging for fundamental oscillators to achieve a wide frequency tuning range with low phase noise and low power consumption at such high frequencies. This chapter discusses and demonstrates mm-Wave and sub-THz oscillators using magnetic tuning techniques, for both coarse and fine frequency tuning, to achieve a wide tuning range, low phase noise and low power consumption.

12.1 Introduction

E/W/D band frequencies ranging from 60 to 170 GHz have various useful and interesting applications including high data rate fixed point-to-point communications, high-resolution vehicular radars and high-sensitivity passive imaging. These applications rely on stringent specifications. These applications rely not only on the transceiver's broadband operation but also on the transmitter's error-vector magnitude and the receiver's signal to noise ratio. These in turn impose challenging requirements on the local oscillator signals in terms of wide tuning range and low phase noise [1].

Frequency tuning method using a switched-capacitor array (SCA) and varactors for coarse and fine frequency tuning, respectively, is popularly used to achieve wide tuning range. However, at mm-Wave frequencies and beyond, the quality factor Q_C of varactors and SCAs becomes much smaller compared to inductor Q_L, and thus the tank Q_T is dominated by the low-varactor Q_C. The tuning

[1]Department of Electronic and Computer Engineering, Hong Kong University of Science and Technology, Clear Water Bay, Hong Kong

range can be enhanced by increasing the size of varactors or SCAs, but it would inevitably further degrade the tank quality factor Q_T and the phase noise performance. To guarantee robust oscillation even with lower Q_C and Q_T, the size of the negative transconductance cell in the oscillators would also need to be increased, which would not only increase parasitic capacitance and thus degrading the tuning range and limiting the maximum oscillation frequency but also increase power consumption. Taking into consideration the phase noise along with the parasitic capacitance introduced by the active devices and buffers, the frequency tuning range of a fundamental VCO operating at around 100 GHz using varactors is currently limited to less than 10% [2], which is not sufficient for most applications, particularly in the presence of process, voltage and temperature (PVT) variations.

In order to avoid low-Q varactors or SCAs, an interpolative phase tuning technique is proposed for mm-Wave oscillators to tune the frequency [3]. However, multiple dedicated phase shifters are required, resulting in high power consumption. A severe trade-off exists between the tuning range and phase noise. Moreover, for these oscillators, the frequency tuning is achieved by intentionally shifting the operation from the peak resonant frequency of the LC tank through injected current, which unfortunately degrades tank Q and results in a steep frequency response.

More recently, magnetic tuning has been used to extend the tuning range and improve the phase noise of radio frequency (RF) and mm-Wave LC oscillators [4–12]. A magnetic-tuning VCO with two extra coupled switched inductors achieves an ultra-wideband tuning of 61.9% at RF range [4]. However, when the same idea was applied to mm-Wave frequencies, it was observed that the frequency tuning range was less than ~28% due to the more dominant parasitic capacitance from the active devices and switches [5–7]. Moreover, these mm-Wave magnetic-tuning oscillators still suffer from reliance on low-Q varactors.

Another way to extend the coarse-tuning range of mm-Wave oscillators is to use high-order LC tanks with dual-resonant modes [13] or three-resonant modes [14]. However, their frequency tuning range is still limited due to the limited coarse-tuning bands. Harmonic extraction [15] or self-mixing [16] oscillators can benefit from a wide tuning range and low phase noise, but at the cost of much lower power output and thus much higher power consumption.

Magnetic tuning by continuously changing the current at the secondary coil of a transformer tank is a good approach to avoid using low-Q varactors or SCAs. As demonstrated in [8], by varying the bias current and thus the transconductance g_m, the impedance from the resonant coil is continuously tuned. However, a small resistor is needed to shunt the resonant inductor, so that linear change of the effective inductance over a wide tuning range can be ensured, resulting in a low tank Q_T. It has been demonstrated in [9] that the currents in the coupling coils are forced to be either in-phase or out-of-phase by a quadrature oscillator topology, and by tuning their relative amplitudes, the oscillation frequency varies. However, as the frequency tuning depends on the changing bias currents, it needs to consume significant power of 30 mW for a wide tuning range. The technique is also leveraged for 100- [10] and 200-GHz [11] oscillators for applications at sub-THz frequencies but with a limited tuning range of 11.2% and 3.5%, respectively, mainly due to the limited current range available without contributing too much parasitic capacitance. Compared to magnetic fine-tuning with continuous control of the current, using a loaded continuously variable resistor does not need to burn as much power for fine-tuning [12]. However, according to [12], only a small tuning range of 16% at 60 GHz was achieved due to the limited frequency tuning bands.

To tackle the above-mentioned issues, mm-Wave and sub-THz oscillators are presented using a magnetic tuning technique, for both coarse and fine frequency tuning, so that high performance is achieved in terms of wide tuning range, low phase noise and low power consumption.

12.2 Magnetic tuning technique

12.2.1 Coarse magnetic tuning

Figure 12.1(a) shows the model of the conventional one-port dual-band (DB) VCO (DB-VCO) with a transformer-based tank [13]. L_1 and L_2 and C_1 and C_2 denote the inductances and capacitances of the primary coil and the secondary coil, respectively, while R_{L1}, R_{L2}, R_{C1} and R_{C2} model the resistance loss in each component. K_{12} is the magnetic coupling coefficient between L_1 and L_2. As illustrated in Figure 12.1(b), the impedance seen from either coils has two distinct frequency

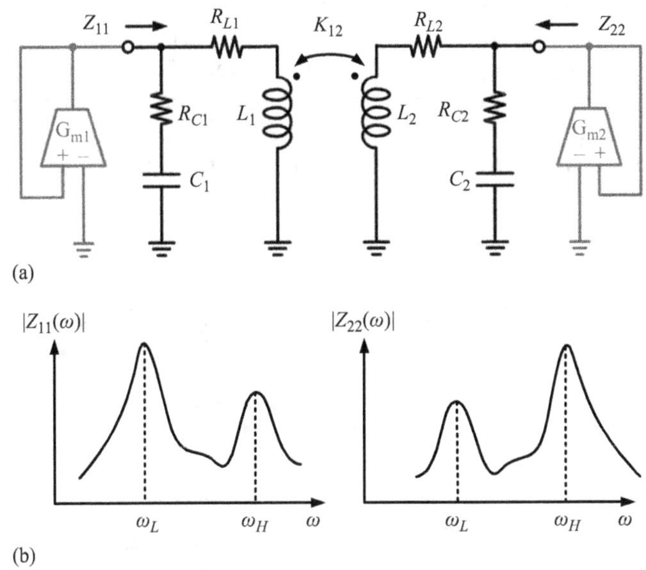

Figure 12.1 Simplified dual-band voltage-controlled oscillator (VCO): (a) model and (b) amplitude response of the tank impedance from the primary coil and secondary coil, respectively

peaks. With negative resistance, the VCO tends to oscillate at peak frequency with higher impedance because it requires less energy.

Assume high component Qs for simplicity (R_{L1}, R_{L2}, R_{C1} and $R_{C2} \rightarrow 0$), so that the oscillation frequency of the DB-VCO can be derived as:

$$\omega_{H,L}^2 = \frac{1}{2}\left(\omega_1^2 + \omega_2^2 \pm \sqrt{\left(\omega_1^2 + \omega_2^2\right)^2 - 4\left(1 - K_{12}^2\right)\omega_1^2\omega_2^2}\right)/\left(1 - K_{12}^2\right) \tag{12.1}$$

where $\omega_1 = (L_1C_1)^{-1/2}$ and $\omega_2 = (L_2C_2)^{-1/2}$. Conventionally, the frequencies of ω_H and ω_L are tuned by varying the capacitors C_1 and C_2, which are usually realized by varactors and SCAs. As discussed earlier, the varactor and SCA Q_Cs decrease dramatically with the frequency and introduce huge losses at mm-Wave and sub-THz frequencies. Fortunately, the resonant frequencies can also be tuned by changing the magnetic coupling coefficient K_{12}, which provides another degree of freedom to tune the frequency. Figure 12.2 shows the relationships between the resonant frequency ω_H (ω_L) and the coupling coefficient K_{12} for different ω_2/ω_1 ratios based on (12.1), assuming that the self-resonant frequencies are kept constant. It is clear that the high resonant frequency ω_H is quite sensitive to the coupling coefficient K_{12}. For instance, with ω_2/ω_1 equal to 1.5, tuning the coupling coefficient K_{12} from 0.1 to 0.35 changes the ratio of ω_H/ω_1 from 1.5 to 1.7, which corresponds to a frequency tuning of more than 10%.

The challenge involved is tuning the coupling coefficient K_{12} while maintaining the self-resonant frequencies ω_1 and ω_2 unchanged. A conventional way to change the coupling coefficient K_{12} is to add a switching coil L_{S1} between the original two coils L_1 and L_2 [17], as shown in Figure 12.3(a). Intuitively, when M_{S1} is turned on, the current i_1 in L_1 induces a current i'_{S1} in L_{S1} and another current i'_2 in L_2, both of which are in the direction opposite to i_1. Since i'_{S1} also induces another current i''_2 in L_2, which tends to cancel i'_2, the effective coupling coefficient $K_{12,\text{eq}}$

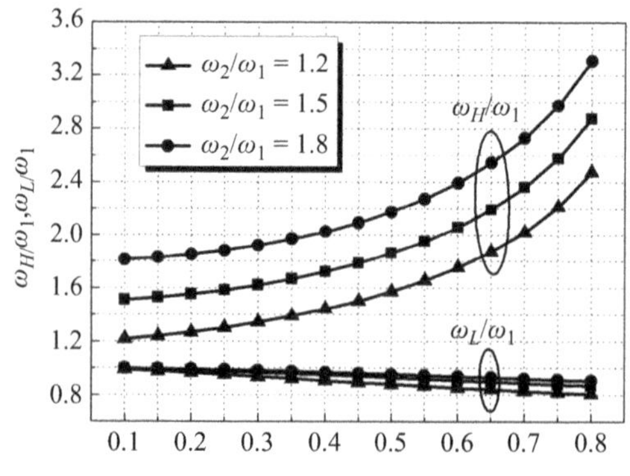

Figure 12.2 Calculated resonant frequencies ω_H and ω_L as functions of the magnetic coupling coefficient K_{12} for different ω_2/ω_1 ratios

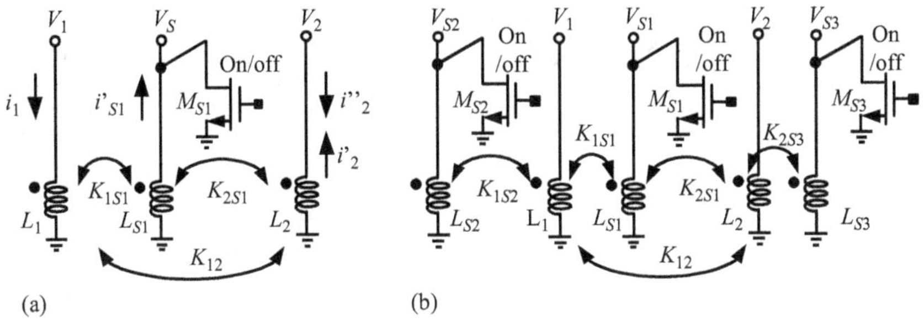

Figure 12.3 Schematics of (a) switched single-shielded transformer and (b) switched triple-shielded transformer

between L_1 and L_2 actually becomes lower than when M_{S1} is turned off. In other words, K_{12} becomes coarsely tunable by switching on/off the M_{S1}.

For the switched single-shielded transformer shown in Figure 12.3(a), because the effective inductances of L_1 and L_2 have also changed when M_{S1} turns on and off [4], the change in magnetic coupling coefficient K_{12} is less significant, and the tuning range is not maximized. As such, in order to extend the tuning range, it is highly desirable to keep L_1 and L_2 constant independent of whether the switch M_{S1} turns on and off. As mentioned in [18], a switched triple-shielded transformer is proposed with two additional shielding coils, L_{S2} and L_{S3}, in series with switches M_{S2} and M_{S3} being added to the left and right sides of L_1 and L_2, respectively, as shown in Figure 12.3(b). When M_{S1} is off, both M_{S2} and M_{S3} are turned on, L_{S1} and L_{S3} introduce counteractive currents in L_1 and L_2 to prevent an increase in inductance for L_1 and L_2. Similarly, when M_{S1} is on, both M_{S2} and M_{S3} are turned off to avoid a decrease in inductance for L_1 and L_2. Consequently, the equivalent inductances of L_1 or L_2 are kept relatively constant when M_{S1} turns on and off to tune K_{12}. Moreover, with the two extra switched coils L_{S2} and L_{S3}, six coarse-tuning bands are created and the oscillator that uses this switched triple-shielded transformer can achieve an ultra-wideband tuning range.

Although the coarse magnetic tuning multimode VCO [18] can offer an ultrawide tuning range of 41.1%, it has only a few coarse-tuning bands and, for fine-tuning, still requires varactors to cover frequency gaps that would inevitably degrade the tank Q, the phase noise and the operation frequency. As previously mentioned, the varactor Q decreases dramatically with frequency. For instance, using a varactor to achieve an ~10% tuning range without consideration of other parasitic capacitance, as shown in Figure 12.4, the simulated worst-case varactor Q drops from 7 at 30 GHz to only 2 at 120 GHz in a standard 65-nm CMOS process. At 100 GHz, with a change in the control voltage from 0 to 1.2 V, the varactor can be tuned from 9 to 4 fF, while the varactor Q varies from 2.5 to 6. Although the inductor Q is very high (~35), the tank Q will be unacceptably small due to the low-Q varactor. To tackle this issue, continuous tuning of the magnetic field can be adopted to fine-tune, so that low-Q varactors are eliminated.

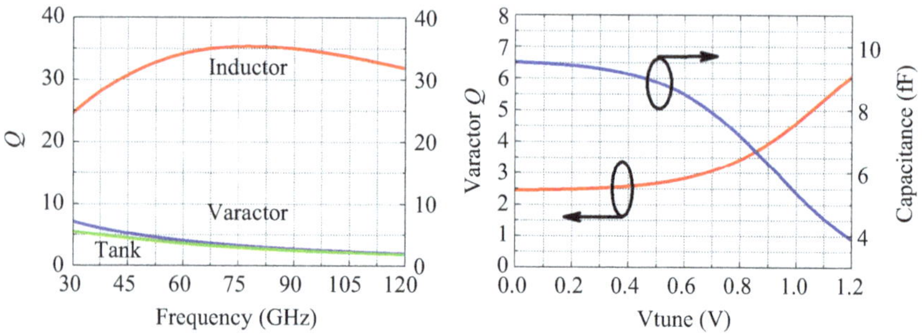

Figure 12.4 Simulated varactor Q with frequency and control voltage at 100 GHz

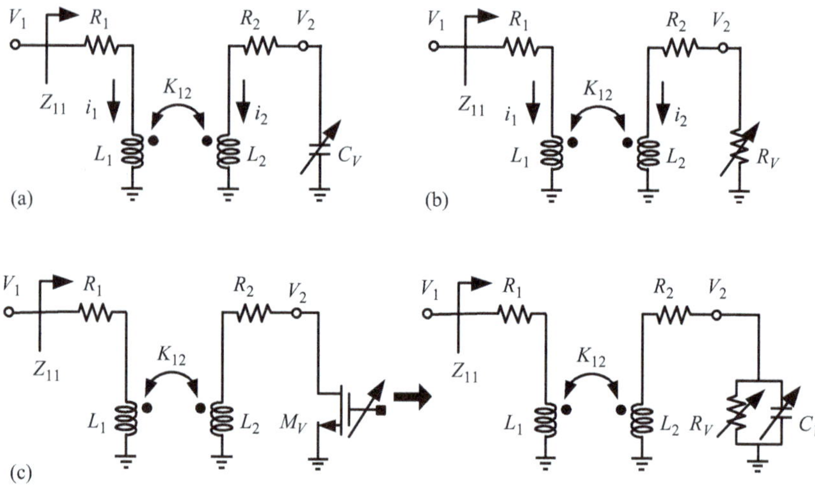

Figure 12.5 Fine magnetic tuning with loaded (a) variable capacitor, (b) variable resistor and (c) variable tuned transistor

12.2.2 Fine magnetic tuning

The realization of fine magnetic tuning by a two-coil transformer is as shown in Figure 12.5. The resonant primary coil L_1 is usually connected with a negative g_m cell for stable oscillation [12]. A loaded component operating as a variable capacitor or resistor is connected in parallel with the secondary coil L_2 for fine magnetic tuning. The loaded component can be a variable capacitor C_V, a variable resistor R_V or a tuned transistor M_V, as shown in Figure 12.5(a), Figure 12.5(b) and Figure 12.5 (c), respectively. Intuitively, voltage V_1 at the primary coil induces voltage V_2 at the secondary coil through magnetic coupling. It follows that two magnetic flux linkages exist: a self-magnetic flux linkage due to the current i_1 through the primary coil L_1 and an injected magnetic flux linkage through mutual magnetic coupling due to the current i_2 through the secondary coil L_2. As a consequence, varying the loaded capacitor C_V or resistor R_V would change the current i_2 and thus the injected magnetic flux linkage, resulting in a variable inductance $L_{1,\mathrm{eff}}$.

12.2.2.1 Fine magnetic tuning with variable capacitor

As shown in Figure 12.5(a), assuming the voltage and current in the primary and secondary coils L_1 and L_2 are (V_1, i_1) and (V_2, i_2), respectively, and R_1 and R_2 model the loss of the two coils, the *I-V* equations at the operation frequency ω can be derived as:

$$V_1(j\omega) = (j\omega L_1 + R_1)i_1 + jK_{12}\sqrt{L_1 L_2}i_2 \tag{12.2}$$

$$V_2(j\omega) = (j\omega L_2 + R_2)i_2 + jK_{12}\sqrt{L_1 L_2}i_1 = -i_2/(j\omega C_V) \tag{12.3}$$

From (12.2) and (12.3), the input impedance Z_{11} can be derived as:

$$Z_{11}(j\omega) = R_1 + j\omega L_1 + \frac{\omega^2 K_{12}^2 L_1 L_2}{R_2 + j\omega L_2 + 1/(j\omega C_V)} \tag{12.4}$$

It follows that the equivalent inductance $L_{1,\mathrm{eff}}$ and equivalent resistance $R_{1,\mathrm{eff}}$ can be derived by calculating the imaginary and real part of the input impedance Z_{11} in (12.4), respectively, as:

$$L_{1,\mathrm{eff}}(C_V) = L_1 - K_{12}^2 \frac{\omega L_1 L_2(\omega L_2 - 1/\omega C_V)}{R_2^2 + (\omega L_2 - 1/\omega C_V)^2} \tag{12.5}$$

$$R_{1,\mathrm{eff}}(C_V) = R_1 + R_2 \frac{\omega^2 K_{12}^2 L_1 L_2}{R_2^2 + (\omega L_2 - 1/\omega C_V)^2} \tag{12.6}$$

The equivalent $Q_{1,\mathrm{eff}}$ can be directly calculated as $Q_{1,\mathrm{eff}} = \omega L_{1,\mathrm{eff}}/R_{1,\mathrm{eff}}$.

The effective $L_{1,\mathrm{eff}}$ and $Q_{1,\mathrm{eff}}$ are calculated and plotted as shown in Figure 12.6, with C_V varying from 10 to 30 fF at 100 GHz, assuming $L_1 = 60$ pH, $L_2 = 90$ pH, $K_{12} = 0.4$, $Q_{L1} = Q_{L2} = 30$, $Q_{CV} = 4$, and the loss of the capacitor C_V is embedded in R_2. The effective inductance $L_{1,\mathrm{eff}}$ is first increased from around 65

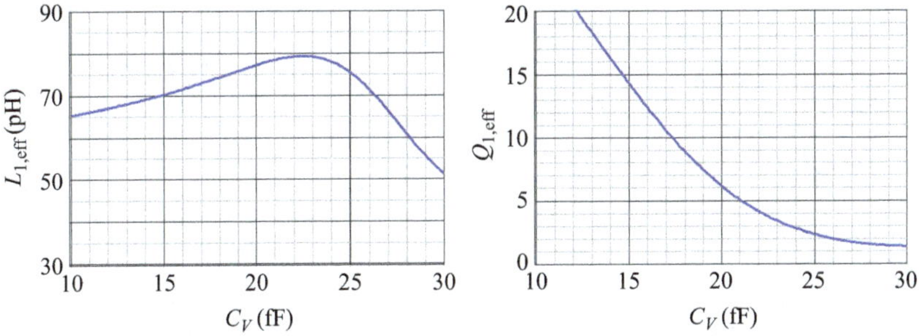

Figure 12.6 Calculated equivalent inductance $L_{1,\mathrm{eff}}$ and equivalent quality factor $Q_{1,\mathrm{eff}}$ with loaded variable capacitor C_V

to 80 pH, with C_V varying from 10 to 23 fF and then begins to decrease with C_V further increasing. On the other hand, the effective $Q_{1,\text{eff}}$ is higher for a small C_V and degrades significantly with increase in C_V. To trade-off between the tuning range and the phase noise performance, C_V should be lesser than 20 fF. However, due to the parasitic capacitance and high loss of the capacitor at high frequencies, the fine magnetic tuning method with a loaded variable capacitor would further degrade the tuning range and tank Q.

12.2.2.2 Fine magnetic tuning with variable resistor

Fine magnetic tuning can also be implemented with a loaded variable resistor R_V, as shown in Figure 12.5(b). Similarly, by deriving the input equivalent impedance Z_{11}, the equivalent inductance $L_{1,\text{eff}}$ and resistance $R_{1,\text{eff}}$ can be obtained as

$$L_{1,\text{eff}}(R_V) = L_1 - \frac{\omega^2 K_{12}^2 L_1 L_2^2}{\omega^2 L_2^2 + (R_2 + R_V)^2} \tag{12.7}$$

$$R_{1,\text{eff}}(R_V) = R_1 + \frac{\omega^2 K_{12}^2 L_1 L_2 (R_2 + R_V)}{\omega^2 L_2^2 + (R_2 + R_V)^2} \tag{12.8}$$

Again at 100 GHz, assuming the same design parameters of the transformer as *above*, the effective $L_{1,\text{eff}}$ and $Q_{1,\text{eff}}$ are calculated and plotted, as shown in Figure 12.7, as the resistor R_V is varied from 0 to 500 Ω. The $L_{1,\text{eff}}$ is continuously tuned from ~50 to 60 pH and the $Q_{1,\text{eff}}$ is more than 8. Compared to the loaded variable capacitor tuning method with the same minimum equivalent Q, the loaded resistor tuning method has a comparable tuning range. Moreover, the loaded tuned resistor contributes less significant parasitic capacitance, offering a more attractive method for fine magnetic tuning.

12.2.2.3 Fine magnetic tuning with variable transistor

A variable resistor can be realized simply using a continuously tuned transistor M_V, as shown in Figure 12.5(c). Modelling the tuned transistor as a resistor R_V parallel

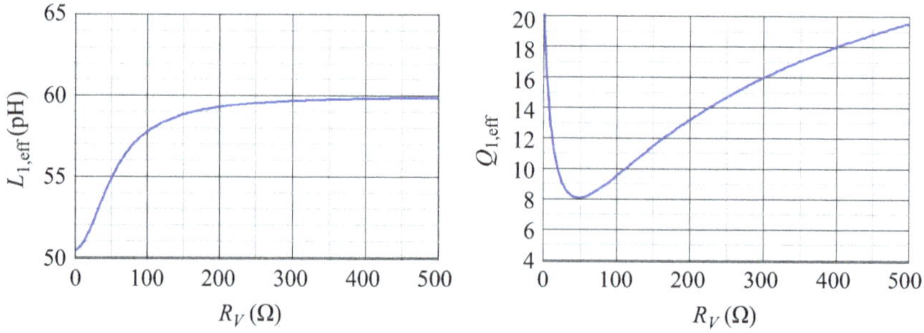

Figure 12.7 Calculated equivalent inductance $L_{1,eff}$ and equivalent quality factor $Q_{1,eff}$ with loaded variable resistor R_V

to a capacitor C_V, the input impedance of the resonant coil L_1 can be derived as:

$$Z_{11} = R_1 + j\omega L_1 + \frac{\omega^2 K_{12}^2 L_1 L_2}{R_2 + j\omega L_2 + R_V/(1 + j\omega R_V C_V)} \tag{12.9}$$

The effective inductance $L_{1,\mathrm{eff}}$ and tank $Q_{1,\mathrm{eff}}$ can then be obtained as:

$$L_{1,\mathrm{eff}}(R_V, C_V) = L_1 - \frac{\omega^2 K_{12}^2 L_1 L_2 (L_2 - \alpha R_V C_V)}{(R_2 + \alpha)^2 + \omega^2 (L_2 - \alpha R_V C_V)^2} \tag{12.10}$$

$$Q_{1,\mathrm{eff}} = \omega \frac{\beta L_1 - \omega^2 K_{12}^2 L_1 L_2 (L_2 - \alpha R_V C_V)}{\beta R_1 + \omega^2 K_{12}^2 L_1 L_2 (R_V + \alpha)} \tag{12.11}$$

where $\alpha = R_V/(1 + \omega^2 R_V^2 C_V^2)$ and $\beta = (R_2 + \alpha^2)^2 + \omega^2 (L_2 - \alpha R_V C_V)^2$. Assuming $L_1 = 60$ pH, $L_2 = 90$ pH, $K_{12} = 0.4$, $Q_1 = Q_2 = 30$, the equivalent inductance $L_{1,\mathrm{eff}}$ and tank $Q_{1,\mathrm{eff}}$ are calculated and plotted as the equivalent resistor R_V is varied up to 500 Ω for both different transistor sizes and control gate voltages. As shown in Figure 12.8, the effective inductance $L_{1,\mathrm{eff}}$ is increased from ~55 to 65 pH with a small transistor and a fixed C_V of 15 fF. On the other hand, $L_{1,\mathrm{eff}}$ is decreased from ~55 to 44 pH with a larger transistor and a fixed C_V of 40 fF. Interestingly, with increase in C_V, the polarity of the denominator in (12.10) is changed, and the change in $L_{1,\mathrm{eff}}$ reverses. As shown in Figure 12.9, the effective $Q_{1,\mathrm{eff}}$ is still above 8 and 5 for fixed C_V of 15 and 40 fF, respectively, as shown in Figure 12.9. The results demonstrate that fine magnetic tuning with a variable transistor is attractive at high mm-Wave frequencies.

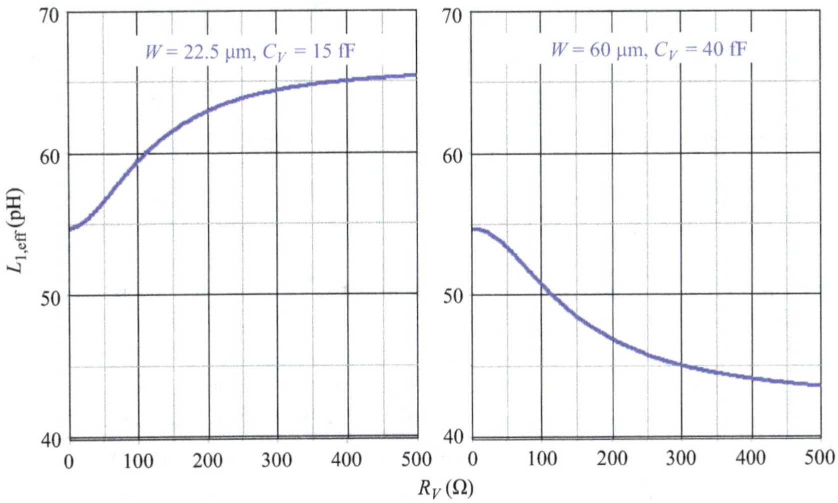

Figure 12.8 Calculated equivalent inductance $L_{1,eff}$ with loaded variable transistor

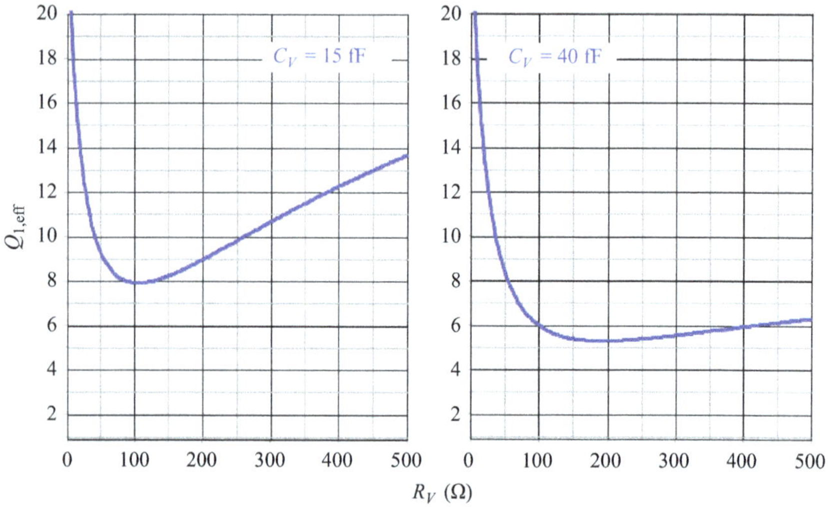

Figure 12.9 Calculated equivalent $Q_{1,eff}$ with loaded variable transistor

Additionally, the coupling coefficient K_{12} can also be tuned to achieve a wider tuning range. Assuming that the quality factors of L_1 and L_2 are sufficiently high and the switch's turn-on resistance is sufficiently small, the equivalent inductance $L_{1,\text{eff}}$ when the tuned transistor M_V is on can be approximated as:

$$L_{1,\text{eff,on}} = L_1\left(1 - K_{12}^2\right) \tag{12.12}$$

When the tuned transistor M_V is off, $L_{1\text{eq,off}}$ can be estimated as:

$$L_{1,\text{eff,off}} = L_1 - \frac{K_{12}^2 L_1}{1 - \left(\omega^2 L_2 C_V\right)^{-1}} \tag{12.13}$$

As a result, the inductance tuning ratio (TR) can be estimated as:

$$\text{TR} = \frac{L_{1,\text{eff,off}} - L_{1,\text{eff,on}}}{L_{1,\text{eff,off}}} \approx K_{12}^2 \tag{12.14}$$

From (12.14), the inductance TR, and thus the tuning range, is highly dependent on the square of coupling coefficient K_{12}. However, a higher coupling coefficient means higher loss and more degradation of the tank Q. In Figure 12.5(c), when the variable resistor R_V is tuned to match the output impedance of the secondary coil L_2, the variable inductor L_1 delivers the maximum output power to the variable resistor R_V, resulting in a minimum quality factor $Q_{1,\text{eff,min}}$ given by [19]:

$$Q_{1,\text{eff,min}} \approx \frac{2}{K_{12}^2} - 1 = \frac{2}{TR} - 1 \tag{12.15}$$

In other words, a clear trade-off exists between the tuning range and the minimum tank Q.

12.2.3 Coarse and fine magnetic tuning

To further improve the tuning range and tank Q, magnetic tuning techniques for both coarse and fine frequency tuning are applied to a DB-VCO as mentioned in [20]. As shown in Figure 12.10(a), a triple-coil transformer is proposed for coarse and fine frequency tuning, while the primary coil L_1 resonates with the parasitic capacitances. A transistor switch M_{SW1} is added in series with the secondary coil L_{SW1} for coarse frequency tuning. Another transistor M_{SW2} operating as a variable resistor is connected in parallel with the third coil L_{SW2} for fine frequency tuning. With the model of the triple-coil transformer as shown in Figure 12.10(b), the effective primary inductance $L_{1,\text{eff}}$ can be derived, which can be coarsely and finely tuned by the two transistors as:

$$L_{1,\text{eff}} = L_1 - \frac{\omega^2 K_{1SW1}^2 L_1 L_{SW1}(L_{SW1} - \alpha_1 R_{SW1} C_{SW1})}{(R_{LSW1} + \alpha_1)^2 + \omega^2 (L_{SW1} - \alpha_1 R_{SW1} C_{SW1})^2} - \frac{\omega^2 K_{1SW2}^2 L_1 L_{SW2}(L_{SW2} - \alpha_2 R_{SW2} C_{SW2})}{(R_{LSW2} + \alpha_2)^2 + \omega^2 (L_{SW2} - \alpha_2 R_{SW2} C_{SW2})^2} \tag{12.16}$$

where $\alpha_i = R_{SWi}/(1 + \omega^2 R_{SWi}^2 C_{SWi}^2)$ for $i = 1, 2$. R_1, R_{LSW1} and R_{LSW2} model the loss of the coupled coils L_1, L_{SW1} and L_{SW2}, respectively; K_{1SW1} and K_{1SW2} are the coupling coefficients between the coils; and the two switches are modelled by R_{SWi} in parallel with C_{SWi} ($i = 1, 2$). For simplicity, ignoring the coupling between L_{SW1} and L_{SW2} due to its indirect effect on the resonant coil L_1 and negligible effect on the equivalent inductance of $L_{1,\text{eff}}$, the equivalent resistance of $R_{1,\text{eff}}$ can be derived as:

$$R_{1,\text{eff}} = R_1 + \frac{\omega^2 K_{1SW1}^2 L_1 L_{SW1}(R_{SW1} + \alpha_1)}{(R_{LSW1} + \alpha_1)^2 + \omega^2 (L_{SW1} - \alpha_1 R_{SW1} C_{SW1})^2} + \frac{\omega^2 K_{1SW2}^2 L_1 L_{SW2}(L_{SW2} + \alpha_2)}{(R_{LSW2} + \alpha_2)^2 + \omega^2 (L_{SW2} - \alpha_2 R_{SW2} C_{SW2})^2} \tag{12.17}$$

From (12.16), the change in the effective inductance $L_{1,\text{eff}}$, and thus the corresponding tuning range, is proportional to the coupling coefficients K_{1SW1} and

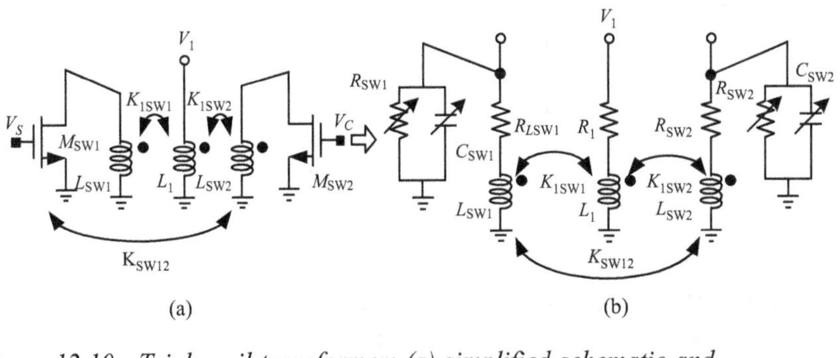

Figure 12.10 Triple-coil transformer: (a) simplified schematic and (b) equivalent model

K_{1SW2}. From (12.17), the loss contribution from the coupled tuned coils L_{SW1} and L_{SW2} increased with K_{1SW1} and K_{1SW2} but decreased with a small R_{1SW1} and R_{1SW2} or large transistor sizes of M_{SW1} and M_{SW2}. Therefore, the design selection of the coupling coefficients and the size of transistor switches are a compromise between the tank Q and tuning range. A larger size switch has smaller loss but larger parasitic capacitance, limiting the frequency tuning range. In contrast, a smaller switch size has smaller parasitic capacitance but larger loss.

Based on (12.16) and (12.17), calculations are made to optimize the design parameters in terms of switch sizes M_{SW1} and M_{SW2} and coupling coefficients K_{1SW1} and K_{1SW2}. In this design, $L_1 = 64$ pH, $L_{SW1} = 35$ pH and $L_{SW2} = 90$ pH. In order to verify the influence of coarse-tuning switch M_{SW1} and coupling coefficient K_{1SW1} on the effective inductance, the size of M_{SW2} is fixed to 60 μm/200 nm and K_{1SW2} is designed to be 0.45. Figure 12.11 shows the calculated effective inductance with different design parameters. In Figure 12.11(a), the two lines of the same colour represent the effective inductances in the low-frequency band (LB) and high-frequency band (HB) by turning off and on the switch M_{SW1}, respectively. The different colour lines maple to different K_{1SW1}, e.g., 0.35 and 0.45, respectively. As we can see from Figure 12.11(a), for the same coarse-tuning switch M_{SW1}, the coarse-tuning range increases with the coupling coefficient K_{1SW1}, as expected. From Figure 12.11, with the same K_{1SW1}, the coarse frequency tuning increases with the coarse-tuning switch M_{SW1}. To ensure enough overlap between the HB and LB, the switch size is chosen to be $M_{SW1} = 30$ μm/60 nm and the coupling coefficient is chosen to be $K_{1SW1} = 0.35$.

After determining the sizes of the coarse-tuning switch M_{SW1} and the coupling coefficient K_{1SW1}, the fine-tuning switch M_{SW2} and the coupling coefficient K_{1SW2} are varied to study their effects on the equivalent inductance and loss contribution. As shown in Figure 12.12(a) and Figure 12.12(b), a larger continually tuned

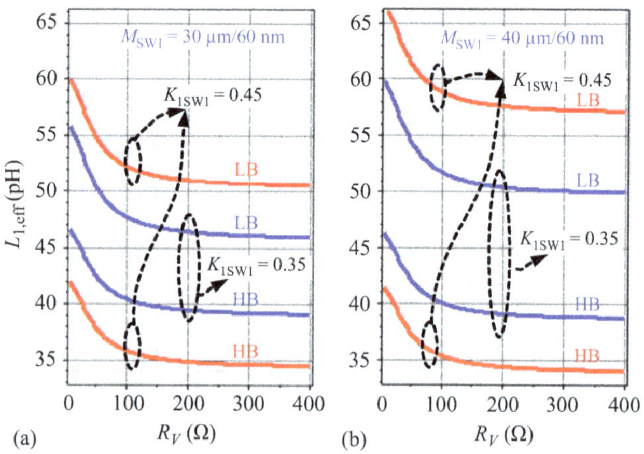

Figure 12.11 Calculated equivalent inductance $L_{1,eff}$ with coarse-tuning switch M_{SW1} and coupling coefficient K_{1SW1}

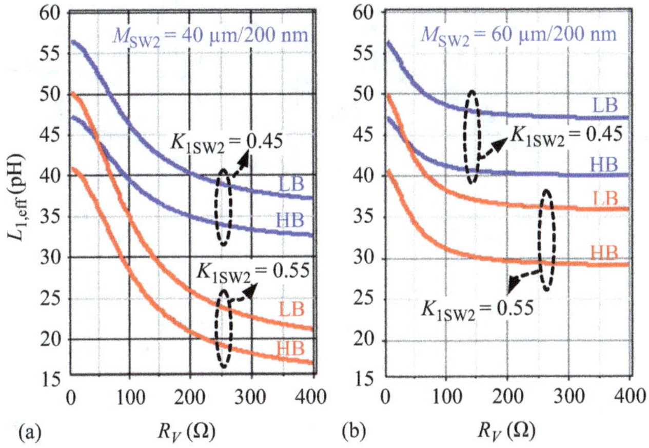

Figure 12.12 Calculated equivalent inductance $L_{1,eff}$ with fine-tuning switch M_{SW2} and coupling coefficient K_{1SW2}

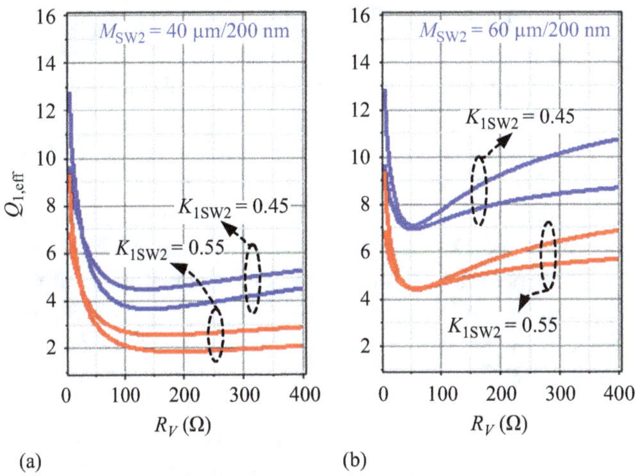

Figure 12.13 Calculated equivalent $Q_{1,eff}$ with fine-tuning switch M_{SW2} and K_{1SW2}

equivalent inductance can be achieved by either increasing the coupling coefficient K_{1SW2} or decreasing the size of the fine-tuning switch M_{SW2}. However, as shown in Figure 12.13(a) and Figure 12.13(b), the equivalent Q would be degraded with increase in K_{1SW2} or decrease in the size of M_{SW2}. As a trade-off between the tuning range and equivalent Q, if the fine-tuning switch size is chosen to be $M_{SW2} = 60$ µm/200 nm and the coupling coefficient is chosen to be $K_{1SW2} = 0.45$, the effective inductance can be tuned from around 40 to 56 pH, while the equivalent Q is larger than 7 across the entire tuning range.

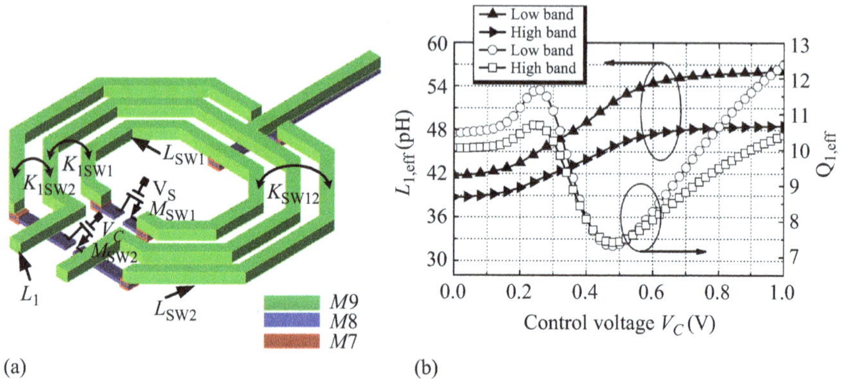

Figure 12.14 The proposed triple-coil transformer: (a) layout and (b) simulated equivalent inductance and tank Q vs. the fine-tuning control voltage

Figure 12.14(a) shows the layout of the proposed triple-coil transformer with an outer diameter of 90 μm. All three coupled coils use the top two thick metal layers to minimize the loss. For propelling the mutual coupling coefficients, L_{SW1} and L_{SW2} are placed inside and outside the primary coil L_1, respectively. The metal widths are 4 μm for all three coupled coils for a small parasitic capacitance. As discussed above, the aspect ratio W/L for the coarse-tuning switch M_{SW1} is designed to be 30 μm/60 nm to minimize its parasitic capacitance C_{SW1}, while the fine-tuning switch M_{SW2} is optimally sized to be 60 μm/200 nm with a large width for low-resistance loss and a long channel for linear equivalent resistance.

Simulated with a 2.5-D electromagnetic (EM) solver (ADS momentum) at 100 GHz, the fitting parameters for the triple-coil transformer are $L_1 = 64$ pH, $L_{SW1} = 35$ pH, $L_{SW2} = 90$ pH, $Q_{L1} = 25$, $Q_{LSW1} = 35$, $Q_{LSW2} = 24.5$, $K_{1SW1} = 0.38$ and $K_{1SW2} = 0.46$. Figure 12.14(b) shows the simulated equivalent inductance and the effective quality factor Q of the tank with the fine-tuning control voltage V_C being tuned from 0 to 1 V. In the HB, the equivalent inductance changes from 39 to 48 pH and the tank Q varies from 7.4 to 10.7. In the LB, the equivalent inductance changes from 42 to 56 pH, and the tank Q varies from 7.3 to 12.4. In other words, the tuning range of the proposed VCO is larger than 10% and the Q of the tank is larger than 7 over the entire frequency range.

12.2.4 Fine magnetic tuning with split transformer

As discussed above, there is a clear trade-off between the magnetic tuning range and the coupling coefficient and thus the loss contribution. The corresponding tuning range of the proposed triple-coil transformer is limited to around 15% due to the limited coarse-tuning sub-bands. It has been proved that with more sub-bands, the tuning range can be improved further while minimizing the tank Q degradation [18]. In order to minimize the Q degradation while extending the tuning range, a split transformer with multiple secondary coils is proposed as the variable inductor,

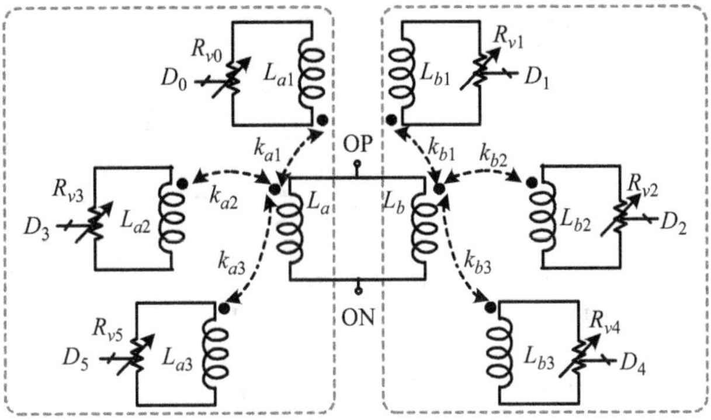

Figure 12.15 Schematic of the split transformer as variable inductor

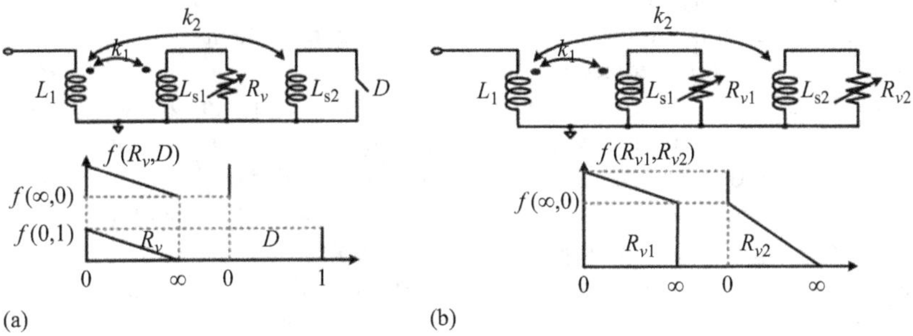

Figure 12.16 Comparison of tuning scheme (a) frequency tuning with a switch and a variable resistor and (b) frequency tuning with multiple variable resistors

as shown in Figure 12.15 [19]. The split transformer consists of two parallel transformers L_a and L_b, each of which is designed to have three parallel secondary coils with a small coupling coefficient k_i to improve the minimum Q_{min}.

The current method is somewhat similar to the above-mentioned coarse-tuning and fine-tuning switched transformer with a variable inductor [20], which creates multiple bands to reduce the tuning range requirement of the variable inductor. However, this split transformer design with multiple variable resistors ensure that there is no frequency gap issue. Figure 12.16 illustrates the comparison between the two tuning schemes, where the vertical axis plots the oscillation frequency, while the horizontal axis shows the tuning of the control signals (D, R_{V1}) and (R_{V1}, R_{V2}) for the two schemes, respectively. For the previous tuning scheme, as shown in Figure 12.16(a), after the frequency is continuously tuned down from $f(R_V = 0, D = 0)$ to $f(R_V = \infty, D = 0)$, the control signals (R_V, D) need to be switched from ($R_V = \infty, D = 0$) to ($R_V = 0, D = 1$) to switch to another band for further tuning. With PVT variations, $f(R_V = \infty, D = 0)$ can typically be higher than $f(R_V = 0, D = 1)$, resulting in a frequency gap. As a result, even more margins for tuning

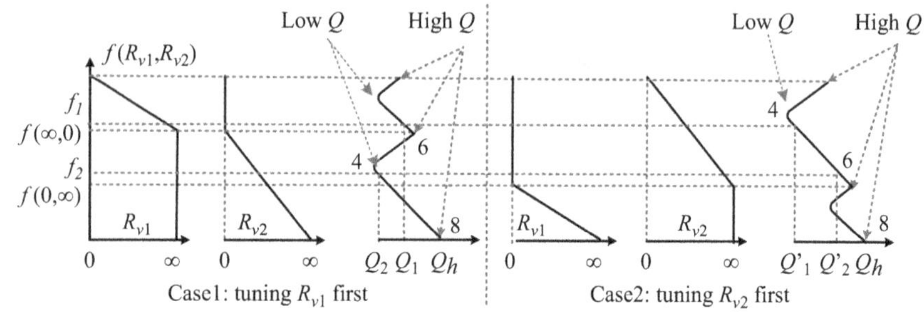

Figure 12.17 Tuning and swapping scheme

range needs to be designed to ensure continuous tuning. However, for the proposed tuning scheme with split transformer, as shown in Figure 12.16(b), after the frequency is continuously tuned down from $f(R_{V1} = 0$ and $R_{V2} = 0)$ to $f(R_{V1} = \infty$ and $R_{V2} = 0)$, R_{V2} is continuously tuned for further tuning, which guarantees no frequency gap between the adjacent tuning bands.

Moreover, in the present scheme, the secondary coils are further designed to be non-uniform, with a progressively scaled coupling coefficient and corresponding tuning range. With such a non-uniform design of secondary coils, it is possible to properly select and swap the tuning order, so that the second coil with the smallest tuning range but with the highest Q is used for the final tuning to avoid operations in the low-Q region. In other words, if the uniform transformer design is used, the tuning range and minimum Q are the same, and the operation in the low-Q region would be unavoidable. On the other hand, with the proposed non-uniform transformer design, the secondary coil with smaller tuning range but higher Q regions can be identified and selected as the last tuning step to reach the target frequency to maximize the Q. Figure 12.17 illustrates the comparison. For simplicity, only two secondary coupled coils are used. The first secondary coil with R_{V1} is designed to have a smaller tuning range but higher minimum Q compared to the second secondary coil with R_{V2}. For a target frequency f_1, tuning R_{V1} first (Case 1) results in a high-Q region associated with R_{V1}, whereas tuning R_{V2} first (Case 2) would operate the tank in a lower Q region associated with R_{V2}. In contrast, for another target oscillation frequency $f_2 < f_1$, the tank can be operated in a higher Q region by swapping the tuning order from Case 1 to Case 2 with R_{V2} being tuned first. In both cases, by swapping the tuning order properly, the coil R_{V1} with higher minimum Q can be assigned to tune in the last tuning step to achieve the target oscillation frequency to maximize the tank Q.

12.3 Design prototypes

12.3.1 Coarse magnetic tuning multi-mode VCO

Figure 12.18 shows the schematic of the coarse magnetic tuning multi-mode VCO [18]. The proposed switched triple-shielded transformer in Figure 12.3(b) is used

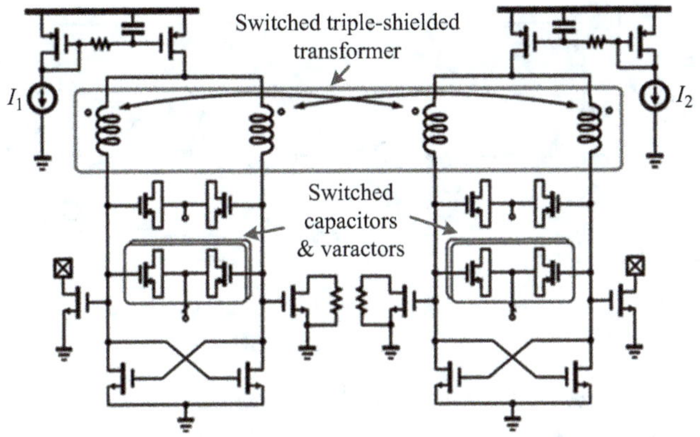

Figure 12.18 Schematic of the coarse magnetic tuning multi-mode VCO

Table 12.1 Control logics and arrangement for mode selection

	Mode 1	Mode 2	Mode 3	Mode 4	Mode 5	Mode 6
$M_{S2}M_{S1}M_{S3}$	101	011	111	010	001	111
I_1	On	On	On	Off	Off	Off
I_2	Off	Off	Off	On	On	On
Bands	Low band			High band		

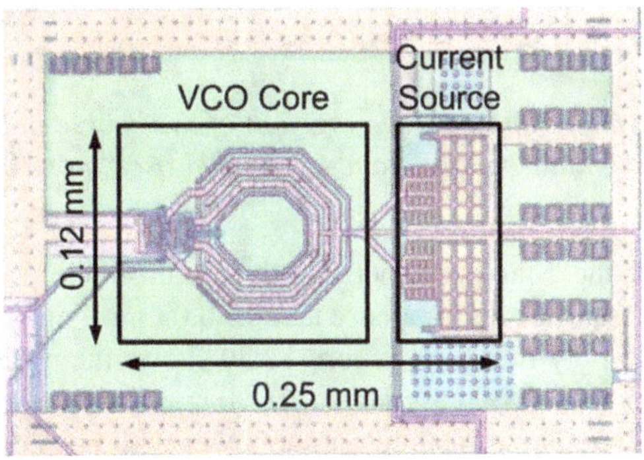

Figure 12.19 Chip photograph of the coarse magnetic tuning multi-mode voltage-controlled oscillator (VCO)

for coarse frequency tuning, and an SCA with controlled binary-weighted varactors and a varactor with continuous analogue control voltage are used for sub-coarse tuning and fine frequency tuning, respectively. By a combination of the bias currents (I_1 and I_2) and the control logics of the switches in the shielded coils, six coarse-tuning modes can be created as shown in Table 12.1. Figure 12.19 shows the

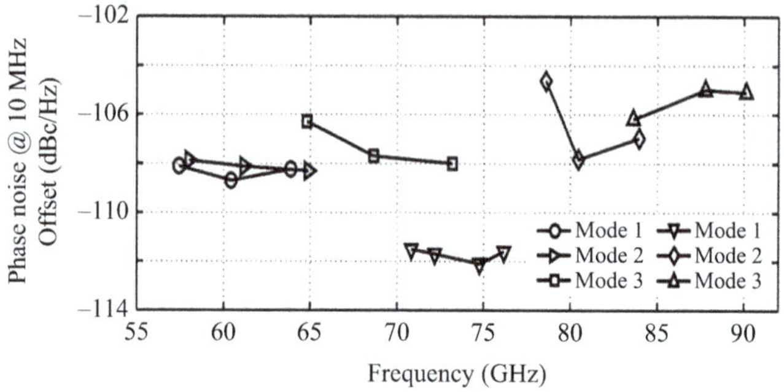

Figure 12.20 Measured phase noise of the coarse magnetic tuning multi-mode voltage-controlled oscillator (VCO) across the tuning range

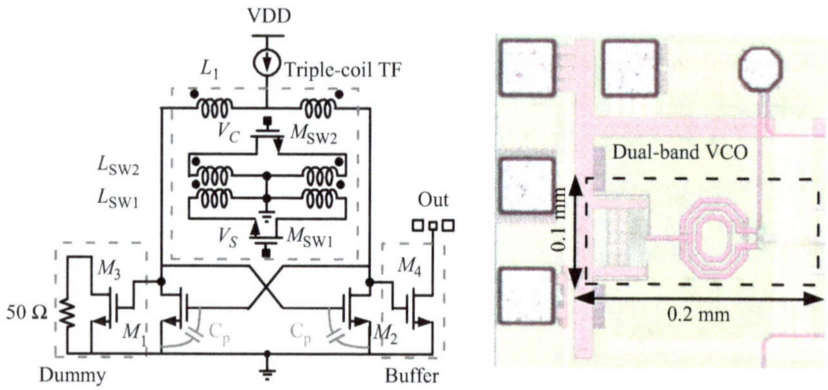

Figure 12.21 Schematic and chip photograph of the coarse and fine magnetic tuning dual-band voltage-controlled oscillator (DB-VCO)

chip photograph of the coarse magnetic tuning multi-mode VCO. Fabricated in 65-nm CMOS process, it occupies a core area of 0.03 mm^2 and consumes a power of 8.4 mW with a supply of 1.2 V. Figure 12.20 shows the measured phase noise varies from −104.6 to −112.2 dBc/Hz at 10-MHz offset across the tuning range from 57.5 to 90.1 GHz. The measured output power with open drain buffer is from −20 to −25 dBm over the tuning range.

12.3.2 Coarse and fine magnetic tuning DB-VCO

Figure 12.21 shows the schematic and chip photograph of the proposed coarse and fine magnetic tuning DB-VCO with open-drain buffers for testing [20]. The VCO core consists of a triple-coil transformer tank proposed in Figure 12.10 and a cross-coupled pair M_1 and M_2 providing negative resistance to compensate for the tank loss. A transistor switch M_{SW1} is added in series with the secondary coil L_{SW1} for coarse frequency tuning, while another transistor M_{SW2} operating as a variable

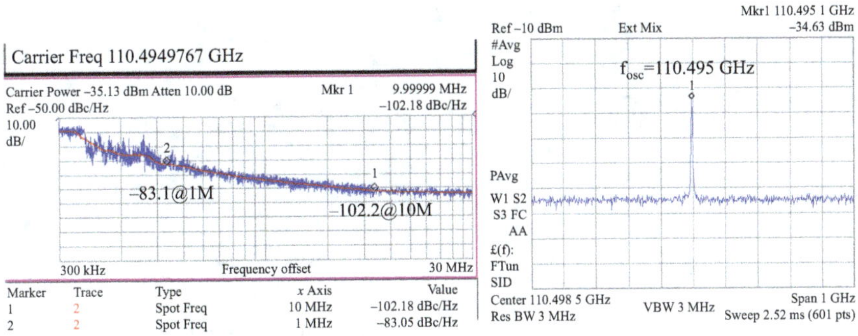

Figure 12.22 Measured phase noise and output spectrum at 110.5 GHz of the coarse and fine magnetic tuning dual-band voltage-controlled oscillator (DB-VCO)

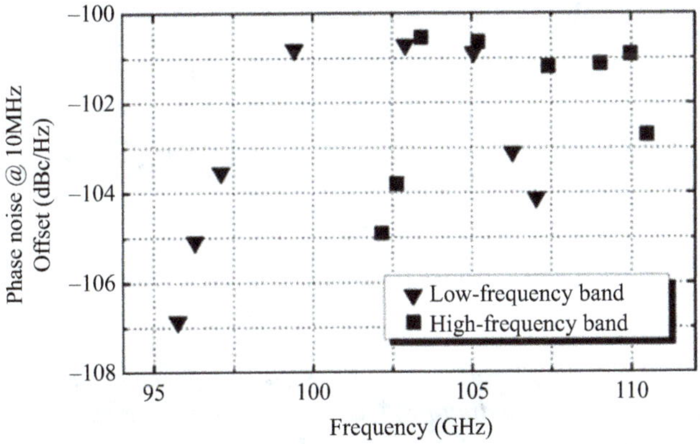

Figure 12.23 Measured phase noise of the VCO across the tuning range

resistor is connected in parallel with the third coil L_{SW2} for fine frequency tuning as discussed in Section 2. As comparison, if varactors were used, the Q value for 10 fF varactors at 100 GHz frequency would be limited to only about 2.

Figure 12.22 shows the phase noise and output spectrum of the DB-VCO at operation frequency of 110.5 GHz, measuring a phase noise of −83.1 dBc/Hz and −102.2 dBc/Hz at 1-MHz and 10-MHz offsets, respectively. Figure 12.23 shows the measured phase noise from −100.6 to −106.9 dBc/Hz at 10 MHz offset across the tuning range from 95.7 to 110.5 GHz. With a supply of 1 V, the power consumption is 6.2 mW, corresponding to figure of merit with tuning range (FoM_T) between −176.0 and −181.7 dBc/Hz.

12.3.3 Fine magnetic tuning split transformer-based VCO

Figure 12.24 shows the schematic and chip photograph of the split transformer-based VCO with the proposed split transformer shown in Figure 12.15. The split

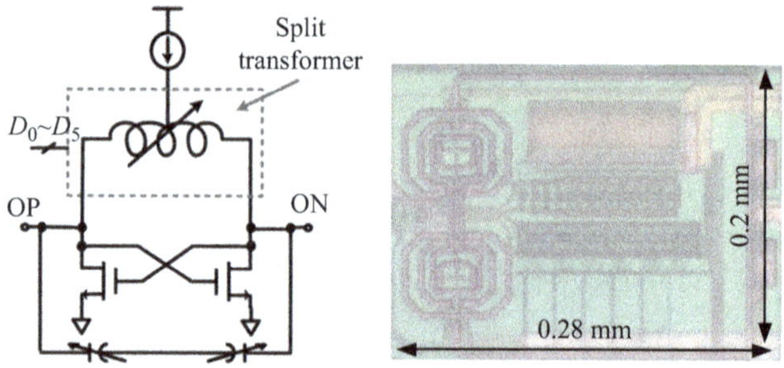

Figure 12.24 Schematic and layout of the split-transformer-based voltage-controlled oscillator (VCO)

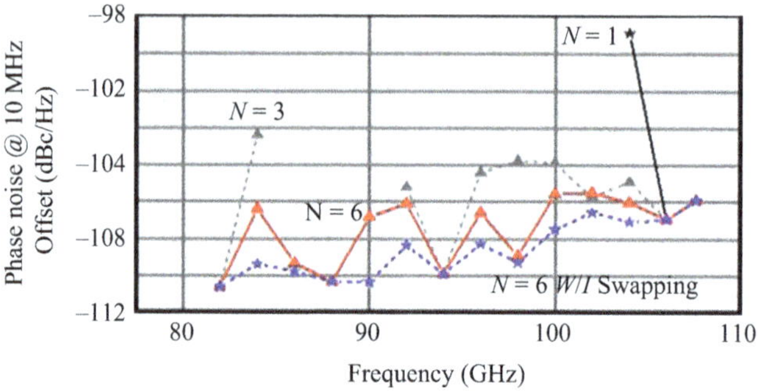

Figure 12.25 Measured phase noise of the split-transformer-based voltage-controlled oscillator (VCO) across the tuning range for different configurations of the split transformer

transformer can be reconfigured into $N = 1$ with the same control signal for $D_{0\sim5}$; $N = 3$ with $D_0 = D_1$, $D_2 = D_3$ and $D_4 = D_5$; and $N = 6$ with different control signal for $D_{0\sim5}$. As shown in Figure 12.25, with $N = 1$, because of severe quality factor degradation, the split-transformer-based VCO fails to oscillate at most of the frequencies. As N is increased, the overall tank Q improves, and the measured phase noise becomes lower. Specifically, for $N = 6$, as compared to $N = 3$, not only does the VCO oscillate over a much wider tuning range (from 21.5% to 27%), but it also has a better Q and phase noise (−106 dBc/Hz instead of −102 dBc/Hz at 84 GHz). By swapping the tuning order, the 10-MHz phase noise is further improved from −106 to −109 dBc/Hz at 84 GHz. The tuning ranges of $D_0 \sim D_5$ are 1.2, 2.95, 4.45, 4.4, 6.2 and 5.8 GHz, respectively, which enables similarly high Q over the entire frequency range. The measured phase noise varies from −106 to −110 dBc/Hz at 10-MHz offset across the tuning range.

Table 12.2 Comparison of mm-Wave and sub-THz oscillators

Ref.	Tuning range (GHz)	Supply Voltage (V)	Power (P_{DC}) (mW)	Phase noise (dBc/Hz)	FoM[(a)] (dBc/Hz)	FoM_T[(b)] (dBc/Hz)	Process
[7]	50.1–59.8 (18%)	1	6.2	−111.9 @10 M	−178.9	−184	65 nm CMOS
[21]	55.1–70.4 (24.6%)	1.2	21.5	−112.2 @10 M	−174.9	−182.6	40 nm CMOS
[22]	62.1–78.3 (22.3%)	1.1	7.7/8.8	−105.8/ −112.0 @10 M	−180.4	−187.4	65 nm CMOS
[23]	87.1–91.7 (5.2%)	1	11	−108.3 @10 M	−176.9	−171.1	65 nm CMOS
[24]	67.8–81.4 (18.2%)	1	13/25	−108/−113 @10 M	−173/ −178	−178/ −183	65 nm CMOS
[25]	98–103.3 (5.2%)	0.8/1.2	12/21	−112.1 @10 M	−178.6	−172.9	65 nm CMOS
[2]	100–110 (9.5%)	1.2	54	−92.8 @10 M	−175.0	−175.5	65 nm CMOS
[26]	93.2–105 (11.9%)	1.5	30	−112.7 @10 M	−177.5	−179.1	65 nm CMOS
[18]	**57.5–90.1 (41.1%)**	**1.2**	**8.4/10.8**	**−104.6/ −112.2 @10M**	**−172/ −180**	**−184.2/ −192.2**	**65 nm CMOS**
[20]	**95.7–110.5 (14.3%)**	**1**	**6.2**	**−100.6/−106.9 @10M**	**−172.9/ −178.6**	**−176/ −181.7**	**65 nm CMOS**
[19]	**82–107.6 (27%)**	**0.8**	**12**	**−106/−110 @10M**	**−175.8/ −177.5**	**−184.4/ −186.1**	**65 nm CMOS**

[(a)] $FoM = PN - 20\log(f_o/\Delta f) + 10\log(P_{DC}/1\ mW)$.
[(b)] $FoM_T = PN - 20\log(f_o/\Delta f) - 20\log(TR[\%]/10) + 10\log(P_{DC}/1\ mW)$.
Tuning range: the frequency range over which a VCO can be adjusted by means of tuning control.
Phase noise: the frequency-domain representation of random fluctuations in the phase of a waveform.
FoM: a numerical quantity based on one or more characteristics (phase noise, power, tuning range) of the VCO that represents a measure of effectiveness.

12.4 Conclusion

Table 12.2 summarizes and compares the performance of the proposed magnetic tuning mm-Wave/sub-THz oscillators with the prior state of the art. The coarse magnetic tuning multi-mode VCO [18] demonstrates the widest tuning range and the best FoM_T. The coarse and fine magnetic tuning VCO [20] achieves the lowest power consumption and the best FoM among 100-GHz oscillators. The split-transformer-based VCO [19] achieves the largest tuning range of 27% with the best worst-case FoM_T among all the existing 100-GHz oscillators. It can be seen that at mm-Wave frequency of around 70 GHz, the coarse magnetic tuning multi-mode VCO can achieve much wider tuning range and comparable phase noise compared to the previous works. With the increase in frequency of above 100 GHz at sub-THz range, the fine magnetic tuning multi-mode VCO has an advantage of wider tuning range and better FoM.

References

[1] K. Okada, N. Li, K. Matsushita, *et al.*, "A 60-GHz 16QAM/8PSK/QPSK/ BPSK direct-conversion transceiver for IEEE802.15.3c," *IEEE J. Solid-State Circuits*, vol. 46, no. 12, pp. 2988–3004, 2011.

[2] M. Adnan and E. Afshari, "A 105-GHz VCO with 9.5% tuning range and 2.8-mW peak output power in a 65-nm bulk CMOS process," *IEEE Trans. Microw. Theory Tech.*, vol. 62, no. 4, pp. 753–762, 2014.

[3] S. Rong and H. C. Luong, "Design and analysis of varactor-less interpolative-phase-tuning millimeter-wave LC oscillators with multiphase outputs," *IEEE J. Solid-State Circuits*, vol. 46, no. 8, pp. 1810–1819, 2011.

[4] M. Demirkan, S. Bruss and R. Spencer, "Design of wide tuning-range CMOS VCOs using switched coupled-inductors," *IEEE J. Solid-State Circuits*, vol. 43, no. 5, pp. 1156–1163, 2008.

[5] M. Vigilante and P. Reynaert, "Analysis and design of an E-Band transformer-coupled low-noise quadrature VCO in 28-nm CMOS," *IEEE Trans. Microw. Theory Tech.*, pp. 1122–1132, 2016.

[6] A. I. Hussein, S. Vasadi and J. Paramesh, "A 50–66-GHz phase-domain digital frequency synthesizer with low phase noise and low fractional spurs," *IEEE J. Solid-State Circuits*, pp. 3329–3347, 2017.

[7] M. Haghi Kashani, A. Tarkeshdouz, R. Molavi, E. Afshari and S. Mirabbasi, "A wide-tuning-range low-phase-noise mm-Wave CMOS VCO with switchable transformer-based tank," *IEEE Solid-State Circuits Letters*, vol. 1, no. 4, pp. 82–85, 2018.

[8] K. Kwok, J. R. Long and J. J. Pekarik, "A 23-to-29GHz differentially tuned varactorless VCO in 0.13μm CMOS," *IEEE Int. Solid-State Circuits Conf. (ISSCC)*, pp. 194–596, 2007.

[9] G. Cusmai, M. Repossi, G. Albasini, A. Mazzanti and F. Svelto, "A magnetically tuned quadrature oscillator," *IEEE J. Solid-State Circuits*, vol. 42, pp. 2870–2877, 2007.

[10] X. Yi, C. C. Boon, J. F. Lin, and W. M. Lim, "A 100 GHz transformer-based varactor-less VCO with 11.2% tuning range in 65nm CMOS technology," *IEEE Eur. Solid-State Circuits Conf. (ESSCIRC)*, pp. 293–296, 2012.

[11] P. Chiang, O. Momeni and P. Heydari, "A 200-GHz inductively tuned VCO with -7-dBm output power in 130-nm SiGe BiCMOS," *IEEE Trans. Microw. Theory Tech.*, pp. 3666–3673, 2013.

[12] T. Lu, C. Yu, W. Chen and C. Wu, "Wide tuning range 60 GHz VCO and 40 GHz DCO using single variable inductor," *IEEE Trans. Circuits Syst. I*, vol. 60, no. 2, pp. 257–267, 2013.

[13] A. Bevilacqua, F. P. Pavan, C. Sandner, A. Gerosa and A. Neviani, "Transformer-based dual-mode voltage-controlled oscillators," *IEEE Trans. Circuits Syst. II*, vol. 54, no. 4, pp. 293–297, 2007.

[14] Z. Safarian and H. Hashemi, "Wideband multi-mode CMOS VCO design using coupled inductors," *IEEE Trans. Circuits Syst. I*, vol. 56, no. 8, pp. 1830–1843, 2009.

[15] B. Sadhu, M. Ferriss, and A. Valdes-Garcia, "A 52 GHz frequency synthesizer featuring a 2nd harmonic extraction technique that preserves VCO performance," *IEEE J. Solid-State Circuits*, vol. 50, no. 5, pp. 1214–1223, 2015.

[16] A. Shirazi, A. Nikpaik, R. Molavi *et al.*, "On the design of mm-Wave self-mixing-VCO architecture for high tuning-range and low phase noise," *IEEE J. Solid-State Circuits*, vol. 51, no. 5, pp. 1210–1222, 2016.

[17] U. Decanis, A. Ghilioni, E. Monaco, A. Mazzanti and F. Svelto, "A low-noise quadrature VCO based on magnetically coupled resonators and a wideband frequency divider at millimeter waves," *IEEE J. Solid-State Circuits*, vol. 46, no. 12, pp. 2943–2955, 2011.

[18] J. Yin and H.C. Luong, "A 57.5–90.1-GHz Magnetically Tuned Multimode CMOS VCO," *IEEE J. Solid-State Circuits*, vol. 48, no. 8, pp. 1851–1861, 2013.

[19] Z. Huang and H. C. Luong, "An 82 GHz-to-107.6 GHz integer-N ADPLL employing a DCO with split transformer and dual-path switched-capacitor ladder and a clock-skew-sampling delta-sigma TDC," *IEEE J. Solid-State Circuits*, vol. 54, no. 2, pp. 358–367, 2019.

[20] X. Liu, C. Chen, J. Ren and H. C. Luong, "Transformer-based varactor-less 96GHz-110GHz VCO and 89GHz-101GHz QVCO in 65nm CMOS," *IEEE Asian Solid-State Circuits Conf. (ASSCC)*, pp. 357–360, 2016.

[21] N. Yanay and E. Socher, "Wide tuning-range mm-Wave voltage controlled oscillator employing an artificial magnetic transmission line," *IEEE Trans. Microw. Theory Tech.*, vol. 63, no. 4, pp. 1342–1352, 2015.

[22] Y. Chao and H. C. Luong, "Analysis and design of wide-band millimeter-wave transformer-based VCO and ILFDs," *IEEE Trans. Circuits Syst. I*, vol. 63, no. 9, pp. 1416–1425, 2016.

[23] T. Xi, S. Guo, P. Gui, D. Huang, Y. Fan and M. Morgan, "Low-phase-noise 54-GHz transformer-coupled quadrature VCO and 76-/90-GHz VCOs in 65-nm CMOS," *IEEE Trans. Microw. Theory Tech.*, pp. 2091–2103, 2016.

[24] L. Wu and Q. Xue, "E-Band multi-phase LC oscillators with rotated-phase-tuning using implicit phase shifters," *IEEE J. Solid-State Circuits*, vol. 53, no. 9, pp. 2560–2571, 2018.

[25] S. Kang and A. M. Niknejad, "A 100GHz active-varactor VCO and a bi-directionally injection-locked loop in 65nm CMOS," *IEEE Radio Freq. Integr. Circuits (RFIC)*, pp. 231–234, 2013.

[26] X. Yi, Z. Liang, G. Feng, *et al.*, "A 93.4-104.8-GHz 57-mW fractional-N cascaded PLL with true in-phase injection-coupled QVCO in 65-nm CMOS technology," *IEEE Trans. Microw. Theory Tech.*, pp. 1–12, 2019.

Chapter 13
Ultra-low phase noise ADPLL for millimeter wave

Zhirui Zong[1] *and Robert Bogdan Staszewski*[2]

Millimeter-wave (mm-wave) frequency synthesizers in complementary metal-oxide-semiconductor (CMOS) suffer from poor phase noise (PN), limited tuning range (TR) and high-power consumption. They are the key subsystems that typically limit the performance of mm-wave transceivers. This chapter presents a new architecture for mm-wave frequency synthesis that improves its PN performance and power efficiency. Various different techniques are introduced and demonstrated in a 60-GHz fractional-N all-digital phase-locked loop (ADPLL).

13.1 Evolution of mm-wave frequency synthesizer architectures

High-performance mm-wave transceivers are the key for various emerging wireless applications. They have set stringent requirements on mm-wave frequency synthesizers. The synthesizers must achieve low PN and wide TR at low-power consumption. In practice, however, physical, circuit- and architecture-level limitations set performance and power-efficiency barriers. Passive devices at mm-wave frequencies feature low-quality (Q) factor, which limits the oscillator's PN performance. In the feedback frequency divider path, the frequency division ratio between the mm-wave output and the input reference clock is large. Design complexity of mm-wave frequency prescalers is significant. These facts pose challenges in designing mm-wave PLLs that are power efficient and of low PN.

Continuous efforts were taken to explore new solutions and implementations for mm-wave frequency synthesis. Several mm-wave frequency synthesizer architectures have been reported in literature. They can be categorized into three groups [1]: a PLL with a fundamental oscillator [1–4], a low-frequency PLL together with a frequency multiplier [5–7], and a PLL with N-push oscillators [8–10]. Figure 13.1 shows these three architectures in the case of 60 GHz frequency generation.

[1]Department of Microelectronics, Delft University of Technology, Delft, The Netherlands
[2]School of Electrical and Electronic Engineering, University College Dublin, Dublin, Ireland

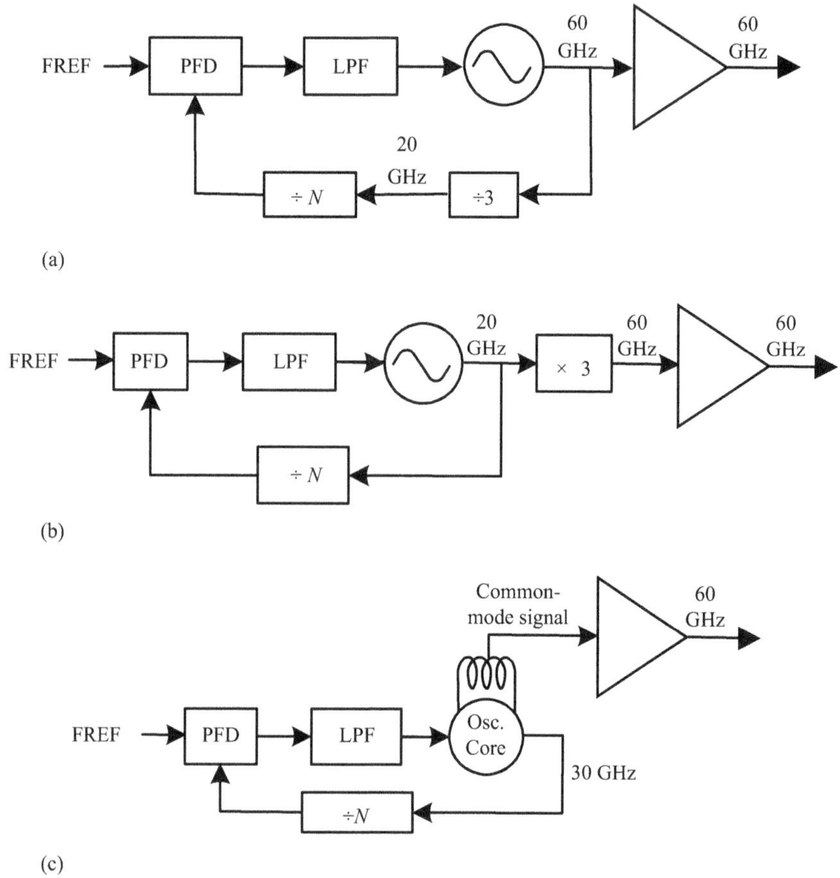

Figure 13.1 Evolution of the mm-wave PLL architectures: (a) PLL with fundamental oscillator, (b) PLL with frequency multiplier, and (c) PLL with push-push oscillator

In the first architecture (Figure 13.1(a)) [1–4], the mm-wave oscillators and high-frequency dividers are the key design challenges [11–16]. The difficulties of 60-GHz oscillators are (1) the parasitic capacitance of active devices takes up a large share of the relatively small tank capacitance, thus limiting the frequency TR and (2) to achieve a TR of >15%, the poor Q factor of the tuning capacitance dominates the Q factor of the mm-wave resonator, thus limiting the achievable PN. Compared to the inductor-capacitor (LC) tank at lower frequencies, the LC resonators at mm-wave frequencies exhibit lower Q factor for the same TR. The 60-GHz frequency dividers must achieve wide locking range to ensure sufficient overlap with the oscillator TR under process, voltage and temperature (PVT) variations. However, there is a strong trade-off between the locking range and power consumption [14–16]. The aforementioned design challenges in the oscillators and

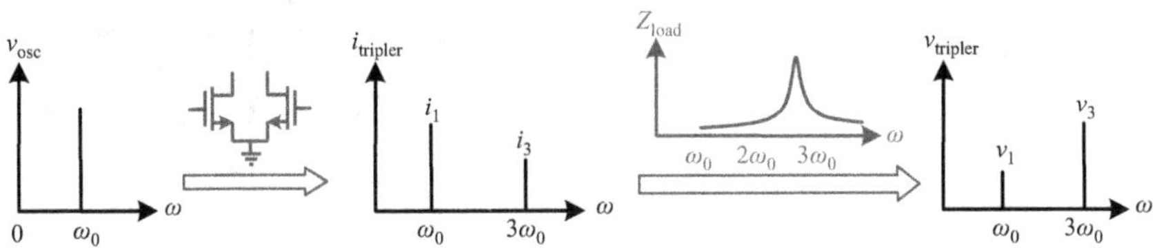

Figure 13.2 Operating principle of a conventional frequency tripler

frequency dividers are relieved in the PLLs based on frequency multipliers (Figure 13.1(b)) [5–7]. This solution can improve the PN performance. However, the 60-GHz frequency multipliers in this architecture typically have limited locking range or consume significant amount of power in order to achieve large locking range. The design challenges remain but are shifted from the mm-wave frequency dividers to frequency multipliers. In PLLs with N-push oscillators (as shown in Figure 13.1(c) for $N = 2$) [8–10], the frequency dividers operate at 60/N GHz, and frequency multipliers are avoided. However, this oscillator type suffers from low output power and mismatches among the N oscillators if $N > 2$. Among this type, push-push oscillators are the most common and easiest to implement. However, the required high common-mode (CM) swing can increase the $1/f$ noise up-conversion [17]. Moreover, the conversion from single-ended CM signal to a differential output may introduce significant phase error [8].

The evolution of mm-wave frequency synthesizers reaches a consensus that the oscillators should operate at low frequencies for good PN. Frequency multipliers remain an obstacle for improving the power efficiency and robustness in frequency synthesizers depicted in Figure 13.1(b). It is desired to have a power-efficient approach to realize frequency multiplication. The operating principle of a conventional frequency tripler is shown in Figure 13.2. It relies on the high-order nonlinearities of the input devices to generate the third harmonic (H3) current. The harmonic current is converted to high voltage swing through an LC tank tuned at the H3 frequency. Since the oscillators undergo large-signal operation, the oscillating current is typically rich in harmonic. In conventional LC oscillators, the harmonic current is filtered out by the high-Q LC tank resonating at the fundamental oscillation frequency. Instead of regenerating the H3 current in the tripler, it will be more power efficient if the harmonic current in the oscillator can be leveraged for frequency multiplication. In the rest of the chapter, a 60-GHz fractional-N ADPLL will be presented, which reuses the readily existing H3 current in a 20-GHz digitally-controlled oscillator (DCO) for implicit frequency tripling.

13.2 Mm-wave ADPLL with implicit frequency tripling

With the aforementioned observations, a new PLL architecture based on a 20-GHz oscillator and an implicit ×3 frequency multiplier [18,19] is introduced in this chapter, as shown in Figure 13.3. The basic concept is to simultaneously generate both 20 GHz and a significant level of its H3 at 60 GHz inside a 20-GHz oscillator.

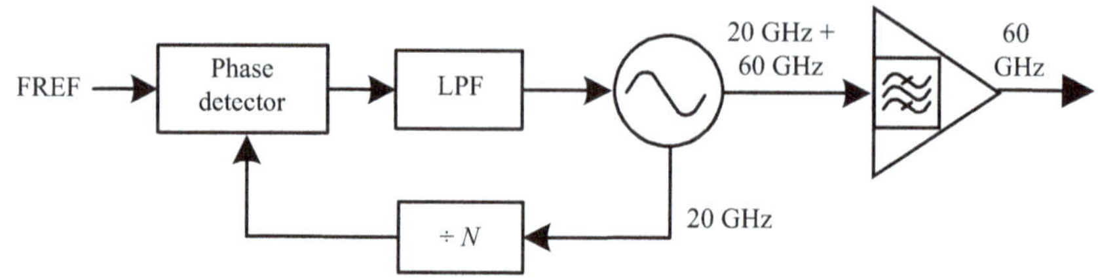

Figure 13.3 Mm-wave PLL architecture with implicit frequency tripler

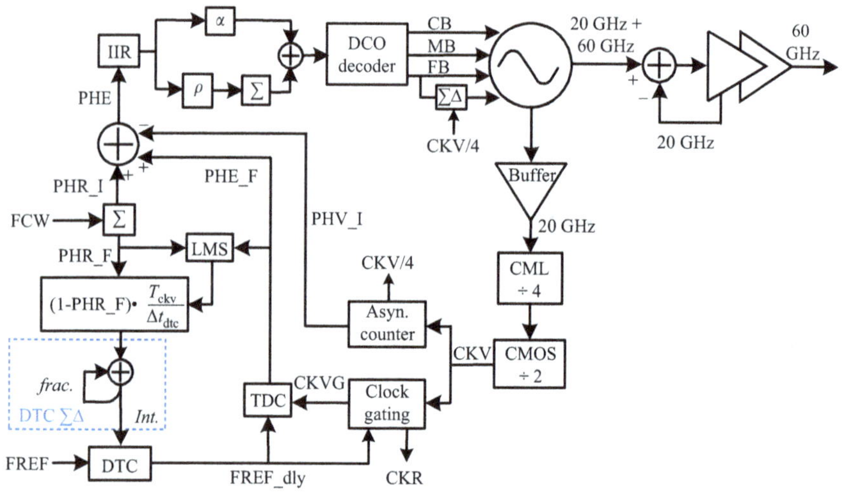

Figure 13.4 System diagram of the 60-GHz digital frequency synthesizer

The generated 60 GHz signal is fed forward to the output buffer, while the 20-GHz signal is fed back for phase detection after further frequency division. The directly available 60-GHz signal from the DCO exempts the demands for further frequency multiplication in the feedforward path. In this way, the conventional 60-GHz frequency dividers or doublers/triplers to 60 GHz are avoided. This simplifies the design and improves the power efficiency of the 60-GHz frequency synthesizer.

Figure 13.4 shows the detailed system diagram of an ADPLL-based 60-GHz digital frequency synthesizer that adopts the new architecture. At its center lies a H3-boosting 20-GHz DCO. The cogenerated H3 is extracted and fed forward to a 60-GHz output buffer, which also cancels out the 20-GHz component. In the feedback path, the 20-GHz signal is prescaled by $\div 4$ current-mode logic and $\div 2$ CMOS dividers to generate variable clock (CKV) for phase detection in the ADPLL. An 8-bit counter digitizes the integer part of the CKV phase (PHV_I[k]). The resulting PHV_I[k] is compared to the integer part of the reference phase PHR_I[k] to generate the integer part of phase error, PHE_I[k]. The reference clock (FREF) clock is delayed by a digital-to-time converter (DTC), whose delay is determined by the fractional part of reference phase (PHR_F[k]). The DTC is controlled by a ΣΔ modulator (ΣΔM) and covers a range that is a bit larger than one CKV period. A least mean square (LMS) algorithm [20] runs in the background to calibrate the DTC gain. The DTC output, FREF_dly, is used to clock gate the CKV to generate CKVG, as well as the retimed clock, CKR, for the digital logic [21]. The time-to-digital converter (TDC) digitizes the phase difference between CKVG and FREF_dly and provides, after normalization, the fractional phase error PHE_F[k]. A small TDC detection range can slow down the transient settling or locking time of the ADPLL in some cases. For a more robust operation, it covers a detection range that is 40% of the CKV period. Furthermore, during the settling process, once TDC overflow or underflow is detected, the TDC output is overwritten by a maximum or minimum value. It facilitates the settling speed [21]. After the PLL is locked, only a few cells in the middle of the TDC chain are active. The composite phase error, PHE [k] = PHE_I[k] + PHE_F[k], is fed to the digital loop filter (LF), which comprises a proportional-integral (PI) controller and a fourth-order infinite-impulse-response (IIR) filter. The LF output is fed to the DCO as its tuning word through a decoder.

In this design, the PLL bandwidth is optimized for low-integrated PN (IPN). The DCO PN is high-pass filtered by the loop. However, unless otherwise wide loop bandwidth is used, the high close-in PN (typically the $1/f^3$ noise) of the oscillators could even dominate the PLLs' in-band PN in many cases. It will be further analyzed in Section 1.3. To have more freedom in optimizing the loop bandwidth, the $1/f^3$ noise of the DCO is targeted to be low in this design Wideband carrier recovery loops in the transceiver baseband can filter out the close-in PN [22]. However, in some applications (such as radars), they may not be available. This work targets low IPN from the PLL itself. To achieve this goal, the noise contribution from FREF, DTC, TDC, and frequency dividers, which are low-pass filtered by the loop, should also be minimized. FREF noise floor is predefined by the crystal oscillator and tends to be conservatively around −150 dBc/Hz for a 40-MHz FREF used in this design. The noise floor of the frequency dividers and DTC are designed to be significantly (about 7 dB) below the FREF noise floor, so that they have negligible contribution to the ADPLL output PN. TDC quantization (Q) noise can be reduced by improving its resolution, typically at the cost of a larger power consumption and degraded integral nonlinearity linearity that will create fractional spurs [23]. However, significant lowering of the TDC Q-noise below that of FREF or DCO's contribution cannot further improve the in-band PN. With the targeting of low $1/f^3$ DCO, its in-band contribution should be less than or comparable to that of FREF. For the optimal power efficiency and linearity, the TDC Q noise is designed at the same level as the FREF noise floor. This corresponds to the TDC resolution of 3 ps with 40 MHz FREF.

The DTC ΣΔM Q-noise-induced PN is given by:

$$L_{dtc}(\Delta f) = \frac{1}{12} \cdot \frac{(\Delta t_{dtc})^2}{f_{ref}} \cdot (2\pi f_o)^2 \cdot \left[2 \sin\left(\frac{\pi \Delta f}{f_{ref}}\right)\right]^2 \cdot |H_{cl}(\Delta f)|^2 \quad (13.1)$$

where Δt_{dtc} is the DTC resolution, f_{ref} is the FREF clock frequency, f_o is the PLL output frequency, and $H_{cl}(\Delta f)$ is the close-loop transfer function of the DTC noise to the output. Besides the PI controller, the IIR filter in $H_{cl}(\Delta f)$ provides extra attenuation to the high-frequency ΣΔM Q noise. The DTC resolution is optimized, so that $L_{dtc}(\Delta f)$ has marginal contribution to the IPN. In this design, the DTC resolution should be finer than 16 ps, so that its degradation on the IPN is less than 0.5 dB. To reduce the TDC dynamic range, fine DTC resolution is desired as well. On the other hand, with better resolution, more delay unit cells are required in the DTC to cover one CKV period. The long DTC delay chain is typically more prone to linearity degradation. As a trade-off, the DTC resolution is chosen as 12 ps in this design.

13.3 Oscillator with implicit frequency multiplication

13.3.1 Third-harmonic boosting

In order to realize frequency tripling with the readily available harmonic current in the oscillator, the LC resonator should preserve the H3 content instead of filtering it out. At mm-wave frequencies, a practical way of implementing it is to introduce an extra LC resonance at the H3 frequency in addition to the fundamental resonance. One possible approach is to use multiple series-connected LC tanks resonating at the fundamental and H3 frequencies [24,25]. Figure 13.5 shows the simplified diagram of such an oscillator. The multiple inductors occupy large area. Since the two resonances have the same phase response and transconductance gain in the oscillation loop, undesired oscillation at the auxiliary resonance could be triggered by increasing the level of H3. Therefore, it appears problematic in our case.

A more compact implementation would be a transformer-based dual-tank resonator. The L and C ratios in primary and secondary windings were optimized in [26] to realize fundamental oscillation and its H3 resonance. However, the H3 generated there is relatively weak (~15% of the fundamental tone), so to make it sufficiently stronger, a high-gain buffering amplifier would be needed. That implies a high power consumption, which would negate the effectiveness of the proposed architecture. In order to reduce the required gain of the following buffer stage, a much stronger H3 must be provided by the oscillator. A harmonic boosting technique is presented and applied to the transformer-based oscillator.

Simplified diagram of the proposed mm-wave oscillator and its operational principle is shown in Figure 13.6. $I_{DH1,3}$, $R_{p1,3}$ and $V_{DH1,3}$ represent the tank's

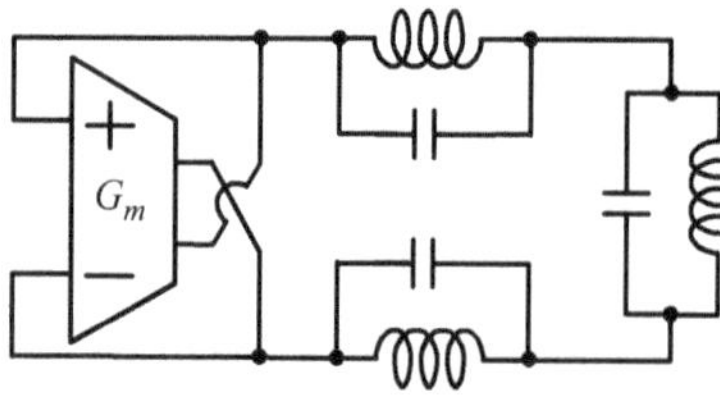

Figure 13.5 A possible harmonic boosting oscillator topology with multiple LC tanks

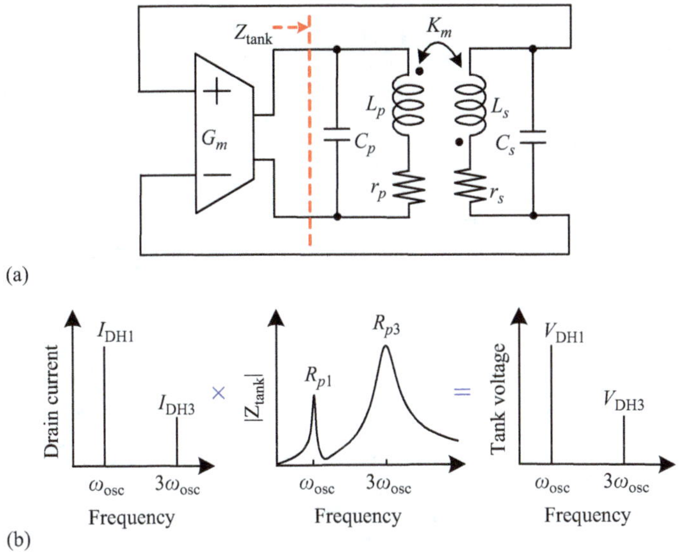

Figure 13.6 (a) Simplified diagram and (b) operational principle of the proposed oscillator

current, equivalent parallel resistance, and voltage, respectively, of the first and H3 components. According to the linear model of oscillators, the oscillation amplitude at the first and H3 is determined by the current component and tank impedance at their respective frequencies. To achieve a larger V_{DH3}/V_{DH1}, there are two possible options: increasing I_{DH3}/I_{DH1} or increasing R_{p3}/R_{p1}. The I_{DH3}/I_{DH1} depends on the nonlinearity of the g_m devices and is typically fixed for a certain type of oscillator (e.g., 0.33 for class-B). It is not straightforward to further increase the I_{DH3}/I_{DH1} ratio without extra efforts. Consequently, larger R_{p3}/R_{p1} is desired. On the other hand, the equivalent Q factor (Q_{eq}) at the two resonant frequencies affect the oscillator performance dramatically. High Q_{eq} at ω_{osc} promotes low PN, while low Q_{eq} at $3\omega_{osc}$ is appreciated for better tolerance to the possible frequency misalignment between the second resonance and $3\omega_{osc}$.

Analysis shows that magnetic coupling coefficient (k_m) affects both the tank impedance (Z_{tank}) and Q_{eq}. The dependence of Z_{tank} and Q_{eq} at ω_{osc} and $3\omega_{osc}$ on k_m are shown in Figure 13.7. As we can see, R_{p1} decreases with smaller k_m, while R_{p3} behaves opposite. Therefore, smaller k_m is desired for larger R_{p3}/R_{p1}. However, larger k_m is required for high Q_{eq} at ω_{osc} and low Q_{eq} at $3\omega_{osc}$. By reducing k_m for larger R_{p3}/R_{p1}, both the PN performance and the tolerance to the possible frequency misalignment between the second resonance and $3\omega_{osc}$ will be degraded. As a trade-off between high H3 and optimal oscillator performance, $k_m = 0.6$ is chosen for $R_{p3}/R_{p1} > 1$ with sufficient Q_{eq}. Simulations show that the H3 to fundamental voltage swing ratio in the DCO (i.e., V_{DH3}/V_{DH1} in Figure 13.2(b)) is 40%.

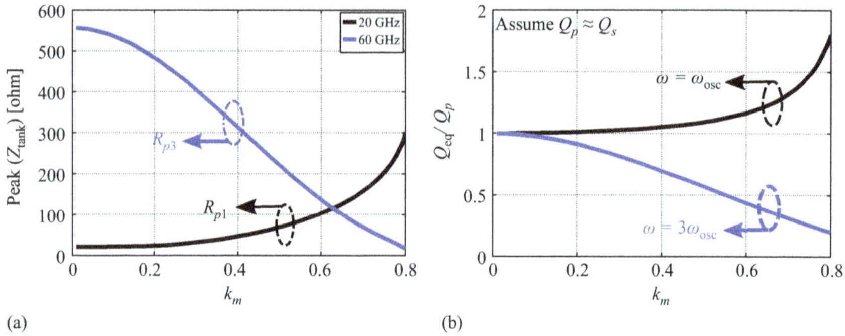

Figure 13.7 Dependency of (a) tank impedance on k_m and (b) Q_{eq} on k_m

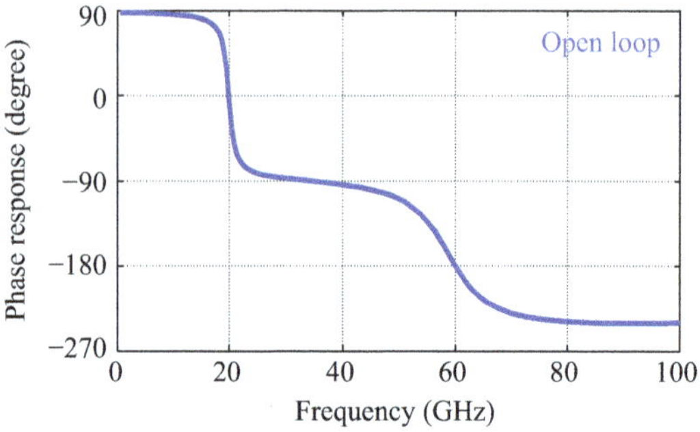

Figure 13.8 Simulated open-loop phase response of the oscillator

A concern might arise that the oscillation could happen at 60 GHz rather than at 20 GHz due to $R_{p3} > R_{p1}$. Start-up conditions are examined to ensure that the oscillation can only happen at 20 GHz, even if $R_{p3} > R_{p1}$: Barkhausen's phase and gain criteria should be satisfied for a stable oscillation. Analysis and simulations show that the open-loop phase response $\measuredangle H_{\text{ol}}(j\omega) = -180°$ at 20 GHz, while $\measuredangle H_{\text{ol}}(j\omega)$ 180° at 60 GHz, as shown in Figure 13.8. The phase criterion is satisfied only at 20 GHz. At 60 GHz, the phase response is $-180°$, which implies a negative feedback. Therefore, only one stable oscillation mode at 20 GHz is possible here. This phenomenon also explains the behavior of Q_{eq} in Figure 13.7. Since the fundamental components at both windings are in phase, its Q_{eq} at 20 GHz benefits from the mutual inductance. However, due to the fact that the H3 at both windings are antiphase, Q_{eq} at 60 GHz decreases with larger k_m.

13.3.2 Suppression of $1/f^3$ noise in oscillators

With the presented PLL architecture, the PN of the oscillator showed significant improvement, especially at 1 MHz offset. However, the $1/f^3$ noise corner is around 1 MHz, which is still quite high. It is even more than 1 MHz in conventional mm-wave oscillators [12,27]. When the loop bandwidth is much smaller than the corner frequency, the $1/f^3$ noise has significant contribution to the output PN. To verify its contribution, the oscillator PN reported in [19] is taken into an s domain linear model of a 60-GHz ADPLL. A 40-MHz FREF clock is used with noise floor of -150 dBc/Hz. The TDC Q noise is assumed to be at the same level as FREF noise. The other nondominant noise sources are neglected for simplicity. The LF is configured for the best PN performance. Figure 13.9 shows the PN result predicted by the model. The DCO's contribution to the in-band PN is much higher than that caused by the TDC and FREF noise floor. Keeping the other loop components and noise sources identical but improving the DCO $1/f^3$ corner to 300 kHz, the DCO PN does not dominate the in-band PN any more, as shown in Figure 13.9. IPN of the PLL with conventional $1/f^3$ DCO is 2.3 dB higher than that with the low $1/f^3$ DCO. Therefore, the currently achievable $1/f^3$ noise corner should be substantially reduced.

Techniques to lower the flicker noise up-conversion have been proposed for single-GHz voltage-biased oscillators [28,29]. However, they are sensitive to parasitic inductances and capacitances between power supply (VDD) and ground (VSS) rails. This effect becomes especially important when the VDD routing is physically away from VSS, effects of which are greatly magnified at mm-wave frequencies. Therefore, those techniques have tight constraints in practical mm-wave designs. Moreover, they are not applicable to current-biased oscillators, i.e., those using a tail current source (e.g., M_T in Figure 13.10), which is helpful for power supply rejection and oscillation amplitude control. In the presence of tail

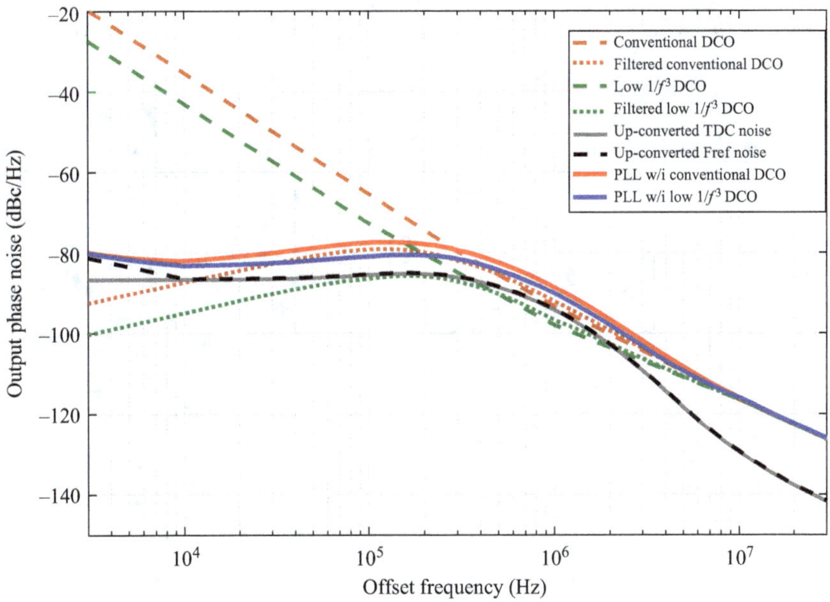

Figure 13.9 A 60-GHz ADPLL output phase noise with two different corners of DCO $1/f^3$ noise corner: 1 MHz and 300 kHz

current, its parasitic capacitance provides an extra medium for the flicker noise of the $M_{A,B}$ transistors to be up-converted to PN [17]. Flicker noise in the tail current source can be up-converted to $1/f^3$ PN. However, with a large-size transistor or its outright replacement by digitally controlled resistors, its contribution can be made marginal. On the other hand, the negative-g_m devices in mm-wave oscillators usually assume the shortest channel length for small parasitic capacitance and generate significant amount of flicker noise. The up-conversion from $1/f$ noise in the negative-g_m devices to PN should be suppressed.

As explored in [30], the up-conversion mechanisms can be categorized into two types: direct and indirect conversions. Fundamental and/or high-order harmonics in v_{gs} of the the g_m devices mixes with the baseband flicker noise and directly up-convert it to around fundamental and harmonic frequencies via the non-linearities of the cross-coupled pair. With any out-of-phase harmonics in v_{gs}, the $1/f$ noise can be converted to phase modulated noise around the oscillation frequency, namely, PN. The up-converted flicker noise at high-order harmonic frequencies in v_{ds} of the the g_m devices can also be indirectly converted down to the oscillation frequency by self-mixing with the periodically varying $g_{ds}(t)$ of the g_m devices. If any high-order harmonic in v_{ds} is out of phase, the $1/f$ noise is indirectly converted to PN. To achieve low $1/f^3$ noise, the out-of-phase harmonic voltage swing in v_{ds} and v_{gs} should be eliminated.

Based on these observations, a generic technique to suppress the $1/f$ noise up-conversion is proposed and applied to the 20-GHz DCO. The schematic and layout implementation are revealed in Figure 13.10. An extra series LC branch (L_{sh} and C_{sh}) is introduced between VDD (transformer's center tap) and a common-source node of $M_{A,B}$ (node *S*). The following two conditions should be satisfied: (1) the

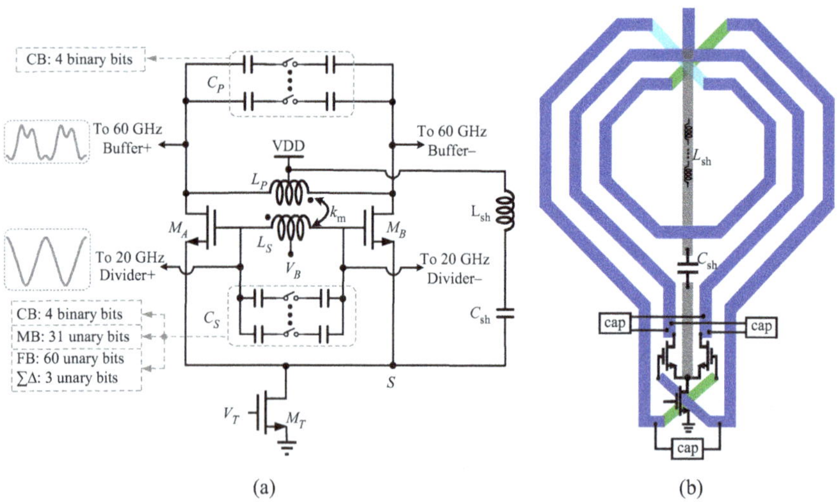

Figure 13.10 (a) Schematic and (b) concept layout of the DCO with the proposed 1/f noise up-conversion suppression technique

impedance of this series LC feed resonates at the second harmonic (H2) frequency and (2) the LC network between the drain of $M_{A,B}$ and node S resonates at H2 in CM. Condition #1 bypasses the practically ill-defined path between VDD and node S with a short-circuit path at H2. Since the H2 current path is explicitly defined by this auxiliary LC branch, it is not subjected to any normal parasitic effects in the VDD/VSS routing. In this way, the H2 component in v_{gs} are ensured to be small due to the short path between node S and center tap at H2. Therefore, condition #1 also prevents the $1/f$ noise up-conversion through direct conversion mechanism. Condition #2 ensures that H2 in v_{ds} is in phase. It reduces the $1/f$ noise indirect up-conversion through self-mixing. This idea provides an extra design freedom for the $1/f$ noise up-conversion reduction at no cost of area or power.

13.3.3 DCO capacitor bank design

The DCO is segmented into four switched-capacitor banks: 4-bit binary coarse tuning (CB), 31-bit unary mid-coarse tuning (MB), 60-bit unary fine-tuning (FB), and 3-bit unary FB for a high-speed ΣΔM. In this DCO, the capacitance in both primary (C_p) and secondary (C_s) sides tunes the oscillation frequency. In order to ensure the second resonance tracks the H3 of the oscillation frequency, C_s and C_p need to maintain a ratio of $C_s/C_p = 2$. The CB is replicated at the primary and secondary windings with 1:2 capacitance ratio, both of which are tuned simultaneously. It covers the largest frequency TR. For the best H3 tracking, it is desired to replicate MB and FB in both windings as well. However, since there are many tuning bits in MB and FB, their replication leads to bulky C_p and C_s in layout. A long interconnect is required to bridge between the inductor and capacitor banks that are far away. It degrades the Q factor and gives rise to parasitics. Fortunately, MB and FB only need to cover a small frequency range (twice the LSB of CB). It is possible to allocate these two banks to only C_s or C_p.

Let's consider the case that MB and FB are allocated to C_s first. Switching the control codes of MB and FB from the middle to the lowest or highest, the oscillation frequency covers ±220 MHz around 20 GHz, which is twice the LSB of CB. The second resonance varies by ±70 MHz. As shown in Figure 13.11, the impedance magnitude and phase at H3 varies by only 0.8% and 0.8°, respectively. If MB and FB are allocated to C_p, the required ΔC_p is 4.7 times larger to achieve the same TR. In this case, the second resonance is tuned by ±6.14 GHz. The impedance magnitude and phase at H3 varies by 52% (from 185 down to 89 Ω) and −46.5°, respectively.

Consequently, MB and FB are only placed at the secondary winding for a simple and compact layout. As revealed above, the TR of MB and FB is small, so that they will not introduce any significant misalignment between the second resonance and H3 frequencies. The three tuning banks are sized for 260 MHz/bit, 35 MHz/bit, and 3 MHz/bit unit steps (i.e., DCO resolution, K_{DCO}) at the 60 GHz carrier, respectively. The MB and FB cover > ±2 × LSB of CB and MB, respectively, to ensure continuous tuning. The high-speed ΣΔM clocked at CKV/4 (600 MHz) reduces the DCO Q noise. It can be programmed as a first or second order. The complete DCO schematic is shown in Figure 13.10(a).

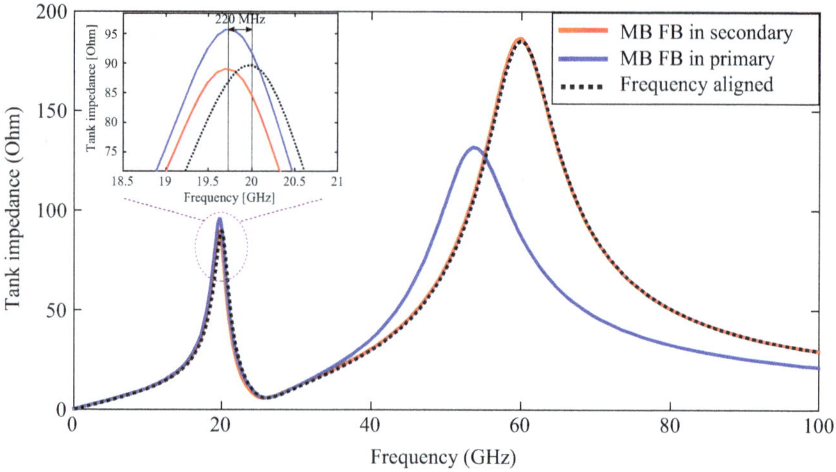

Figure 13.11 Simulated tank impedance with MB and FB in primary and secondary sides

13.4 The 20-GHz component suppression

This ADPLL is intended for a wireless transceiver. In such a case as in many others, the undesired 20 GHz tone should be sufficiently attenuated to lie well below the spectrum mask of a transmitter or to meet the blocker tolerance of a receiver [31]. To get a "pure" spectrum at the 60-GHz carrier output, the 20-GHz fundamental DCO tone should be suppressed in the 60-GHz buffer path. Ideally, the buffer would only react to the H3 component, while discarding the fundamental component. To achieve this effect, a cancellation scheme of this 20-GHz component is proposed. Figure 13.12(a) reveals the core idea in which the source of buffer's input transistor (M_1) follows the DCO's 20-GHz component but sees the ground at 60 GHz. In other words, M_1 senses only the H3 component between its gate and source (i.e., v_{gs}). This requires a high impedance at 20 GHz, but a very low impedance (ideally zero) at 60 GHz. A parallel LC tank that resonates at 20 GHz is therefore placed between the source of M1 and ground. Figure 13.12(b) shows the schematic of the soft-cancellation technique in a 60-GHz buffer with LC output matching network. At 20 GHz, M_1 is source degenerated by a high impedance provided by the 20-GHz tail tank. No 20-GHz current flow is allowed in M_1, thus no g_m gain. The LC tank has a very low impedance (<10 Ω) at 60 GHz, hence the desired 60-GHz input signal experiences high g_m. Since the capacitive source degeneration at high frequencies in M_1 may cause stability concerns, neutralization capacitor C_1 is added differentially between the gate and source to ensure differential-mode stability. A dampening resistor $R_T = 10\,\Omega$ is inserted between the common node of the tail tank capacitors C_T and ground to ensure CM stability. A simulated stability factor shows that this buffer is unconditionally stable.

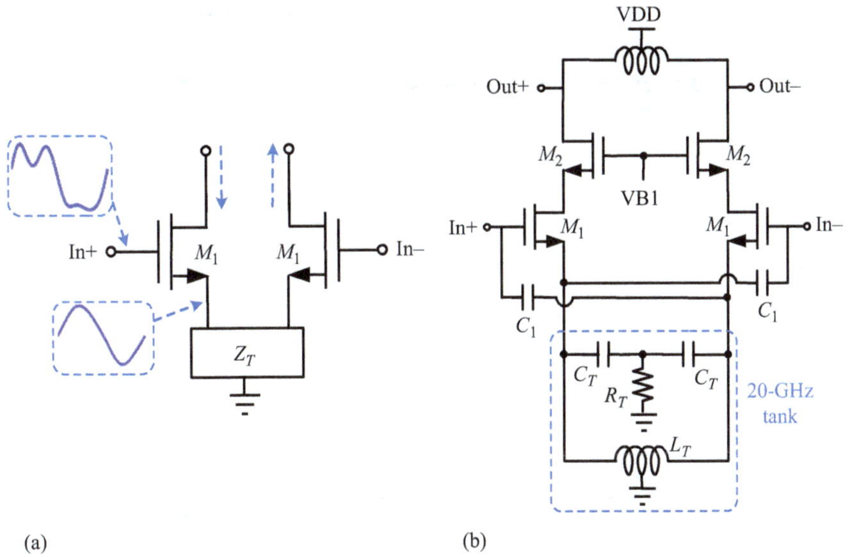

Figure 13.12 (a) Working principle and (b) schematic of the subharmonic soft-cancellation 60-GHz buffer with LC output matching network

Compared to the conventional buffers and notch-filter solutions [32], the proposed suppression technique offers several advantages. It achieves a better 20-GHz component rejection, provides more gain at 60 GHz, and prevents the generation of other harmonic contents (i.e., 40 and 80 GHz). With the same circuit parameters and input signal as that used in this design, the performance of proposed subharmonic soft-cancellation buffer (Figure 13.12(b)) is simulated and compared against the conventional and notch-filter buffers.

In the proposed buffer, the 20-GHz signal gain is:

$$\frac{G_m \cdot Z_{\text{load}}(j\omega_0)}{1 + G_m \cdot Z_{\text{tail}}(j\omega_0)} \tag{13.2}$$

where G_m is the transconductance of M_1. The effective transconductance gain at 20 GHz is decreased by a ratio of $1 + G_m Z_{\text{tail}}(j\omega_0)$. Figure 13.13 shows the simulated drain current in different buffers. With source degeneration by the 20-GHz LC tank, the 20-GHz drain current in the proposed buffer is reduced by 10–23 dB when compared to the other buffers. It validates the effectiveness of the proposed subharmonic soft-cancellation technique.

The subharmonic soft-cancellation technique presented here also provides more gain to the signal of interest at 60 GHz. The input of the buffer is the 60-GHz signal accompanied by a high 20-GHz tone. It is well known that a large blocker signal can desensitize the low-noise amplifiers in the receivers. The large 20-GHz tone can also desensitize the buffer in a similar way. The input transistor (M_1) has nonlinearities with a high signal excitation. In the conventional buffer and the

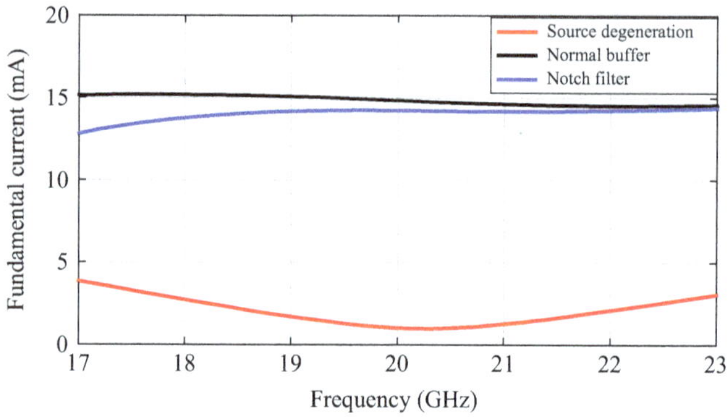

Figure 13.13 Simulated drain current at 20 GHz in different buffers

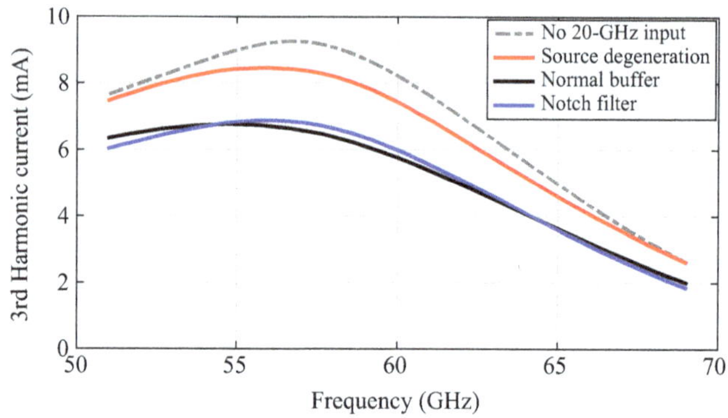

Figure 13.14 Simulated drain current at 60 GHz in different buffers

buffer with notch filter, the 20-GHz component directly presents in the v_{gs} of M_1. It decreases the 60-GHz output current in these buffers due to the third-order nonlinearity [33], while it is intact in the presented buffer. Figure 13.14 shows the simulated drain current at 60 GHz in different buffers. Compared to the conventional buffer and buffer with notch filter, the proposed solution delivers the highest drain current at 60 GHz. It is further compared against the case with 20-GHz tone removed from the input (dash line in Figure 13.14). In the conventional and notch-filter buffers, the drain current at 60 GHz is decreased by 2.6 and 2.8 dB, respectively. With the proposed technique, the reduction is limited to less than 0.8 dB.

The reduction in the 20-GHz component in v_{gs} of M_1 also prevents the generation of other harmonic components (such as 40 and 80 GHz), which can result from the second-order nonlinear distortion of the 20-GHz tone or harmonic mixing effects between 20 and 60 GHz. The simulated drain currents at 40 GHz in different buffers are shown in Figure 13.15. The proposed technique reduces the 40-GHz output current by 12–27 dB across the frequency range.

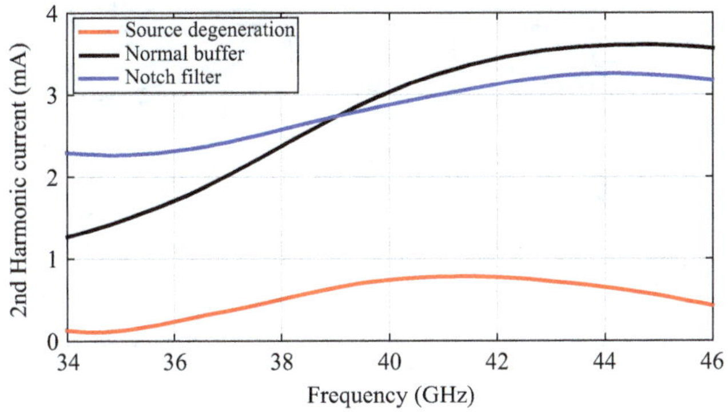

Figure 13.15 Simulated drain current at 40 GHz in different buffers

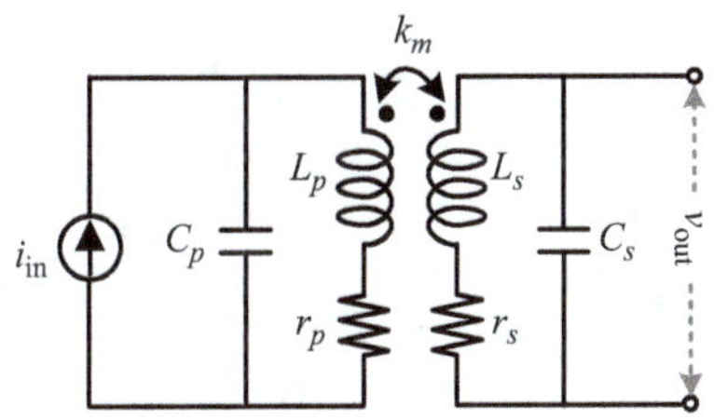

Figure 13.16 Transformer-based matching network

Besides the subharmonic soft-cancellation technique, the matching network of the buffer also provides extra filtering to the undesired 20-GHz tone. The matching network can have two possible implementations: a single LC tank (as in Figure 13.12(b)) and transformer-based LC network (Figure 13.16). The transformer solution prevails at mm-wave frequencies. Benefiting from the mutual inductance, it can achieve better Q factor and less insertion loss. The transimpedance of the transformer matching network can be expressed as:

$$Z_{\text{trans}}(j\omega) \cong \frac{j\omega M}{(1 + j\omega\tau_L - \omega^2/\omega_L^2)(1 + j\omega\tau_H - \omega^2/\omega_H^2)} \tag{13.3}$$

where ω_L and ω_H are the frequencies of the two conjugate pole pairs. With a high k_m, the second conjugate pole pair occurs at frequency that is much higher than the operating frequency of the transformer matching network (i.e., $\omega_L << \omega_H$). A high-k_m transformer matching network can be approximated as a single LC tank with one conjugate pole pair at the frequency of interest. Due to the enhanced Q by the mutual inductance, it provides more filtering capability to the out-of-band signal. With a low-k_m transformer matching network, the two conjugate pole pairs can be designed to be close to each other (i.e., $\omega_L \cong \omega_H$) [34]. If the two pole pairs are infinitely close, the suppression to the out-of-band signal is twice that of the single LC tank. Figure 13.17 shows the normalized filter response of the single LC

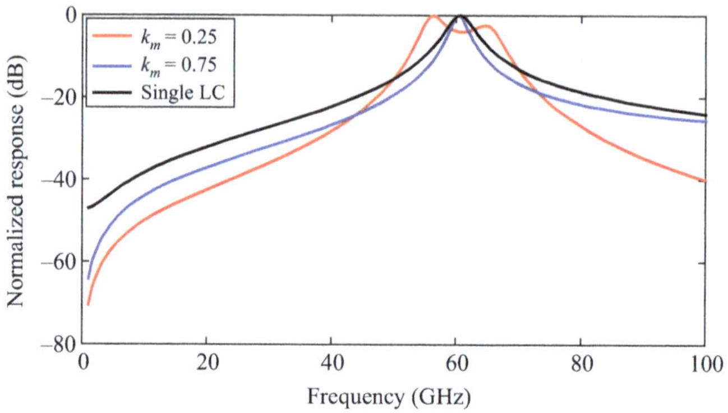

Figure 13.17 Normalized filter response of the single LC tank, low-k_m, and high-k_m matching network

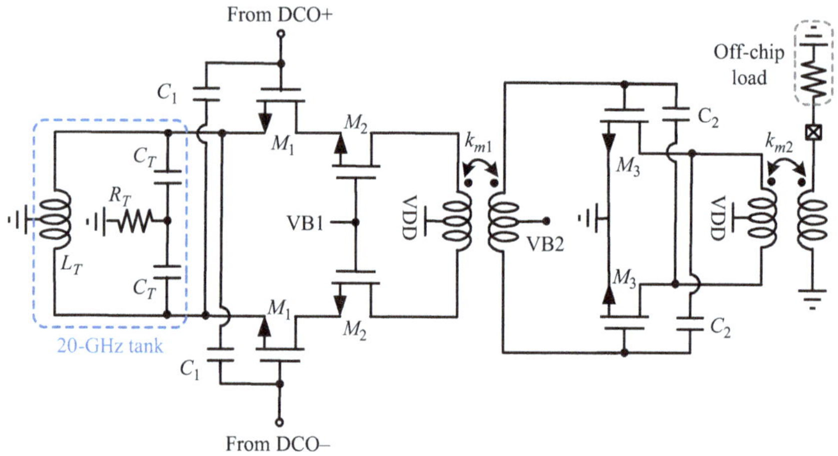

Figure 13.18 Schematic of the subharmonic soft-cancellation 60-GHz buffer with low-k_m transformer matching network and output driver

tank, high-k_m ($k_m = 0.75$) and low-k_m ($k_m = 0.25$) transformer matching network (assuming the loading and Q factor of the self-inductance in each winding are the same as that in the single LC tank). These three matching networks are all centered at 60 GHz. The low-k_m matching network exhibits 5 and 10 dB more attenuation to 20 GHz signal than the high-k_m and single LC counterparts. Meanwhile, it also provides the widest bandwidth at 60 GHz.

In this design, a loosely coupled transformer ($k_{m1} = 0.25$) is used for the matching network of the 60-GHz buffer, as shown in Figure 13.18. The simulated transfer function is shown in Figure 13.19. With the subharmonic soft-cancellation technique and low-k_m transformer matching network, this 60-GHz buffer provides a 50-dB suppression at 20 GHz. Across the frequency range, the buffer stage alone provides −43 to −51 dB suppression of the 20-GHz component and 6–7 dB gain of

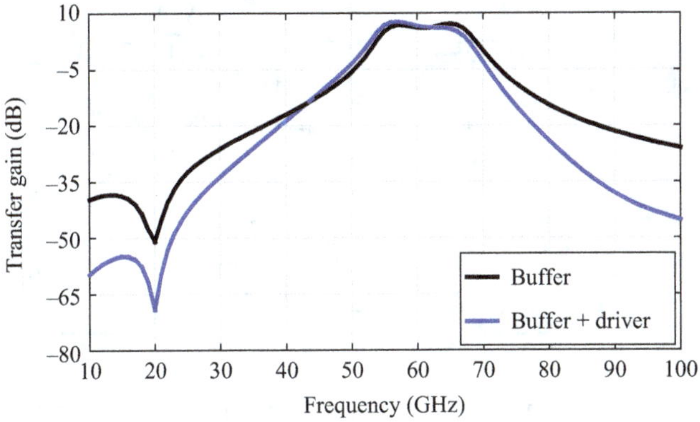

Figure 13.19 Simulated transfer gain of the 60-GHz buffer stage and combined buffer and driver stage

the 60-GHz input, while achieving 13-GHz (22%) bandwidth. The buffer is followed by an output driver to deliver >0 dBm for measurements. Including the output driver, the two stages provide −70 dB suppression at 20 GHz, as also shown in Figure 13.19.

13.5 Phase detection

In this design, an 8-bit variable-phase high-speed counter, PHV_I, and a noise-shaping DTC-assisted TDC act as the phase detector. There are two design options for a high-speed counter: asynchronous and synchronous operations among different bits. As a trade-off between power consumption and design complexity, the 2-LSB integer bits are handled by an asynchronous counter to support high speed at low power. The remaining 6-MSB bits are handled by a synchronous counter. Timing for reading out the combined counter output is critical due to its asynchronous and high-speed nature. Due to the staggered settling of the counter's LSB bits, extreme care must be taken to guarantee proper sampling of PHV_I by the CKR clock edge, especially over PVT variations. Any misalignment between various bits or meta-stability will result in catastrophic phase jumps and thus failure in locking.

To ensure a robust operation, an asynchronous sampling scheme is used in the high-speed counter readout. Figure 13.20 shows the circuit diagram and timing sequence. A specific case is also demonstrated in Figure 13.20(b), in which the counter output bits c[1:7] are not yet settled at the CKR rising edge t_1. In the proposed solution, the CKV-to-Q delay for each asynchronous bit is added onto the respective sampling clock. The falling edge of CKV samples CKR and generates the sampling clock CKS_0 for bit c[0]. The delay between the rising edge of CKS_0 and c[0] is exactly half the CKV period, if the *D* flip-flops have the same delay. The rising edge of CKS_0 samples VDD and generates CKS_1, the sampling clock for bit c[1]. The sampling clocks for all the other asynchronous bits are generated in this manner. In this way, the delay in asynchronous bits also propagates in the respective sampling clocks. It ensures that the rising edge of each sampling clock always appears half CKV period after the

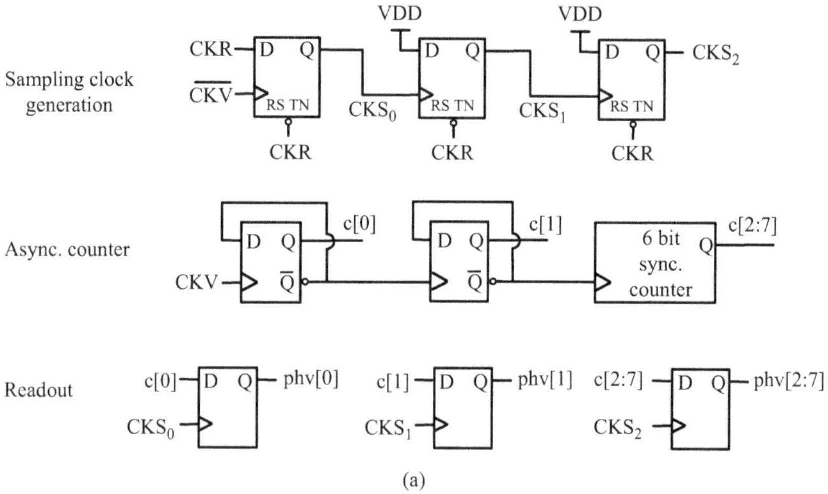

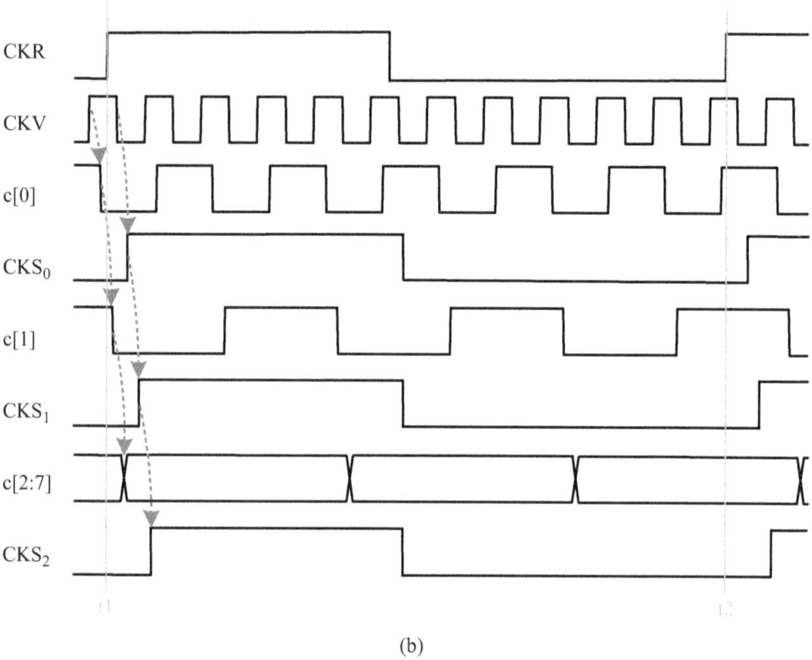

Figure 13.20 (a) Block diagram and (b) timing sequence of the asynchronous high-speed counter

corresponding counter bit is settled. The counter output can be robustly read out in each cycle with this asynchronous sampling technique.

The diagram of the DTC-TDC subsystem is shown in Figure 13.21. A 64-stage DTC with 12 ps resolution is based on current-starved delay cells. The DTC helps to reduce the required detection range of the TDC. A 64-stage 2.7 ps fine-resolution TDC uses a vernier line and ensures that its Q noise level is comparable to the FREF noise floor. To avoid a dead-zone issue in the TDC, the TDC output is encoded in mid-rise code. The TDC dynamic range covers about $0.4 \cdot T_{CKV}$. During the locking or settling process, the TDC input can exceed its TDC dynamic range. When overflow or underflow happens, the TDC output is overwritten by a maximum or minimum value (+64 or −64). In this way, the loop bandwidth is adaptively increased to facilitate the settling speed. After the PLL is locked, only 7–8 unit cells in the middle of the TDC chain are effectively operating.

The DTC gain inaccuracy can introduce spurious tones at the ADPLL output. An LMS algorithm [20] is used and implemented on chip to automatically calibrate the DTC gain (K_{DTC}) in the background. An inaccurate K_{DTC} causes extra phase error at the TDC output. The induced phase error is proportional to PHR_F[k]. The cross-correlation between PHE_F[k] and PHR_F[k] provides an indication of the inaccuracy in K_{DTC}. By iteratively updating the estimated K_{DTC}, an accurate DTC gain can be obtained when the cross-correlation between PHE_F[k] and PHR_F[k] is not detectable. Figure 13.22 shows the block diagram of the DTC gain

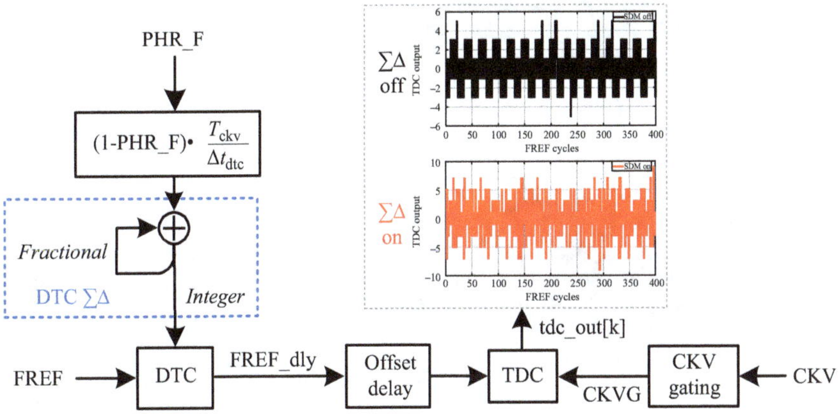

Figure 13.21 Block diagram of the DTC-TDC subsystem

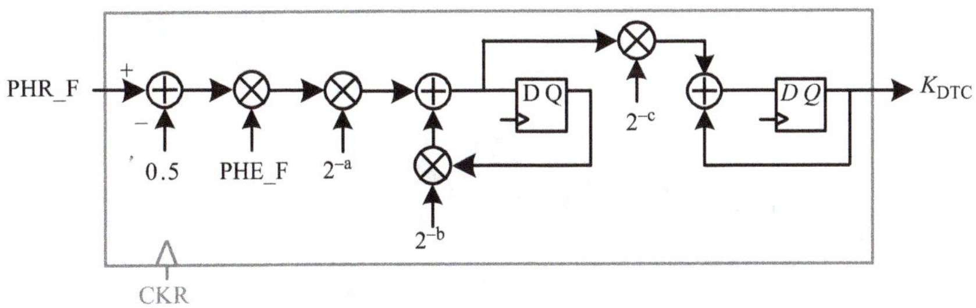

Figure 13.22 Diagram of the LMS algorithm for DTC gain calibration

calibration algorithm. An IIR filter attenuates the high-frequency components in the cross-correlation input. The IIR filter output then updates the K_{DTC} with proper gain. The transient settling behavior of the LMS algorithm is simulated and shown in Figure 13.23. With 20% initial error in the DTC gain, the calibration converges in 25 μs. The residual gain error is only 0.4%.

It is well known that the Q error and nonlinearity of DTC and TDC can introduce spurious tones. An error-feedback ΣΔ modulation is introduced to the DTC, as highlighted in Figure 13.21. It eliminates the spur induced by the Q error from the DTC-TDC and by the TDC nonlinearity. ΣΔM shapes the DTC Q noise to high frequencies, which is then filtered by the LF. Moreover, the TDC input is scrambled by the preceding ΣΔM-DTC such that the TDC Q error and nonlinearity will not induce spurious tones.

With the DTC ΣΔM on/off, the respective TDC outputs are read out in the PLL close-loop simulations. Figure 13.24 shows the corresponding frequency spectra.

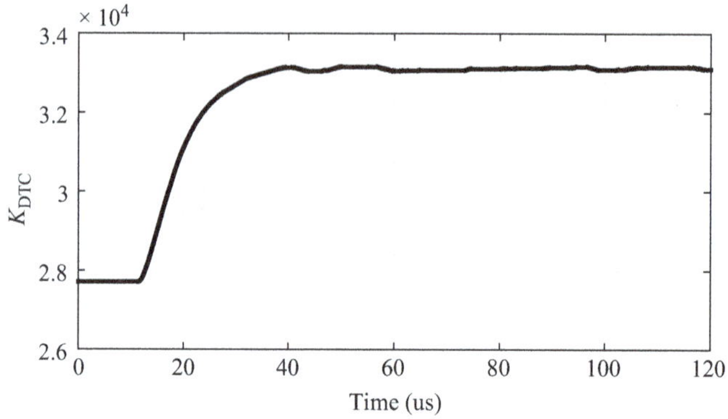

Figure 13.23 Simulated transient converging process of the DTC gain calibration algorithm with 20% initial error

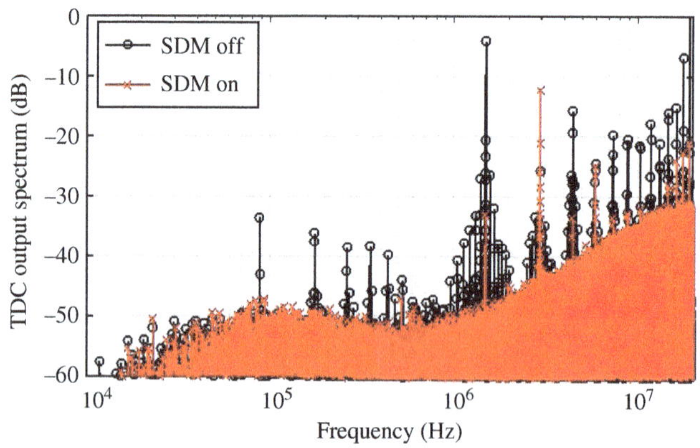

Figure 13.24 Simulated frequency spectrum of the TDC output with DTC ΣΔM on/off

As we can see, the low-frequency tones at the TDC output disappear or are suppressed below the system noise when the ΣΔM is on. The remaining spurs are caused by the nonlinearity in the DTC and limited cycle output of the ΣΔM. They are attenuated by the following LF.

13.6 Digital LF

The digital LF determines the dynamic behavior and filtering capability to different noise sources in the ADPLL. In this work, the LF comprises a cascaded four-stage IIR filter and a PI controller with gear-shifting mechanisms. All the parameters (λ_{1-4}, α, and ρ) are programmable. The integral path and each IIR stage can be switched on/off. When all the IIR filter stages and the integral path are enabled, type-II and sixth-order ADPLL is achieved. The gear-shifting mechanisms enable dynamic loop bandwidth in the locking process. Upon initiation, wide loop bandwidth is used to acquire the desired frequency in short time. Once the loop is locked, it is switched to the optimal loop bandwidth for low IPN performance.

The IIR filter is intended to attenuate the Q noise of the DTC ΣΔM at high frequencies. It also help to suppress the spurs at high offset frequencies. The DTC ΣΔM Q noise is shaped to high frequencies. The filtering capability provided by the PI controller may be not sufficient to attenuate the Q noise below the other noise sources. In this work, the IIR filter bandwidth is configured to be 2–3 MHz, which is much higher than the loop bandwidth (300 kHz). With the assistance of the IIR filter, the DTC DSM noise is attenuated below the DCO PN. In this way, the degradation on the IPN caused by the DTC DSM is limited to only 0.3 dB.

13.7 Experimental results

The presented 60-GHz digital fractional-N frequency synthesizer is prototyped in TSMC 28-nm 1P9M LP CMOS. Figure 13.25 shows the chip micrograph. The design occupies a core area of 0.38 mm^2, while the total chip area (including pads) is 1.1 mm^2. The PN and spectra of the generated 60-GHz carrier are measured via on-wafer probing. Outputs of the 2.5-GHz frequency dividers could also be conveniently monitored.

13.7.1 Open-loop test

The ADPLL is first set to an open-loop mode for the DCO performance characterization using a R&S FSUP50 Signal Source Analyzer. The TR of the DCO is 57.5–67.2 GHz, covering all four IEEE 802.11ad channels (58.32, 60.48, 62.64, and 64.8 GHz) with sufficient margin. The DCO draws 19 mA from a 0.9-V supply. PN of the free-running DCO is measured after the frequency dividers (CKV point in Figure 13.4), as shown in Figure 13.26. The PN referred to the 66.88-GHz carrier is −98 dBc/Hz at 1 MHz offset. Across the TR, the $1/f^3$ noise corner varies between 300 and 400 kHz, as shown in Figure 13.27. The slightly higher $1/f^3$ corner at lower oscillation frequencies is expected to be caused by the possible

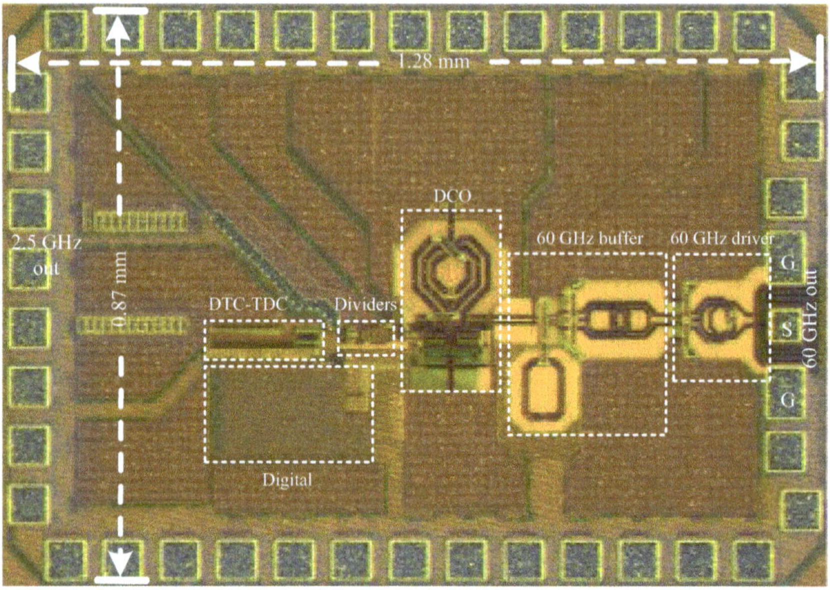

Figure 13.25 Chip micrograph of the proposed 60 GHz ADPLL

misalignment between the auxiliary resonances at H2/H3 and the fundamental frequencies. With a more accurate EM modeling, it can be improved further.

13.7.2 Close-loop test

The ADPLL is separately measured with different external crystal reference clocks of $f_R = 40$ and 100 MHz, at which it, respectively, consumes 28.6 and 31 mW (17 mW for DCO and 9.5 mW for frequency dividers). The 60-GHz buffer and 50-Ω load driver stage consume 10.5 and 11.5 mW, respectively. All the circuit blocks (except for DCO) are supplied at 1.05 V (i.e., the nominal voltage for this technology). The ADPLL PN is measured with R&S FSUP50 with an extension V-band harmonic mixer for down-conversion. The loop bandwidth is programmed to 200–300 kHz for the lowest RMS jitter. PN at an integer-N channel (65.28 GHz) and a fractional-N channel (65.411 GHz) is shown in Figure 13.28. The RMS jitter integrated from 10 kHz to 30 MHz is 237 fs at 65.28 GHz (integer-N) and 268 fs at 65.411 GHz (fractional-N). With the 100 MHz reference, the in-band PN and integrated jitter are substantially improved (see Figure 13.29). Figure 13.30 shows the spectrum at 65.411 GHz with DTC ΣΔM on (in red) and off (in blue). When the ΣΔM is engaged, the fractional spurs are significantly attenuated and the highest spur level observed at this carrier frequency is −51 dBc.

The measured RMS jitter across the fractional frequency-command word (FCW) offsets away from the 60-GHz integer-N channel, with 40 and 100 MHz references, which are summarized in Figure 13.31(a). With $f_R = 40$ MHz, the measured RMS jitter is 236–266 and 236–316 fs, respectively, across the integer-N

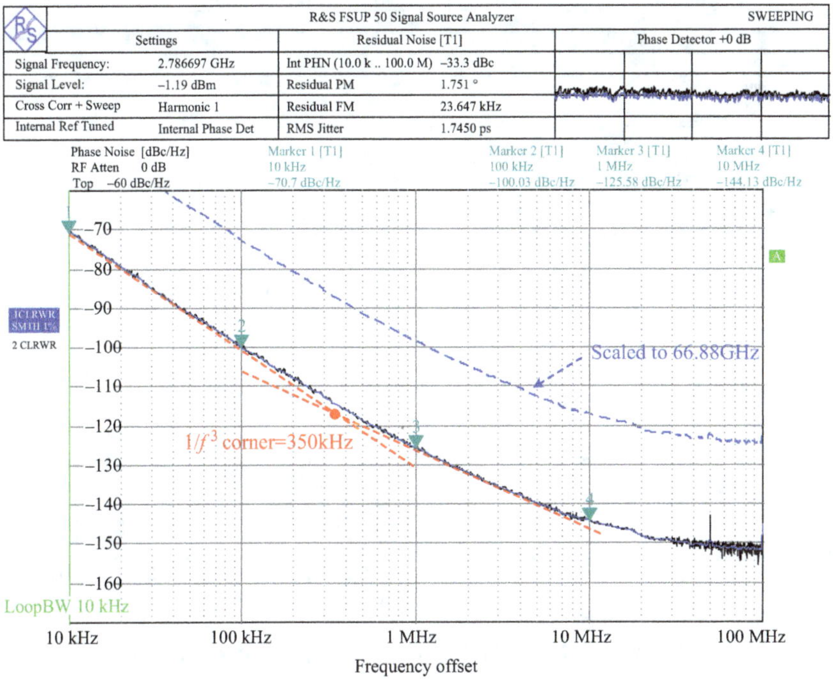

Figure 13.26 Measured DCO phase noise at dividers' output and scaled up to 66.9 GHz

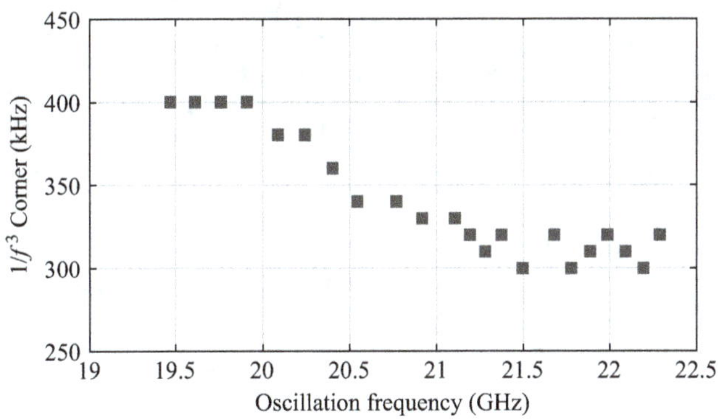

Figure 13.27 Measured DCO $1/f^3$ noise corner over TR

(all swept) and fractional-N settings. With $f_R = 100$ MHz, the RMS jitter is improved to 213–241 and 213–277 fs across the respective channels. Figure 13.31 (a) reveals that the RMS jitter improves at some special fractional FCWs (such as 0.125, 0.25, and 0.5), with $f_R = 40$ MHz, while this reduction becomes smaller with $f_R = 100$ MHz. With these fractional FCWs, the TDC Q error is not uniformly

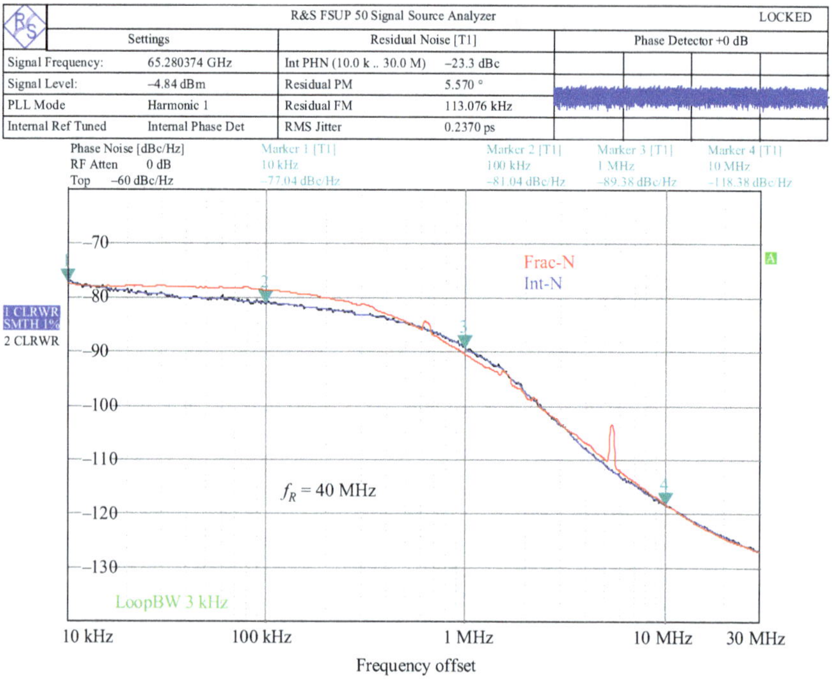

Figure 13.28 Measured ADPLL PN with $f_R = 40$ at integer-N and fractional-N

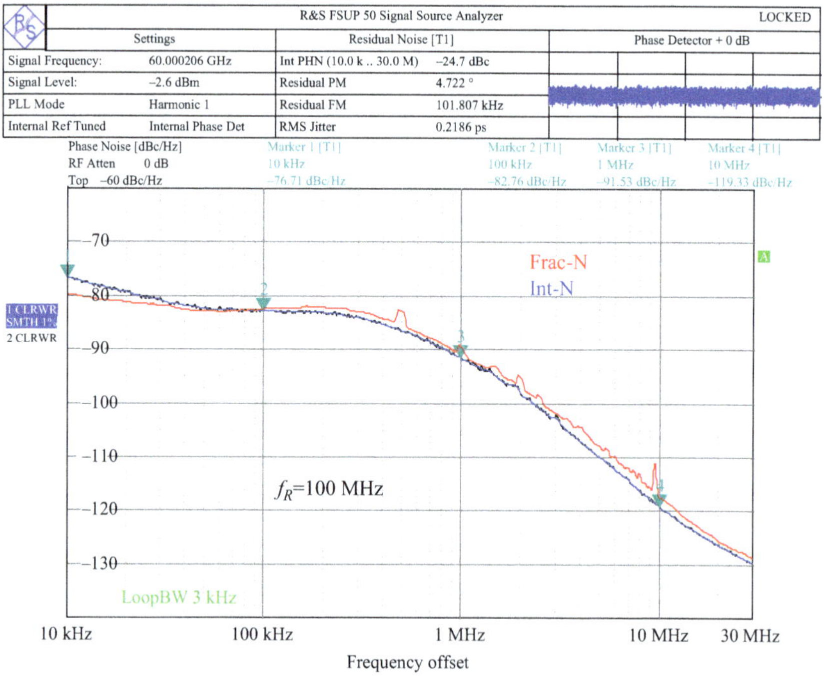

Figure 13.29 Measured ADPLL PN with $f_R = 100$ MHz at integer-N and fractional-N

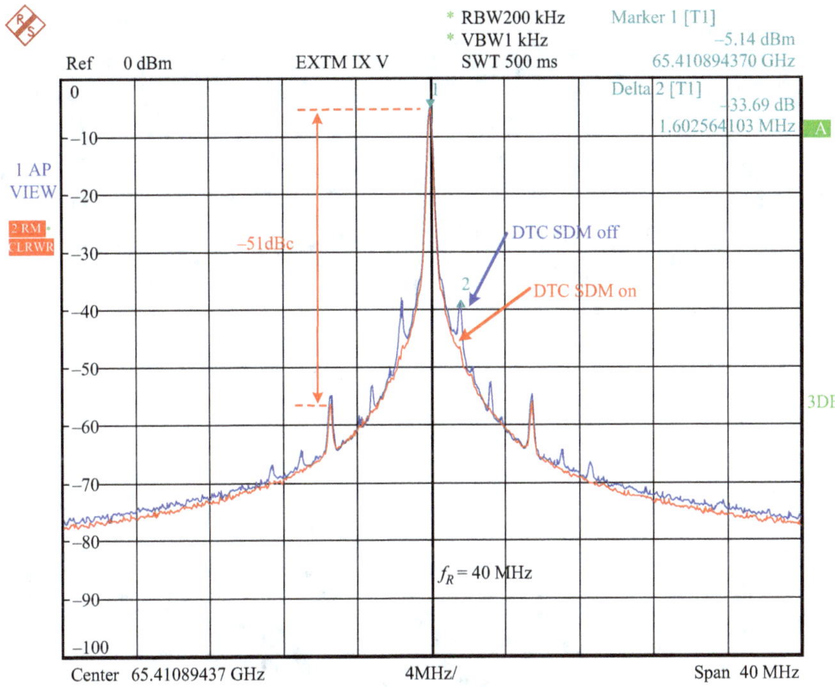

Figure 13.30 Measured ADPLL output spectrum with DTC ΣΔM on/off

distributed, and its contribution to the ADPLL PN decreases. This causes the drop in RMS jitter in case of $f_R = 40$ MHz. Since the uniformly distributed TDC Q noise with $f_R = 100$ MHz is already a marginal contributor (4 dB below the FREF noise) to the ADPLL PN, the aforementioned effect exhibits less impact on the RMS jitter.

Fractional spurs are measured at the 2.5-GHz divider output with DTC ΣΔM enabled and scaled to the corresponding 60-GHz frequencies. The fractional spurs across different fractional settings are summarized in Figure 13.31(b). At 2.5-GHz divider output, the fractional spurs are measured <−57 dBc. When referring to 60-GHz carrier frequencies, they are below −30 dBc. It should be noted that only a few papers report fractional spur levels at mmW frequencies. In [35], the spur levels appear to have been incorrectly taken from PN plots, without accounting for the resolution bandwidth. In [36], with sophisticated DTC nonlinearity calibration, a lower fractional spur of −38 dBc (normalized to 60 GHz) was achieved. There is no DTC/TDC nonlinearity calibration done in our design, but it can be applied if the spurs need further suppression in some applications. Random dithering in ΣΔM [37] can also reduce part of the spurs.

Measurements have been done to verify the transient settling performance of this ADPLL. After the PLL is locked with 40 MHz FREF, a step ΔFCW is added to the original FCW. The step response of the ADPLL is monitored at the 2.5-GHz divider output. As shown in Figure 13.32, with a 100-MHz step at 60 GHz, the ADPLL settles within 5 μs. With smaller frequency step or higher FREF frequency, the settling is even faster.

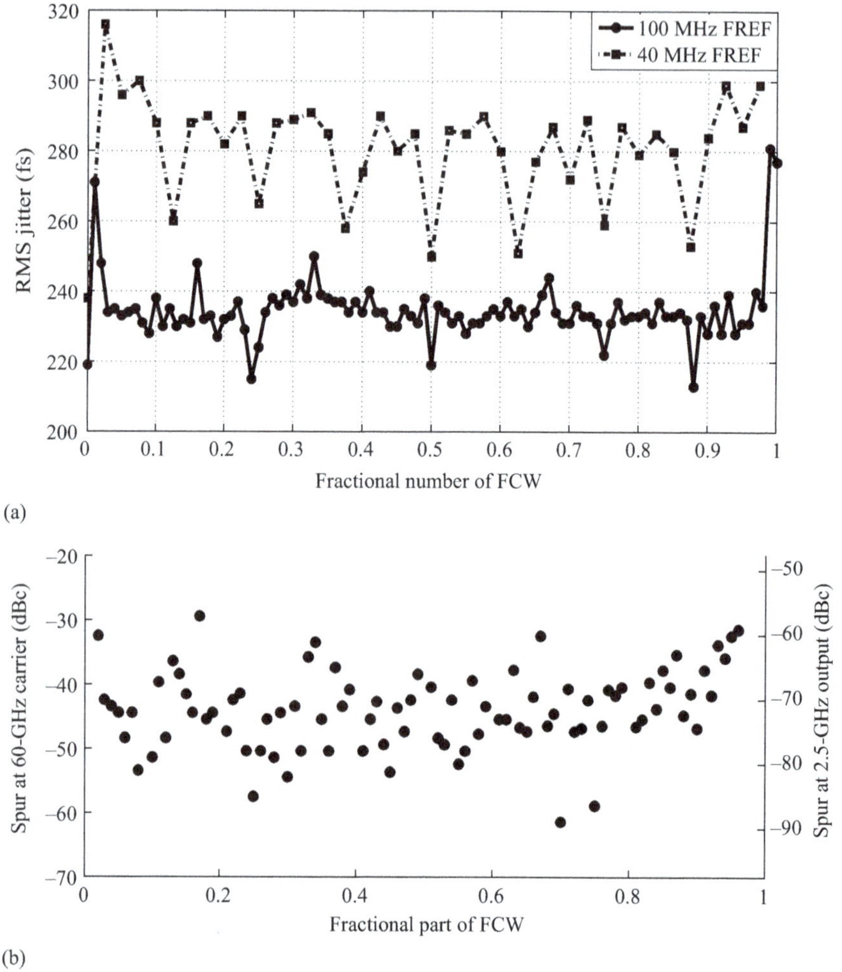

Figure 13.31 *Measured ADPLL: (a) RMS jitter and (b) worst-case fractional spurs, across fractional offsets from 60.0-GHz integer-N*

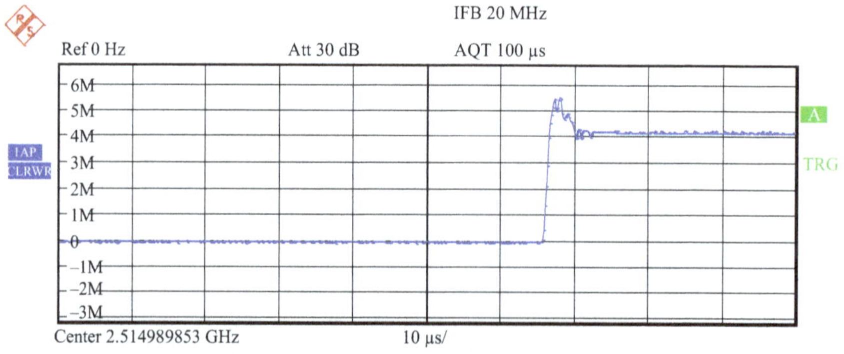

Figure 13.32 *Measured ADPLL settling behavior with 100 MHz frequency step at 60 GHz*

The output power levels at the 60-GHz carrier and the leaked 20-GHz tone are measured to validate the effectiveness of the 20-GHz tone suppression technique. After de-embedding cable losses, the driver delivers 1 dBm at 60 GHz to the external 50-Ω load, with ±1 dB variation across the TR (see Figure 13.33). The 20-GHz residual level is within −51 to −57 dBm across the TR. The achieved harmonic rejection ratio is 20 dB better than the notch-filter solution in [32]. The 40-GHz tone, which is caused by the nonlinear effects in the buffer stage, is measured at −44 to −64 dBm across the TR.

Table 13.1 summarizes the performance of the proposed 60-GHz ADPLL and compares it with the relevant state-of-the-art PLLs at 60 GHz or more [2–5,35,38–40].

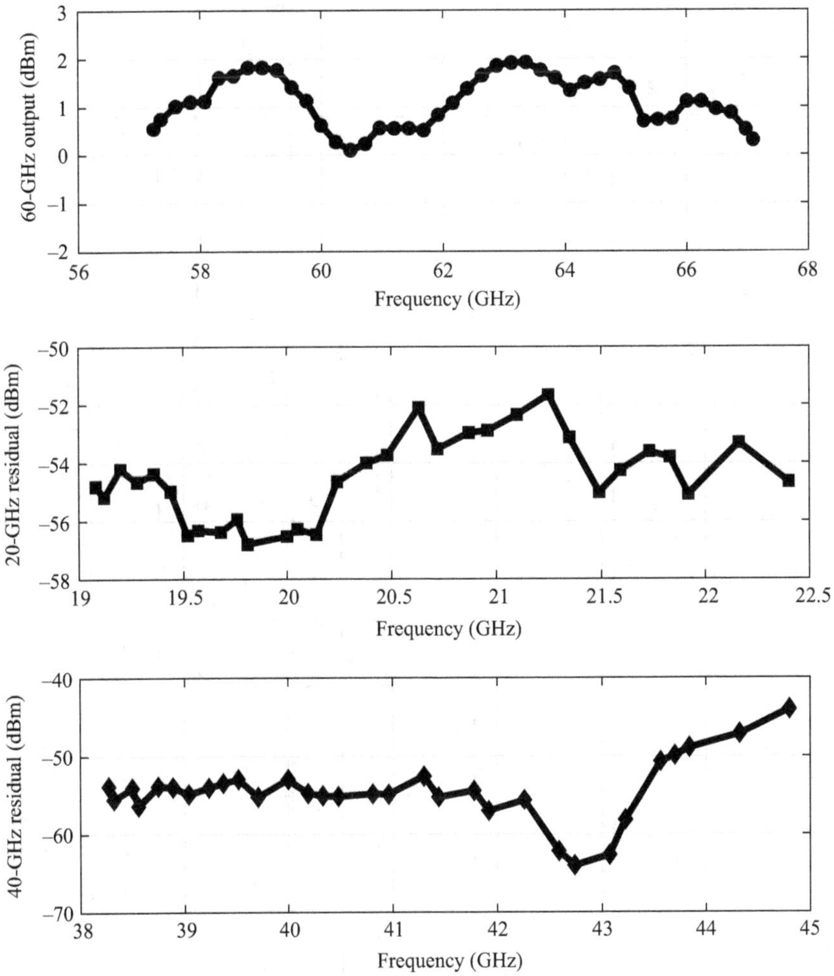

Figure 13.33 Measured power across the tuning range of 60-GHz carrier and 20-GHz residual

Table 13.1 Comparison of the 60-GHz fractional-N ADPLL with relevant state-of-the-art

		ISSCC'09 Scheir	JSSC'11 Musa	JSSC'16 Siriburanon	ISSCC'13 Yi	JSSC'14 Wu	JSSC'15 Szortyka	ISSCC'18 Huang	JSSC'17 Hussein	This Work	
Architecture		Analog PLL	20G PLL + 60G QILO	20G SSPLL + 60G QILO	Analog PLL	ADPLL	SSPLL	ADPLL	ADPLL	ADPLL	
Type		INT-N	INT-N	INT-N	INT-N	FRAC-N	INT-N	INT-N	FRAC-N	FRAC -N	
Output frequency (GHz)		57–66 (14.6%)	58–63 (8.3%)	58.3–64.8 (10.5%)	58–68.5 (8.3%)	56.5–63.5 (11.6%)	53.8–63.3 (16.2%)	82–107.6 (27%)	50.2–66.5 (28%)	57.5–67.2 (15.6%)	
Reference frequency (MHz)		100	36	40	135	100	40	125	100	40	100
PN (dBc/Hz)	In-band	–70	–60	–75	NA	–75	–88– –92	–84– –87	–79– –83	–77– –81	–81– –84
	1MHz	–75	–95	–92	–89.8– –91.5	–90	–88– –92	–81	–88– –94.5	–89– –92	
RMS jitter (fs)	Integer channel	NA	NA	290	238	590.2	200	276~328	223–302.5	236–266	213–241
	Fractional channel								962–1540	236–316	213–277
V_{DD} (V)		1.1	1.2	1	1.2	1.2	1	1.2/0.8	1	1.05	
P_{DC} (mW)		78	80	32	24.6	48	42	35.5	46	28.6 + 10.5 *	31 + 10.5 *
FoM # (dB)	Integer channel	NA	NA	–235.7	–238.5	–227.8	–237.7	–234–235.7	–233.8–236.4	–235.6–236.6 *	–236.2–237.2 *
	Fractional channel								–219.6–223.7	–234–236.6 *	–235–237.2 *
Reference spur (dBc)		–42	<–47	–38	–54.5	–74	<–40	–34~ –52	NA	–62	–65
Area (mm^2)		0.82	1.68	1.09	0.19 **	0.48 **	0.16 **	0.36 **	0.45 **	1.1 (core area is 0.38)	
CMOS technology (nm)		65	65	65	65	65	40	65	65	28	

*: include the power consumption of the 60-GHz output buffer stage (10.5 mW)

**: only include the core chip area

#: $FoM = 10 \cdot \log_{10}[(\frac{Jitter_{rms}}{1s})^2 \cdot \frac{P_{DC}}{1mW}]$

In literature, only two fractional-N PLLs at 60 GHz or above are found. The integrated jitter of our ADPLL is the best compared to the other fractional-N counterparts, while consuming the lowest power. The figure of merit improves the state-of-the-art by 3 dB at the fractional-N operation, even by including the power consumption of the 60-GHz buffer stage.

References

[1] Kang S, Chien JC, and Niknejad AM. A W-Band Low-Noise PLL with a Fundamental VCO in SiGe for Millimeter-Wave Applications. IEEE Transactions on Microwave Theory and Techniques. 2014;62(10):2390–2404.

[2] Wu W, Staszewski RB, and Long JR. A 56.4-to-63.4 GHz Multi-Rate All-Digital Fractional-N PLL for FMCW Radar Applications in 65 nm CMOS. IEEE Journal of Solid-State Circuits. 2014;49(5):1081–1096.

[3] Yi X, Boon CC, Liu H, Lin JF, and Lim WM. A 57.9-to-68.3 GHz 24.6 mW Frequency Synthesizer with In-Phase Injection-Coupled QVCO in 65 nm CMOS Technology. IEEE Journal of Solid-State Circuits. 2014;49(2): 347–359.

[4] Szortyka V, Shi Q, Raczkowski K, Parvais B, Kuijk M, and Wambacq P. A 42 mW 200 fs-Jitter 60 GHz Sub-Sampling PLL in 40 nm CMOS. IEEE Journal of Solid-State Circuits. 2015;50(9):2025–2036.

[5] Musa A, Murakami R, Sato T, Chaivipas W, Okada K, and Matsuzawa A. A Low Phase Noise Quadrature Injection Locked Frequency Synthesizer for Mm-Wave Applications. IEEE Journal of Solid-State Circuits. 2011;46(11): 2635–2649.

[6] Deng W, Siriburanon T, Musa A, Okada K, and Matsuzawa A. A Sub-Harmonic Injection-Locked Quadrature Frequency Synthesizer with Frequency Calibration Scheme for Millimeter-Wave TDD Transceivers. IEEE Journal of Solid-State Circuits. 2013;48(7):1710–1720.

[7] Li A, Zheng S, Yin J, Luo X, and Luong HC. A 21–48 GHz Subharmonic Injection-Locked Fractional-N Frequency Synthesizer for Multiband Point-to-Point Backhaul Communications. IEEE Journal of Solid-State Circuits. 2014;49(8):1785–1799.

[8] Sadhu B, Ferriss M, and Valdes-Garcia A. A 46.4–58.1 GHz Frequency Synthesizer Featuring a 2nd Harmonic Extraction Technique That Preserves VCO Performance. In: 2014 IEEE Radio Frequency Integrated Circuits Symposium. IEEE; 2014. p. 173–176.

[9] Chao Y, Luong HC, and Hong Z. Analysis and Design of a 14.1-mW 50/100-GHz Transformer-Based PLL with Embedded Phase Shifter in 65-nm CMOS. IEEE Transactions on Microwave Theory and Techniques. 2015;63(4):1193–1201.

[10] Catli B, and Hella MM. Triple-Push Operation for Combined Oscillation/Division Functionality in Millimeter-Wave Frequency Synthesizers. IEEE Journal of Solid-State Circuits. 2010;45(8):1575–1589.

[11] Cao C, O K. Millimeter-Wave Voltage-Controlled Oscillators in 0.13-mm CMOS Technology. IEEE Journal of Solid-State Circuits. 2006;41(6):1297–1304.
[12] Li L, Reynaert P, and Steyaert MS. Design and Analysis of a 90 nm Mm-Wave Oscillator Using Inductive-Division LC Tank. IEEE Journal of Solid-State Circuits. 2009;44(7):1950–1958.
[13] Yin J, and Luong HC. A 57.5–90.1-GHz Magnetically Tuned Multimode CMOS VCO. IEEE Journal of Solid-State Circuits. 2013;48(8):1851–1861.
[14] Wong YH, Lin WH, Tsai JH, and Huang T-W. A 50-to-62GHz Wide-Locking-Range CMOS Injection-Locked Frequency Divider with Transformer Feedback. In: 2008 IEEE Radio Frequency Integrated Circuits Symposium. IEEE; 2008. p. 435–438.
[15] Gu Q, Xu Z, Huang D, *et al.* A Low Power V-Band CMOS Frequency Divider with Wide Locking Range and Accurate Quadrature Output Phases. *IEEE Journal of Solid-State Circuits*. 2008;43(4):991–998.
[16] Rong S, Ng AW, and Luong HC. 0.9 mW 7GHz and 1.6 mW 60GHz Frequency Dividers with Locking-Range Enhancement in 0.13 μm CMOS. In: 2009 IEEE International Solid-State Circuits Conference-Digest of Technical Papers. IEEE; 2009. p. 96–97.
[17] Rael J, and Abidi AA. Physical Processes of Phase Noise in Differential LC Oscillators. In: Proceedings of the IEEE 2000 Custom Integrated Circuits Conference (Cat. No. 00CH37044). IEEE; 2000. p. 569–572.
[18] Zong Z, Babaie M, and Staszewski RB. A 60 GHz 25% Tuning Range Frequency Generator with Implicit Divider Based on Third Harmonic Extraction with 182 dBc/Hz FoM. In: 2015 IEEE Radio Frequency Integrated Circuits Symposium (RFIC). IEEE; 2015. p. 279–282.
[19] Zong Z, Babaie M, and Staszewski RB. A 60 GHz Frequency Generator Based on A 20 GHz Oscillator and An Implicit Multiplier. IEEE Journal of Solid-State Circuits. 2016;51(5):1261–1273.
[20] Zhuang J, and Staszewski RB. Gain Estimation of A Digital-to-Time Converter for Phase-Prediction All-Digital PLL. In: 2013 European Conference on Circuit Theory and Design (ECCTD). IEEE; 2013. p. 1–4.
[21] Liu YH, Van Den Heuvel J, Kuramochi T, *et al.* An Ultra-Low Power 1.7-2.7 GHz Fractional-N Sub-Sampling Digital Frequency Synthesizer and Modulator for IoT Applications in 40 nm CMOS. IEEE Transactions on Circuits and Systems I: Regular Papers. 2017;64(5):1094–1105.
[22] Okada K, Kondou K, Miyahara M, *et al.* Full Four-Channel 6.3-Gb/s 60-GHz CMOS Transceiver with Low-Power Analog and Digital Baseband Circuitry. IEEE Journal of Solid-State Circuits. 2013;48(1):46–65.
[23] Levantino S, Marzin G, and Samori C. An Adaptive Pre-Distortion Technique to Mitigate the DTC Nonlinearity in Digital PLLs. IEEE Journal of Solid-State Circuits. 2014;49(8):1762–1772.
[24] Wu L, and Luong HC. Analysis and Design of a 0.6 V 2.2 mW 58.5-to-72.9 GHz Divide-by-4 Injection-Locked Frequency Divider with Harmonic Boosting. IEEE Transactions on Circuits and Systems I: Regular Papers. 2013;60(8):2001–2008.

[25] Kim H, Ryu S, Chung Y, Choi J, and Kim B. A Low Phase-Noise CMOS VCO with Harmonic Tuned LC Tank. IEEE Transactions on Microwave Theory and Techniques. 2006;54(7):2917–2924.

[26] Babaie M, and Staszewski RB. A Class-F CMOS Oscillator. IEEE Journal of Solid-State Circuits. 2013;48(12):3120–3133.

[27] Wu W, Long JR, and Staszewski RB. High-Resolution Millimeter-Wave Digitally Controlled Oscillators with Reconfigurable Passive Resonators. IEEE Journal of Solid-State Circuits. 2013;48(11):2785–2794.

[28] Murphy D, Darabi H, and Wu H. A VCO with Implicit Common-Mode Resonance. In: 2015 IEEE International Solid-State Circuits Conference-(ISSCC) Digest of Technical Papers. IEEE; 2015. p. 1–3.

[29] Shahmohammadi M, Babaie M, and Staszewski RB. A 1/f Noise Upconversion Reduction Technique for Voltage-Biased RF CMOS Oscillators. IEEE Journal of Solid-State Circuits. 2016;51(11):2610–2624.

[30] Zong Z, Chen P, and Staszewski RB. A Low-Noise Fractional-*N* Digital Frequency Synthesizer With Implicit Frequency Tripling for mm-Wave Applications. IEEE Journal of Solid-State Circuits. 2019;54(3):755–767.

[31] Zong Z, and Staszewski RB. Effects of Subharmonics in LO Generation on RF Transceivers. In: 2018 IEEE MTT-S International Microwave Workshop Series on 5G Hardware and System Technologies (IMWS-5G). IEEE; 2018. p. 1–3.

[32] Kuo CN, and Yan TC. A 60 GHz Injection-Locked Frequency Tripler with Spur Suppression. IEEE Microwave and Wireless Components Letters. 2010;20(10):560–562.

[33] Razavi B. RF Microelectronics. Prentice Hall PTR; 1998. Available from: http://books.google.nl/books?id=TQZTAAAAMAAJ.

[34] Thyagarajan SV, Niknejad AM, and Hull CD. A 60 GHz Drain-Source Neutralized Wideband Linear Power Amplifier in 28 nm CMOS. IEEE Transactions on Circuits and Systems I: Regular Papers. 2014;61(8): 2253–2262.

[35] Hussein AI, Vasadi S, and Paramesh J. A 50–66-GHz Phase-Domain Digital Frequency Synthesizer With Low Phase Noise and Low Fractional Spurs. IEEE Journal of Solid-State Circuits. 2017;52(12):3329–3347.

[36] Cherniak D, Grimaldi L, Bertulessi L, Samori C, Nonis R, and Levantino S. A 23-GHz Low-Phase-Noise Digital Bang–Bang PLL for Fast Triangular and Sawtooth Chirp Modulation. *IEEE Journal of Solid-State Circuits*. 2018;53(12):3565–3575.

[37] Markulic N, Raczkowski K, Martens E, *et al.* A DTC-Based Subsampling PLL Capable of Self-Calibrated Fractional Synthesis and Two-Point Modulation. *IEEE Journal of Solid-State Circuits*. 2016;51(12): 3078–3092.

[38] Scheir K, Vandersteen G, Rolain Y, and Wambacq P. A 57-to-66GHz Quadrature PLL in 45nm Digital CMOS. In: *2009 IEEE International Solid-State Circuits Conference-Digest of Technical Papers. IEEE*; 2009. p. 494–495.

[39] Siriburanon T, Kondo S, Katsuragi M, *et al.* A Low-Power Low-Noise Mm-Wave Subsampling PLL Using Dual-Step-Mixing ILFD and Tail-Coupling Quadrature Injection-Locked Oscillator for IEEE 802.11 ad. *IEEE Journal of Solid-State Circuits*. 2016;51(5):1246–1260.

[40] Huang Z, and Luong HC. An 82-to-108GHz- 181dB-FOM T ADPLL Employing a DCO with Split-Transformer and Dual-Path Switched-Capacitor Ladder and a Clock-Skew-Sampling Delta-Sigma TDC. In: *2018 IEEE International Solid-State Circuits Conference-(ISSCC). IEEE*; 2018. p. 260–262.

Chapter 14

DTC-based subsampling PLLs for low-noise synthesis and two-point modulation

Nereo Markulic[1]*, Jan Craninckx*[1]*, and Piet Wambacq*[1]

The analog integer-N subsampling phase lock loop (PLL) [1] introduced in 2009 is an architecture that enables frequency synthesis with extremely low integrated phase noise (i.e., root mean square (RMS) jitter) at low power consumption. Its attractive performance parameters are enabled by high gain within the phase-error detection subsampling core and operation without a (power consuming) divider in the feedback path of the loop. More than 10 years from its original publication [1], the simple subsampling loop (implemented in an old and forgotten 180 nm node) still holds one of the record figure of merits in the art of frequency synthesis.

There is unfortunately another side of the coin when it comes to subsampling PLL-based synthesis. In its original form, the PLL is incapable of fractional synthesis or frequency modulation. This makes the loop inadequate for the modern wireless standards where communication channels are densely filling with the available spectrum (which is a scarce resource today). Moreover, the loop's attractive characteristics cannot be applied in the context of frequency/ phase modulation, limiting its use in numerous applications where power-efficient polar transmission is desired.

This chapter reveals techniques for fractional-N synthesis and phase/frequency modulation using the subsampling loop. A digital-to-time converter (DTC) is applied at the input of the system to enable compensation of fractional/modulation phase residue. We will discuss how the original characteristics of a subsampling PLL can be preserved by careful DTC design and power-efficient predistortion. The concepts explored are proven with measured performance of a prototype implemented in bulk 28 nm complementary metal–oxide–semiconductor (CMOS). The PLL operates with almost no performance gap (176/198 fs RMS jitter) between the integer and fractional operation, achieving up to −247.6 dB FOM. The designed loop is also capable of two-point, 10 Mbit/s Gaussian minimum shift keying (GMSK) modulation with measured −40.5 dB error vector magnitude (EVM) around a carrier of nearly 10 GHz. This chapter is based on the material reported in [2–4].

[1]imec, Leuven, Belgium

14.1 Fractional-N subsampling PLL enabled by a DTC

The subsampling PLL relies on phase-error detection based on differential sampling of the voltage-controlled oscillator (VCO) sinusoidal output with a repetition rate set by the reference frequency (Figure 14.1). During integer-N synthesis, which is analyzed first, the DTC is transparent, i.e., it does not modulate the phase of the input signal. In a phase-locked state, the sampling happens precisely at the zero crossings of the differential sinewave. Any output phase deviation will result in a non-zero voltage being sampled, which in turn is converted to a correction current (by a transconducting stage G_M) fed to the loop filter. The loop filter re-adjusts VCO control voltage correcting for the existing phase/frequency error. Since the sampling events occur only at the edges of the reference, the VCO zero crossings are aligned to these edges and the VCO produces a frequency that is an exact integer-N multiple of the reference.

The DTC is needed to ensure that the sampling edge falls at the VCO zero crossing even when the ratio between the input and the output frequency is a noninteger. A simple example of fractional-N multiplication (that differs from integer-N by 0.25) is depicted in Figure 14.1. In the first cycle, the sampling edge is aligned with the VCO zero crossing. In the second cycle, the sampling edge is

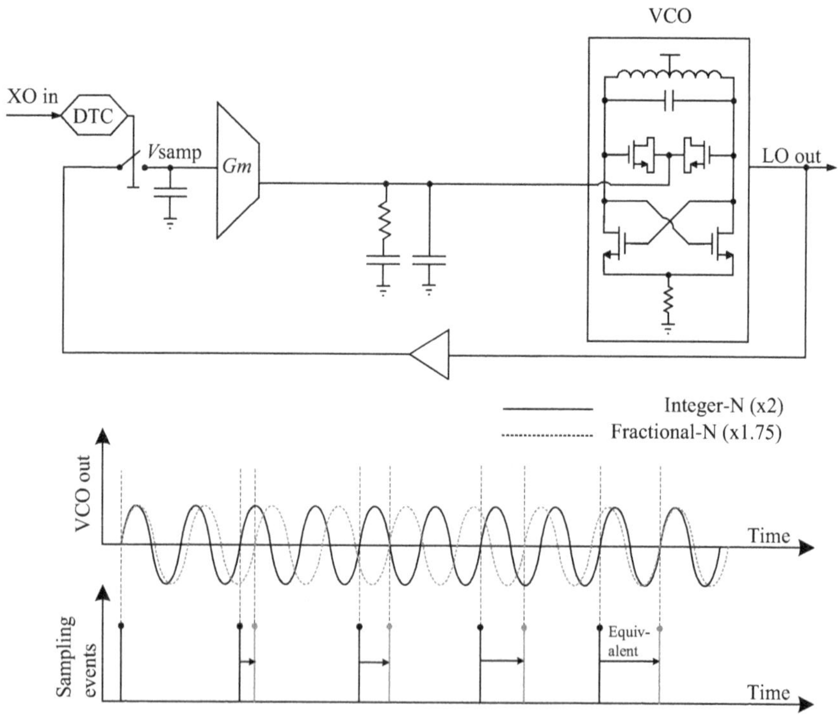

Figure 14.1 Integer (dash line) and fractional-N (full line) operation example

delayed by $0.25 \cdot T_{VCO}$ (where T_{VCO} is the period of the VCO's oscillation) beyond the reference edge. In the following third cycle, the sampling edge is delayed by $0.5 \cdot T_{VCO}$ and then by $0.75 \cdot T_{VCO}$. Finally, in the fifth cycle, the sampling could happen at a delay of a full VCO period; however, simply skipping a VCO cycle and sampling at the original reference edge yields the same effect. Since the required PLL frequency and the reference frequency are known, it is possible to calculate the position of any following zero crossings with absolute precision. The DTC should cover at least one VCO period of delay for functional fractional synthesis.

14.1.1 Limitations introduced by the DTC and their mitigation on the system level

A fractional-N subsampling PLL based on an *ideal* DTC would have the same performance as an integer-N subsampling PLL. This lies in stark contrast to traditional mixed signal ΔΣ PLL, where an unavoidable penalty in phase noise (or spurious content) is associated with the modulation through the divider. In practice, however, DTC design necessitates careful considerations not to impose limitations to the loop performance.

The implemented digital computation path of the DTC input code is depicted in Figure 14.2. First the difference between the multiplication factor *N.f* and the

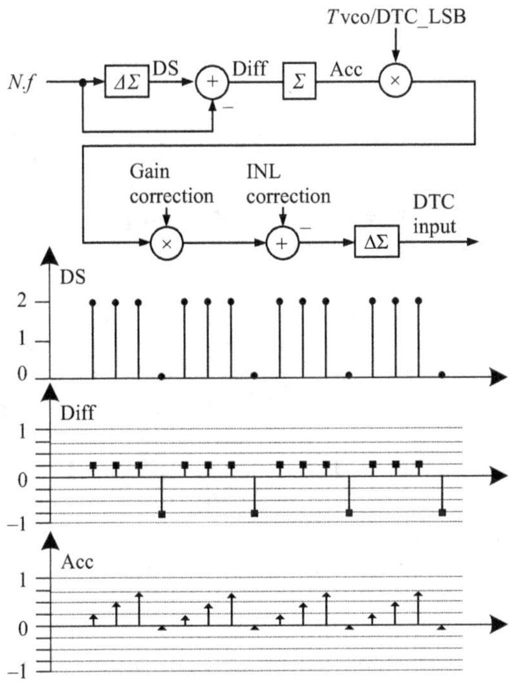

Figure 14.2 Digital computation of the DTC input code. Note that gain correction and INL correction equal 1 and 0, respectively, for a DTC that does not operate with gain imbalance or nonlinearity

integer-quantized value is extracted. A $\Delta\Sigma$ modulator is used to generate the integer quantization of *N.f*, so that the signal *Diff* is a zero-mean stream accumulated with the desired "phase wrapping" behavior, as explained above. The accumulated value *Acc* is then appropriately scaled on the available DTC input quantization range. Gain and integral nonlinearity (INL) correction are initially set at 1 and 0, respectively, not to influence the signal propagation to the DTC input (discussed later).

14.1.1.1 DTC resolution

A DTC, just like any data converter, has limited resolution. To scale the accumulated phase to a digital value, representing the DTC input tuning code, the output of the accumulator *Acc* in the Figure 14.2 needs to be multiplied by a factor $T_{REF}/LSB_{DTC}{\cdot}N_{FRAC}$. The sampling moment hence occurs with accuracy limited by the DTC and the resulting error is fed into the low-pass filter (LPF), thereby instantaneously modulating the VCO, creating quantization noise.

System-level simulations show that choosing a DTC least significant bit (LSB) of 0.5 ps ensures that the quantization noise appears well below other loop noise, even in a low-noise subsampling PLL environment. Moreover, a second $\Delta\Sigma$ modulator (Figure 14.2) is used in front of the DTC to shape the associated quantization noise beyond the PLL bandwidth. Thanks to the fact that the stream is perfectly accurate on average, the average PLL frequency is also accurate, with no visible modulation. Stringent mathematical analysis yields the following expression for the resulting DTC quantization noise power spectral density (expressed as sideband phase noise):

$$\mathcal{L} = 10\log_{10}\left\{\left(\frac{2\pi}{\sqrt{12}}\cdot\frac{LSB}{T_{VCO}}\right)^2\frac{1}{F_{REF}}[2\sin(\pi f T_{REF})]^2\right\}\left[\frac{\text{dBc}}{\text{Hz}}\right] \qquad (14.1)$$

at the in-band PLL output, where *LSB* represents the DTC LSB delay, T_{VCO} is the VCO period, F_{REF} and T_{REF} are the crystal reference frequency and period, respectively, and f is the frequency offset. The term $[2\sin(\pi f T_{REF})]^2$ comes from first-order $\Delta\Sigma$ shaping at reference rate. If the DTC quantization error is set as 0.5 ps (as suggested earlier), then the in-band (500 kHz offset) phase noise at 3.6 GHz output (40 MHz reference) appears at −148 dBc/Hz. This value indeed suggests that quantization noise can easily be kept under the thermal noise floor in a DTC-based (subsampling) PLL.

In the context of DTC-based PLL design and related to quantization noise limitations imposed by a DTC, it is interesting to briefly discuss DTC versus a time-to-digital converter (TDC)-based phase-error detection. Digital PLLs that exploit TDC rather than voltage-domain phase-error processing during frequency synthesis have undoubtedly shaken the field of LO synthesis (see Part II of this book). Digital PLLs reveal new paths toward the goal of jitter vs power optimization and may, even more importantly, permit cost-effective designs (minimization of area consuming loop filter). However, the quantization error of the TDC used for phase-error detection in a digital, fractional-N PLL limits the spectral purity of the system (similar to a DTC in a DTC-based PLL). The minimum time-quantization step of the TDC is typically related to the minimum gate delay defined by the physical technology used. Even in the advanced 28-nm CMOS, it is approximately 10 ps, which is enough to limit the PLL phase noise performance. A DTC, on the other hand, can easily operate with a resolution on the order of 100 fs [5] which is fundamentally superior to a TDC in the same technology.

14.1.1.2 DTC gain error

DTC gain can be defined as the amount of delay per LSB code. DTC is analog in its nature and susceptible to process voltage temperature (PVT) variations, hence its absolute gain is unknown and varies with time and temperature. Gain error in the delay steps introduces problematic noise/spurs in the spectrum of the PLL. Automatic background calibration that tracks the gain variations becomes therefore necessary.

An automatic DTC gain calibration based on an least mean square (LMS) algorithm can be designed as depicted in Figure 14.3. The sign of the instantaneously sampled voltage (i.e., phase error) is extracted in every cycle and correlated with the change in direction of the DTC word. Intuitively, if the modulator instructs the DTC to sample later, but due to a gain error in the system "early" samples get consecutively detected – it is possible to deduct that the DTC gain is too low. After accumulation of the average error sign and code change product, the correction word can be applied as a scaling factor to the computation path, as depicted in Figure 14.2. When the correction loop converges, no penalty on the phase noise is observed if the LMS update factor (α) is low enough. Figure 14.4 shows a simulation result where a 5% gain error was applied to the DTC. This error introduces a large ripple in the sampled voltage, which in turn results in significant noise and spurious content at the

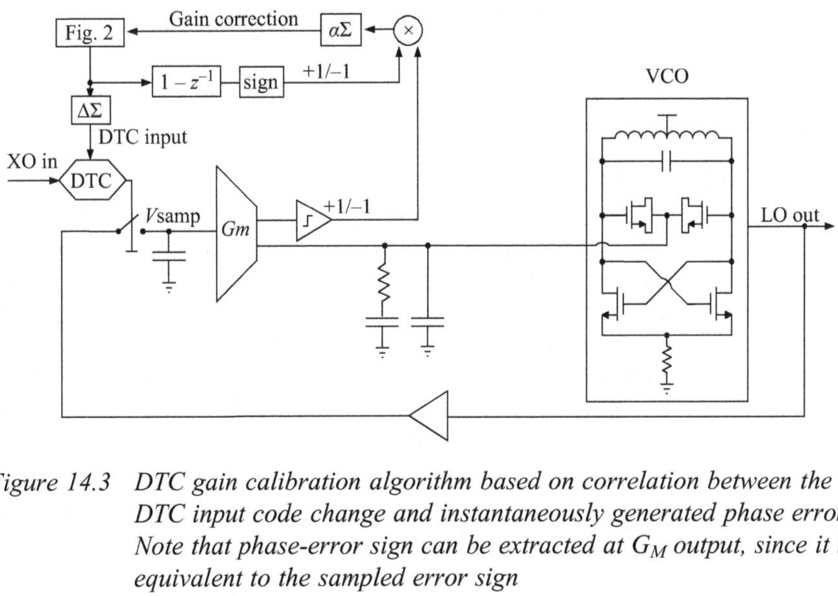

Figure 14.3 DTC gain calibration algorithm based on correlation between the DTC input code change and instantaneously generated phase error. Note that phase-error sign can be extracted at G_M output, since it is equivalent to the sampled error sign

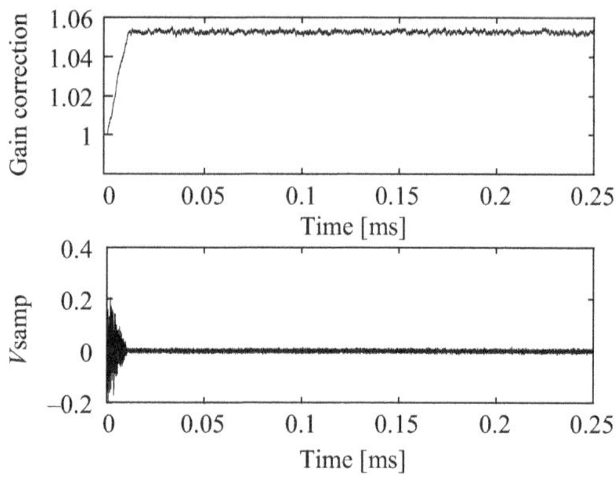

Figure 14.4 Example of the DTC gain error calibration algorithm convergence in the presence of 5% initial error

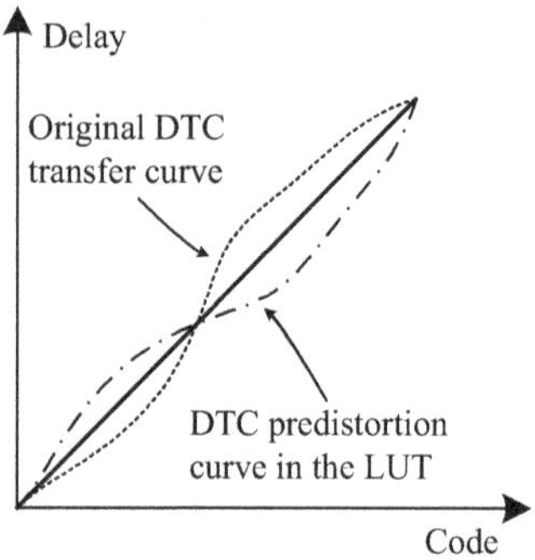

Figure 14.5 Principle of predistortion based on a LUT

PLL output. After the DTC gain is corrected, the sampled voltage converges back to zero.

14.1.1.3 DTC nonlinearity

Any nonlinearity in the phase-error comparison path of a PLL will induce fractional spurs and noise folding. In the context of analog PLLs, this problem typically originates from a nonlinear charge pump. Similarly, in classical digital PLLs where an analog phase-frequency detector is replaced by a TDC, the TDC causes similar issues. In DTC-based PLLs, this problem arises due to the nonlinearity of the DTC, since the delay through the DTC affects the phase-error detection. Intuitively, if the DTC has a pronounced INL, by application of a periodic signal at its input during fractional synthesis to compensate for the periodically accumulated fractional residue (see Figure 14.2), the DTC will create periodical errors. The periodicity is related to the fractional multiplication number and results in spurs at the fractional residue frequency and its harmonics that are rather pronounced when they fall in PLL band.

Instead of directing design effort into analog DTC linearization for low spur content at the PLL output, we propose digital self-calibration. If the DTC INL is *known* (measured *a priori*), predistortion can be used, as depicted in Figure 14.5, to remove the predictable errors. A look-up-table (LUT) could for instance store the INL value related to every DTC code. At a given cycle, the appropriate INL error can simply be subtracted from the original DTC input code. With this correction to the nonlinear DTC, the sampler correctly samples near the VCO zero crossing, restoring spur-free PLL operation. For accurate predistortion, the DTC INL needs to be measured. This is not a trivial task, especially for small DTC time steps (typically in the order of 100 fs). Moreover, the DTC transfer curve is very sensitive to process and environmental variations, meaning that the correct predistortion values will drift over time. It is therefore mandatory to have a digital self-calibration that runs in the background while the PLL operates normally.

To ensure linearity of phase-error detection path, we use DTC nonlinearity calibration loop (Figure 14.6). The calibration is based on observation of the instantaneous phase-error sign. This 1-bit information (+1/−1) about the phase error (see error extraction depicted in Figure 14.3) is random and zero on average in a linear system (we assume DTC gain error is calibrated out and that the gain correction algorithm is disabled); but in the presence of DTC INL, it becomes "colored" by its nonlinearity. Essentially, we exploit the correlation between the error sign signal and the particular DTC input code, which induced it, to restore the

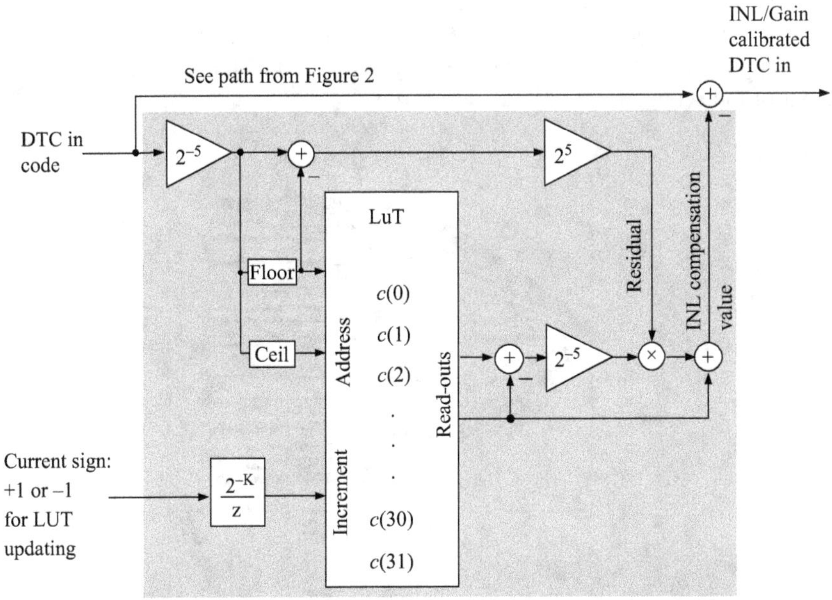

Figure 14.6 DTC INL background calibration based on a look-up-table approach

INL curves and precompensate for them (note that the approach is rather similar to DTC gain error correction, which is disabled during DTC INL calibration). A LUT with a set of coefficients $c(0{:}k-1)$ is used to approximate the N-bit DTC INL curve (k is in this example set as 32 to cancel nonlinearity of an exemplified 10-bit DTC). In every clock cycle, the input code addresses two neighboring LUT coefficients that linearly approximate piece-wise the expected, instantaneous INL error. The INL compensation value is simply subtracted from the original code, which ideally forces the DTC to produce no error (zero mean error sign) for the given code. The LUT correction coefficients are updated gradually by integrating the scaled error sign value to the appropriate address (defined with the input code) in every cycle. When the calibration is initialized, the LUT is reset to zero. While the PLL operates, the coefficients $c(0{:}k-1)$ slowly change toward their optimal calibrated values that cancel the INL error. At this moment, the coefficient updating can be disabled. The algorithm convergence speed is determined by the tap gain G, where a typical value of 2^{-13} results in ~10 ms approximate calibration time. The DTC INL calibration loop is enabled after the gain error is corrected. An important detail is that the offset in the extracted error sign needs to be digitally compensated for the loop to converge. An example of the DTC INL convergence is depicted in Figure 14.7, for an INL error of approximately 3 LSBs as exemplified in Figure 14.8. Once when the coefficients converge, they only dither around the final value compensating for the present nonlinearity. The dithering level depends on the algorithm gain, which can be set, such that there is no influence on the output spectral purity.

The algorithm convergence speed is not of great importance if the background calibration can track temperature variation, which typically happens with time constants expressed in seconds. Moreover, initializing coefficients to a previously acquired estimate significantly reduces calibration time. Gear-shifting techniques can be applied as well to optimize convergence speed of the algorithm, but these are out of the scope of the presented material.

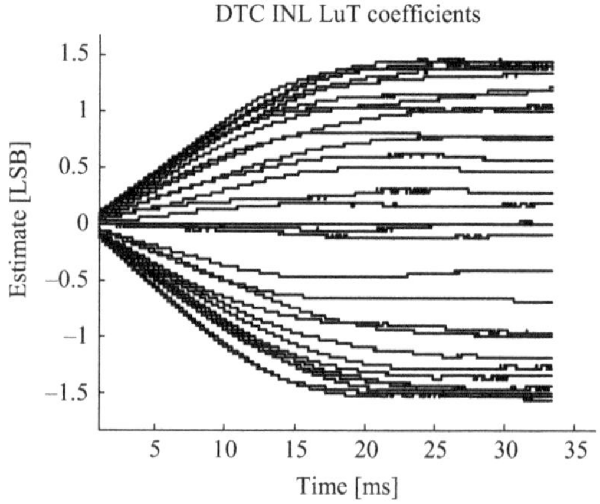

Figure 14.7 Example of the DTC INL error calibration algorithm convergence in the presence of INL error depicted in Figure 14.8

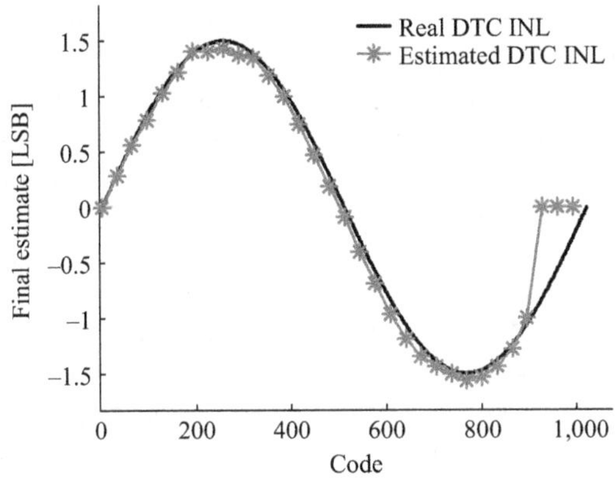

Figure 14.8 Final INL correction coefficients stored within the LUT

14.1.1.4 DTC phase noise

In this chapter, we propose a solution to enhance an integer-N subsampling PLL by placing a phase modulator (DTC) in the reference path. Unfortunately, the phase noise contribution of the DTC adds directly to the phase noise of the reference. The in-band phase noise performance of the subsampling PLL is influenced by the phase noise of both the reference crystal oscillator and the DTC, since both pass the system in the same way. Note that the phase noise floor of the DTC in fact gets multiplied by N^2 in the process of transfer to the PLL in-band output (filtered out of band). Therefore, great care must be taken to minimize the contribution of DTC to phase noise, otherwise the unique phase noise advantages of the subsampling architecture could be lost. Hence, Section 14.3.1 highlights the DTC phase noise optimization procedure.

14.2 Subsampling PLL-based phase modulator enabled by a DTC

Fractional-N PLLs are the fundamental building block of a polar transmitter often used for polar modulation. This section explains system-level enhancements to a DTC-based subsampling PLL that enables wideband and accurate phase/frequency modulation using the core loop.

Typical phase modulators based on fractional-N PLLs with a programmable divider in the feedback path of the loop operate based on phase/frequency injection through division ratio modulation. The desired central frequency of the PLL is then continuously modulated (by changing the desired division factor in the feedback path of the loop) to achieve the desired phase accumulation pattern at the PLL output. This approach is, however, strongly limited in bandwidth by the PLL's loop filter. The modulation signal injected through the divider is, namely, low-pass filtered by the loop by transferring to the output. Even in a low-in-band noise subsampling PLL environment that permits wide PLL bandwidth, realistic modulation bandwidth rises only up to 1 MHz (otherwise significant amount of signal distortion can be expected).

Similar principle could be applied in the context of a fractional-N subsampling PLL. Phase modulation can be achieved by direct modulation of the DTC. The DTC can force controlled "phase errors" (modulation signal) into the loop by purposely moving away from the VCO zero crossings. The modulation signal is then transferred to the PLL output, but, again, due to the presence of the loop filter, the achievable data rate is limited by the loop bandwidth.

Higher data rates could be enabled by predistortion of the modulation signal in front of the DTC. However, the predistortion transfer function is not easily determined since it depends on inaccurate analog components. Furthermore, it can initiate clipping in the PD: the linear range of the subsampling phase detection (PD) is valid only for small phase errors.

14.2.1 Two-point modulation

Two-point modulation technique overcomes the typical bandwidth limitations described above. We present here a two-point modulator in the context of a DTC-based subsampling PLL as shown in Figure 14.9. The basic principle is to inject the modulation signal simultaneously in front of the VCO (point two) and in the reference path, through the DTC (point one). The VCO frequency instantaneously changes with respect to the injected modulation signal. In the time domain, this means that the VCO instantaneously changes its period and the amount of accumulated phase over one reference period changes. The DTC receives the same modulation signal (expressed in phase) on top of the regular fractional residue compensation signal. It must delay the sampling event by an amount that will keep the VCO zero-crossing sampling unperturbed. Importantly, there are no phase errors in the PD induced by the modulation. The modulation signal transfers unfiltered to the output, i.e., the injection between the two points is canceled within the sampler and the loop does not "sense" the modulation data.

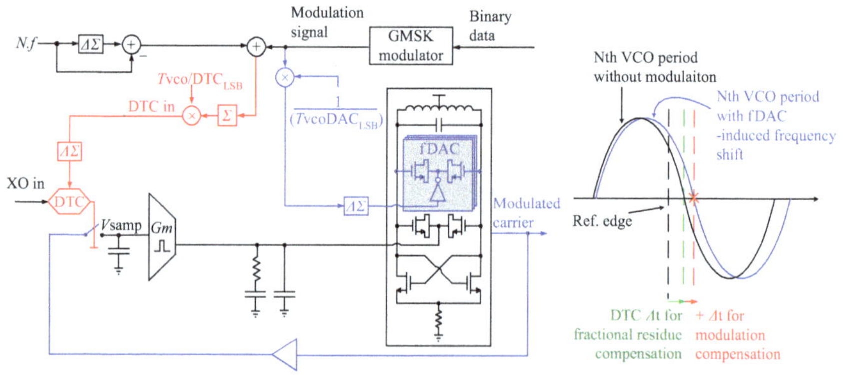

Figure 14.9 Two-point modulation principle using a DTC-based subsampling PLL

Two-point injection can be analyzed in frequency domain. Injection point two has a high-pass profile for the PLL output, while point one implements low pass. An *all-pass* transmission of the modulation data is available for the PLL output if the bandwidths of the two profiles match. Under the same conditions, there is a *no pass* of the data into the loop, i.e., the loop filter of the PLL does not experience any perturbations related to the injection of the modulation signal.

14.2.2 Limitations of the technique and their mitigation on the system level

The quality of the modulated output can be significantly degraded in presence of gain errors (often referred to as gain mismatch) or nonlinearities in the digital-to-modulated phase conversion that occurs in both the DTC and the modulating point two, through the VCO. The DTC gain/INL calibration as explained in Section 14.1.1 is readily reused here. Indeed, during the two-point modulation the zero-crossing subsampling condition still holds, and the PD works within its linear range. The DTC input code and phase error remain correlated in the presence of gain/INL errors in the same way as in the fractional-N synthesizer mode. The modulator's injection point one (DTC) is already calibrated in the background. We show next that a similar algorithm can be implemented for calibration of gain/INL errors in injection point two.

Nonlinear transfer of the modulation signal through point two to the output of the system typically arises from several nonidealities: mismatch/nonlinearity of the frequency DAC/varactor used in conversion, nonlinear capacitance-to-frequency conversion in the LC tank or similar. At this point, we assume a frequency modulation DAC (fDAC) is used for instantaneous VCO frequency tuning. We again use background estimated predistortion to linearize the modulator behavior. Note that all the effects of nonlinear data-to-modulated phase transfer are addressed simultaneously.

In the presence of nonlinearity, the instantaneous frequency shift induced by the fDAC will be wrong, i.e., the modulated PLL output period will be larger or smaller than expected. This leads to the DTC sampling away from the ideal zero crossing (late or early) and, consequently, to the transconductor injecting an error current into the loop filter. This current and its extracted sign is strongly correlated with the fDAC input code and can be exploited for fDAC self-calibration (Figure 14.10). To linearize the frequency modulation, we use again a LUT-based approach. In this design, as discussed later with building block implementation details, the frequency modulating DAC is of 8-bit resolution, i.e., 256 unique fDAC input codes are used. To correctly correlate the PD error with the fDAC input code, it is necessary to track the derivative of the PD error since the VCO integrates the phase and the modulation signal appears in the frequency domain. The correction coefficients (stored within the LUT) are again gradually increased/decreased (depending on the PD sign) at appropriate addresses, until they converge to positions that cancel out the fDAC nonlinearity, i.e., where the PD output becomes a zero-mean stream per particular fDAC input code. Since DTC and fDAC INL calibration loops cancel uncorrelated static errors independently of each other, they can run simultaneously in the background.

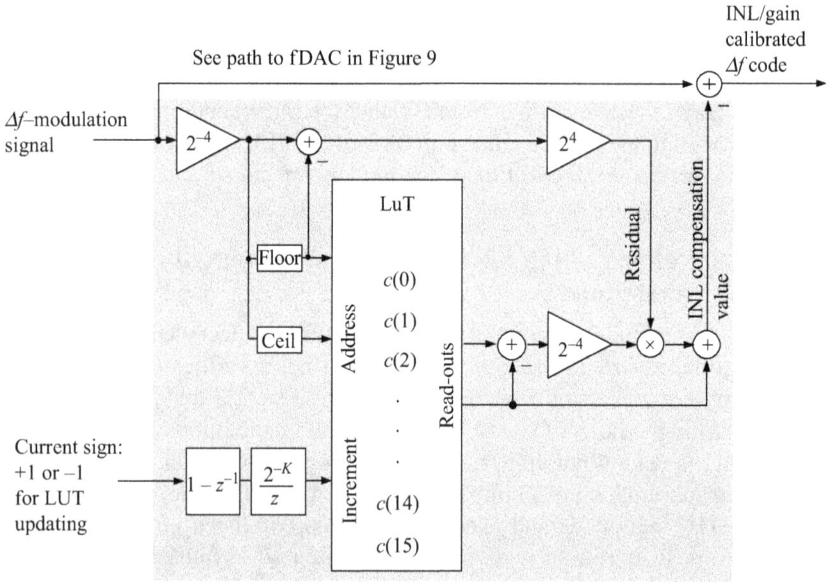

Figure 14.10 Frequency modulation in point two or fDAC INL background calibration based on a look-up-table approach. The algorithm is very similar to the proposed technique in Figure 14.6

This finalizes the overview of the DTC-based subsampling phase modulator. Note that the proposed environment efficiently addresses cancellation of *all* the relevant analog nonidealities that generate distortion at the system level. The proposed techniques are built in the digital domain, where the power-efficient CMOS can be used. Note that the proposed environment as depicted in Figure 14.9 on the left differs from the integer-N subsampling PLL only by addition of the DTC and fDAC in the signal path (ignoring the digital signal processing necessary that is not part of the core loop). After linearization, the DTC-based subsampling phase modulator in fact operates with performance directly set by the *integer*-N subsampling PLL: the output integrated noise is not limited by DTC/fDAC quantization noise (both of these data converter's LSBs can be set well below the thermal noise floor) and the two-point modulation technique allows optimal loop filtering set completely independent of the desired modulation bandwidth.

14.3 Analog circuits

14.3.1 Digital-to-time converter

A DTC lies at the input of the system and serves as a part of the PLLs' phase-error comparison path. As already discussed in previous sections, its resolution and linearity are essential for the overall system performance. Compromising these potential results in noise folding and spurious content at the PLL output, by design the resolution is set *low enough* and DTC nonlinearity is calibrated in the background.

DTC random phase noise performance is equally important. In fact, since the block is directly at the PLL input, its output random phase noise is multiplied by N^2 by transferring to the PLL output (where N is the PLL multiplication number) which potentially introduces a bottleneck for the in-band phase noise of the loop (low-pass PLL filtering profile compresses DTC-originated phase noise out of band).

Another key performance metric in the design is noise in the supply. Care must be taken that any noise (coming from, e.g., digital switching in the PLL) does not modulate delay generated by DTC. Since DTC is implemented as a digital delay line (as discussed here, an inverter with a tunable load), any supply noise becomes potentially part of the delay signal. This holds in the other direction as well, i.e., it is equally important that the DTC itself does not disturb the power supply (there should be no code-dependent power consumption) to avoid influence on the sensitive analog subsampling path.

Architectural overview of the DTC is given in Figure 14.11. The input buffer drives the delay control block, which is realized as an inverter with a dedicated resistor in the NMOS discharging path loaded by a controllable capacitor C_L. The delay stage output transition speed (discharging at the top plate of C_L) is dependent on the load capacitance and the resistor size (ignoring parasitics, assuming a steep input slope and assuming that the resistance of the NMOS is much smaller than the resistance of R). The discharging is exponential (RC settling) but introduced delay is a linear function of C_L for a fixed comparator threshold (in the following stage). We concentrate on the relevant high-to-low transition at the delay stage output, because this transition induces the sample taking. Note that capacitance discharging could be implemented by an NMOS transistor only (without a resistor), however, this would lead to an excessive $1/f$ noise contribution, which would dominate the

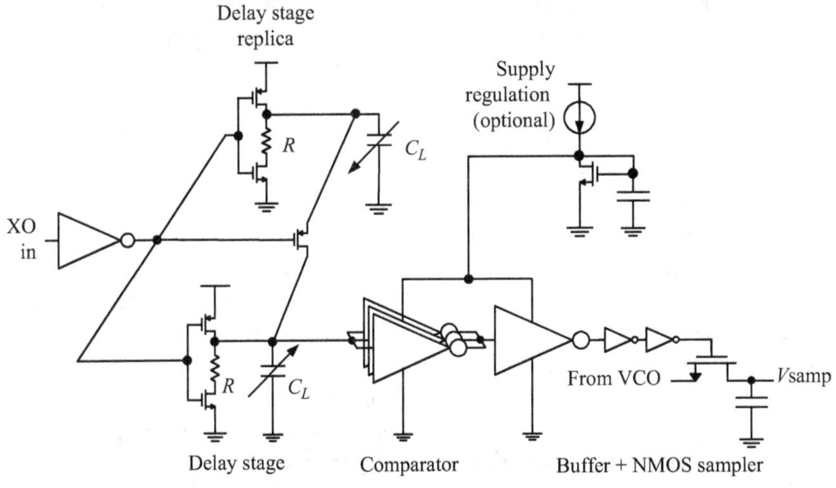

Figure 14.11 Architectural overview of the DTC

PLL in-band phase noise. By ensuring that the resistor sets the discharging, we also ensure that its contribution is the dominant source to phase noise. This is desired because a resistor generates no $1/f$ noise. The NMOS transistor can have minimal length and operate merely as a switch, introducing no $1/f$ noise ($1/f$ contribution is relevant only when the NMOS is used as a current source dominantly discharging C_L). Furthermore, any supply ripple coming from the preceding buffer only modulates the NMOS switch resistance, which is an order of magnitude smaller than the discharging resistor and does not affect delay. The phase noise induced by the delay stage (dominant phase noise source in the path) is proportional to

$$\mathcal{L} \sim 10\log_{10}\left(f_{\text{out}}\frac{kTC_LR^2}{VDD^2}\right) \sim 10\log_{10}\left(f_{\text{out}}\frac{kT\tau_{\text{delay}}R}{VDD^2}\right)\left[\frac{\text{dBc}}{\text{Hz}}\right] \tag{14.2}$$

where R is the value of the resistance, VDD is the supply voltage level, f_{out} is the clock input frequency and τ_{delay} is the delay induced by the block.

Note that the DTC uses a delay-stage replica that produces complementary delay with respect to the main path. In this way, we equalize the amount of charge pulled from the supply in every transition, avoiding any kind of dependence between code and supply ripple. Care must also be given to the fact that regeneration of the RC-delayed slope is most vulnerable to supply modulation. A tunable regulated supply shown in Figure 14.11 is used to protect the supply of the comparator and the following buffer. The regulated supply consists of a constant current source biasing a diode-connected transistor. A capacitor of several picofarads is used for additional decoupling of the regulated node. At the moment of toggling, charge is instantaneously pulled from the capacitor and not from the VDD. The dip in the regulated supply voltage is suppressed by the gain of the current source before reaching the top supply. The dynamic charge flow is in this way kept within the structure itself, protecting the rest of the PLL. In the same way, any noise from the overall supply does not couple in the delay signal path. Two small and fast inverters (steep slopes insensitive to noise) are then used in the following stages to ensure robust square wave drive of the sampling switch.

14.3.2 Subsampling loop

The subsampling loop consists of a VCO buffer, a sampler, and a pulsed transconductor (G_M). Additionally, the DTC provides the required phase modulation, as discussed in the previous sections. The overview of the complete loop is given in Figure 14.12.

The VCO buffer implemented as a source follower is required to reduce potential kickback from the sampler to the VCO. The sampler is built around an NMOS switch and a small metal-oxide-metal (MOM) capacitor (note that implementation is differential but for simplicity shown as single ended). Sampler kT/C noise can be neglected (in comparison to other loop noise generators) since it is suppressed by the large phase-error detection gain in the loop. The implemented sampler uses an auxiliary sampler operating in inverted phase to the primary sampler in order to reduce load variability of the VCO. Since the implemented VCO can operate from the input-output (IO) voltage, the tuning voltage also has a range larger than the core voltage. Therefore, the output

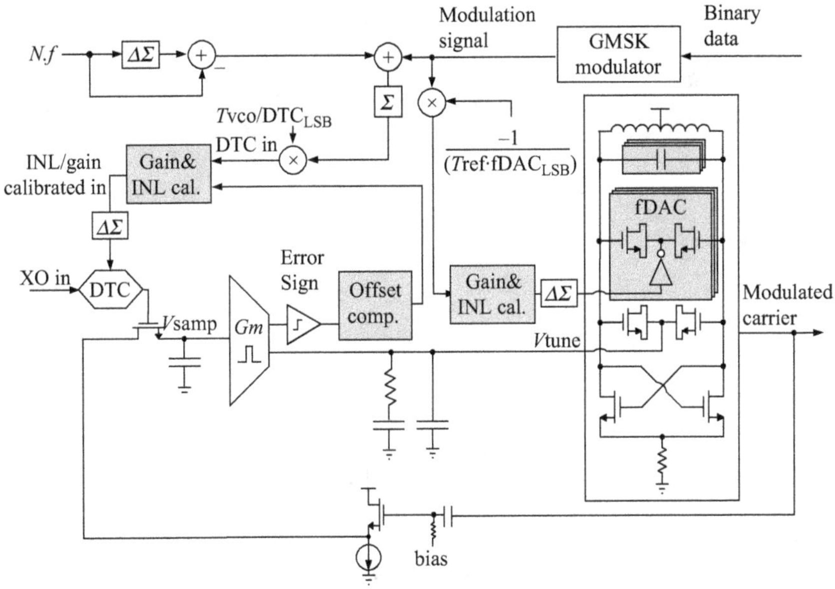

Figure 14.12 Overview of the subsampling loop for fractional synthesis and phase modulation

stage of the G_M needs to provide translation from the low-voltage domain of the sampler to the high-voltage domain of the LPF and the VCO. This translation is done in current domain between the first and the second stage (Figure 14.13). The first stage is a simple differential pair providing the necessary transconductance, whereas the second stage implements a charge pump-like output. The detection gain is so large that duty cycling is required in the output stage of the transconductor. Pulsing is done with a simple digital pulse generator that opens the output switches of the G_M.

Figure 14.13 shows details of the G_M stage implementation. Note that the calibration loops depicted in Figures 14.3, 14.6, and 14.10 necessitate information on the instantaneous phase-error measured in every sample. Instead of tracking the sampled voltage, we extract sign of the *current* pumped into the loop filter in every cycle. Indeed, in case of any mismatch in the phase detection circuitry (VCO buffers, samplers, G_M), the PLL adjusts phase lock, so that the output current of the G_M is zero on average and not necessarily the differentially sampled voltage. The slightest imbalance of current in the output branch results in a large swing of voltage at the sign extraction output node. A dynamic comparator is used to detect the sign of this voltage. Any offset in the comparator is removed digitally in post processing.

14.3.3 VCO implementation

The VCO from this design [6] is used in the class B operation mode. A digital varactor provides 6-bit coarse frequency tuning, an analog thick-gate-oxide varactor provides fine-tuning, while for modulation a simple 8-bit fDAC is used as depicted in Figure 14.14. The cross-coupled transistor M_C are implemented in thick

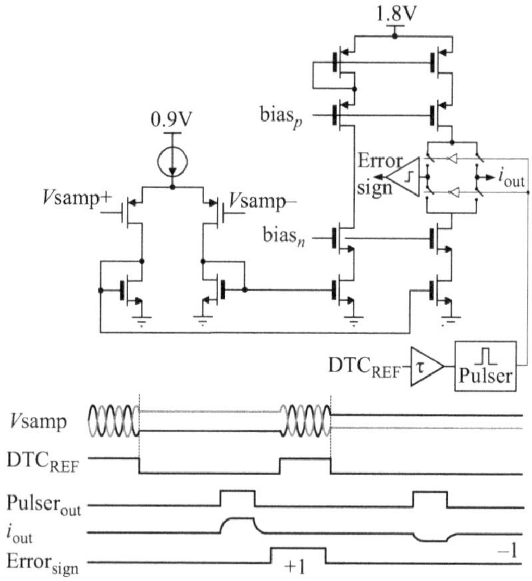

Figure 14.13 Transconductor schematic. The input pair receives differentially sampled voltage. The output current is duty cycled in the flow to the loop filter

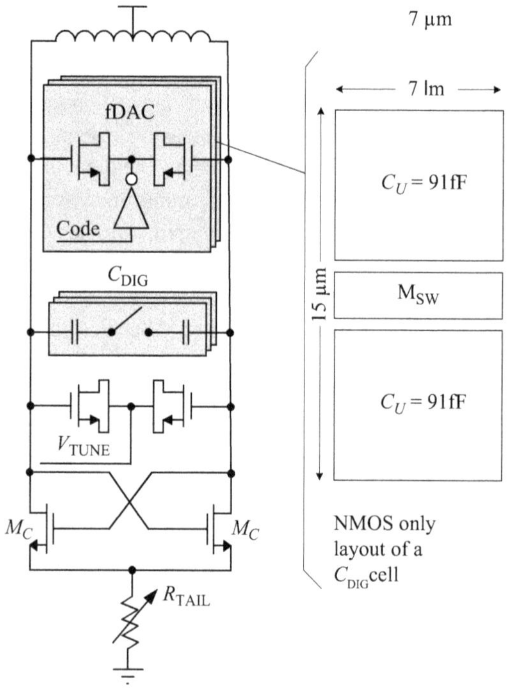

Figure 14.14 VCO schematic (in class B) and layout floor plan of the NMOS-only digital varactor cell as shown in Figure 14.15 (left)

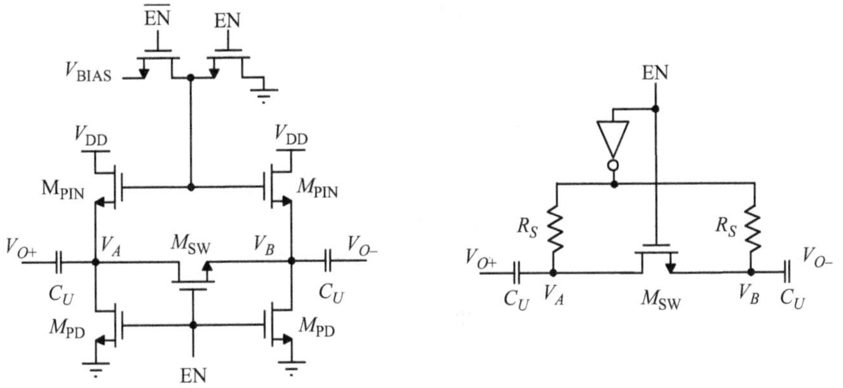

Figure 14.15 Digital varactor cell proposed in this design (left) in comparison to a conventional cell (right)

oxide for reliability since they "see" full VCO swing, and the VCO can operate from the nominal 0.9 V supply to 1.8 V in low-noise mode. A digitally tunable resistor in the tail is used to trade power for phase noise performance.

The digital varactor unit cell used in this VCO is depicted in Figure 14.15 (left). Compared to the widely used conventional cell shown on the right-hand side of the same figure, the proposed structure has several advantages, particularly in the context of nanoscale CMOS technology used. In the on-state (EN = 1), the proposed switched capacitor cell operates similarly to the conventional cell: the switch M_{SW} differentially shorts nodes V_A and V_B and the (MOM) capacitors C_U add to the overall tank capacitance. In the off-state (EN = 0), the cell behaves differently than the conventional cell. The M_{PIN} transistors provide a "bottom-pinning" functionality, setting the DC bias levels of the cell such that voltage stress on the devices can be minimized. Additionally, this structure naturally produces the highest off-state Q possible, since all leakage is dynamically compensated by the pinning transistors. The proposed cell also benefits from its compact, NMOS-only implementation as exemplified in layout from Figure 14.15 (right-hand side).

The layout can be realized as a single-composite NMOS block placed between the two-unit capacitors. By comparison, the conventional cell uses both PMOS and NMOS transistors and a polysilicon resistor, which cannot be abutted.

14.3.4 Frequency acquisition

A subsampling PLL is incapable of distinguishing frequencies separated by $\pm F_{REF}$, where F_{REF} is the operating frequency brought to the input of the PLL, typically by a crystal oscillator. To mitigate this issue, a frequency acquisition loop runs in parallel to the subsampling loop to acquire the desired frequency. The implementation of the acquisition loop is completely overtaken from the design presented in [1]. Importantly, this loop can operate at wide bandwidth to acquire the frequency as fast as possible. The performance of the auxiliary loop in terms of power consumption, phase noise, or nonlinearity is irrelevant since it is automatically disabled after the frequency lock (using a dead-zone detection mechanism).

14.4 Experimental results

The integrated circuit (IC) prototype (Figure 14.16) was fabricated in 28 nm bulk digital CMOS, with an active area of approximately 0.25 mm^2. It operates on 0.9 and 1.8 V supplies (IO interface and the G_M stage). The VCO can be put in a high-power/-performance mode at a 1.4-V supply if desired.

The total power consumption of the frequency synthesizer is 5.6 mW, of which 1.8 mW is for the loop components, 2.7 mW for the VCO, and 1.1 mW for digital circuitry. The VCO has a 20% tuning range around a central frequency of 11.25 GHz.

The measured phase noise profile at the output of the PLL at a close-to-integer fractional 11.72 GHz channel depicted in Figure 14.17 (left for a low-power VCO). The in-band phase noise is always below −107.9 dBc/Hz. The measured RMS jitter is 198 fs, with integration ranging from 10 kHz to 40 MHz, which includes all the spurs. In integer-N mode, the RMS jitter improves by only 22 fs. This proves that the system's nonidealities were well identified and successfully mitigated in this design. In other words, fractional spurs are no longer an issue, thanks to the proposed background DTC calibration.

The measured spurious content (worst measured spur) at the PLL output is shown in Figure 14.18, across different fractional channels. The background calibration suppresses spurs by > 15 dB, and the worst-case fractional spur appears at −56.6 dBc. The reference spur is at −69 dBc. The PLL achieves a competitive

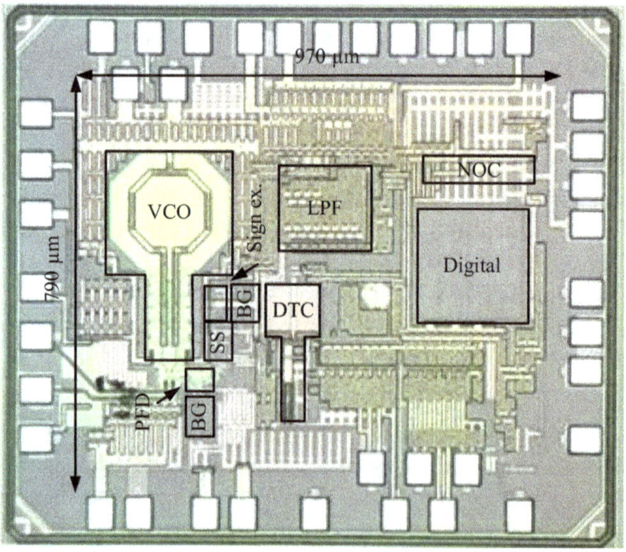

Figure 14.16 Chip microphotograph

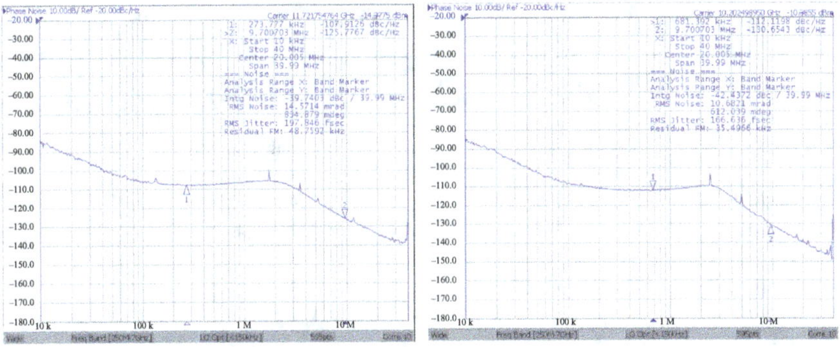

Figure 14.17 Phase noise profile at the PLL output in low-power VCO mode (left) and high-power VCO (right). The integration range includes all the spurs (integer and fractional)

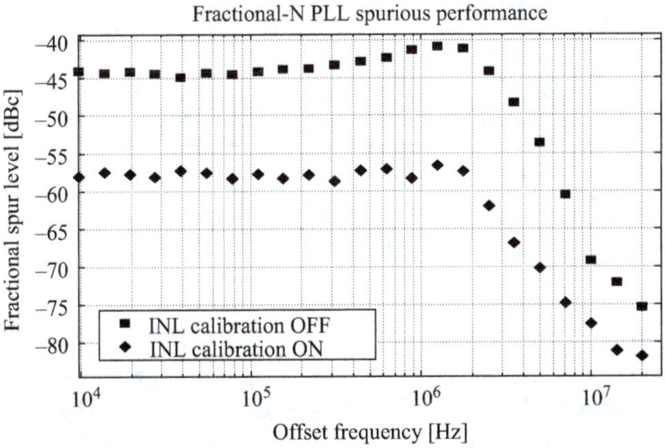

Figure 14.18 Spurious performance of the PLL at different synthesized fractional channel without and with background calibration

−246.6/−247.6 dB FOM (worst-case/best-case fractional-N mode), while with digital circuitry disabled in integer-N mode, the FOM is at −248.5 dB. In the high-performance mode, the VCO can operate at 1.4 V supply with up to 22.4 mW power consumption. The RMS jitter (see Figure 14.17 right) is then between 167 and 147 fs for the worst-case fractional and integer-N mode. The small difference again proves the effectiveness of the background calibrated environment. A summary of the PLL performance is given in Table 14.1. Thanks to the efficient background calibration algorithms and a DTC-based subsampling environment, this PLL achieves one of the highest FOMs reported up to date.

The VCO is operated in high-power mode during modulation since the fDAC achieves ~5× larger LSB than the targeted 50 kHz/LSB in the nominal mode at

Table 14.1 PLL performance summary and comparison to the state of art

	This work	Chen ISSCC'15	Huang ISSCC'14	Gao ISSCC'16
Architecture	Subsampling analog	Subsampling digital	Subsampling analog	Sampling digital
Reference (MHz)	40	49.15	48	40
Output (GHz)	10.1–12.4	2.6–3.9	2.2–2.4	2.7–4.3
Tuning range (%)	20.4	40.0	8.7	46
Bandwidth (kHz)	1,800	700	500	500[3]
In-band PN (dBc/Hz)[1]	−107.9	−97.8	−97.9	−99.4[3]
Ref. spur (dBc)	−69	−60	−55	−78
Worst fractional spur (dBc)	−56.6	−62	−53	−54
RMS jitter (fs) frac-N best/worst	176/197 (10 k–40 M)	226/240 (1 k–100 M)	266/400 (10 k–30 M)	159/n.a. (10 k–40 M)
Power (mW)	5.6	11.5	17.3	8.2
FOM (dB)[2] **best/worst**	Frac-N: −247.6/−246.6; Int-N: −248.5	Frac-N: −242.3/−241.8; Int-N: n.a.	Frac-N: −239.1/−235.6; Int-N: n.a.	Frac-N: −246.8/n.a.; Int-N: n.a.
Area (mm^2)	0.25	0.23	0.75	0.3
Process (nm)	28	65	180	28

[1]Scaled to 11.72 GHz.

[2]$FOM_{PLL} = 10\,log_{10}\left[\left(\frac{\sigma^2_{t,PLL}}{1s}\right)\cdot\left(\frac{P_{PLL}}{1mW}\right)\right]$

[3]Estimated from the source analyzer measurement.

low-power and at low VCO swing (due to a design mistake). By running the VCO in high-power mode (6.4–22.4 mW), the LSB reduces to 113–83 kHz/LSB. This is due to an increased swing at the VCO output that minimizes the effective ΔC [fF] in the modulating fDAC. In the high-power mode, the fDAC quantization noise does not limit the EVM performance. Figure 14.19 shows the measured 10 Mb/s GMSK spectrum and constellation with all the background calibration algorithms enabled, around a fractional carrier close to 10.24 GHz. The modulated output has an undistorted spectral profile with bandwidth larger than 10 MHz even though the PLL bandwidth (BW) is approximately 2.5 MHz. The measured EVM with DTC/fDAC calibration is −40.5 dB, which is only 1.9 dB above the fundamental limit of the total integrated phase noise in the high-power synthesizer mode. Note that EVM scaled with $20\log_{10}(N)$, where N is the frequency multiplication factor, which means that the modulator would provide −49.6 dB EVM at 3.6 GHz, after division.

To prove the importance of calibration, we show EVM at different fractional offsets (Figure 14.20) with and without calibration. Using only gain correction results in average in 8 dB EVM improvement. The DTC and fDAC INL calibration results in additional 7 dB EVM improvement (6 dB after DTC calibration and then 1 dB after fDAC calibration).

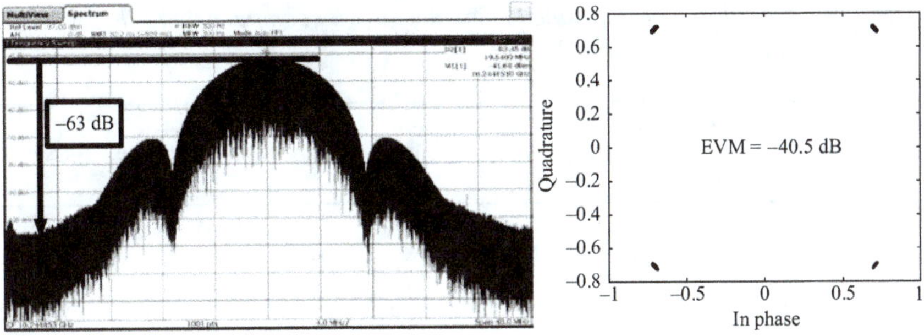

Figure 14.19 Modulated output spectrum and measured constellation at GMSK 10 Mb/s close to a 10.24-GHz fractional carrier

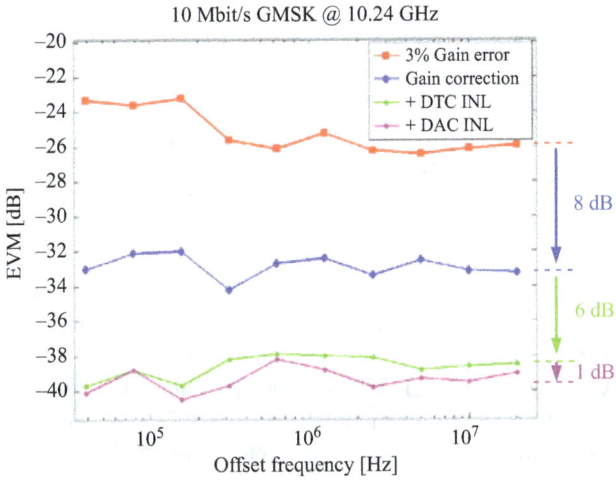

Figure 14.20 Performance comparison with different numbers of background calibration loops enabled

The modulator performance summary is given in Table 14.2, together with a comparison to the similar work in the state of art. The modulator achieves an order of magnitude EVM improvement compared to the similar work in the field.

14.5 Conclusion

The presented chapter describes a DTC-based subsampling PLL capable of fractional synthesis and phase modulation. With the combination of the extremely low-noise subsampling error-detection core and a DTC, which is used for fractional residue compensation, the PLL achieves up to −247.6 dB FOM and compares favorably to the state of art. The DTC induces insignificant degradation in the

Table 14.2 Modulator performance summary and its comparison to the state of art

	This work (low VCO power)	**This work (high VCO power)**	**Marzin JSSC'12**	**Li IWS'12**	**Shanan ISSCC'09**	**Xu JSSC'14**
Architecture	Direct FM		Direct FM	Direct FM	Direct FM	Direct FM
Clock frequency (MHz)	40		40	n.a.	52	26
Output (GHz)	9.2–12.4		2.9–4.0	1.5–1.8	2.4	1.7–2.1
Power dissipation (mW)	8.1	25.4	5	10	16	6.9
Modulation type	GMSK		GMSK	GMSK	GMSK	GMSK
Data rate (Mb/s)	10		10	10	2	1.08
Energy/bit (nJ/bit)	0.8	2.5	0.5	1	8	6.4
Total integrated phase noise (dBc)[1]	−41.7	−42.4	−29.9	−13.8	n.a.	−20.3
EVM RMS (dB)[1]	−37.4	−40.5	−26.9	n.a.	−5.4	−13.9
Out-of-band emission (dBr)	−55	−63	−56	n.a.	n.a.	n.a.
Area (mm^2)	0.25		0.5	0.5	1.1	0.49
Process (nm)	28		65	65	180	65

[1]Scaled to 10.24 GHz.

phase-error comparison path, thanks to its high-resolution and low-noise operation. Its nonlinearity is compensated in the background by efficient digital predistortion.

Moreover, we propose a two-point modulation technique is the context of the DTC-based subsampling loop. The modulation signal is directly applied through the VCO, while the DTC phase modulates in the reference path for maintained zero-crossing subsampling. The gain imbalance/nonlinearities in the digital-to-modulated conversion are background corrected for a 10-MHz BW, 10-Mb/s GMSK modulation with <−40 dB EVM at 10 GHz. The achieved performance surpasses performance achieved by similar work in the field, proving the effectiveness of the proposed approach.

References

[1] X. Gao, E. A. M. Klumperink, M. Bohsali, and B. Nauta, "A low noise subsampling PLL in which divider noise is eliminated and PD/CP noise is not multiplied by N^2," *IEEE Journal of Solid-State Circuits*, vol. 44, no. 12, pp. 3253–3263, 2009.

[2] N. Markulic, K. Raczkowski, E. Martens *et al.*, "A DTC-based subsampling PLL capable of self-calibrated fractional synthesis and two-point modulation," *IEEE Journal of Solid-State Circuits*, vol. 51, no. 12, pp. 3078–3092, 2016.

[3] N. Markulic, K. Raczkowski, E. Martens *et al.*, "9.7 A self-calibrated 10 Mb/s phase modulator with −37.4 dB EVM based on a 10.1-to-12.4 GHz, −246.6 dB-FOM, fractional-N subsampling PLL," in *2016 IEEE International Solid-State Circuits Conference (ISSCC)*, 2016, pp. 176–177.
[4] K. Raczkowski, N. Markulic, B. Hershberg, and J. Craninckx, "A 9.2–12.7 GHz wideband fractional-N subsampling PLL in 28 nm CMOS with 280 fs RMS jitter," *IEEE Journal of Solid-State Circuits*, vol. 50, no. 5, pp. 1203–1213, 2015.
[5] N. Markulic, K. Raczkowski, P. Wambacq, and J. Craninckx, "A 10-bit, 550-fs step Digital-to-Time Converter in 28 nm CMOS," in *European Solid State Circuits Conference (ESSCIRC), ESSCIRC 2014-40th*, 2014, pp. 79–82.
[6] B. Hershberg, K. Raczkowski, K. Vaesen, and J. Craninckx, "A 9.1–12.7 GHz VCO in 28 nm CMOS with a bottom-pinning bias technique for digital varactor stress reduction," in *European Solid State Circuits Conference (ESSCIRC), ESSCIRC 2014-40th*, 2014, pp. 83–86.

Chapter 15

Hybrid two-point modulation with 1b high-pass modulation and embedded FIR filtering

Ni Xu[1]

This chapter presents a hybrid two-point modulator architecture based on the semi-digital phase-locked loop (PLL). By utilizing a 1-bit ΔΣ modulation (DSM) with embedded finite-impulse response (FIR) filtering, the high-pass modulation path does not suffer from the digitally controlled oscillator (DCO) gain nonlinearity. The digital FIR filter in the high-pass modulation path not only suppresses quantization noise but also reduces noise coupling with time-interleaved switching of partitioned capacitors. In addition, the proposed two-path DCO modulation enables a flexible gain partition for the high-pass and low-pass modulation path within the DCO. To further enhance the linearity of the fractional-N PLL and the modulator, a hybrid FIR filtering method is used for the low-pass modulation path. The chapter gives two design examples to validate the proposed two-point modulation: A 1.8-GHz two-point modulator based on a semi-digital PLL implemented in 65 nm complementary metal-oxide-semiconductor transistor (CMOS and the 2.4-GHz two-point modulator implemented in 0.18 μm CMOS). The proposed architecture is useful for low-cost phase modulator design without requiring complex calibration algorithm or an advanced CMOS technology, while achieving comparable modulation performance.

15.1 Introduction

The ΔΣ PLL enables digital phase modulation without requiring digital-to-analog converters (DACs) and radio-frequency (RF) up-converters, which significantly simplifies the overall transmitter architecture such as polar transmitter. Considering the typical PLL bandwidth is not wide enough to meet the required modulation symbol rate, a digital compensation method [1] or a two-point modulation method is usually used [2,3] to overcome the bandwidth limitation. In the digital compensation method, the transfer function of the digital compensation filter needs to be matched well with that of the PLL. However, it is difficult to have perfect match

[1]Credo Semiconductor, San Jose, CA, USA

because the loop dynamics of the PLL is highly sensitive to process and temperature variations, which makes it less attractive for on-chip modulation.

Different from the digital compensation method, the two-point modulator provides an all-pass transfer function by combining high-pass and low-pass modulation paths in the ΔΣ fractional-N PLL. Hence, the PLL bandwidth can be optimized for phase noise performance regardless of the data rate [2,3]. To achieve an all-pass transfer function with the high-pass and the low-pass modulation paths, the transfer function of the high-pass modulation must be accurately scaled by the voltage-controlled oscillator (VCO) gain. In practice, the VCO gain is highly sensitive to process-voltage-temperature (PVT) variations and is one of the most critical parameters to determine the overall performance of the two-point modulator. In addition, the VCO nonlinearity also degrades the out-of-band spectral emission of the transmitter, resulting in poor performance of the error–vector magnitude (EVM). The all-digital PLL (ADPLL) with the DCO provides a robust way of calibrating the absolute DCO gain [4,5], but DCO nonlinearity calibration is still challenging. The use of the full thermometric capacitor banks improves the DCO linearity but suffers from a large number of elements with substantial area occupation, causing tuning range reduction in the DCO. To reduce the DCO area, a multicapacitor bank topology is proposed with the all-digital background calibration method [6], but it requires a complex algorithm to accurately calibrate the intrabank nonlinearity.

In this chapter, we introduce a hybrid-loop two-point modulator with 1-bit high-pass modulation to mitigate the DCO nonlinearity [7]. The digital FIR filter and the hybrid FIR filter are utilized in the high-pass and low-pass modulation paths, respectively, to improve the linearity and reduce high-frequency quantization noise. In the proposed two-point modulator, only the absolute DCO gain needs to be calibrated and the complex DCO nonlinearity calibration is not required. This can substantially simplify the two-point modulator architecture while achieving good linearity.

The chapter is organized as follows. Section 15.2 describes an overall architecture of the hybrid two-point modulator, the 1-bit DCO modulation method, and the modeling for the design. Section 15.3 provides a design example and detailed circuit implementation. Experimental results are presented in Section 15.4, followed by conclusion in Section 15.5.

15.2 Hybrid two-point modulation with embedded FIR filter

15.2.1 Semi-digital PLL with hybrid FIR filtering

The ADPLL-based modulation is considered as a robust method to implement the two-point modulator with the digital gain/delay mismatch calibration [5,6]. However, the high-performance time-to-digital converter (TDC) is required in the ADPLL design that is power hungry and requires advanced CMOS technology. Moreover, the TDC nonlinearity becomes another source of the fractional spur and degrades in-band phase noise of the ADPLL.

The TDC-less semi-digital PLL architectures achieving linear loop dynamics without a large integration capacitor have been proposed in [8–13]. Figure 15.1 shows a simplified block diagram of the semi-digital PLL with hybrid FIR filter. The DCO/VCO is used to accommodate both analog and digital controls. For the proportional-gain path, a conventional analog control with the PFD and the charge pump followed by a second-order passive loop filter is used. Since the capacitance values for high-order poles are not large, a small-area loop filter can be used. As for the integral path, digital implementation is done with a bang-bang phase detector (BBPD) and a finite-state machine (FSM) to compensate the limited frequency tracking capability of the proportional-gain path. The ΔΣ modulator is used to provide a high-frequency resolution as in the ADPLL. As illustrated in Figure 15.1, the integral path provides continuous frequency tracking with high resolution, which significantly reduces the effective analog control voltage range. Such a small control voltage range apparently relaxes the linearity requirement of the analog gain of the VCO and the charge pump.

In addition, the hybrid FIR filter is helpful to further improve the charge pump linearity. Basically, the charge pump nonlinearity is another source of the in-band noise and the fractional spurs in the ΔΣ PLL. Figure 15.2 demonstrates the simulation results of the PLL closed loop output spectrum and the charge pump output voltage waveform with and without the proposed FIR filtering. In frequency domain, the embedded FIR filter reduces the ΔΣ modulation quantization noise a lot. And in time domain, the peak error voltage at the charge pump output with the hybrid.

FIR filter is much smaller than that of the conventional one. Since the up/down current mismatch due to large output voltage swing is one of the main reasons for charge pump nonlinearity [14], by reducing the charge pump output swing, the hybrid FIR could partially improve the charge pump linearity.

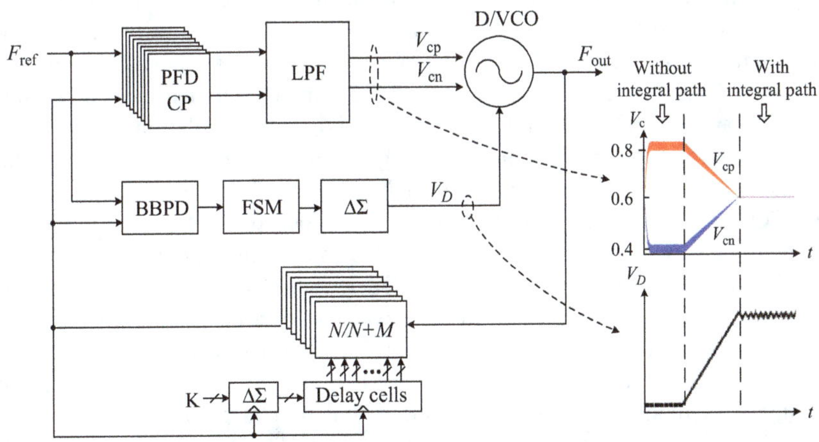

Figure 15.1 Functional block diagram of the semi-digital phase-locked loop (PLL) with hybrid finite-impulse response (FIR) filter

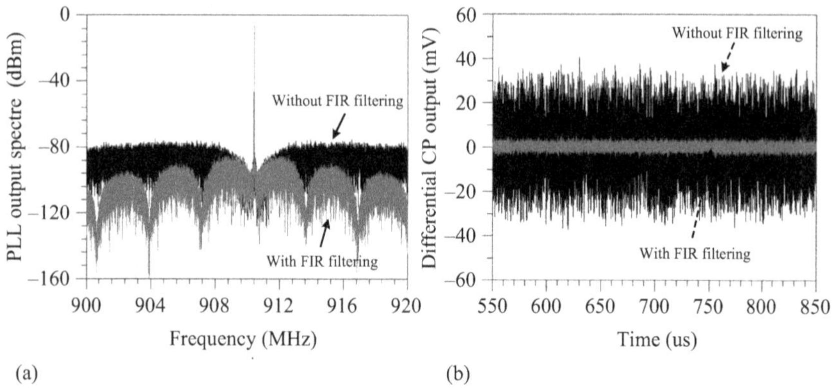

Figure 15.2 Simulation results of the hybrid phase-locked loop (PLL) with and without finite-impulse response (FIR) filtering: (a) PLL output spectrum. (b) Differential charge pump (CP) output voltage swing

15.2.2 Dedicated high-pass modulation path with FIR-embedded 1-bit control

The low-pass and high-pass modulation bits are combined at the input of the DCO to control the capacitor array in the conventional ADPLL-based two-point modulation. To achieve high-frequency resolution with minimum quantization noise, each capacitance of the capacitor array needs to be small. To meet the required frequency range, the capacitor array needs to be huge, resulting in large parasitic capacitance, which significantly degrades the DCO linearity. Several methods to improve the DCO nonlinearity have been proposed, such as using a frequency-to-voltage converter for VCO linearization [2], adaptive gain compensation for DCO gain calibration [6], and second-order approximation scheme in the DCO gain calibration [15]. However, those methods significantly increase design complexity.

Considering the capacitor array of the DCO can be easily partitioned, we split the DCO capacitor into two parts: the one for high-pass modulation and the other for low-pass modulation (frequency tracking). It makes the design much flexible, and the two paths can be optimized separately. Since the frequency tracking path of the DCO does not move much when the loop is locked, the linearity is not that critical for the overall performance. In the meantime, the high-pass modulation path plays the dominant role in frequency modulation, which requires high linearity within the frequency range. Wide frequency range is needed for high data rate. Therefore, the EVM and out-of-band phase noise performance are more sensitive to the nonlinearity of the high-pass modulation pass. To solve the nonlinearity issue in the high-pass modulation pass, the 1-bit modulation is proposed as shown in Figure 15.3. As the high-pass modulation is implemented by a single capacitor, the high-pass modulation of the DCO does not suffer from the nonlinearity, which is similar to the $\Delta\Sigma$ ADC with a 1-bit quantizer. On the other hand, the capacitance of

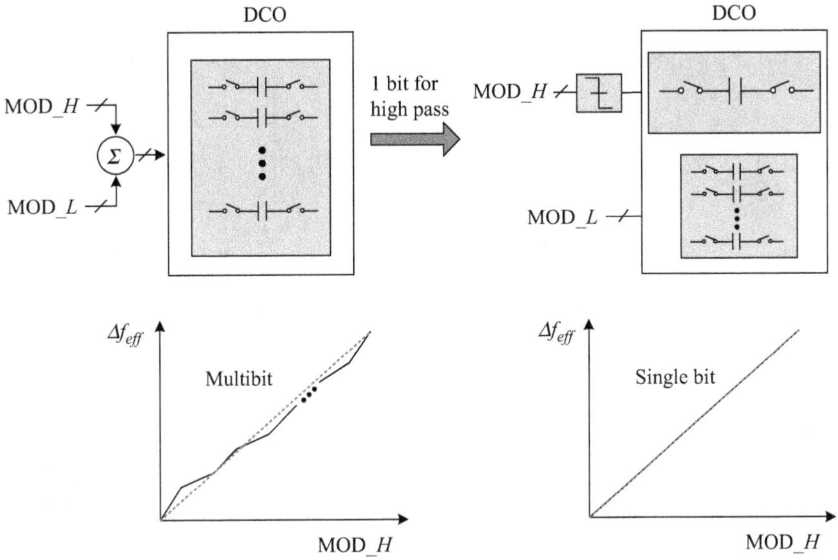

Figure 15.3 Dedicated high-pass modulation with 1-bit control

the single capacitor needs to be large enough to cover the required frequency deviation for the target data rate.

The 1-bit modulation with the single capacitor indeed overcomes the DCO nonlinearity. However, the single large capacitor in the DCO causes substantial quantization noise and makes the DCO extremely sensitive to the noise coupling due to the high DCO gain. To solve these problems effectively, an FIR filtering is used for the DCO [16]. Figure 15.4 illustrates the implementation. The large varactor is partitioned into small varactors, and each varactor is controlled by bit-shifting operation to perform the FIR filtering. The proposed FIR filter offers several advantages. First, it reduces the quantization noise in high frequencies due to the 1-bit modulation. Second, the FIR-based operation provides a time-interleaved modulation with uniformly partitioned small varactors in the high-pass modulation path, which significantly reduces the noise coupling sensitivity for the given time. Different from the traditional multibit high-pass modulation, the proposed 1-bit modulation with the FIR filtering architecture does not have the nonlinearity issue, even if the small varactor array is utilized. It will be analyzed and discussed later.

Figure 15.5 shows the design flowchart of the proposed 1-bit DCO modulation. The required minimum frequency deviation $\Delta f_{\min}$ can be calculated from the given modulation and data rate. For example, for a 1-Mb/s Gaussian minimum shift keying (GMSK) modulation with a modulation index of 0.5, the required minimum frequency deviation $\Delta f_{\min}$ is 500 kHz [17]. Generally, the 1-bit modulation can be realized by a first-order multi-stage noise shaping (MASH) modulator (MASH1) or

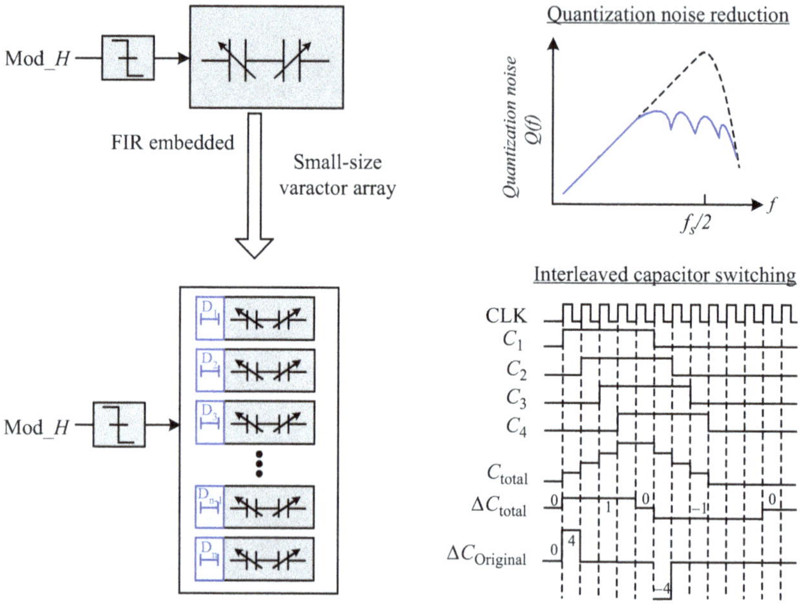

Figure 15.4 One-bit digitally controlled oscillator (DCO) modulation with finite-impulse response (FIR) embedded

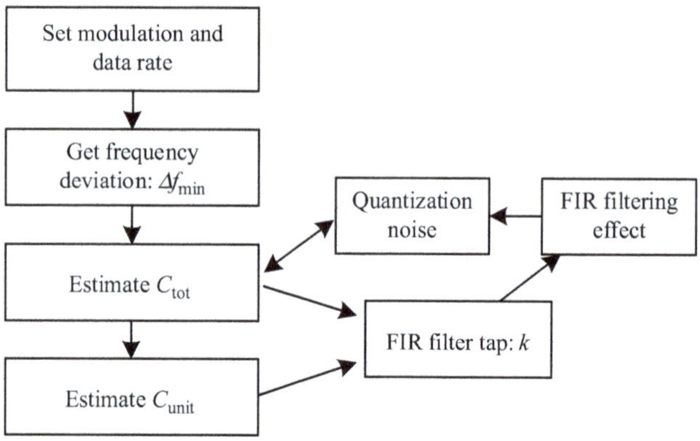

Figure 15.5 Design flow for the 1-bit finite-impulse response (FIR)-embedded digitally controlled oscillator (DCO) modulator

a second-order single-loop ΔΣ modulator (SLDSM2). Compared with MASH1, the SLDSM2 achieves better noise reduction, but its input dynamic range is limited by the internal logic implementation [9]. In this work, the MASH1 is used for the 1-bit modulation to avoid the input dynamic range problem and reduce power consumption. With enough margin over PVT variations, the total capacitance C_{tot}

required for the 1-bit modulation is estimated from Δf_{min}. When the k-tap FIR filter is used, the C_{tot} is divided into k unit capacitances C_{unit}. Since the minimum capacitance C_{min_unit} is limited by the given CMOS technology, the maximum number of the FIR taps k_{max} for the required C_{tot} is set by $C_{to}t/C_{min\_unit}$. When the number of the FIR taps k increases, the FIR filter provides better low-pass filtering for the modulation spectrum in high frequencies due to higher order of FIR filter. However, with a fixed C_{unit}, increasing k means increased total capacitance, leading to larger quantization noise. Thus, a trade-off was observed between the FIR filtering and the quantization noise. Therefore, it is important to choose the optimum number of the FIR filter taps for the 1-bit DCO modulation.

To verify the effectiveness of the proposed 1-bit high-pass modulation, behavioral simulations are performed with the 1 Mb/s GMSK modulation signal. The simulation results of the root-mean-square (RMS) EVM and the quantization noise with different FIR taps are demonstrated in Figure 15.6. For each FIR filter design, the C_{unit} is selected based on the instruction given in Figure 15.5. The 1-bit DCO open-loop and the PLL close-loop simulations are performed to demonstrate the EVM contribution from the DCO as shown in Figure 15.6(a). As shown in Figure 15.6(b), the higher-order FIR filter results in higher quantization noise with increased total capacitance when the C_{unit} is fixed. But it can support higher data rate in practice. Considering the required frequency deviation for the 1-Mb/s data rate, the quantization noise, and the EVM performance, we select the 16-tap FIR filter for the 1-Mb/s GMSK modulator.

To verify the linearity improvement from the proposed architecture, three different DCO modulations are compared: the conventional DCO modulation, the 1-bit modulation without the FIR filter, and the 1-bit modulation with the 16-tap FIR. The DCO with the 16-bit thermometer-coded control is modeled by a nonideal

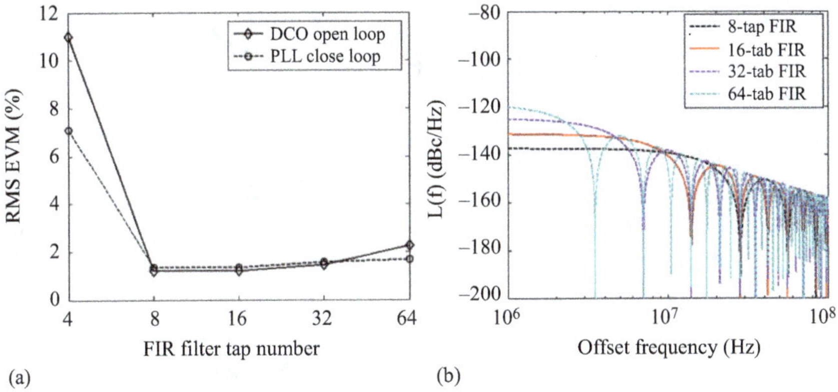

Figure 15.6 Error–vector magnitude (EVM) and spectrum comparison with different finite-impulse response (FIR) filter taps. (a) Root-mean-square (RMS) EVM with 4–64 taps FIR filtering. (b) Quantization noise with different tap FIR filtering

DAC and an ideal VCO for both conventional and 1-bit DCO modulation. The integral nonlinearity is controlled from 0% to 5% to simulate the DCO nonlinearity. The modulation signal is digitalized into 12-bit digital signal. As shown in Figure 15.7(a), the 8 bits in the least significant bit (LSB) part of the 12-bit control are fed to an 8-bit MASH1 in the conventional modulation. Then the 1-bit output of the MASH1 is added to the 4 bits in the MSB part. For the 1-bit DCO modulation, the 12-bit MASH1 is used to generate the 1-bit output for the FIR or the DCO control. Figure 15.7(b) shows that the EVM performance with the 1-bit high-pass modulation is superior to that with the multibit high-pass modulation in the presence of DCO nonlinearity. If the FIR filtering method is used, the mismatch between the partitioned varactors slightly degrades the EVM performance, but the 1-bit high-pass modulation still exhibits much better performance than the multibit high-pass modulation in spite of the mismatch in the FIR filter. As shown in Figure 15.7, when there is no nonlinearity or <2% nonlinearity, the DCO with the FIR filter has slightly better EVM compared to that without the FIR with the 1-bit high-pass modulation case. In practice, the use of the FIR filter is important to meet the spectrum mask and to reduce DCO sensitivity.

15.2.3 Linear model of the hybrid loop two-point modulator

A linear model of the proposed two-point modulator is shown in Figure 15.8, where G_H is the gain parameter in the high-frequency modulation that sets a compensated gain before the DCO/VCO input, G_L is the gain parameter in the low-frequency modulation that sets a normalized frequency deviation for the divider modulation, τ_L and τ_H are the delay control for low- and high-frequency modulation paths, respectively, N is the division ratio, β is the integral gain coefficient, K_{VCO} is the gain of the analog-controlled oscillator, and K_{BBPD} is the gain of the BBPD. $H_a(s,z)$ and $H_d(z)$ are the transfer functions of the analog modulation path and the digital modulation path, respectively. The overall transfer function $A(s,z)$ of the semi-digital PLL is given by:

$$A(s,z) = \frac{H_a(s,z)\cdot K_{\mathrm{VCO}} + H_d(z)\cdot K_{\mathrm{DCO}}}{N}\frac{1}{s} \tag{15.1}$$

where $H_a(s,z)$ and $H_d(z)$ are:

$$H_a(s,z) = \frac{I}{2\pi}\cdot H_{\mathrm{LPF}}(s)\cdot\frac{(1+z^{-1}+\cdots+z^{-7})}{8} \tag{15.2}$$

$$H_d(z) = K_{\mathrm{BBPD}}\cdot\frac{\beta\cdot z^{-1}}{1-z^{-1}} \tag{15.3}$$

The hybrid FIR filter shown in (15.2) is clocked by the divider output with a corner frequency of 1.5 MHz, which is much higher than the bandwidth of the PLL. Therefore, (15.2) can be simplified to;

$$H_a(s) \approx \frac{I}{2\pi}\cdot H_{\mathrm{LPF}}(s) \tag{15.4}$$

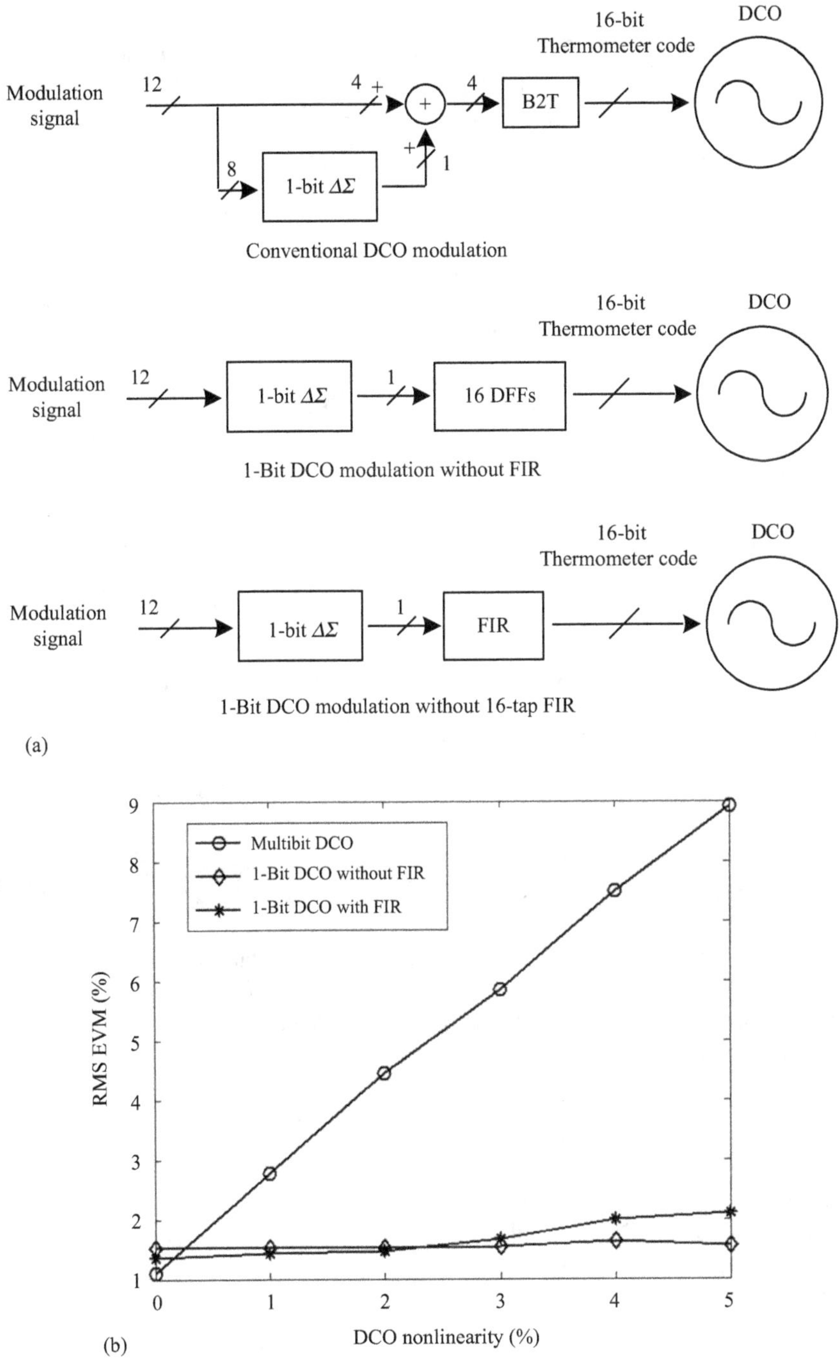

Figure 15.7 *(a) Schematics with multibit control nonlinear digitally controlled oscillator (DCO), 1-bit control DCO without finite-impulse response (FIR), and 1-bit control DCO with FIR. (b) Simulated error–vector magnitude (EVM) performance for each case*

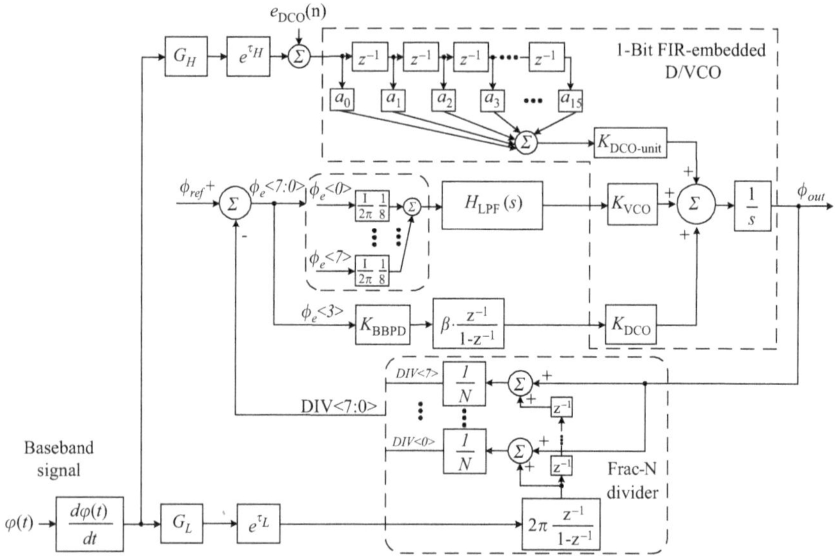

Figure 15.8 Linear model of the proposed two-point modulator

Then the transfer functions for the high-frequency modulation $H_H(s,z)$ and the low-frequency modulation $H_L(s,z)$ can be written as:

$$H_H(s,z) = \frac{G_H \cdot e^{s\tau_H}}{1 + A(s,z)} \cdot \left(a_0 + a_1 z^{-1} + \cdots + a_{15} z^{-15}\right) \cdot K_{\mathrm{DCO_unit}} \cdot \frac{1}{s} \cdot s \quad (15.5)$$

$$H_L(s,z) = \frac{G_L \cdot e^{s\tau_L} \cdot A(s,z)}{1 + A(s,z)} \frac{2\pi \cdot z^{-1}}{1 - z^{-1}} \cdot s \quad (15.6)$$

where a_0–a_{15} are the coefficients of the FIR filter. The 16-tap FIR filter in the DCO modulation path is clocked by one eighth of the DCO/VCO frequency, having a corner frequency of about 6 MHz which is much wider than the bandwidth of the baseband signal. Then (15.5) can be simplified to:

$$H_H(s,z) \approx \frac{G_H \cdot e^{s\tau_H}}{1 + A(s,z)} \cdot (a_0 + a_1 + \cdots + a_{15}) \cdot K_{\mathrm{DCO_unit}} \cdot \frac{1}{s} \cdot s \quad (15.7)$$

where $K_{\mathrm{DCO\_unit}}$ is the unit gain of the small varactor in the DCO modulation path. The 1-bit gain of the DCO modulation path is $(a_0 + a_1 + \cdots + a_{15})$. $K_{\mathrm{DCO\_unit}}$ is defined as K_{DCOM}. The 16-tap FIR filter is realized by setting $\{a_0, a_1, \ldots, a_{15}\}$ to be 1. By applying the s-domain transform with the reference clock period T as following:

$$\frac{z^{-1}}{1 - z^{-1}} \approx \frac{1}{j2\pi fT} = \frac{1}{sT} \quad \text{for } f << \frac{1}{T} \quad (15.8)$$

Then we obtain:

$$H_H(s,z) + H_L(s,z) = \frac{G_H \cdot e^{s\tau_H} \cdot K_{\mathrm{DCOM}}}{1 + A(s,z)} + \frac{G_L \cdot e^{s\tau_L} \cdot A(s,z)}{1 + A(s,z)} \frac{2\pi}{T}. \tag{15.9}$$

If we assume $\tau = \tau_L$, $G_H = 1/K_{\mathrm{DCOM}}$, and $G_L = T/2\pi$ with the reference clock period T, then:

$$|H_H(s,z) + H_L(s,z)| = 1 \tag{15.10}$$

proving that the all-pass transfer function is also valid for the hybrid-loop two-point modulator.

15.2.4 Gain and delay mismatch analysis and calibration

In the two-point modulation, the delay and gain mismatches between the high-pass and the low-pass modulation path, and the gain nonlinearity of the DCO/VCO are critical to the modulator performance. Thanks to the 1-bit DCO modulation structure, only the 1-bit absolute gain needs to be calibrated, which can be done in a simpler way such as the just-in-time algorithm [17], the consecutive counting operation [18], or the least-mean-square-based gain calibration [19]. In this work, the manual calibration is done with the assistance of the field programmable gate array (FPGA). In the calibration mode, the FIR filter input is set to a constant value (1 or 0) to detect the 1-bit DCO gain. Then the output from the FPGA could adjust the baseband signal to match the gain between the two paths.

The delay mismatch effect is also analyzed in the hybrid-loop two-point modulator. Figure 15.9 shows the simulated EVM performance with the delay mismatch. The x-axis is the delay mismatch measured in the reference clock period. The y-axis is the relative RMS EVM value. In practice, it is easy to control the

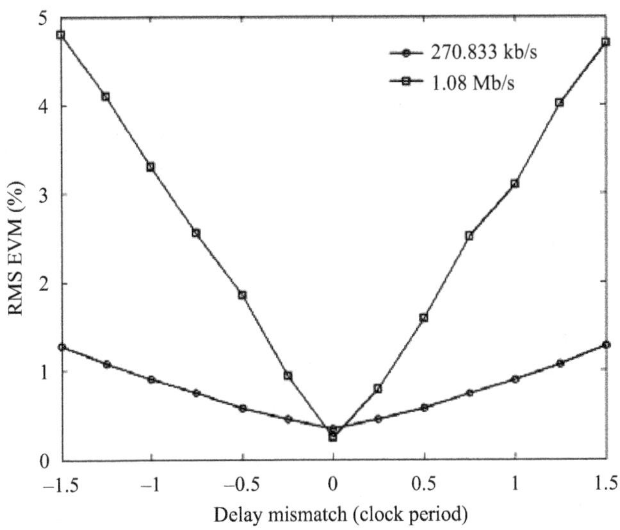

Figure 15.9 Effect of delay mismatch between two paths

delay mismatch with the time resolution of the DCO/VCO DSM clock period in the high-pass modulation path. In this design, the time resolution is 8 times the DCO/VCO period, which means the RMS EVM contribution from the delay mismatch can be well controlled to be less than 0.4% for the 1.08 Mb/s GMSK signal.

15.3 Circuit implementation

Block diagram of the proposed 1.08 Mb/s GMSK hybrid-loop two-point modulator [7] is shown in Figure 15.10. The BBPD with the FSM replacing the large integration capacitor in the loop filter works as frequency tracking [20]. In the low-pass modulation path, an 8-tap hybrid FIR filter is utilized to suppress the DSM quantization noise [14]. On the other hand, the hybrid FIR filtering significantly reduces the peak phase deviation caused by the ΔΣ modulator, which further improves the linearity of the charge pump and the varactor of the DCO/VCO. In the high-pass modulation path, the digital FIR filter is used after the first-order ΔΣ modulator to partition a single large capacitor into 16 small capacitors as discussed in Section 15.2. The digital baseband modulation is performed by an FPGA through GMSK/GFSK lookup table, in-phase/quadrature (IQ)-to-phase modulation (PM) signal translation, and differentiation of the PM signal.

Figure 15.11 shows the detailed block diagram of the digital integral path. It contains a BBPD as the phase detector and an FSM as the phase integrator followed by the ΔΣ modulator. The FSM works at the reference clock frequency of 26 MHz. It contains a 21-bit accumulator and a 1-bit selector that is controlled by the output of the BBPD. The FSM output increases or decreases by a constant value that is programmed to change the gain of the digital integral path during each clock period.

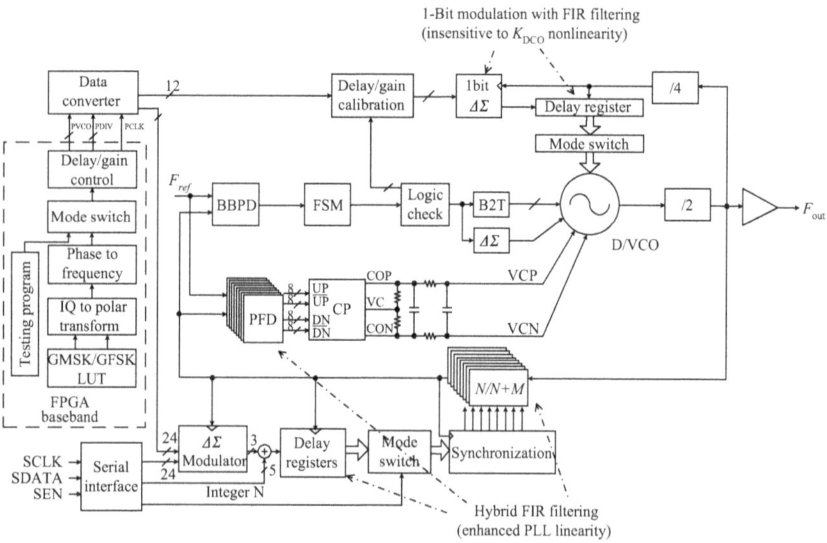

Figure 15.10 Block diagram of the proposed 1.08 Mb/s GMSK modulator

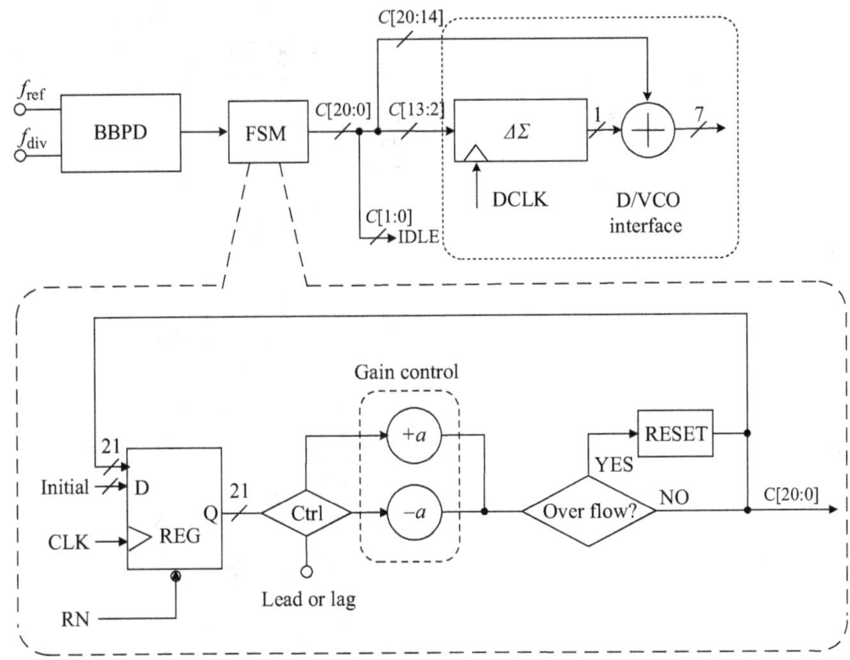

Figure 15.11 Block diagram of integral path

The bandwidth of the digital path is designed to be much smaller than that of the analog path, so that the digital integral path mostly performs frequency acquisition like a large signal bias in the overall system, while the analog proportional-gain path performs phase tracking as a small signal path in the linear system.

Figure 15.12 shows the details of the 1-bit high-pass modulation path, including a 12-bit first-order ΔΣ modulator, 16 DFFs, and 16 varactor pairs of the same size. The ΔΣ modulator and the DFFs are clocked at one eighth of the DCO/VCO output frequency. The quantization noise is generated by the DCO/VCO with the first-order DSM and the 16-tap FIR filter. The FIR filter is given by:

$$H(z) = \frac{\Delta f_1 \cdot z^{-1} + \Delta f_2 \cdot z^{-2} + \Delta f_3 \cdot z^{-3} + \cdots + \Delta f_{16} \cdot z^{-16}}{\Delta f} \tag{15.11}$$

where Δf_1 to Δf_{16} stand for the frequency deviation of each pair of the varactors, and Δf is the total frequency deviation. Δf_1 to Δf_{16} are assumed to be identical to the same unit varactor size. However, the mismatch in the layout can cause frequency nonlinearity. As shown in Figure 15.7, the nonlinearity of up to 5% is shown to be tolerable in the proposed architecture. With the clock frequency of 225 MHz, the 3-dB bandwidth of the FIR filter is about 6 MHz, which is large enough for the 1.08 Mb/s GMSK/GFSK modulation.

Figure 15.13 shows the DCO/VCO schematic. To achieve good power supply rejection, a current bias and a cross-coupled pair are designed with P-channel metal

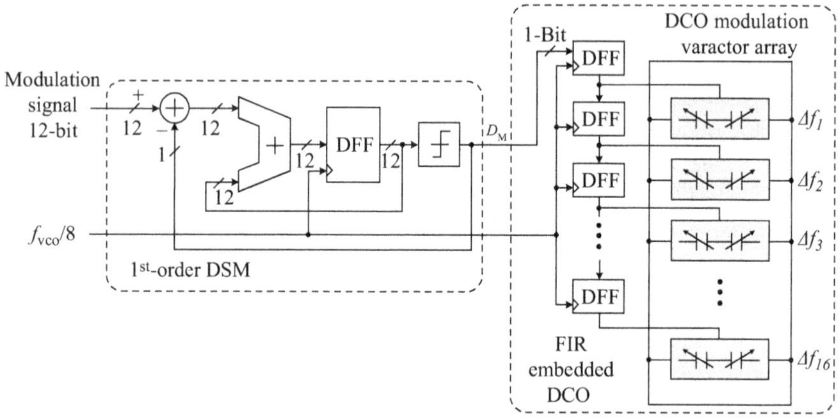

Figure 15.12 Detailed diagram of 1-bit high-pass modulation path. DFF: digital flip-flops

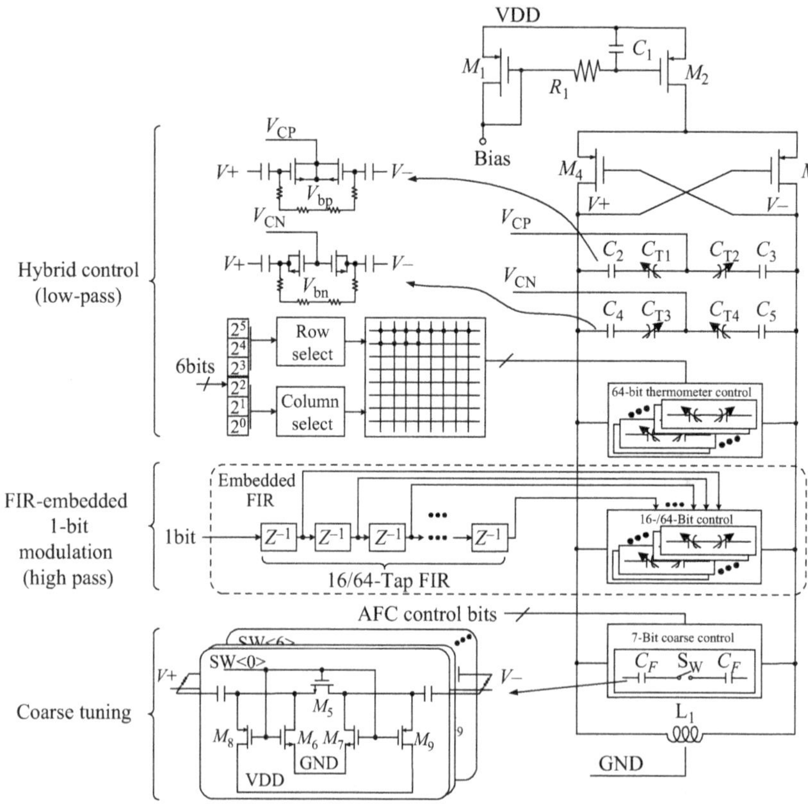

Figure 15.13 Schematic of digitally controlled oscillator (DCO)/voltage-controlled oscillator (VCO)

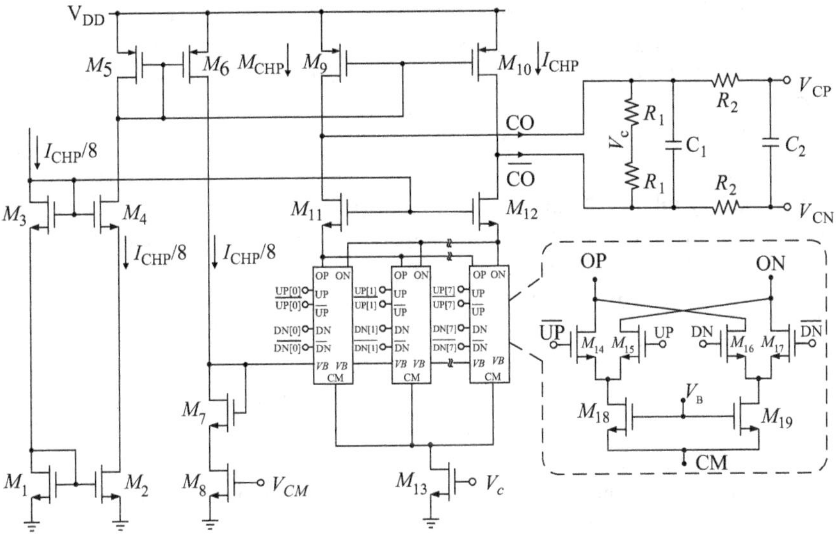

Figure 15.14 Schematic of differential charge pump with loop filter

oxide semiconductor (PMOS) transistors. The DCO/VCO input consists of a hybrid input for the PLL control and a 1-bit control for the high-pass data modulation, and seven binary-coded control bits for coarse tuning. Since the fully differential charge pump and loop filter are used, differential input varactors, C_{T1} to C_{T4}, are designed for the analog control path. The digital FIR filter has 16 taps and can be configured to 64 taps as well. However, a large tap number increases the total capacitance for the minimum unit capacitor C_{unit}. Take 1.08 Mb/s GMSK modulation for example, wherein the total capacitance of 19.2 fF (= 16 × 1.2 fF) is designed with the 16-tap FIR filter to have a frequency deviation of 1.1 MHz for the 1-bit control. The 16-tap digital FIR filter also reduces the high-frequency quantization noise by about 13 dB. The sensitivity of the analog control path is 22 MHz/V and the digital frequency step is 140 kHz with 64-bit thermometer code for the hybrid PLL control. With the differential spiral inductor of 2 nH, the tuning range of 1.66–2.03 GHz is achieved.

Figure 15.14 shows the schematic of the charge pump followed by a differential loop filter. To minimize the supply and substrate noise coupling and enhance the linearity of the charge pump output, the differential loop filter is used. The charge pump contains 8-tap switches to control the current outputs from 40 to 320 μA in differential loop with the hybrid FIR [14]. The 8-tap currents are added directly to the charge pump output. N-channel metal oxide semiconductor (NMOS) transistors M8 and M13 are in triode region for common-mode feedback to set the common-mode voltage V_{CM} of the differential charge pump.

15.4 Implement results

A prototype of 1.08 Mb/s GMSK/GFSK two-point modulator chip was fabricated in 65 nm CMOS process [7]. The active area occupied by the semi-digital PLL and the digital modulation circuits is 0.49 mm^2. An external analog loop filter is used to control the PLL bandwidth. By implementing the analog filter on the chip, the area will be only 0.027 mm^2 since the loop filter does not have the integration capacitor. The GMSK/GFSK modulated data are generated by the FPGA and fed to the low-pass and high-pass modulation paths of the ΔΣ PLL.

With a 26-MHz reference clock, the spur level of −79.9 dBc is measured with the PLL bandwidth of 80 kHz. The in-band phase noise varies from −92.8 to −96.7 dBc/Hz with the PLL bandwidth range of 75–85 kHz over the entire locking range, while the out-of-band phase noise varies from −115 to −118 dBc/Hz at 400 kHz offset frequency. When the phase noise is integrated from 10 kHz to 100 MHz, the integrated phase error is 0.45°rms and 0.65°rms at 832 and 1014 MHz, with the corresponding RMS jitter of 1.5 and 1.63 ps, respectively. Figure 15.15 shows the measured phase noise performance with an output frequency of 913.2 MHz. The measured in-band phase noise is −94.2 dBc/Hz at 40 kHz offset frequency.

The third pole of the LPF is removed to clearly demonstrate the effect of the hybrid FIR filtering. As shown in Figure 15.16, the out-of-band phase noise between 2 and 10 MHz offset is significantly reduced, proving that the hybrid FIR filter is functional and can improve the linearity of the charge pump with a normal loop configuration as discussed previously. The measured integrated noise with and without the FIR filter are 0.5238 rms and 0.5412 rms, with the corresponding RMS jitter of 1.59 and 1.65 ps, respectively. To see the DCO FIR filtering effect in the high-pass modulation, the modulated spectra with the 64-tap and the 16-tap FIR filter are compared in Figure 15.17. The spectrum with the 16-tap FIR filter achieves lower out-of-band noise due to the smaller total capacitance, showing that the use of 16 taps is optimal among 8, 16, 32, and 64 taps as expected from the simulation.

Figure 15.18 shows the measured output spectra, the constellations, the EVM values, and the eye diagrams with the 1.08 Mb/s GMSK modulations. The modulator consumes 6.9 mW from a 1 V supply where the DCO/VCO consumes 4.2 mW and the digital blocks consume 0.9 mW, achieving 1.79% RMS EVM.

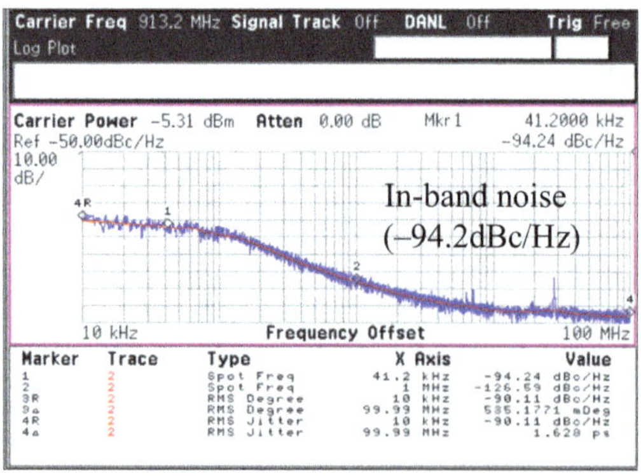

Figure 15.15 Measured phase noise performance at 913.2 MHz output

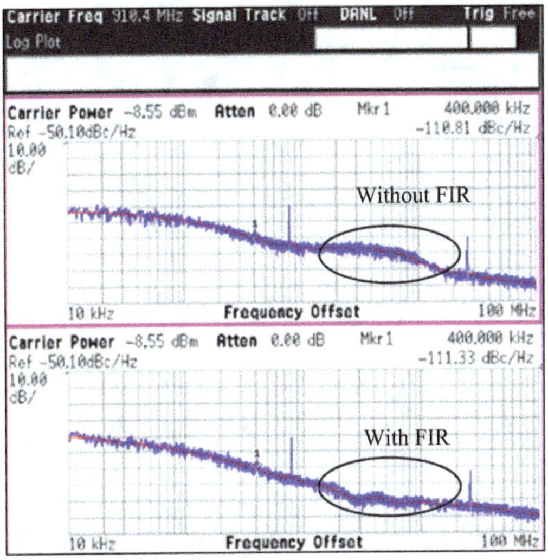

Figure 15.16 Measured hybrid finite-impulse response (FIR) filtering effect; third pole is removed for clear demonstration

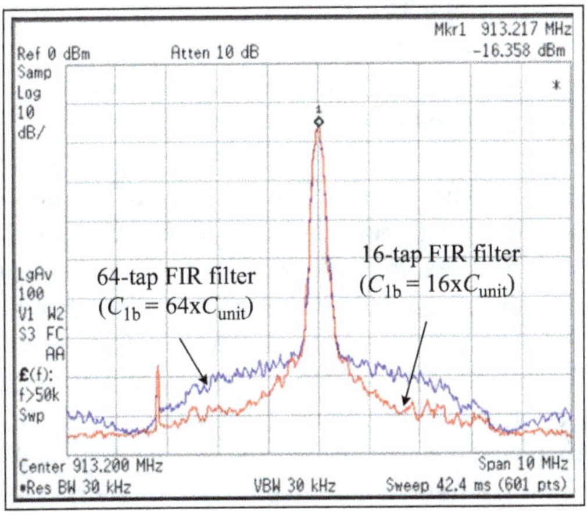

Figure 15.17 Measured finite-impulse response (FIR) filtering effect in high-pass modulation

The measurement results show that the proposed two-point modulator based on the FIR-embedded 1-bit high-pass modulation method is useful to simplify the TDC-less semi-digital PLL or ADPLL-based two-point modulator design.

Thanks to the proposed hybrid modulation with 1-bit high-pass modulation architecture, a digital-intensive modulator can be designed without using complex calibration algorithm or an advanced CMOS technology. Ref. [3] implements a 2-Mb/s two-point modulator with 0.18 μm CMOS with the proposed hybrid FIR

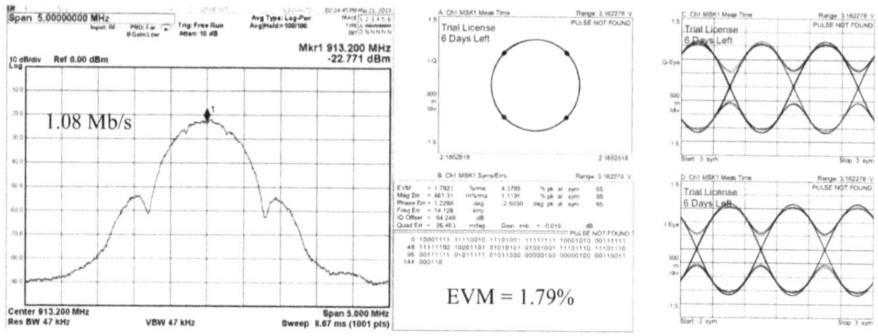

Figure 15.18 Measured spectrum, error–vector magnitude (EVM), and eye diagram with 1.08 Mb/s GMSK modulation

filtering. The digital control of the low-pass modulation path consists of a DAC followed by a low-pass filter and the analog varactor since it is difficult to design a small-size varactor array for fine frequency resolution in 0.18 μm CMOS. The analog control of the low-pass modulation path is the normal analog varactor. In the high-pass modulation path, the digital FIR filter clocked by one eighth of the DCO/VCO frequency is used after the first-order ΔΣ modulator to partition a single-large capacitor into small capacitors. The active areas occupied by the semi-digital PLL and the digital modulation circuit are 1.37 and 0.28 mm^2, respectively.

Figure 15.19 shows the measurement performance of the semi-digital PLL at 2.005 GHz output with the reference frequency of 50 MHz. The phase noise of −132 dBc/Hz at 3 MHz offset is measured and the integrated phase noise is about 1.29°rms, with the PLL bandwidth of about 100 kHz. The reference spur of −80 dBc is measured. The fractional spur at 1.3 MHz offset is −71.9 dBc, with the digital integral path enabled, exhibiting nearly 20 dB improvement in the fractional spur with the reduced analog control voltage range or enhanced charge pump linearity. Figure 15.19 also shows the measured 2 Mb/s GFSK signals with the 32-tap and the 64-tap FIR-embedded 1-bit high-pass modulation modes. Another high-pass multibit modulation by combining the 1-bit modulator is also compared. Since a baseband modulation signal from the FPGA is oversampled with 6.25 MHz clock, some harmonic tones are observed in the spectrum. In the measurement, the spectrum with the 64-tap FIR filtering mode shows an optimum performance.

Figure 15.20 shows the EVM performance with the multi-bit and the 64-tap FIR-embedded 1-bit high-pass modulation modes. The RMS EVM with the 64-tap FIR-embedded 1-bit modulation is about 2.03%, while the RMS EVM with the other mode is about 3.36%, proving that the proposed FIR-embedded 1-bit modulation improves the modulation linearity. The 2 Mb/s modulator in 0.18 μm CMOS draws a current of 28 mA from a 1.8-V supply, where 8 and 3.6 mA are consumed by the digital phase modulation and the DCO/VCO, respectively. Compared to [7] in 65 nm CMOS, comparable modulation performance is achieved with 0.18 μm CMOS technology, thanks to the hybrid modulation with 1-bit high-pass modulation architecture and the use of the DAC-based DCO/VCO control.

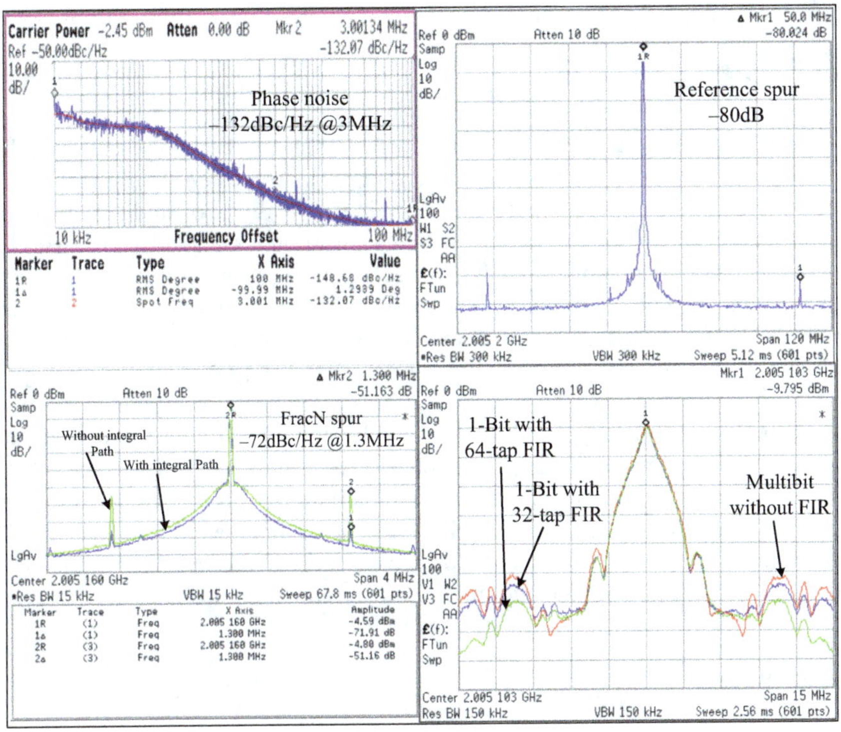

Figure 15.19 Measured finite-impulse response (FIR) filtering effect in high-pass modulation

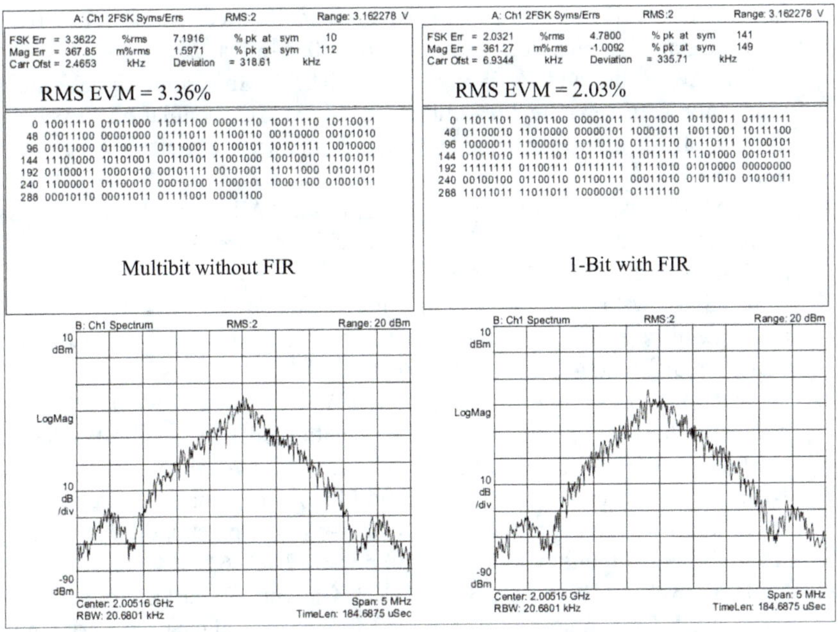

Figure 15.20 Measured finite-impulse response (FIR) filtering effect in high-pass modulation

15.5 Conclusion

In this chapter, we present the hybrid-loop two-point modulation architecture by utilizing a 1-bit high-pass modulation that is immune to the nonlinear effect of the DCO. It significantly simplifies the DCO nonlinearity calibration and only requires the absolute gain calibration. To reduce the quantization noise and DCO sensitivity due to the 1-bit modulation, an embedded FIR filtering along with the 1-bit high-pass modulation is also introduced. The digital FIR filter in the high-pass modulation path not only suppresses quantization noise but also reduces noise coupling with time-interleaved switching of partitioned capacitors. The proposed 1-bit high-pass modulation method can be also extended to the ADPLL-based two-point modulator. Moreover, the modulator uses the mixed-mode loop control and the hybrid FIR filtering method in the low-pass modulation path to achieve the enhanced linear loop dynamics. The proposed two-point modulation architecture is implemented in both 65 nm CMOS and 0.18 μm CMOS technology. The experimental results show that the proposed architecture is useful for moderate data rate modulation without requiring complex calibration algorithm or advanced CMOS technology.

References

[1] M. H. Perrot, T. T. Tewksbury, and C. G. Sodini, "A 27-mW CMOS fractional-N synthesizer using digital compensation for 2.5-Mb/s GFSK modulation," IEEE J. Solid- State Circuits, vol. 32, no. 12, pp. 2048–2060, 1997.

[2] M. Youssef, A. Zolfaghari, B. Mohammadi, H. Darabi, and A. A. Abidi, "A low-power GSM/EDGE/WCDMA polar transmitter in 65-nm CMOS," IEEE J. Solid-State Circuits, vol. 46, no. 12, pp. 3061–3074, 2011.

[3] N. Xu, W. Rhee, and Z. Wang, "A 2 GHz 2 Mb/s semi-digital+-point modulator with separate FIR-embedded 1-bit DCO modulation in 0.18μm CMOS," IEEE Microwave and Wireless Components Letters, vol. 25, no. 4, pp. 253–255, 2015.

[4] R. B. Staszewski, J. Wallberg, S. Rezeq, *et al.*, "All-digital PLL and transmitter for mobile phones," IEEE J. Solid-State Circuits, vol. 40, no. 12, pp. 2469–2482, 2005.

[5] R. B. Staszewski, K. Waheed, F. Dulger, and O. E. Eliezer, "Spur-free multirate all-digital PLL for mobile phones in 65 nm CMOS," IEEE J. Solid-State Circuits, vol. 46, no. 12, pp. 2904–2919, 2011.

[6] G. Marzin, S. Levantino, C. Samori, and A. L. Lacaita., "A 20Mb/s phase modulator based on a 3.6GHz digital PLL with -36dB EVM at 5mW power," IEEE J. Solid-State Circuits, vol. 47, no. 12, pp. 2974–2988, 2012.

[7] N. Xu, W. Rhee, and Z. Wang, "A hybrid loop two-point modulator without DCO nonlinearity calibration by utilizing 1-bit high-pass modulation," IEEE J. Solid-State Circuits, vol. 49, no. 10, pp. 2172–2186, 2014.

[8] M. H. Perrott, Y. Huang, R. Baird, B. W. Garlepp, D. Pastorello, and E. T. Kin, "A 2.5-Gb/s multi-rate 0.25-μm CMOS clock and data recovery circuit

utilizing a hybrid analog/digital loop filter and all-digital referenceless frequency acquisition," IEEE J. Solid-State Circuits, vol. 41, no. 12, pp. 2930–2944, 2006.

[9] Y. Sun, N. Xu, M. Wang, *et al.*, "A 1.75 mW 1.1 GHz semi-digital fractional-N PLL with TDC-less hybrid loop control," IEEE Microwave Wireless Comp. Lett., vol. 22, no. 12, pp. 654–656, 2012.

[10] P.-Y. Wang, J.-H. Zhan, H.-H. Chang, and H.-M. Sherman Chang, "A digital intensive fractional-N PLL and all-digital self-calibration schemes," IEEE J. Solid-State Circuits, vol. 44, pp. 2182–2192, 2009.

[11] W. Yin, R. Inti, A. Elshazly, M. Talegaonkar, B. Young, and P. K. Hanumolu, "A TDC-less 7mW 2.5Gb/s digital CDR with linear loop dynamics and offset-free data recovery," in IEEE Int. Solid-State Circuits Conf. (ISSCC), pp. 440–441, 2011.

[12] A. Sai, T. Yamaji, and T. Itakura, "A 570fsrms integrated-jitter ring-VCO-based 1.21GHz PLL with hybrid loop," in IEEE Int. Solid-State Circuits Conf. (ISSCC) Dig. Tech. Papers, pp. 98–99, 2011.

[13] Y. Sun, J. Qiao, X. Yu, W. Rhee, B.-H. Park, and Z. Wang , "A continuously tunable hybrid LC-VCO PLL with mixed-mode dual-path control and bi-level delta-sigma modulated coarse tuning," IEEE Trans. Circuits and Systems-I, vol. 58, no. 9, pp. 2149–2158, 2011.

[14] X. Yu, Y. Sun, L. Zhang, and Z. Wang, "An FIR-embedded noise filtering method for $\Delta\sum$ fractional-N PLL clock generators," IEEE J. Solid-State Circuits, vol. 44, no. 9, pp. 2426–2436, 2009.

[15] H. Kobayashi, S. Kousai, and Y. Yoshihara, "An all-digital 8-DPSK polar transmitter with second-order approximation scheme and phase rotation-constant digital PA for bluetooth EDR in 65 nm CMOS," in IEEE Int. Solid-State Circuits Conf. (ISSCC) Dig. Tech. Papers, pp. 174–176, 2011.

[16] R. He, C. Liu, X. Yu *et al.*, "A low-cost, leakage-insensitive semi-digital PLL with linear phase detection and FIR-embedded digital frequency acquisition," in Proc. IEEE Asian Solid-State Circuits Conf. (A-SSCC), pp. 197–200, 2010.

[17] R. Staszewski, D. Leipold, and P. T. Balsara, "Just-in-time gain estimation of an RF digitally controlled oscillator for digital direct frequency modulation," IEEE Trans. Circuits Syst. II, vol. 50, no. 11, pp. 887–892, 2003.

[18] H. K. Ahn, K. S. Lee, H. Yu, H.-S. Oh, D. Keum, and B. H. Park., "VCO gain calibration technique for GSM/EDGE polar modulated transmitter," in Proc. IEEE Radio Freq. Integr. Circuits Symp. (RFIC), pp. 383–386, 2008.

[19] R. B. Staszewski, J. Wallberg, C. M. Hung, G. Feygin, M. Entezari, and D. Leipold, "LMS-based calibration of an RF digitally controlled oscillator for mobile phones," IEEE Trans. Circuits Syst. II, vol. 53, no. 3, pp.225–229, 2006.

[20] M. H. Perrott, M. D. Trott, and C. G. Sodini, "A modeling approach for $\Delta\sum$ fractional-N frequency synthesizers allowing straightforward noise analysis," IEEE J. Solid-State Circuits, vol. 37, no. 8, pp. 1028–1038, 2002.

Part IV

Clock-and-data recovery and clocking

Chapter 16

An overview of CDR in ultra-high-speed wireline transceivers

Jay Im[1] *and Hongtao Zhang*[1]

A survey of recent developments of clock data recovery (CDR) solutions in ultra-high-speed wireline transceivers. Design trade-offs between performance, power, and silicon area. CDR design considerations for non-return-to-zero (NRZ) and pulse–amplitude–modulation (PAM-4) signal modulation schemes and silicon examples.

16.1 Introduction

Clock recovery, or clock data recovery (CDR), plays a crucial role in most wireline transceivers in production. By adaptively and dynamically placing the data sampling clock edge at the optimal phase within two adjacent symbol boundaries, usually at the center of one unit-symbol interval (UI), CDR reduces the timing margin loss attributed to the accumulated jitter on the received symbol edges that arrive at the data quantizer [or the track-and-hold (T&H) switch in analog-to-digital converter (ADC)-based receivers (RXs)] input and on the sampling clock edge itself. As the need for transceivers capable of high data throughput and low latency rises, keeping in pace with the ongoing rapid increase in the worldwide data traffic within and across data centers, mobile communication terminals, Internet of things (IoT) devices, and upcoming autonomous vehicles, the role CDR plays has become more critical to maximize the symbol rate that a wireline link can operate at, without increasing the bit error rate (BER) beyond the system-level tolerance.

In this chapter, after a brief high-level discussion of CDR fundamentals in high-speed digital input-output (IO)s using serial interface, or serializer/deserializer (SerDes), a survey of CDR implementations in ultra-high-speed wireline RXs from recently published circuit conferences is discussed in the context of their requirements and design trade-offs. The next two sections discuss select topics on CDR for binary modulation scheme (NRZ) and then on CDR for four-level pulse–amplitude–modulation (PAM-4) signaling that has become currently the most adopted modulation scheme at data rates of 56 Gb/s or higher. The PAM-4 CDR is

[1]Xilinx, Inc., 2100 Logic Drive, San Jose, CA, USA

relatively new to the industry and is dedicated a short theoretical discussion on its unique aspects in contrast to NRZ CDR, mostly derived from the presence of multiple transition modes between adjacent PAM-4 symbols. Subtleties of the baud-rate phase detection (BR-PD) scheme in PAM-4 CDR are noted, as the BR-PD architecture has become a de facto necessity for high-speed PAM-4 links primarily due to its reduced power consumption. Lastly, recently published 56 Gb/s and 112 Gb/s PAM-4 RXs are described, which adopted an ADC-based RX analog front-end (RX AFE) to digitize incoming symbols, followed by a digital block that performs intersymbol interference (ISI) equalization in the digital domain using a digital signal processing (DSP) unit. Effects of CDR loop latency on performance, especially in the context of the ADC-/DSP-based digital RX architecture, and published examples adopting a low-latency CDR with dedicated path are noted.

16.2 CDR in SerDes

Figure 16.1 illustrates a simplified block diagram representing one lane of transceiver including a transmitter (TX) serializer and an RX deserializer, converting parallel data bits to serial data stream and vice versa. In this architecture of SerDes, no explicit clock is being transmitted together with the data as in source-synchronous system. The clock phase information due to data transitions is embedded in the serial TX data signal itself.

A multitude of protocols are designed for different applications. SONET, USB, Ethernet, and peripheral component interconnect (PCI) Express are a few well-known serial interfaces. Each of these serial interfaces can be sorted into one of two groups, synchronous or asynchronous, depending on whether the reference clock used to generate TX serial waveforms is also used for the RX that receives the transmitted data stream. As illustrated in Figure 16.2(a), a synchronous serial

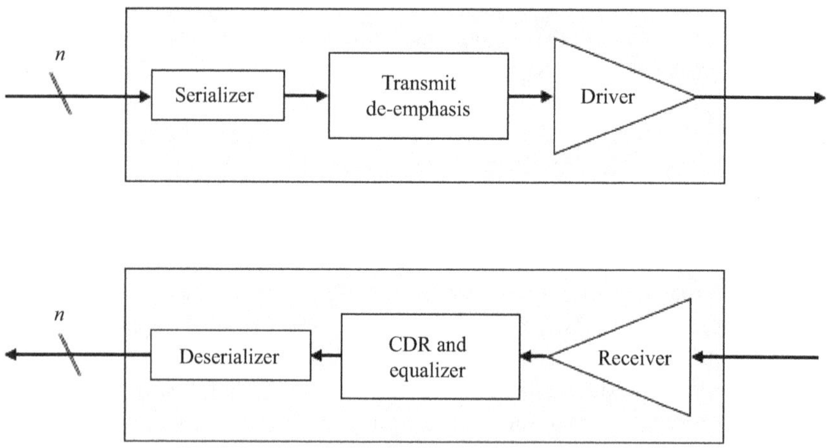

Figure 16.1 One pair of TX (top) and RX (bottom) in a SerDes lane

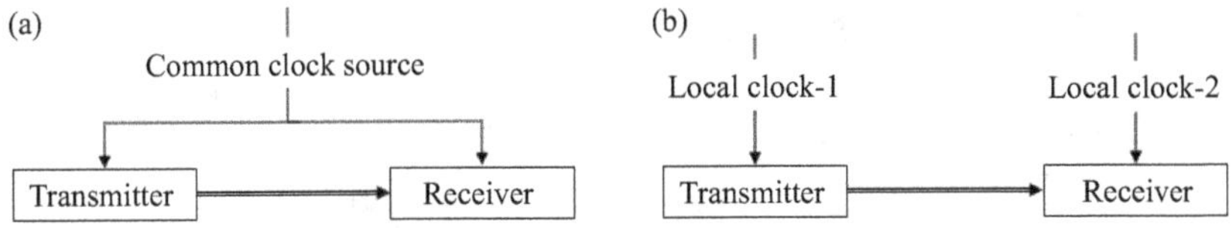

Figure 16.2 (a) Synchronous and (b) asynchronous (pleisiochronous) clocking architectures

interface involves both the TX and the RX, which share a common reference clock. That is, the normalized reference clock frequency is identical to both TX and RX. Over time, the synthesizer output clock frequency might drift due to environment changes such as temperature or voltage, but the frequency seen on the RX side follows that seen on the TX side. The CDR in the RX would only need to track the instantaneous phase variations between the TX and the RX due to noise and jitter.

Asynchronous serial interfaces in Figure 16.2(b), on the other hand, imply that the TX and the RX use different reference clocks that could differ by some small values usually characterized in units of parts per million (ppm, defined as the difference between the normalized frequencies of TX and RX divided by the normalized frequency). If the reference clocks for TX and RX originate from two independent onboard clock synthesizers, the system is considered asynchronous or plesiochronous. The RX clock frequency cannot be guaranteed to be the same as the TX clock frequency even for a short period of time because of the allowed error tolerance of reference clock sources (usually realized using crystal quartz oscillators). Temperature, voltage, or process variations between the two clock synthesizers result in an average ppm offset between the two clock frequencies. For protocols that require spread-spectrum clocking (SSC) modulation of reference clock frequencies, the RX CDR needs to not only track the instantaneous phase difference but also compensate for the time-varying frequency offset between TX and RX.

A phase-locked loop (PLL) in SerDes is used to synchronize the local clock frequency with the reference clock frequency and to multiply the reference clock frequency up for the actual data rate. In a synchronous system, the RX PLL and the TX PLL receive the same reference clock, whereas in an asynchronous system, the reference clock frequencies for RX and TX PLLs are only guaranteed to be within a certain tolerance specification. One approach for clock recovery, as illustrated in Figure 16.3(a), is to have the phase-frequency detector (PFD) in the loop to directly adjust the frequency and phase of the voltage-controlled oscillator (VCO) output. When no data are received to lock the VCO, the loop normally uses a reference clock to maintain a frequency and phase lock to a phase location in close proximity to that of expected data stream. Some CDR designs remove the need for such reference clocks altogether and operate in a "referenceless" mode. Once the SerDes RX starts to receive incoming data edges, the PLL clock is compared with the data edges in frequency and phase in a PFD to generate the error signal, which is then

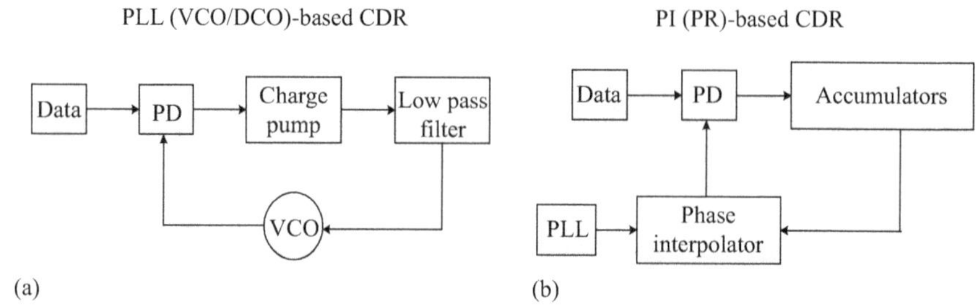

Figure 16.3 (a) Phase-locked loop (PLL) based vs. (b) phase interpolator (PI) (or phase rotator)-based clock data recovery (CDR)

used to adjust the VCO, or digitally controlled oscillator (DCO), in such a way that the incoming data are sampled around the symbol center.

Another commonly used approach, illustrated in Figure 16.3(b), is to use a phase interpolator (PI) to adjust the RX sampling clock phase from a local PLL to track and match the TX frequency and phase.

The phase detector (PD) implementations vary in terms of linear *vs.* sign-sign (or bang-bang) types for different architectures. The frequency of phase error detection is typically either (1) twice the data of the symbol rate (hence 2× over-sampled), quantizing the data and the associated edge (zero crossing or reference crossing) or (2) the same as the symbol rate (BR), typically sampling the residual error at the cursor height of the waveform at the same phase as the data quantizer.

The commonly used CDR feedback loops can be categorized into either first order or second order. A first-order CDR can track phase offsets within its loop bandwidth and gain ranges but would result in a phase error if a finite frequency offset exists in the system. The larger the frequency offset is, the larger the phase error becomes. A second-order CDR, on the other hand, can track the frequency offset without phase errors within its tracking range.

For an asynchronous system, the frequency offset ppm must be well controlled if a first-order CDR is desired for more efficient design. If the system margin is small, the ppm offset must be guaranteed to be small enough for a given link speed, ISI residual budget, and system jitter budget; otherwise, a second-order CDR is needed to reduce the margin loss.

PI-based second-order CDR architecture is typically modeled as shown in Figure 16.4. In the model diagram, G_p represents the first-order or proportional path gain and G_f represents the second-order or integral path gain. If G_f is 0, this architecture reduces to a first-order CDR. The system behavior may be analyzed in a straightforward manner if the PD behavior can be approximated to be linear. If the CDR loop is in a locked state, the PD output, which represents the phase error between sampling clock and incoming RX data, ideally averages to 0.

With a small frequency ppm offset, Δf, the PD output in the steady state does not average to 0, if the CDR loop is only of the first order. The resulting phase error, P_e, is proportional to the ppm offset magnitude and is inversely proportional

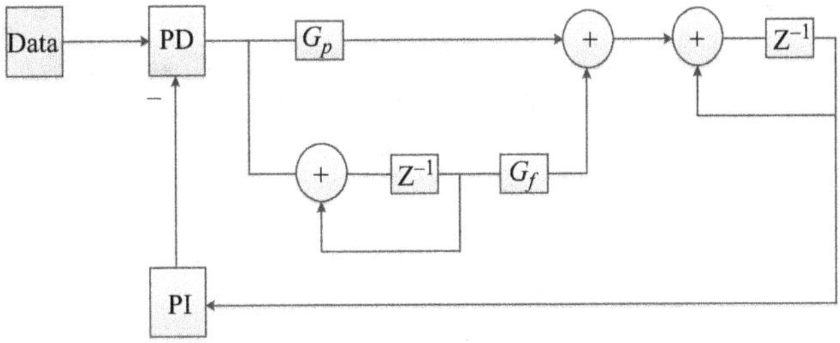

Figure 16.4 System model diagram of second-order clock data recovery (CDR) loop

to G_p and G_{pi} (the PI gain) as shown in (16.1). On the other hand, the phase error of a second-order CDR can be reduced to a smaller value than the PI resolution, as shown in (16.2):

$$P_e^{\text{1st-order}} \sim \frac{\Delta f}{G_{pi} \times G_p} \tag{16.1}$$

$$P_e^{\text{2nd-order}} \sim 0 \tag{16.2}$$

In practice, PD cannot track an arbitrarily large frequency offset due to its phase wrapping-around nature, which results in an incorrect PD output. CDR loop will not converge properly even with a second-order path if given an incorrect PD output. In other words, if frequency offset is out of the CDR capture range, CDR can no longer lock onto the incoming data and therefore results in a closed eye. This behavior can be seen from the time-domain simulation results plotted in Figure 16.5 based on the simple architecture model depicted in Figure 16.4.

Another insight that can be derived from the simple model is about the jitter tolerance (JTOL). It is well known that CDR can track low-frequency jitter up to its loop bandwidth. A second-order CDR loop can be especially useful with a large amplitude but very-low-frequency jitter because of the second pole that the frequency path introduces in its jitter transfer function (Figure 16.5). In addition, a first-order CDR JTOL can be easily compromised in an asynchronous system with finite frequency offset, while a second-order CDR JTOL is barely affected, as can be observed in either a time-domain simulation or a small signal analysis in Figure 16.6(a) and (b), respectively.

Although a simple architecture model may be used to predict, to a large degree, the most fundamental behaviors of a CDR in the presence of a frequency offset and a sinusoidal jitter, a time-domain simulation using an elaborate system model with cycle-matching accuracy and analog delays is desired for high-confidence results. Such models are extensively used to account for convergence complexities and interactions among various adaptation loops and CDR in SerDes systems. This is usually addressed by input/output buffer information specification algorithmic modeling interface (IBIS-AMI) simulation framework as described in [1].

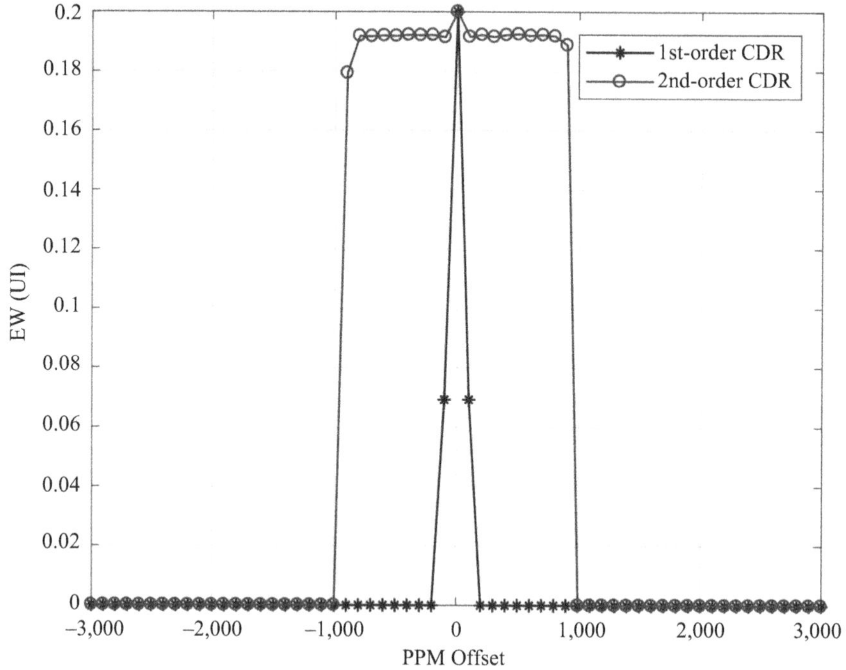

Figure 16.5 Simulated eye width vs. parts per million (ppm) offset for a first-order and a second-order clock data recovery (CDR) loops

16.3 Requirements and approaches for CDR in ultra-high-speed wireline transceivers

The main application of ultra-high-speed transceivers is in relatively short reach channels (e.g., see optical inter-networking forum common electrical I/O (OIF-CEI) specifications for XSR, VSR, MR, and LR [2] or IEEE 802.3 standards pertaining to electrical subsystems for wired Ethernet [3]), where a multitude of contiguous transceiver lanes are embedded on a silicon die to realize a very-high-aggregate bit transfer throughput per die edge, per chip, and/or per ball grid array (BGA) bump, often reaching over 4 TB/s per chip from hundreds of TX/RX lanes [4]. Since the electric energy cost and the system cooling cost contribute significantly to the operation overhead of modern data centers, power efficiency is one of the highest priority considerations in designing such highly integrated and high data rate transceivers. In extreme cases, the area heat dissipation density itself can become the limiting factor in determining the maximum number of transceivers per area that can be integrated in a very large scale integration (VLSI) package, for a given system-level cooling capability.

Meanwhile the continuously increasing wafer production cost in advanced process design kit (PDK)s poses a strong financial incentive toward using a smaller silicon area

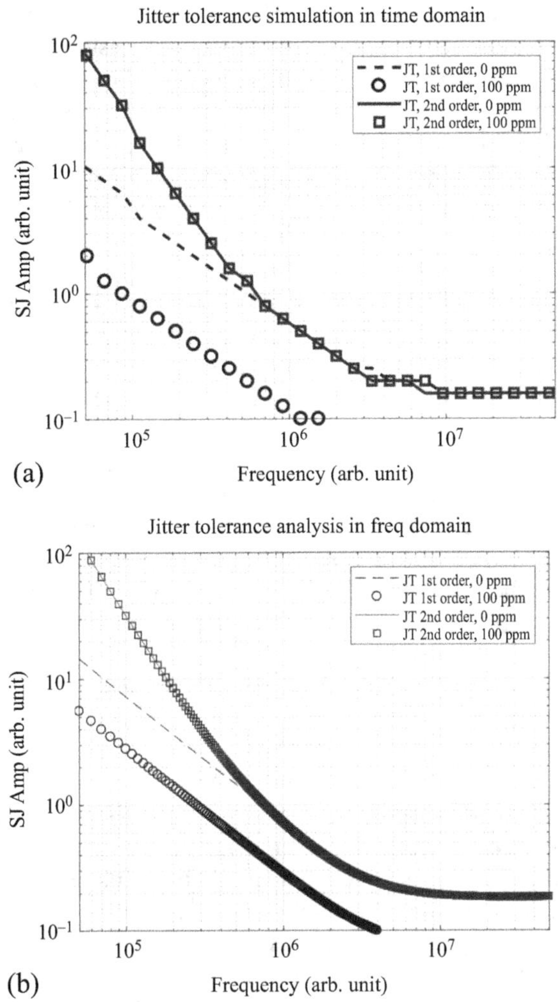

Figure 16.6 Jitter tolerance (JTOL) analysis in (a) time domain and (b) frequency domain

for circuit designs, given a choice of multiple implementation possibilities. With the advancement of the device scaling that enables more digital functions to be realized in a smaller silicon area than were feasible in older PDKs, moving more circuit functions from analog to digital domain becomes a natural choice for highly integrated sophisticated designs. A more compact circuit layout may also enable more power efficient designs due to a reduced on-die wire parasitic capacitance needed to interconnect circuit components. Overall, the power and area efficiency have become the primary focus of the contemporary design optimization to realize transceiver front-end and back-end circuits and to choose and implement CDR architectures.

High-speed clock generation and distribution often consume a major portion of the power budget in advanced high-speed transceivers, easily taking up to or even beyond 50% of the total power consumption (see references in Table 16.1). Therefore, multiphase clocks required for performing an oversampling phase error detection (e.g., Alexander type PD) are inherently less power efficient. Moreover, as the target line rate increases and the required ISI correction becomes more sophisticated and power demanding, providing an additional data path hardware just for symbol edge samples may incur a prohibitively high power consumption.

On the other hand, BR-CDR architecture, as usually implemented using Mueller–Muller (MM) PD scheme [5], in principle adds no additional hardware to realize it in most SerDes RXs, because the error slicer output needed for its implementation is also needed for ISI cancellation [feed-forward equalizer (FFE), decision feedback equalizer (DFE), and continuous-time linear equalizer (CTLE) peaking coefficients and pole/zero locations] and automatic gain control and off-set correction (AGC/OS) loops in the first place. This enables a very power-efficient CDR solution.

Performance of BR-CDR, as discussed in more detail in PAM-4 CDR subchapter, heavily relies on fine details of the equalized single-bit response (SBR) pulse shape, such as the peak position with respect to pre- and post-cursors and the slope of SBR near the sampling phase. For example, a low loss channel that results in a near flat waveform around the lock point can suffer from a phase wander due to the extremely low PD gain.

Despite these challenges, MM-PD-based BR-CDR with single-phase sampling clock per UI (for data-slicer clock or rank-1 T&H clock in analog- or ADC-based RX AFE, respectively) is the favored choice for power-competitive ultra-high-speed RXs. At lower speeds, especially NRZ RXs with an AFE, where multiphase clock generation costs are more acceptable, the oversampling CDR architecture is preferred over BR type for better jitter performance (see Table 16.1).

From a system-performance perspective, the jitter filtering by CDR, as commonly specified quantitatively by JTOL masks, is one of the two key variable components (the other being PLL to filter out the reference clock jitter) that determine the end-to-end link budget for a given reference clock noise profile and TX and RX nonidealities, such as deterministic jitter (DJ), periodic jitter (PJ), duty cycle distortion (DCD), residual ISI, ADC quantization noises, device noise, and cross-talk noise.

The loop latency impact on CDR performance is approximately given by a phase margin degradation of (360 degree)*(f_bw)*(t_delay), where f_bw represents the loop bandwidth and t_delay the loop latency. For example, for a 56-Gb/s PAM-4 RX with one unit symbol interval of 1 UI = 1/(28GS/s) = 35.7 ps, an extra loop latency of 3*64 UI from three pipelining stages in DSP running at 875 MHz and 1 ns of total clock tree delay, in a 4-MHz bandwidth CDR loop would result in a phase-margin degradation of about 11 degree (= (360 degree)*(3*64*35.7 ps + 1 ns)*(4 MHz)), a nontrivial burden in keeping jitter peaking from causing JTOL margin loss at the sinusoidal jitter (SJ) frequency around the loop bandwidth.

Major contributors to the CDR loop latency include (1) deserialization [including time-interleaving (TI) stages in a TI-ADC-based AFE] and data-path

Table 16.1 *A survey of CDR architectures in ultra-high-speed receivers published in recent conferences*

Modulation	Data rate (Gb/s)	PDK	CDR architecture				RX-AFE		Reference
			Baud-rate		Over-sampling		ADC	Analog	
			VCO	PI	VCO	PI			
PAM-4	112	10 nm	v				v		Intel, VLSI 2019
	112	16 nm		v			v		Xilinx, VLSI 2018
	100	14 nm		v				v	IBM, ISSCC 2019
	100	16 nm	v				v		Inphi, ISSCC 2019
	64	28 nm FDSOI				v		v	STMicro, ISSCC 2018
	60	7 nm		v			v		Huawei, ISSCC 2019
	56	7 nm		v			v		TSMC, VLSI 2019
	56	7 nm				v		v	TSMC, VLSI 2019
	56	7 nm	v				v		eSilicon, ISSCC 2019
	56	7 nm		v			v		MediaTek, ISSCC 2019
	56	7 nm				v		v	Huawei, ISSCC 2019
	56	16 nm		v			v		Xilinx, ISSCC 2018
	56	65 nm			v			v	TAMU, CICC 2018
	56	16 nm		v				v	Xilinx, ISSCC 2017
	56	40 nm				v		v	NTU, ISSCC 2017
	56	16 nm		v			v		Xilinx, VLSI 2016
	56	40 nm			v			v	NTU, VLSI 2015
	52	40 nm			v			v	HKUST, VLSI 2019
	50	28 nm	v				v		Inphi, ISSCC 2016
	32	65 nm		v			v		TAMU, CICC 2018
	32	65 nm				v		v	Peking U., ISSCC 2018
	32	28 nm		v			v		Broadcom, ISSCC 2016
	28	65 nm			v		v		U. Alberta, CICC 2017

(Continues)

Table 16.1 (Continued)

Modulation	Data rate (Gb/s)	PDK	CDR architecture				RX-AFE		Reference
			Baud-rate		Over-sampling		ADC	Analog	
			VCO	PI	VCO	PI			
NRZ	60	14 nm				v		v	IBM, VLSI 2017
	60	65 nm	v					v	UCB, ISSCC 2017
	56	14 nm				v		v	IBM, ISSCC 2018
	56	16 nm		v				v	Xilinx, ISSCC 2018
	56	28 nm	v					v	Fujitsu, ISSCC 2016
	56	20 nm		v				v	Fujitsu, VLSI 2014
	40	65 nm			v			v	Tsinghua U., CICC 2015
	40	28 nm				v		v	Rambus, VLSI 2014
	36	28 nm	v					v	U. Toronto, ISSCC 2019
	36	20 nm				v		v	Fujitsu, VLSI 2014
	32.75	16 nm				v		v	Xilinx, VLSI 2016
	32.75	20 nm				v		v	Xilinx, ISSCC 2015
	32	28 nm	v					v	U. Toronto, ISSCC 2017
	32	22 nm		v				v	Intel, ISSCC 2014
	30	28 nm			v			v	U. Toronto, CICC 2019
	28.3	28 nm				v		v	Socionext, VLSI 2016
	28.05	14 nm			v			v	Rambus, ISSCC 2018
	28.05	32 nm SOI				v		v	IBM, VLSI 2017
	28	7 nm				v		v	Xilinx, VLSI 2018
	28	28 nm				v		v	U. Toronto, ISSCC 2017
	28	28 nm				v		v	Broadcom, ISSCC 2015
	28	28 nm				v		v	Hitachi, ISSCC 2015
	28	28 nm				v		v	LSI, ISSCC 2014
	28	40 nm				v		v	Broadcom, ISSCC 2014
	25	28 nm				v		v	Hitachi, ISSCC 2016
	25	32 nm SOI		v			v		IBM, ISSCC 2016
	25	32 nm SOI				v		v	IBM, ISSCC 2015

pipelining in parallel data domain to satisfy timing closure, (2) accumulated clock tree delay, and (3) the time delay from CDR PD output change to the phase change in the sampling high-speed clock at the sampler. In PI-based CDR, (3) includes PI latency given by the time delay from the PI code update to high-speed clock phase change. For a digital implementation of CDR loop, the total cycle-based delays accrued from pipelining stages in DSP and CDR (for both PD and accumulators) clocked at the parallel data bus speed is the dominant latency source.

A clever use of partial-response-based RX architecture has been demonstrated to reduce the required slicer counts. In this scheme, the signal at slicer input needs to be shaped to be $1 + 0.5$ D by TX-FFE and CTLE, so that CDR can share the speculative DFE slicers' outputs, considerably reducing the number of high-speed slicers to achieve a very power-efficient solution. The technique has been shown to be applied to both NRZ [6] and PAM-4 [7] analog RXs. Applications of the scheme, however, may be limited to special cases of channels because RX CTLE (and possibly TX-FFE) needs to be tuned to satisfy the partial-response constraint at the slicer input, i.e., $h1/h0 = 0.5$, rather than to minimize the overall end-to-end ISI for optimal performance.

Table 16.1 summarizes the recently published NRZ and PAM-4 complementary metal-oxide-semiconductor (CMOS) RX silicon results from major circuit conferences (CICC, ISSCC, and VLSI) from 2014 to 2019. The tabulation includes only those RXs with a closed-loop CDR, either integrated on chip or implemented off chip using a field-programmable gate array (FPGA) or a micro-controller, with a max data rate of 25 Gb/s or higher. Data points in the table are sorted by the modulation schemes (PAM-4 and NRZ) and then by the highest reported data rates.

Most NRZ RXs adopt analog-based RX AFE architectures, along with a PI-based oversampled CDR, especially below 32 Gb/s, where sharing of the power- and area-costly PLL among multiple channels is beneficial. An ADC-based solution for NRZ at this speed would be less power/area efficient. The popularity of oversampled CDR in 25–28 Gb/s range is likely motivated by the more stringent JTOL mask and loop bandwidth (BW) that the supported protocols require. For scalable data rate transceivers, switching between BR and oversampled CDR is an option for rate-based power-efficiency optimization [8].

For PAM-4 RXs, the trend of adopting ADC-based AFE architectures in conjunction with BR phase error sampling and PI-based phase correction is rather clear, consistent with the mandate of using more digital-heavy and power-efficient ISI cancellation and CDR solutions. All but one ADC-based PAM-4 RXs use BR CDR.

All but one report in the table adopt sign-sign PDs (BB-PD).

16.4 CDR for NRZ modulation scheme

For 28 Gb/s NRZ RXs designed to address high-volume market demand for various widely adopted protocols such as Ethernet (50/100/200/400GE, etc. [3]) and PCI Express [9], the favored CDR circuit choice is 2× oversampling by a BB-PD driving either a VCO or PIs. For multilane applications, the PI-CDR is usually preferred over analog CDR in consideration of (1) the shared PLL allowing less amortized silicon area per channel, (2) more robust loop performance over PVT variations, and (3) easier design migration between different PDKs. For multi-protocol transceivers designed to support multiple line rates and protocol-specific compliance items, the main circuit challenge is to develop power- and area-optimized solutions that satisfy the union of all supported protocol requirements. For example, PCI Express electrical compliance typically requires a CDR JTOL specification of 10 MHz or higher nominal loop BW while subjected to a frequency offset of up to 5,000 ppm time-varying (30–33 kHz) component for SSC in addition to a maximum of 600 ppm relative offset from the allowed tolerance of independent reference clock sources for RX and TX. In order to clear such stringent JTOL masks using a digital CDR loop, PD and accumulators need to run at as high a speed as the digital timing closure permits. Running CDR loop at high speed helps to (1) reduce the loop latency for less jitter peaking and (2) reduce the amount of PI code changes per CDR cycle (thereby suppressing PJ induced by abrupt phase change) that is necessary to null out the frequency offset between the input data and the RX PLL.

In optical link applications, upcoming next-generation passive optical networks (PONs), such as 100G-EPON standard (IEEE 802.3ca), drive the need for 25 Gb/s PHY IP that can receive bursts of packets sent from optical network unit RXs to deserialize and relay them to a packet processing unit in the same application-specific integrated circuit (ASIC) or FPGA that the SERDES resides. In such a use model, fast CDR relocking after a long idle state between packet bursts is of critical importance because CDR receives no data signal edges for its PD to infer and correct phase error accumulated during the idle state. Without a fast locking mechanism, the clock phase difference between TX and RX may drift far enough from the locking point, so that the link error may persist beyond PON training symbols inserted in the header of the incoming packet burst.

Reference [10] is an example of multilane 28 Gb/s NRZ transceivers implemented in 7 nm FinFET CMOS PDK optimized for power and area, taking advantage of the technology scaling, especially for digital circuits. This category of multiprotocol flexible-reach wireline transceivers has become a crucial IP component in FPGA and system on a chip (SoC) devices to address demands from data centers and telecommunication infrastructures. A compact transformer-based wide-range inductor-capacitor phase-locked loop (LC-PLL) (8–16 GHz) sends a single-phase clock to TX/RX lanes. Each lane generates eight-phase clocks using an injection-locked ring oscillator for four slices of CMOS PI [11], namely, RX-Data, RX-Edge, RX-Eyescan, and TX.

Figure 16.7 illustrates the RX block diagram. The CTLE is implemented by a single-stage current-mode logic (CML) amplifier with programmable capacitor arrays at both the source degeneration node and the output node, realizing a max peaking of 16 dB with a fixed DC gain. The long-tail cancellation [12] is performed on the same stage by providing a mid-frequency [MF ($\sim$300 MHz)] and a low-frequency ($\sim$35 MHz) negative feedback paths. In this design, a large resistor value is used to implement the resistor-capacitor (RC) filters to significantly reduce capacitor area. Two cascaded CML stages with source degeneration are used to realize a programmable gain amplifier (PGA) following the CTLE, which

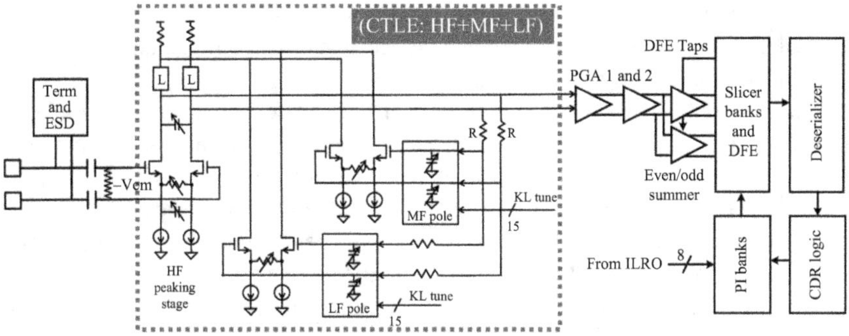

Figure 16.7 Block diagram of a 28-Gb/s NRZ multiprotocol receiver

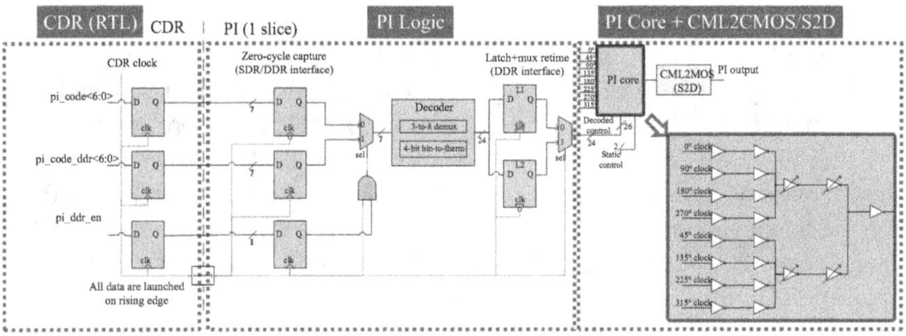

Figure 16.8 Clock data recovery (CDR) to phase interpolator (PI) logic interface and CMOS PI block diagram

subsequently drives two half-rate DFE summing buffers. The output common modes of the three CML stages (CTLE, PGA1, and PGA2) are independently controlled through a shared digital feedback control loop. The half-rate DFE stage provides a 15-tap postcursor ISI cancellation with speculative *h*1.

A 2× oversampling second-order CDR loop with a BBPD and digital CMOS PIs is used in this design. As illustrated in Figure 16.8, in the single data rate PI mode, the CDR logic updates the PI phase code upon every rising edge of the CDR clock. In the double data rate (DDR) PI mode, the CDR logic transfers one half of the intended PI code steps at the falling edge of the CDR clock and the other half of the code steps at the rising edge of the CDR clock. The DDR PI mode reduces the PJ magnitude of the output clock caused by an abrupt PI output clock phase change upon PI code changes.

To address the challenging BW requirement, latency reduction in all sub-blocks of the CDR loop is given a high priority. The CDR logic (PD, accumulators, adders, multipliers, and dividers) is implemented using an automated place-and-route (PnR) flow. CDR outputs interface with the PI logic via a zero-cycle data transfer mode (Figure 16.8). CDR PnR block is clocked at the highest possible speed and the number of inserted pipe-lining stages and the clock tree delay are minimized. A careful circuit and layout optimization is performed in the custom-designed PI logic and core to achieve fast response from input code change to output clock phase change.

The 28-Gb/s PRBS-31 performance over a 30-dB channel with fully adapted CTLE and DFE is summarized in Figure 16.9, showing (a) an eye diagram with own TX driving RX over the channel, (b) the corresponding horizontal BER bathtub curve, and (c) JTOL measured using bit error rate tester (BERT). The channel between TX and RX consists of a 27.2-dB device under test (DUT) trace, a pair of DC-blocking caps, three pairs of 26” radio frequency (RF) cables, and 3 dB loss from directional coupler (for cross-talk noise injection tests). A horizontal margin of >0.4 UI is achieved at <1E−10 BER.

The JTOL test is performed with a fixed LC-PLL reference clock and by adding a SJ to PRBS-31 RX input driven by BERT via a 27-dB back-plane channel. The measured JTOL satisfies the OIF-CEI mask depicted by the solid line. BERT is set to add 1,000 ppm SSC (33 kHz) to the sweeping SJ and BER is set at 1E−10. Two different CDR settings (#1 and #2 in Figure 16.9(d)) are tested to demonstrate the bandwidth and gain trade-off. In setting #1, BW is set high (14 MHz), so that the link has less margin at the injected SJ higher than 10 MHz but more margin below 10 MHz than #2. In contrast, setting #2 with BW = 6.8 MHz shows 0.4 UI more margin above 7 MHz but closer to the JTOL mask below 6 MHz. Note that the BERT limits the maximum amplitude of sweeping SJ below 5 MHz for high-BW setting (#1) and below 200 kHz for low-BW setting (#2). Due to this equipment limitation, the plotted JTOL data points below CDR BW show only a narrow SJ frequency range with larger than 20 dB/dec roll-off, as expected from a second-order loop.

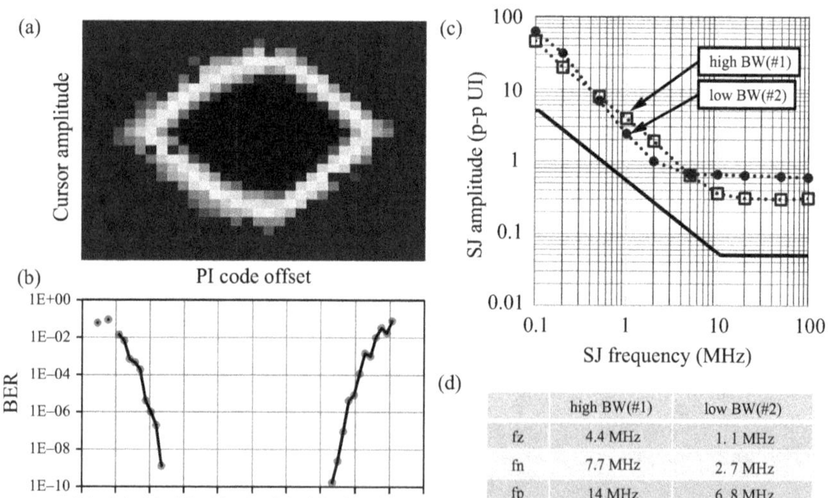

(d)

	high BW(#1)	low BW(#2)
fz	4.4 MHz	1. 1 MHz
fn	7.7 MHz	2. 7 MHz
fp	14 MHz	6. 8 MHz
ζ	0.89	1.25

Figure 16.9 28.21 Gb/s 30 dB channel OIF-stressed (a) RX eye, (b) bit error rate (BER) bathtub, and (c) jitter tolerance (JTOL) (at 1E−10 BER)

Figure 16.10 plots measured CDR BB-PD output *vs.* data sampling phase varied by sweeping PI code. The test is done with the RX receiving 25 Gb/s PRBS-31 data stream driven by BERT. The normalized up-vote counts (VoteUP) and down-vote counts (VoteDN) are plotted by accumulating several thousands of PD output samples per each PI code. From a bit PI modulating half-rate DFE clock, 1 UI corresponds to 64 steps of PI code. CDR achieves a stable lock condition at around code 32 in this setup with approximately symmetric left and right timing margins. Meta-stable lock conditions exist around PI code of 0 and 64, where VoteUP and VoteDN balance out evenly. If CDR starts from these states, it may take a long time for some disturbance to assist CDR to move away from the meta-stable region and to eventually reach the stable lock position.

To prevent the meta-stable region from slowing down the CDR re-locking process in a burst-mode operation, the CDR logic includes an algorithm to detect meta-stable regions based on PD outputs. The detection logic monitors a simultaneous occurrence of nonzero VoteUP and nonzero VoteDN in a CDR cycle. The thresholds of nonzero vote counts and the number of consecutive CDR cycles for the detection logic to trigger can be programmable. If the detection logic triggers, CDR is deduced to be either around a meta-stable region (Figure 16.11(a)) or near a stable locking point (Figure 16.11(b)). Regardless, the CDR logic will execute a series of predetermined steps (Figure 16.12): (1) break the loop by disconnecting PD output from the accumulator and (2) then fast ramp the PI code (PI walk) in the direction determined by the sign of voteUP–voteDN, by a programmable number of steps. Subsequently, the CDR will close the loop back and naturally acquire a stable lock. This scheme enables adjusting data sampling phase quickly, if needed, upon the link wakeup from a long idle state to minimize the CDR lock time.

Notice that the detection scheme based on voteUP and voteDN cannot distinguish a stable lock condition from a meta-stable region. As a consequence, CDR re-locking from near a stable lock point may actually take longer if the meta-stable region detection scheme is turned on. As shown in Figure 16.10, however, the range of PI codes across which VoteUP and VoteDN are not unanimous is much

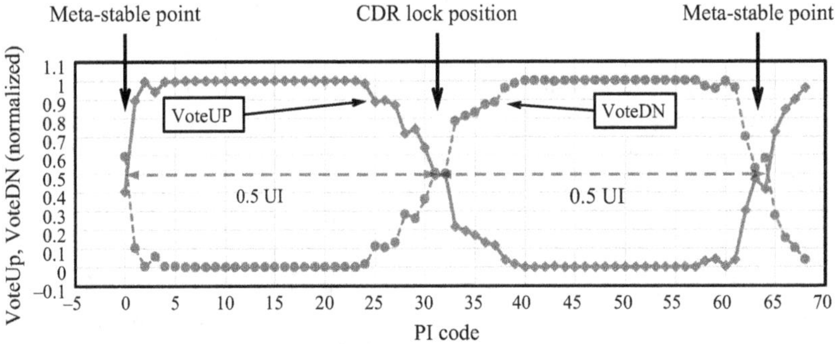

Figure 16.10 Normalized bang-bang phase detector (BBPD) output (VoteUP and VoteDN) vs. phase interpolator (PI) code

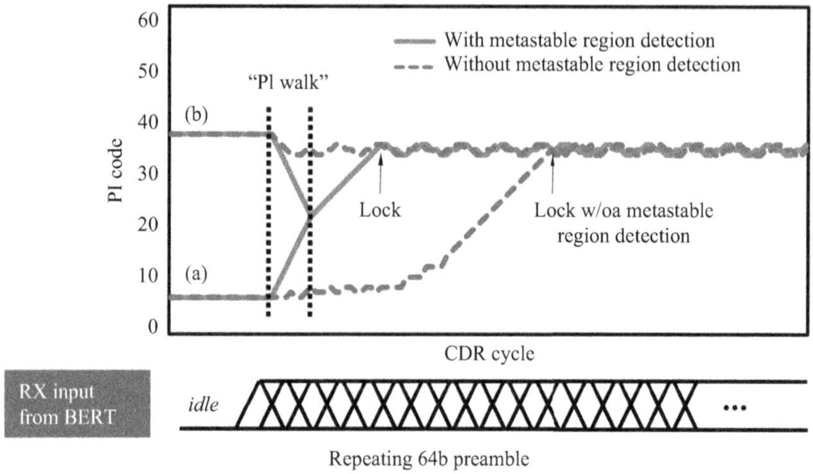

Figure 16.11 Burst-mode clock data recovery (CDR) re-locking starting (a) from a meta-stable region and (b) near a stable lock position

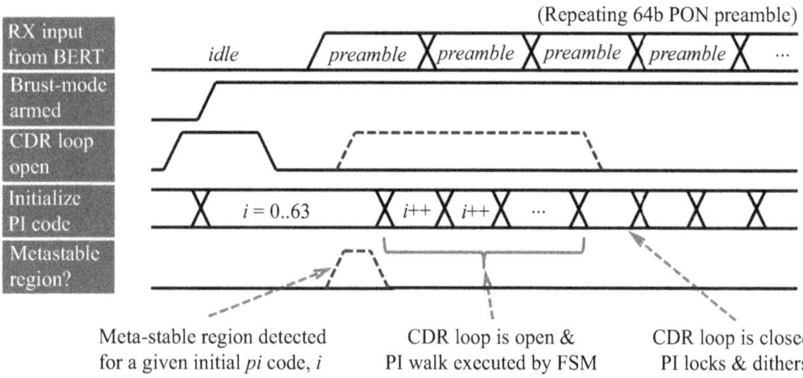

Figure 16.12 Burst-mode clock data recovery (CDR) test sequence

wider for a stable lock condition than for meta-stable regions. Therefore, fast PI walking by a fixed step of, e.g., 10, in this particular case, could effectively eliminate the possibility of a very long lock time for those cases where CDR starts re-locking at the proximity of a meta-stable point, at the cost of a small increase in the lock time for those cases where CDR starts near a stable lock position (Figure 16.11(b)).

Reduction in CDR lock times with and without the designed meta-stable region detection scheme is tested. BERT is configured to synchronously send a series of zeros first, emulating a long PON idle state, followed by a repeating 64-bit PON training patterns at 25 Gb/s rate to the RX via a VSR channel (Figure 16.12). The CDR lock time is measured by the total time required (in UI) from the first preamble to the first 64-bit training pattern received without any error. The test is repeated over all possible initial values of the PI code to emulate random clock phase drift in burst-mode operations, testing multiple runs with identical settings per each initial PI code to obtain statistics.

As plotted in Figure 16.13, the long tail distribution of CDR lock time without the meta-stable region detection reaches up to 650 UI in this sample size. When the detection and fast-PI-walk mode is turned on, the maximum lock time is reduced to 350 UI, demonstrating the implemented digital CDR achieves less than 14 ns lock time in a synchronous burst-mode operation.

Similar fast locking burst-mode CDR solutions by direct PI controlling are reported for 25 Gb/s [13] and 56 Gb/s NRZ optical RXs [14].

The next NRZ example [15] is for synchronous interfaces targeting chip-to-chip communication across a very short printed circuit board (PCB) trace represented by CEI-56G-XSR-NRZ standard. Eight 56 Gb/s lanes sharing an LC-PLL and a bias generation block (Figure 16.14) provide 448 Gb/s aggregate throughput in each direction. The target test channel loss at 28 GHz is measured to be 8 dB from BGA to BGA plus 3 dB additional loss from package traces. Although the TX and RX circuits can be relatively simpler due to the low channel loss, a per-lane de-skewing CDR scheme is still needed to compensate the propagation delay variations among multiple lanes.

Figure 16.15 shows the RX subsystem architecture. The single-stage CTLE comprises a source-degenerated CML stage providing up to 12 dB of high-frequency (HF) peaking at the 28 GHz Nyquist frequency. The CTLE input common mode is dynamically adjusted to ensure that the tail device is adequately saturated for reliable operation but still provides maximum headroom for the differential pair. The CTLE output drives five StrongArm-based slicers [16] that are clocked by four phases of a 14-GHz clock. Four of the slicers are data slicers to recover the 56 Gb/s data. The fifth slicer is an error slicer that samples the peak of

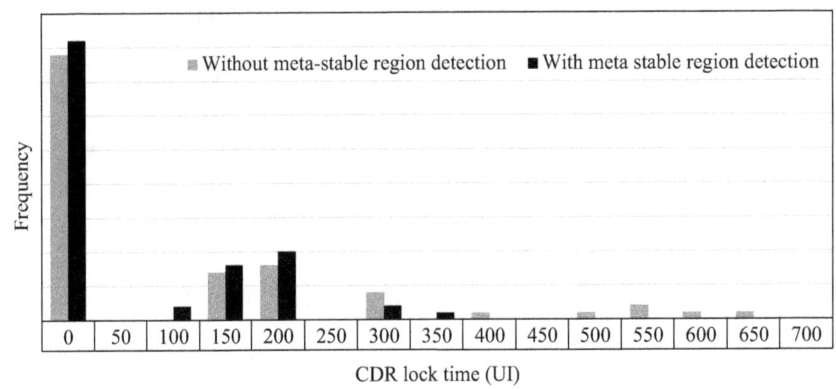

Figure 16.13 Measured burst-mode clock data recovery (CDR) lock time histogram

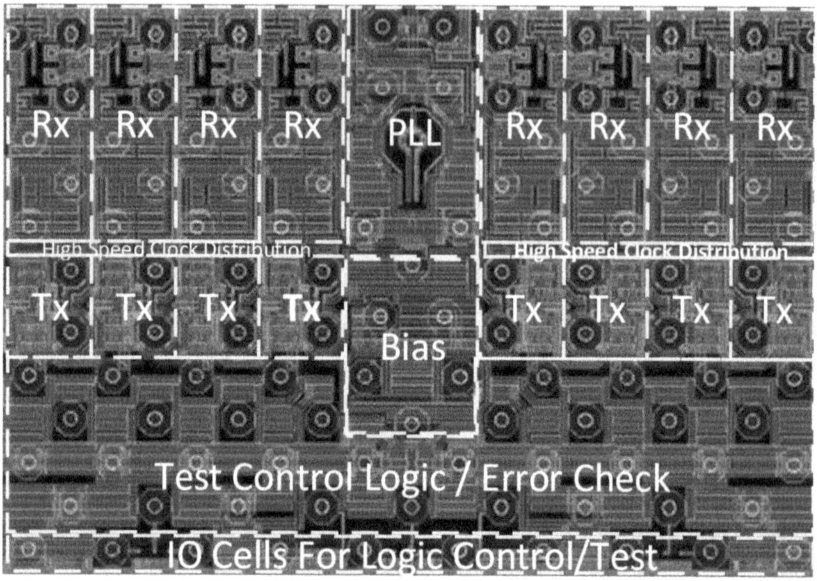

Figure 16.14 Eight channels of 56 Gb/s XSR NRZ transceiver sharing phase-locked loop (PLL) and bias

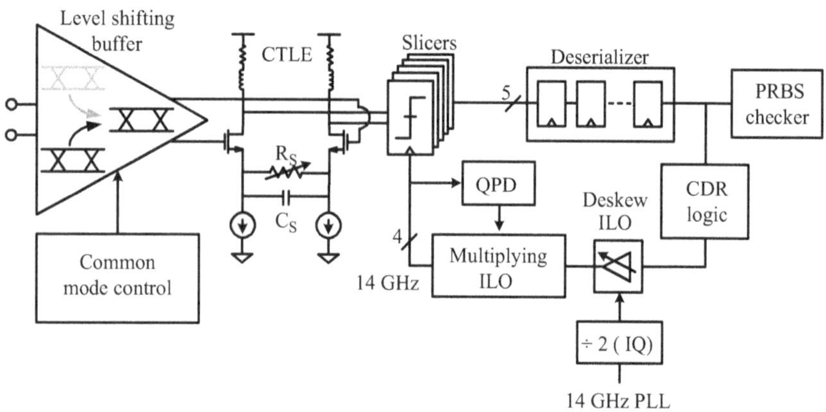

Figure 16.15 Receiver subsystem block diagram

the input signal and is used in conjunction with the data samples to perform a BR-CDR control loop.

The CDR loop (Figure 16.16) positions the slicer sampling clocks to their optimal phases by digitally controlling a delay adjustment block implemented using a de-skewing injection-locked oscillator (ILO) [17]. In order to maximize the delay adjustment range, the de-skewing ILO is designed to operate at 3.5 GHz.

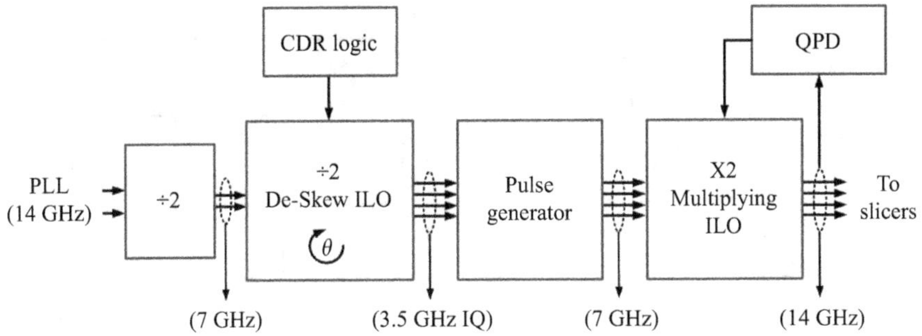

Figure 16.16 RX clocking

The de-skewing ILO output is frequency multiplied by a factor of four by a multiplying ILO (M-ILO) to generate the final four-phase 14-GHz sampling clocks. The phases 0 and 180 of M-ILO outputs are re-timed at 14 GHz by phase shift transfer and the phase 90 and 270 outputs are determined by an implicit interpolation via quadrature PD (QPD) loop that controls relative spacing between 0 and 90 (or 180 and 270). The four-phase sampling clocks at the slicers are sensed and processed by a QPD circuit. The QPD circuit controls the supply level of the M-ILO to ensure a free running oscillation frequency that is matched to the frequency of the injected clocks to minimize in-phase and quadrature-phase (IQ) error [17].

Figure 16.17 shows the architecture of the proposed de-skewing ILO circuit. It consists of a four-stage pseudo-differential ring oscillator with transmission gate switches that are used to short the output of the first stage when clk is high and that of the third stage when clk_b is high. This symmetric injection sequence ensures division by two and maintains the IQ accuracy [17]. Additionally, division by two doubles the time of de-skew range for a given amount of phase de-skew range, which is limited by the injection strength [18]. The injection strength is controlled by digitally setting the effective resistance of the injecting switch. The outputs of the second and fourth delay cells are used as the IQ outputs of the ILO. To perform time de-skewing, the free running frequency of the ILO is detuned by using two digitally controlled resistor DACs, which control the supply and the effective ground of the ILO. Each DAC has 6b for de-skew control and 4b for PVT calibration. A one-time PVT calibration is performed by observing the frequency of the free running oscillator and altering the calibration bits to bring it close to the injected frequency.

The design implements an LC-based PLL (Figure 16.18) with a VCO oscillating at 28 GHz. A ring-based injection-locked frequency divider (ILFD) divides the VCO clock by 2, generating four phases of 14 GHz clocks. The ILFD design provides a low-power solution to HF clock division in the PLL. It drives the CMOS-based feedback divider in the PLL and the four-phase 14-GHz output clocks. Pseudo-differential inverters with regulated supply are used to distribute the four-phase 14-GHz output clocks to the TX/RX channels.

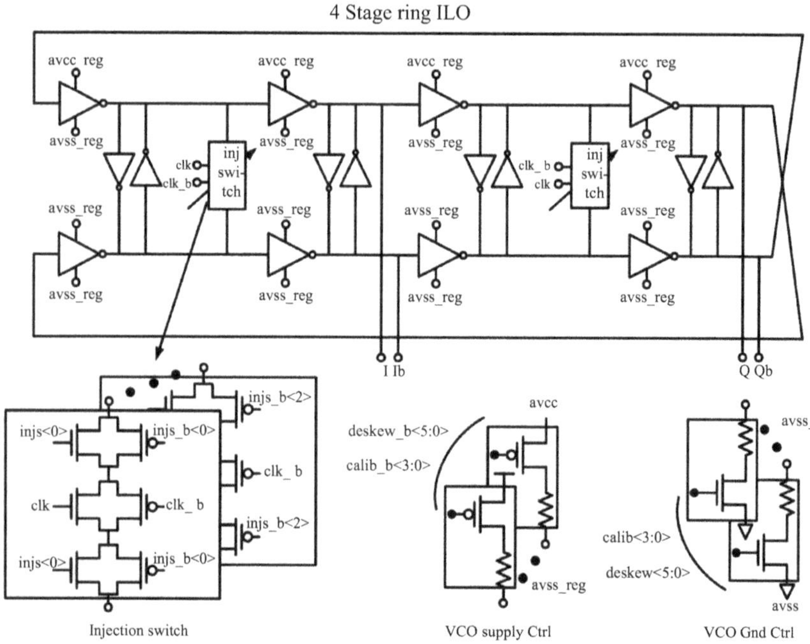

Figure 16.17 De-skewing injection-locked oscillator (ILO) used as phase DAC in clock data recovery (CDR) loop

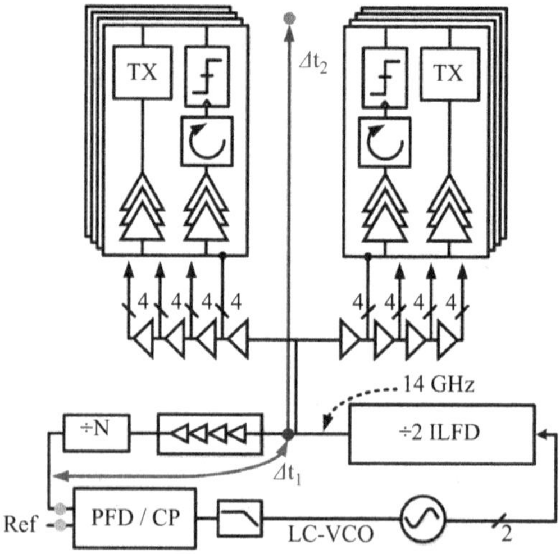

Figure 16.18 Clock delay matching between TX and RX (Δt2) and between channel and phase-locked loop (PLL) loop (Δt1)

After the initial CDR lock, the incoming data may drift relative to the RX sampling clocks as the TX and RX silicon temperature changes. Since the de-skewing ILO has a limited delay adjustment range, the drift must be sufficiently small over a wide temperature range. In order to de-sensitize the leaf nodes of the RX and TX clock trees to temperature change, the insertion delay on RX and TX clock paths ($\Delta t2$ in Figure 16.18) is matched as closely as possible to the first order. Additionally, an inverter delay chain is added to the PLL feedback ($\Delta t1$), such that the total PLL feedback delay is approximately the same as the average TX/RX clock tree delay across all eight lanes. Matching $\Delta t1$ and $\Delta t2$ ensures that TX data phase and RX sampling clock phase track the common reference clock phase to the first order regardless of temperature drift.

A BR MM CDR may suffer from phase wandering when the SBR at data slicer input has a very shallow slope near the sampling point as may be the case for low loss channels or low data rate operations. RXs adopting integrating front end are inherently more robust in this regard because the integrated RX signal peaks at the desired sampling point.

The NRZ transceiver in [19] implements a half-rate integrating front end with two-tap FFE and thee-tap DFE to realize a 60-Gb/s RX using 65 nm CMOS process. The BR-CDR driving a voltage DAC-controlled VCO detects the sampling phase that results in the largest integrated signal. The optimal sampling phase is obtained by modulating the sampling clock phase by a small amount, $\Delta\phi[n]$, and accumulating the correlation between $\Delta\phi[n]$ and the instantaneous error slicer output, $e[n]$, to adjust the VCO phase.

16.5 CDR for PAM-4 modulation scheme

As shown in Figure 16.19(a), an NRZ symbol is represented by one of two levels for each symbol defined as (−1, +1) or as (0, 1). On the other hand, a PAM-4 symbol is represented by one of the four levels for each symbol defined as (−3, −1, +1, +3) in Figure 16.19(b). The four PAM-4 levels are sometimes denoted as (−1, −1/3, +1/3, +1) or (0, 1, 2, 3). The alternate notations will be interchangeably used in the rest of this sub-chapter depending on the context whenever doing so does not cause confusion.

The main difference between PAM-4 CDR and NRZ CDR is in their PD. Two most commonly used PDs for SerDes are 2× oversampled PD and BR-PD, notably MM-PD [5]. Both can be used for either NRZ signaling or PAM-4 signaling.

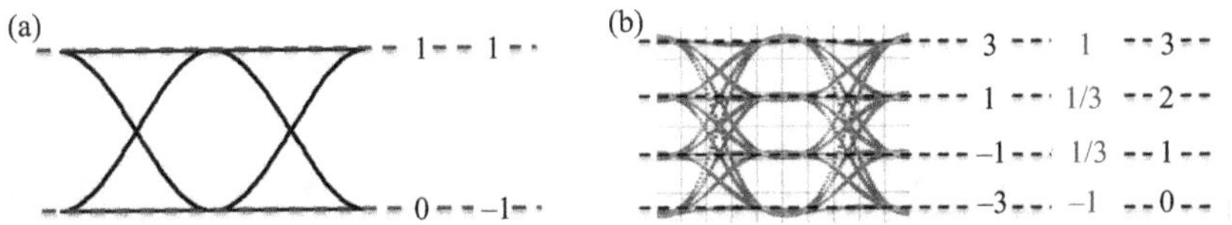

Figure 16.19 Symbol-level conventions for (a) NRZ and (b) PAM-4 signaling

For 2× oversampled PD, the received signal is sampled at twice the frequency of the signal BR or symbol rate. That is, two samples are used in PD for each transmitted symbol. If one sample is taken at the center of the symbol, then the other sample falls on the edge of the symbol. The sampling phase information can be extracted from an edge sample and its two adjacent data samples, if the two data values are different (i.e., whenever a data transition occurs). As shown in Figure 16.20 for the NRZ case, if the locking point is ideal, the data sample (marked as asterisks) would be at the symbol center while the edge sample (marked as circular dots) would be 0. If the locking point is too early, for a $+1 \rightarrow -1$ transition, the edge sample would be positive; while for a $-1 \rightarrow +1$ transition, the edge sample would be negative. Similarly, if the locking point is too late, for a $+1 \rightarrow -1$ transition, the edge sample would be negative; while for a $-1 \rightarrow +1$ transition, the edge sample would be positive.

For NRZ signaling, $0 \rightarrow 0$ or $1 \rightarrow 1$ transitions are not considered as valid in PD and do not provide any phase information for CDR. Only $0 \rightarrow 1$ or $1 \rightarrow 0$ data sequences are considered as transition data patterns and provide phase information as described. For a random data input, each of the four data permutation sequences has an equal probability, so the valid transition density or the effective voting density for a random NRZ pattern is close to 50%.

For PAM-4 signaling, 16 different data permutation sequences are available for two adjacent symbols. Among them, $-3 \rightarrow -3$, $-1 \rightarrow -1$, $+1 \rightarrow +1$, or $+3 \rightarrow +3$ do not provide any phase information. As such, out of the 16 sequences, only 12 transitions are valid and can be used to deduce phase information. Furthermore, the PAM-4 edge sample values at the ideal locking point are more complicated than in NRZ. As shown in Figure 16.21, at the ideal locking point at ±0.5 UI, the edge sample value is 0 for transitions $+1 \rightarrow -1$, $-1 \rightarrow +1$, $-3 \rightarrow +3$, and $+3 \rightarrow -3$; 1 for $-1 \rightarrow +3$ and $+3 \rightarrow -1$; 2 for $+1 \rightarrow +3$ and $+3 \rightarrow +1$; -1 for $+1 \rightarrow -3$ and $-3 \rightarrow +1$; and -2 for $-1 \rightarrow -3$ and $-3 \rightarrow -1$.

The detection of early or late sampling phase for PAM-4 is similar to that for NRZ. For example, for a $+3 \rightarrow +1$ transition, if the edge sample is greater than 2, PD will generate an early vote; while if the edge sample is less than 2, then PD will generate a late vote. In another example, for a $-3 \rightarrow +1$ transition, if the edge sample is greater than -1, PD will generate a late vote; while if the edge sample is less than -1, PD will generate an early vote. The voting logic is summarized in Table 16.2.

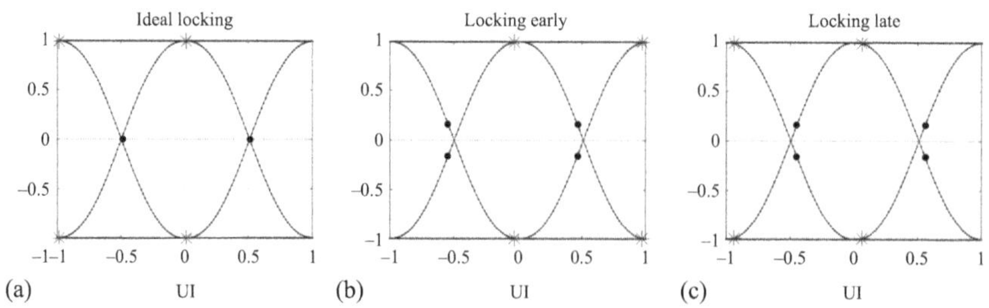

Figure 16.20 Oversampling CDR PD output for (a) locked, (b) early, and (c) late clock phase in NRZ

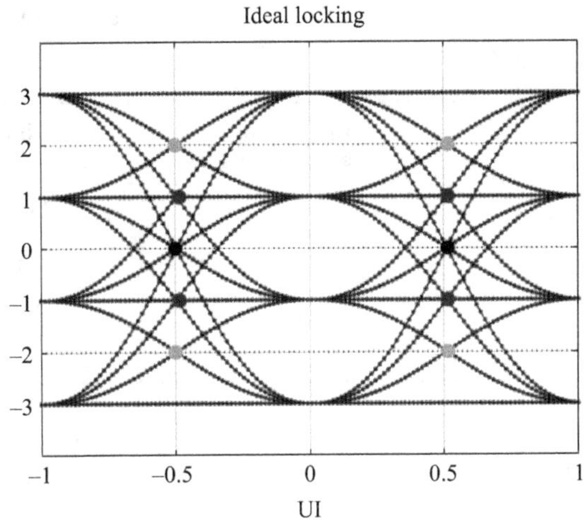

Figure 16.21 Edge sample values vary depending on the adjacent PAM-4 symbols

Table 16.2 Phase detection logic for 2× oversampling PAM-4 clock data recovery (CDR)

Previous data sample	**Edge sample**	**Current data sample**	**PD vote**
−3	>−2	−1	Late
−3	<−2	−1	Early
−3	>−1	+1	Late
−3	<−1	+1	Early
−3	>0	+3	Late
−3	<0	+3	Early
−1	>−2	−3	Late
−1	<−2	−3	Early
−1	>0	+1	Late
−1	<0	+1	Early
−1	>+1	+3	Late
−1	<+1	+3	Early
+1	>−1	−3	Late
+1	<−1	−3	Early
+1	>0	−1	Late
+1	<0	−1	Early
+1	>+2	+3	Late
+1	<+2	+3	Early
+3	>0	−3	Late
+3	<0	−3	Early
+3	>+1	−1	Late
+3	<+1	−1	Early
+3	>+2	+1	Late
+3	<+2	+1	Early

Table 16.3 Oversampling phase detector (PD) voting density for random PAM-4 data

Valid transitions	Edge sample value at ideal locking	Voting density from a random data sequence
$+1 \rightarrow -1$, $-1 \rightarrow +1$, $-3 \rightarrow +3$, $+3 \rightarrow -3$	0	0.25
$-1 \rightarrow +3$, $+3 \rightarrow -1$	+2	0.125
$+1 \rightarrow +3$, $+3 \rightarrow +1$	−2	0.125
$+1 \rightarrow -3$, $-3 \rightarrow +1$	+1	0.125
$-1 \rightarrow -3$, $-3 \rightarrow -1$	−1	0.125

As noted earlier, ideal edge samples have different values for different transitions. Furthermore, different transitions occur with different probabilities. The voting density from different transitions for a random PAM-4 data sequence is summarized in Table 16.3.

As shown in Table 16.3, multiple comparator levels are needed to achieve a high voting density. To achieve a similar voting density as NRZ, i.e., 50%, at least two extra slicing levels are needed in addition to the one at 0 level, at either +2 or −2 (also known as data slicing levels because data bits are determined by these slicer outputs at the data sampling phase) or +1 or −1 (also known as error slicing levels because their outputs at the data sampling phases indicate the amount of deviation between the received and the expected data values). If all five levels are used for the PD, PAM-4 signaling can provide a higher voting density (0.75) than NRZ due to its richness in data transitions, albeit at the cost of more power and area to implement the needed slicers and clock phases. Typically, for an analog-based RX design, not all five levels are assigned corresponding comparators for all edges [20]. The less the number of assigned comparators, the lower the PD voting density. A CDR having lower PD voting density becomes more sensitive to the data pattern and add more undesirable jitter. To reduce the pattern sensitivity, the PD sometimes alternatively chooses different slicing levels for the edge samples or uses different slicing levels for different clock phases. In a half-rate RX architecture where the data are sampled by even/odd clocks (clock phase 0 and 180), the edges are sampled by the quadrature clock phases (clock phase 90 and 270). The edge samples from the quadrature clock phases can use different slicing levels. For example, the clock 90 comparators use slicing levels +2, −2, and 0, whereas the clock 180 comparators use slicing levels +1, −1, and 0.

The 2× oversampling PD does not always guarantee that the locked data sampling phase is at the center or at the peak of the eye. This is particularly true for PAM-4. Figure 16.22(a) plot illustrates a case where the data sampling phase at 0 UI (half a UI away from the two adjacent crossings by convention) is not necessarily at the center of the eye due to an asymmetry of the eye, particularly for the top and the bottom PAM-4 eyes. If the data slicing levels for top and bottom eyes move down and up, respectively, to trade-off between the horizontal eye width and the vertical eye height, the eye width margin on the left and right side of the

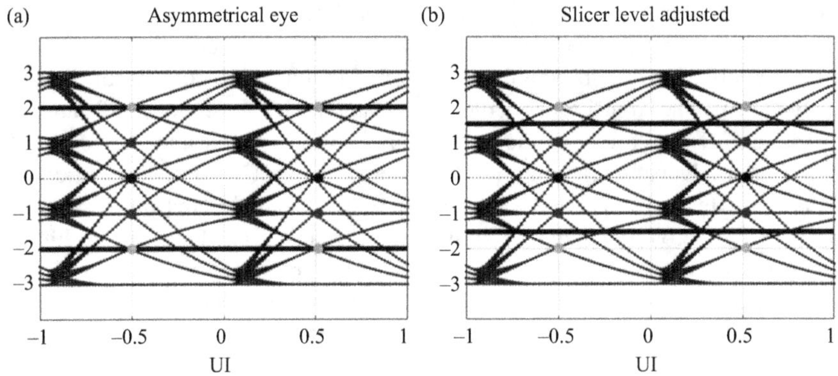

Figure 16.22 PAM-4 data slicing levels for asymmetrical eyes adjusted for optimal horizontal margin

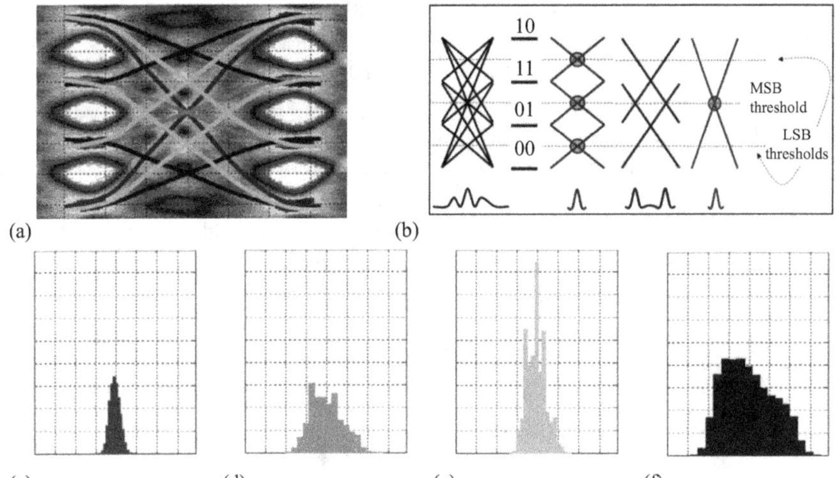

Figure 16.23 Crossing distributions for various types of PAM-4 symbol transitions

locked data sampling phase can be rebalanced, as shown in Figure 16.22(b). A fixed skew between the data clock and the edge clock may be added to re-center the data sampling point for optimal margins.

The quality of different types of transition in realistic systems can be very different from that of a simple ideal model prediction [21]. Due to the differences in slopes for different PAM-4 transitions, coupled with circuit impairments, such as noise and nonlinearity, the crossing distributions at its corresponding slicing levels usually are quite different from the ideal model. This in turn leads to different PD voting gains from different crossing [22]. Figures 16.23(c)–(f) plot crossing histograms for four types of transition obtained from a model simulation. Notice that the major transitions between +3 and −3 levels in Figure 16.23(c) result in the tightest crossing distribution due to higher slew rate and better circuit linearity near level 0. The transitions between +3 and +1 or between −3 and −1 (Figure 16.23(f)) give rise to the widest distributions, which may lead to CDR-induced jitter adversely affecting link BER performance. Crossing distribution also affects the effective linearized gain of the BBPD. A wide crossing distribution reduces the effective PD gain and limits the achievable CDR loop bandwidth.

As noted previously, 2× oversampling CDR may not implement all five comparators for the edge samples due to the implementation cost, especially for an analog-based design. The reduction in edge sampling directly affects the PD voting density and hence the CDR bandwidth. Some novel voting schemes [23,24] can be used for PAM-4 signaling to improve the CDR tracking behavior. For example, Figure 16.24 shows the case where only three slicing levels are used for edge samples at −2, 0, and +2. The transition sampled at ±0.5 UI is used for the regular locking, whereas samples obtained slightly away from ±0.5 UI can be used with a selective voting logic to enhance acquisition and jitter performance.

Table 16.4 shows the voting logic for the transitions marked by the crossing samples taken at around ±0.65 and ±0.35 UI in Figure 16.24. During an initial locking process when the sampling phase is far away from the ideal locking phase at ±0.5 UI, the added logic essentially increases the PD voting density, accelerating the CDR settling behavior and extending its acquisition range. In another case, if a system has large jitter at frequencies slightly beyond the CDR tracking range based only on the votes from the conventional voting logic, the added logic may effectively increase the PD voting density and hence increases the CDR bandwidth.

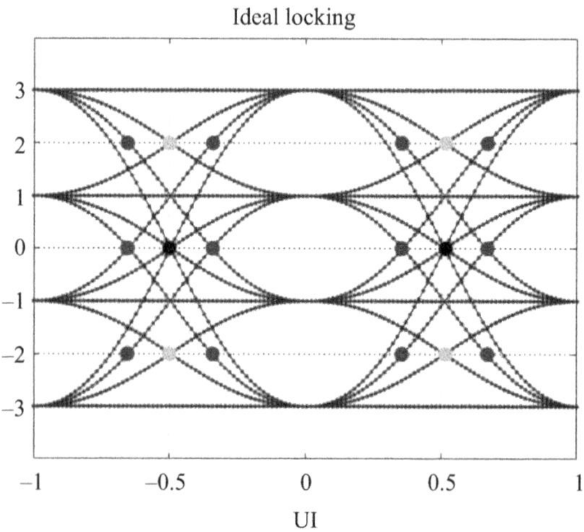

Figure 16.24 Transitions at ideal phases (±0.5 UI) vs. aggressive phases (∼ ±0.65 and ±0.35 UI)

Table 16.4 Voting logic for aggressive phase error decision

Previous data sample	Edge sample	Current data sample	Phase detector vote
−3	<−2	+1	Early
−3	>0	+1	Late
−1	<0	+3	Early
−1	>2	+3	Late
+3	>2	−1	Early
+3	<0	−1	Late

BR-CDR does not need an extra sampling phase for the edge samples and hence is widely adopted for SerDes RXs at data rates of 56 Gb/s and beyond (see Table 16.1), especially for ADC-based SerDes designs where adding extra clock phases could significantly increase power consumption.

The BR-CDR PD for PAM-4 is fundamentally the same as for NRZ. The locking point is determined where the system SBR has the same magnitude at $h(-1)$ and $h(1)$. One of the commonly used equations for the phase-settling behavior is described in (16.3).

$E\{\cdot\}$ represents the expectation value of a random variable, realized in real systems as temporal average of a measurement. Data and corresponding error sampled at time n are denoted as $d(n)$ and $e(n)$, respectively. For NRZ, $d(n)$ can assume +1 or −1. For PAM-4, $d(n)$ can be −3, −1, and +1 or +3:

$$E\{e(n) \times [d(n+1) - d(n-1)]\} = 0 \tag{16.3}$$

The error signal comes from uncancelled ISI and noise, $N(n)$. Simplifying the model by limiting ISI only to $h(1)$ and $h(-1)$, the error is expressed as (16.4), ignoring all other residual ISI from higher order terms. In (16.4), $N(n)$ represents the sampled noise:

$$e(n) = h_{-1} \times d(n+1) + h_1 \times d(n-1) + N(n) \tag{16.4}$$

The expectation formula (16.3) can then be rewritten as (16.5) below:

$$\begin{aligned} E\{e(n) \times [d(n+1) - d(n)]\} &= E\{[h_{-1} \times d(n+1) + h_1 \times d(n-1) + N(n)] \\ &\quad \times [d(n+1) - d(n-1)]\} \\ &= h_{-1} \times E\left\{d(n+1)^2 - d(n+1) \times d(n-1)\right\} \\ &\quad - h_1 \times E\left\{d(n-1)^2 - d(n+1) \times d(n-1)\right\} \\ &\quad + N(n) \times E\{[d(n+1) - d(n-1)]\} \end{aligned} \tag{16.5}$$

For a random NRZ sequence, each data symbol has an equal probability of being −1 and +1. Therefore, on average, (16.5) is reduced to:

$$E\{e(n) \times [d(n+1) - d(n)]\} = (h_{-1} - h_1) \tag{16.6}$$

For a random PAM-4 sequence, each data symbol has an equal probability of being −3, −1, +1 and +3, so (16.5) is reduced to:

$$E\{e(n) \times [d(n+1) - d(n)]\} = 5 \times (h_{-1} - h_1) \tag{16.7}$$

Driving the expectation value in (16.6) or (16.7) to 0 means that $h_{-1} = h_1$ at the sampling point in locked condition, as illustrated in the SBR plotted in Figure 16.25(a). The three black dots mark the precursor tap h_{-1}, the main cursor tap h_0, and the postcursor tap h_1. The corresponding PAM-4 eye is plotted in Figure 16.25(b).

If the clock phase deviates from this point, the expectation becomes:

$$\begin{aligned} E\{e(n) \times [d(n+1) - d(n)]\} = {} & 5 \times (\Delta h_{-1} - \Delta h_1) + \Delta h_0 \\ & \times E\{d(n) \times [d(n+1) - d(n-1)]\} \end{aligned} \tag{16.8}$$

In (16.8), Δh_0, Δh_1, and Δh_{-1} represent the changes in main-, post-, and pre-cursor taps, respectively. Again, for a random pattern, the expectation value of the last term in (16.8) is zero.

From the channel SBR, if locking point drifts to the left, as illustrated in Figure 16.26(a), Δh_{-1} typically becomes more negative and Δh_1 becomes more positive, moving the expectation value to a negative value. Similarly, if the locking point drifts to the right, as illustrated in Figure 16.26(b), Δh_{-1} and Δh_1 becomes more positive, moving the expectation value to a positive value.

Usually only the sign of the error signal is extracted from the error slicers to reduce design complexity and power consumption, even for ADC-based RXs, where the magnitude of *e*(*n*) is readily available if needed. The BR-CDR loop then drive the below convergence equation to zero:

$$E\{\text{sign}[e(n)] \times [d(n+1) - d(n-1)]\} = 0 \tag{16.9}$$

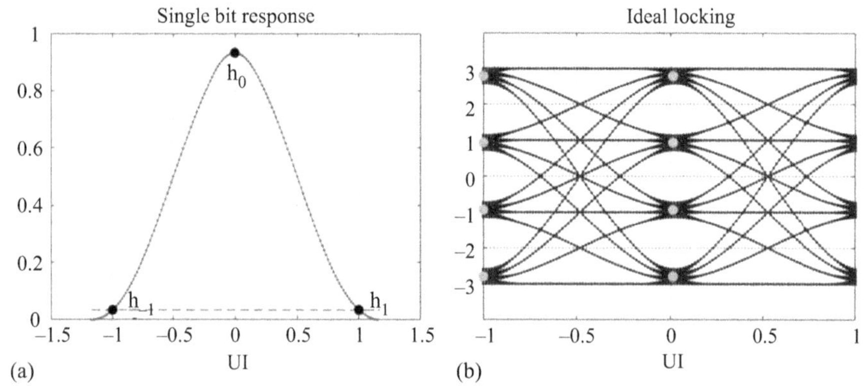

Figure 16.25 Baud-rate clock data recovery (CDR) locking phase at 0 UI in (a) single-bit response (SBR) and (b) eye diagram

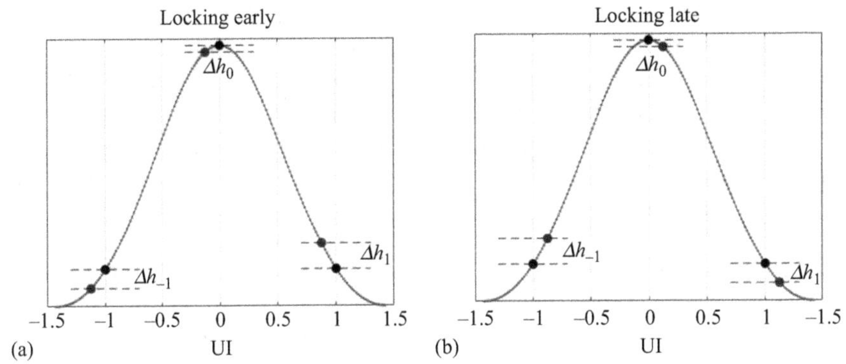

Figure 16.26 Locking point drift (a) early or (b) late causing nonzero expectation value

(16.9) can be rewritten as:

$$E\{\text{sign}[e(n)] \times [d(n+1) - d(n)]\}$$
$$= E\{\text{sign}[h_{-1} \times d(n+1) + h_1 \times d(n-1) + N(n)] \times [d(n+1) - d(n-1)]\} \quad (16.10)$$

If $d(n+1) = -d(n-1)$, then the convergence equation, (16.10), drives the locking phase to satisfy $h_{-1} = h_1$ condition. For a random NRZ pattern, $d(n)$ can be either +1 or −1 with equal probability and the voting density is 50%. This voting density is the same as that for a 2× oversampling CDR. For a random PAM-4 pattern, $d(n)$ can be −3, −1, +1, or +3, so if we limit the selection of the valid patterns to between +3 and −3 or between +1 and −1, the voting density is only 25%. In the presence of an equalization in the system where h_1 is close to 0, the locking position will be where $h_{-1} = h_1 = 0$. In this case, as long as $d(n+1)$ and $d(n-1)$ have opposite signs, the votes would still be valid (assuming the noise and residual ISI is sufficiently small or averaged to zero) and the voting density is improved to 50%, the same level as NRZ. Note that for an analog-based RX architecture, not all error slicers are implemented for power efficiency reasons. Analogous to what have been discussed for 2× oversampling CDR, the less the number of used comparators is, the lower the voting density and the resulting CDR bandwidth. ADC-based RXs are typically not subjected to this constraint due to the relatively lower implementation cost for digital slicers.

Now assuming CDR has achieved a lock condition, i.e., $h_{-1} = h_1$, whenever the sampling clock phase deviates from the ideal locking phase, the voting logic output becomes:

$$E\{\text{sign}[e(n)] \times [d(n+1) - d(n)]\} = E\{\text{sign}[\Delta h_{-1} \times d(n+1) + \Delta h_1 \times d(n-1) + \Delta h_0 \times d(n) + N(n)] \times [d(n+1) - d(n-1)]\} \quad (16.11)$$

In (16.11), the sign of an error sample, sign[$e(n)$], depends not only on the previous and the next symbols but also on the current symbol. Typically, Δh_0 is small relative to Δh_{-1} or Δh_1 and can be safely ignored in NRZ CDR. For PAM-4 CDR, however, Δh_0 may not be negligible due to the multiplication factor from the PAM-4 data. Therefore, some kind of pattern filtering schemes are usually adopted to reduce the jitter induced by CDR voting logic. For example, certain votes can be ignored when $d(n)$ equals 3 or -3.

For asymmetrical channel SBRs, which are typically the case in real systems, corresponding eye diagrams are also asymmetrical, as illustrated in Figure 16.27. The data sampling phases may be either at the peak of the eye or at the horizontal center of the eye. Analogous to the oversampling case, the eye height can be traded off with the eye width to achieve a minimal BER.

Several PAM-4 RXs capable of data rates of 56 Gb/s or higher have been realized in silicon and published in recent years (Table 16.1). Majority of them adopt an ADC-based front-end architecture, followed by FFE and DFE stages performing ISI cancellation in digital domain using a dedicated DSP block. A 56 Gb/s silicon example in [25] demonstrates a fully integrated and adaptive 19–56 Gb/s PAM-4 (9.5–28 Gb/s in NRZ mode) transceiver implemented in 16 nm FinFET technology.

Figure 16.28 depicts the RX block diagram showing its sub-systems. The RX AFE consists of a three-stage CTLE and one variable gain amplifier (VGA) stage. In addition to the two HF boosting CTLE stages, an MF peaking stage is inserted for long-tail cancellation [12], which reduces the total power consumption by reducing the number and range of postcursor FFE taps, while also improving the signal to noise ratio at the ADC input.

The 28GS/s 32-way TI asynchronous successive approximation register ADC (TI-SAR-ADC) consists of a four-way (e.g., 7 GHz for 56 Gb/s) front-end interleaving stage followed by an eight-way (875 MHz for 56 Gb/s) interleaving stage [26,27]. Since the ADC input signal level can be higher than the maximum allowed voltage for thin-oxide devices, a T&H stage with a boot-strapped n-type

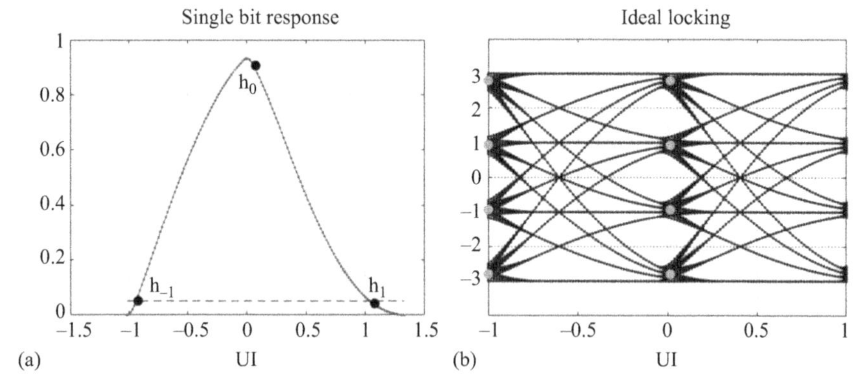

(a) (b)

Figure 16.27 Asymmetrical (a) single-bit response (SBR) and (b) corresponding eye diagram in baud-rate clock data recovery (CDR)

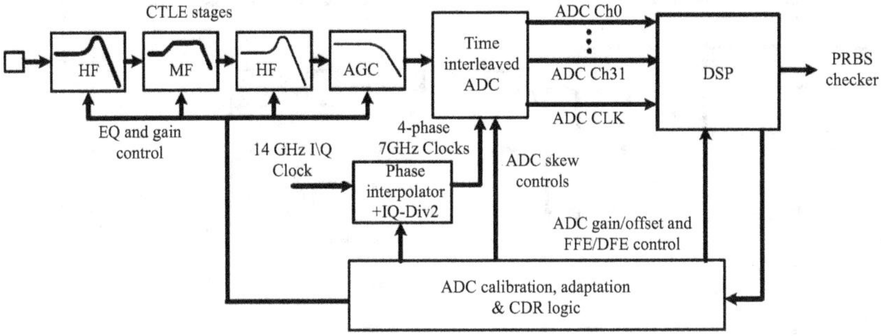

Figure 16.28 Sub-systems of ADC-DSP-based 56 Gb/s PAM-4 receiver

metal-oxide-semiconductor (NMOS) switch is used to meet the long-term reliability requirements and to maintain good linearity. The SAR ADC can be configured from maximum 7b operation down to 3b operation by inserting a resolution control logic block. Taking advantage of the sequential nature of asynchronous SAR ADCs, the resolution control is performed by simply gating the asynchronous clock after the desired number of conversions is completed. The lower ADC resolution mode is used to opportunistically save ADC and DSP power when the RX is used on a clean, low-loss channel.

The digitized RX signal from ADC output is re-timed and sent to DSP for signal processing and slicing. The data from the TI-ADC are deserialized by a factor of 2 to allow the DSP to operate at a frequency optimal for timing closure and power efficiency. Fully integrated equalization adaptation, ADC calibration, and BR-CDR logic concurrently and dynamically adjust CTLE/VGA settings, TI-ADC front-end sampling phases, ADC gain/offset, and FFE/DFE coefficients in the DSP. An architectural view of DSP, autoadaptive ISI equalization and ADC calibration, and CDR are shown in Figure 16.29.

The ADC offset and gain are detected and corrected digitally in the DSP, while the ADC timing skew is corrected in the analog domain using a clock skew correction block driving TI-ADC front-end interleaving stage. The calibration is done in the background from live data for PVT variation.

A 14-tap FFE is used to equalize four precursor taps (hm4–hm1) and ten postcursor taps ($h2$–$h11$). A speculative one-tap DFE is used for dominant first postcursor ($h1$) equalization to minimize noise and cross-talk amplification. A bank of digital slicers slices the FFE-corrected digitized data into speculative PAM-4 data and error samples. The speculative samples are fed to the DFE multiplexer which in turn performs one-tap decision feedback data selection. The resolutions and the coefficient ranges of the DFE and FFE taps are carefully chosen to balance between the DSP power consumption and the quantization noise. The FFE coefficients are adapted using the sign-sign least mean squares (LMS) algorithm.

The BR-CDR logic modulates the output clock of a 1-UI/64 resolution RX PI, which is subsequently divided by two to generate four phases of clocks. The four-phase clocks are then skew calibrated by four digitally controlled variable delay

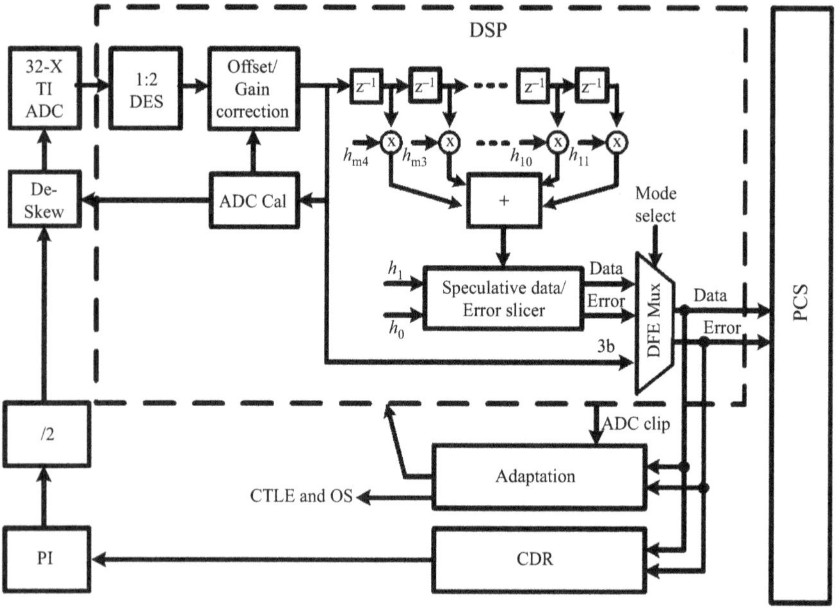

Figure 16.29 Digital signal processing (DSP), ADC calibration, equalization adaptation, and clock data recovery (CDR)

cells to drive the TI-ADC front-end interleaving stage. CDR is based on a modified MM algorithm, so that the optimal sampling phase can be fine-tuned.

The fabricated silicon tested in 7 bit resolution mode over a channel of 32 dB IL at 14 GHz achieves less than 1E−12 BER without explicitly added cross talk and less than 1E−6 with 2 mV-rms cross talk injected using a random noise generator with a crest factor of 10. This is a sufficient margin to meet the OIF-CEI-56G-LR-PAM-4 standard [2], which requires a pre-FEC BER <1E−4. Figure 16.30 plots JTOL measured over several parts across supply and temperature variations. The RX clears the OIF-CEI-56G-LR JTOL mask with the BR-CDR tracking a 200 ppm frequency offset.

A similar ADC-DSP architecture using BR digital CDR is implemented to realize a 112 Gb/s PAM-4 RX in [28]. With a 2× extended tap ranges (31-tap FFE and 1-tap speculative DFE) to account for the halved UI compared to the 56 Gb/s RX, the designed silicon achieves 2E−5 BER, with its own on-chip TX transmitting PAM-4 PRBS-31 signal over a channel of 20 dB loss at 28 GHz. Figure 16.31 illustrates a 112 Gb/s post-DSP eye diagram clearly showing PAM-4 levels centered at the CDR lock point (PI code offset = 0).

As discussed earlier in the chapter, potential performance degradation of PAM-4 RXs due to CDR loop latency is inherently more pronounced in ADC-DSP-based architecture than in analog-based ones [29]. Several ADC-based BR-CDR RXs published adopt a parallel post-ADC data path approach, decoupling the CDR loop from the main data path, so that the loop latency can be reduced at the cost of

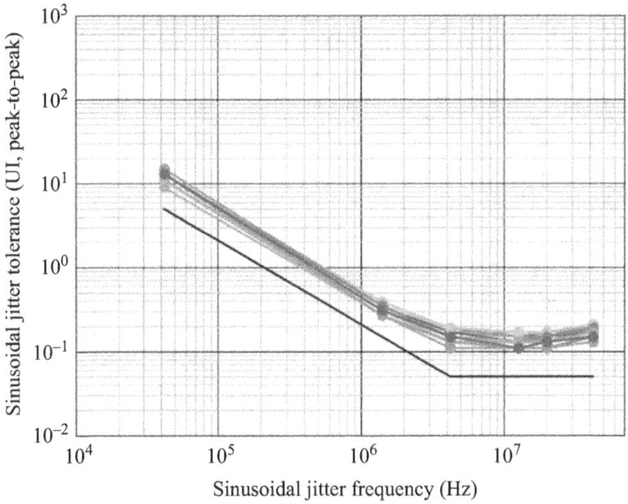

Figure 16.30 Jitter tolerance (JTOL) of 56 Gb/s LR PAM-4 with 200 ppm (OIF-CEI-56G-LR mask)

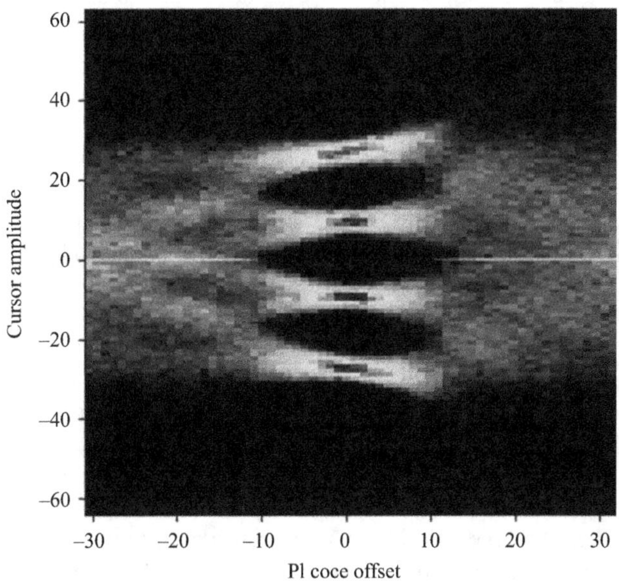

Figure 16.31 Equalized 112 Gb/s PAM-4 eye over a 34-dB channel

more power consumption and more residual ISI for the PD. This approach was applied in ADC-based RXs for both NRZ and PAM-4. For example, reference [30] reports a 25-Gb/s ADC-based NRZ RX utilizing quarter rate 5b flash ADCs followed by a primary DSP performing 8-tap FFE and 8-tap DFE for the main data path. The BR-CDR PD receives a data stream from a secondary low-latency 2-tap FFE and 1-tap DFE path. A majority of the recently published high-speed ADC-based PAM-4 RXs in Table 16.1 adopt a parallel path approach. Others achieve latency reduction by teeing off partially equalized data for CDR, e.g., from RX FFE output before the DFE stage that follows [31]. Even more aggressive CDR loop latency reduction has been reported in a few other publications [32,33], where the ADC output is directly applied to CDR PD before digital filtering for ISI reduction in DSP.

The impact of residual ISI resulting from using low-latency path may include (1) limiting the CDR lock acquisition range, (2) more residual jitter in steady state, and (3) sub-optimal locked sampling clock phase. The performance trade-off between lower latency *vs.* lower residual ISI for a given channel characteristics profile is not comprehensibly studied and published at this time.

16.6 Conclusion

The crucial role CDR plays in SerDes system has become more important and challenging in ultra-high-speed wireline RXs at a line rate of 25 Gb/s or higher. The main driving force of circuit development being power and area efficiency, especially for multi-lane SerDes needed for highly integrated applications, CDR solutions for such high data rates require a careful design trade-off evaluation in terms of circuit architecture choice and implementation details that can vary widely depending on speed, system-level constraints, and signal modulation schemes. Although most topics of wireline CDR have been well studied and understood in terms of theories and general design trade-offs, robust silicon realizations suitable for high-volume production of ASIC, FPGA, CPU, GPU, and SoCs are still an active research and development area with a room for innovation.

References

[1] Zhang, H., Rao, F., Wu, D. *et al.*, "IBIS-AMI Modeling of Asynchronous High Speed Link Systems," DesignCon, 2017.
[2] Optical Internetworking Forum (OIF) web site (https://www.oiforum.com/technical-work/current-work/).
[3] IEEE 802.3 Ethernet Working Group web site (http://www.ieee802.org/3/).
[4] Xilinx, Inc. web site (https://www.xilinx.com/applications/industrial/instrumentation.html).
[5] Mueller, K., and M. Muller, "Timing Recovery in Digital Synchronous Data Receivers," IEEE Transactions on Communications, vol. COM-24, no. 5, pp. 516–531, 1976.
[6] Shibasaki, T., Danjo, t., Ogata, Y. *et al.*, "A 56Gb/s NRZ-Electrical 247mW/lane Serial-Link Transceiver in 28nm CMOS," ISSCC, pp. 64–65, 2016.

[7] Cevrero, A., Ozkaya, I., Francese, P. A. *et al.*, "A 100Gb/s 1.1pJ/b PAM-4 RX with Dual-Mode 1-Tap PAM-4/3-Tap NRZ Speculative DFE in 14nm CMOS FinFET," ISSCC, pp. 112–113, 2019.

[8] Jaussi, J., Balamurugan, G., Hyvonen, S. *et al.*, "A 205mW 32Gb/s 3-Tap FFE/6-Tap DFE Bidirectional Serial Link in 22nm CMOS," ISSCC, pp. 440–441, 2014.

[9] PCI-SIG web site (https://pcisig.com/specifications/pciexpress/).

[10] Im, J., Chen, S., Freitas, D. *et al.*, "A 0.5-28Gb/s Wireline Transceiver with 15-Tap DFE and Fast-Locking Digital CDR in 7nm FinFET," Symposium on VLSI Circuits, 2018.

[11] Chen, S., Zhou, L., Zhuang, I. *et al.*, "A 4-to-16GHz Inverter-Based Injection-Locked Quadrature Clock Generator with Phase Interpolators for Multi-Standard I/Os in 7nm FinFET," ISSCC, pp. 390–391, 2018.

[12] Savoj, J., Hsieh, K., Upadhyaya, P. *et al.*, "Design of High-Speed Wireline Transceivers for Backplane Communications in 28nm CMOS," CICC, 2012.

[13] Rylyakov A., Proesel, J., Rylov, S. *et al.*, "A 25Gb/s Burst-Mode Receiver for Rapidly Reconfigurable Optical Networks," ISSCC, pp. 400–401, 2015.

[14] Ozkaya, I., Cevrero, A., Francese, P. A. *et al.*, "A 56Gb/s Burst-Mode NRZ Optical Receiver with 6.8ns Power-On and CDR-Lock Time for Adaptive Optical Links in 14nm FinFET CMOS," ISSCC, pp. 266–267, 2018.

[15] Erett, M., Carey, D., Hudner, J. *et al.*, "A 126mW 56Gb/s NRZ Wireline Transceiver for Synchronous Short-Reach Applications in 16nm FinFET," ISSCC, pp. 274–275, 2018.

[16] Montanaro, J., Witek, R. T., Anne, K. *et al.*, "A 160-MHz, 32-b, 0.5-W CMOS RISC Microprocessor," JSSC, 1996.

[17] Raj, M., Saeedi, S., Emami, A. *et al.*, "A 4-to-11GHz Injection-Locked Quarter-Rate Clocking for an Adaptive 153fJ/b Optical Receiver in 28nm FDSOI CMOS," ISSCC, pp. 404–405, 2015.

[18] O'Mahony, F., Shekhar, S., Mansuri, M. *et al.*, "A 27Gb/s Forwarded-Clock I/O Receiver Using an Injection-Locked LC-DCO in 45nm CMOS," ISSCC, pp. 452–453, 2008.

[19] Han, J., Lu, Y., Sutardja, N. *et al.*, "A 60Gb/s 288mW NRZ Transceiver with Adaptive Equalization and Baud-Rate Clock and Data Recovery in 65nm CMOS Technology," ISSCC, 2017.

[20] Peng, P.-J., Li, J.-F., Chen, L.-Y. *et al.*, "A 56Gb/s PAM-4/NRZ Transceiver in 40nm CMOS," ISSCC, pp. 110–112, 2017.

[21] Zhang, H., Jiao, B., Liao, Y. *et al.*, "PAM4 Signaling for 56G Serial Link Applications – A Tutorial," DesignCon, 2016.

[22] Tang, L., Gai, W., Shi, L. *et al.*, "40Gb/s 74.9 mW PAM4 Receiver with Novel Clock and Data Recovery," IEEE ISCAS, 2017.

[23] Toi, T., Wadatsumi, J., Kobayashi, H. *et al.*, "A 25.6Gb/s Uplink-Downlink Interface Employing PAM-4 Based 4-Channel Multiplexing and Cascaded CDR Circuits in Ring Topology for High-Bandwidth and Large-Capacity Storage Systems," ISSCC, pp. 478–479, 2019.

[24] Shahramian, S., Dehlaghi, B., Liang, J. *et al.*, "A 1.41pJ/b 56Gb/s PAM-4 Wireline Receiver Employing Enhanced Pattern Utilization CDR and Generic Adaptation Algorithms in 7nm CMOS," ISSCC, pp. 482–483, 2019.

[25] Upadhyaya, P., Poon, C. F., Lim, S. W. *et al.*, "A Fully Adaptive 19-58 Gb/s PAM-4 and 9.5-29 Gb/s NRZ Wireline Transceiver With Configurable ADC in 16nm FinFET," JSSC, 2019.

[26] Frans, Y., Shin, J., Zhou, L. *et al.*, "A 56-Gb/s PAM4 Wireline Transceiver Using a 32-Way Time-Interleaved SAR ADC in 16-nm FinFET," IEEE Journal of Solid-State Circuits, vol. 52, no. 4, pp. 1101–1110, 2017.

[27] Frans, Y., Elzeftawi, M., Hedayati, H. *et al.*, "A 56Gb/s PAM4 Wireline Transceiver Using a 32-way Time-Interleaved SAR ADC in 16nm FinFET," IEEE Symposium on VLSI Circuits, pp 1–2, 2016.

[28] Hudner, J., Carey, D., Casey, R. *et al.*, "A 112GB/S PAM4 Wireline Receiver Using a 64-Way Time-Interleaved SAR ADC in 16NM FinFET," IEEE Symposium on VLSI Circuits, pp. 47–48, 2018.

[29] Im, J., Freitas, D., Roldan, A. *et al.*, "A 40-to-56 Gb/s PAM-4 Receiver with 10-tap Direct Decision-Feedback Equalization in 16 nm FinFET," in ISSCC Dig. Tech. Papers, pp. 114–115, 2017.

[30] Rylov, S., Beukema, T., Toprak-Deniz, Z. *et al.*, "A 25Gb/s ADC-Based Serial Line Receiver in 32nm CMOS SOI," in ISSCC Dig. Tech. Papers, pp. 56–57, 2016.

[31] Krupnik, Y., Perelman, Y., Levin, I. *et al.*, "112 Gb/s PAM4 ADC Based SERDES Receiver for Long-Reach Channels in 10nm Process," IEEE Symposium on VLSI Circuits, pp. 266–267, 2019.

[32] Gopalakrishnan, K., Ren, A., Tan, A. *et al.*, "A 40/50/100Gb/s PAM-4 Ethernet Transceiver in 28nm CMOS," in ISSCC Dig. Tech. Papers, pp. 62–63, 2016.

[33] Kiran, S., Cai, S., Luo, Y. *et al.*, "A 32 Gb/s ADC-Based PAM-4 Receiver with 2-bit/Stage SAR ADC and Partially-Unrolled DFE," CICC, 2018.

Chapter 17

Clock and data recovery for optical links

Wei Zen Chen[1], Ming-Chiuan Su[2], and Yuan-Sheng Lee[3]

Optical links played a main role as Internet backbones for ultra-long-haul telecommunication networks for many decades and have been widely deployed in the metropolitan and local area networks. Recently, they have become more and more popular in short-reach interconnects such as active optical cables that aggregate multiple electrical channels for an extended distance at tens of gigabit/s operation. Compared to the electrical interconnects, they provide the merits of a wider bandwidth with a lower electromagnetic interference (EMI) effect. Thus broadband data communication can be achieved with lower equalization, lower latency, and lower power consumption. With the booming applications of cloud services, optical links also find their niches in the data center that bridge servers and storage devices. According to the report in [1], global data centers consumed about 416 terawatts in 2016. The projected power consumption is doubled every 4 years, among which over 20% is consumed by the networking devices and the high-speed inputs/outputs (IOs). Thus, energy-efficient networking is highly demanding to reduce the cost and power consumption of data centers [2].

Power saving of the high-speed IOs can be improved in two ways. One direction is to improve the energy efficiency in the active mode during data transmission. As the ratio of network utilization is low during light load, it is necessary to shape the traffic in burst-mode (BM) data packets and then turn the network into power-saving mode during the idle periods. In this context, the network must resume data link after activation in a short period of time without much power, timing, and hardware overhead. From this perspective, BM clock and data recovery (CDR) (BMCDR) techniques, which have been widely utilized in passive optical networks, also become more and more attractive in high-speed I/Os for consumer electronics to improve the energy efficiency.

This chapter focuses on the design of energy-efficient CDR circuits for high-speed optical links. In the first part, a BMCDR is introduced [3]. It incorporates a selective gating ring oscillator in a phase-locked loop (PLL). The selective gating

[1]Department of Electronics Engineering and Institute of Electronics, National Chiao Tung University, Hsinchu, Taiwan
[2]Mediatek, Hsinchu, Taiwan
[3]Realtek, Hsinchu, Taiwan

oscillator is running at $1/n$ data rate, which enables high-speed operation with low-power consumption. Meanwhile, the proposed BMCDR manifests both the advantages of instant phase locking and data jitter suppression in contrast to conventional gating-only or PLL-only CDRs. In the second part, a fully integrated optical receiver consisting of an integrating-type receiver front-end and a new baud-rate CDR is described [4]. A hybrid loop filter is proposed to facilitate programmable bandwidth with a constant phase margin. Both the experimental prototypes have small form factors and high energy efficiency. They demonstrate strong potentials to be applied in consumer electronics and high-performance computing platforms.

17.1 Prior art of BMCDR

BMCDR circuits have been widely deployed in passive optical networks. In the Ethernet Passive Optical Network system, an optical line terminal (OLT) in the central office (CO) communicates with multiple optical network units (ONUs) under the subscribers' control. The ONU transmits BM data to the OLT using a time-division multiple-access scheme. In the CO, the BM data received from subscribers are phase asynchronous relative to the CO's clock. It is required to recover data burst within tens to hundreds of bit time.

Conventionally, BMCDRs can be categorized into three folds: (1) phase interpolator (PI)-based [5,6], (2) gating voltage-controlled oscillator (GVCO)-based [7,8], or (3) injection-locked oscillator (ILO)-based architectures [9,10]. As is shown in Figure 17.1(a), PI-based CDRs rapidly extract the weighting coefficients of the PI through a BMPD and then synthesize a clock waveform (CK_S) that is closest to the data edge. Incorporating with a binary search engine [5] or a sample and hold circuit [6], agile phase locking is achieved through a direct phase synthesis from the I/Q phases of the clock signals (CK_I, CK_Q). On the other hand, the GVCO-based BMCDR is another common approach that is similar to the ILO-based counterparts, which adjust the sampling phase through gating or injection locking the oscillator with the input data. Figure 17.1(b) illustrates the GVCO-based BMCDR architecture, which consists of a PLL for voltage-controlled oscillator (VCO) frequency initialization and two-gated VCOs (VCO1 and VCO2) controlled by serial input data for phase synchronization [7]. To adjust the sampling phase properly, the free running frequency of the GVCO must be brought into the locking range of the gating oscillator initially. The sampling clock is synthesized by recombining the clock phases of the two GVCOs. As GVCO1 and GVCO2 are enabled alternatively by the input data, the start-up time of each oscillator limits the operating speed of the recovered clock and the data rate. Furthermore, the inevitable frequency mismatch between the master VCO and the two slave GVCOs leads to an accumulated phase error in the long run of consecutive identical digital (CID) bits input data. It would severely degrade the timing margin for sampling clock and also limit the bit error rate (BER) performance. The frequency mismatch issue between the master and slave oscillators in Figure 17.1(b) can be circumvented by using a single-gated VCO, as is shown in Figure 17.1(c) [11].

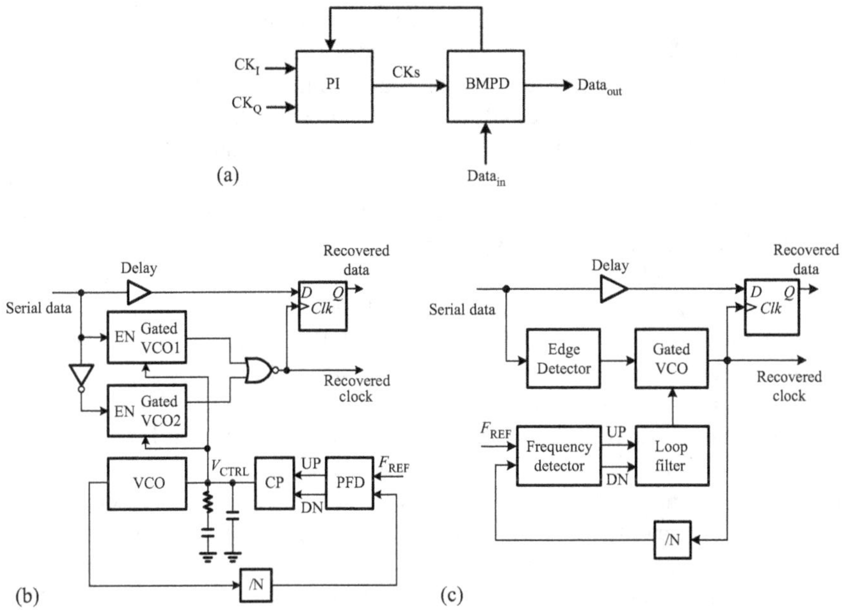

Figure 17.1 (a) PI-based BMCDR, (b) BMCDR with replica gated VCO, And (c) BMCDR with single-gated VCO

Additionally, to minimize the phase error accumulation when receiving a long run of CID bits, frequency mismatch between the transmitter and the receiver should be eliminated by continuously tracking the frequency of the input data.

17.2 BMCDR with jitter filtering

The shortcomings of the conventional BMCDRs can be summarized as follows: (1) Limited data jitter suppression capability due to the feed forward and open-loop operation. In BMCDRs, sampling clock is resynchronized with the data edge instantly. In this context, the data jitter translates to the clock output without much filtering. It deteriorates the clock jitter performance, which is a common drawback for most BMCDRs. (2) Lack of frequency offset tracking capability. (3) In GVCO-based architectures, the start-up time of the gating oscillator limits the operating speed. To overcome the aforementioned issues, a BMCDR is proposed to achieve instant phase locking, free from frequency offset problem, and provide jitter suppression capability simultaneously [3]. Moderated by a lock detector (LD), the strategy is to adopt an open-loop (feed forward) operation for instant phase-locking response at the beginning, while resume to a close-loop (feedback) operation during phase tracking for noise filtering. Meanwhile, an $1/n$ rate GVCO with selective gating is adopted to facilitate high-speed (10 Gbps) operation with a high energy efficiency (1.24 pJ/bit).

Figure 17.2 shows the BMCDR architecture. It consists of a 1/5-rate selective-gating VCO (SGVCO), a 1/5-rate phase detector (PD), an LD, and a dual-mode PLL with a gating controller. As the clock rate is much lower than the data rate, a special gating technique is required to properly modulate the SGVCO at different stages through the input data. In contrast to the conventional full-rate or half-rate BMCDR architectures [7,12], the SGVCO and PD frequencies are substantially reduced to 1/5 the data rate. It greatly relaxes the timing constraint of the SGVCO recovery time and facilitates a high-speed operation with a lower power consumption. Meanwhile, 1:5 demultiplexing is accomplished after data recovery, which also alleviates the design challenge of the succeeding demultiplexer.

The detailed operation of the proposed BMCDR is described as follows. Under the start-up condition, both the *lock* (=0) and *gate* (=0) control signals are low. Frequency initialization is achieved by using a PLL with a local reference. The *lock* indicator would be flagged (=1) once the frequency initialization is completed. It ensures that the GVCO's frequency is preset at the vicinity of the targeting frequency (1/5 data rate). When the received signal strength indicator asserts that the data bursts in, the gating pulse (*gate*) would be flagged (=1). The BMCDR is then switched to the data gating mode for instant phase locking. During this mode, the gating controller would pass the input data to the corresponding stages of SGVCO that has the closest transition phases. Meanwhile, the CDR is open loop by

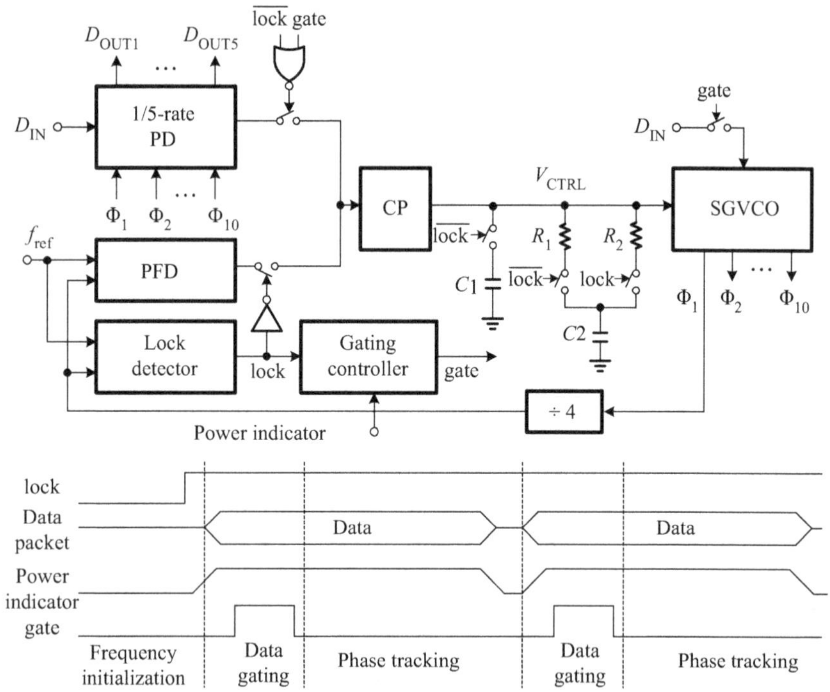

Figure 17.2 Burst-mode clock and data with jitter filtering

disabling the phase frequency detector (PFD) and the 1/5-rate PD to prevent the SGVCO from being mutually pulled by the PD loop during the gating process.

Typically, the GVCO can be synchronized with the input data within several unit-interval (UI), thanks to its open-loop operation. In this architecture, the data gating time is programmable by the gating pulse generator, which is inversely proportional to the data rate. After the phase gating process, the proposed BMCDR then resumes its close-loop operation by resetting *gate* (=0) signal and is reconfigured as a second-order, 1/5 data rate phase tracking CDR. Thus, it can provide frequency tracking and jitter suppression capability while tolerating the long-run CID bits input data. Another BMCDR with noise filtering is proposed in [13], where gating and close-loop PLL operation are working concurrently. The jitter transfer bandwidth is adjusted by changing the data injection rate. It must be treated with caution to avoid mutual pulling between the ILO and PLL during the phase tracking process while setup the jitter tolerance (JTOL) bandwidth.

17.3 Burst mode CDR circuit implementation

17.3.1 Lock detector

The LD is utilized to monitor the frequency accuracy after frequency initialization by the PLL. Figure 17.3(a) shows the detailed architecture. When the PLL approaches phase locking, the feedback signal at the divider output f_{div} would dangle around the edge of f_{ref}. The locked condition is asserted when the timing difference (Φ_e) between f_{div} and f_{ref} is less than a predefined timing window (T_d) over a continuous period of time (64 reference cycles). To detect this situation, f_{div} is sampled by both f_{ref} and its delay version and vice versa.

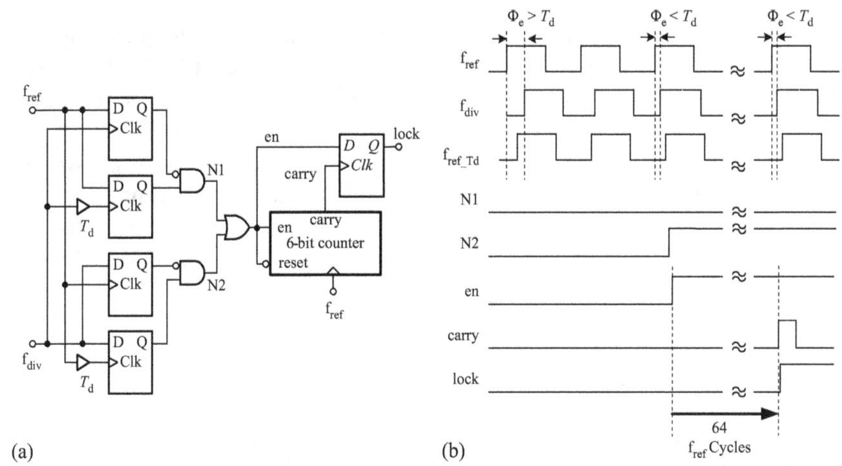

Figure 17.3 (a) Proposed lock detector and (b) proposed lock detector timing diagram

Figure 17.3(b) illustrates the scenario where the rising edge of $f_{\rm div}$ falls into the timing window of $f_{\rm ref}$, where $N1$ or $N2$ would be raised to high. The confident level of frequency locked is ensured by a 6-bit counter, whose output would then issue the locked condition. Assuming that the locking window is T_d with a variation of ΔT, under the locked condition, it can be shown that:

$$\left|\frac{1}{f_{\rm div}} - \frac{1}{f_{\rm ref}}\right| \times 64 \le 2 \times (T_d + \Delta T) \tag{17.1}$$

$$\frac{1}{\frac{1}{f_{\rm ref}} + \frac{T_d+\Delta T}{32}} \le f_{\rm div} \tag{17.2}$$

Thus, the percentage of frequency error can be derived as:

$$\left|\frac{\Delta f}{f_{\rm ref}}\right| = \frac{\left(\frac{T_d+\Delta T}{32}\right)f_{\rm ref}}{1+\left(\frac{T_d+\Delta T}{32}\right)f_{\rm ref}} \cong \left(\frac{T_d+\Delta T}{32}\right)f_{\rm ref} \tag{17.3}$$

According to (17.3), it can be seen that the issue of locking window variation can be circumvented by the confident counter. With a 20% variation ($\Delta T = 0.2\times T_d$), we arrive at the frequency accuracy of 0.037% in the locked state. Thus, it can tolerate up to 72 CID bits in the phase gating mode.

17.3.2 Reconfigurable loop filter

In the frequency initialization mode, the SGVCO is brought to the vicinity of 1/5 data rate through a third-order PLL. Figure 17.4(a) shows the third-order PLL linear model. The PLL loop bandwidth ($\mathrm{PLL_{LPBW}}$) can be approximated as:

$$\mathrm{PLL_{LPBW}} \cong \frac{I_{\rm CP} \times K_{\rm VCO} \times R_1}{N} \tag{17.4}$$

During frequency initialization, major design concerns are loop stability and minimizing clock jitter to ensure that the clock edge would fall into the locking window. In the phase tracking mode, the BMCDR is reconfigured as a second-order bang-bang PLL [14]. Assuming that the phase difference between the data edge and the clock edge is small compared to the data period, and the CDR loop bandwidth is much smaller than the data rate, the loop behavior can be approximated using a linear model. Figure 17.4(b) shows the second-order CDR linear model. Its loop bandwidth ($\mathrm{CDR_{LPBW}}$) and damping factor (ξ) can be approximated as:

$$\mathrm{CDR_{LPBW}} \cong K_{\rm PD} \times K_{\rm VCO} \times I_{\rm CP} \times R_2 \times D_{\rm T} \tag{17.5}$$

$$\xi = \frac{R_2}{2}\sqrt{D_{\rm T} \times I_{\rm CP} \times C_2 \times K_{\rm PD} \times K_{\rm VCO}} \tag{17.6}$$

where D_T is the data transition density and is about 0.5 for a random input data.

During the CDR mode, the major issues on loop filter design are to provide an adequate loop bandwidth and damping factor for jitter suppression. As the loop bandwidths and damping factors in the PLL and CDR modes are different, the loop

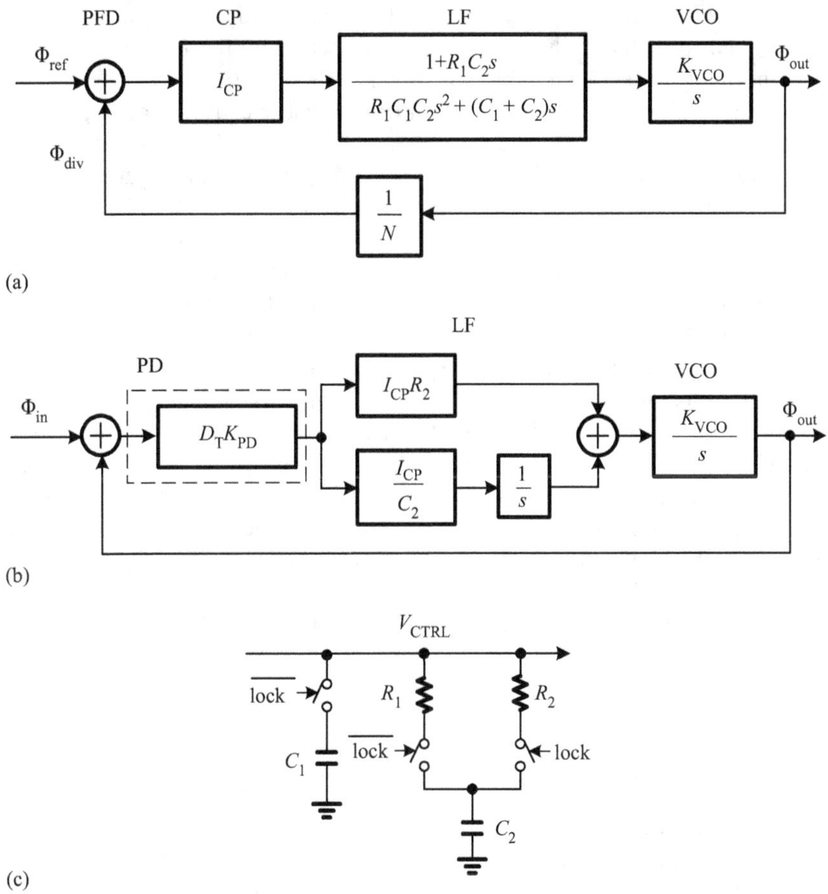

Figure 17.4 (a) A third-order PLL behavioral model, (b) a second-order CDR behavioral model, and (c) reconfigurable loop filter

filter must be reconfigured to fit the target specifications. Figure 17.4(c) shows the circuit schematic of the reconfigurable loop filter, which is controlled by the *lock* signal. The loop bandwidths in these two modes can be optimized by properly choosing R_1 and R_2, respectively. Besides, the charge pump currents are also programmable to provide more design flexibility. As the CDR loop bandwidth varies under different D_T, R_2 is adjustable by using tunable resistor string to accommodate D_T variations.

17.3.3 Selective gating VCO

Conventionally, GVCO-based BMCDRs are operated in a full-rate or half-rate architecture [7–12]. To enable high-speed operation, the oscillation frequency of the GVCO can be running at the $1/n$ data rate to mitigate the start-up problem during data gating mode [3,8–15]. On the other hand, multiphase sampling requires careful layout to maintain the timing matching between different signal traces to minimize timing offset. To comply with an 1/5-rate operation, the GVCO is composed of a five-stage ring oscillator accompanied by a selectively gating generator, as is shown in Figure 17.5(a). The gating generator selectively passes the data edge onto the different stages of the GVCO, so as to perform phase gating properly (SGVCO). The concept of SGVCO can be easily applied to BMCDRs with different decimation factor *n*.

Figure 17.5(b) illustrates the timing of the 1/5-rate selective gating process, where the sampling period is divided into five zones (GE1, GE3, GE5, GE7, and GE9). During the data gating mode, the input data continuously samples the five zones (GE1, GE3, GE5, GE7, and GE9) to pick the corresponding SGVCO stage that has the closest transition edges with the D_{IN}. The gating generator then provides the gating pulses (GS1, GS3, GS5, GS7, and GS9) to the corresponding stages of the SGVCO. Guiding by these gating pulses, D_{IN} is capable of adjusting the SGVCO output phases at the 1/5 data rate, which drastically mitigates the speed and power requirement of the BMCDR. When the SGVCO is locked, D_{IN} will be aligned with Φ_1, Φ_3, Φ_5, Φ_7, and Φ_9. Meanwhile, Φ_2, Φ_4, Φ_6, Φ_8, and Φ_{10} can be utilized for data recovery. As the input data are retimed by the 1/5-rate SGVCO, it also provides 1:5 demultiplexing at the BMCDR output after data resample.

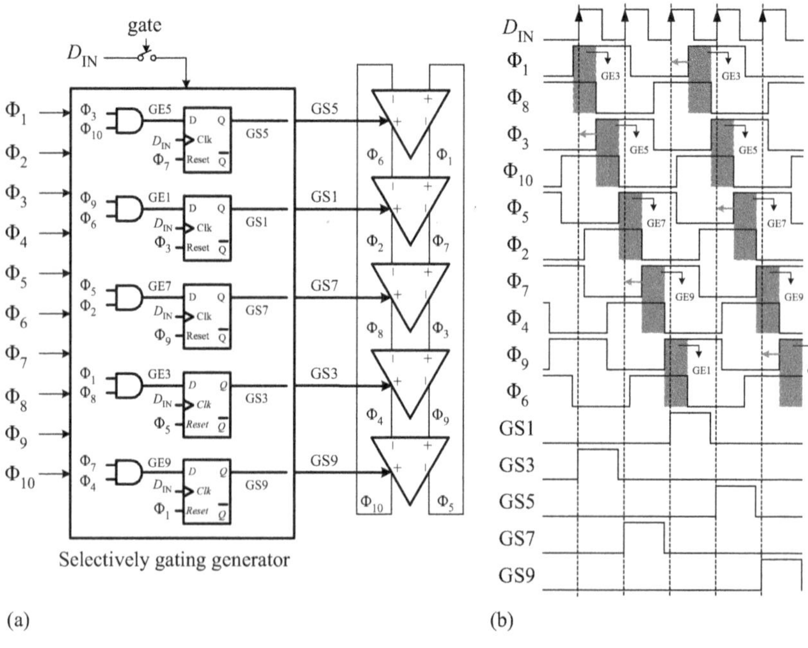

Figure 17.5 (a) SGVCO architecture and (b) timing diagram

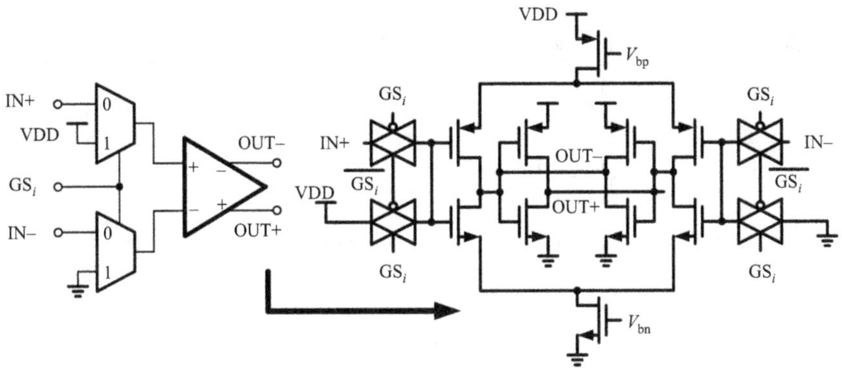

Figure 17.6 SGVCO's differential delay cell

Figure 17.6 depicts the circuit schematic of the SGVCO's delay cell. The multiplexer (MUX)-based structure provides dual-mode operation controlled by GS_i. When $GS_i = 1$ ($i = 1, 3, 5, 7$, and 9), the delay cell is operated in gating mode whose output will be reloaded to high/low for instant phase resynchronization. On the contrary, when $GS_i = 0$, the SGVCO is configured as a closed loop oscillator whose output is toggled by its input. A cross-coupled latch is placed in the middle of the delay cell to accelerate switching speed and reduce phase noise. On the other hand, it provides nearly rail-to-rail output swing, so as to have a higher immunity to external noise disturbance.

According to the SGVCO architecture in Figure 17.5, the intrinsic phase mismatch between the SGVCO output and data edge includes the CK-to-Q delay for GS_i generation and the transmission gate delay in the SGVCO. It amounts to about 0.36 UI at 10 Gbps operation. This issue is circumvented by inserting a dummy buffer at the data path of the 1/5-rate phase detector.

17.3.4 An 1/5-rate phase detector

A binary phase detector is commonly used in a high-speed CDR to generate the phase lead-lag information. By extending the Alexander-type phase detector [16] to a 5-way time-interleaved architecture, an 1/5-rate PD can be implemented to generate UP/DN signals out of 10 sampling points. Figure 17.7(a) illustrates the PD circuit schematic, and the corresponding timing diagram is shown in Figure 17.7 (b). Here the circles and solid dots on D_{IN}, respectively, represent the sampling points by (Φ_1, Φ_3, Φ_5, Φ_7, and Φ_9) and (Φ_2, Φ_4, Φ_6, Φ_8, and Φ_{10}). As the input data are sampled sequentially, they are realigned using retiming registers before extracting the lead/lag information through the XOR gates. The UP/DN information for charge pump is then derived by tallying the UP_i/DN_i ($i = 1, 2, 3, 4$, and 5) collected from the five sub-PDs. In the locked state (Φ_1, Φ_3, Φ_5, Φ_7, and Φ_9) are aligned with the data edge. Meanwhile (Φ_2, Φ_4, Φ_6, Φ_8, and Φ_{10}) are served as the sampling phases to recover and demultiplex the input data.

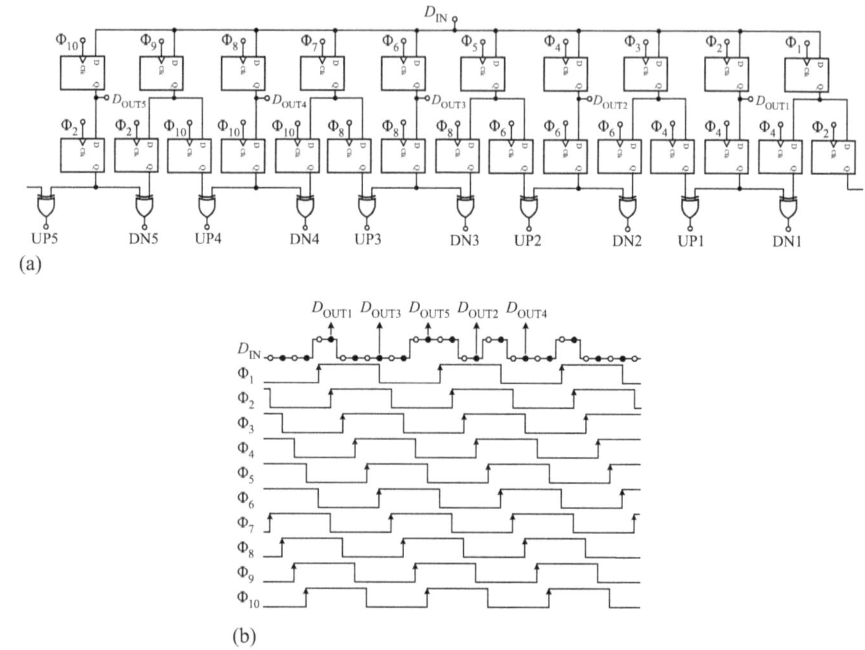

Figure 17.7 (a) An 1/5-rate phase detector and (b) timing diagram

17.3.5 Dual-mode loop filter

Figure 17.8 shows the dual-mode loop filter combined with charge pump and phase detector. Here the UP and DN pulses are generated from either the PFD or the 1/5-rate PD, which are steered by the *lock* and *gate* control signals through the select logic. During the frequency initialization mode, *lock* and *gate* signals are both low and the charge pump current I_{CP1} is dictated by the PFD. Meanwhile, the loop filter is configured as a second-order filter. Once frequency initialization is completed, both the *lock* and *gate* signals are raised high and the BMCDR then steps into the data gating mode.

During the data gating mode, the UP and DN signals are both low and the output voltage V_{CTRL} is kept in the hold mode. The PLL is open loop to avoid confusing the phase detector during data gating. According to the simulation results, the leakage current of the loop filter is about 10 nA, which results in about 1.2 kHz frequency drift during the open-loop operation. It corresponds to about 0.6 ppm frequency drift and can be easily recovered by the gating process. Definitely, the PLL can also be implemented in an all-digital (AD) manner (ADPLL) to fully circumvent the frequency drift issue without departing from the operation principles.

After the data gating process, which lasts for couples of bit time, the control signal's *lock* remains high and *gate* returns to low, and the BMCDR would then

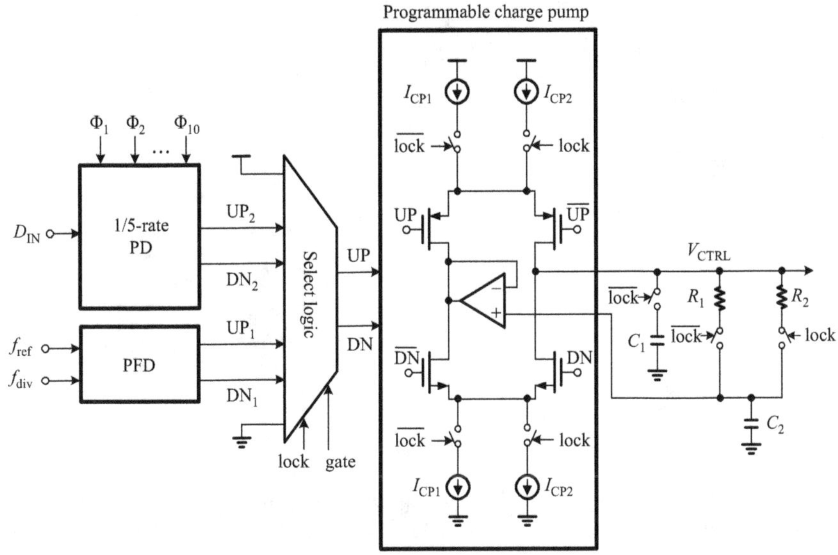

Figure 17.8 Dual-mode loop filter combined with charge pump and phase detector

resume its phase tracking process. During this mode, the charge pump with tail current I_{CP2} is then handed over to the 1/5-rate PD, and the loop filter is reconfigured as a first-order filter by disconnecting C_1. For the dual-mode operation, the charge pump current and the resistors R_1 and R_2 are programmable to meet different loop bandwidth requirement. A unity gain buffer is utilized to mitigate charge sharing during the ON/OFF switching of the charge pump [17]. The frequency jump associated with voltage glitch can also be alleviated by using a dummy switch. The mode switching is completed during the preamble of each data packet.

17.3.6 Burst mode CDR experimental results

An experimental prototype of a 10 Gbps, 1/5-rate BMCDR is fabricated in a 55-nm complementary metal-oxide-semiconductor (CMOS) technology. Figure 17.9 shows the chip micrograph. The chip area is pad limited, and the core area is only $200 \times 150\ \mu m^2$. The SGVCO and the 1/5-rate PD are placed close to each other to mitigate sampling phase mismatches. In this experimental prototype, the reference clock frequency is 500 MHz and the frequency multiplication factor is 4, which generates a 2-GHz sampling clock during frequency initialization. The BER performance is verified using Anritsu pattern generator (MU181020A) and Anritsu error detector (MU181040A). By applying a 2^7-1 PRBS, jitter stressed input data at 10 Gbps to the BMCDR, the measured recovered clock spectrum at 2 GHz is shown in Figure 17.10(a). The measured phase noise performance from a 2-GHz carrier is shown in Figure 17.10(b), which is about −105.68 dBc/Hz at 1 MHz offset.

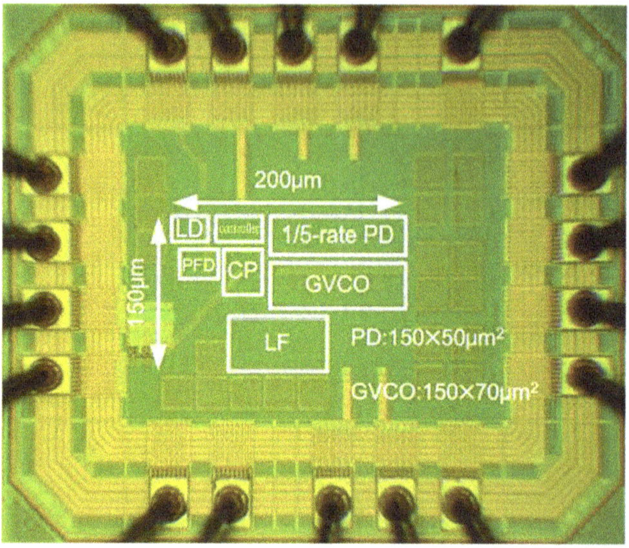

Figure 17.9 Chip micrograph

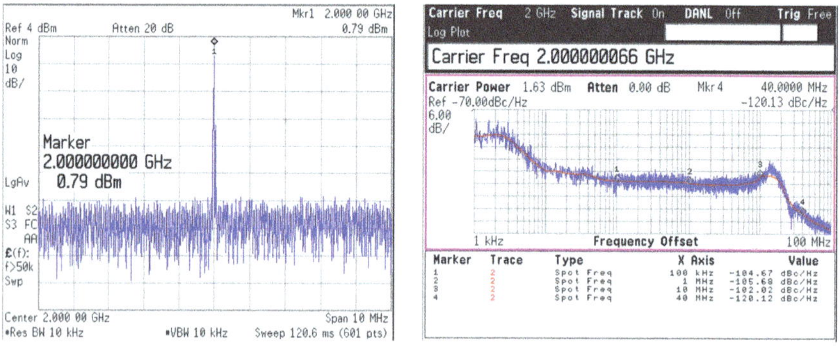

(a) (b)

Figure 17.10 Measured (a) recovered clock spectrum and (b) phase noise @ 2 GHz

To verify the jitter suppression capability, the 10 Gbps input data without and with jitter stressed are applied to the BMCDR, as shown in Figure 17.11(a) and (b), respectively. The corresponding eye diagrams at the demultiplexer output are shown in Figure 17.11(c) and (d) for comparison. Without jitter stressed, the measured data jitter at the 10 Gbps input is about 15.8 ps_{pp}, while its demultiplexed data jitter at the 2 Gbps output is about 42.6 ps_{pp}. On the other hand, with a 4-MHz, 0.22 UI_{pp} sinusoidal jitter, the data jitter at the 10 Gbps input is about 44 ps_{pp}, and

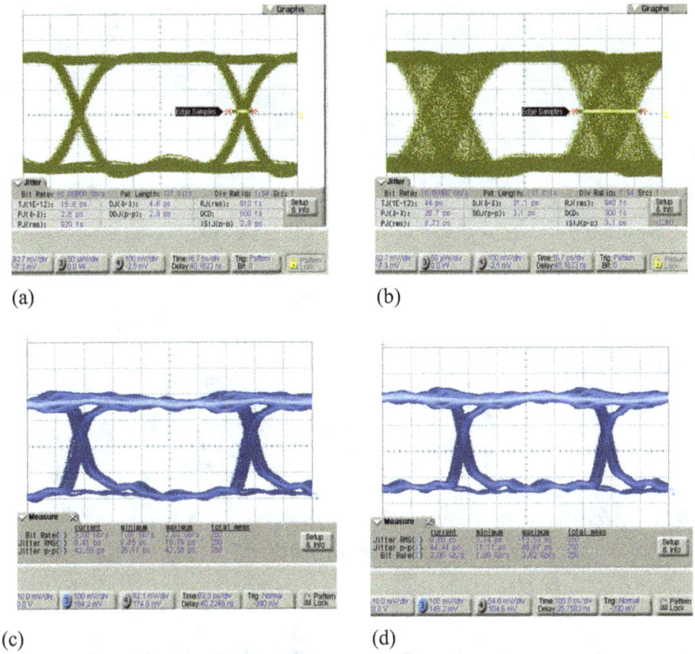

Figure 17.11 Measured 10 Gbps input data eye diagram (a) without and (b) with 4 MHz, 0.22 UI_{pp} sinusoidal jitter. Demultiplexed 2 Gbps output data eye diagram (c) without and (d) with sinusoidal jitter at 10 Gbps input data

its data jitter at the 2 Gbps output is about 44 ps_{pp}. Note that all the measured jitter includes the intrinsic noise of the oscilloscope and other noise sources associated with the tapered buffer. The data jitters at the CDR output are relatively constant despite the jitters at the data input. Figure 17.12 shows the measured recovered clock and its jitter histogram with a jitter-stressed input data (4 MHz, 0.22 UI_{pp}). The clock jitter is 2.94 ps_{rms} (24.89 ps_{pp}) at 2 GHz. It can be seen that the input data jitter has been significantly suppressed at the output clock.

With a D_{IN} amplitude of 0.4 $V_{p\text{-}p}$, the BMCDR combining with a 1:5 demultiplexer achieves a BER of less than 10^{-12}. Figure 17.13(a) shows the JTOL test under the target BER ($< 10^{-12}$). The JTOL test reveals that it passes the 10G-PON JTOL mask with a 2^7-1 PRBS input pattern [18]. Figure 17.13(b) shows the locking behavior for a 10-Gbps input data. With a 10-Gbps repeated data sequence of (00111 00100 11111 11101 11100 11011 11011 00101), one of the 1:5 demultiplexed data at 2 Gbps is observed as (10100110 ...). The recovered data are captured at the first input data edge. Thus, the lock time is less than 1 UI. The latency including the serial to parallel data conversion and I/O buffer is about 5 ns. Excluding the I/O buffers, the 10-Gbps BMCDR and 1:5 demultiplexer consume only 12.4 mW. It accomplishes a BMCDR and 1:5 demultiplexer with a high

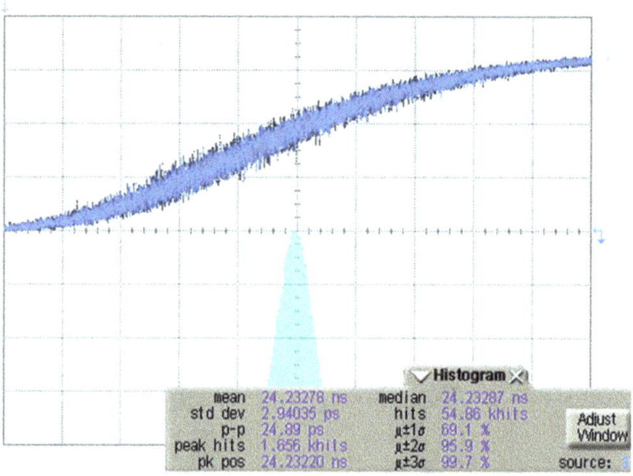

Figure 17.12 Measured recovered clock jitter and its histogram @ 2 GHz (with 4 MHz, 0.22 UI_{pp} sinusoidal jitter at 10 Gbps input data)

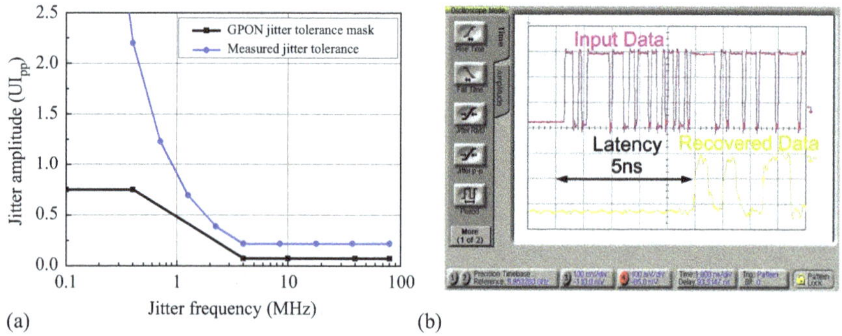

Figure 17.13 (a) Measured jitter tolerance with 10 G-PON jitter tolerance mask (10 Gbps, 2^7–1 PRBS) and (b) locking behavior

energy efficiency of 1.24 pJ/bit. On the other hand, the proposed architecture is also highly area efficient, which only occupies 0.03 mm^2.

17.4 Fully integrated optical receiver with baud-rate CDR

Conventionally, an optical receiver is composed of multi-chips in different technologies. For example, the transimpedance amplifier (TIA), limiting amplifier, and clock and data recovery circuits are fabricated in BiCMOS or CMOS technologies, respectively. In this session, a single-chip optical receiver, which consists of a current boost preamplifier, a baud-rate CDR, and a 1:4 demultiplexer, is presented [4]. Monolithic integration of the receiver front-end significantly reduces the form factor and enables high-density optical IOs. Additionally, it circumvents the signal-integrity issue associated with the electrostatic discharge (ESD) capacitance and boding inductance for the integration of multi-chips or multi-packages. Implemented in a 40-nm CMOS technology, the core area of the chip is only 0.09 mm^2. The experimental prototypes demonstrate a high input sensitivity of −8.7 dBm with a high energy efficiency of 2.1 pJ/bit at 25 Gb/s operation (the PD responsivity is 0.53 A/W).

17.5 Architecture of optical receiver with baud-rate CDR

Recently, integrating-type optical receiver front-ends have drawn many research attentions [19,20] and demonstrated superior energy and area efficiency compared to conventional TIA-based architectures [21,22]. Baud-rate CDRs are commonly incorporated with the integrating-type receivers. Compared to conventional 2X CDRs that require edge samples for clock timing adjustment, no quadrature phase samples are needed for clock recovery in a baud-rate CDR. It reduces the number of sampling phases by half and also saves the power and area for clock tree distributions [23,24]. Moreover, in contrast to conventional Mueller–Müller baud-rate CDRs that may encounter ambiguous locking point problems for ideal square wave input signals [25], this issue is circumvented by incorporating an integration-type receiver frond-end in this design. Palermo *et al.* [26] proposed a baud-rate phase detector by comparing the integrated signal levels for specific 4-bit patterns. The major demerits are that the voltage headroom of integrators would limit the run length of receiving CID bits. Meanwhile, the integrating voltage is susceptible to noise disturbance for the retention of long-run data. Raj *et al.* [27] utilized a source synchronized receiver with ILO for timing recovery. The unavoidable phase offset between the clock path and the data path leads to a residual timing error and sub-optimal sampling in data recovery. Musa and Carusone [23] proposed a timing recovery circuit based on sign-sign minimum mean square error algorithm with a slope detector. However, it becomes very challenging to generate slope and error signals at a high-speed operation. Han *et al.* [25] proposed a baud-rate PD with a dithered slope detector. The timing step control is critical for high-speed operation. Kim *et al.* [28] proposed a sub-baud-rate CDR by combining direct voltage sampler and integrator sampler. The timing matching between these two paths is challenging to maintain a low BER.

To overcome the aforementioned issues, the proposed optical receiver is composed of a periodically reset integrating-type receiver incorporating with a new baud-rate CDR [4]. It is capable of receiving a data stream with a long run of CID bits without suffering from the overflow issues. A longer reset time during each sample reduces the residual on the integrator and the inter-symbol interference (ISI) effect on PD. It also reduces the integration time and the conversion gain of the receiver front-end. On the other hand, as the data edge is also eliminated by the periodically reset operation, a new baud-rate phase detector and a CDR circuit are proposed for the timing recovery.

Figure 17.14(a) shows the overall receiver architecture, and Figure 17.14(b) illustrates the timing diagram. It is composed of a current preamplifier [19]

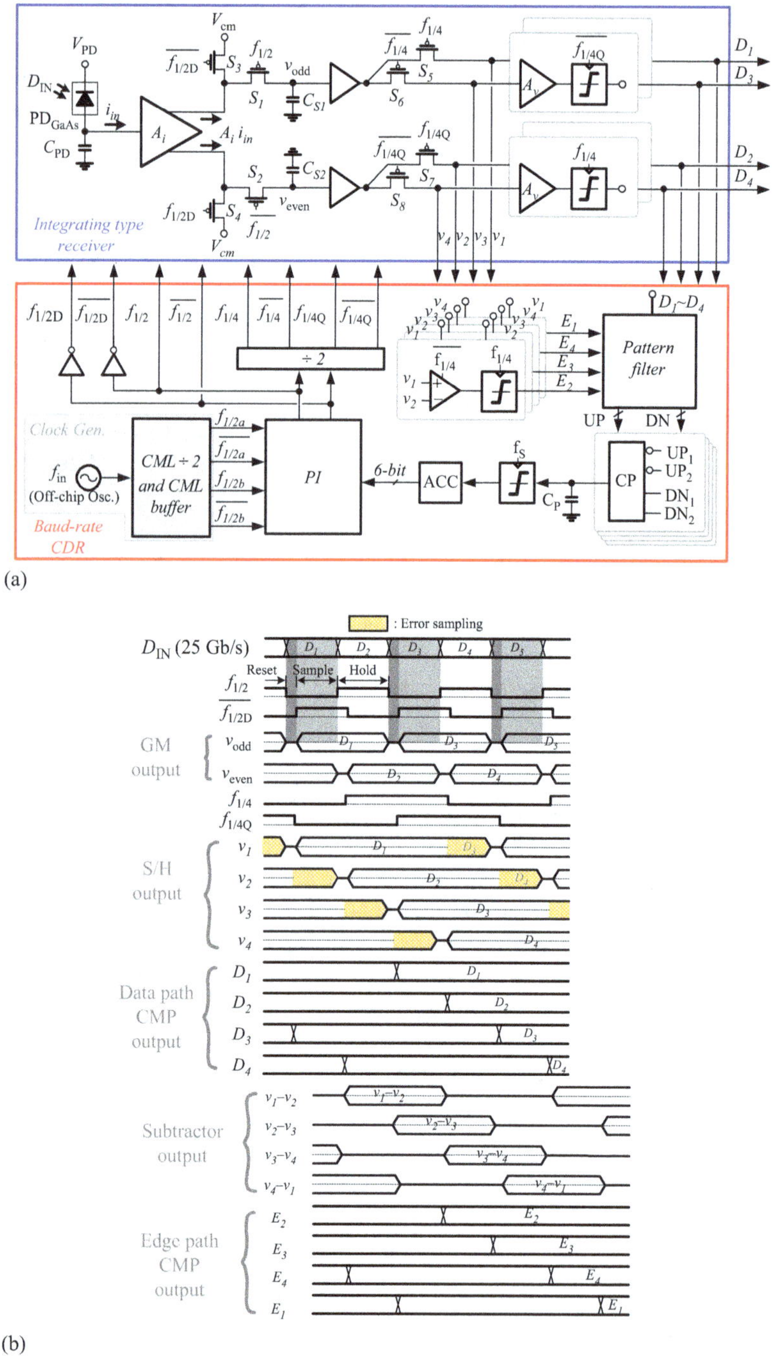

Figure 17.14 (a) Proposed time-interleaved integrating-type optical receiver with baud-rate CDR and (b) its timing diagram

followed by a two-stage cascaded 1:2 demultiplexer and a baud-rate clock and data recovery circuit. The photo current (i_{in}) of the data input D_{IN} is first enlarged by the current amplifier (A_i), whose output is then demultiplexed by 2 and integrated on C_{S1} and C_{S2} for current-to-voltage conversion. The integrating output voltages (v_{odd}, v_{even}) are further demultiplexed as ($v_1 - v_4$) to drive four slicers, so as to generate the demultiplexed digital output ($D_1 - D_4$). Meanwhile, the analog voltages ($v_1 - v_4$) are also applied to four differential voltage amplifiers to generate voltage differences, which are subsequently quantized as edge information ($E_i, i = 1.2.3.4$). Moderated by the digital output ($D_1 - D_4$), the edge information then passes through a pattern filter to generate UP/DN signals, which are then fed into a hybrid loop filter to adjust the sampling phase of the integrating-type receiver.

As the UP/DN signals are generated at baud rate, it imposes severe design challenges to implement an all-digital loop filter. To overcome this design challenge, the proposed hybrid loop filter is composed of a charge pump filter in the first stage, whose output is then decimated to drive a digital accumulator in the second stage. The digital filter output is utilized to control a 6-bit PI for the sampling phase adjustment over 2 UI (half-rate sampling clock). The loop dynamics of the CDR is designed based on the PD characteristic, which depends on the statistics of UP/DN signals.

The oscillator (f_{in}) of the optical receiver is running at a full-rate clock (25 GHz) and is applied externally, which is then subdivided by 2 and 4 to provide the half-rate ($f_{1/2}$) and the quarter-rate ($f_{1/4}$) clocks for the receiver front-end and the baud-rate CDR. The sampling phase is tuned by interpolating the half-rate clock. Thus, only a single PI is needed, and both the half- and quarter-rate (I/Q) sampling phases can be rotated simultaneously. Additionally, to comply with the JTOL specifications, the CDR loop bandwidth is programmable through the hybrid loop filter.

17.6 Circuit implementation of optical receiver with baud-rate CDR

17.6.1 Current amplifier

Figure 17.15 shows the circuit schematic of the current amplifier, which is composed of a trans-impedance gain stage followed by dual-output trans-conductance gain stage [19]. The current amplifier drives two integrators for the first 1:2 demultiplexer. A negative impedance converter is used in the G_M stage to increase its output impedance. It provides the advantage of a high sensitivity with low-power consumption. The integrator output is periodically reset through switches ($M_{s1} - M_{s3}$). Thus, a long run of CID bits is tolerable. In contrast to the conventional TIAs that are designed with a bandwidth of about 0.7 X data rate, the bandwidth of the current amplifier is chosen as about 0.4 X data rate to maximize the receiver signal to noise ratio (SNR) under a constant gain bandwidth product constraint. Typically, the input sensitivity of the optical receiver directly compromises with its 3 dB bandwidth. By reducing its bandwidth from 0.7 X to 0.4 X data rate, the input sensitivity improves by about 2.43 dB with a manageable ISI.

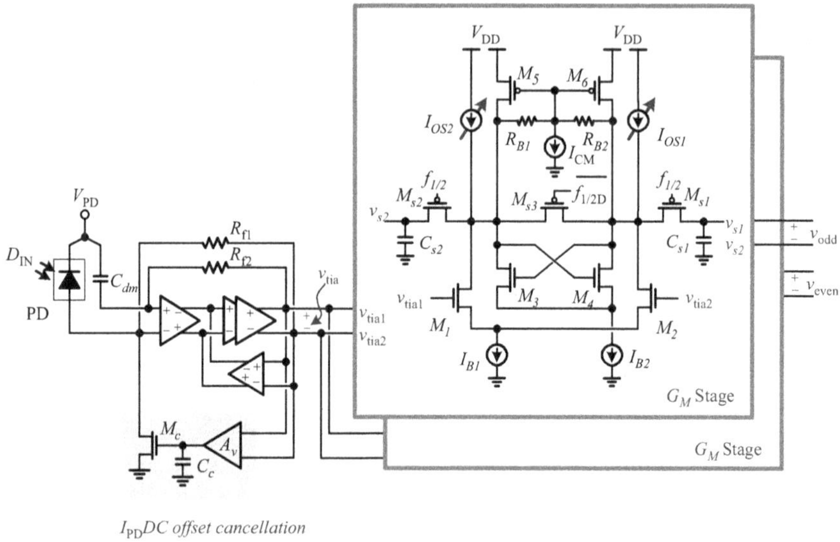

Figure 17.15 Dual-output current amplifier with current integrator

17.6.2 Comparator

The integrated output voltages are applied to preamplifiers and clocked comparators to generate the data output (D_1-D_4) and the edge (E_1-E_4) information. As the eye opening before the data slicers is critical to the overall receiver sensitivity, active-feedback preamplifiers are utilized to further boost up the integrating voltage swing and alleviate the kickback noise, as is shown in Figure 17.16. In addition, the effective voltage gain of the dynamic comparator can be expressed as:

$$A_{\text{CMP}} \approx \exp\left(\frac{g_m T_R}{C_L}\right) \tag{17.7}$$

where g_m, T_R, and C_L, respectively, represent the trans-conductance of the dynamic comparator, the regeneration time of the latch, and the parasitic capacitance at the comparator output.

According to (17.7), A_{CMP} can be increased exponentially by enlarging g_m and T_R. Incorporating with the quarter-rate CDR, it can provide a higher gain while relaxing the bandwidth requirement of the preamplifier and the clocked comparator.

17.6.3 Baud-rate PD

The operation principle of the proposed baud-rate CDR is described in Figure 17.17. A half-rate receiver front-end with a two-way time-interleaved architecture is first considered for the simplicity of illustration. In contrast to the oversampling-based CDRs that extract the edge information of input clock by

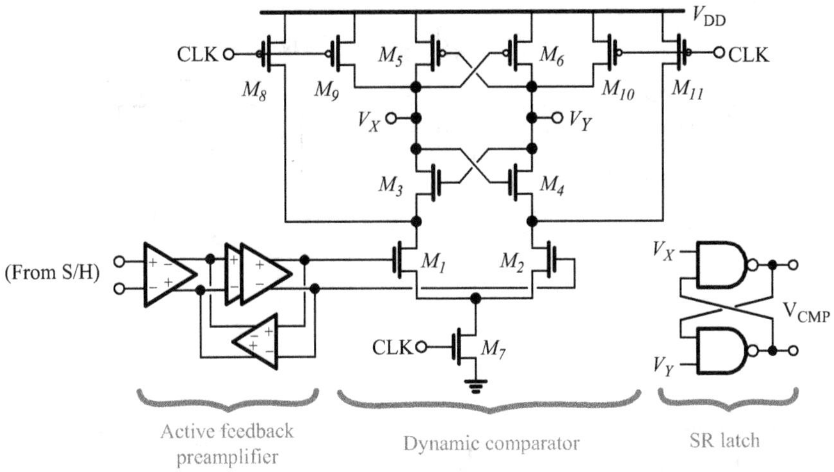

Figure 17.16 Preamplifier and dynamic comparator

detecting data transitions between 2 bits (1,0) and (0,1), the clock edges (E_n) are detected based on specific 3-bit patterns. Here the DN signals are detected through the input patterns (D_{n-1}, D_n, and D_{n+1}), which comprise a data transition followed by two identical bits such as (1,0,0) or (0,1,1). On the other hand, the UP signals are detected through the input patterns that consist of two identical bits followed by a data transition such as (1,1,0) or (0,0,1). The truth table for UP/DN detection is shown in Figure 17.17(b), which is generated based on the detection of the integration voltages (v_{n-1}, v_n, v_{n+1}) from (D_{n-1}, D_n, D_{n+1}). Here a pattern filter is implemented to activate the UP/DN pulses when a valid input pattern is received. Assuming that the sampling clock edge is leading the data edge, the integration timing window of 1 unit interval (UI) would overlap (D_{n-1}, D_n) and (D_n, D_{n+1}), as shown in Figure 17.17(c). Thus, the integrating voltages v_n would become smaller than v_{n+1} with the input pattern of (0,1,1). Contrarily, v_n would become larger than v_{n+1} with the input pattern of (1,0,0).

Define:

$$E_n = \mathrm{Sgn}[v_{n-1} - v_n] = \mathrm{Sgn}[\Delta v_n] = \mathrm{Sgn}[\alpha \times \Delta\theta] \tag{17.8}$$

$$E_{n+1} = \mathrm{Sgn}[v_n - v_{n+1}] = \mathrm{Sgn}[\Delta v_{n+1}] = \mathrm{Sgn}[\alpha \times \Delta\theta] \tag{17.9}$$

where α is the coefficient of the integrator and $\Delta\theta$ is the phase difference at the PD input. The Boolean function for the DN signal can then be described as:

$$\mathrm{DN} = D_{n-1} \cdot \overline{D_n} \cdot \overline{D_{n+1}} \times E_{n+1} + \overline{D_{n-1}} \cdot D_n \cdot D_{n+1} \times \overline{E_{n+1}} \tag{17.10}$$

Similarly, the Boolean function for the UP signal can be described as:

$$\mathrm{UP} = D_{n-1} \cdot D_n \cdot \overline{D_{n+1}} \times E_n + \overline{D_{n-1}} \cdot \overline{D_n} \cdot D_{n+1} \times \overline{E_n} \tag{17.11}$$

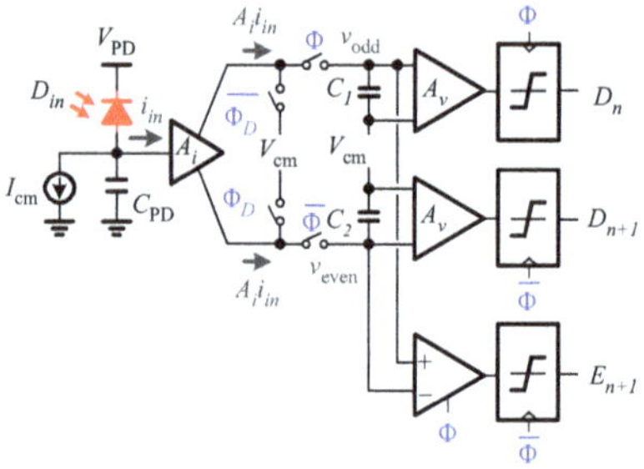

(a)

D_{n-1}	D_n	D_{n+1}	E_n	E_{n+1}	DN	UP
1	0	0	x	1	1	0
0	1	1	x	0	1	0
1	1	0	1	x	0	1
0	0	1	0	x	0	1

(b)

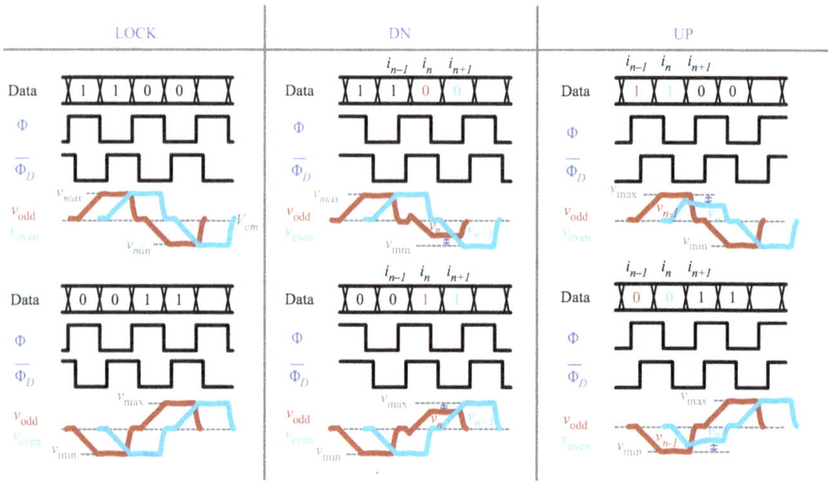

(c)

Figure 17.17 Concept of the baud-rate phase detector (a) schematic, (b) truth table, and (c) integrating voltage waveforms

Conventional MM-CDRs rely on a reference-level V_{REF} for UP/DN (early/late) detection [24,29]. For it to function properly, the input signal level must be adjusted through an automatic gain control (AGC) loop to be close to V_{REF}, which in general takes ~kUI to converge [29]. In the proposed PD architecture, the reference level for phase error detection is automatically generated from the consecutive identical bits (such as 011 or 110). It facilitates the convergence of the CDR loop without depending on another (low-speed AGC) loop. On the other hand, the edge detection based on 3-bit patterns reduce the probability of transition detection, which would be taken into account for loop bandwidth design. When the CDR is receiving an ISI stressed input data, the probability distribution of UP/DN signals may also vary, which changes the effective loop gain of the CDR. The probability density of the proposed baud-rate PD under different phase shift ($\Delta\theta$) conditions is shown in Figure 17.18. As is shown in Figure 17.18(a), the integrating voltages (v_{odd} or v_{even}) at the receiver front-end depend on the phase offset ($\Delta\theta$). They

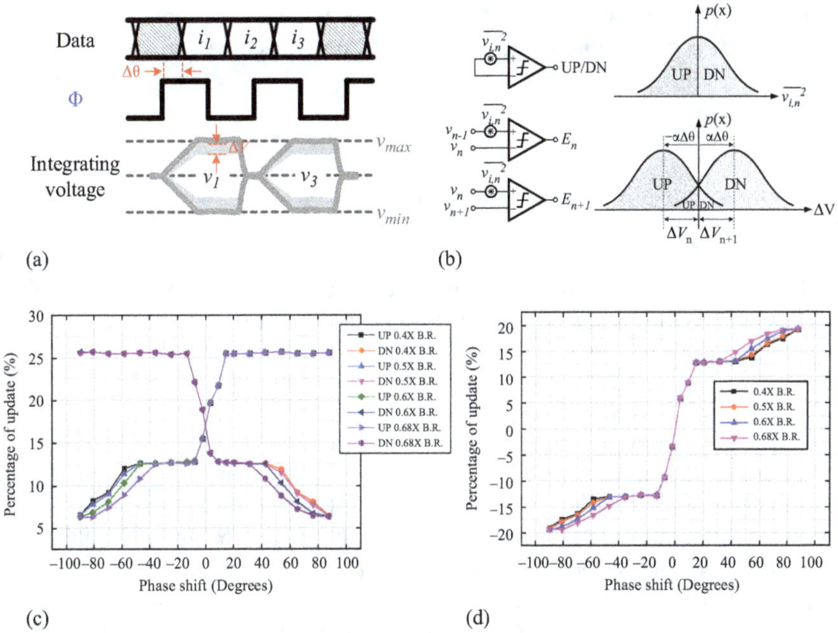

Figure 17.18 Proposed baud-rate PD (a) integrating voltage, (b) probability density function of the comparator input noise, (c) probability density of UP/DN pulses with a limited receiver bandwidth, and (d) simulated PD gain curve

reach the maximum when $\Delta\theta = 0$ and decreases along with the increment of $\Delta\theta$ as $0 < \Delta\theta < 2\pi$. Assume that the input data pattern is random and generated with an equal probability, and the input referred noise of the data slicers is approximated as a Gaussian noise. Figure 17.18(b) illustrates the probability density function $p(x)$ of UP/DN at the slicer output. Since the UP signal will be generated with the input patterns of (1,1,0) or (0,0,1), the probability of generating these two patterns is 1/4. If $\Delta\theta = 0$, the average of Δv_n (or Δv_{n+1}) would be close to zero. Thus, the probabilities of UP and DN pulses are equal (50%) and can be calculated as:

$$P_{\mathrm{UP/DN}}(0) = \frac{1}{4}\int_0^{\infty} p(x)dx = 12.5\% \tag{17.12}$$

If the sampling clock is leading the data edge ($2\pi > \Delta\theta > 0$), the probability of generating DN pulses based on (17.9) becomes:

$$P_{\mathrm{DN}}(\Delta\theta) = \frac{1}{4}\int_{-|\Delta v_{n+1}|}^{\infty} p(x)dx \tag{17.13}$$

For $|\Delta v_{n+1}| \gg \sigma$, where σ is the standard deviation of the input referred noise of the slicer, the probability of generating DN pulses becomes:

$$P_{\mathrm{DN}}(\Delta\theta) \approx \frac{1}{4} = 25\% \tag{17.14}$$

It is noteworthy that in the phase-leading case ($\Delta\theta > 0$), false-UP signal might be generated from the input patterns ($D_{n-2}, D_{n-1}, D_n, D_{n+1}$) of (1110) or (0001). In these cases, $v_{n-1} = v_n$ when $\Delta\theta > 0$. As the probability of receiving (1110) or (0001) signals is equal to 1/8, the probability of generating false-UP in the phase leading cases due to the input noise of the slicers becomes:

$$P_{\mathrm{fasle-UP}}(\Delta\theta) = \frac{1}{8}\int_0^{\infty} p(x)dx = 6.25\% \tag{17.15}$$

Similarly, the probability density of generating UP (and false-DN) pulses in the $\Delta\theta < 0$ cases are symmetric with respect to those for $\Delta\theta > 0$. For $|\Delta v_n|$ or $|\Delta v_{n+1}| \sim \sigma$, the integration function in (17.13) linearizes the probability distribution of UP/DN signals versus the phase shift ($\Delta\theta$). Thus, the phase detector performs as a binary-quantized linear phase detector [14]. In order to verify the PD model and illustrate the PD characteristics under different bandwidths of the receiver front-end, Figure 17.18(c) shows the simulated probability density functions of UP/DN pulses under different receiver bandwidth in terms of bit rate. The corresponding PD gain is shown in Figure 17.18(d). Similar to a bang-bang PD, the baud-rate PD gain is desensitized with a large phase offset. Note that the false-UP and false-DN probabilities increase due to the ISI effect as the receiver bandwidth decreases. Given that the bandwidth of the receiver front-end is larger than 0.4 X data rate, the variations in UP/DN probabilities are negligible as the phase error is within $\pm 10°$. It implies a stable operation of the baud-rate PD over a wide bandwidth variation of the front-end receiver (0.4 X–0.7 X data rate).

17.6.4 Charge mode logic subtractor

As is described in the previous session, the UP/DN information is detected by comparing the integrating voltages (v_{n-1}, v_n, and v_{n+1}). In this design, the subtractors for phase detection are implemented in a charge mode logic [30] and operated at quarter rate to reduce the DC power consumption. Figure 17.19 illustrates the operation of the charge mode logic in the (a) precharge ($\overline{f_{1/4}} = 0$) mode and (b) evaluation mode ($\overline{f_{1/4}} = 1$). The detailed timing diagram is shown in Figure 17.19(c). Here the quarter-rate clock signal $\overline{f_{1/4}}$ is enabled when both the integrating voltages (v_1 and v_2) are ready. The subtractor output is then applied to a dynamic comparator for the RZ to NRZ conversion. The architecture of the dynamic comparator is the same as that shown in Figure 17.16. The rail-to-rail UP/DN information is then applied to a pattern filter before being fed into the hybrid loop filter for the sampling phase adjustments.

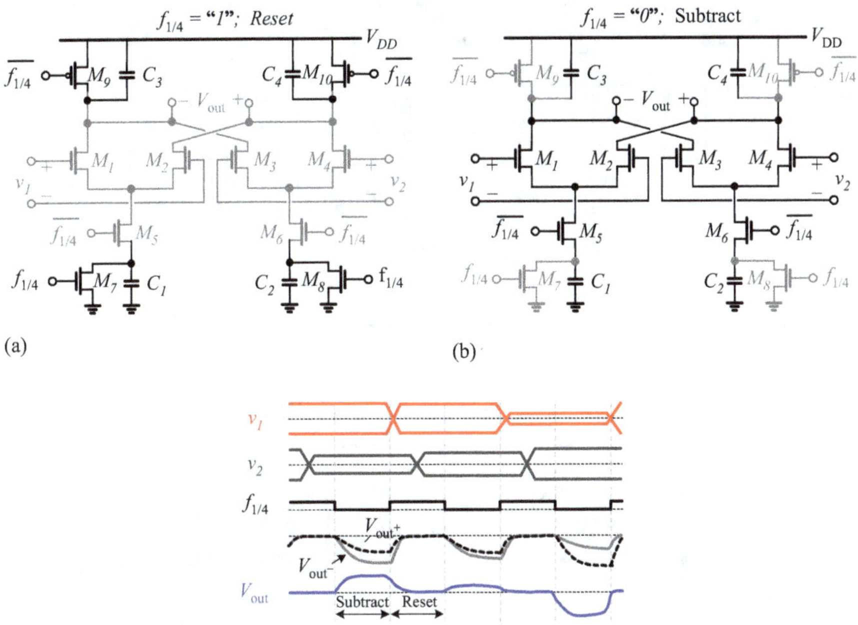

(a) (b) (c)

Figure 17.19 Charge mode logic subtractor (a) when clock is "1," (b) when clock is "0," and (c) timing diagram

17.6.5 Hybrid loop filter and phase interpolator

Conventionally, the digital loop filter can be categorized into two types. The first-type generates UP/DN signals from the demultiplexed parallel data [31] and then performs summation and decimation in the digital domain, as is shown in Figure 17.20(a). The loop latency, which consists of demultiplexer, accumulator, and decimator delay, should be well managed to avoid degrading JTOL performance. The second type uses a decimation ripple counter after a baud-rate PD [32,33]. The major drawback is its high-power consumption due to the high toggle rate of the ripple counter. It also imposes severe design challenges to be implemented in an all-digital circuit manner. To circumvent this design issue, a hybrid loop filter composed of a charge pump-based integrator followed by a digital accumulator is utilized, as is shown in Figure 17.20(b). In this design, a current steering charge pump is implemented to perform the same function as a tally circuit and decimation filter [4,34], which manifests sufficient speed in this technology. Besides, cascode current sources are implemented to mitigate current mismatch effects. The pattern filter is also embedded in the four-way time-interleaved loop filter, which activates the UP/DN pulses based on two-specific input patterns, respectively. Thus, both the UP/DN pulses are composed of eight combinations

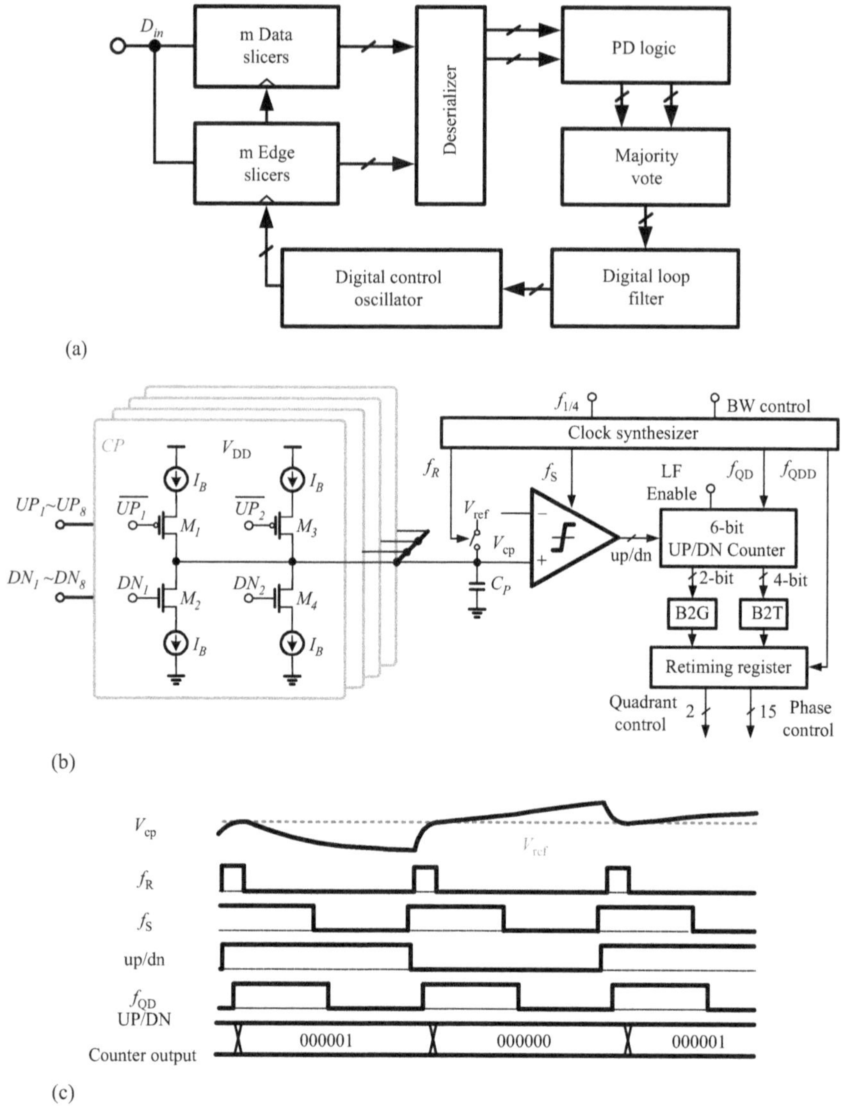

Figure 17.20 (a) Conventional all-digital loop filter, (b) schematic of the proposed hybrid loop filter, and (c) hybrid loop filter timing diagram

($UP_{1–8}$, $DN_{1–8}$). The four paths, quarter-rate UP/DN signals, are first integrated through an analog integrator whose output are periodically quantized and reset to generate the decimated digital output signals. Figure 17.20(c) illustrates the timing diagram. Here f_R is the reset clock of the charge pump, f_s denotes the sampling clock of the comparator, and f_{QD} represents the toggling clock of the UP/DN

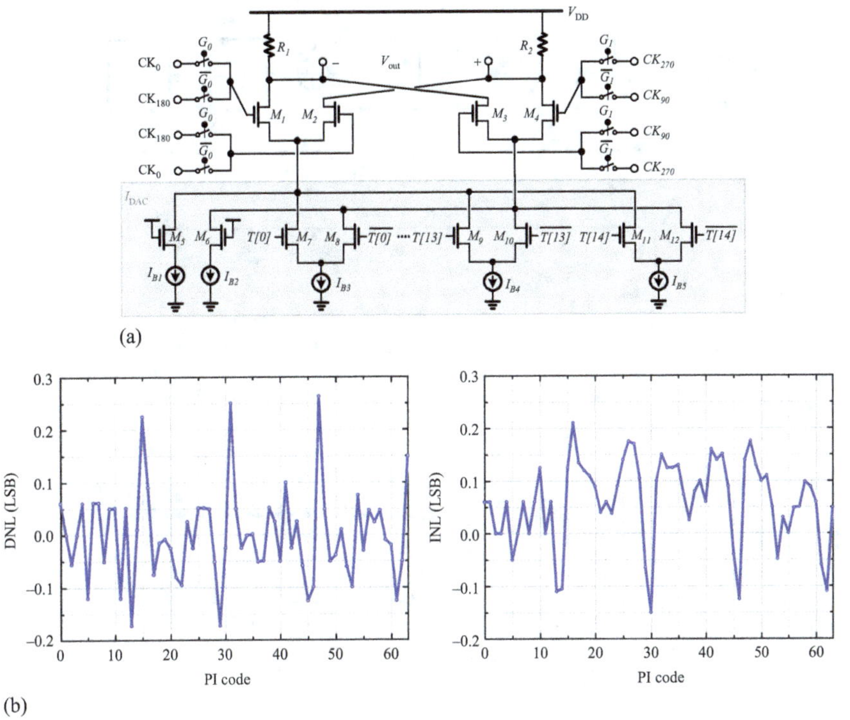

Figure 17.21 (a) The schematic of the phase interpolator (PI) and (b) simulated DNL and INL of the PI output

counter, which are all running at 97 MHz for the loop bandwidth to be equal to 4 MHz at 25 Gbps (f_b) operation. Figure 17.21(a) illustrates the circuit schematic of the current steering PI. Here G_0 and G_1 control the phase rotation in four quadrants, and the weighting coefficients of PI is adjusted by a 4-bit DAC controlled by the thermometer code. The sampling phase is adjusted across 2 UI (half rate) with a 6-bit resolution. Figure 17.21(b) shows the simulated differential nonlinearity (DNL) and integral nonlinearity (INL) at the PI output, which are less than ±0.3 UI across the four quadrants.

Figure 17.22(a) shows the equivalent circuit model of the first-order baud-rate CDR. Let

- K_{PD}: PD gain
- T_b: Bit time
- T_S: Integration time window of the charge pump
- $\frac{I_{CP}}{2\pi C_P}\frac{1-e^{-sT_S}}{s}$: Boxcar CP filter
- $\frac{4T_b}{T_S}$: Decimation factor
- K_q: 1 bit quantizer gain

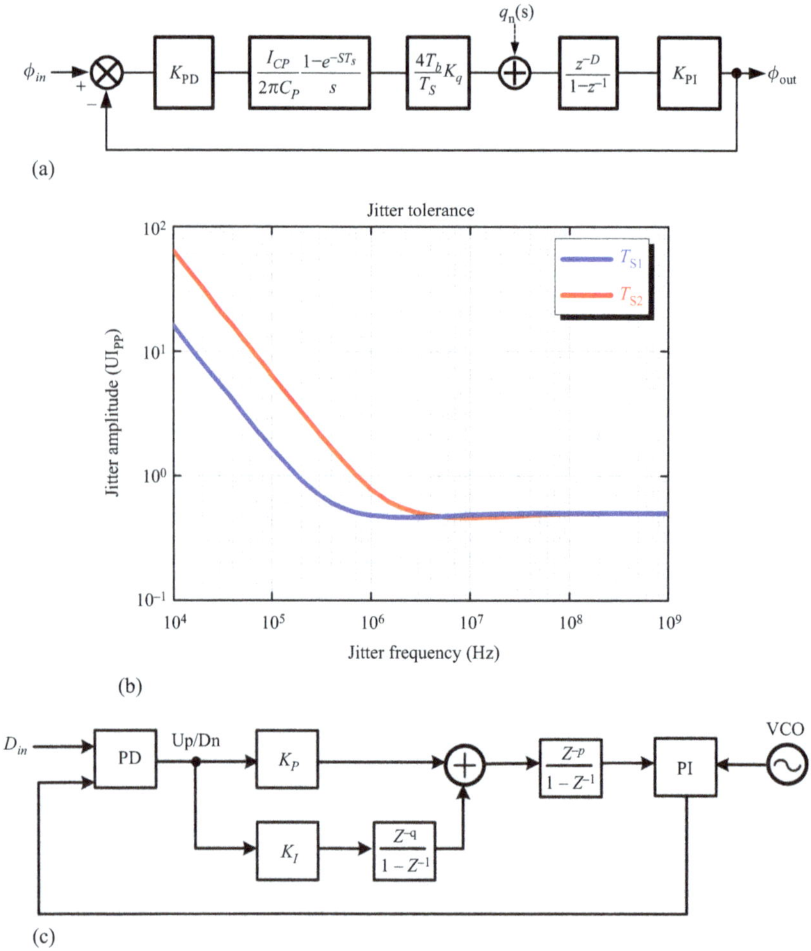

Figure 17.22 (a) The first-order CDR model, (b) simulated jitter tolerance curve, and (c) second-order CDR

- $q_n(s)$: Quantization noise due to 1 bit quantizer
- $\frac{z^{-D}}{1-z^{-1}}$: UP/DN accumulator and loop delay
- K_{PI} : PI phase step

The PD is toggled at the quarter-rate ($1/4T_b$), and the integrate-and-dump charge pump filter performs as a boxcar filter with an integration time window of T_S, where T_S ($1/f_s$) represents the sampling period of the quantizer and the succeeding UP/DN accumulator. As the charge pump output voltage is digitized by a single-bit quantizer, the decimation factor of ($4T_b/T_S$) leads to a reduced loop gain. By setting the sampling rate ω_s to be much higher than the CDR loop bandwidth ω_u (25 X in this design), the loop gain $L(s)$ of the CDR can be rewritten by using bilinear transform.

For a first-order DLL, the unity gain bandwidth (ω_u) is determined by K_{loop} and is scalable by adjusting T_S. Meanwhile, the quantization noise $q_n(s)$ is attenuated by the CP prefilter, which is negligible within the loop bandwidth:

$$
\begin{aligned}
L(s) &= \frac{K_{\text{PD}} \times I_{\text{cp}} \times 4T_b \times K_q \times K_{\text{PI}}}{2\pi C_p} \times \frac{1}{T_S} \times \frac{1}{s} \times \left(\frac{1 - \frac{sT_S}{2}}{1 + \frac{sT_S}{2}} \right)^D \\
&= \frac{K_{\text{PD,effective}} \times K_{\text{PI}}}{T_S} \times \frac{1}{s} \times \left(\frac{1 - \frac{s}{\omega_Z}}{1 + \frac{s}{\omega_p}} \right)^D \\
&= \frac{K_{\text{loop}}}{s} \times \left(\frac{1 - \frac{s}{\omega_Z}}{1 + \frac{s}{\omega_p}} \right)^D
\end{aligned}
\tag{17.16}
$$

The filter output generates a 6-bit control signal to the PI to adjust the sampling phase. $K_{\text{PI}} = 80 \times 10^{-12}/64$ (s), $K_{\text{PD,effective}} = 6.64 \times 10^{10}$ (1/s). The loop delay $D = 1.1$ clock cycles, where 1 clock cycle delay for data retiming, and 0.1 clock cycle delay for PI current switching. By adjusting the sampling rate from $T_{S1} = 2^{10} \times 40$ ps to $T_{S2} = 2^8 \times 40$ ps, the simulated JTOL curves are shown in Figure 17.22(b). The corresponding loop latency in the 4-MHz loop bandwidth case is 10.3 ns, while that in the 1-MHz loop bandwidth case is 40.1 ns. In both cases, the phase margin is about 87°, which are always stable. The CDR is based on a first-order delay locked loop, which has limited frequency offset tolerance. This range can be extended by using a high-order PI filter, as is shown in Figure 17.22(c).

17.6.6 Experimental results of optical receiver with baud-rate CDR

The optical receiver comprising a font-end amplifier, a baud-rate CDR, and a 1:4 demultiplexer is implemented in a 40-nm bulk-CMOS process. The receiver chip is integrated with a PD (Albis PDCS32T-GS) on a FR-4-printed circuit board through bonding wire. The core area of the chip size is 0.09 mm^2 only. The PD bandwidth is about 17 GHz with responsivity of 0.53 A/W. An 850 nm vertical-cavity surface-emitting laser (VCSEL) light source with an extinction ratio of 6 dB is applied for optical test. When coupling the PD with an OM3-MMF, the power loss is about 3.87 dB due to the limited aperture size. It is de-embedded in the sensitivity measurement. A digital variable optical attenuator is adopted for the sensitivity test. To demonstrate the CDR's capability of tolerating long-run CID bits, a 2^{31}-1 PRBS test pattern is applied for the input-sensitivity measurement. The measured timing jitters at the 1:4 demultiplexer output with 20 Gb/s and 25 Gb/s test data are about 1.73 and 1.69 ps-rms, respectively. Figure 17.23 shows the measured input sensitivities of the receiver front-end, which are about −13.8 and −8.7 dBm, respectively, for the 20 and 25 Gb/s operations for BER of less than 10^{-12}.

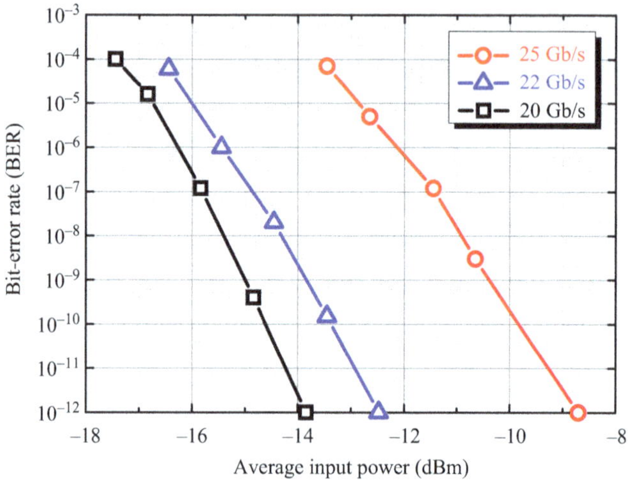

Figure 17.23 Measured input sensitivity

Define the figure of merit (FoM) of the optical receiver as:

$$\text{FoM} = \frac{\text{Power}[\text{mW}] \times i_{\text{Sensitivity}}[\mu\text{A}_{\text{PP}}]}{D_R[Gb/s] \times C_{\text{PD}}[fF]} \tag{17.17}$$

The proposed receiver manifests superior performance in terms of integration level (with CDR), chip area, and energy efficiency compared to the prior art (FoM = 1.8).

17.7 Conclusion

BMCDRs using gating oscillators are known for instant phase-locking response. The feed forward operation and wide bandwidth nature deteriorates its jitter filtering capability, as a direct compromise between locking speed and jitter transfer. To circumvent the design trade-off, a BMCDR that is switchable from data gating mode (feed forward) to phase tracking mode (feedback) operation is described. Moderated by an LD, it achieves instantaneous phase locking and jitter suppression simultaneously. Moreover, the sub-baud-rate CDR incorporates a selectively gating GVCO, which paves the way for a wide-range operation with a scalable power consumption. The experimental prototype demonstrates a superior energy efficiency of 1.24 pJ/bit at 10 Gbps operation using a 55-nm CMOS technology.

To further integrate the receiver front-end with the CDR, an optical receiver comprising a font-end amplifier, a PI-based baud-rate CDR, and a 1:4 demultiplexer is described. A new baud-rate PD is proposed to tolerate long run of CID bits, and its dynamic behavior is investigated in detail. As no edge samples are required in the baud-rate CDR, it greatly reduces the power consumption of the clock distribution networks. Fabricated in a 40-nm CMOS process, the footprint of the receiver core is only 0.09 mm^2. The input sensitivities and energy efficiency of the entire receiver are about −13.8 dBm, −8.7 dBm, and 2.4 pJ/bit, 2.1 pJ/bit, respectively, for 20 Gb/s and 25 Gb/s operations. Combing the selectively gating and baud-rate CDR techniques, it demonstrates strong potentials for versatile optical links applications.

References

[1] "Why Energy is a Big and Rapidly Growing Problem for Data Centers", https://www.forbes.com/sites/forbestechcouncil/2017/12/15/why-energy-is-a-big-and-rapidly-growing-problem-for-data-centers/#4bc056a55a30

[2] C. Kachris, K. Bergman, and I. Tomkos. *Optical Interconnects for Future Data Center Networks*. Springer, 2013.

[3] M.-C. Su,W.-Z. Chen, P.-S. Wu, Y.-H. Chen, C.-C. Lee and S.-J. Jou. "A 10-Gb/s, 1.24 pJ/bit, Burst-Mode Clock and Data Recovery with Jitter Suppression," *IEEE Transactions on Circuits and Systems-I*, vol. 62, no. 3, pp. 743–751, 2015.

[4] Y.-S. Lee, W.-H. Ho, and W.-Z. Chen, "A 25-Gb/s, 2.1 pJ/bit, Fully Integrated Optical Receiver with a Baud-Rate Clock and Data Recovery," *IEEE Journal of Solid-State Circuits*, vol. 54, no. 8, pp. 2243–2254, August, 2019.

[5] W.-Z. Chen, C.-Y. Wei, and J.-W. Chen, "A Wide-Range Burst Mode Clock and Data Recovery Circuit", *IEEE Asian Solid-State Circuits Conference*, Nov. 2006, pp. 403–406.

[6] B. Abiri, R. Shivnaraine, A. Sheikholeslami, H. Tamura, and M. Kibune, "A 1-to-6 Gb/s Phase-Interpolator-Based Burst-Mode CDR in 65 nm CMOS," *IEEE ISSCC Digest of Technical Papers*, Feb. 2011, pp. 154–155.

[7] M. Banu and A. Dunlop, "A 660 Mb/s CMOS Clock and Data Recovery Circuit with Instantaneous Locking for NRZ Data and Burst-mode Transmission," *IEEE ISSCC Digest of Technical Papers*, Feb. 1993, pp. 102–103.

[8] Y.-H. Chen and W.-Z. Chen, "A 0.6-7 Gbps, 1/7 Rate, Burst Mode Clock and Data Recovery Circuit and Demultiplexer," *IEEE Radio Frequency Integrated Circuits Symposium*, Jun. 2012, pp. 531–534.

[9] K. Maruko, T. Sugioka, H. Hayashi, *et al.*, "A 1.296-to-5.184 Gb/s Transceiver with 2.4 mW(Gb/s) Burst-Mode CDR Using Dual-Edge Injection-Locked Oscillator," *IEEE ISSCC Digest of Technical Papers*, Feb. 2010, pp. 364–365.

[10] J. Lee and M. Liu, "A 20-Gb/s Burst-Mode Clock and Data Recovery Circuit Using Injection-Locking Technique," *IEEE Journal of Solid-State Circuits*, vol. 43, no. 3, pp. 619–630, 2008.

[11] J. Terada, K. Nishimura, S. Kimura, H. Katsurai, N. Yoshimoto, and Y. Ohtomo, "A 10.3 Gb/s Burst-Mode CDR Using a ΔΣ DAC," *IEEE Journal of Solid-State Circuits*, vol. 43, no. 12, pp. 2921–2928, 2008.

[12] M. Nogawa, K. Nishimura, S. Kimura, *et al.*, "A 10 Gb/s burst-mode CDR IC in 0.13 μm CMOS," *IEEE ISSCC Digest of Technical Papers*, Feb. 2005, pp. 228–229.

[13] W. S. Choi, T. Anand, G. Shu, and P. K. Hanumolu, "A Fast Power-on 2.2 Gb/s Burst-mode Digital CDR with Programmable Input Jitter Filtering," *IEEE Symposium on VLSI Circuits*, Jun. 2013, pp. 280–281.

[14] J. Lee, K. S. Kundert, and B. Razavi, "Analysis and Modeling of Bang-bang Clock and Data Recovery Circuits," *IEEE Journal of Solid-State Circuits*, vol. 39, no. 9, pp. 1571–1580, 2004.

[15] S. L. J. Gierkink, "A 2.5 Gb/s Run-Length-Tolerant Burst-Mode CDR Based on a 1/8th-Rate Dual Pulse Ring Oscillator," *IEEE Journal of Solid-State Circuits*, vol. 43, no. 8, pp. 1763–1771, 2008.

[16] J. D. H. Alexander, "Clock Recovery from Random Binary Signals," *Electronics Letters*, vol. 11, no. 22, pp. 541–542, 1975.

[17] I. A. Young, J. K. Greason, and K. L. Wong, "A PLL Clock Generator with 5 to 110 MHz of Lock Range for Microprocessors," *IEEE Journal of Solid-State Circuits*, vol. 27, no. 11, pp. 1599–1607, 1992.

[18] 10-Gigabit-Capable Passive Optical Networks (XG-PON): Physical Media Dependent (PMD) Layer Specification, ITU-T G.987.2 (2010).

[19] S.-H. Huang and W.-Z. Chen, "A 25 Gb/s 1.13 pJ/b −10.8 dBm Input Sensitivity Optical Receiver in 40 nm CMOS," *IEEE Journal of Solid-State Circuits*, vol. 52, no. 3, pp. 747–756, 2017.

[20] M. H. Nazari and A. Emami-Neyestanak, "A 24-Gb/s Double-Sampling Receiver for Ultra-low-power Optical Communication," *IEEE Journal of Solid-State Circuits*,vol. 48, no. 2, pp. 344–357, 2013.

[21] T. Takemoto, H. Yamashita, T. Yazaki, N. Chujo, Y. Lee, and Y. Matsuoka, "A 4 × 25-to-28 Gb/s 4.9 mW/Gb/s –9.7 dBm High-sensitivity Optical Receiver Based on 65 nm CMOS for Board-to board Interconnects," *IEEE ISSCC Digest of Technical Papers*, Feb. 2013, pp. 118–119.

[22] T. Shibasaki, Y. Tsunoda, H. Oku, *et al.*, "4 × 25.78 Gb/s Retimer ICs for Optical Links in 0.13 μm SiGe BiCMOS," *IEEE ISSCC Digest of Technical Papers*, Feb. 2015, pp. 412–414.

[23] F. Musa and A. C. Carusone, "A Baud-rate Timing Recovery Scheme with a Dual-function Analog Filter," *IEEE Transaction on Circuits Systems II*, vol. 53, no. 12, pp. 1393–1397, 2006.

[24] R. Dokania, A. Kern, M. He *et al.*, "A 5.9 pJ/b 10 Gb/s Serial Link with Unequalized MM-CDR in 14 nm Tri-gate CMOS," *IEEE ISSCC Digest of Technical Papers*, Feb. 2015, pp. 184–185.

[25] J. Han, Y. Lu, N. Sutardja *et al.*, "A 60 Gb/s 288 mW NRZ Transceiver with Adaptive Equalization and Baud-rate Clock and Data Recovery in 65 nm CMOS Technology," *IEEE ISSCC Digest of Technical Papers*, Feb. 2017, pp. 112–113.

[26] S. Palermo, A. Emami-Neyestanak, and M. Horowitz, "A 90 nm CMOS 16 Gb/s Transceiver for Optical Interconnect," *IEEE Journal of Solid-State Circuits*, vol. 43, no. 5, pp. 1235–1246, 2008.

[27] M. Raj, S. Saeedi, and A. Emami-Neyestanak, "A 4-to-11 GHz Injection-locked Quarter-rate Clocking for an Adaptive 153 fJ/b Optical Receiver in 28 nm FDSOI CMOS," *IEEE ISSCC Digest of Technical Papers*, Feb. 2015, pp. 404–406.

[28] D. Kim, W. Choi, A. Elkholy *et al.*, "A 15 Gb/s 1.9 pJ/bit Sub-baud-rate Digital CDR," *IEEE Custom Integrated Circuits Conference*, Apr. 2018, pp. 8–11.

[29] F. Spagna, L. Chen, M. Deshpande *et al.*, "A 78 mW 11.8 Gb/s Serial Link Transceiver with Adaptive RX Equalization and Baud-Rate CDR in 32 nm CMOS," *IEEE ISSCC Digest of Technical Papers*, Feb. 2010, pp. 366–367.

[30] J. W. Jung and B. Razavi, "A 25-Gb/s 5-mW CMOS CDR/deserializer," *IEEE Journal of Solid-State Circuits*, vol. 48, no. 3, pp. 684–697, 2013.

[31] J. L. Sonntag and J. Stonick, "A Digital Clock and Data Recovery Architecture for Multi-gigabit/s Binary Links," *IEEE Journal of Solid-State Circuits*, vol. 41, no. 8, pp. 1867–1875, 2006.

[32] I.-F. Chen, R.-J. Yang, and S.-I. Liu, "Loop Latency Reduction Technique for All-digital Clock and Data Recovery Circuits," *IEEE Asian Solid-State Circuits Conference*, 2009, pp. 309–312.

[33] Z.-H. Hong, Y.-C. Liu, and W.-Z. Chen, "A 3.12 pJ/bit, 19–27 Gbps Receiver with 2-Tap DFE Embedded Clock and Data Recovery," *IEEE Journal of Solid-State Circuits*, vol. 50, no. 11, pp. 2625–2634, 2015.

[34] H. J. Jeon, R. Kulkarni, Y. C. Lo, J. Kim, and J. S. Martinez, "A Bang-Bang Clock and Data Recovery Using Mixed Mode Adaptive Loop Gain Strategy," *IEEE Journal of Solid-State Circuits*, vol. 48, no. 6, pp. 1398–1415, 2013.

Chapter 18

Digital clock and data recovery circuits

Saurabh Saxena[1] and Pavan Kumar Hanumolu[2]

18.1 Introduction

Clock and data recovery unit performs the key function of recovering clock and incoming data bits in all serial communication systems. Figure 18.1 shows block diagram of an analog clock and data recovery unit. Phase error, Φ_E, between the recovered clock phase, Φ_{RCK}, and input phase, Φ_{IN}, is detected by a linear phase detector (PD). PD outputs signals UP and DN whose pulse widths are modulated by the input phase error. UP and DN pulses add an average charge $q(= I_{CP}T \cdot \Phi_E/2\pi)$ to the loop filter, thereby varying control voltage, V_{CTRL}, of a voltage controlled oscillator (VCO). VCO's output frequency, $f_{RCK} = \omega_{RCK}/2\pi$, is proportional to V_{CTRL}, that is, $\omega_{RCK} = \omega_{free} + K_{VCO} \cdot V_{CTRL}$, where ω_{free} is free-running frequency of the oscillator and K_{VCO} is its gain. Feedback loop ensures that VCO's output phase (and therefore frequency) is locked to the input bit stream such that $d\Phi_E/dt = 0$. In steady state, small phase/frequency perturbations, random or deterministic, of input or output signals are corrected by the feedback loop. The small-signal loop gain of the CDR is given by:

$$\mathrm{LG}(s) = K_{PD} \cdot I_{CP} \cdot \frac{1 + s\mathrm{RC}_1}{sC_1} \cdot \frac{K_{VCO}}{s} \tag{18.1}$$

where K_{PD} represents phase detector gain. Loop gain and bandwidth of the CDR loop are dictated by the jitter transfer function (JTRAN) and jitter generation (JGEN) specifications. Jitter transfer function quantifies the amount of input noise filtering provided by the CDR and is defined as:

$$\mathrm{JTRAN} = \frac{\Phi_{RCK}(s)}{\Phi_{IN}(s)} = \frac{\mathrm{LG}(s)}{1 + \mathrm{LG}(s)} \tag{18.2}$$

A low −3 dB bandwidth, $\omega_{-3dB} \approx K_{PD}K_{VCO}I_{CP}$, and a large phase margin, $\Phi_M = \tan^{-1}(\omega_u \mathrm{RC}_1)$ are required to filter input noise without excessive input noise peaking, respectively. On the other hand, jitter generation is defined as jitter added to the recovered clock by the CDR's loop components. Each noise source, as shown

[1]Department of Electrical Engineering, Indian Institute of Technology Madras, Chennai, India
[2]Electrical and Computer Engineering, University of Illinois Urbana–Champaign, Champaign, IL, USA

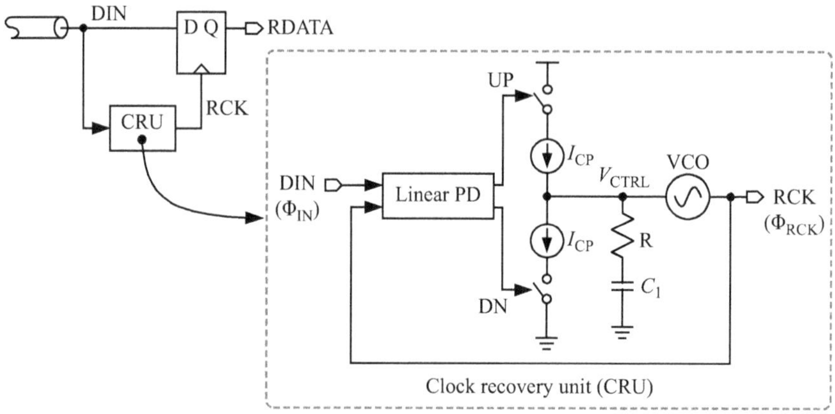

Figure 18.1 Block diagram of an analog clock and data recovery

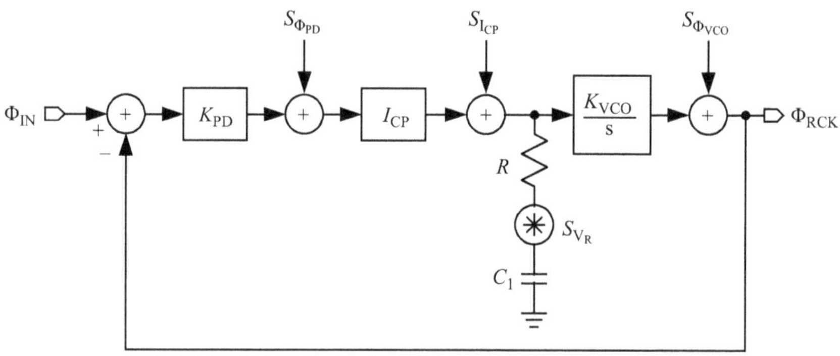

Figure 18.2 Small signal model of clock recovery unit (CRU) for noise analysis

in Figure 18.2, contributes to output noise, Φ_{RCK}, through noise transfer functions (NTFs) as defined below:

$$\mathrm{NTF_{IN}} = \frac{\Phi_{RCK}(s)}{\Phi_{IN}(s)} = \frac{\mathrm{LG}(s)}{1 + \mathrm{LG}(s)} \tag{18.3}$$

$$\mathrm{NTF_{PD}} = \frac{\Phi_{RCK}(s)}{\Phi_{PD}(s)} = \frac{1}{K_{PD}} \cdot \frac{\mathrm{LG}(s)}{1 + \mathrm{LG}(s)} \tag{18.4}$$

$$\mathrm{NTF_{CP}} = \frac{\Phi_{RCK}(s)}{I_{CP}(s)} = \frac{1}{K_{PD}I_{CP}} \cdot \frac{\mathrm{LG}(s)}{1 + \mathrm{LG}(s)} \tag{18.5}$$

$$\mathrm{NTF}_R = \frac{\Phi_{RCK}(s)}{V_R(s)} = \frac{K_{VCO}}{s} \cdot \frac{1}{1 + \mathrm{LG}(s)} \tag{18.6}$$

$$\mathrm{NTF_{VCO}} = \frac{\Phi_{\mathrm{RCK}}(s)}{\Phi_{\mathrm{VCO}}(s)} = \frac{1}{1 + \mathrm{LG}(s)} \tag{18.7}$$

The output noise power spectral density (PSD) is obtained by summing the contribution of each noise source filtered by their respective NTF as evaluated below:

$$\begin{aligned} S_{\Phi_{\mathrm{RCK}}}^{\mathrm{TOTAL}} = \; & S_{\Phi_{\mathrm{IN}}} \cdot |\mathrm{NTF_{IN}}|^2 + S_{\Phi_{\mathrm{PD}}} \cdot |\mathrm{NTF_{PD}}|^2 + S_{I_{\mathrm{CP}}} \cdot |\mathrm{NTF_{CP}}|^2 \\ & + S_{V_R} \cdot |\mathrm{NTF}_R|^2 + S_{\Phi_{\mathrm{VCO}}} \cdot |\mathrm{NTF_{VCO}}|^2 \end{aligned} \tag{18.8}$$

Integrating $S_{\Phi_{\mathrm{RCK}}}^{\mathrm{TOTAL}}$ over frequency gives variance of recovered clock jitter. Hence, long-term absolute recovered clock jitter is given by:

$$\Phi_{\mathrm{RCK}}^{\mathrm{TOTAL}} = \sqrt{\int_0^{\infty} S_{\Phi_{\mathrm{OUT}}}^{\mathrm{TOTAL}}(f)\,df} \;\; [\mathrm{rad}] \tag{18.9}$$

$$\mathrm{Jitter}(\sigma_j) = \frac{T_{\mathrm{RCK}}}{2\pi} \sqrt{\int_0^{\infty} S_{\Phi_{\mathrm{OUT}}}^{\mathrm{TOTAL}}(f)\,df} \;\; [s] \tag{18.10}$$

Figures 18.3(a) and 18.3(b) show noise transfer functions and output noise PSD (S_X^{RCK}) for different noise sources in a 2.5 Gb/s full-rate clock recovery unit. In this example, a linear phase detector with gain $K_{\mathrm{PD}} = 1/2\pi$ and a 2.5 GHz VCO with spot phase noise −120 dBc/Hz at 1 MHz are employed. CDR's loop parameters I_{CP}, R, and C_1 are calculated to be 1.26 µA, 5 kΩ, and 2.76 nF, respectively, for achieving 2 MHz JTRAN bandwidth and 0.1 dB JTRAN peaking. Noise analysis for CRU reveals that input, phase detector, and charge pump noise are low

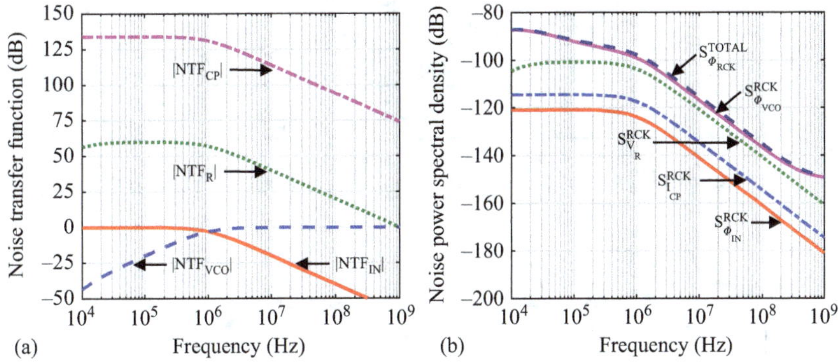

Figure 18.3 (a) Noise transfer functions for noise sources in Figure 18.2 and (b) noise power spectral density for noise sources in 2.5 Gb/s full-rate analog CDR

pass filtered, resistor noise is bandpass filtered, and VCO's phase noise is high pass filtered.

Analog CDRs require a large loop filter capacitor in the order of at least a few hundred pico-farads to meet the noise and (<0.1 dB) peaking requirements. With a capacitor density of 1–2 fF/μm^2, loop filter capacitor occupies a large area. MOS capacitor has a higher density, but their leakage current severely degrades jitter performance of a CDR.

A digital equivalent of an analog RC-loop filter alleviates the problem of a large loop filter capacitor. Figure 18.4 shows a block diagram of a CDR with an analog phase error detector and a digital loop filter (DLF). The digital loop filter mimics resistor R with a proportional path gain, K_P, and realizes integration with a digital accumulator, ACC_I. However, the DLF has to be preceded by a time-to-digital converter (TDC) to digitize pulse width modulated PD's output and later followed by a digital-to-analog converter (DAC) to control the voltage controlled oscillator. The TDC digitizes input phase error and feeds its output, D_{TDC}, to the digital loop filter. In this CDR architecture, TDC adds quantization noise and incurs large power penalty while improving TDC's resolution. Well known data conversion techniques such as delta-sigma modulation (DSM) can be used to achieve high resolution. One such example was reported in [1] wherein a 1-bit continuous-time DSM was used to digitize PD output with high resolution. However, this approach requires a low bandwidth low pass filter to suppress DSM quantization error, thus severely limited both JTOL and JTRAN. The bandwidth constraint can only be relaxed by eliminating phase quantization error in the proportional path as shown in the CDR architecture illustrated in Figure 18.5 [2]. The pulse width modulated linear PD's output controls current controlled oscillator (CCO) through a current DAC, PDAC, without introducing any quantization error. Integral path uses quantized phase, D_{TDC}, but the impact of quantization error is greatly reduced by the low bandwidth integral path.

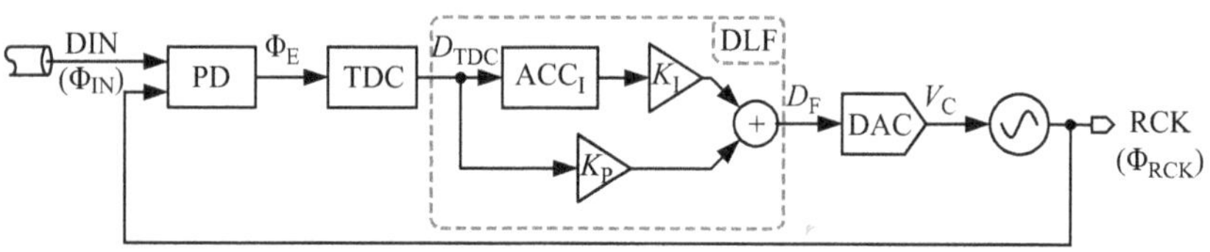

Figure 18.4 Clock recovery unit with a digital loop filter

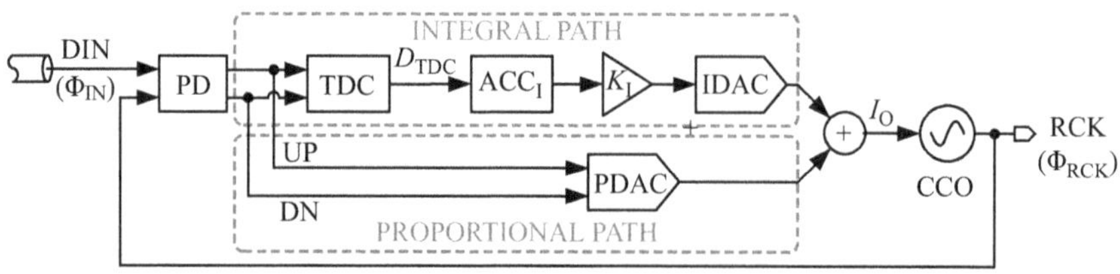

Figure 18.5 Clock recovery unit with analog proportional and digital integral path

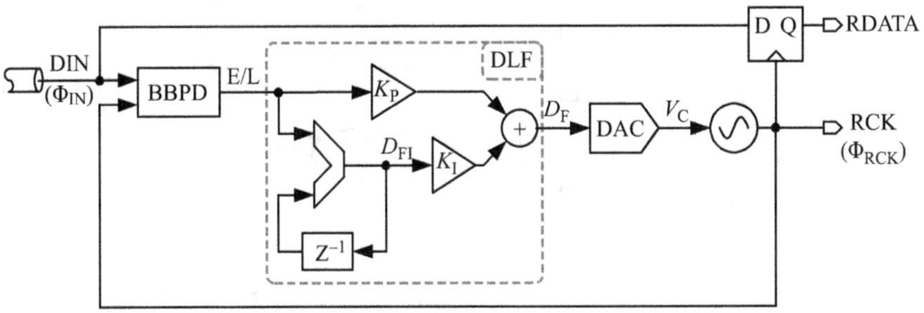

Figure 18.6 Block diagram of a standard digital clock and data recovery unit

The linearity and speed bottlenecks associated with the analog proportional path can be relaxed by using a bang-bang PD (BBPD) in the digital proportional path as depicted in Figure 18.6. In Section 18.2, we analyze the BBPD-based digital CDR and discuss its design challenges. The design techniques overcoming these problems are examined in Section 18.3. We incorporate the design improvement techniques and present a case study of a digital CDR with decoupled JTRAN and JTRACK bandwidth in Section 18.4. The design issues and possible solutions for a wide range of digital clock and data recovery are illustrated with another case study in Section 18.5.

18.2 Design and analysis of digital CDR

In the commonly used BBPD-based digital CDR (Figure 18.6), BBPD measures the sign of the phase error between received bits, DIN, and the recovered clock. It outputs E and L signals that signify early and late arrival of the received bit stream concerning the recovered clock. E/L signals are filtered by the digital loop filter and the resulting output digital control word, D_F, is used to control oscillator's phase/frequency through a voltage/current DAC. Operation of all the buildings block and their small-signal models needed to analyze loop dynamics are discussed below.

18.2.1 Bang-bang phase detector (BBPD)

Figure 18.7 shows schematic of a BBPD. Incoming data are sampled twice per UI, generating samples S_2 and S_3. Using these two samples along with previous data sample, S_1, E/L signals are generated as per the conditions described in the figure. In case of no bit transitions ($S_1 = S_3$), both *early* and *late* signals remain low. In the locked state, CRU aligns RCK's rising and falling edges to bit-center and bit-transition, respectively. The falling-edge triggered DFF's output, S_4, is randomized due to timing jitter on the sampling clock, which in turn dithers *early/late* signals. During bit transitions, BBPD operation can be captured using a single D flip-flop (DFF) as shown in Figure 18.8. In the presence of phase error, 'Δt', between the input signal and sampling clock, an average output of BBPD is evaluated below:

$$\overline{D_{\mathrm{PD}}} = \alpha_T \times [1 - 2P(t_j < -\Delta t)] \tag{18.11}$$

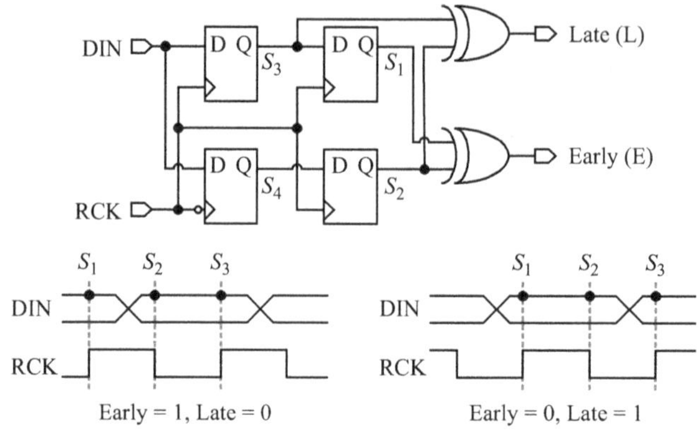

Figure 18.7 Schematic diagram of a BBPD with signal waveforms

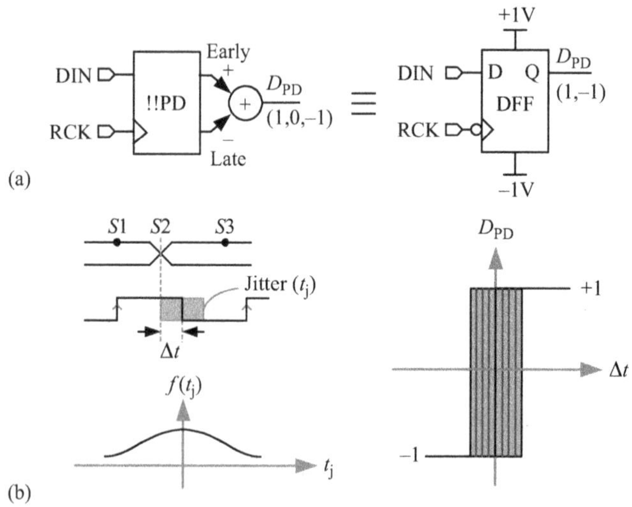

Figure 18.8 (a) Simplified model of a BBPD. (b) Signal waveforms in BBPD

where α_T is the input transition density of the received bit stream. Under Gaussian noise approximation for the clock jitter $t_j \sim \mathcal{N}(0, \sigma_j^2)$, BBPD's small signal gain can be evaluated as:

$$K_{PD} = \frac{d\overline{D_{PD}}}{d\Phi_E} = \frac{d\Delta t}{d\Phi_E} \times \frac{d\overline{D_{PD}}}{d\Delta t} = \alpha_T \sqrt{\frac{2}{\pi\sigma_{\Phi_j}^2}} \exp\left(-\frac{\Delta t^2}{2\sigma_{\Phi_j}^2}\right) \tag{18.12}$$

where $\sigma_{\Phi_j}^2$ is variance of the sampling clock jitter. In CRU's locked state, BBPD's gain $K_{PD}(\Phi_E = 0) = \alpha_T \sqrt{2/\pi\sigma_{\Phi_j}^2}$ can be used to perform small-signal analysis.

18.2.2 Digital loop filter (DLF)

BBPD's output, D_{PD}, is filtered by using a digital filter whose output is given by:

$$D_F[nT] = K_P \times D_{PD}[nT] + K_I \times (D_{FI}[\overline{n-1}T] + D_{PD}[nT]) \tag{18.13}$$

where $1/T$ is the update of the DLF and is equal to 1/UI in the full-rate implementation. For lower frequency (ω) perturbations in the loop, the transfer function can be simplified by using approximations $|j\omega/f_{\text{RCK}} \ll 1|$ and $z = e^{s/f_{RCK}} \approx 1 + s/f_{\text{RCK}}$:

$$D_F(z) = K_P D_{\text{PD}}(z) + \frac{K_I}{z-1} D_{\text{PD}}(z) \tag{18.14}$$

$$D_F(s) \approx K_P D_{\text{PD}}(s) + \frac{K_I f_{\text{RCK}}}{s} D_{\text{PD}}(s) \tag{18.15}$$

Because accumulator clocked at the data rate consumes significant power, DLF's update rate is reduced to a fraction of the data rate in most practical implementations at the cost of increased latency.

18.2.3 Digital-to-analog converter

Digital-to-analog converter (DAC) acts as an interface between DLF's output and voltage or current controlled oscillator. Denoting its gain by K_{DAC} [V/LSB], VCO's control voltage equals $V_C[nT] = K_{\text{DAC}} D_F[nT]$. Typically, DACs are implemented using Nyquist-rate converters or oversampling converters [3].

18.2.4 Voltage controlled oscillator

DAC's output controls oscillator's frequency and phase according to the mathematical relations given below:

$$\omega_{\text{RCK}}[nT] = \omega_{\text{free}} + K_{\text{VCO}} \cdot K_{\text{DAC}} D_F[nT] \tag{18.16}$$

$$\Phi_{\text{RCK}}(t) = \int_0^t \omega_{\text{RCK}}(\tau) d\tau \tag{18.17}$$

For small perturbations in control voltage, ΔV_c, change in the output phase, $\Delta\Phi_{\text{RCK}}$, can be simplified as:

$$\Delta\Phi_{\text{RCK}}(s) = \frac{K_{\text{VCO}}}{s} \Delta V_c(s) \tag{18.18}$$

Using above linearized gain models, a small-signal model of the clock and data recovery loop can constructed as shown in Figure 18.9. Loop gain of the CDR loop equals:

$$\text{LG}(s) = K_{\text{PD}} \left(K_P + \frac{K_I f_{\text{RCK}}}{s} \right) \frac{K_{\text{DAC}} K_{\text{VCO}}}{s} \tag{18.19}$$

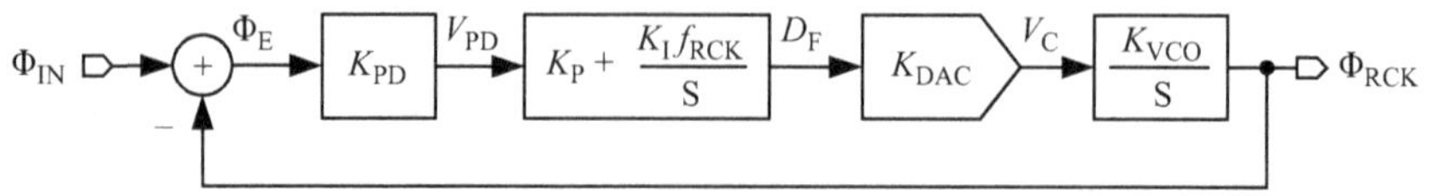

Figure 18.9 Small signal model of a standard digital CDR

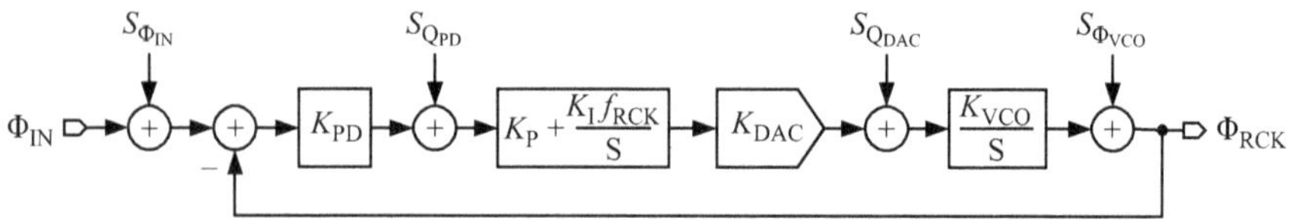

Figure 18.10 Small signal noise model of a standard digital CDR

18.2.5 Noise analysis of a digital CDR

Bandwidth of a CDR loop is dictated by jitter transfer bandwidth and jitter tolerance or jitter tracking bandwidth requirements. For a given target bandwidth and phase margin, recovered clock jitter is limited by the input jitter tracked by the CDR loop. Jitter tracking is defined with the help of a jitter tracking transfer function JTRACK as defined below:

$$\text{JTRACK} = \frac{\Phi_E(s)}{\Phi_{\text{IN}}(s)} = \frac{1}{1 + \text{LG}(s)} \tag{18.20}$$

If $\delta \cdot T$ is the peak-to-peak sinusoidal jitter tracked at $\omega_{-3\text{dB}}$ bandwidth by the clock recovery loop with BER $\leq 10^{-12}$, then:

$$\Phi_E \leq \frac{2\pi}{T}(T - 14\sigma_j) \tag{18.21}$$

where $0 \leq \delta \leq 1$, σ_j is the standard deviation of the recovered clock jitter, and $\pm 7\sigma_j$ timing margin is required for BER $\leq 10^{-12}$. Substituting Φ_E in the above expression yields an upper bound on the recovered clock jitter [4]:

$$\frac{\Phi_{\text{IN}}}{1 + \text{LG}(s)} \leq \frac{2\pi}{T}(T - 14\sigma_j) \tag{18.22}$$

$$\begin{aligned} &\Rightarrow \frac{\delta T}{T - 14\sigma_j} \leq 1 + \text{LG}(s) \\ &\Rightarrow \sigma_j \leq \frac{1}{14}\left(T - \frac{\delta T}{1 + \text{LG}(s)}\right) \end{aligned} \tag{18.23}$$

Now, we can include all noise sources in the CDR loop and optimize their contribution to the output jitter. Figure 18.10 shows small-signal model illustrating all the different noise sources. Noise power spectral density of these sources are listed in Table 18.1. Here, α_T is data transition density and DAC_{LSB} is quantization

Table 18.1 Noise parameters in a standard digital CDR

Noise source	Noise type	Noise PSD
Input	Phase noise	$S_{\Phi_{\text{IN}}}$
BBPD	Quantization noise	$S_{Q_{\text{PD}}} = \left(\alpha_T - \frac{2}{\pi}\alpha_T{}^2\right)$
DAC	Quantization noise	$S_{Q_{\text{DAC}}} = \frac{2\text{DAC}_{\text{LSB}}^2}{12 f_{\text{RCK}}}$
VCO	Phase noise	$S_{\Phi_{\text{VCO}}}$

Table 18.2 Performance specifications for digital CDR design example

Parameter	Value
Input jitter	0.01 UI rms
α_T	0.5
Jitter generation	4 ps rms
Jitter transfer bandwidth	< 2 MHz
Jitter peaking	< 0.1 dB
Jitter tolerance (at f > 1 MHz)	> 0.15 UI

error in voltage/current controlling the oscillator. Using noise transfer functions associated with each of the noise source, total output noise power spectral density is given below. Integrating over frequency gives variance of the recovered clock jitter:

$$
\begin{aligned}
S_{\Phi_{\text{RCK}}}^{\text{TOTAL}} = \;& S_{\Phi_{\text{IN}}} \cdot |\text{NTF}_{\text{IN}}|^2 + S_{Q_{\text{PD}}} \cdot |\text{NTF}_{\text{PD}}|^2 \\
& + S_{Q_{\text{DAC}}} \cdot |\text{NTF}_{\text{DAC}}|^2 + S_{\Phi_{\text{VCO}}} \cdot |\text{NTF}_{\text{VCO}}|^2
\end{aligned}
\tag{18.24}
$$

18.2.5.1 Design example

Find loop parameters for a full-rate digital CDR at 2.5 Gb/s with specifications are given in Table 18.2.

Solution: Loop gain of a type-II digital CDR is:

$$
\text{LG}(s) = \frac{K_{\text{PD}} K_P K_{\text{DAC}} K_{\text{VCO}}}{s^2} \cdot \left(s + \frac{K_I f_{\text{RCK}}}{K_P}\right) \tag{18.25}
$$

Here, oscillator's gain is $K_{\text{DCO}} = K_{\text{DAC}} K_{\text{VCO}}$ [Hz/LSB]. For the above loop gain, peaking in closed loop response can be approximated by [5]:

$$
\text{Jitter peaking[in dB]} = \frac{8.686}{\omega_{-3dB}/\omega_z} \tag{18.26}
$$

where $\omega_{-3\text{dB}}$ is the closed loop -3 dB bandwidth and $\omega_z = K_I f_{\text{RCK}}/K_P$ is zero frequency. From desired peaking and bandwidth, ω_z can be calculated to be 14.467 Mrad/s. To achieve the desired jitter, we can start with noise performance of a

Table 18.3 Performance of a sample oscillator

Parameter	Value
K_{VCO}	4 GHz/V
K_{DAC}	1 mV/LSB
SSB phase noise @ 10 MHz	−114 dBc/Hz
Flicker noise corner frequency	200 kHz
Noise floor	−150 dBc/Hz

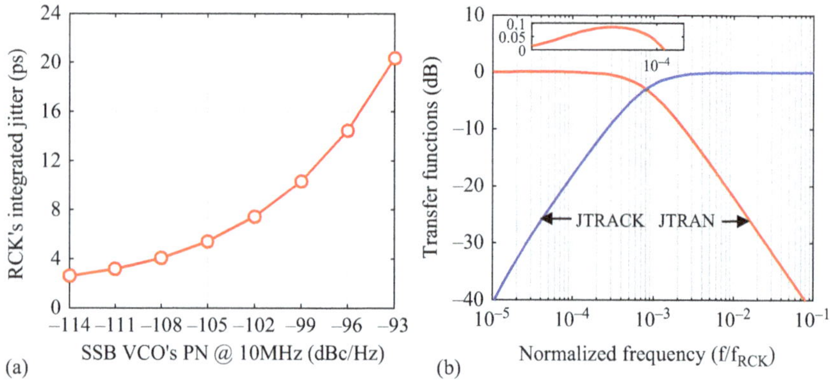

Figure 18.11 (a) Variation in the recovered clock jitter with VCO's phase noise and (b) JTRAN and JTRACK transfer functions for desired bandwidth and jitter peaking

sample design of an oscillator whose design parameters are listed in Table 18.3. In general, oscillator's noise performance can be improved by −3 dB at the cost of doubling the power consumed in the oscillator. Figure 18.11(a) shows jitter generated in CDR loop for different amounts of VCO phase noise, 2 MHz JTRAN bandwidth, and 0.1 dB jitter peaking. It is to be noted that gain of BBPD is inversely proportional to output jitter and is therefore updated with output jitter, iteratively. Using oscillator with −108 dBc/Hz @ 10 MHz phase noise, other loop parameters are calculated to be: $K_{\mathrm{PD}} = 4.456\ \mathrm{V/rad}$, $K_P = 0.1119$, $K_I = 6.4757 \times 10^{-6}$, and integrated recovered clock jitter $\sigma_j = 4.07$ ps. Jitter transfer function and jitter tracking function for the CDR are shown in Figure 18.11(b). It confirms <0.1 dB peaking and 2 MHz bandwidth. Small signal JTOL is higher than required performance.

We can make the following observations from the above design exercise.

1. The digital loop filter is clocked at the data rate. It increases power consumption in phase error processing due to the high frequency of n-bit arithmetic operation involved in the digital loop filter.
2. The digital loop filter with proportional and integral path gains $K_P' = K_P K_{\mathrm{DCO}} = 447.6\ \mathrm{kHz/LSB}$ and $K_I' = K_I K_{\mathrm{DCO}} = 25.9\ \mathrm{Hz/LSB}$, respectively, can be

implemented as shown in Figure 18.12. Here, a three-level PDAC with gain, $K_{PDAC} = 0.1118\ \text{mV/LSB}$ and n-bit IDAC with gain, $K_{IDAC} = 1\ \text{mV/LSB}$ are used. A low K_{IDAC} is required for smaller DAC quantization noise whereas large K_{IDAC} is needed for controlling large frequency range of the oscillator with a lesser number of bits.

3. To reduce jitter generated in a CDR loop in power efficient manner, CDR loop bandwidth can have a different optimum as compared to JTRAN bandwidth requirements. Figure 18.13 shows variation in the recovered clock jitter with unity gain frequency of CDR's loop for the same noise sources used in the example. For low JTRAN bandwidth, output jitter is dominated by oscillator's phase noise, thereby increasing power consumption in the oscillator.
4. In the above CDR design, JTRAN and JTRACK bandwidths are tightly coupled, which is a problem for CDRs in repeater applications. Here, recovered clock with low jitter is required to re-transmit the recovered data, which is achieved with low JTRAN bandwidth and low jitter generated within the loop. However, a large JTRACK bandwidth is needed to track the input jitter and low BER.

Next, we will discuss modifications to standard digital CDR that address the above issues.

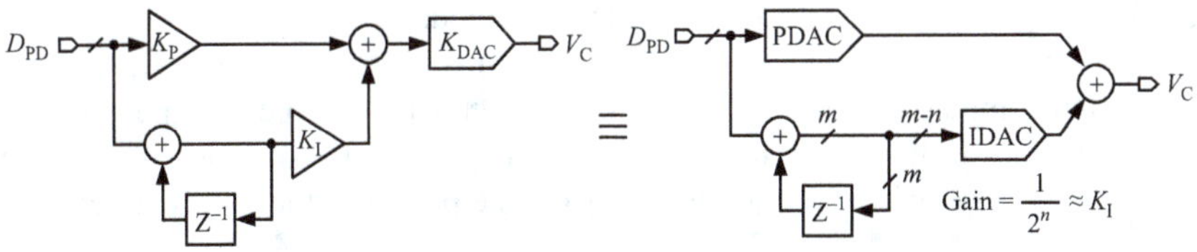

Figure 18.12 Block level implementation of a digital loop filter

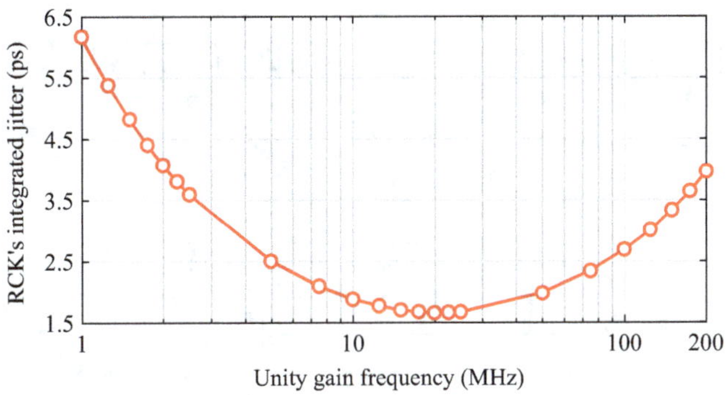

Figure 18.13 Variation in RCK*'s integrated jitter with CDR's unity gain frequency*

18.3 Design techniques for improving standard digital CDR

In this section, we will discuss design techniques to lower the clocking speed of digital loop filter, generate high speed clock source with low jitter and low power, and decouple JTRAN and JTRACK bandwidth.

18.3.1 Low-speed digital loop filter

Clocking speed of a digital loop filter can be reduced in multiple ways as shown in Figure 18.14. Sub-rate CDRs employ *2P* clock phases at f_{DR}/P in BBPD (Figure 18.14(a)) while providing *P* early/late samples for digital processing at rate f_{DR}/P. Typically, *P* is 2 and 4 for half-rate and quarter-rate CDRs, respectively. Higher *P* is not feasible due to implementation problems in generating multiple low-jitter clock phases with precise phase matching. On the other hand, deserialization of BBPD's output by a factor of *P* by using the recovered clock at f_{RCK} and its divided frequencies is much simpler to reduce DLF's clocking rate as shown in Figure 18.14(b). 1:P de-multiplexor (DMUX) outputs *P* early/late signals at f_{RCK}/P. However, loop filter has *p*-bit input to process every $1/f_{ACC} = P/f_{RCK}$ period. DLF's input bit-width can be reduced by decimating BBPD's output as shown in Figure 18.14(c). Decimation is typically performed by realizing either majority voting (MV) or boxcar filtering. Block diagram of a majority voting scheme to decimate E/L signals by a factor of 4 is shown in Figure 18.15. Majority voting is a nonlinear operation with gain, K_V dependent on input bit-pattern and jitter. In contrast, a finite impulse response (FIR) boxcar filter with a typical output as given below decimates BBPD's output (D_{PD}) to smaller bit-length word at a lower clock frequency, P/T_{RCK}, where $P > 0$ and T_{RCK} is time period of the recovered clock:

$$D_C[k \cdot \mathrm{PT_{RCK}}] = \frac{1}{P}\sum_{i=1}^{P} D_{PD}(\overline{k-1} \cdot PT_{RCK} + i \cdot T_{RCK}) \tag{18.27}$$

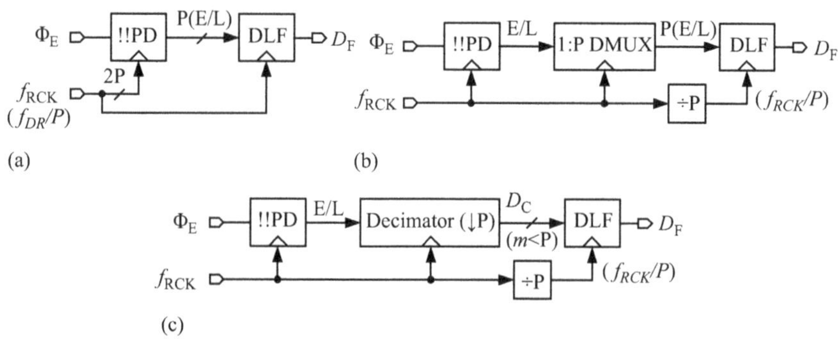

Figure 18.14 Digital filtering options for (a) sub-rate CDRs, (b) 1P deserialized BBPD's output, and (c) decimated BBPD's output

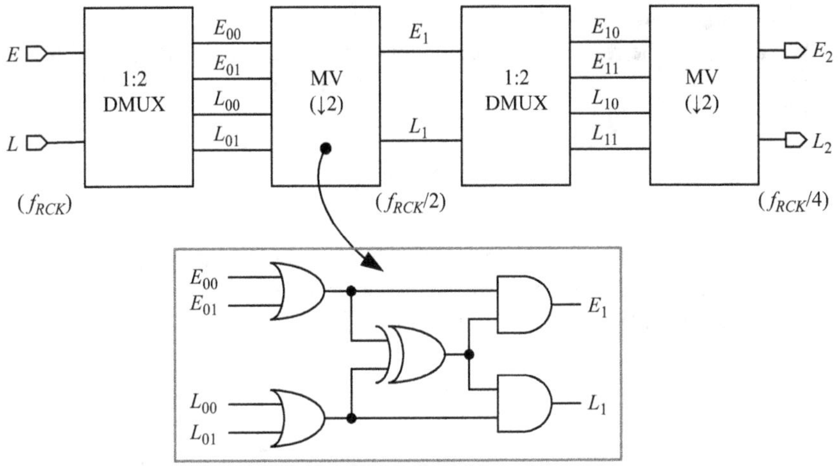

Figure 18.15 Decimation by majority voting

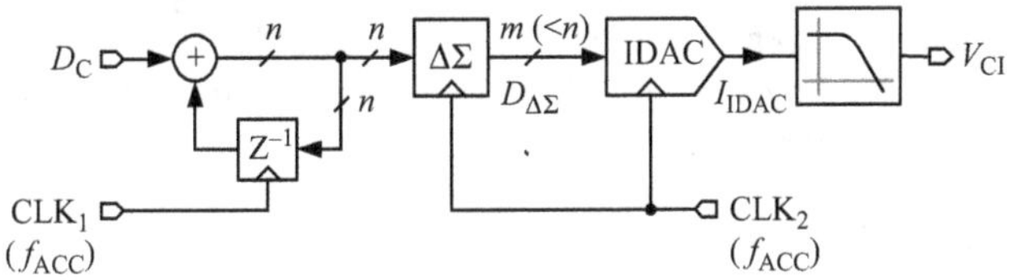

Figure 18.16 Block diagram of an integral path using digital delta sigma modulator

Note that deserialization is inherent to both MV and boxcar filtering. With decimation, digital loop filter can be clocked at $f_{ACC} = f_{RCK}/P$ thereby reducing power consumption of the accumulator. However, this adds a loop delay of $\geq T_d = 1/f_{ACC}$, which adversely affects phase-margin, jitter peaking, and JTOL performance of the CDR [6]. Hence, there exists a tradeoff between CDR's performance metrics and power consumption.

18.3.2 Digital frequency control in oscillators

In the locked state of the CDR loop, frequency of the oscillator is defined by integral path control word D_I and resolution of voltage/current DAC, IDAC. Because D_I dithers by a few LSBs in steady state, a smaller IDAC LSB and large frequency control word length are required to lower quantization error while achieving wide operating range. It is difficult to meet these requirements using a Nyquist-rate DAC, especially in the presence of mismatch between DAC cells. Figure 18.16 shows an implementation of a high-resolution integral path using a digital delta-sigma modulator, low-resolution Nyquist DAC, and a lowpass filter. ΔΣ truncates n-bit accumulator output to m-bits with $m \leq n$. A m-bit Nyquist DAC outputs current, which is fed into a low pass filter that suppresses ΔΣ's quantization noise. In contrast, proportional path can be easily implemented using Nyquist-DAC clocked at f_{ACC} as the resolution requirements are typically in the range of few bits. With a decimator and above digital loop filter, the new loop gain of the CDR can be simplified as below:

$$\mathrm{LG}(s) = K_{\mathrm{PD}} \cdot K_V \cdot \left(K_P + \frac{K_I f_{\mathrm{ACC}}}{s}\right) \cdot \frac{K_{\mathrm{VCO}}}{s} \cdot e^{-sT_d} \tag{18.28}$$

where K_P and K_I include voltage gain of DAC in proportional and integral paths, respectively. K_V is the gain of the decimator and T_d is delay in the loop, mainly due to deserialization and buffering of digital bits.

18.3.3 Digitally controlled oscillator (DCO)

Using a low power and poor phase noise oscillator in the CDR requires a wide bandwidth to filter its phase noise, which increases jitter generation. Another approach to realizing a low phase noise clock source involves a two-step process shown in Figure 18.17. A low jitter and high-frequency reference clock, $f_{\mathrm{OUT,REF}}$, nominally equal to f_{OUT}, is generated separately using a phase-locked loop. This PLL's bandwidth is independent of CDR's bandwidth. Multiple clock phases, $I/Q/\overline{I}\overline{Q}$, at $f_{\mathrm{OUT,REF}}$ frequency are interpolated based on the control word, D_{PI}, to produce f_{OUT}. Output phase can be expressed as:

$$\Phi_{\mathrm{OUT}}[kT_{\mathrm{PI}}] = \Phi_{\mathrm{OUT,REF}}[kT_{\mathrm{PI}}] + K_{\mathrm{PI}} D_{\mathrm{PI}}[kT_{\mathrm{PI}}] \tag{18.29}$$

where $T_{\mathrm{PI}}(= 1/f_{\mathrm{PI}})$ is the period at which PI's control word is changed and K_{PI} is PI's gain in rad/LSB. Since digital loop filter output is accumulated to generate PI's control word, PI's output accumulates phase proportional to DLF's output as evaluated below:

$$\Phi_{\mathrm{OUT}}[kT_{\mathrm{PI}}] = \Phi_{\mathrm{OUT,REF}}[kT_{\mathrm{PI}}] + K_{\mathrm{PI}}(D_{\mathrm{PI}}[\overline{k-1}T_{\mathrm{PI}}] + D_F[kT_{\mathrm{PI}}]) \tag{18.30}$$

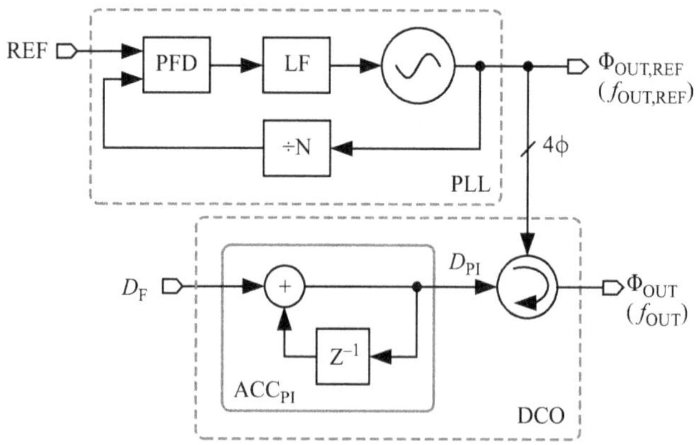

Figure 18.17 Block diagram of a PI-based digitally controlled oscillator

Any change in D_F is translated to output phase with the following transfer function:

$$\begin{aligned} \frac{\Phi_{\text{OUT}}(z)}{D_F(z)} &= \frac{K_{\text{PI}}}{1 - z^{-1}} \\ \Rightarrow \frac{\Phi_{\text{OUT}}(z)}{D_F(z)} &\approx \frac{K_{\text{PI}} f_{\text{PI}}}{s} = \frac{K_{\text{DCO}}}{s} \end{aligned} \tag{18.31}$$

A phase interpolator together with an accumulator operate as a digitally controlled oscillator with gain $K_{\text{DCO}} = K_{\text{PI}} f_{\text{PI}}$.

18.3.4 Decoupling JTRAN and JTRACK bandwidth in CDR

In a standard digital CDR, JTRAN $(= \Phi_{\text{RCK}}/\Phi_{\text{IN}})$ and JTRACK $(= 1 - \Phi_{\text{RCK}}/\Phi_{\text{IN}})$ bandwidth are coupled as they depend on the same signal transfer function, $\Phi_{\text{RCK}}/\Phi_{\text{IN}}$. One way to de-couple this tradeoff is by generating two separate clocks namely recovered clock (RCK) and sampling clock (SCK). Both clock signals should have the same frequency, that is, proportional to data rate, but RCK meets JTRAN requirement while SCK satisfies JTRACK. Figure 18.18 shows an adaptation of a standard digital CDR to generate the two clocks. Phase shift in the output clock due to proportional control is realized by adding phase shift in the feedback path. Phase added due to proportional control in the two CDR architectures are given by:

$$\Phi_{\text{RCK,PROP}}[kT] = \Phi_{\text{RCK,PROP}}[\overline{k-1}T] + 2\pi K_{\text{DCO}} K_P D_F[\overline{k-1}T] \tag{18.32}$$

$$\Phi_{\text{SCK,PROP}}[kT] = \Phi_{\text{SCK,PROP}}[\overline{k-1}T] + K_\phi D_F[\overline{k-1}T] \tag{18.33}$$

To keep the same loop dynamics for Φ_{RCK} and Φ_{SCK}, $K_\phi = 2\pi K_{\text{DCO}} K_P T_{\text{ACC}}$ where $T_{\text{ACC}} = 1/f_{\text{ACC}}$ is the update rate of the loop filter. In addition, we have a recovered clock RCK that is devoid of perturbations in the proportional path. JTRAN and JTRACK transfer functions for the modified CDR loop are given below:

$$\text{JTRAN}(s) = \frac{K_{\text{PD}} K_D K_I K_{\text{DCO}} f_{\text{ACC}}}{s^2 + s K_{\text{PD}} K_D K_\phi f_{\text{ACC}} + K_{\text{PD}} K_D K_I K_{\text{DCO}} f_{\text{ACC}}} \tag{18.34}$$

$$\text{JTRACK}(s) = 1 - \frac{K_{\text{PD}} K_D f_{\text{ACC}} (s K_\phi + K_I K_{\text{DCO}})}{s^2 + s K_{\text{PD}} K_D K_\phi f_{\text{ACC}} + K_{\text{PD}} K_D K_I K_{\text{DCO}} f_{\text{ACC}}} \tag{18.35}$$

Jitter transfer bandwidth, ω_{JTRAN} and jitter tracking bandwidth, ω_{JTRACK} can be approximated as $K_I K_{\text{DCO}}/K_\phi$ and $K_{\text{PD}} K_D K_\phi f_{\text{ACC}}$, respectively. Hence, JTRAN and JTRACK bandwidth can be regulated independently to meet CDR specifications in repeaters. We will undertake case studies of two different CDR architectures and discuss their design considerations.

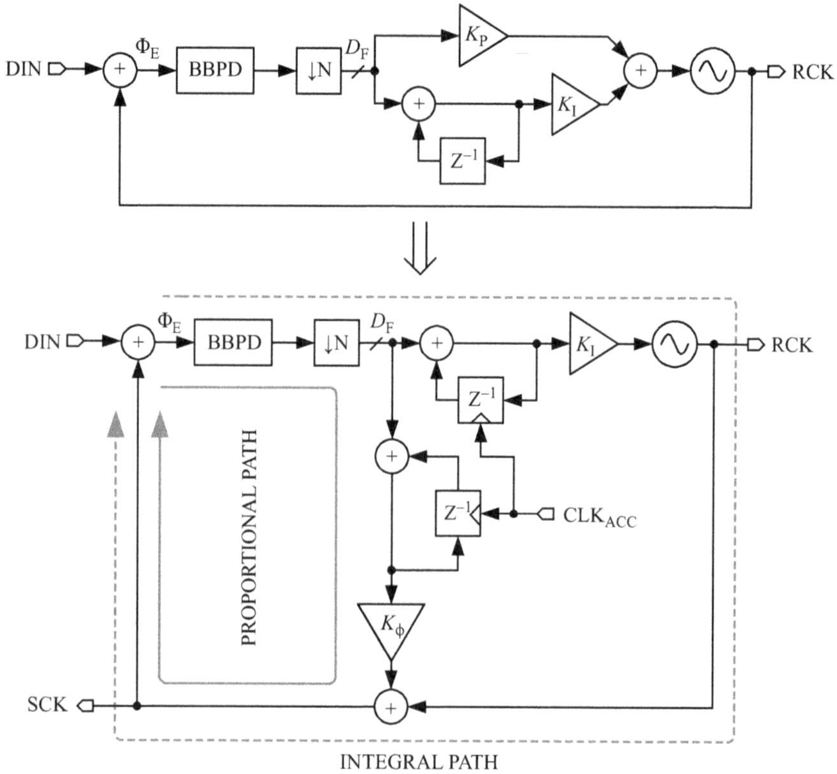

Figure 18.18 Evolution of a digital CDR with decoupled JTRAN and JTRACK

18.4 Case study I: digital CDR with decoupled JTRAN and JTRACK

The digital CDR architecture derived in Figure 18.18 decouples JTRAN and JTRACK bandwidth by separating the recovered clock from the sampling clock and adding faster phase tracking to the sampling clock. Figure 18.19 shows different options to add phase to the recovered clock. A digitally controlled delay line (DCDL) whose output phase is controlled by a digital control word, DCW as shown in Figure 18.19(a), can be used to add phase to RCK. Phase shift provided by the delay line can be expressed as below:

$$\Phi_{SCK} = \Phi_{RCK} + K_{DCDL} \cdot DCW + \Phi_{OS} \tag{18.36}$$

where Φ_{OS} is a fixed offset and K_{DCDL} is gain of the delay line. The maximum delay or phase shift offered by the DCDL is limited and its gain has to be calibrated across PVT variations. DCDL's design suffers from tradeoffs between range, resolution, and power while retaining sharp output clock edges needed for sampling [7].

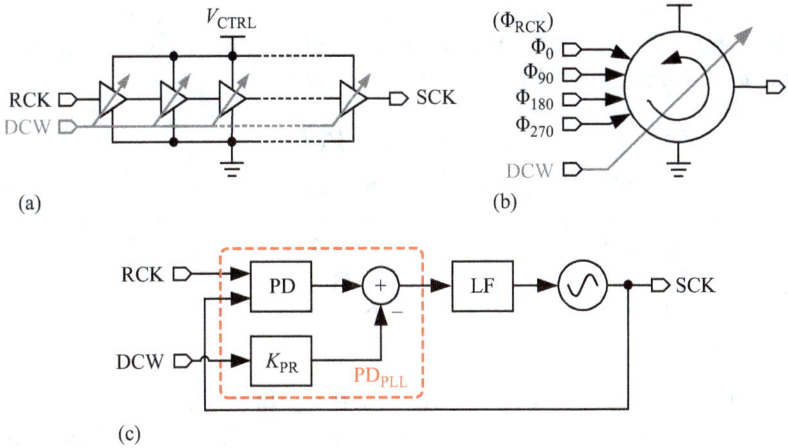

Figure 18.19 Phase addition to the recovered clock by using (a) digitally controlled delay line, (b) phase interpolator, and (c) phase rotating PLL

In sub-rate CDRs, number of RCK's clock phases and delay lines required are directly proportional to the number of SCK's phases employed.

In contrast, a phase interpolator–based phase shifting requires a minimum of four RCK's clock phases independent of the number of SCK clock phases used during sampling (Figure 18.19(b)). However, the number of phases interpolators double with each additional sampling phase. The output phase shift can be expressed by (1.37), where K_{PI} [rad/LSB], that is, PI's gain replaces K_{DCDL}. Unlike a delay line, a PI can provide infinite phase shift, which makes it a viable option for plesiochronous links. Both, delay line and PI-based phase-shifting techniques are open-loop phase addition techniques and their performance is limited by their quantization noise, that is, fixed phase-step and digital-to-phase conversion non-linearity [8,9]. Several noise-filtering, resolution improvement, and nonlinearity suppression techniques have been introduced to improve the performance of the above phase-addition techniques.

On the other hand, a closed-loop phase addition technique with the help of a phase rotating PLL (PRPLL) in Figure 18.19(c) helps to overcome some of the above drawbacks [10]. Phase quantization noise and nonlinearity during phase addition are suppressed by the local negative feedback loop. Multiple sampling clock phases are generated either directly by an oscillator in the PRPLL or by dividing oscillator's output. Noise and bandwidth of the PLL are optimized for minimum jitter generation while meeting jitter tracking requirements. Depending on the data rate and sub-rate CDR architecture, a choice between delay line, phase interpolator, and PRPLL can be made to achieve the same performance with minimum power consumption. For example, open-loop phase addition methods are used at high data rates. Delay line with lower quantization noise and nonlinearity is preferred for limited data rate and full/half-rate CDRs. PI is preferred for wide-range clock recovery. Closed-loop phase addition is useful at low-to-medium data rates and in sub-rate CDRs. We will now discuss the design of a 5 Gb/s half-rate digital CDR using a PRPLL.

18.4.1 A 5 Gb/s digital CDR using PRPLL

Figure 18.20 shows block diagram of a digital CDR that uses a PRPLL in the feedback loop. Here, phase error is accumulated using an accumulator, ACC_I, whose output, D_F, controls frequency of the oscillator that feeds RCK into the PRPLL. Stability of the above CDR loop, its JTRAN/JGEN behavior and decoupled JTRAN/JTRACK can be analyzed by using small-signal analysis as described below.

18.4.1.1 Linear analysis of PRPLL-based digital CDR

A linear model of above CDR is shown in Figure 18.21. Loop gain of the CDR, $\mathrm{LG_{CDR}}(s)$ is given by:

$$\mathrm{LG_{CDR}}(s) = K_{\mathrm{BBPD}} \cdot \left(K_P K_{\mathrm{PR}} \frac{f_{\mathrm{ACC}}}{s} + \frac{K_I K_{\mathrm{DCO}}}{s} \frac{f_{\mathrm{ACC}}}{s} \right) \cdot \frac{\mathrm{LG_{PRPLL}}(s)}{1 + \mathrm{LG_{PRPLL}}(s)} \tag{18.37}$$

$\mathrm{LG_{PRPLL}}$ is loop gain of the PRPLL and is given below:

$$\mathrm{LG_{PRPLL}}(s) = K_{\mathrm{PD}} \cdot LF(s) \cdot \frac{K_{\mathrm{VCO2}}}{s} \tag{18.38}$$

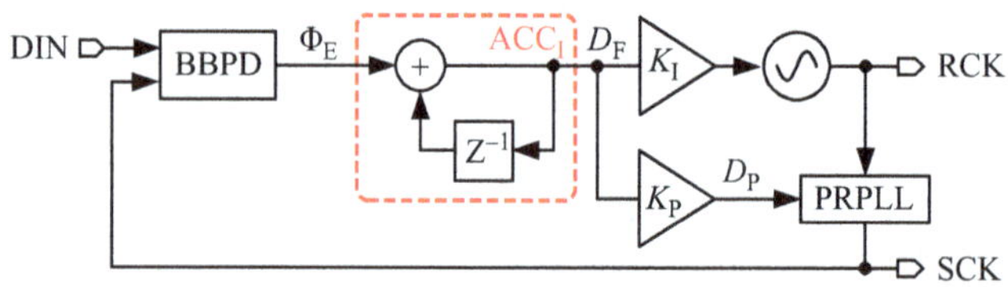

Figure 18.20 Block diagram of a digital CDR with PRPLL in feedback path

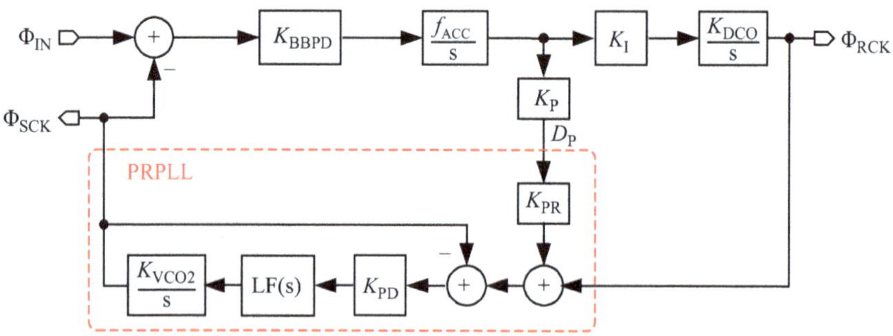

Figure 18.21 Small signal diagram of a digital CDR with PRPLL in feedback path

where K_{PD} is gain of the phase error detector, $LF(s)$ is loop filter transfer function, and K_{VCO2} is gain of the VCO used in the PRPLL. The PRPLL is designed with closed loop -3 dB bandwidth larger than the unity gain frequency of CDR's loop gain. Hence, an approximation $|LG_{PRPLL}(s)/(1+LG_{PRPLL}(s))| \approx 1$ holds true in the frequencies of interest. CDR's loop gain can be simplified as below:

$$LG(s) = K_{BBPD}\frac{f_{ACC}}{s}\cdot\left(K_P K_{PR} + \frac{K_I K_{DCO}}{s}\right) \tag{18.39}$$

The closed-loop jitter transfer function from input-to-recovered clock, H_{IN2RCK} is evaluated below:

$$H_{IN2RCK} = \frac{\Phi_{RCK}(s)}{\Phi_{DIN}(s)} = \frac{K_{BBPD}K_I K_{DCO}f_{ACC}}{s^2 + sK_{BBPD}K_P K_{PR}f_{ACC} + K_{BBPD}K_I K_{DCO}f_{ACC}} \tag{18.40}$$

Above transfer function has two poles and there is no zero. Hence, input jitter peaking due to zero present in the forward path of a standard type-II CDR loop is eliminated and the peaking will be limited by quality factor of two closed-loop poles, which are given by:

$$\omega_{p1,p2} = \frac{K_{BBPD}K_P K_{PR}f_{ACC}}{2}\left[-1 \pm \sqrt{1 - \frac{4K_{BBPD}K_I K_{DCO}f_{ACC}}{(K_{BBPD}K_P K_{PR}f_{ACC})^2}}\right] \tag{18.41}$$

$$\begin{aligned} &= \frac{K_{PP}}{2}\left[-1 \pm \sqrt{1 - \frac{4K_{INT}}{K_{PP}^2}}\right] \\ &\approx \frac{K_{PP}}{2}\left[-1 \pm \left(1 - \frac{2K_{INT}}{K_{PP}^2}\right)\right] \end{aligned} \tag{18.42}$$

Under the approximation, the poles are widely separated and equal to $-K_{PP}$ and $-K_{INT}/K_{PP}$. Lower of the two poles restricts JTRAN bandwidth which is typically K_{INT}/K_{PP}. In contrast, the sampling clock SCK tracks input jitter with the following transfer function:

$$H_{IN2SCK} = \frac{\Phi_{SCK}(s)}{\Phi_{DIN}(s)} = \frac{sK_{BBPD}K_P K_{PR}f_{ACC} + K_{BBPD}K_I K_{DCO}f_{ACC}}{s^2 + sK_{BBPD}K_P K_{PR}f_{ACC} + K_{BBPD}K_I K_{DCO}f_{ACC}} \tag{18.43}$$

In addition to the same two poles as those of H_{IN2RCK}, the above transfer function has a zero, $\omega_z = -K_{INT}/K_{PP}$. Under approximation $\omega_{p1} = \omega_z$, the jitter tracking bandwidth is extended to $|\omega_{p2}| = K_{PP}$. The loop gain analysis confirms the decoupling of jitter transfer bandwidth and jitter tracking bandwidth or JTOL. Though we have simplified the analysis of above CDR similar to the standard type-II CDR, the operation and stability of the former is complicated by a PRPLL nested in the CDR loop. The impact of the PRPLL on CDR's stability can be minimized by making the slew-rate of phase tracking in the CDR to be much smaller than slew-rate of the PRPLL and the condition can be stated as below:

$$\frac{K_P K_{PR} + K_I K_{DCO} T_{ACC}}{2\pi T_{ACC}} \ll f_{PRPLL} \tag{18.44}$$

where f_{PRPLL} is −3 dB bandwidth of the PRPLL and T_{ACC} is the update period of the accumulator. A stability factor ξ is given below and it must be chosen greater than 1 for stable operation [11]:

$$\text{Stability factor,} \quad \xi = \frac{K_P K_{PR}}{K_I K_{DCO}} \tag{18.45}$$

In the above design example, $f_{VCO} = 2.5$ GHz, $K_{PR} = 2\pi/64$ rad/s, $K_{DCO} = 2\pi \times 10^5$ rad/s/LSB, $K_P = 0.5$, $K_I = 0.25$, and $f_{ACC} = f_{VCO}/4$. The lower bound on f_{PRPLL} is about 5 MHz and the upper bound is roughly 1/10th of the DCO's frequency, that is, 250 MHz. Having discussed the basic operation of above CDR, we will discuss implementation details of important blocks next.

The detailed block diagram of the CDR in Figure 18.20 with important implementation-level features is shown in Figure 18.22 [12]. Input DIN is terminated with 50 Ω impedance and amplified using a limiting amplifier to drive a half-rate BBPD. BBPD's outputs are decimated by a factor of 4 while using a majority voting, thereby reducing the update rate in CDR to 1/8th-data rate. The fast phase error correction and low jitter clock recovery are accomplished through a proportional and integral path, respectively. In CDR's proportional path, ACC_P accumulates the decimated phase error and updates the sampling clock phases with 6-bit control of the PRPLL. The first two bits are used to select two 90° separated clock phases from VCO2's multiple output phases. The last four bits are used to find a

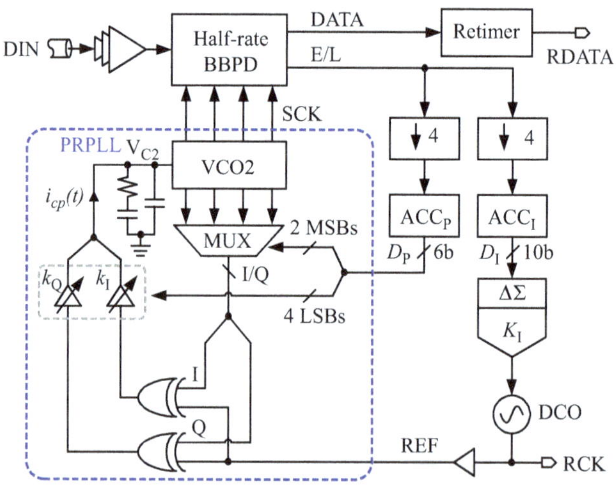

Figure 18.22 Detailed block diagram of a 5 Gb/s half-rate digital CDR

weighted mean of phase errors $\Delta\Phi_I$ and $\Delta\Phi_Q$ between phases I and Q and PRPLL's reference clock, REF, which is same as the recovered clock. The variable averaging weights, k_I and k_Q, are implemented through CML-based combined PD and CP. In steady state:

$$\Phi_{E,\text{PRPLL}} = k_I \cdot \Delta\Phi_I + k_Q \cdot \Delta\Phi_Q = 0 \;\forall\; \Delta\Phi_I,\;\; \Delta\Phi_Q \;\in [0, \pi/2] \tag{18.46}$$

Depending on weights k_I and k_Q, CDR locks to an intermediate phase between I and Q. The weighted average is $i_{\text{cp}}(t)$, which is proportional to $\Phi_{E,\text{PRPLL}}$, and filtered by using a second-order passive low-pass filter. The filtered voltage, V_{C2} controls VCO2's output frequency, thereby correcting for the phase error in CDR's proportional path. Here, k_I's and k_Q's resolution limits the dithering jitter in sampling clock phases. PRPLL is designed with 200 MHz bandwidth to filter VCO2's noise, which is often the dominant noise source. In the integral path, the phase error is accumulated using ACC_I to control the frequency of the digitally controlled ring oscillator, DCO. ACC_I's 10-bit output is truncated to 8-bits by using a digital delta–sigma modulator to reduce the linearity, matching, and area requirement for Nyquist-DAC controlling the oscillator. Finally, an 8-bit current DAC controls DCO's frequency with a resolution of 400 kHz/LSB. Assuming 0.01 UI_{rms} input jitter and an input transition density of 0.5, JTRAN bandwidth of 2 MHz mandates DCO's phase noise to be −100 dBc/Hz at 1 MHz frequency offset for input jitter and DCO's phase noise to contribute equally to the recovered clock jitter. The frequency acquisition range of the CDR can be extended by using an additional path correcting for large frequency errors.

Incorporating the design techniques discussed above, a 5 Gb/s CDR was realized in TSMC 90 nm CMOS process. The long-term recovered clock jitter, and sampling clock jitter with jitter histograms for the prototype is shown in Figure 18.23. A 5 ps r.m.s jitter for the recovered clock and 14.9 ps r.m.s jitter for the sampling clock was achieved with a higher value for the later due to PRPLL's phase quantization error. The decoupled JTRAN bandwidth (2 MHz) and JTOL bandwidth (16 MHz) for the CDR architecture are demonstrated by corresponding measurement results in Figure 18.24(a) and Figure 18.24(b), respectively.

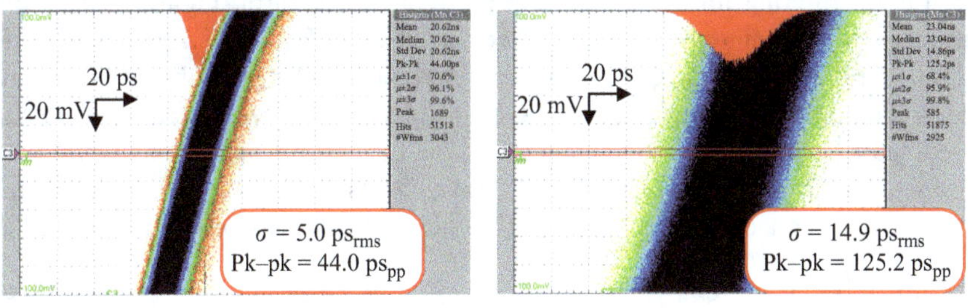

(a) (b)

Figure 18.23 Measured RCK and SCK jitter with PRBS31 input data: (a) RCK jitter, (b) SCK jitter

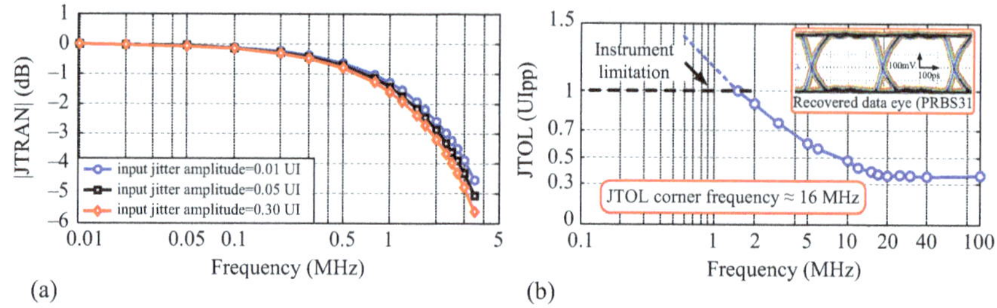

Figure 18.24 (a) Measured JTRAN with different jitter input jitter amplitudes and (b) measured jitter tolerance with a BER threshold of 10^{-12} and PRBS7 input data

At 5 Gb/s, the CDR dissipated 18.6 mW power with a significant portion (37%) for the ring oscillator. An LC-based oscillator can reduce the power consumption and improve jitter performance at the expense of the area. Also, applications with wide frequency range would require oscillators with wide tuning range, which is difficult to achieve for a standalone LC oscillator. A choice between a widely tunable ring oscillator and low power and low jitter LC oscillator is difficult when we want both characteristics. This problem will be addressed in the next design example, which will incorporate a widely tunable, low power, and low jitter digitally controlled oscillator in a digital CDR with decoupled JTRAN and JTOL. Also, the example will discuss the use of digitally controlled delay line (DCDL) with low quantization noise while separating JTRAN and JTOL bandwidth.

18.5 Case study II: A wide range digital clock and data recovery

A wide range digital CDR requires a widely tunable digitally controlled oscillator. A ring oscillator is an obvious choice for a wide range, but large power consumption, power versus phase noise tradeoff, and jitter generation versus jitter transfer bandwidth requirements overshadow its tuning range benefits. Separating phase noise filtering for oscillator from its frequency control, Figure 18.25 shows frequency generation options to substitute the DCO. In Figure 18.25(a), a low jitter and high-frequency clock, CLK_{IN}, is generated independently with low power in a closed-loop operation. CLK_{IN}'s I and Q clock phases are interpolated using a phase interpolator to generate output clock, CLK_{OUT}. Earlier, we fixed PI's control word D_{PI} that produced a fixed phase offset. Now, D_{PI} is updated at a fixed rate while using a roll-over accumulator, ACC_{PI}. The output clock phase can be expressed as:

$$\Phi_{OUT}[kT_{ACC}] = \Phi_{OUT}[\overline{k-1}\,T_{ACC}] + K_{PI}D_F[\overline{k-1}\,T_{ACC}] \tag{18.47}$$

If D_F is fixed, the average output clock frequency is given by:

$$f_{OUT}(t) = f_{IN}(t) + \frac{K_{PI} \cdot D_F}{T_{ACC}} \tag{18.48}$$

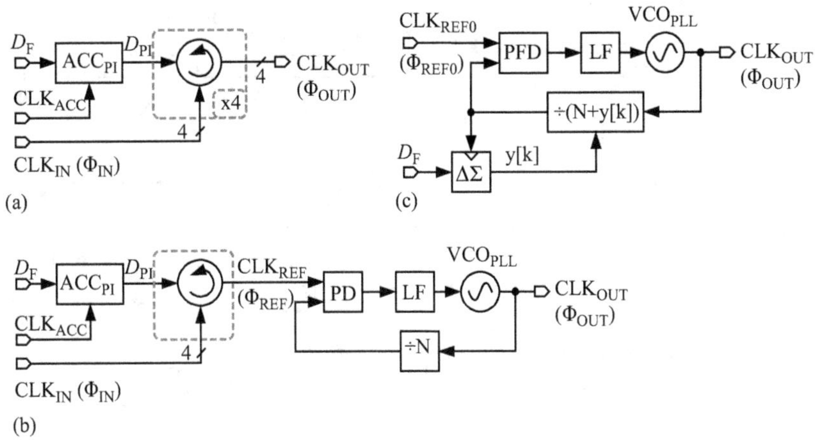

Figure 18.25 Block diagrams of DCOs using (a) phase interpolator, (b) phase interpolator and PLL, and (c) fractional-N PLL

where f_{IN} is the frequency of the input clock. For multiple sampling clock phases that are required in sub-rate CDRs, more than one PIs are used. The output clock jitter is limited by PI's quantization error, that is, fixed phase interpolation step. To overcome above drawbacks, PI-based oscillator can be used to generate a reference clock that is later multiplied to generate the output clock with desired frequency and multiple clock phases (Figure 18.25(b)) [13]. CLK_{IN} is already a low jitter clock, and VCO_{PLL}'s phase noise is filtered by high bandwidth for the PLL. PI's quantization error is rejected beyond PLL's bandwidth. Unlike the PI-based DCO, Figure 18.25(c) depicts a digital frequency control by controlling a frequency control word (FCW) of a fractional-N PLL (FNPLL). The fractional-N PLL has relatively lower reference frequency. The PLL-based frequency generation produces a low jitter clock in reduced power budget, but it comes with increased latency in the CDR loop. Since FNPLL is nested in the CDR loop, FNPLL's bandwidth should be larger than the desired CDR's bandwidth. In the following design example, FNPLL-based DCO is incorporated for 4–10 Gb/s wide range digital clock and data recovery. The design details are discussed below.

18.5.1 A 4–10 Gb/s digital clock and data recovery

Figure 18.26 shows a block diagram of a wide range, 2–5 GHz, half-rate digital clock and data recovery with decoupled JTOL and JTRAN bandwidth. The received signal, V_{RX} is amplified using a limiting amplifier to feed full-scale signal, DIN, to the clock recovery unit. The CDR implements proportional path by adding phase shift to DIN through DCDL and corrects for frequency error by changing frequency control word of FNPLL-based DCO. In the proportional path, the phase error between phase shifted input, DIN_{Φ} and recovered clock is determined by BBPD. The *early/late* signals are accumulated using accumulator, ACC_P whose

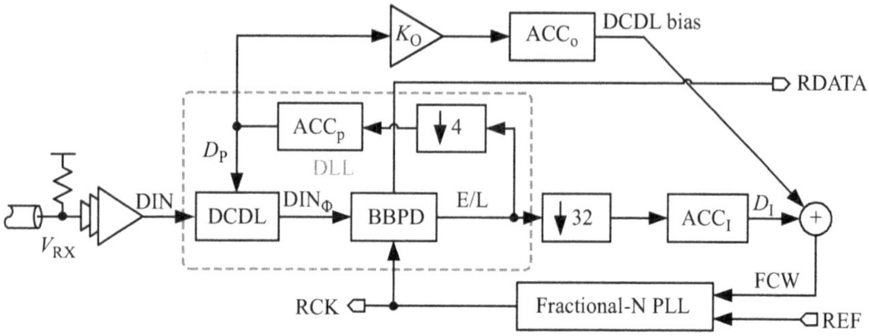

Figure 18.26 Block diagram of a wide range digital CDR using FNPLL-based DCO

output D_P is used to add or subtract phase/delay to DIN. BBPD's output is deserialized by a factor of 4 and 32 for fast proportional path and slow integral path, respectively. A lower accumulation rate is preferred to save power consumption in digital accumulators. The higher accumulation rate, $f_{RCK}/4$ reduces latency in the proportional path with sufficient phase margin for the loop stability and higher JTOL bandwidth.

Once DCDL's output saturates with minimum or maximum delay, continuous *early/late* signals quickly enforce a frequency shift through ACC$_O$ and phase recovery with zero average at ACC$_O$'s output in steady state. The DCDL is biased around its mid-delay point with the help of ACC$_O$ and maximum possible delay range is $\pm$100 ps. On the other hand, the CDR loop can withstand comparatively large propagation delay in the low gain integral path while saving energy during accumulation. The bandwidth and phase margin of FNPLL-based DCO should be optimized for desired CDR's loop bandwidth and stability. A separate frequency locked loop (FLL) can be used to bring the DCO within the looking range of type-II digital CDR. The recovered clock frequency is not specifically tied to FNPLL's reference clock frequency. Hence it acts like a referenceless digital CDR. The implementation details of important blocks in the above CDR are discussed below.

18.5.2 Fractional-N PLL-based DCO

Figure 18.27 shows block diagram of an analog fractional-N PLL employed as a DCO in the CDR loop [14]. It consists of a three-state phase frequency detector (PFD), a charge pump, an analog loop filter, a widely tunable ring oscillator (2–5.5 GHz), a 4-to-15 multimodulus divider, and a digital delta–sigma modulator. A 17-bit frequency control word from CDR is generated at an arbitrary delay offset from FNPLL's sampling instants. Hence, it is synchronized with FNPLL's divided clock, CLK$_{DIV}$, before truncated to 4-bit divider control, N_{DIV}. The metastability issues during synchronization can be avoided if synchronization rate is higher than twice the update rate for FCW. The gain from CDR's frequency control word to RCK's output phase is expressed below:

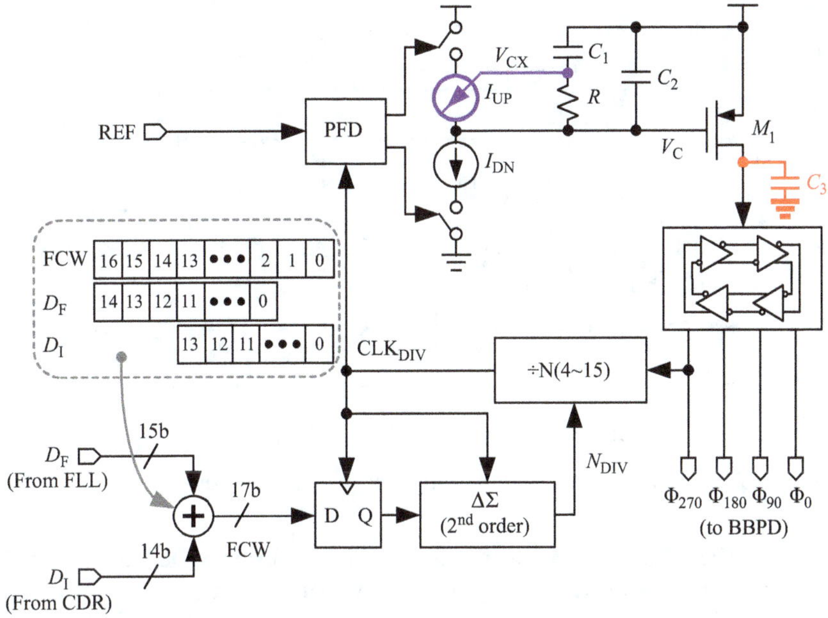

Figure 18.27 Simplified schematic diagram of a FNPLL-based DCO

$$\frac{\Phi_{RCK}(s)}{FCW(s)} = \frac{2\pi}{s} \times \frac{LG_{FNPLL}(s)}{1 + LG_{FNPLL}(s)} \tag{18.49}$$

$$LG_{FNPLL}(s) = \frac{I_{CP} R K_{VCO}}{2\pi s^2 (C_1 + C_2) N_{nom}} \times \frac{1 + sRC_1}{\left(1 + \frac{sRC_1C_2}{C_1+C_2}\right)} \tag{18.50}$$

$LG_{FNPLL}(s)$ is open-loop gain of the fractional-N PLL and N_{nom} is the average clock frequency multiplication factor. With FNPLL's −3 dB bandwidth much larger than the frequencies of interest within the CDR loop, we can mimic oscillator's transfer function as $\Phi_{RCK}(s)/FCW(s) \approx 2\pi/s$. As the ring oscillator spans a wide frequency range, its control voltage V_C varies over a wide range thereby causing mismatch between current sources I_{UP} and I_{DN}. The mismatch can be minimized by using a local feedback loop and settled oscillator's control voltage V_{CX} as a reference voltage for charge pump current regulation. In the divider path, the quantization noise during 17-bit FCW to 4-bit N_{DIV} quantization is first noise shaped by a second-order ΔΣ noise-shaping transfer function and later filtered by a low pass transfer function for the PLL. Noise contribution from FNPLL's blocks is optimized for the required recovered clock jitter with minimum power.

18.5.3 Digitally controlled delay line

Figure 18.28 shows a schematic diagram of a pseudo-differential digitally controlled delay line. The full-scale differential signals DIP and DIN drive the 16-stage

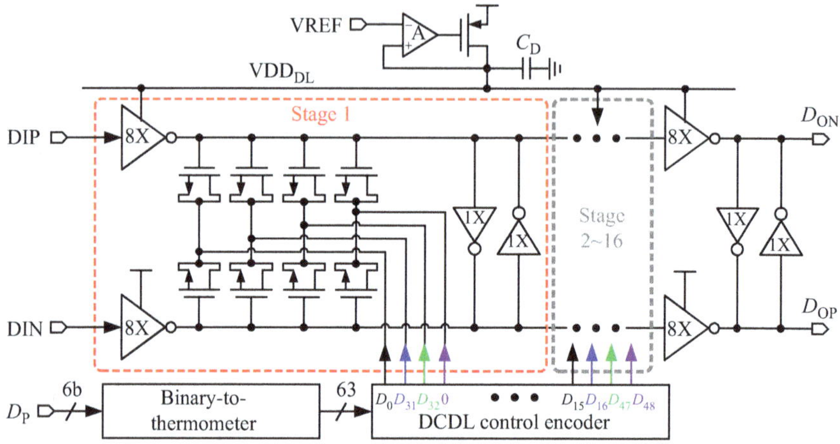

Figure 18.28 Simplified schematic diagram of a digitally controlled delay line

delay line where the delay of each stage is varied by using MOS capacitors. The delay line control is distributed in such a way that each delay stage contributes roughly equal delay to the output stage differing only by 1 LSB delay. This minimizes the nonlinearity during digital control bits to output delay conversion. The delay line has a finite bandwidth, and it adds to ISI while processing the wide band received signal. The ISI can be reduced by maximizing the bandwidth for each delay stage that comes with increased power consumption. The overall delay of the delay line can be varied by controlling its supply voltage, $\mathrm{VDD_{DL}}$, with the help of a low dropout regulator. Also, the LDO reduces the supply noise sensitivity of the delay line. It is important to note that using a DCDL in the data path to add phase-shift and decouple JTOL and JTRAN bandwidth is accompanied by bandwidth and power consumption tradeoff. DCDL in the clock path would have also decoupled JTOL and JTRAN bandwidth but at the increased cost of two delay lines for I/Q phases in half-rate CDR. Having understood the operation of basic building blocks of the wide range CDR, we will analyze the complete CDR loop next.

18.5.4 Linear analysis of a wide range CDR

The small-signal model of above wide range CDR is depicted in Figure 18.29. The jitter tracking in the CDR loop is given by:

$$
\begin{aligned}
\frac{\Phi_E(s)}{\Phi_{\mathrm{IN}}(s)} &= \frac{1}{1 + \dfrac{K_{\mathrm{BBPD}} K_{\mathrm{DCDL}} K_P}{4s/f_{RCK}} + \dfrac{2\pi K_{\mathrm{BBPD}} K_I f_{\mathrm{REF}}}{32s^2/f_{RCK}} \times \dfrac{LG_{\mathrm{FNPLL}}(s)}{1 + LG_{\mathrm{FNPLL}}(s)}} \\
&\approx \frac{1}{1 + \dfrac{K_{\mathrm{BBPD}} K_{\mathrm{DCDL}} K_P}{4s/f_{RCK}} + \dfrac{2\pi K_{\mathrm{BBPD}} K_I f_{\mathrm{REF}}}{32s^2/f_{RCK}}}, \\
&\text{for } \left| \frac{\mathrm{LF_{FNPLL}}(s)}{1 + \mathrm{LG_{FNPLL}}(s)} \right| = 1
\end{aligned}
\tag{18.51}
$$

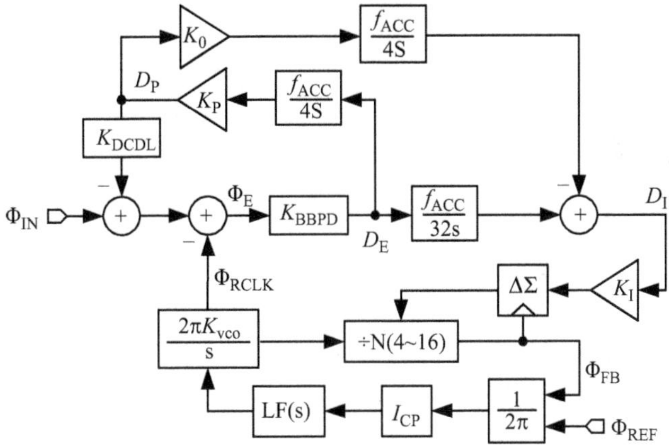

Figure 18.29 Small signal model of a wide range CDR

Accumulators in proportional and integral paths are clocked at $f_{RCK}/4$ and $f_{RCK}/32$, respectively. Under the assumption of widely separated poles, the loop has two zeros as $\omega_{z1} = \omega_{z2} = 0$ and two poles as $\omega_{p1} = -K_{BBPD}K_{DCDL}K_P f_{RCK}/4$ and $\omega_{p2} = -2\pi K_I f_{REF}/8K_{DCDL}K_P$. Since the proportional path gain dominates the integral path gain for stability reasons, the JTRACK bandwidth is given by higher of the two poles, that is, $\omega_{-3dB} = -\omega_{p1}$. On the other hand, the input jitter is transferred to the recovered clock jitter with transfer function as given below:

$$\frac{\Phi_{RCK}(s)}{\Phi_{IN}(s)} = \frac{\dfrac{2\pi K_{BBPD}K_I f_{REF}}{32s^2/f_{RCK}} \times \dfrac{LG_{FNPLL}(s)}{1+LG_{FNPLL}(s)}}{1 + \dfrac{K_{BBPD}K_{DCDL}K_P}{4s/f_{RCK}} + \dfrac{2\pi K_{BBPD}K_I f_{REF}}{32s^2/f_{RCK}} \times \dfrac{LG_{FNPLL}(s)}{1+LG_{FNPLL}(s)}}$$
$$\approx \frac{2\pi K_{BBPD}K_I f_{REF} f_{RCK}}{32s^2 + 8sK_{BBPD}K_{DCDL}K_P f_{RCK} + 2\pi K_{BBPD}K_I f_{REF} f_{RCK}} \qquad (18.52)$$

The JTRAN bandwidth can be approximated as $\omega_{-3dB} = 2\pi K_I f_{REF}/8K_P K_{DCDL}$. Hence, the JTRAN and JTRACK bandwidths are decoupled.

Above CDR was designed and fabricated in TSMC 65 nm CMOS process [14]. Figure 18.30 shows measured long-term absolute jitter of the recovered clock when the CDR operates with PRBS31 input. The jitter is 2.9 ps_{rms}/25.1 ps peak-to-peak at 4 Gb/s and 2.2 ps_{rms}/24.1 ps peak-to-peak at 10 Gb/s. Independently, the fractional-N PLL has a long-term absolute integrated jitter varying from 1.41 to 2.3 ps_{rms} depending on the fractional frequency control word for frequencies around 5 GHz. At 5 GHz output, the FNPLL consumes 5 mW from 1 V supply voltage. As the jitter due to oscillator phase noise is greatly suppressed by wide bandwidth FNPLL, a very low JTRAN bandwidth can be chosen to suppress the input jitter. Figure 18.31(a) shows JTRAN transfer functions for different input jitter amplitudes. The measured JTRAN bandwidth is about 0.2 MHz and independent of input

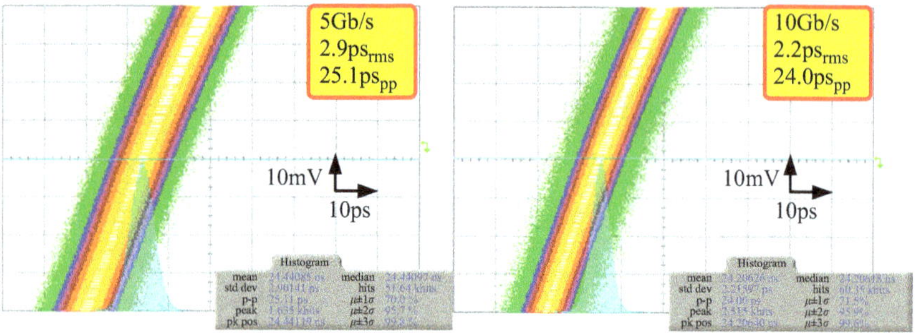

Figure 18.30 Recovered clock jitter histograms at different data rates

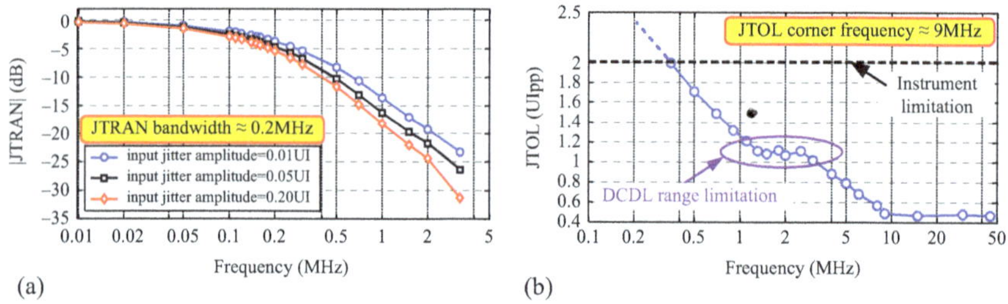

Figure 18.31 (a) Measured JTRAN with different jitter amplitudes. (b) Measured JTOL with PRBS7 input data at 10 Gb/s

jitter even while using BBPD. The decoupled JTRACK and JTOL bandwidth can be observed from jitter tolerance measurement shown in Figure 18.31(b). Measured JTOL's corner frequency is 9 MHz at 10 Gb/s. JTOL is limited by DCDL range in 1.1-to-2.5 MHz frequency band at 10 Gb/s.

In the above design example, we could recover clock over a wide range of frequencies with moderate recovered clock jitter and low power consumption. The jitter performance of the design is limited by the jitter generated within the FNPLL, which can be improved with modified FNPLL implementation techniques as used in [7,15]. It is difficult to implement but an LC oscillator with 2X tuning range can help to increase FNPLL's bandwidth and reduce output jitter. The degradation in JTOL's performance due to the limited tunable range of DCDL can be minimized with increased delay range at the cost of increased power or ISI. ISI can be eliminated by adding phase shift to the recovered clock through DCDLs instead of adding a delay in the data path but it increases power consumption due to multiple delay lines for multiple sampling clock phases in sub-rate CDRs.

Recovering a wide range of clock frequencies from the received signal is desirable for both electrical and optical communication networks. However, CDRs with conventional linear or bang-bang phase detectors suffer from limited frequency acquisition range. Typically, a separate frequency detector operates simultaneously with regular phase detector in the CDR loop while extending the range of data rates recovered [16–19]. A divider-based stochastic reference clock generator in [20,21] extends the frequency acquisition range but makes it depend on input data transition density. A continuous rate and wide-range frequency acquisition in digital CDRs and independent of input transition density is demonstrated in [14].

In this chapter, we mostly discussed 2X oversampling digital CDRs, which required multiple high-frequency clock phases for data and edge sampling. In contrast, baud-rate CDRs employ a single clock phase for data and clock recovery, thereby reducing the power consumed in multiphase clock generation and distribution circuits. Interested readers can read more about baud-rate CDRs, their benefits and drawbacks in [22–26].

References

[1] Perrott MH, Huang Y, Baird RT, *et al.* "A 2.5-Gb/s multi-rate 0.25-μm CMOS clock and data recovery circuit utilizing a hybrid analog/digital loop filter and all-digital referenceless frequency acquisition." *IEEE J Solid-State Circuits*. 2006;41(12):2930–2944.

[2] Yin W, Inti R, Elshazly A, *et al.* "A TDC-Less 7 mW 2.5 Gb/s digital CDR with linear loop dynamics and offset-free data recovery." *IEEE J Solid-State Circuits*. 2011;46(12):3163–3173.

[3] Pavan S, Schreier R, and Temes GC. *Understanding Delta-Sigma Data Converters*. 2nd edn. Baker RJ, editor. Hoboken, NJ: Wiley-IEEE Press; 2017.

[4] Sonntag JL, and Stonick J. "A digital clock and data recovery architecture for multi-gigabit/s binary links." *IEEE J. Solid-State Circuits*. 2006;41(8):1867–1875.

[5] De Vito L. "A versatile clock recovery architecture and monolithic implementation." In: Razavi B, editor. *Monolithic Phase-Locked Loops and Clock Recovery Circuits: Theory and Design*. Wiley-IEEE Press; 1996. pp. 405–420.

[6] Park M, and Kim J. "Pseudo-linear analysis of bang-bang controlled timing circuits." *IEEE Transactions on Circuits and Systems—Part I: Regular Papers*. 2013;60(6):1381–1394.

[7] Elkholy A, Saxena S, Shu G, *et al.* "Low-jitter multi-output all-digital clock generator using DTC-based open loop fractional dividers." *IEEE J Solid-State Circuits*. 2018;53(6):1806–1817.

[8] Santiccioli A, Samori C, Lacaita AL, *et al.* "Power-jitter trade-off analysis in digital-to-time converters." *Electron. Lett.* 201753(5):306–308.

[9] Sidiropoulos S, and Horowitz MA. "A semidigital dual delay-locked loop." *IEEE J Solid-State Circuits*. 1997 Nov;32(11):1683–1692.

[10] Toifl T, Menolfi C, Buchmann P, *et al.* "A 0.94-ps-RMS-jitter 0.016-mm^2 2.5-GHz multiphase generator PLL with 360° digitally programmable phase shift for 10-Gb/s serial links." *IEEE J Solid-State Circuits*. 2005;40 (12):2700–2712.

[11] Hanumolu PK, Wei G, and Moon U. "A wide-tracking range clock and data recovery circuit." *IEEE J Solid-State Circuits*. 2008;43(2):425–439.

[12] Shu G, Saxena S, Choi W, *et al*. "A reference-less clock and data recovery circuit using phase-rotating phase-locked loop." *IEEE J Solid-State Circuits*. 2014;49(4):1036–1047.
[13] Saxena S, Shu G, Nandwana RK, *et al*. "A 2.8 mW/Gb/s, 14 Gb/s serial link transceiver." *IEEE J Solid-State Circuits*. 2017;52(5):1399–1411.
[14] Shu G, Choi W, Saxena S, *et al*. "A 4-to-10.5 Gb/s continuous-rate digital clock and data recovery with automatic frequency acquisition." *IEEE J Solid-State Circuits*. 2016;51(2):428–439.
[15] Elkholy A, Saxena S, Nandwana RK, *et al*. "A 2.05.5 GHz wide bandwidth ring-based digital fractional-N PLL with extended range multi-modulus divider." *IEEE J Solid-State Circuits*. 2016;51(8):1771–1784.
[16] Pottbacker A, Langmann U, and Schreiber H. "A Si bipolar phase and frequency detector IC for clock extraction up to 8 Gb/s." *IEEE J Solid-State Circuits*. 1992;27(12):1747–1751.
[17] Dalton D, Kwet Chai, Evans E, *et al*. "A 12.5-Mb/s to 2.7-Gb/s continuous-rate CDR with automatic frequency acquisition and data-rate readback." *IEEE J Solid-State Circuits*. 2005;40(12):2713–2725.
[18] Kratyuk V, Hanumolu PK, Mayaram K, *et al*. "A 0.6GHz to 2GHz digital PLL with wide tracking range." In: *Proceedings of the IEEE Custom Integrated Circuits Conference*. 2007. pp. 305–308.
[19] Kocaman N, Fallahi S, Kargar M, *et al*. "An 8.511.5Gbps SONET transceiver with referenceless frequency acquisition." In: *Proceedings of the IEEE Custom Integrated Circuits Conference*; 2012. pp. 1–4.
[20] Inti R, Yin W, Elshazly A, *et al*. "A 0.5-to-2.5 Gb/s reference-less half-rate digital CDR with unlimited frequency acquisition range and improved input duty-cycle error tolerance." *IEEE J Solid-State Circuits*. 2011;46(12):3150–3162.
[21] Han J, Yang J, and Bae H. "Analysis of a frequency acquisition technique with a stochastic reference clock generator." *IEEE Transactions on Circuits and Systems—Part II: Express Briefs*. 2012;59(6):336–340.
[22] Mueller K, and Müller M. "Timing recovery in digital synchronous data receivers." *IEEE Trans on Communications*. 1976;24(5):516–531.
[23] Spagna F, Chen L, Deshpande M, *et al*. "A 78 mW 11.8Gb/s serial link transceiver with adaptive RX equalization and baud-rate CDR in 32 nm CMOS." In: *IEEE International Solid-State Circuits Conference (ISSCC) Digest of Technical Papers*; 2010. pp. 366–367.
[24] Francese PA, Toifl T, Buchmann P, *et al*. "A 16 Gb/s 3.7 mW/Gb/s 8-tap DFE receiver and baud-rate CDR with 31 kppm tracking bandwidth." *IEEE J Solid-State Circuits*. 2014;49(11):2490–2502.
[25] Rahman W, Yoo D, Liang J, *et al*. "A 22.5-to-32-Gb/s 3.2-pJ/b referenceless baud-rate digital CDR with DFE and CTLE in 28-nm CMOS." *IEEE J Solid-State Circuits*. 2017;52(12):3517–3531.
[26] Kim D, Choi W, Elkholy A, *et al*. "A 15-Gb/s sub-baud-rate digital CDR." *IEEE J Solid-State Circuits*. 2019;54(3):685–695.

Chapter 19
Spread spectrum clock generator: a low-cost EMI solution

Minyoung Song[1] *and Chulwoo Kim*[2]

In this chapter, spread spectrum clock generation (SSCG)—an advanced clock generation solution for electromagnetic interference (EMI)—is presented. EMI becomes critical as the clock speed and the degree of integration are increased. Various EMI suppression techniques are introduced in this chapter. Among EMI suppression techniques, SSCG is a simple but cost-effective solution since it requires no external components and can be implemented using existing clock generators. Several SSCG design approaches are described. Similar to other clock generators, SSCG also faces implementation issues, for example, noise, non-linearity, and mismatch. Finally, recent spread spectrum clocking (SSC) modulation profiles are presented and discussed because the SSC modulation profile dominates the EMI performance.

19.1 Methods for electromagnetic interference (EMI) reduction

In synchronous digital systems, for example, CPU and DSP, the clock signal contributes highly to the electromagnetic interference (EMI). Owing to its periodicity, most of the clock energy is concentrated in its fundamental frequency and harmonics, having a narrow bandwidth. Therefore, the concentrated electromagnetic energy exceeds the regulatory limits for the electronic interference. Since the adjacent circuits near EMI sources are directly affected by EMI radiation, one of the best solutions is to shield either the EMI aggressors or victims susceptible to EMI. However, this is difficult to apply for interchip applications and expensive owing to the addition of external bulky components.

Instead of shielding solutions, we may consider the reduction of EMI from the source itself. EMI can be regarded as coupled noise. Since the coupled noise effect increases when the signal frequency or power increases, decreasing the power of high-frequency harmonics would be good for EMI reduction. In digital

[1]imec-Netherlands, Eindhoven, the Netherlands
[2]School of Electrical Engineering, Korea University, Seoul, Republic of Korea

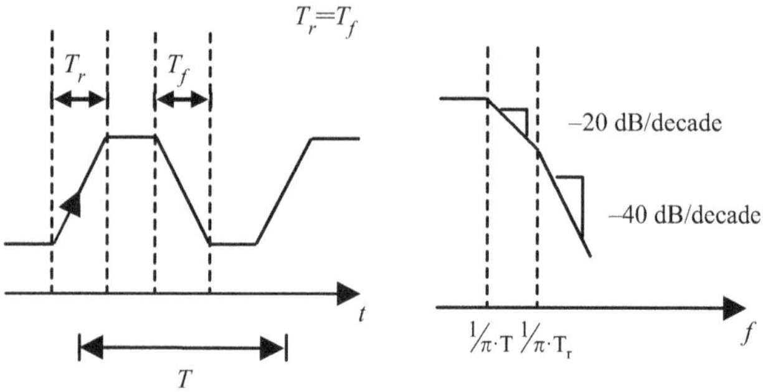

Figure 19.1 Trapezoidal waveform and its harmonic amplitudes

communications, the shape of the clock signal can be modeled with trapezoidal waveforms, which have short rise/fall times. Figure 19.1 shows a trapezoidal waveform and spectrum. The magnitude of the trapezoidal spectrum can be expressed as:

$$| C_n |= A\frac{\tau}{T}\text{sinc}\left(n\cdot\omega_0\frac{\tau}{2}\right)\text{sinc}\left(n\cdot\omega_0\frac{\tau_r}{2}\right) \tag{19.1}$$

where $f_0 = \frac{1}{T}$ is the signal frequency, τ is the pulse width, and τ_r is the rise or fall time. The power of high-frequency harmonics is reduced as the rise/fall times of the trapezoidal waveform increase. Therefore, the use of low-pass filtering to increase the rise/fall times of the digital signal can reduce EMI. However, in high-speed applications, it is difficult to increase the rise/fall times of a signal since the timing margin is critical. Moreover, the increased rise/fall times only decrease the power of signal harmonics and not that of fundamental.

Low-voltage differential clocking, which decreases the signal level, can be applied at the cost of level conversion and complex routing. Moreover, the reduced signal results in a lower signal-to-noise ratio (SNR), which limits the communication budget.

19.2 Spread spectrum clock generation (SSCG): a cost-effective EMI solution

The EMI receiver, that is, the victim, is band-limited. Therefore, instead of decreasing the total signal power, it is sufficient to decrease the power level within the EMI receiver band to reduce EMI. Spread spectrum clocking (SSC) is used to spread the clock spectrum so that it is wider than the bandwidth of the EMI receiver, as illustrated in Figure 19.2(a). Owing to SSC, the peak power level within the EMI receiver bandwidth is reduced while the total signal power is unchanged (see Figure 19.2(b)). When measuring EMI, the bandwidth of the EMI receiver is shaped as the resolution bandwidth (RBW) of the spectrum analyzer [1]. A spectrum analyzer with a finite RBW operates as a band-pass filter for the signal,

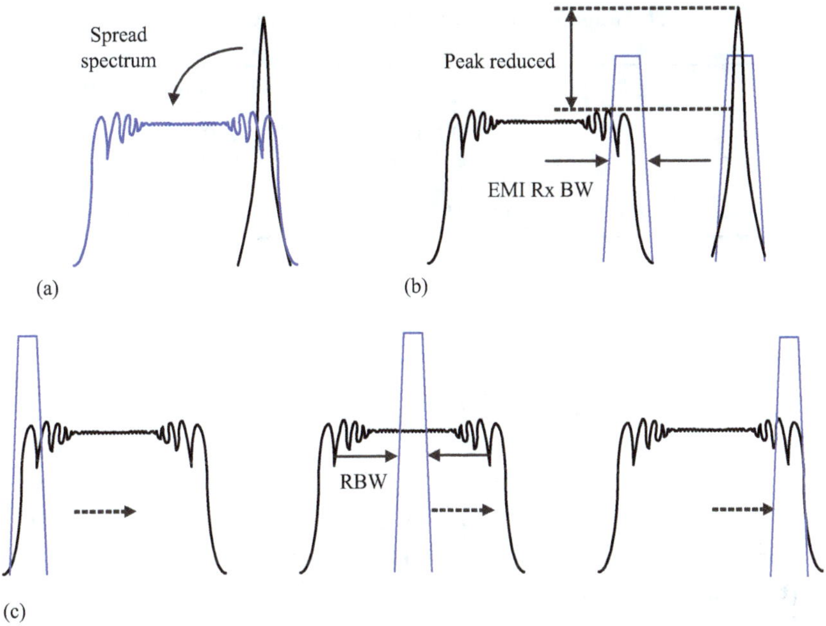

Figure 19.2 (a) SSC, (b) peak level reduction in the bandwidth of the EMI receiver, and (c) measurement method of the spectrum analyzer

as shown in Figure 19.2(c). Therefore, the measured signal power spectrum depends on the RBW and the bandwidth of measured signal.

SSC uses a broadband signal whose bandwidth exceeds the RBW, and the subharmonics of SSC are distributed more widely than the RBW. Unlike a narrowband signal whose bandwidth is narrower than the RBW, the measured SSC spectrum power level relies on the RBW since the subharmonics are "moving-averaged" within the RBW. The "moving-averaged" subharmonics form ripples on top of the spectrum. As the RBW becomes wider, the spectrum power level increases and the ripples become smoother.

Therefore, the RBW needs to be specified to evaluate the EMI performance properly. The minimum RBW is specified to reflect the actual bandwidth of the EMI receiver for each frequency band for EMI measurement, as summarized in Table 19.1 [1].

Figure 19.3 shows the SSC modulation profile (triangular) and its modulated spectrum. The SSC modulation profile can be defined by the frequency deviation (δ) and the modulation frequency (f_m). As δ increases, the bandwidth becomes wider, and the peak power of the spectrum is reduced. f_m determines the distance between two adjacent subharmonics and the frequency of the ripples in an SSC spectrum.

As f_m decreases, the frequency spacing of each subharmonic also decreases; therefore, the signal power is distributed more densely within the same signal

Table 19.1 Regulatory situation for EMI measurement [1]

Frequency band		Minimum RBW
Band A	9–150 kHz	220 Hz
Band B	150 kHz–30 MHz	9 kHz
Band C	30 MHz–1 GHz	120 kHz
Beyond band C	Over 1 GHz	1 MHz

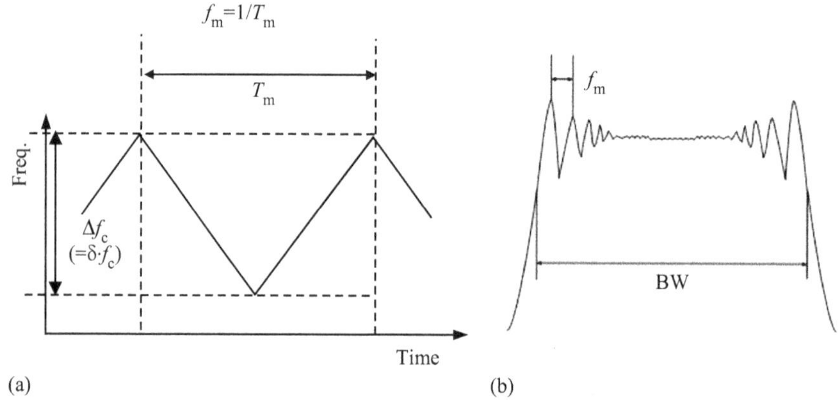

Figure 19.3 Relationship between (a) the modulation profile (triangular) and (b) its spectrum

bandwidth. Consequently, the peak power of the spectrum is reduced. With these characteristics of δ and f_m, the peak reduction can be approximated [2] as:

$$P\,|_{\text{reduction}} \propto 10 \cdot \log\left(\frac{\delta \cdot f_c}{f_m}\right) \tag{19.2}$$

To consider the RBW effect simply, a postulation can be introduced; the frequency components are uniformly distributed in the SSC spectrum, as shown in Figure 19.4. Since the spectrum is flat, the integrated power in the RBW can be approximated as:

$$P\,|_{\text{measured}} = P \cdot \left(\frac{\text{RBW}}{\Delta f_c}\right) \tag{19.3}$$

where Δf_c is $\delta \cdot f_c$. As the RBW becomes wider, the amount of EMI reduction is reduced.

Using (19.2) and (19.3), the amount of EMI reduction can be estimated as:

$$P\,|_{\text{reduction}} = 10 \cdot \log\left(\alpha \cdot \frac{\delta \cdot f_c}{f_m} \cdot \frac{\Delta f_c}{\text{RBW}}\right) \tag{19.4}$$

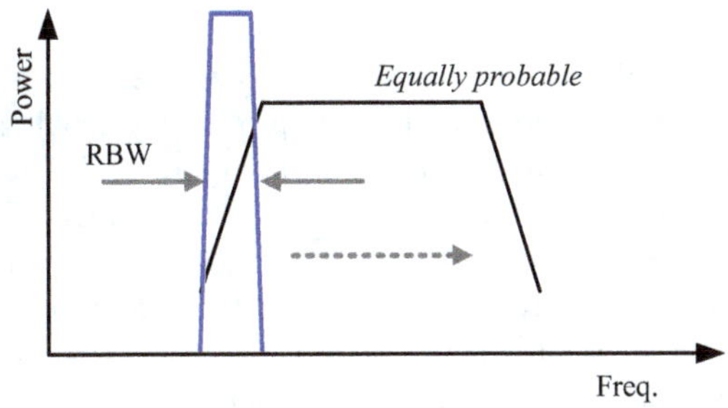

Figure 19.4 Effects of the RBW on broadband spectrum measurement

where α is a variable for other conditions such as the shape of the SSC modulation profile. Although f_m and δ are guided in many applications [3,4] to achieve a certain level of EMI, the EMI can be further reduced with the SSC modulation profile. The effect of the SSC modulation profile is explained in Section 19.4.

Indeed, SSC is a cost-effective EMI solution, as it requires no external devices, for example, shielding or capacitors. Moreover, SSC is also a powerful EMI solution for high-speed communications since it allows the signal power level to be preserved—and thus the SNR and rise/fall times of the signal for EMI reduction.

19.3 Design approaches for SSCG

To reduce the power within the EMI receiver, the frequency spacing of each subharmonic (the modulation frequency, f_m) should be substantially less than the RBW. Therefore, SSCG needs to change the clock frequency slightly. A phase-locked loop (PLL)-based clock generator with output frequency modulation can be a good solution for low-frequency modulation. A delay-line-based clock generator, unlike a PLL, could be also applied for SSCG. However, it is difficult to implement finer frequency control with low power dissipation. A frequency-locked loop (FLL) can provide a simple SSCG implementation, but it suffers from frequency accuracy.

19.3.1 PLL-based SSCG

Since SSC is used to modulate the clock frequency, a PLL-based clock generator is usually used for SSCG. Analog PLL (APLL)-based SSCG is popular since it allows a simple implementation. The subblocks of a PLL, for example, a voltage-controlled oscillator (VCO), charge pump (CP), and/or divider, are controlled to modulate the frequency. However, the APLL-based SSCG suffers from a mismatch that leads to over-/under-SSC.

Digital PLLs (DPLLs) have become attractive in advanced nodes and can be also an alternative for SSCG to reduce the mismatch since they are highly adaptable for digital calibration. On the other hand, the additional noise contributions, for example, quantization noise, limit the SSC performance.

19.3.1.1 Analog PLL

APLLs have been widely used for SSCG owing to their robustness and simple implementation. To apply an SSC modulation profile, the subblocks of APLLs are exploited to modulate the frequency, as shown in Figure 19.5. The SSC modulation profile can be directly inputted into the VCO [5]. An auxiliary CP current is controlled by the dedicated SSC modulation profile so that it directly modulates the VCO input, as shown in Figure 19.6. The VCO direct control method is straightforward to implement with frequency modulation since the VCO has a linear relationship between the input voltage and the output frequency. However, it distorts the SSC modulation profile when the modulation frequency f_m is less than the bandwidth of the PLL, as the VCO input is high-pass-filtered by the PLL.

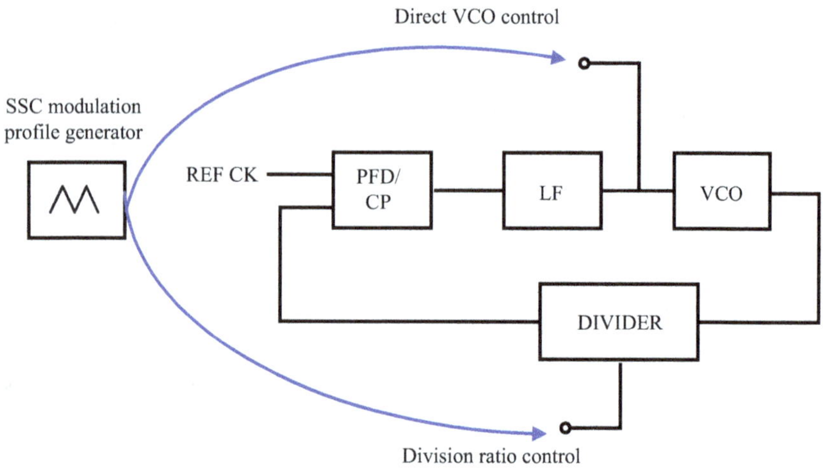

Figure 19.5 Frequency modulation methods for APLLs

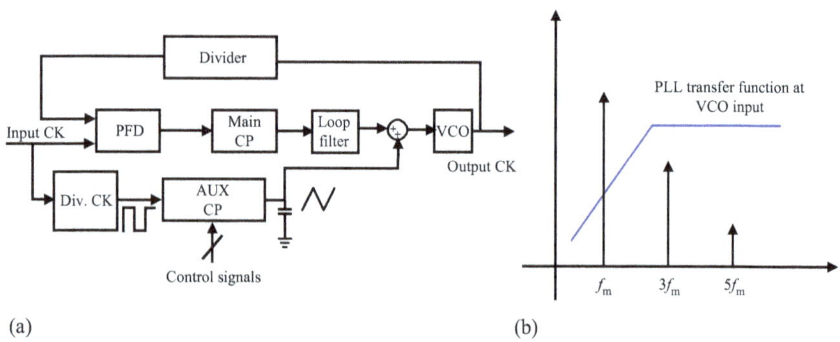

Figure 19.6 (a) Direct VCO control method [5] and (b) SSC distortion by a PLL

As described in Section 19.2, this limits the EMI reduction. Moreover, this method is also prone to PVT variations, which also distort the SSC modulation profile.

There is another method for controlling the division ratio of the programmable feedback divider, as shown in Figure 19.7. For fine frequency tuning such as SSC modulation, a ΔΣ fractional-*N* PLL is adopted. The ΔΣ modulator between the SSC profile generator and the divider is used to filter the quantization noise [6–8]. Although the ΔΣ modulator realizes fine frequency control, it still has to deal with the quantization noise, which significantly decreases the clock purity. To minimize the quantization noise, the PLL loop bandwidth should be sufficiently narrow to filter it. However, the narrow loop bandwidth requires a larger loop filter (LF). Unlike the VCO direct control method, the divider input is low-pass filtered by a PLL. Therefore, SSC modulation profile is distorted when the PLL loop bandwidth is narrower than f_m.

A higher-order or multibit ΔΣ modulator for shaping the quantization noise further can take advantage of the reduction in the quantization noise while retaining the PLL loop bandwidth. However, it also requires a higher-order LF to filter the out-of-band quantization noise or multimodulus divider. Compensation for the PLL loop bandwidth in the SSC profile to break the trade-off between the ΔΣ quantization noise and the SSC profile distortion is further discussed in Section 19.2.

19.3.1.2 Digital PLL

As process technology is scaled down, analog circuit design becomes challenging owing to the increase in short-channel effects, device mismatch, and a low supply voltage. Digitally assisted analog circuit design, for example, a DPLL can maintain analog functionalities during this downscaling trend. Moreover, new technology helps to maximize the digital performance with faster-speed operation and lower cost-effectiveness. To gain the benefits from the downscaling trend while maintaining PLL performance, APLLs can be replaced with DPLLs.

A DPLL usually comprises a digitally controlled oscillator (DCO), a digital loop filter (DLF), and either a time-to-digital converter (TDC) [9] or bang-bang phase detector [10]. DPLLs can be usually classified by the phase detector type, that is, a TDC or bang-bang phase detector, as shown in Figure 19.8.

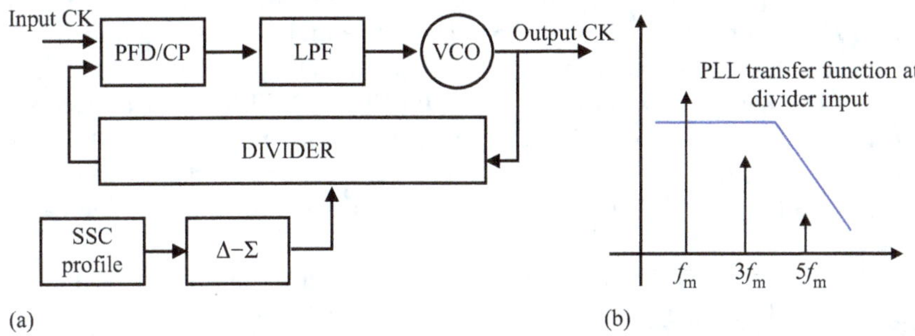

Figure 19.7 (a) Division ratio control method and (b) SSC distortion by a PLL

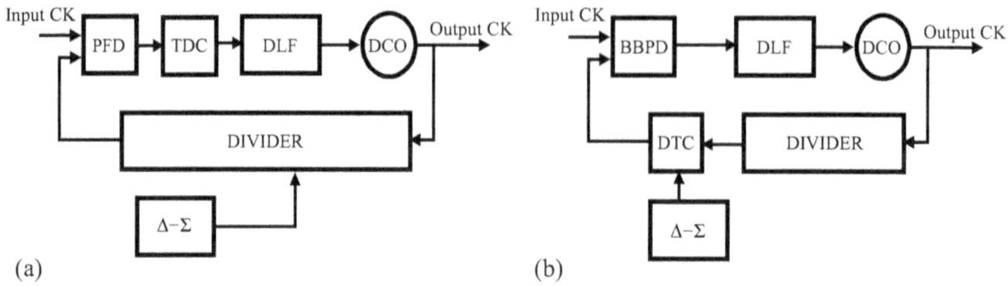

Figure 19.8 Common DPLLs: (a) TDC-DPLL and (b) BB-DPLL

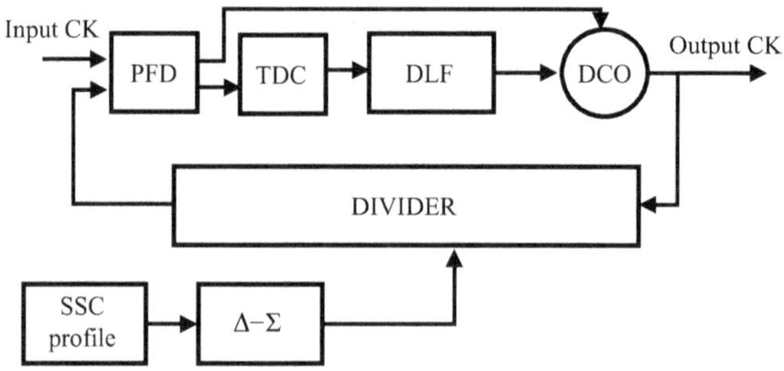

Figure 19.9 TDC-DPLL-based SSCG [11]

A TDC-based DPLL (TDC-DPLL) has linear behavior for phase tracking, similar to APLLs. However, an additional noise contribution, for example, the TDC quantization error, arises. The in-band noise floor of a TDC-DPLL is determined by the TDC resolution and division factor of the divider, N. The TDC-induced phase noise is scaled up by $20\log_{10}N$. High in-band noise can degrade the SNR; hence, it degrades the transfer of the small in-band signal such as the SSC modulation profile. Moreover, the limit cycle noise additionally deteriorates the DCO frequency modulation such as the SSC modulation profile since the discrete operation of a DPLL can have a longer loop latency than that of an APLL. TDC-DPLL-based SSCG [11] exploits the PFD to control the DCO directly, as illustrated in Figure 19.9. As a result, the limit cycle noise can be minimized by minimizing the loop latency on the proportional path. To reduce the TDC noise contribution of the TDC-DPLL, Figure 19.10 shows a subsampling TDC-DPLL that has no feedback divider within the PLL for SSCG [12]. Since the subsampling TDC-DPLL is a divider-less approach, the TDC noise is no longer scaled up by the division factor and also free from the ΔΣ quantization noise; thus, clean SSC modulation can be obtained.

Unlike the TDC-DPLL, it is difficult to implement a bang-bang-based digital PLL (BB-DPLL) for the fractional-N PLL, for example, SSCG due to its nonlinear operation, but it is more area- and power-effective than a TDC-DPLL [10]. BB-DPLL-based SSCG [13] achieves SSC modulation with a phase rotator and

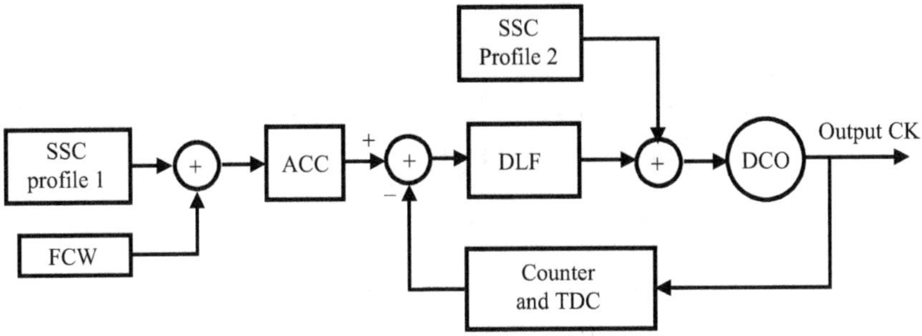

Figure 19.10 Subsampling TDC-DPLL-based SSCG [12]

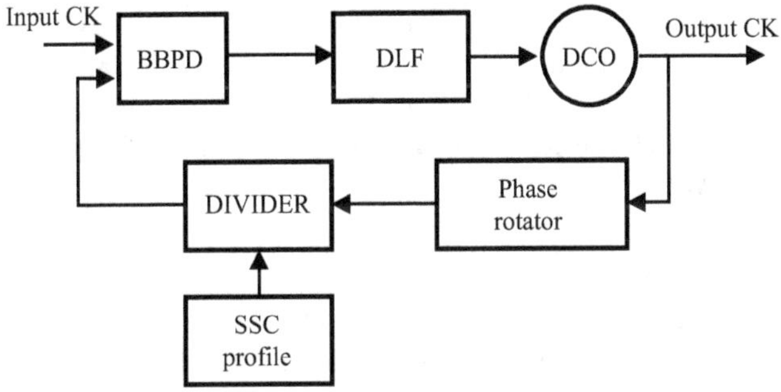

Figure 19.11 BB-DPLL-based SSCG [13]

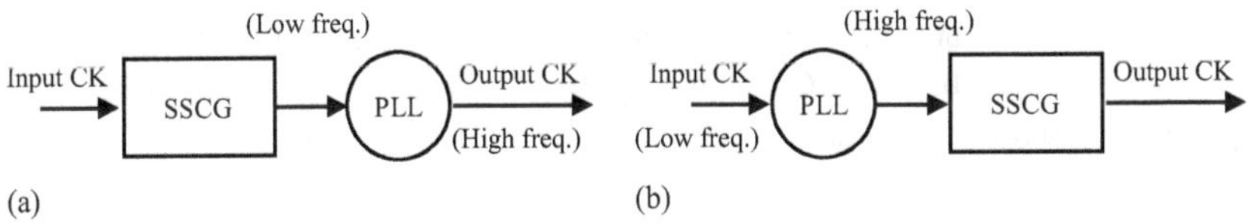

Figure 19.12 Delay-line-based SSCG: (a) preprocessing and (b) postprocessing

integer-N bang-bang PLL, as shown in Figure 19.11. Since it is based on integer-N operation, there is also no $\Delta\Sigma$ quantization noise issue.

19.3.2 Delay-line-based SSCG

Instead of frequency modulation with a PLL, phase modulation with a delay line can be also an option for SSC. Compared to frequency modulation with a PLL, phase modulation with a delay line can be easily implemented and more robust against random period jitter by the PLL because it is carried out after (post-processing) [14,15] or before (preprocessing) the PLL [16] (see Figure 19.12). The postprocessing method has advantages in terms of distortion in the noise and

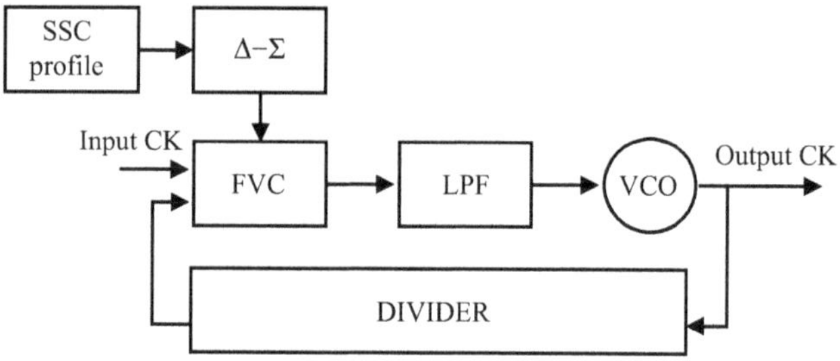

Figure 19.13 FLL-based SSCG [17]

modulation profile over preprocessing, but it consumes more power for high-frequency operation.

The quality of SSC modulation depends on the delay-line resolution in delay-line-based SSCG, which limits the clock frequency and its deviation. For instance, a frequency deviation of 5,000 ppm at 1 GHz equivalently requires a period deviation of 5 ps, which is too small to achieve with common delay cells in modern CMOS technology.

19.3.3 FLL-based SSCG

Compared to a PLL, an FLL can reduce the amount of area used with better stability. Therefore, it can be an area-effective alternative to filter ΔΣ quantization noise from the fractional-*N* approach. Figure 19.13 shows FLL-based SSCG [17]. It consists of an FLL and SSC modulation generator. To reduce the quantization error, the frequency-to-voltage-converter (FVC) can be directly controlled instead of the divider. Direct FVC control realizes more linear and finer frequency control than division ratio control. However, since the FLL cannot avoid the constant frequency offset, FLL-based SSCG is not applicable when an accurate frequency is required. Although the frequency offset can be allowed, it should be less than the SSC frequency range ($\Delta f_c = \delta \cdot f_c$).

19.4 Methodologies for the SSC modulation profile

As discussed in Section 19.2, the SSC modulation profile determines how to spread the spectrum and thus the EMI reduction. The frequency deviation (δ) and modulation frequency (f_m) of the modulation profile influence the spectral bandwidth and subharmonic distance, respectively. Moreover, the shape of the modulation profile also contributes to the EMI reduction since it forms the distribution of the power spectrum in the signal bandwidth. In this section, several SSC modulation profiles are introduced.

19.4.1 Sinusoidal modulation profile

Various modulation profiles for SSC have been presented in the literature. As shown in Figure 19.14, a sinusoidal modulation profile [18] can modulate the

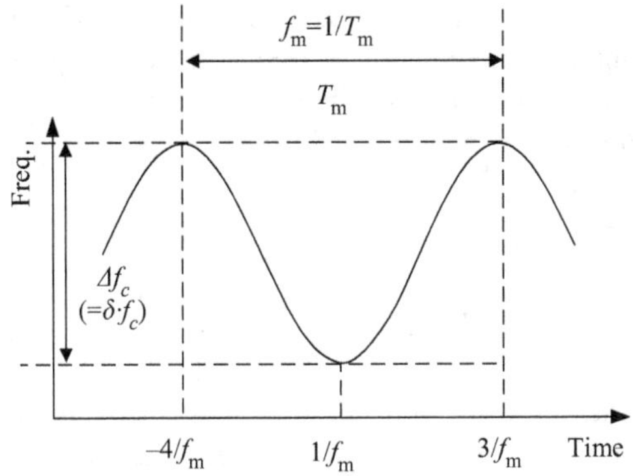

Figure 19.14 Sinusoidal modulation profile

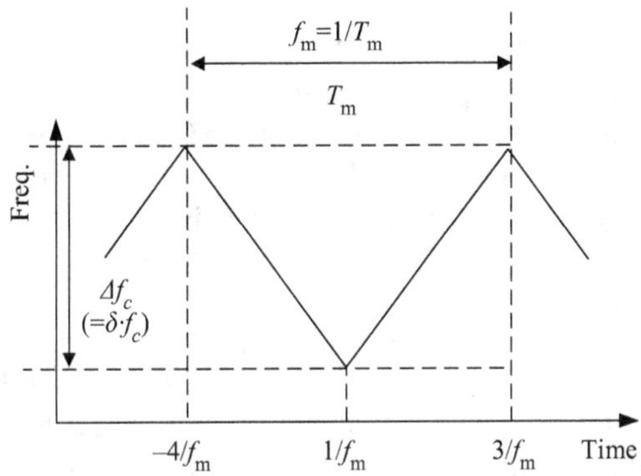

Figure 19.15 Triangular modulation profile

frequency with a simple implementation since the sinusoidal signals can be defined with the amplitude and frequency:

$$\omega_{\mathrm{SSC,sin.}}(t) = \Delta f_c \cdot \sin(2\pi f_m \cdot t) - \Delta f_c \tag{19.5}$$

However, the sinusoidal modulation profile has poor EMI performance. It cannot generate a flat spectrum to maximize the peak reduction since the rate of change in the frequency at the maximum/minimum of the modulation profile is low compared to other modulation profiles; accordingly, the power is concentrated on the sides of the spectrum.

19.4.2 Triangular modulation profile

Figure 19.15 shows a triangular modulation profile [5–8,11,14–16,19,20], which is widely used in many applications [3,4] since it is license-free and guarantees sufficient EMI performance. Its spectrum is flatter than that of the sinusoidal profile since the frequency components are evenly distributed in the modulation profile. Moreover, the triangular modulation profile can be implemented with simple digital blocks such as a counter or an analog integrator with a clock [5]. The triangular modulation profile can be defined as:

$$\omega_{\mathrm{SSC,Tri.}}(t) = \begin{cases} 4\Delta f_c \cdot f_m \cdot t - \Delta f_c, & -\dfrac{1}{4f_m} \le t \le \dfrac{1}{4f_m}, \\ -4\Delta f_c \cdot f_m \cdot \left(t - \dfrac{1}{2f_m}\right) - \Delta f_c, & \dfrac{1}{4f_m} \le t \le \dfrac{3}{4f_m} \end{cases} \tag{19.6}$$

However, the measured spectrum by triangular modulation profile is not sufficiently flat according to the relationship between the RBW and f_m. A spectrum analyzer acts as a moving-average filter, as explained in Section 19.2, whose bandwidth is the RBW, and it distorts the modulation profile at the rapid frequency transition points, for example, the maximum and minimum of the profile.

19.4.3 PLL-bandwidth-compensated triangular modulation

The triangular modulation profile still suffers from the band-limited implementation of the clock source, for example, the PLL. The PLL loop bandwidth is a limiting factor for the modulation profile accuracy. A ΔΣ-modulator-based fractional-*N* PLL usually requires a narrow loop bandwidth to filter the quantization noise of the ΔΣ modulator sufficiently. However, the narrow loop bandwidth distorts the modulation profile and reduces the frequency deviation (δ); consequently, it decreases the peak levels at the minimum and maximum frequencies, as illustrated in Figure 19.16. To eliminate the effect of the PLL loop bandwidth on the modulation profile, the two-point modulation technique [21] has been adopted, as shown in Figure 19.17. The output of the triangular modulation profile feeds two blocks: the ΔΣ-modulator divider and VCO. Since the divider and VCO inputs are low- and high-pass-filtered with the same bandwidth, respectively, the two-point modulation technique can realize all-pass behavior for the triangular modulation

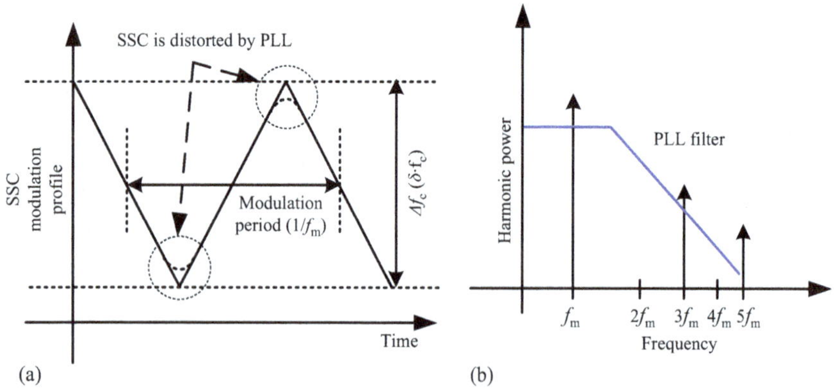

Figure 19.16 Profile distortion a by PLL: (a) time and (b) frequency domains

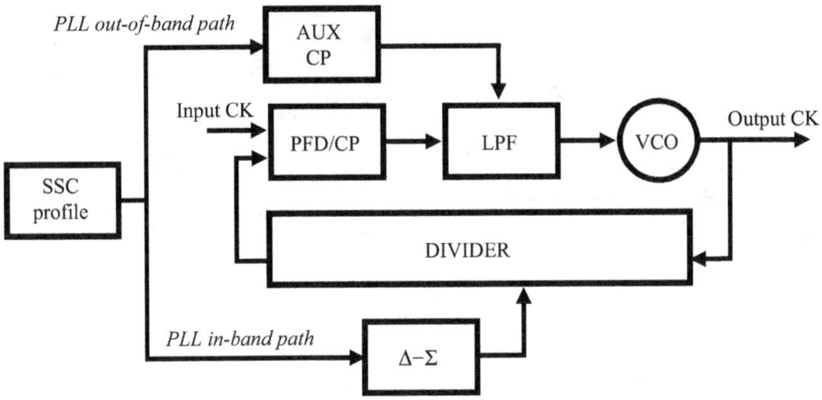

Figure 19.17 SSCG using two-point modulation [21]

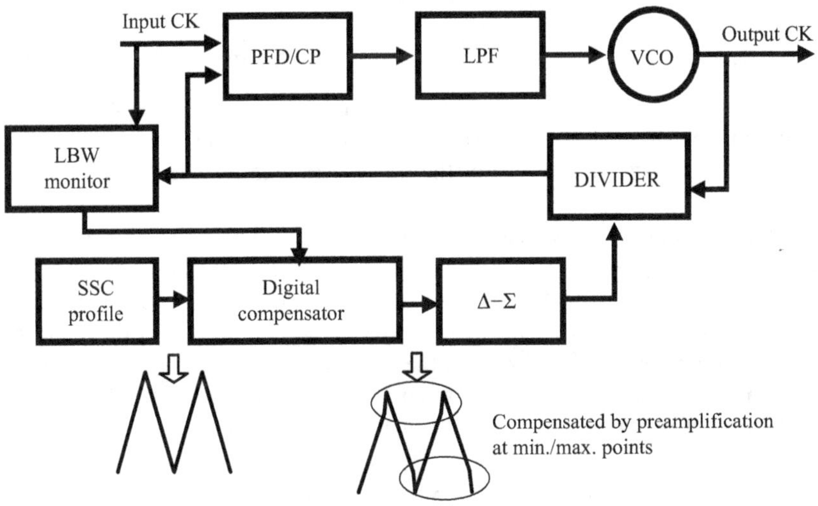

Figure 19.18 SSCG with digital compensation [22]

profile. However, the mismatch between two points is unavoidable. For instance, the VCO gain and CP current can easily vary with PVT variations. This mismatch can increase the frequency deviation and exceed the jitter.

A digital compensator before the ΔΣ modulator can also mitigate the effect of the PLL loop bandwidth [22] (see Figure 19.18). Compared to two-point modulation, the digital-compensator-aided modulation profile is more robust to PVT variations. DPLL-bandwidth-compensated SSCG comprises a PLL, digital compensator, and PLL bandwidth monitor. To compensate for the effect of PLL loop bandwidth on the SSC profile, the bandwidth of the digital compensator should be matched to the PLL loop bandwidth, as detected by the PLL loop bandwidth monitor. The digital

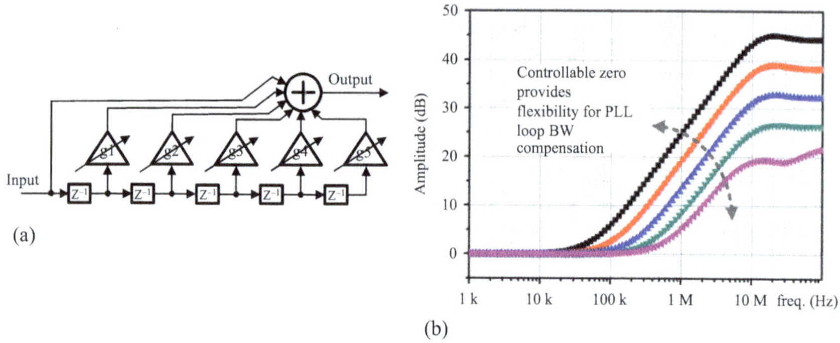

Figure 19.19 (a) Block diagram and (b) transfer function of the digital compensator [22]

compensator is based on the finite impulse response (FIR), as shown in Figure 19.19, and its transfer function, $H_{\mathrm{COMP}}(z)$ is expressed as:

$$H_{\mathrm{COMP}}(z) = 1 + g_1 z^{-1} + g_2 z^{-2} + g_3 z^{-3} + g_4 z^{-4} + g_5 z^{-5} \tag{19.7}$$

where g_1–g_5 are the gains of each delayed input and, except for the 4th gain (g_4), the gains are approximated to a power of 2 to reduce the hardware complexity. The FIR type has advantages in terms of the power and area over the infinite impulse response (IIR).

Since the digital compensator is used to compensate for the distortion in the SSC modulation profile by the PLL, the location of its zero should be flexible to overcome the variation in the bandwidth of the PLL. The PLL loop bandwidth suffers from PVT variations in the charge pump current (I_{CP}) and the VCO gain (K_{VCO}) and the passive components of the LF. The location of its zero varies from 54 kHz to 1.35 MHz and is sufficient to cover the variation in the bandwidth of the PLL because a common PLL loop bandwidth in many applications is 54 kHz–1.35 MHz [3,4].

The PLL loop bandwidth can be evaluated by the PLL locking time, although the exact PLL locking time is determined by the loop bandwidth and stability condition, for example, the phase margin. When the PLL loop bandwidth is much less than 1/10 of the reference clock frequency, it can be approximated by:

$$f_{\mathrm{BW}} = \frac{4}{T_{\mathrm{LOCK}}} \tag{19.8}$$

where f_{BW} and T_{LOCK} are the bandwidth and locking time of the PLL, respectively [23].

The PLL loop bandwidth monitor is organized with a lock detector with a small capture range, a counter, and a division factor (N) controller. The PLL loop bandwidth is automatically compensated with the detected PLL loop bandwidth from the monitor, as shown in Figure 19.20. The foreground method [24] is sufficient to compensate for the PLL loop bandwidth, although the background method [25] is more reliable [22].

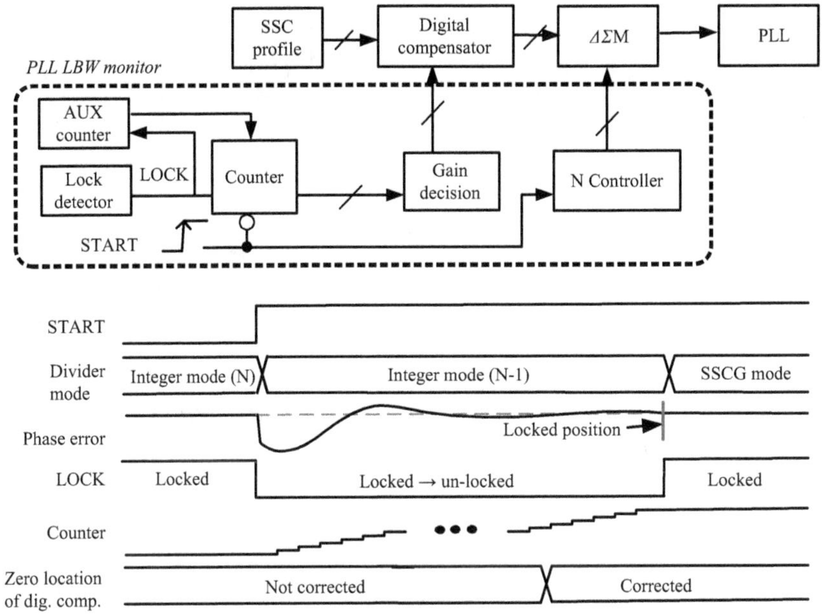

Figure 19.20 Foreground automatic calibration method [22]

To minimize the nonlinear effect of the PLL on the locking time, the PLL is first locked at the target frequency. Then, the smallest frequency change, for example, the reference clock frequency, is applied by shifting the division ratio by 1 when lock-time detection is started. The lock detector flags that the PLL is locked, and the transient response of the PLL is finished when the phase error induced by the frequency change is less than the detector's capture range. However, since the lock detection algorithm exploits the frequency step response while it monitors the phase information, the lock detector can generate a false alarm that appears when the phase error is within capture range, but the frequency error is still far from the locked condition. Therefore, an auxiliary counter for buffering the false alarm is required. The auxiliary counter is fully accumulated as long as the lock state of the PLL is consecutively buffered. During the transition period of the PLL, the counter counts the lock time until the auxiliary counter is fully accumulated. Finally, the locking time of the PLL can be determined on the basis of two counter values and utilized to match the bandwidth of the digital compensator to that of the PLL. Figure 19.21 shows the measurement results of SSCG with digital compensation.

19.4.4 Hershey–Kiss modulation profile

As introduced in Section 19.4.2, the triangular modulation profile is distorted by the RBW. The well-known Hershey–Kiss modulation profile for mitigating the RBW effect has been presented [26]. Since the modulation profile is mostly

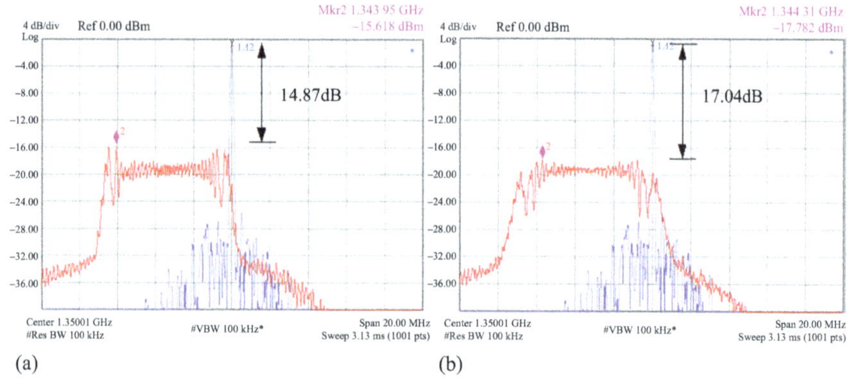

Figure 19.21 Measurement results of SSCG (a) without and (b) with digital compensation [22]

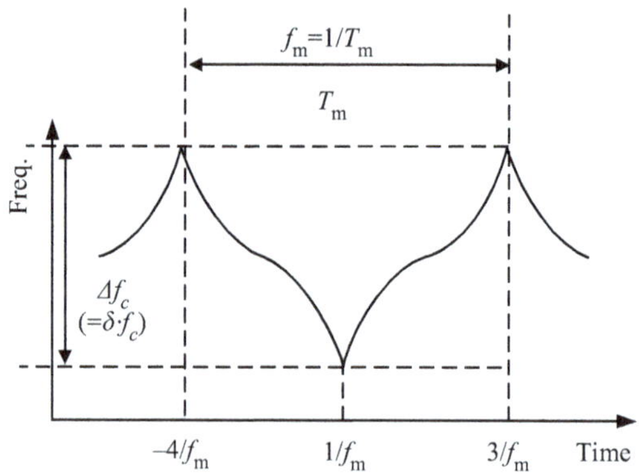

Figure 19.22 Hershey–Kiss modulation profile [27]

distorted at the maximum and minimum points, the Hershey–Kiss modulation profile compensates for the distortion by predistortion to boost the power of the harmonics, increasing the slopes at these points, as illustrated in Figure 19.22.

The Hershey–Kiss modulation profile can be defined as:

$$\omega_{\mathrm{SSC,H-K}}(t) = \begin{cases} \mathrm{K} \cdot (0.45t^3 + 0.55t) - \Delta f_c, & -\dfrac{1}{4f_m} \le t \le \dfrac{1}{4f_m}, \\ \mathrm{K} \cdot \left(-0.45\left(t - \dfrac{1}{2f_m}\right)^3 + 0.55\left(t - \dfrac{1}{2f_m}\right)\right) - \Delta f_c, & \dfrac{1}{4f_m} \le t \le \dfrac{3}{4f_m} \end{cases} \tag{19.9}$$

where $K = \frac{\Delta f_c}{0.45\frac{1}{4^3 f_m{}^3} + 0.55\frac{1}{4f_m}}$. Δf_c is $\delta \cdot f_c$, as defined in Section 19.2.

The Hershey–Kiss modulation profile [27,28] can reduce the effects of the RBW on the maximum/minimum points of the profile owing to the steep signal slopes at both the maximum and minimum. Thus, the Hershey–Kiss modulation profile can compensate for the RBW distortion. However, the implementation of the modulation profile is complex because the profile shape is nonlinear. Moreover, the Hershey–Kiss modulation profile is optimized to the RBW for band C in Table 19.1 [26]. It is difficult to guarantee the optimization of the EMI reduction at other RBW criteria.

19.4.5 Newton–Raphson modulation profile

Although the Hershey–Kiss modulation profile maximizes the EMI reduction by virtue of a cubic-order profile formula, it requires additional expensive and complex solutions, for example, a look-up table (LUT) [27], SRAM [14], or a dual ΔΣ modulator [29]. The LUT and SRAM consume a large amount of power, require a large die area, and even require extra semiconductor process steps that significantly increase the manufacturing cost. The dual ΔΣ modulator suffers from large quantization noise to increase clock period jitter during SSCG operation. To reduce the hardware complexity, the Newton–Raphson method, used to find a square-root value (also known as Heron's method), is applied to mimic the Hershey–Kiss modulation profile [17] at a low cost. The original Newton–Raphson equation is:

$$Y[n,\ i+1] = \frac{1}{2}\left(Y[n,i] + \frac{X[n]}{Y[n,i]}\right) \approx \sqrt{X[n]} \tag{19.10}$$

To find a precise value, this equation should be iteratively calculated in terms of the index i. However, the iteration process can be eliminated to achieve a square-root value when the Newton–Raphson method starts from a certain point [30].

Therefore, (19.10) can be simplified as a first-order mathematical function with coefficients A and B:

$$Y[n,\ i] = \frac{X[n]}{A} + B \tag{19.11}$$

where A and B are used to set up the starting points. Substituting (19.11) into (19.10), (19.10) can be expressed as:

$$Y[n] = \frac{1}{2}\left(\frac{X[n]}{A} + B + \frac{X[n]}{\frac{X[n]}{A} + B}\right) \tag{19.12}$$

Finally, the Newton–Raphson equation is:

$$Y[n] = \frac{\beta}{2A}\left(X[n] + AB + S\left\lfloor\frac{A^2X[n]}{X[n] + AB}\right\rfloor\right) \tag{19.13}$$

by adding a scaling factor β, which is equal to $2A$, to remove unnecessary operations and to match the Hershey Kiss profile [27]. A, B, and S are properly chosen to achieve to Hershey–Kiss profile with a certain f_m and δ. Note that the dividing operation in (19.13) is round down. Figures 19.23 and 19.24 show the coincidence between the Newton–Raphson and Hershey–Kiss profiles according to the profile shape and spectrum.

The Newton–Raphson profile generator is shown in Figure 19.25. A coefficient selector (CS) contains the coefficients in (19.13). A and S are expressed as a power of 2 to reduce the complexity of the digital hardware. A profile counter (PC) generates $X[n]$ for the square-root approximation (SRA), expressed as:

$$X[n] = \begin{cases} \mathrm{STEP} \cdot \mathrm{CNT}[n] \text{ for } 0 < n \cdot T_{N-R} < \dfrac{T_m}{4}, \\ \mathrm{CNT_{MAX}} - \mathrm{STEP} \cdot \mathrm{CNT}[n] \text{ for } \dfrac{T_m}{4} < n \cdot T_{N-R} < T_m \end{cases} \tag{19.14}$$

where STEP is the step size of the PC, T_m is the period of the modulation profile, $T_{N\text{-}R}$ is the period of the profile generator clock, and n is a nonnegative integer (0, 1, 2, ...).

STEP determines the slope of the profile and thus the modulation frequency f_m. Equation (19.12) is calculated by SRA. Finally, the profile switch (PS) turns half of

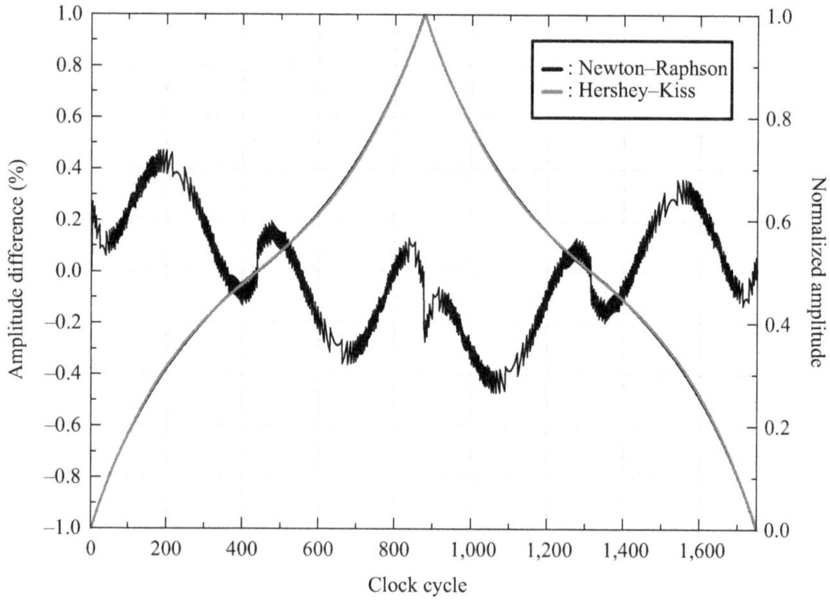

Figure 19.23 Difference between the Newton–Raphson and Hershey–Kiss profiles [17]

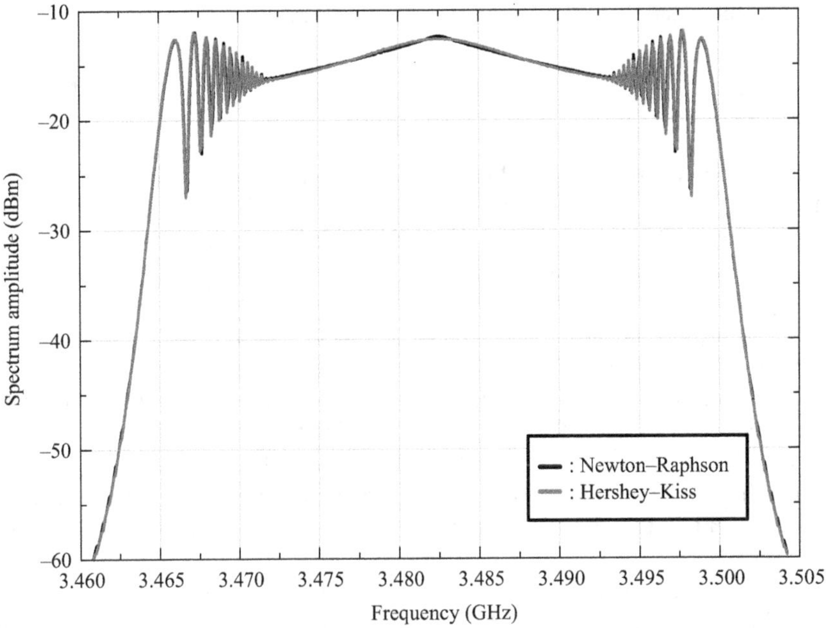

Figure 19.24 Simulated spectra of the Newton–Raphson and Hershey–Kiss profiles [17]

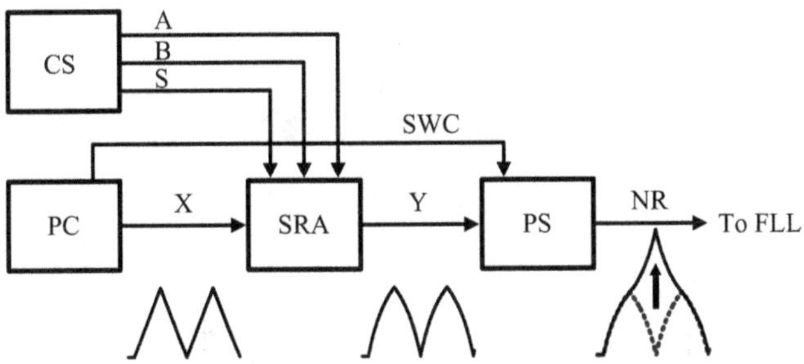

Figure 19.25 Block diagram of the Newton–Raphson profile generator [17]

$Y[n]$ upside down when the SWC in Figure 19.25 is high. The Newton–Raphson profile is finally expressed as:

$$NR[n] = \begin{cases} Y[n] \text{ for } 0 < n \cdot T_{N-R} < \dfrac{T_m}{4},\ \dfrac{3}{4}T_m < n \cdot T_{N-R} < T_m, \\ (2^{16} - 1) - Y[n] \text{ for } \dfrac{T_m}{4} < n \cdot T_{N-R} < \dfrac{3}{4}T_m \end{cases} \tag{19.15}$$

Therefore, the Newton–Raphson profile $NR[n]$ is generated without any memory.

19.4.6 Piecewise-linear modulation profile

As discussed in the previous sections, a trade-off between the hardware burden and EMI performance is observed for the choice of the SSC modulation profile. The Hershey–Kiss modulation profile can reduce EMI further than the triangular modulation profile on the effect of the RBW with expensive hardware solutions. Although the Newton–Raphson method in Section 19.4.5 can alleviate the hardware cost for the generation of the Hershey–Kiss profile, it still requires nonlinear digital processing, for example, divider operation.

The piecewise linear (PWL) modulation profile [2] can overcome this trade-off, as shown in Figure 19.26. The PWL modulation profile consists of linear signals that have two different slopes. By changing the slope and time interval of each linear signal, the EMI reduction can be optimized. Since the PWL modulation profile only consists of linear signals, it allows for a simple implementation.

The PWL modulation profile comprises several linear signals that have different slopes (α_1 and α_2). Higher EMI reduction can be achieved with steeper slopes at the minimum and maximum points; however, more gradual slopes are at the zero-crossing points in the profile, as in the Hershey–Kiss modulation profile.

The period of the PWL profile is the sum of each linear signal, T_1 and T_2, as shown in Figure 19.26, computed from:

$$\frac{1}{f_m} = T_m = 4T_1 + 2T_2 \tag{19.16}$$

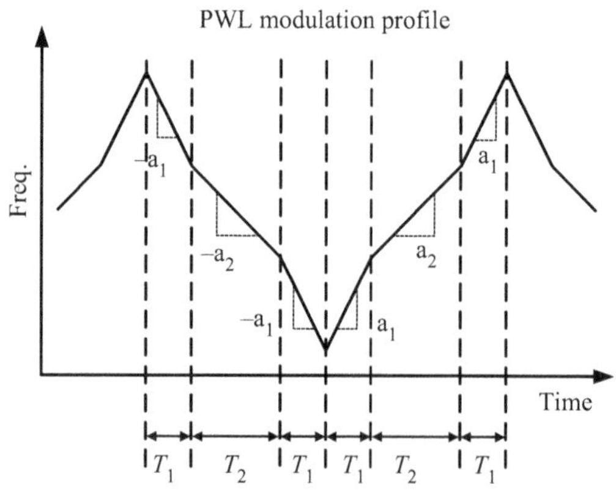

Figure 19.26 Piecewise linear (PWL) modulation profile [2]

The PWL modulation profile can approach Hershey–Kiss modulation using the ratio between the time sections (T_1 and T_2) and the slopes of the linear signals (α_1 and α_2). The PWL modulation profile in Figure 19.26 can be defined with the following six sections with different linear signals:

- Section 1: $0 \le t < \frac{\rho \cdot T_m}{4}$,
- Section 2: $\frac{\rho \cdot T_m}{4} \le t < \frac{(2-\rho) \cdot T_m}{4}$,
- Section 3: $\frac{(2-\rho) \cdot T_m}{4} \le t < 0.5 \cdot T_m$,
- Section 4: $0.5 \cdot T_m \le t < \frac{(2+\rho) \cdot T_m}{4}$,
- Section 5: $\frac{(2+\rho) \cdot T_m}{4} \le t < \frac{(4-\rho) \cdot T_m}{4}$,
- Section 6: $\frac{(4-\rho) \cdot T_m}{4} \le t < T_m$

Therefore, the PWL modulation profile can be derived for each section i as:

$$\omega_{\mathrm{SSC,PWL}}(t) = K_i \cdot t + B_i, \quad i = 1, 2, \cdots, 6, \tag{19.17}$$

where:

$$\begin{cases} K_1, K_3 = -\alpha_1 = -K_4, -K_6, \\ \quad K_2 = \dfrac{4 \cdot \Delta f_c - \alpha_1 \cdot T_m \cdot \rho}{(\rho - 1) \cdot T_m} = -\alpha_2 = -K_5 \end{cases}$$

and

$$\begin{cases} B_1 = \Delta f_c, \\ B_2 = \Delta f_c - \dfrac{\alpha_1(\rho \cdot T_m)}{4} - \dfrac{\rho \cdot (4 \cdot \Delta f_c - \alpha_1 \cdot T_m \cdot \rho)}{4 \cdot (\rho - 1)}, \\ B_3 = -\Delta f_c + 0.5 \cdot T_m \cdot \alpha_1, \\ B_4 = -\Delta f_c - 0.5 \cdot T_m \cdot \alpha_1, \\ B_5 = -\Delta f_c + \dfrac{\alpha_1(\rho \cdot T_m)}{4} + \dfrac{(\rho + 2) \cdot (4 \cdot \Delta f_c - \alpha_1 \cdot T_m \cdot \rho)}{4 \cdot (\rho - 1)}, \\ B_6 = \Delta f_c - \alpha_1 \cdot T_m \end{cases}$$

The shape of the PWL modulation profile can be modified with two variables: α_1 and ρ. α_1 is the slope of the signal during T_1. ρ is the ratio between T_1 and T_2:

$$\rho = \frac{2T_1}{2T_1 + T_2}, \quad 0 < \rho < 1 \tag{19.18}$$

To approximate the Hershey–Kiss modulation profile, several constraints are required. First, the crossing point needs to be greater than zero:

$$\Delta f_c - \alpha_1 \cdot \left(\frac{\rho \cdot T_m}{4} \right) > 0 \tag{19.19}$$

α_1 should also be greater than α_2 in order to increase the slopes at the maximum and minimum points and decrease the signal slope at the zero-crossing point $(\alpha_1 > \alpha_2)$:

$$\alpha_1 > \frac{4 \cdot \Delta f_c - \alpha_1 \cdot T_m \cdot \rho}{(\rho - 1) \cdot T_m} \tag{19.20}$$

As shown in Figure 19.27, the optimum values of α_1 and ρ which make the PWL profile most similar to the Hershey–Kiss profile can minimize the maximum peak in the spectra, thus maximizing the EMI reduction.

The PWL modulation profile is beneficial in terms of its flexibility. Unlike the Hershey–Kiss modulation profile, which has a fixed shape, the PWL modulation profile is adaptable to different RBW criteria by controlling α_1 and ρ. Moreover, the flexibility of the PWL modulation profile can be easily expanded to have more

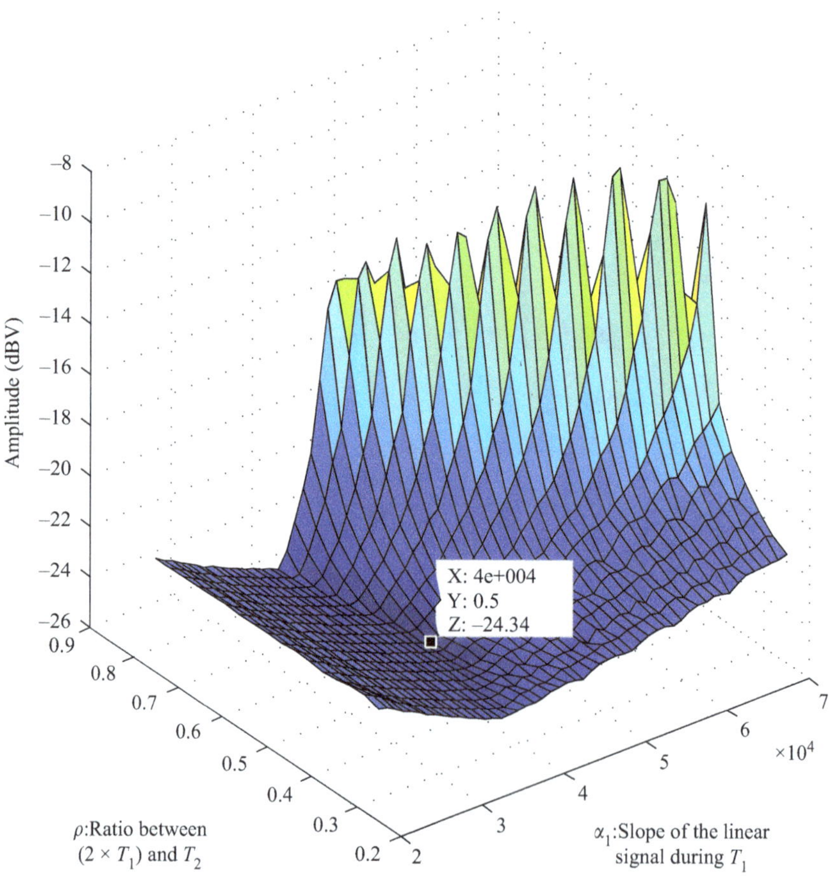

Figure 19.27 Maximum peak amplitudes in the spectra [2]

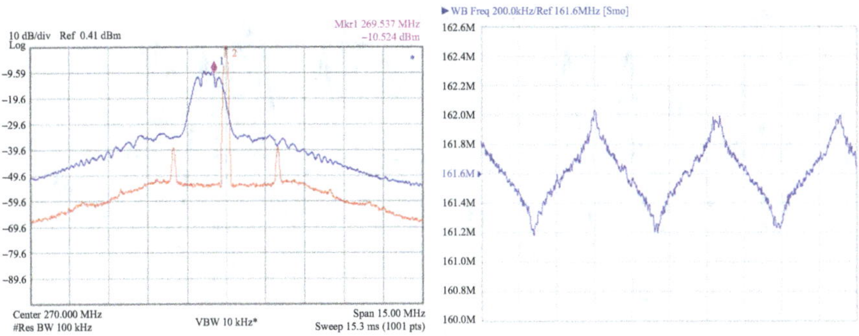

(a) (b)

Figure 19.28 Measurement results of SSCG with the PWL profile: (a) clock spectrum and (b) time trend [2]

slopes than 2 without requiring much more hardware [2]. Figure 19.28 shows the measurement results of SSCG with the PWL modulation profile.

19.4.7 Random modulation profile

As introduced in Section 19.2, further EMI reduction is achieved when the modulation frequency (f_m) is decreased since it densely spreads the clock spectrum. A random modulation profile can minimize the modulation frequency and thus the peak power of the spectrum. The chaotic phase-amplitude modulation (PAM) method [31] was first reported as a random modulation profile, but it requires a complex analog circuit and is not suitable for higher frequency deviations owing to the steep frequency transitions in the profile. Another approach for the random modulation profile [12] is based on pseudo-random bit sequence (PRBS), which is simpler than that for chaotic PAM. Although random modulation profiles are attractive, they are impractical since f_m should be beyond the band of human hearing (20–20 k) to avoid the audible interference.

19.4.8 Nested modulation profile

As discussed in Section 19.2, when the RBW is sufficiently small to distinguish the subharmonics of the clock, the EMI reduction can be improved by either increasing δ or decreasing f_m because the range of the frequency distribution increases and the frequency spacing of the power spectrum decreases. Moreover, other issues arise in addition to the EMI reduction, for example, the cycle-to-cycle jitter of the interface clock or data resulting from an increase in δ. To increase the EMI suppression further without increasing δ and decreasing f_m, the nested modulation profile has been introduced [32].

Figure 19.29 shows the spread spectrum profiles of a conventional triangular profile and the nested profile. EMI can be further reduced by either increasing δ or decreasing f_m and accordingly increasing the modulation area (A). However, the

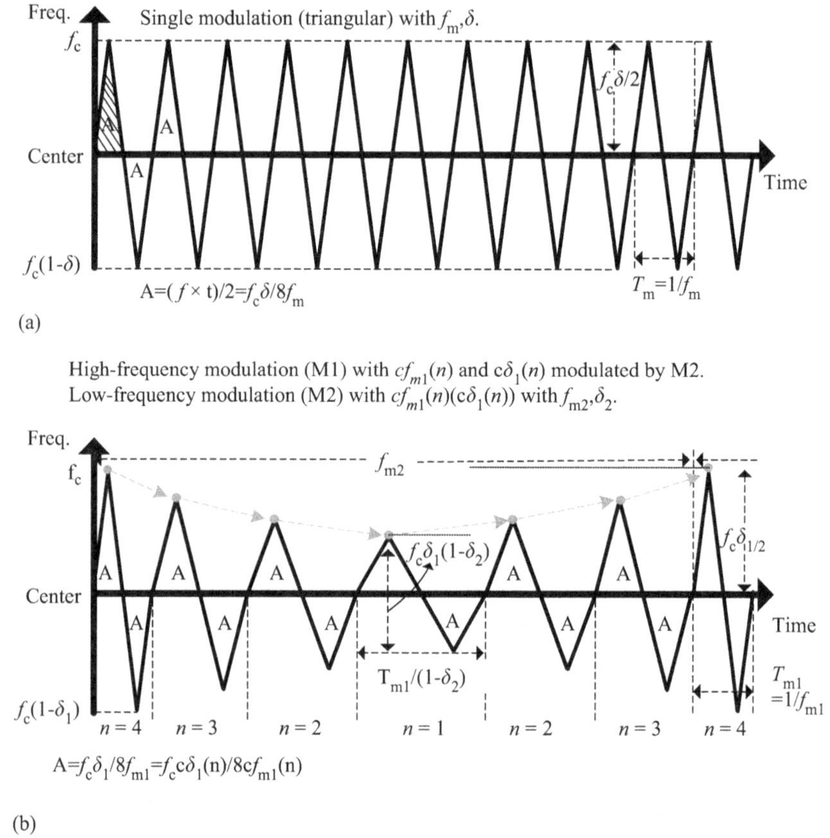

Figure 19.29 (a) Triangular and (b) nested (ex. four steps) modulation profiles [32]

accumulated jitter is increased by the modulation area. A first-in-first-out (FIFO) interface is generally required to eliminate the degradation in the data integrity due to the accumulated jitter [12]. The size of the FIFO interface becomes larger as the modulation area increases. To reduce EMI further but maintain the modulation area, the nested modulation profile comprises two modulations, M1 and M2. The modulation frequency and frequency deviation of the high-frequency modulation profile (M1) periodically change in response to the low-frequency modulation profile (M2). Meanwhile, the ratio between the instant frequency deviation ($c\delta_1(n)$) and the modulation frequency ($cf_{m1}(n)$) of M1 is constant to maintain the modulation area.

Since the frequency distributions of the nested profile vary at the rate of the modulation frequency of M2 (f_{m2}), the frequency spacing of the power spectrum for the modulated carrier signal in the sideband decreases to f_{m2} when $f_{m2} \ll f_m$. Therefore, the power spectrum of the nested profile signal can spread in a narrower space with a smaller amplitude compared to that of the conventional method.

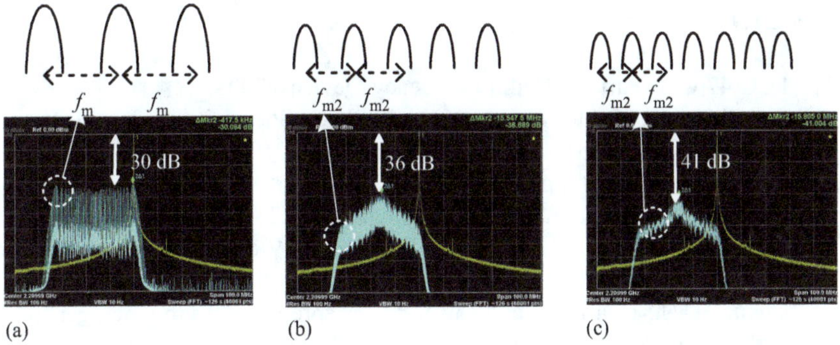

Figure 19.30 Power spectra of the (a) triangular, (b) nested, and (c) nested with random (RBW = 100 Hz) profiles [34]

From the perspective of the size of the FIFO interface, the FIFO area for nested modulated SSC is the same as that for triangular modulated SSC because they have the same modulation area. Moreover, the maximum cycle-to-cycle jitter of the nested modulation profile is also the same as that of the triangular modulation profile.

Since the EMI can be further reduced by decreasing f_{m2}, the random modulation profile can be considered as M2 to minimize f_{m2}, as shown in Figure 19.30(c). Note that f_{m2} should be higher than the band of human hearing, as explained in Section 19.4.7. As shown in Figure 19.30, a significant EMI reduction can be achieved by the nested modulation profile compared to the triangular modulation profile.

The nested modulation profile has several interesting benefits. A significant EMI reduction can be achieved without sacrificing clock purity. In addition, the nested profile can be easily implemented by adding a digital counter to the conventional triangular profile [32].

19.5 Conclusions

In this chapter, EMI reduction by SSCG and its implementation were presented. Without any external components, SSCG can reduce EMI by spreading the clock spectrum itself. The EMI performance of SSCG is affected by the frequency deviation, the modulation frequency of the modulation profile, and the RBW. As a result, we can find the optimum modulation profile to maximize the EMI reduction with a certain RBW criteria. On the basis of these observations, we overviewed several SSC modulation profiles introduced in the literature. The benefits and drawbacks of each modulation profiles were explained. Since SSCG is a type of clock generator, it faces the similar issues on analog circuit design, for example, noise, nonlinearity, and mismatch. Several implementation examples of SSCG architectures were presented and discussed to overcome these issues.

References

[1] FCC 47 CFR Part 16, Radio Frequency Devices, 2008.

[2] Song M., Ahn S., Jung I., Kim Y., and Kim C. "Piecewise linear modulation technique for spread spectrum clock generation." *IEEE Trans. VLSI Circuits*. 2013, vol. 41, no. 8, pp. 1234–1245.

[3] Serial ATA International Organization, Serial ATA Revision 2.6, 2007.

[4] Display Port Standard Version 1, Revision 2, Video Electronics Standard Association, 2010.

[5] Chang H.-H., Hua I.-H., and Liu S.-I., "A spread-spectrum clock generator with triangular modulation." *IEEE J. Solid-State Circuits*. 2003, vol. 38, no. 4, pp. 673–676.

[6] Sugawara M., Ishibashi T., Ogasawara K., *et al.* "1.5 Gbps, 5150 ppm spread spectrum SerDes PHY with a 0.3 mW, 1.5 Gbps level detector for serial ATA." In: *IEEE Symposium on VLSI Circuits, Digest of Technical Papers*. 2002, pp. 60–63.

[7] Aoyama M., Ogasawara K., Sugawara M., *et al.* "3Gbps, 5000ppm spread spectrum SerDes PHY with frequency tracking phase interpolator for serial ATA." In *IEEE Symposium on VLSI Circuits, Digest of Technical Papers*. 2003, pp. 107–110.

[8] Kokubo M., Kawamoto T., Oshima T., *et al.* "Spread-spectrum clock generator for serial ATA using fractional PLL controlled by ΔΣ modulator with level shifter." in *IEEE International Solid-State Circuits Conference, Digest of Technical Papers*. 2005, pp. 160–161.

[9] Song M., Jung I., Pamarti S., and Kim C. "A 2.4 GHz 0.1-fref-bandwidth all-digital phase-locked loop with delay-cell-less TDC." *Trans. Circuits Syst. I*. 2013, vol. 60, no. 12, pp. 3145–3251.

[10] Song M., Kim T., Kim J., Kim W., Kim S.-J., and Park H. "A 0.009mm2 2.06mW 32-to-2000MHz 2nd-order ΔΣ analogous bang-bang digital PLL with feed-forward delay-locked and phase-locked operations in 14nm FinFET technology." in *IEEE International Solid-State Circuits Conference, Digest of Technical Papers*. 2015, pp. 266–267.

[11] Lin S.-Y., and Liu S.-I. "A 1.5 GHz all-digital spread-spectrum clock generator." *IEEE J. Solid-State Circuits*. 2009, vol. 44, no. 11, pp. 3111–3119.

[12] Dalt N.D., Pridnig P., and Grollitsch W. "An all-digital PLL using random modulation for SSC generation in 65nm CMOS." in *IEEE International Solid-State Circuits Conference, Digest of Technical Papers*. 2013, pp. 252–253.

[13] Yang J., Lee J.-Y., Lim S.-J., and Bae H.-M. "Phase-rotator-based all-digital phase-locked loop for a spread-spectrum clock generator." *IEEE Trans. Circuits Syst. II*. 2014, vol. 61, no. 11, pp. 880–884.

[14] Caro D.D., Romani C.A., Petra N., Strollo A.G.M., and Parrella C. "A 1.27GHz, all-digital spread spectrum clock generator/synthesizer in 65nm CMOS," *IEEE J. Solid-State Circuits*. 2010, vol. 45, pp. 1048–1060.

[15] Kim J., Kam D.G., Jun P.J., and Kim J. "Spread spectrum clock generator with delay cell array to reduce electromagnetic interference." *IEEE Trans. Electromagn. Compat.* 2005, vol. 47, no. 11, pp. 908–920.

[16] Damphousse S., Ouici K., Rizki A., and Mallinson M. "A 1.5 GHz all-digital spread-spectrum clock generator." *IEEE J. Solid-State Circuits*. 2009, vol. 42, no. 1, pp. 145–150.

[17] Hwang S., Song M., Kwak Y.-H., Jung I., and Kim C. "A 3.5-GHz spread-spectrum clock generator with a memoryless Newton–Raphson modulation profile." *IEEE J. Solid-State Circuits*. 2012, vol. 47, no. 5, pp. 1199–1208.

[18] Lin F., and Chen D.Y. "Reduction of power supply EMI emission by switching frequency modulation." *IEEE Trans. Power Electron.* 1994, vol. 9, no. 1, pp. 132–137.

[19] Cheng K.-H., Hung C.-L., and Chang C.-H. "A 0.77 ps RMS Jitter 6-GHz spread-spectrum clock generator using a compensated phase-rotating technique." *IEEE J. Solid-State Circuits*. 2011, vol. 46, no. 5, pp. 1198–1213.

[20] Hsieh Y.-B., and Kao Y.-H. "A fully integrated spread-spectrum clock generator by using direct VCO modulation." *IEEE Trans. Circuits Syst. I.* 2008, vol. 55, no. 7, pp. 1845–1853.

[21] Kao Y.-H., and Hsieh Y.-B. "A low-power and high-precision spread spectrum clock generator for serial advanced technology attachment applications using two-point modulation." *IEEE Trans. Electromagn. Compat.* 2009, vol. 51, no. 2, pp. 245–254.

[22] Bae S.-G., Hwang S., Song J., Lee Y., and Kim C. "A ΔΣ-modulator-based spread-spectrum clock generator with digital compensation and calibration for phase-locked loop bandwidth." *IEEE Trans. Circuits Syst. II*, In press.

[23] Shu K., and Sanchez-Sinencio E. CMOS PLL *Synthesizers: Analysis and Design*. Boston, Springer, 2005.

[24] Shin J., and Shin H. "A 1.9–3.8 GHz ΔΣ fractional-N PLL frequency synthesizer with fast auto-calibration of loop bandwidth and VCO frequency." *IEEE J. Solid-State Circuits*. 2012, vol. 47, no. 3, pp. 665–675.

[25] Kuan T.-K., and Liu S.-I. "A loop gain optimization technique for integer-N TDC-based phase-locked loops." *IEEE Trans. Circuits Syst. I.* 2015, vol. 62, no. 7, pp. 1873–1882.

[26] Hardin K.B., Fessler J.T., and Bush D.R. "Spread spectrum clock generation for the reduction of radiated emissions." in *Proc. IEEE Int. Symp. Electromagnetic Compatibility*. 1994, pp. 227–231.

[27] Hardin K., Oglesbee R.A., and Fisher F. "Investigation into the interference potential of spread-spectrum clock generation to broadband digital communications." *IEEE Trans. Electromagn. Compat.* 2003, vol. 45, no. 1, pp. 10–21.

[28] Caro D.D., Romani C.A., Petra N., Strollo A.G.M., and Parrella C. "A 1.27 GHz, all-digital spread spectrum clock generator/synthesizer in 65 nm CMOS." *IEEE J. Solid-State Circuits*. 2010, vol. 45, no. 5, pp. 1048–1060.

[29] Park H.-M, Jin H.-B., and Kang J.-K. "SSCG with Hershey-Kiss modulation profile using dual sigma-delta modulators." *IEICE Electron. Express.* 2010, vol. 7, no. 18, pp. 1349–1353.
[30] Allie M., and Lyons R. "A root of less evil." *IEEE Signal Process. Mag.* 2005, vol. 22, no. 2, pp. 93–96.
[31] F. Pareschi, G. Setti, and R. Rovatti, "A 3-GHz serial ATA spread-spectrum clock generator employing chaotic PAM modulation." *IEEE Trans. Circuits Syst. I.* 2010, vol. 57, no. 10, pp. 2577–2587.
[32] Jun J., Bae S.-G., Lee Y., and Kim C. "A spread spectrum clock generator with nested modulation profile for a high-resolution display system." *IEEE Trans. Circuits Syst. II.* 2018, vol. 65, no. 11, pp. 1509–1513.

Chapter 20

High-performance CMOS clock distribution

Xunjun Mo[1], Nijwm Wary[1] and Tony Chan Carusone[1]

20.1 Introduction

In modern communication systems, the distribution of high-frequency precision clocks across integrated circuits is a significant challenge. For example, in wireline transceivers, multiple parallel lanes share a single clock source (a PLL) whose output is distributed to the lanes as shown in Figure 20.1. Clock jitter introduces timing uncertainty into the link, thus limiting the maximum achievable data rate [1]. Thus, as data rates increase, jitter specifications tighten. For example, the OIF-CEI VSR standards for NRZ and 4-PAM links at 56 Gb/s and 112 Gb/s, respectively, call for random clock jitter below 180 fs-rms. Naturally, some of this jitter will arise in the PLL. However, since the power consumption of the PLL is amortized over the many parallel links sharing it, considerable power can be allocated to reducing its jitter. By contrast, the power required to distribute a clock across N lanes is generally proportional to N, and is therefore greatly constrained. As a result, the jitter contributed by the clock distribution can become predominant. For instance, in [2], a PLL having only 85 fs rms random jitter (RJ) is distributed across two transmitters, and the 28 Gbps NRZ transmitter output has 185 fs rms RJ.

Considering its importance in determining the performance of modern transceivers, it is not surprising that clock distribution consumes a significant fraction of the total power in modern high-speed communication ICs. For example, in [2], the clocking circuitry consumes 48% of the total 785 mW power consumed in each transceiver lane. Similarly, in [3], the power consumed by clock buffers including duty-cycle correction (DCC) and quadrature error correction (QEC) occupies 70% of the total front-end transmitter power when operating at 112 Gb/s (4-PAM). Therefore, it is important to properly model and simulate sources of jitter in the distribution of clocks and come up with design methodologies that minimize the jitter introduced in global clock distribution.

[1]Department of Electrical & Computer Engineering, University of Toronto, Toronto, Canada

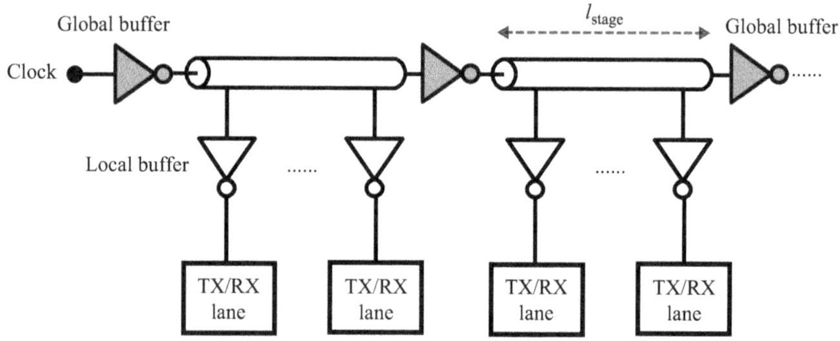

Figure 20.1 Global clock distribution for high-speed links

20.1.1 Sources of jitter in clock distribution

There are three major jitter sources in global clock distribution: power supply noise, intrinsic device noise, and the amplification of input jitter [1]. Supply noise mainly originates from transient current drawn by on-chip circuits, including analog transceiver circuits and digital circuits. Broadband transient currents drawn from the power supply flow through a power distribution network (PDN) consisting of parasitics on the printed circuit board, IC package, and on-die interconnect. Thus, the PDN impedance develops transient voltage drops across it resulting in fluctuation of the internal local supply voltage. The amplitude of the supply noise is therefore dependent on the frequency content of the transient current and the frequency-dependent impedance of the PDN [4]. PDN impedance typically resonates at around tens to hundreds of MHz [5] due to the resonance of typical package parasitic inductance and on-chip decoupling capacitance. Supply noise, in turn, modulates the delay of clock distribution buffers resulting in power supply induced jitter (PSIJ). Since, in principle, PSIJ can be predicted given perfect knowledge of the PDN and all currents, it is considered deterministic jitter. Intrinsic device noise, including the thermal and flicker noise of MOSFETs in clock distribution circuits, give rise to random jitter (RJ). In contrast to PSIJ, it is truly random and unbounded. Finally, jitter on the clock source can be amplified as it goes through clock buffers. Jitter amplification is defined as the ratio of jitter at the output of a buffer chain to the jitter at its input. We will see that jitter amplification arises from the finite bandwidth of the clock path, even if there is no supply noise or device noise.

20.1.2 CMOS clocking for broadband clock distribution

Many applications must accommodate a wide range of data rates, and thus a wide range of clock frequencies. It is therefore desirable to have wideband clock distribution scheme. Examples include the wireline transceivers in high-performance FPGAs [6] and high-performance computing systems [7]. In [6], transceivers are expected to support 40 protocols with data rates ranging from 0.6 to 13.1 Gb/s. Similarly, in [7], the per-lane data rate is scalable from 2 to 16 Gb/s, and the clock frequency is scalable from 0.5 to 4 GHz. In high-performance computing systems, the demanded I/O bandwidth can vary dramatically over time, so it is desirable to save power by lowering clock frequencies when bandwidth demand is low.

CMOS clocking is a popular method for broadband clock distribution in highly scaled technologies. It essentially relies on the basic CMOS inverter to buffer a full-swing clock to the capacitive loads presented by interconnect and other buffers and logic. Although current mode logic (CML) buffers are less susceptible to PSIJ and have high bandwidth, they consume static power which makes it difficult to scale their power consumption with clock frequency. They also suffer from swing versus bandwidth trade-offs. CML performance degrades at reduced power supply voltages and with variations in output impedance at high temperature [8]. Clocking schemes based on LC resonance have superior jitter performance [9,10]. However, the lumped inductor occupies large on-chip area and has limited tuning range. Clock distribution based on distributed oscillators can reduce the distance from the clock generator to the load and thus reduces the number of buffer stages required [11,12]. However, it is also a narrowband with limited tuning range because its resonant frequency depends on transmission line geometry. By contrast, the simple CMOS buffer is inherently broadband and scaling-friendly. Although its sensitivity to PSIJ is a major drawback, we will see that PSIJ is proportional to rise time, and in advanced nodes rise time gets very small so PSIJ becomes less problematic. Moreover, integrated linear voltage regulators are often used to mitigate supply noise and PSIJ [13]. Therefore, this chapter focuses on CMOS clocking for broadband clock distribution in highly scaled CMOS technologies.

There are many examples of modern wireline transceivers achieving state-of-the-art performance using CMOS clock distribution. For example, transceivers operating at 56 GS/s included 14 GHz CMOS clock distribution [3,14,15]. Broadband CMOS clock distribution was demonstrated in 7 nm CMOS operating from 4 to 16 GHz in [8].

Research on CMOS clock distribution is active, including efforts to combine clever frequency planning and co-optimization of on-die transmission line and termination schemes. For example, a hierarchical clock distribution architecture was reported in [7] wherein a lower frequency clock is first passively distributed along a long unterminated transmission line to several "bundles" of transceivers. Within each bundle, the received clock is multiplied to a higher frequency and passively distributed to all transceivers. Circuit techniques can also afford benefits. For example, a small series capacitor inserted between a CMOS buffer and the capacitive load of a transmission line in [16] extends the buffer bandwidth. It also lowers the clock swing, thus reducing dynamic power consumption. In [17], a long unterminated on-die transmission line results in reflections of the rising and falling clock transitions. The induced incident and reflected waves superimpose resulting in clock waveforms with zero skew all along the transmission line. All these techniques make use of the CMOS inverter, so our study begins there.

20.2 Power supply induced jitter (PSIJ) in the CMOS inverter

In this section, we study how power supply noise is converted into clock jitter through the elemental buffer circuit in CMOS clock distribution: the simple inverter. Supply noise refers to the deviation of local supply and ground voltages from their ideal (nominal) values. The deviations are defined $\Delta V_{\mathrm{DD}} = V'_{\mathrm{DD}} - V_{\mathrm{DD}}$ and $\Delta V_{\mathrm{SS}} = V'_{\mathrm{SS}} - V_{\mathrm{SS}}$, where V'_{DD} and V'_{SS} are the local supply and ground voltages, respectively, and V_{DD} and V_{SS} are their nominal values. In general, supply noise induces jitter in the CMOS inverter by affecting the inverter's threshold voltage, its voltage swing ($V'_{\mathrm{DD}} - V'_{\mathrm{SS}}$), and its output edge rate.

20.2.1 Jitter sensitivity

Consider an inverter with nominal delay t_d, but whose delay changes by Δt_d due to supply noise. The sensitivity of the buffer delay to supply noise is [1,18–20]:

$$\frac{\Delta t_d}{t_d} = K\frac{\Delta V_{\mathrm{DD}}}{V_{\mathrm{DD}}} \tag{20.1}$$

Here, K is a constant for a given inverter design. Rearranging (20.1):

$$\Delta t_d = K\frac{\Delta V_{\mathrm{DD}}}{V_{\mathrm{DD}}}t_d \propto K\Delta V_{\mathrm{DD}}t_d \tag{20.2}$$

Equation (20.2) affords an important insight: to reduce the PSIJ Δt_d, the total gate delay in the clock path t_d should be minimized. Moreover, a combination of supply regulation and decoupling can be used to reduce supply voltage variations ΔV_{DD}. There have also been attempts to reduce Δt_d by reducing K [19].

For a chain of series-connected CMOS inverters, K can be derived analytically with a few simplifying assumptions [21]. As shown in Figure 20.2, the input is assumed to be a step and the buffer delay is defined as the time taken to charge the output capacitance C_L from a clean voltage V_{SS} to 0.5 V'_{DD}. During the transition, PMOS transistor M_P is in saturation, so assuming a square-law model, the nominal delay t_d can be expressed as:

$$t_d = \frac{Q}{I} = \frac{C_L V_{\mathrm{DD}}/2}{\frac{1}{2}\mu C_{\mathrm{ox}}\frac{W}{L}(V_{\mathrm{GS}} - V_{\mathrm{th}})^2} = \frac{C_L V_{\mathrm{DD}}/2}{\frac{1}{2}\mu C_{\mathrm{ox}}\frac{W}{L}(V_{\mathrm{DD}} - V_{\mathrm{th}})^2} \tag{20.3}$$

Taking the derivative of t_d with respect to V_{DD}, we have:

$$\frac{dt_d}{dV_{\mathrm{DD}}} = \frac{C_L/2}{\frac{1}{2}\mu C_{\mathrm{ox}}\frac{W}{L}(V_{\mathrm{DD}} - V_{\mathrm{th}})^2}\left(1 - \frac{2V_{\mathrm{DD}}}{V_{\mathrm{DD}} - V_{\mathrm{th}}}\right) = \frac{t_d}{V_{\mathrm{DD}}}\left(1 - \frac{2V_{\mathrm{DD}}}{V_{\mathrm{DD}} - V_{\mathrm{th}}}\right) \tag{20.4}$$

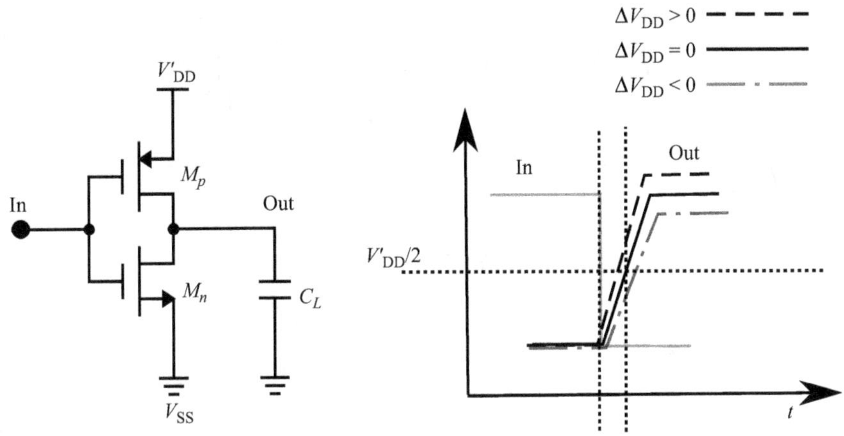

Figure 20.2 The step response of a CMOS inverter in the presence of supply voltage variations

The delay variation Δt_d due to finite variation in the supply ΔV_{DD} is approximated by $\Delta t_d = \frac{dt_d}{dV_{DD}} \Delta V_{DD}$. Substitution into (20.4) and rearranging results in:

$$\frac{\Delta t_d}{t_d} = -\frac{V_{DD} + V_{th}}{V_{DD} - V_{th}} \frac{\Delta V_{DD}}{V_{DD}} \tag{20.5}$$

Comparing (20.1) and (20.6), we have:

$$K = -\frac{V_{DD} + V_{th}}{V_{DD} - V_{th}} \tag{20.6}$$

So, higher V_{DD} reduces K while higher threshold increases it.

Though intuitive, the preceding derivation is an over-simplification. For example, zero input rise time neglects the significant impact of ΔV_{DD} on the inverter threshold (V_M), which in turn impacts its delay. Nevertheless, the linear relationship in (20.1) is observed in practice, and, K can be obtained more accurately through simulation. Using the test bench shown in Figure 20.3(a), a 10 GHz clock is passed through a chain of identical buffers in a 65 nm CMOS technology, where every buffer sees the same supply noise. To vary the nominal gate delay, the buffers sizes are swept from 8 to 20 figures where each finger of the NMOS and PMOS devices are $\frac{500\ \text{nm}}{60\ \text{nm}}$ and $\frac{200\ \text{nm}}{60\ \text{nm}}$, respectively. The capacitor sizes are swept from 6 fF to 10 fF. A 200 MHz sinusoid having 40 mV peak-to-peak amplitude is injected onto V_{DD}. The delay is measured with respect to a crossing point halfway between the noisy supply rails. Thus, the crossing point varies sinusoidally along with the supply noise. The nominal delay t_d and delay variation Δt_d of INV3 (averaged over both rising and falling transitions) are plotted in Figure 20.3(b). Evidently, the delay variation is proportional to nominal delay. The straight line is described by the equation $\Delta t_d = 1.45\ (40\ \text{mV}/1\ \text{V}) t_d$. By comparison with (20.2), we see that $K \approx 1.45$ for this chain of buffers in 65 nm CMOS technology.

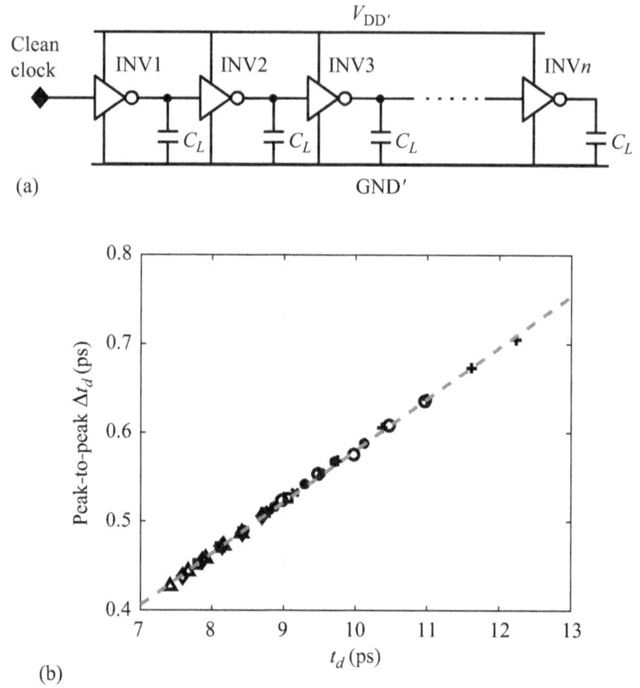

Figure 20.3 (a) Test bench for simulating delay sensitivity, K and (b) peak-to-peak delay variation in the presence of 40 mV peak-to-peak supply variation with different input rise/fall-time, buffer size and C_L

20.2.2 Jitter induced by common-mode and differential supply noise

In certain cases, there is a correlation between the noise on the local supply and ground of a buffer. When these correlations exist, supply noise can be decomposed into common-mode (CM) and differential mode (DM) components to simplify the analysis [22]. The CM supply noise (ΔV_{CM}) is the noise component that causes a local supply and ground to go up and down together whereas DM supply noise (ΔV_{DM}) causes a local supply and ground to vary in a complementary fashion. So V'_{DD} and V'_{SS} can be expressed in terms of ΔV_{DM} and ΔV_{CM} as:

$$V'_{DD} = V_{DD} + \Delta V_{CM} + \Delta V_{DM}/2 \tag{20.7}$$

$$V'_{SS} = V_{SS} + \Delta V_{CM} - \Delta V_{DM}/2 \tag{20.8}$$

where:

$$\Delta V_{CM} = \frac{V'_{DD} + V'_{SS}}{2} - \frac{V_{DD} + V_{SS}}{2} \tag{20.9}$$

$$\Delta V_{DM} = (V'_{DD} - V'_{SS}) - (V_{DD} - V_{SS}) \tag{20.10}$$

The CM supply noise reflects how far the average of the local supply and ground deviates from its nominal value whereas DM supply noise reflects how far the difference of the supply and ground deviates from its nominal value.

The mode (CM vs. DM) and correlation of the supply noise seen by multiple buffers is affected by the PDN, including the packaging technology [4,22]. In wire-bonded packaging, supply noise is predominantly due to ac supply currents flowing through the shared bond wire inductance in the PDN. As a result, supply noise over the whole chip is highly correlated and DM noise predominates [22]. However, in flip-chip packaging, different supply bumps deliver current to different blocks located on different parts of the chip, so the supply noise between them may be less correlated [22]. Moreover, ac V_{DD} currents may flow through V_{SS} return paths in a neighboring block, and as a result, supply noise can have a large CM component. More generally, CM supply noise arises whenever V_{DD} and V_{SS} currents flow through unbalanced PDN impedances. For instance, often all V_{SS} bumps are connected together, but V_{DD} bumps are isolated to create separate supply voltage "islands." In such cases, the impedance of the ground PDN is lower than the supply PDN.

The effect of CM supply noise ΔV_{CM} on PSIJ is illustrated in Figure 20.4. When ΔV_{CM} of adjacent buffers are different, the threshold voltage V_M of the buffer is shifted with respect to its input. As illustrated in Figure 20.4(a), when $\Delta V_{CM} > 0$ for the input clock, the falling transition of the input takes longer to reach V_M compared to the case where $\Delta V_{CM} = 0$. As a result, gate delay gets larger. On the other hand, gate delay gets smaller for the input rising transition since it reaches V_M

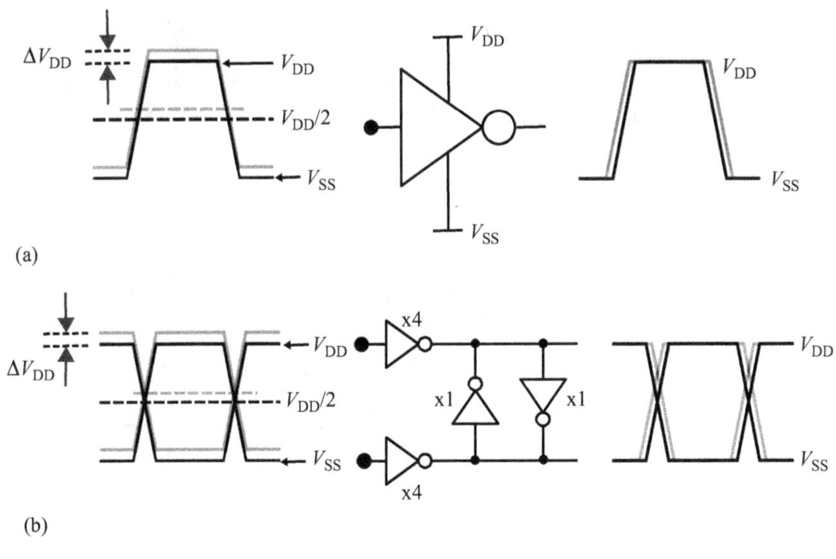

Figure 20.4 PSIJ from CM supply noise for (a) a single-ended inverter and (b) a pseudo-differential inverter

sooner. So, the duty cycle of the output clock changes leading to duty cycle distortion (DCD). The delay variation is approximately:

$$\Delta t_d = 2\frac{\Delta V_{\text{CM}}}{V_{\text{DD}}/(t_r/2)} = t_r\frac{\Delta V_{\text{CM}}}{V_{\text{DD}}} \tag{20.11}$$

where t_r is 0%–100% input transition time assuming a linear rising/falling transition waveform. Since only the difference in ΔV_{CM} between adjacent buffers results in jitter, no jitter is induced if all buffers share the same ΔV_{CM}. For a pseudo-differential inverter, CM noise does not result in DCD. As illustrated in Figure 20.4(b), although the transition of one edge starts Δt_d later, the transition of its differential counterpart starts Δt_d sooner, so the differential zero crossing point is shifted in voltage, but not in time. Therefore, CM noise does not affect pseudo-differential inverters if only the differential zero-crossing is of the concern.

The PSIJ of DM supply noise, ΔV_{DM}, is illustrated in Figure 20.5. When $\Delta V_{\text{DM}} > 0$, the output voltage change required to reach the logic threshold increases by $\Delta V_{\text{DM}}/2$ which tends to increase delay. However, at the same time, V_{GS} of the transistors also increases by $\Delta V_{\text{DM}}/2$ which provides more current to charge and discharge the buffer output, tending to reduce delay. The later effect dominates resulting in a smaller delay. This is illustrated in Figure 20.5(a). For a pseudo-differential inverter illustrated in Figure 20.5(b), ΔV_{DM} has the same effect on both edges, and delay variation is almost the same as the single-ended signal.

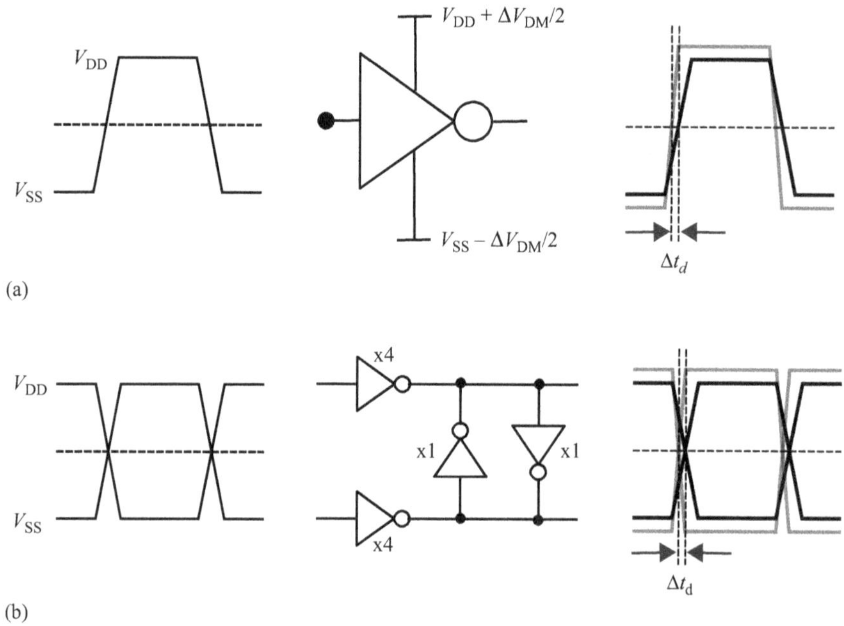

Figure 20.5 PSIJ of DM supply noise for (a) a single-ended inverter and (b) a pseudo-differential inverter

20.2.3 PSIJ accumulation

The way PSIJ accumulates in a chain of series-connected clock buffers depends on the type of noise (i.e., CM and/or DM) and the correlation of supply noise seen by adjacent buffers [23]. Since gate delay is affected by input edge rate, in a chain of buffers, supply noise modulating the rise/fall time of a buffer stage also modulates the delay of next buffer. This effect either adds constructively or destructively to the PSIJ between buffer stages depending on the phase of supply noise seen by them. Thus, for buffers only affected by DM supply noise, Δt_d is:

$$\Delta t_d = -(K_{\mathrm{DM0}}\Delta V_{\mathrm{DM0}} + K_{\mathrm{DM1}}\Delta V_{\mathrm{DM1}} + K_{\mathrm{DM2}}\Delta V_{\mathrm{DM2}} + \cdots + K_{\mathrm{DM}k}\Delta V_{\mathrm{DM}k}) \tag{20.12}$$

where ΔV_{DM0} and $\Delta V_{\mathrm{DM}k}$ are the DM supply noise of the current buffer and kth preceding buffer, respectively, and K_{DM0} and $K_{\mathrm{DM}k}$ are corresponding supply noise sensitivity constants which may be characterized by simulation. For example, $K_{\mathrm{DM0}} \approx 0.46$, $K_{\mathrm{DM1}} \approx 0.69$ and $K_{\mathrm{DM2}} \approx 0.22$ in a 65 nm CMOS process. The values of $K_{\mathrm{DM}k}$ are quite small for $k > 3$ and can be ignored.

When the supply noise frequency is low and the buffers are close together, then the supply noise seen by different buffers is highly correlated and the total delay of the buffer chain is likely much smaller than the period of the supply noise. In such cases, the PSIJ introduced by each inverter is roughly the same, and PSIJ accumulates linearly with the number of buffers in the chain. Quantitatively, $\Delta V_{\mathrm{DM0}} = \Delta V_{\mathrm{DM1}} = \Delta V_{\mathrm{DM2}} \ldots = \Delta V_{\mathrm{DM}k} = \Delta V_{\mathrm{DM}}$ and $K_{\mathrm{DM}} = K_{\mathrm{DM0}} + K_{\mathrm{DM1}} + K_{\mathrm{DM2}} \cdots + K_{\mathrm{DM}k}$ in (20.12), and if PSIJ_1 is the PSIJ from one inverter, and n is the number of buffer stages, we have:

$$\mathrm{PSIJ}_1 = \Delta t_d = -K_{\mathrm{DM}}\Delta V_{\mathrm{DM}} \tag{20.13}$$

$$\mathrm{PSIJ}_{\mathrm{total}} = n \times \mathrm{PSIJ}_1 \tag{20.14}$$

From simulations, $K_{\mathrm{DM}} \approx 1.4$ in 65 nm CMOS technology. An eight-stage pseudo-differential inverter chain with a DM peak-to-peak supply noise of 40 mV at a frequency of 200 MHz is simulated. The absolute jitter (also called the time interval error, TIE) at the input, and at the output of the 2nd, 4th, and 8th inverter stages are shown in Figure 20.6(a). The TIE accumulates linearly with the number of stages, and has the opposite polarity, as predicted by (20.13) and (20.14). In Figure 20.6(b), peak-to-peak TIE jitter is plotted against the number of stages, highlighting the linear accumulation of PSIJ.

PSIJ accumulates more slowly if the DM supply noise at different stages of a buffer chain is uncorrelated. Assume all buffers see the same supply noise, a buffer chain has a total delay of t_d, and every buffer stage is identical. The effect of supply noise on the PSIJ of the entire buffer chain is filtered in the time domain by a rectangular window with width t_d and equivalently in the frequency domain by a sinc function:

$$\mathrm{PSIJ}(f_{\mathrm{sup}n}) \propto t_d \sin c(\pi t_d f_{\mathrm{sup}n}) \tag{20.15}$$

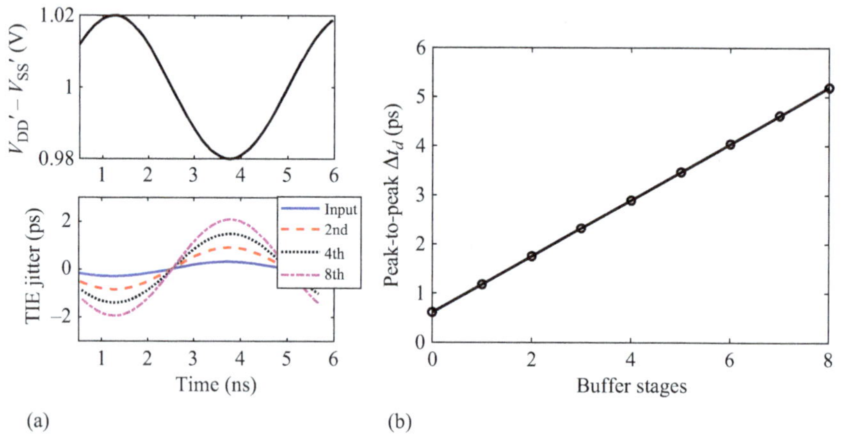

Figure 20.6 PSIJ accumulation along a buffer chain under highly correlated DM supply noise: (a) sinusoidal DM supply variation (top) and the resulting TIE jitter at the output of various stages (bottom) and (b) accumulation of peak-to-peak PSIJ along the buffer chain

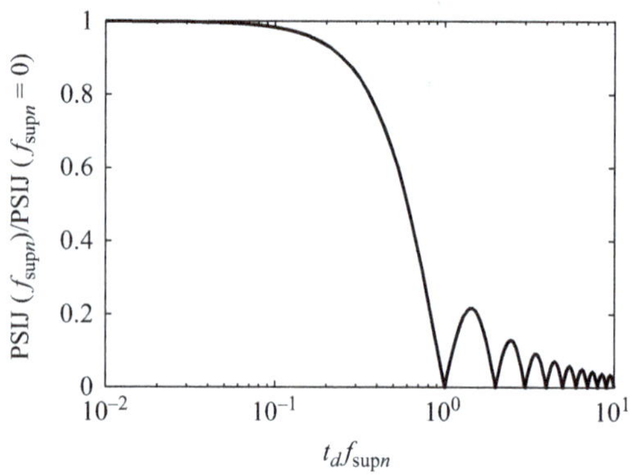

Figure 20.7 PSIJ sensitivity versus noise frequency and clock path delay

As plotted in Figure 20.7, the effect of supply noise on jitter starts to drop as the supply noise frequency $f_{supn} \geq 1/(10t_d)$ and drops to zero at $f_{supn} = 1/t_d$. Intuitively, when t_d is comparable to the supply noise period, as a clock edge propagates through the clock chain, different buffers see different phases of the supply noise. As a result, delay increases for some buffers and decreases for others, and the net effect gets canceled out. PSIJ(f_{supn}) is reported in [5,24] as PSIJ sensitivity: the induced peak-to-peak jitter versus supply noise frequency. For example, in [5] a DDR3 interface with a maximum clock frequency of 1 GHz has a PSIJ sensitivity that starts to drop at around 200 MHz reaching zero at around 1 GHz.

However, this low pass characteristic of jitter sensitivity becomes practically irrelevant as clock frequency increases. For example, consider a clock frequency of 10 GHz. Equation (20.15) indicates that the jitter sensitivity of a buffer chain is reduced to 90% of its dc value at 200 MHz if the chain has a total delay of $t_d = 1.25\ \text{ns} = 12.5T_{\text{clk}}$. However, this would imply an impractically huge number of buffers in the chain. Even modest decoupling capacitance effectively filters supply noise above 1 GHz. Hence, the jitter sensitivity of a typical buffer chain is essentially flat over the range of frequencies typical of supply noise. In rare cases where t_d is 10 or more clock periods, the buffers are separated by significant supply impedances. Thus, DM supply noise seen by different buffers in the chain is less correlated and their PSIJ does not add constructively.

As explained in Section 2.2, CM supply noise only induces DCD in single-ended inverters, and only if the CM supply noise ΔV_{CM} seen by neighboring buffers is different. Moreover, due to the inverting nature of CMOS buffers, supply noise-induced DCD accumulates constructively only if neighboring inverters have ΔV_{CM} with opposing polarity. Such a scenario is unlikely in practice.

20.3 Random jitter generation in the CMOS inverter

Random jitter (RJ) due to thermal noise in a CMOS inverter arises in two ways. When the inverter is not transitioning, either the NMOS or PMOS transistor is in triode is connecting the inverter output to a supply voltage and it contributes kT/C voltage noise to the output node. This introduces an initial random voltage at the start of the next transition. For a rising transition, assuming a constant current I_D charging a capacitance C_L from 0 to V_{DD} with a rise time t_r, the slope of the transition waveform is $I_D/C_L = V_{\text{DD}}/(t_r)$. The jitter due to the initial random noise voltage can be obtained by dividing it by transition slope of the rising voltage waveform:

$$\sigma_{j1}^2 = \frac{\sigma_{V_{n1}}^2}{(I_D/C_L)^2} = \frac{kTC_L}{I_D^2} = \frac{kT}{I_D V_{\text{DD}}} t_r \tag{20.16}$$

Secondly, during a transition, white current noise with a PSD of $4kT\gamma g_m$ is superimposed on the charging/discharging current. Voltage noise at the $V_{\text{DD}}/2$ crossing can be obtained by integrating white current noise over time window $t_r/2$ and divide it by load capacitor. The integration results in a factor of $t_r/4$ [25], so we have:

$$\sigma_{V_{n2}}^2 = 4kT\gamma g_m \frac{t_r}{4} \frac{1}{C_L^2} = \frac{kT\gamma g_m t_r}{C_L^2} \tag{20.17}$$

The jitter is obtained by dividing $\sigma_{V_{n2}}^2$ by the slope at transition:

$$\sigma_{j2}^2 = \frac{\sigma_{V_{n2}}^2}{(I_D/C_L)^2} = \frac{kT\gamma g_m t_r}{I_D^2} = \frac{2kT\gamma}{I_D(V_{\text{DD}} - V_{\text{th}})} t_r \tag{20.18}$$

In this expression, the last equality assumes the MOSFET square-law model and $V_{GS} = V_{DD}$. From (20.16) and (20.18), it is observed that both jitter components are directly proportional to $\frac{t_r}{I_D}$. So, jitter variance from a single inverter increases with decreasing transition current and decreases with decreasing rise time or gate delay.

The random jitter, RJ, can be simulated using either transient noise analysis or a combination of periodic steady state (PSS) and periodic noise (PNoise) analysis. Using the transient noise method, device noise is updated with a maximum time step of $1/(2f_{\max})$ where $f_{\max}$ is a simulation parameter. For accurate RJ simulation, $f_{\max}$ should be set large enough to allow for several time steps during clock transitions (e.g., $f_{\max} \geq \frac{1}{10t_r}$). To obtain accurate statistical information about the clock jitter (e.g., its RMS value), a sufficient number of clock cycles is needed. Unfortunately, the combination of large $f_{\max}$ and a large number of clock cycles leads to a long simulation time.

Alternatively, with PSS and PNoise simulations, results may be obtained more quickly. First, the periodic steady state of the clock waveform is computed using a PSS analysis. Then, the jitter spectrum (i.e., phase noise) at the logic threshold crossing can be obtained by a linear PNoise analysis. An accurate RMS value can be obtained by integrating the jitter spectrum from DC up to $f_{\text{clk}}/2$. This is because noise is sampled at either the rising or falling edge of a clock at a rate of f_{clk}, and the Nyquist frequency is $f_{\text{clk}}/2$. Since this approach is statistical, simulation time is usually shorter. However, unlike transient noise analysis which takes into account all the nonlinear effects of transistors, this analysis assumes device noise is small enough that it does not affect the periodic steady state of the clock. For RJ simulation in a chain of CMOS clock buffers, device noise is usually small, so PSS and PNoise analysis may generally be used for faster and accurate simulation.

20.3.1 Random jitter trend

Since the device noise of each inverter is inherently random and uncorrelated, mean-squared jitter accumulates linearly along a buffer chain. For example, a PSS and PNoise simulation is performed at the 1st and the 8th nodes of an eight-stage inverter chain. The RJ of rising and falling edges is shown in Figure 20.8. It shows that in a 65 nm CMOS technology, each four-finger inverter cell contributes 16.3 fs RMS jitter. This property of RJ accumulation is independent of input RJ as long as there is no jitter amplification. Therefore, an eight-stage inverter chain adds $\sqrt{8} \cdot 16.3 = 46$ fs RMS jitter to a clean clock. If a clock with 100 fs RMS RJ is applied to the input of the eight-stage buffer chain, the output will have $\sqrt{46^2 + 100^2} = 110$ fs RMS RJ.

To verify that RJ accumulation is inversely proportional to buffer current, I_D in (20.16) and (20.18), the buffer size is varied and the resulting RJ variance for one stage, σ_j^2, is plotted in Figure 20.9(a). It verifies that σ_j^2 is inversely proportional to transistor size, which is in turn proportional to I_D. Similarly, to verify that RJ accumulation is proportional to signal rise time, the buffer size is kept fixed to maintain constant I_D and C_L is varied to change the rise time. The variance of RJ for

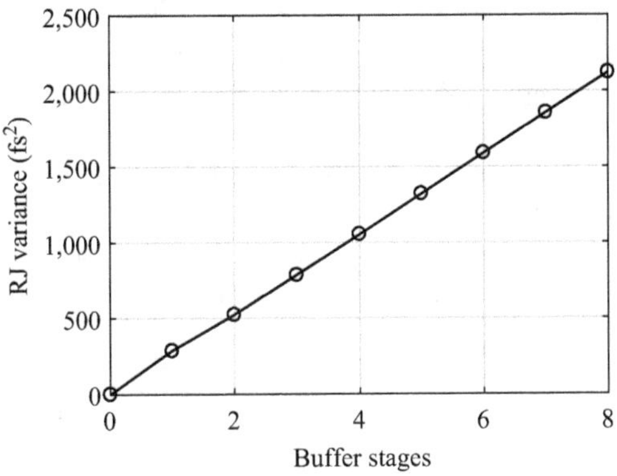

Figure 20.8 RJ variance accumulates linearly along a chain of buffers

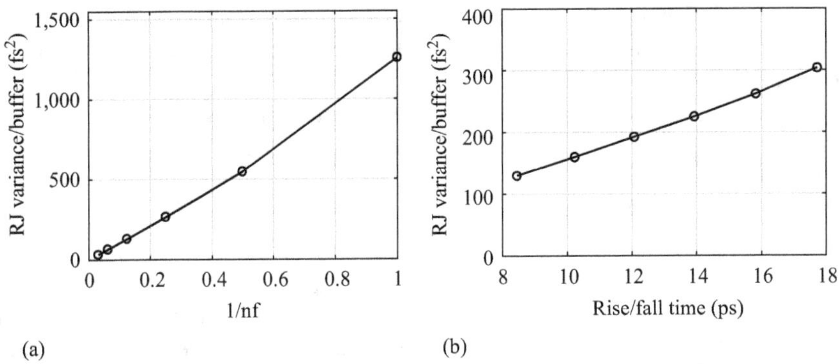

Figure 20.9 RJ variance accumulation per stage (a) with buffer size and (b) with rise/fall time

one inverter stage is plotted against the 10%–90% rise/fall time in Figure 20.9(b). It illustrates that σ_j^2 is directly proportional to t_r. In conclusion, RJ in an inverter decreases with faster rise and fall times and larger buffer sizes. Thus, RJ arising from thermal noise can be reduced by burning more power.

20.4 Jitter generation in global clock distribution

Jitter in local clocks depends on both the jitter accumulated along the global clock distribution buffer chain and the jitter added by any local buffers. Each of those, in turn, has RJ and PSIJ components. The RMS RJ of the global buffer chain is $\sqrt{n}$ times the RJ of each stage. The PSIJ of the global buffer chain, assuming supply noise along the chain is highly correlated, is n times the PSIJ of each stage. Both the RJ and PSIJ contributed by the global clock distribution buffers decrease with faster rise and fall times. The jitter contributed by the local clock buffer also depends significantly on the rise and fall times at its input. Thus, jitter is minimized by designing the global clock distribution to reduce n while still maintaining fast rise and fall times throughout the chain, including at all points where local buffers are connected.

In summary, to distribute a clock across a fixed distance, nl_{stage}, using fewer buffers n will result in less jitter unless rise and fall times suffer. Once rise and fall times become much slower, jitter can be expected to increase. To explore this tradeoff, we must model the propagation of clocks along the global clock interconnect (transmission line).

20.4.1 Model of a CMOS inverter driving a low-loss transmission line

Our model for clock propagation along global clock interconnect is shown in Figure 20.10. The transmission line is loaded by the input capacitance C_{in} of the next-stage buffer. For each rising and falling transition, the buffer launches an incident wave that is reflected at the far end by the capacitive load. The clock waveform at each point along the line is the superposition of the incident and the reflected wave.

The amplitude of the incident wave depends on the output impedance of the driver R_{drv} and the characteristic impedance of the wire Z_0:

$$V_{\text{inc}}(x=0) = V_{\text{in}}\frac{Z_0}{Z_0 + R_{\text{drv}}} \tag{20.19}$$

For simplicity, R_{drv} is assumed to be a linear resistance, whereas in reality, it is nonlinear with a value that depends on V_{in}. So, the incident wave amplitude increases with increases in Z_0, and decreases with increases in R_{drv}. At a distance x from the driver, the incident wave amplitude decreases by $e^{-\alpha x}$ where α is the interconnect's attenuation constant. Thus, at the far end, the incident wave is attenuated by a factor of $e^{-\alpha l}$, where l is the length of the wire. With capacitive termination, the clock undergoes exponential settling with a time constant $\tau = Z_0 C_{\text{in}}$ [26]:

$$V_{\text{clk}}(x=l) = 2V_{\text{inc}}(x=0)e^{-\alpha l}(1 - e^{-t/(Z_0 C_{\text{in}})}) \tag{20.20}$$

Hence, C_{in} significantly degrades rise and fall times at the far end if τ is much greater than the launch transition time, t_r. As the reflected wave travels back toward

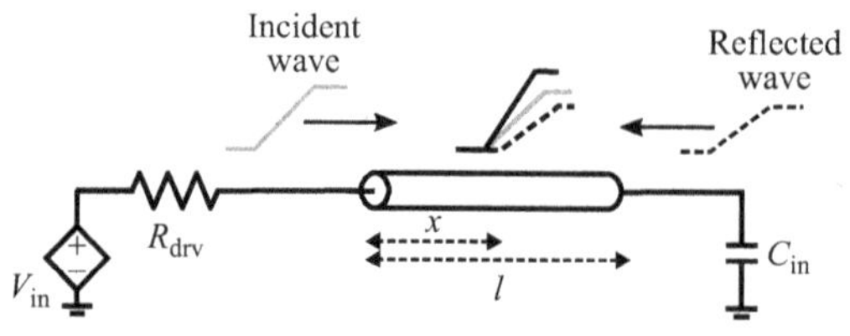

Figure 20.10 Model for CMOS inverter driving long RLC wire

the near end, at position x, it is attenuated by a total factor of $e^{-\alpha(2l-x)}$ relative to incident wave amplitude. Assuming $\tau \ll t_r$, the reflected wave at distance x from the near end has an amplitude:

$$V_{\text{ref}}(x) = V_{\text{inc}}(x = 0)e^{-\alpha(2l-x)}. \tag{20.21}$$

For a propagation velocity along the interconnect v_p, it takes $t_f = 2(l - x)/v_p$ between the moment the incident wave reaches the point x and the moment the reflected wave reaches the same point and superimposes with the incident wave. At any point x where t_f is comparable to the input clock transition time t_r, the observed clock settling will be slowed while waiting for the reflected wave to arrive. Note that t_f is smallest at the far end and largest at the near end, so this effect degrades clock edges at the near end (small x).

20.4.2 Design of CMOS clock distribution with low jitter generation

To reduce both PSIJ and RJ of clock buffers, it is desired for clock waveforms to rise and fall rapidly between ground and V_{DD}, thus turning buffer transistors fully on and providing the maximum possible current for charging and discharging their outputs. From (20.19), the edge rate of the incident wave launched by a global buffer improves with a sharper edge on the input clock V_{in}, and a larger ratio $\frac{Z_0}{Z_0+R_{\text{drv}}}$. Also, from (20.20), we see that the far end clock doubles the amplitude of the attenuated incident wave. Thus, to ensure fast edge rates all along the global clock distribution, two situations are to be avoided by designers.

First, it is undesirable to design for $R_{\text{drv}} \gg Z_0$. In this case, the launched incident wave has an amplitude much less than V_{DD}, and multiple reflections are required before each clock edge will fully settle. Thus, l_{stage} must be made very small to ensure fast settling, which in turn requires large n to cover a fixed total global clock distribution distance nl_{stage}. We have already seen that a large number of series-connected buffers n implies larger accumulation of RJ and PSIJ, as well as potentially high power consumption.

Second, a design where l_{stage} is so large that the round-trip time-of-flight exceeds the clock transition time is undesirable. As discussed at the end of the Section 20.4.1, at the transmission line near end, $x = 0$, the reflected wave will return $t_d = 2l_{\text{stage}}/v_p$ after the incident wave is launched. For a short wire, the reflected wave returns back before the incident clock edge rises to $\frac{V_{\text{DD}}}{2}$, thus providing a fast edge rate all along the transmission line. However, as l_{stage} increases, the reflected wave experiences more attenuation and takes more time to return. At locations closer to the near end, late arrival of reflected waves reduces the superposition effect between incident and reflected waves. As a result, clock waveforms may rise quickly to a fraction of V_{DD} due to the incident wave, but then stall awaiting the attenuated reflected wave. Figure 20.11(a) shows an example clock waveforms of the global clock at different distances x from the near end in a 65 nm CMOS technology. Note that in advanced CMOS technologies, high clock frequencies and lower t_r are targeted, so degradation of edge rates due to reflections

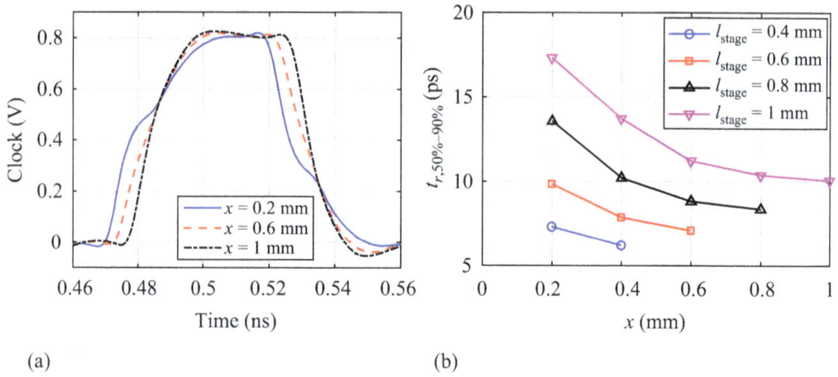

Figure 20.11 (a) Global clock waveforms at different distances from near end and (b) global clock $t_{r,50-90\%}$ at different distances from near end for different l_{stage}

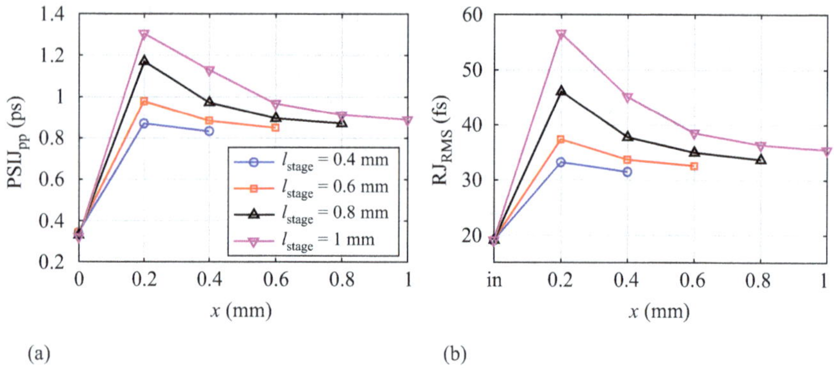

Figure 20.12 Jitter of local clock at different distances from the near end for different l_{stage}: (a) $PSIJ_{pp}$ and (b) RJ_{RMS}

occurs even at small l_{stage}. Figure 20.11(b) shows the variation in $t_{r,50\%-90\%}$ for global clocks versus distance from the near end x for different l_{stage}. Notice that the global clock always has larger $t_{r,50\%-90\%}$ at the near end than at the far end due to the reflections, and as l_{stage} increases, the effect is exaggerated. Since the effect is caused by reflections that are fundamentally limited by the propagation delay of the interconnect, this issue can not be mitigated by designing larger buffers.

When tapped off by local clock buffers, the slower edge rates at the transmission line near end increase the PSIJ and RJ introduced by the local buffer. The PSIJ_{pp} and RJ introduced by a local clock buffer at different tapping points are plotted in Figure 20.12. In this analysis, sinusoidal supply noise with a peak-to-peak amplitude of 40 mV (DM) was introduced, and PSS and PNoise analyses were used to find the RJ.

A typical design choice is to select $R_{drv} \approx Z_0$. We have already seen that $R_{drv} \gg Z_0$ is undesirable. On the other hand, $R_{drv} \ll Z_0$ can cause undesirable overshoot and ringing in the clock waveforms due to the resulting negative reflections. The choice of $R_{drv} \approx Z_0$ is impacted by load capacitances including the input capacitance of global and local buffers (C_{Load}). The time constant $Z_0 C_{Load}$ must be limited to maintain fast edge rates. Additionally, C_{Load} adds to the capacitance of the transmission line, decreasing the equivalent impedance $Z_{0,eqv} = \sqrt{\frac{L_w}{(C_w+C_{Load})}}$. Lower $Z_{0,eqv}$ requires a larger global buffer and thus higher power consumption to maintain $R_{drv} \approx Z_0$. In addition, increasing C_{Load} tends to increase transmission line delay (proportional to $\sqrt{\frac{L_w}{C_w+C_{Load}}}$), so l_{stage} may need to be reduced to prevent the problem of reflections illustrated in Figures 20.11 and 20.12. Although reducing C_{Load} is in many ways desirable, we have already seen that the minimum size of clock buffers is limited by their RJ contribution.

20.4.3 *Example: jitter generation in 16 nm CMOS clock distribution*

Simulations in a 16 nm CMOS technology of a 10 GHz clock distributed across a total distance of 2-mm tapped by ten local buffers with uniform 200-μm spacing exemplify the preceding analysis. The number of global buffers *n* is varied from 2 to 10. Global buffers are sized to have R_{drv} slightly larger than Z_0 to maintain a sharp edge at the far end without significant overshoot. The corresponding number of fins (nfin) and the number of fingers (nf) are 4 and 36, respectively, with minimum length. For local buffers, four times the minimum device size (nfin = 4 and nf = 4) is used as a compromise between low RJ generation and low capacitive loading on the transmission line.

20.4.3.1 PSIJ

Supply noise of 40 mV peak-to-peak amplitude (DM) is introduced. All buffers share the same supply noise, the worst case scenario for DM supply noise. While varying the number of global buffers *n*, the PSIJ of all ten local clocks at the output of each local clock buffer is plotted in Figure 20.13(a). When $n = 2$ or 3, the corresponding l_{stage} results in higher PSIJ at the near end of each transmission line segment because reflections degrade $t_{r,50\%-90\%}$ of the clock waveforms at the tapping points. Nevertheless, the peak PSIJ among all the ten lanes is less than the case for larger values of *n* (more buffers) because PSIJ is introduced at fewer global buffers. By contrast, when $n = 5$ or 10 the peak PSIJ is higher because it accumulates over more global buffers, and the last lane has the highest PSIJ.

In the same simulation environment, global clock buffer size is varied from 28 to 36 fingers. In each case, predrivers are inserted so that the same input clock waveform is maintained at the input to the global buffer chain. The worst PSIJ_{pp} among the ten tapping points is plotted versus the number of buffers inserted, *n*, in Figure 20.13(b). The PSIJ_{pp} reduces with fewer and larger global buffers.

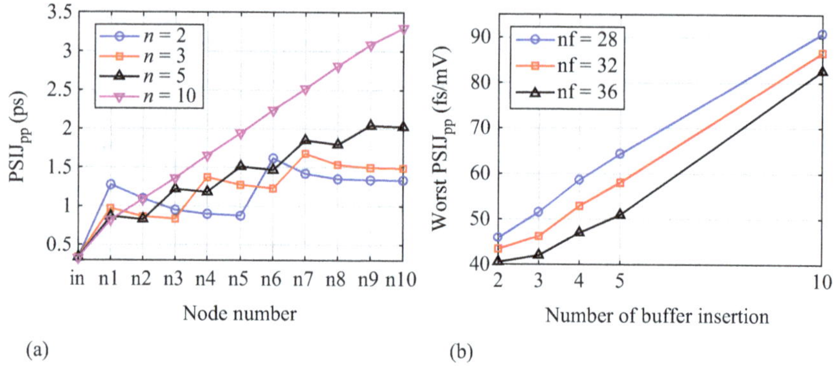

Figure 20.13 (a) PSIJ at the output of local clock buffers at different points along a 2-mm global clock distribution line with a different number of buffers inserted, n, each sized with nf = 36 NMOS and PMOS fingers and (b) The worst PSIJ among all local clocks in the 2-mm clock distribution with in the 2-mm clock distribution buffer sizes

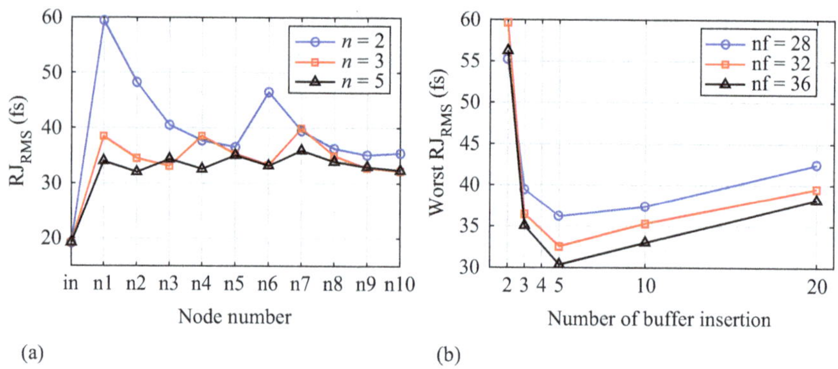

Figure 20.14 (a) Variation in RJ of local clocks with n and (b) worst RJ in local clocks with variation in buffer size

However, this simulation assumes all buffers share the same supply noise, which may not be true for global buffers spanning 2 mm. When the correlation of DM supply noise among buffers reduces, PSIJ accumulates more slowly with n, which may suggest the use of somewhat more buffers.

20.4.3.2 Random jitter

With the global buffers sized so that $R_{drv} \approx Z_0$, RJ accumulation among the global buffers is relatively small. Unlike PSIJ accumulation, the RJ introduced by each buffer is generally uncorrelated so it accumulates only in proportion to $\sqrt{n}$. On the other hand, fewer buffers imply larger l_{stage} and slower edge rates. Figure 20.14(a) shows RJ in the local clocks at different tapping points with a different number of global buffers. When $n = 2$, large RJ can be observed at the near end due to reflections causing a poor $t_{r,50\%-90\%}$ in the global clocks. As n increases, l_{stage} decreases and $t_{r,50\%-90\%}$ is improved which eventually reduces the variation of RJ along the clock distribution.

Varying the global buffer size from 28 to 36 fingers while maintaining the same input clock waveform and excluding RJ from any predrivers, as before, peak RJ among the ten lanes is plotted versus the number of global buffers n while varying the global buffer sizes in Figure 20.14(b). In this case, $n = 5$ results in the lowest RJ, where each global buffer drives a 0.4-mm long transmission line segment tapped by two local buffers. Smaller n and larger l_{stage} suffer from a reduced edge rate due to the reflection effects described in Section 20.4.2 and, thus, additional RJ contributed by the local buffers tapping there. Increasing n beyond 5 does not improve edge rate much; rather it results in the accumulation of more RJ while also consuming more power.

20.5 Jitter amplification

Whereas Sections 20.2–20.4 were concerned with the generation of jitter by clock distribution circuits, we here consider the impact of clock distribution circuits on jitter that is already present at their input. Jitter amplification arises when a clock does not settle sufficiently between transitions. For example, it has been observed that jitter is amplified when a clock is passed through lossy PCB channels in a source-synchronous link [27,28], and through on-chip clock distribution when the clock path has insufficient bandwidth [1]. Two models have been developed to characterize this effect: the JIR and JTF model [29], and the MER model [30]. Each will be described in turn this section.

20.5.1 Jitter impulse response (JIR) and jitter transfer function (JTF)

When impulsive jitter is injected into clock distribution circuits (i.e., a small incremental time delay or advance applied to an individual clock edge), it results in jitter in multiple subsequent edges in the output clock.The jitter impulse response (JIR) is defined as a sequence of TIE jitter observed when a jitter impulse is injected at input clock [1]. The circuit's response to a random sequence of TIE jitter, $\{h_i\}_{\text{in}}$, is an output jitter sequence:

$$\{h_i\}_{\text{out}} = \{h_i\}_{\text{in}} * \text{JIR} \tag{20.22}$$

Since this assumes the circuit's jitter response is linear, a jitter transfer function (JTF) can be obtained by taking a Fourier transform of the JIR. The JTF describes the amplification of input jitter as a function of jitter frequency. A simple simulation example is shown in Figure 20.15 to demonstrate this concept wherein each buffer drives a lumped RC load. A 10 GHz clock is passed through the clock path with and without a 5 ps jitter impulse injected and the resulting clock edge crossing times are compared to find the JIF. With insufficient RC-limited bandwidth in the buffers, the

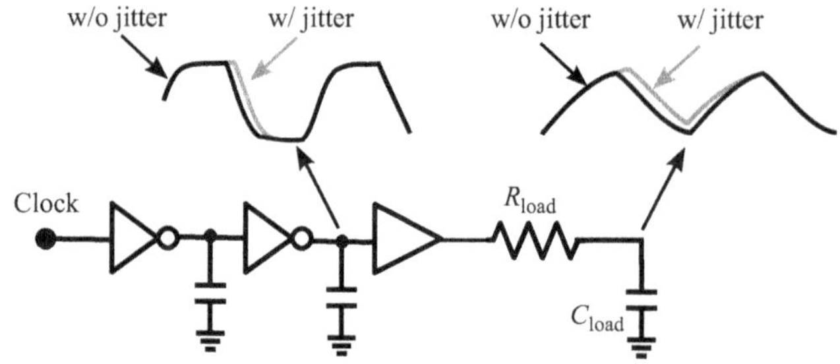

Figure 20.15 Test bench: single-stage buffer drive lumped RC load

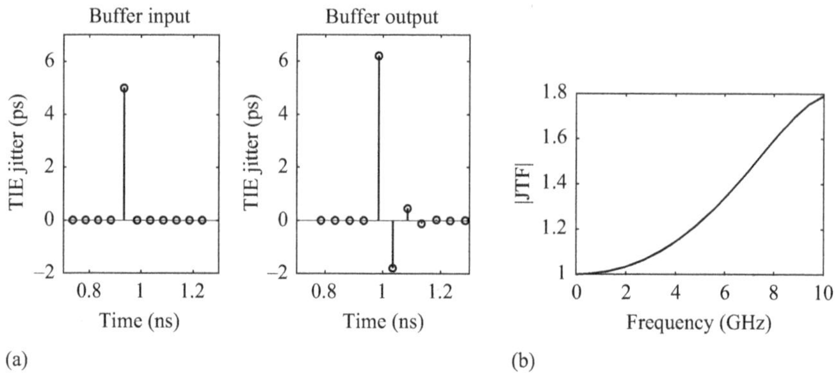

Figure 20.16 JIR and JTF for single buffer driving lumped RC load (a) jitter impulse at input and JIR and (b) JTF

output clock cannot settle sufficiently between transitions. The resulting JIR is plotted in Figure 20.16(a) for a single buffer. It shows that input jitter on multiple consecutive edges will all contribute to the jitter on each output edge. The corresponding JTF is shown in Figure 20.16(b). There is almost no amplification for low-frequency jitter, whereas high-frequency jitter (especially DCD, equivalent to jitter at 10 GHz in this case) is amplified up to 1.8 times. In other words, the JTF of a bandwidth-limited low-pass buffer has a high-pass characteristic. It has been shown that circuits with high-pass and band-pass transfer functions have a low-pass JTF [1].

20.5.2 JIR and JTF for a cascade of clock buffers

The analysis of jitter amplification from the previous section can be extended to a cascade of buffers such as arise in global clock distribution. The JTF of the entire chain of n buffer stages is the product of the JTF of each stage:

$$\mathrm{JTF_{chain}} \simeq \prod_{i=1}^{n} \mathrm{JTF}_i \tag{20.23}$$

where JTF_i is the JTF of each stage. A 24-stage buffer chain shown in Figure 20.17 is used to illustrate this model. Each stage uses equally sized buffers to drive wire segments of same length, so the JTF of each stage is the same, $\text{JTF}_{\text{stage}}$, and $\text{JTF}_{\text{chain}}$ is $(\text{JTF}_{\text{stage}})^{24}$. A small jitter impulse is injected into the clock chain, and JIR at each buffer stage is recorded. The magnitude of the jitter impulse in this simulation must be small enough so that jitter amplification remains linear throughout the buffer chain. We obtain $\text{JTF}_{\text{stage}}$ by dividing the Fourier transforms of the measured JIRs after the second and third stages, as shown in Figure 20.18(a). We can also find $\text{JTF}_{\text{chain}}$ directly by taking the Fourier transform of the JIR after the 24th stage. In Figure 20.18(b), $(\text{JTF}_{\text{stage}})^{24}$ is plotted alongside $(\text{JTF}_{\text{chain}})$ with good agreement.

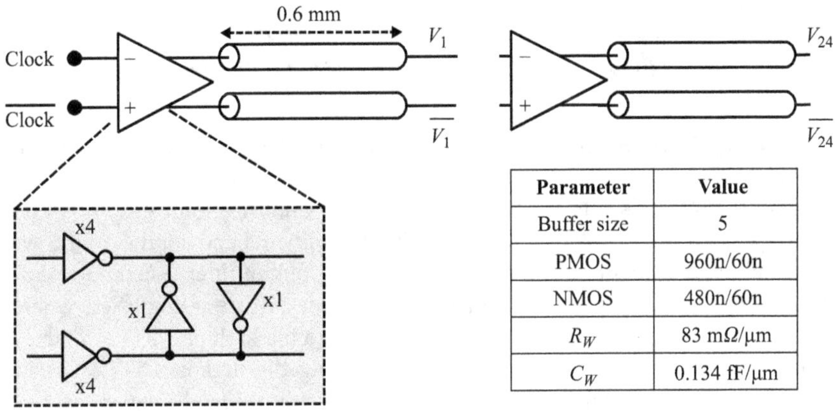

Parameter	Value
Buffer size	5
PMOS	960n/60n
NMOS	480n/60n
R_W	83 mΩ/μm
C_W	0.134 fF/μm

Figure 20.17 Test case for the analysis of jitter amplification in a clock distribution buffer chain. Each stage comprises a differential buffer driving $l_{stage} = 0.1$ mm interconnect

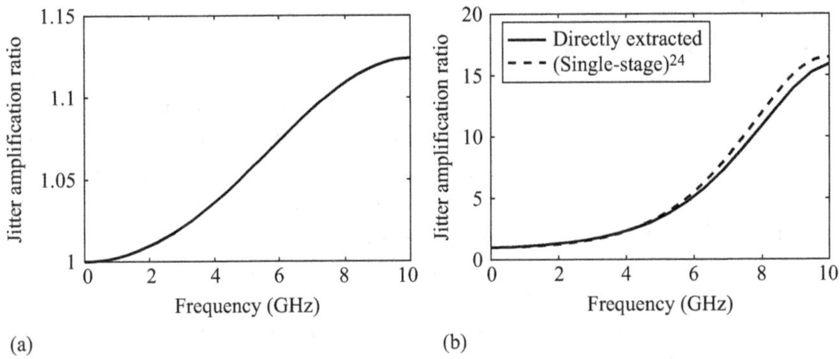

Figure 20.18 JTF of (a) a single buffer stage JTF_{stage} and (b) a 24-stage clock distribution, obtained directly and inferred from that of an individual stage $(JTF_{stage})^{24}$

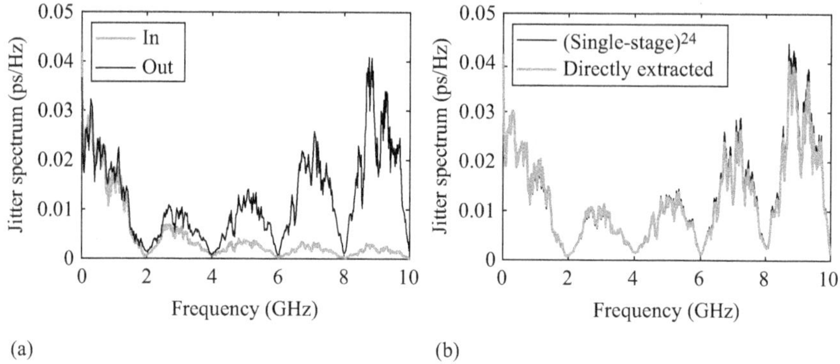

Figure 20.19 (a) Spectrum of RJ at input and output and (b) output RJ from simulation and from prediction based on two methods

Thus, a linear model for jitter amplification is appropriate (for small enough jitter amplitudes).

A similar analysis can be used to estimate the amplification of RJ. For example, low-pass RJ is generated by lowpass filtering white RJ and applied to the same 24-stage clock distribution. The resulting input and output jitter spectra are shown in Figure 20.19(a). It shows that jitter up to 2 GHz is barely amplified whereas high-frequency jitter is significantly amplified due to the high pass JTF. In fact, the output RJ may be accurately predicted by applying the highpass $\text{JTF}_{\text{chain}}$ to the input spectrum, as shown in Figure 20.19(b). Since jitter amplification increases exponentially with the number of buffer stages, even a small gain can become huge for a large number of stages, as illustrated in this example.

20.5.3 Design of CMOS clock distribution with low jitter amplification

Jitter amplification occurs when a clock waveform does not settle between transitions, which typically occurs in one of two situations. First, when there is a slow RC settling time constant as shown in Figure 20.20(a). For a lowpass clock path with a single time constant τ, insufficient settling occurs when τ is comparable to one-half the clock period, $T_{\text{clk}}/2$. Note that some clock distribution interconnect, such as a top metal layer on-chip transmission line, has low loss and cannot be accurately modeled by RC single time constants. However, RC settling behavior provides qualitative insight and occurs even for low-loss transmission lines when the buffer size is too small, in which case the buffer output resistance R_{drv} (being much larger than the interconnect characteristic impedance Z_0) determines the settling time constant. The second possibility arises when the interconnect has low series resistance, and the buffer is sized large enough so that $R_{\text{drv}} \approx Z_0$. In such cases, the clock path is no longer RC-limited, but incomplete settling still occurs when the round trip delay of the transmission line t_{round} is comparable to $T_{\text{clk}}/2$ as shown in Figure 20.20(b). As described in Section 20.4, proper settling of the clock

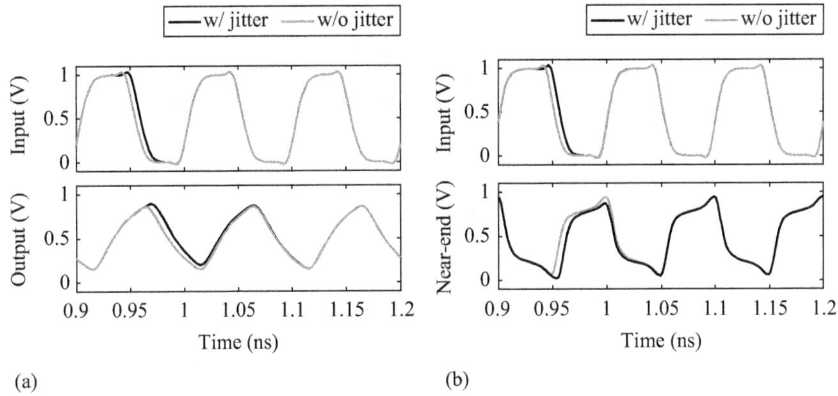

Figure 20.20 Unsettled clock due to (a) low BW of RC-limited clock path and (b) low Z_0 and long unterminated wire

waveform must await the return of the reflected clock edge, which takes a particularly long time at the near end. The time taken to settle, t_{settle}, is roughly the sum of the round-trip delay, t_{round}, and the rise time, t_r, of the attenuated reflected wave:

$$t_{\text{settle}} = t_r + t_{\text{round}} \tag{20.24}$$

For proper settling, $t_{\text{settle}} < T_{\text{clk}}/2$. Moreover, t_{round} depends on the propagation velocity and line segment length:

$$t_{\text{round}} = 2\frac{l_{\text{stage}}}{v_p} = 2l_{\text{stage}}\sqrt{L_w(C_w + C_{\text{load}})} \tag{20.25}$$

where C_{load} is the capacitance per unit length due to the input of local buffers tapping onto the transmission line. To prevent excessive jitter amplification, the maximum wire length is therefore limited by:

$$l_{\text{stage}} < \frac{1}{2}\frac{T_{\text{clk}}/2 - t_r}{\sqrt{L_w(C_w + C_{\text{load}})}} \tag{20.26}$$

This maximum wire length decreases with higher f_{clk} and larger local buffers. For example, in the clock distribution example of Section 20.4.3 with a f_{clk} of 10 GHz and 14 GHz and a global buffer size of 32 fingers, RJ amplification factor* is plotted against wire length per stage in Figure 20.21. In both cases, it is evident that jitter amplification is absent up to some maximum wire length, beyond which jitter amplification increases dramatically. The maximum wire length is lower at 14 GHz than 10 GHz, as suggested by (20.26).

We have seen that both jitter amplification and jitter generation (discussed in Section 20.4.2) degrade as l_{stage} increases, particularly at the transmission line near

*The RJ at the input and output are measured in RMS.

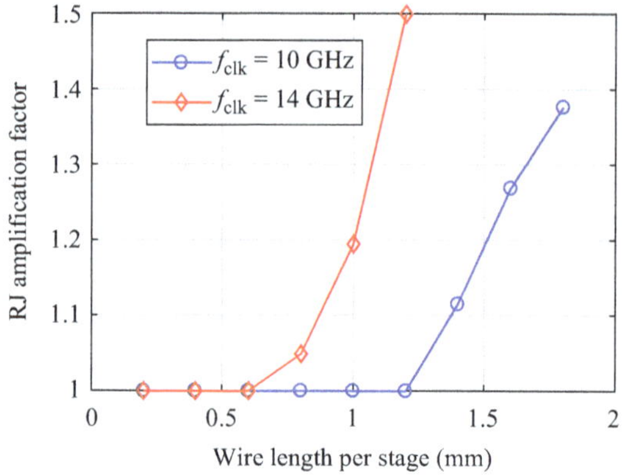

Figure 20.21 RJ amplification versus l_{stage}

end, so we next consider which presents a stricter limit on l_{stage}. When f_{clk} is low, the available settling time $T_{\text{clk}}/2$ is much larger than the input clock transition time, t_r. Thus as shown in Figure 20.20(b), jitter amplification occurs only for relatively large l_{stage} and t_{round}. However, in such cases even at shorter l_{stage}, the transition time of global clock waveforms being tapped by local buffers increases significantly. Thus, jitter generation increases significantly even before there is noticeable jitter amplification. This is the case for f_{clk} = 10 GHz, where jitter amplification occurs as $l_{stage} > 1.2$ mm as shown in Figure 20.21 whereas jitter generation at the transmission line near end degrades significantly even at $l_{stage} =$ 1 mm. However, at higher f_{clk} where $T_{\text{clk}}/2$ is comparable to the input clock transition time t_r, even a slight increase in the global clock transition times may cause it to approach the available settling time $T_{\text{clk}}/2$. For example, at f_{clk} = 14 GHz, $T_{\text{clk}}/2 \approx 30$ ps, and jitter amplification is incurred at $l_{\text{stage}} > 0.6$ mm, whereas jitter generation increases mildly.

Process and temperature variations may further limit l_{stage} due to their impact on buffer R_{drv}. In a slow process corner at high temperature, R_{drv} increases and clock waveforms may become RC limited, resulting in jitter amplification. For the same design, in a fast corner at low temperature, R_{drv} decreases and a large incident wave may result in overshoot. Such variations can be tracked and the strength of clock buffers tuned accordingly using either analog [31] and digital [32] means.

20.6 Summary

In this chapter, we provided background on three major jitter sources in high-performance CMOS clock distribution: power supply induced jitter (PSIJ), random jitter (RJ), and jitter amplification. We discussed how PSIJ is introduced in the CMOS inverter and its accumulation along a chain of buffers depending on the type of supply noise and its variation along the buffer chain. We also reviewed the analysis of RJ in the CMOS inverter. We described design tradeoffs to minimize both PSIJ and RJ in global clock distribution. We described linear models of jitter amplification, including the jitter impulse response (JIR) and jitter transfer function (JTF). Jitter amplification for buffers driving transmission-line interconnect was analyzed quantitatively, and simulations were used to develop insight. Design guidelines are also given for both cases. Finally, we discussed design considerations for jitter amplification in CMOS clock distribution. With the increasing use of CMOS circuits for high-performance clock distribution in advanced CMOS technologies, we believe the methods and guidelines in this chapter will prove ever more useful.

References

[1] Casper B, and O'Mahony F. Clocking analysis, implementation and measurement techniques for high-speed data links: a tutorial. *IEEE Trans. Circuits Syst. I: Regular Papers*. 2009;56(1):17–39.

[2] Upadhyaya P, Savoj J, An FT, *et al.* A 0.5-to-32.75 Gb/s flexible-reach wireline transceiver in 20 nm CMOS. In: *2015 IEEE International Solid-State Circuits Conference - (ISSCC) Digest of Technical Papers*; 2015. pp. 1–3.

[3] Kim J, Balankutty A, Dokania R, *et al.* A 112 Gb/s PAM-4 transmitter with 3-Tap FFE in 10 nm CMOS. In: *2018 IEEE International Solid - State Circuits Conference - (ISSCC)*; 2018. pp. 102–104.

[4] Shahramian S, Dehlaghi B, Beerkens R, *et al.* Chip-level power integrity methodology for high-speed serial links. DesignCon, San Francisco, CA, 2018, pp. 1–18.

[5] Schmitt R, and Lan H. Design and characterization of the power supply system for a high speed 1600 Mbps DDR3 interface in wirebond package. DesignCon, 2012, pp. 1–25.

[6] Savoj J, Hsieh K, Upadhyaya P, *et al.* Design of high-speed wireline transceivers for backplane communications in 28 nm CMOS. In: *Proceedings of the IEEE 2012 Custom Integrated Circuits Conference*; 2012. pp. 1–4.

[7] Mansuri M, Jaussi JE, Kennedy JT, *et al.* A scalable 0.128-1 Tb/s, 0.8-2.6 pJ/bit, 64-lane parallel I/O in 32-nm CMOS. *IEEE J. Solid-State Circuits*. 2013;48(12):3229–3242.

[8] Chen S, Zhou L, Zhuang I, *et al.* A 4-to-16 GHz inverter-based injection-locked quadrature clock generator with phase interpolators for multi-standard I/Os in 7 nm FinFET. In: *2018 IEEE International Solid - State Circuits Conference - (ISSCC)*; 2018. pp. 390–392.

[9] Poulton J, Palmer R, Fuller AM, *et al.* A 14-mW 6.25-Gb/s transceiver in 90-nm CMOS. *IEEE J. Solid-State Circuits*. 2007;42(12):2745–2757.

[10] Kocaman N, Ali T, Rao LP, *et al.* A 3.8 mW/Gbps quad-channel 8.5-13 Gbps serial link with a 5 tap DFE and a 4 tap transmit FFE in 28 nm CMOS. *IEEE J. Solid-State Circuits*. 2016;51(4):881–892.

[11] Wood J, Edwards TC, and Lipa S. Rotary traveling-wave oscillator arrays: a new clock technology. *IEEE J. Solid-State Circuits*. 2001;36(11):1654–1665.

[12] O'Mahony F, Yue CP, Horowitz MA, *et al.* A 10-GHz global clock distribution using coupled standing-wave oscillators. *IEEE J. Solid-State Circuits*. 2003;38(11):1813–1820.

[13] Frans Y, McLeod S, Hedayati H, *et al.* A 40-to-64 Gb/s NRZ transmitter with supply-regulated front-end in 16 nm FinFET. In: *2016 IEEE International Solid-State Circuits Conference (ISSCC)*; 2016. pp. 68–70.

[14] Menolfi C, Braendli M, Francese PA, *et al.* A 112 Gb/S 2.6 pJ/b 8-Tap FFE PAM-4 SST TX in 14 nm CMOS. In: *2018 IEEE International Solid - State Circuits Conference - (ISSCC)*; 2018. pp. 104–106.

[15] Erett M, Carey D, Hudner J, *et al.* A 126 mW 56 Gb/s NRZ wireline transceiver for synchronous short-reach applications in 16 nm FinFET. In: *2018 IEEE International Solid - State Circuits Conference - (ISSCC)*; 2018. pp. 274–276.

[16] Song YH, Yang HW, Li H, *et al.* An 8-16 Gb/s, 0.65-1.05 pJ/b, voltage-mode transmitter with analog impedance modulation equalization and sub-3 ns power-state transitioning. *IEEE J. Solid-State Circuits*. 2014;49(11): 2631–2643.

[17] Dickson TO, Liu Y, Rylov SV, *et al.* A 1.4 pJ/bit, power-scalable 16x12 Gb/s source-synchronous I/O with DFE receiver in 32 nm SOI CMOS technology. *IEEE J. Solid-State Circuits*. 2015;50(8):1917–1931.

[18] Sidiropoulos S, Liu D, Kim J, *et al.* Adaptive bandwidth DLLs and PLLs using regulated supply CMOS buffers. In: *2000 Symposium on VLSI Circuits. Digest of Technical Papers (Cat. No.00CH37103)*; 2000. pp. 124–127.

[19] Mansuri M, and Yang CKK. A low-power adaptive bandwidth PLL and clock buffer with supply-noise compensation. *IEEE J. Solid-State Circuits*. 2003;38(11):1804–1812.

[20] O'Mahony F, Mansuri M, Casper B, *et al.* A low-jitter PLL and repeaterless clock distribution network for a 20 Gb/s link. In: *2006 Symposium on VLSI Circuits, 2006. Digest of Technical Papers*; 2006. pp. 29–30.

[21] N D Dalt AS. *Understanding Jitter and Phase Noise*. Cambridge University Press; 2018.

[22] Chen LH, Marek-Sadowska M, and Brewer F. Buffer delay change in the presence of power and ground noise. *IEEE Trans. Very Large Scale Integr. Syst.* 2003;11(3):461–473.

[23] Figueiredo M, and Aguiar RL. Dynamic jitter accumulation in clock repeaters considering power and ground noise correlations. In: *2011 IEEE International Symposium of Circuits and Systems (ISCAS)*; 2011. pp. 2565–2568.

[24] Wong KL, Rahal-Arabi T, Ma M, *et al.* Enhancing microprocessor immunity to power supply noise with clock-data compensation. *IEEE J. Solid-State Circuits*. 2006;41(4):749–758.

[25] Abidi AA. Phase noise and jitter in CMOS ring oscillators. *IEEE J. Solid-State Circuits*. 2006;41(8):1803–1816.

[26] Bakoglu HB. *Circuits, Interconnection, and Packaging for VLSI.* Addison-Wesley; 1990.
[27] Chaudhuri S, Anderson W, Bryan J, *et al.* Jitter amplification characterization of passive clock channels at 6.4 and 9.6 Gb/s. In: *2006 IEEE Electrical Performance of Electronic Packaging*; 2006. pp. 21–24.
[28] Madden C, Chang S, Oh D, *et al.* Jitter amplification considerations for PCB clock channel design. In: *2007 IEEE Electrical Performance of Electronic Packaging*; 2007. pp. 135–138.
[29] Beyene W. Modeling and analysis techniques of jitter enhancement across high-speed interconnect systems. In: *2007 IEEE Electrical Performance of Electronic Packaging*; 2007. pp. 29–32.
[30] Ren J, and Oh KS. Multiple edge responses for fast and accurate system simulations. *IEEE Trans. Adv. Packaging.* 2008;31(4):741–748.
[31] Wilson JM, Turner WJ, Poulton JW, *et al.* A 1.17 pJ/b 25 Gb/s/pin ground-referenced single-ended serial link for off- and on-package communication in 16 nm CMOS using a process- and temperature-adaptive voltage regulator. In: *2018 IEEE International Solid - State Circuits Conference - (ISSCC)*; 2018. pp. 276–278.
[32] Mahfuzul IAKM, Tsuchiya A, Kobayashi K, *et al.* Variation-sensitive monitor circuits for estimation of Die-to-Die process variation. In: *2011 IEEE ICMTS International Conference on Microelectronic Test Structures*; 2011. pp. 153–157.

Part V

Advanced clock/frequency generation

Chapter 21
Sub-sampling PLL techniques

Xiang Gao[1], Eric Klumperink[2] and Bram Nauta[2]

The phase-locked-loop (PLL) is a ubiquitous component in modern ICs due to its versatility. It can, for instance, be used for clock generation, frequency synthesis, frequency modulation and demodulation, clock and data recovery, synchronization, and spread spectrum signal generation, in applications like high-performance analog-to-digital converters, wireline and optical transceivers, and wireless radio transceivers. Of the many known PLL architectures, the one shown in Figure 21.1(a) is perhaps the most widely used which we call the "classical PLL" architecture. It consists of a voltage-controlled oscillator (VCO) locked to a reference clock Ref by a feedback loop with the following "loop components": a phase detector (PD), a charge pump (CP), a loop filter (LF), and a frequency divider with ratio N (÷N). Jitter (or phase noise) and power are often the two critical performance matrix for many PLL applications. With the trend of higher data-rate, higher carrier frequency and higher order of modulation, the jitter or phase noise requirement becomes more and more demanding while the power budget is still limited. The PLL's jitter performance for a given power can be evaluated with the PLL jitter and power figure-of-merit (FOM) [1]. In the classical PLL, the PD and CP noise is multiplied by N^2 and often dominates the in-band phase noise, thus limiting the achievable PLL FOM.

The sub-sampling PLL (SSPLL) proposed in [2] uses a PD that sub-samples the high-frequency VCO output with the reference clock. With this architecture, it is shown that the PD and CP noise is greatly attenuated by the high phase-detection gain and it is possible to remove the N^2 noise amplification factor, leading to lower in-band phase noise and better PLL FOM. This chapter reviews the development of the sub-sampling PLL techniques and their applications in recent PLL architectures.

21.1 Classical charge pump PLL

A linear, phase domain model for the classical PLL and its noise are shown in Figure 21.1(b), where K_d is the PD/CP detection gain, $F_{LF}(s)$ the loop filter trans-impedance transfer function and K_{VCO} the VCO tuning gain. $1/f$ noise has been

[1]Institute of VLSI, Zhejiang University, Hangzhou, China
[2]EEMCS, University of Twente, Enschede, The Netherlands

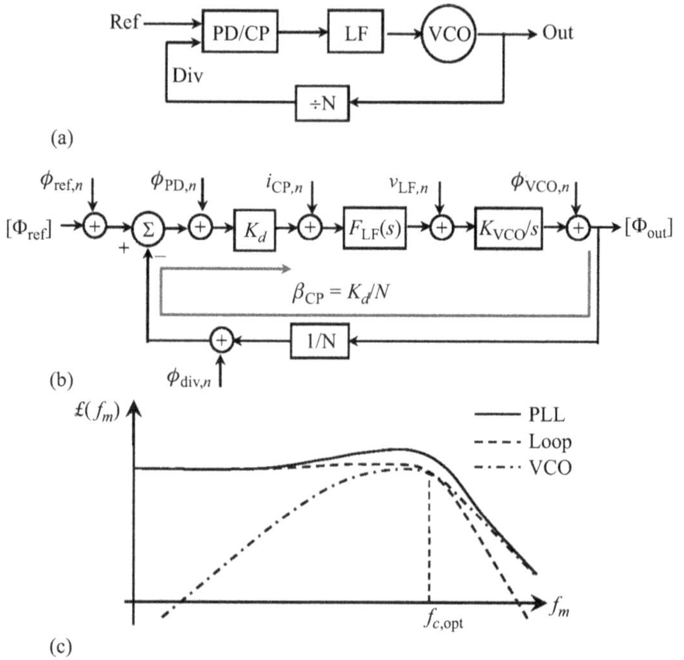

Figure 21.1 Classical PLL: (a) architecture, (b) phase domain model, and (c) phase noise spectrum (1/f noise neglected)

neglected to simplify the analysis so that we can focus on the fundamental limitations due to thermal noise.

Defining a CP feedback gain β_{CP} as the gain from the PLL output to the CP output current, the closed-loop CP noise transfer function can be calculated as:

$$H_{CP}(s) = \frac{1}{\beta_{CP}} \cdot \frac{\beta_{CP} \cdot F_{LF}(s) \cdot K_{VCO}/s}{1 + \beta_{CP} \cdot F_{LF}(s) \cdot K_{VCO}/s} = \frac{1}{\beta_{CP}} \cdot \frac{G(s)}{1 + G(s)} \tag{21.1}$$

where $G(s)$ is the PLL open-loop transfer function. Inside the PLL bandwidth, $G(s) \gg 1$ and the PLL in-band phase noise contributed by the CP can be approximated as:

$$\mathcal{L}_{\text{in-band,CP}} \approx \frac{1}{2} \cdot S_{iCP,n} \cdot |H_{CP}(s)|^2 \approx \frac{S_{iCP,n}}{2\beta_{CP}^2} \tag{21.2}$$

where the phase noise is expressed with the often used single sideband noise power to carrier power ratio $\mathcal{L}$ and $S_{iCP,n}$ is the power spectral density of the CP current noise.

Equation (21.2) indicates that the CP noise is suppressed by $(\beta_{CP})^2$ when transferred to the PLL output. A larger β_{CP} is thus desired for more CP noise suppression. In a classical PLL design, the three-state PFD/CP in Figure 21.2 is often used. The VCO output is firstly divided down so that the divider output Div

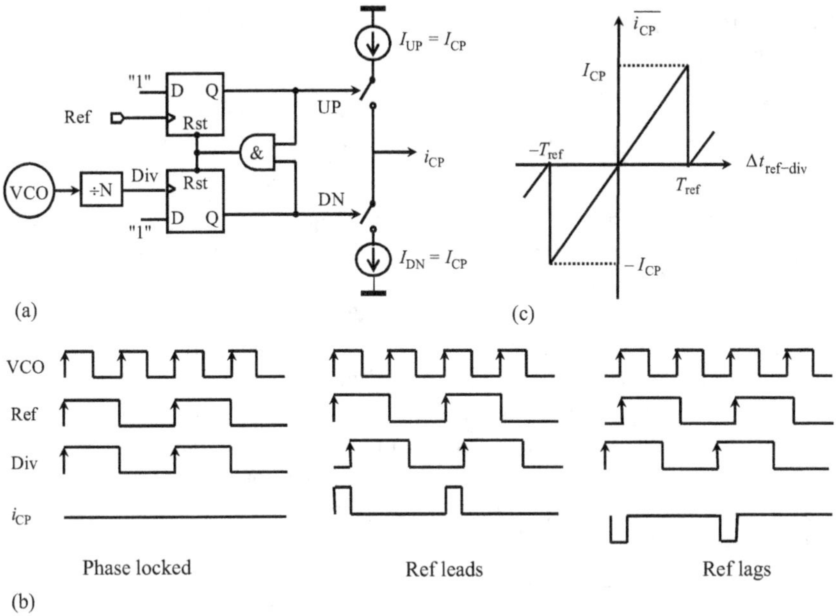

Figure 21.2 Three-state PFD/CP: (a) schematic, (b) timing diagram, and (c) characteristic

has similar frequency as Ref. The timing of Div and Ref are then compared and the CP outputs a current pulse with width equal to the amount of timing error. The CP feedback gain for a given CP bias current I_{CP} is:

$$\beta_{CP,PFD} = \frac{\Delta\overline{i_{CP}}}{\Delta\phi_{VCO}} = \frac{I_{CP}}{2\pi}\cdot\frac{1}{N} \tag{21.3}$$

We can see that $\beta_{CP,PFD}$ is reduced by N and that is why the CP noise (in power) is multiplied by N^2 in a classical PLL.

The reduction of $\beta_{CP,PFD}$ by N can be understood in the time domain. The VCO timing error Δt is directly transferred to the divider output with a gain of 1. The PFD/CP detects this error and outputs a current pulse with width Δt over one Ref period T_{ref}. If we increase N for the same f_{VCO}, f_{ref} becomes lower and T_{ref} larger. Consequently, the mean CP output current $I_{CP}\cdot\Delta t/T_{ref}$ corresponding to the same Δt becomes smaller, resulting in a lower $\beta_{CP,PFD}$.

21.2 PLL figure-of-merit

A benchmarking FOM that takes into account the fundamental tradeoffs between the key design parameters can be useful in comparing the relative merits of different designs and stimulating the development of power-efficient high-performance circuits. Two widely used examples are the ADC FOM and the VCO FOM. The VCO FOM is defined as [3]:

$$\mathrm{FOM}_{\mathrm{VCO}} = 10\log\left[\mathcal{L}_{\mathrm{VCO}}(f_m)\cdot\left(\frac{f_m}{f_{\mathrm{VCO}}}\right)^2\cdot\frac{P_{\mathrm{VCO}}}{1\ \mathrm{mW}}\right] \tag{21.4}$$

where $\mathcal{L}_{\mathrm{VCO}}(f_m)$ is the phase noise at an offset frequency f_m, f_{VCO} is the operation frequency, and P_{VCO} the power consumption. The unit of $\mathrm{FOM}_{\mathrm{VCO}}$ is dBc/Hz. A more negative $\mathrm{FOM}_{\mathrm{VCO}}$ corresponds to a better VCO design.

When transferred to the PLL output, the VCO noise is high-pass filtered and dominates the out-of-band phase noise. Another important part of the PLL phase noise is the noise from the loop components like reference clock buffer, PD, CP, and Divider. The loop noise is low pass filtered and dominates in-band. Assuming all the fundamentally required power for a timing circuit is dynamic (related to the switching events) and neglecting $1/f$ noise, [1] showed that the in-band loop noise at the PLL output $\mathcal{L}_{\mathrm{loop,in\text{-}band}}$ is proportional to the consumed power P_{loop} and the square of the output frequency f_{out}:

$$\mathcal{L}_{\mathrm{loop,in\text{-}band}} \propto N^2\cdot f_{\mathrm{ref}}\cdot\frac{f_{\mathrm{ref}}}{P_{\mathrm{loop}}} = \frac{f_{\mathrm{out}}^2}{P_{\mathrm{loop}}} \tag{21.5}$$

which is independent on f_{ref} because using a larger f_{ref} not only reduces the loop phase noise but also increases the dynamic power consumed by the loop. A FOM for PLL loop design is thus proposed in [1] as:

$$\mathrm{FOM}_{\mathrm{loop}} = 10\log\left[\mathcal{L}_{\mathrm{loop,in\text{-}band}}\cdot\left(\frac{1\ \mathrm{Hz}}{f_{\mathrm{out}}}\right)^2\cdot\frac{P_{\mathrm{loop}}}{1\ \mathrm{mW}}\right] \tag{21.6}$$

With (21.4) and (21.6), we have now separate FOMs for the loop and VCO, while we would like a combined PLL FOM. In many PLL applications, minimizing the total integrated phase noise or jitter is the design target, where a tradeoff exists between the VCO and loop noise filtering. It can be shown that an optimum loop bandwidth $f_{c,\mathrm{opt}}$ exists [1,4]. It is approximately where the spectrum density of the VCO and the loop noise intersect as shown in Figure 21.1(c). In that case, the VCO and the loop contribute equal jitter. For a given PLL power budget P_{PLL}, it would be best to spend it equally between the loop and VCO [1]: $P_{\mathrm{loop}} = P_{\mathrm{VCO}} = P_{\mathrm{PLL}}/2$. Under such optimized conditions, the minimum achievable jitter can be calculated as [1]:

$$\sigma_{t,\mathrm{PLL,min}}^2 = \frac{1}{(1\ \mathrm{Hz})^2}\cdot\frac{1\ \mathrm{mW}}{P_{\mathrm{PLL}}}\cdot\frac{1}{\pi}\cdot 10^{\frac{\mathrm{FOM}_{\mathrm{loop}}+\mathrm{FOM}_{\mathrm{VCO}}}{20}} \tag{21.7}$$

which says that for a given quality of VCO and loop design (constant $\mathrm{FOM}_{\mathrm{VCO}}$ and $\mathrm{FOM}_{\mathrm{loop}}$), the minimum achievable PLL jitter variance is inversely proportional to the consumed power. In [1], we thus proposed a PLL benchmarking FOM as:

$$\mathrm{FOM_{PLL}} = 10\log\left[\left(\frac{\sigma_{t,\mathrm{PLL}}}{1\,\mathrm{s}}\right)^2 \cdot \frac{P_{\mathrm{PLL}}}{1\,\mathrm{mW}}\right] \tag{21.8}$$

A smaller $\mathrm{FOM_{PLL}}$ corresponds to a better PLL design.

Comparing (21.7) and (21.8), we get:

$$\mathrm{FOM_{PLL}} = \frac{\mathrm{FOM_{loop}} + \mathrm{FOM_{VCO}}}{2} + 10\log\frac{1}{\pi} \tag{21.9}$$

Therefore, the design quality of the loop and VCO is equally important in improving PLL FOM. This is intuitive since the loop and the VCO have equal contribution to both power and jitter in an optimized PLL design. Figure 21.3 shows the FOM for PLL designs with integrated LC oscillator in recent years' literature.

Note that there is no f_{ref} and f_{VCO} involved in the PLL FOM definition because the derivations in [1] assumed the optimum case: all the loop power is dynamic (scale with f_{ref}) and is fundamentally needed to meet the noise performance. In reality, some power will not scale with f_{ref}, for example, the CP DC bias current. Also, noise performance is not always defining the required power. Some power may be needed to achieve sufficient speed, for example, in a high-frequency divider or clock buffer driving certain load. Therefore, in practice, it might be easier to

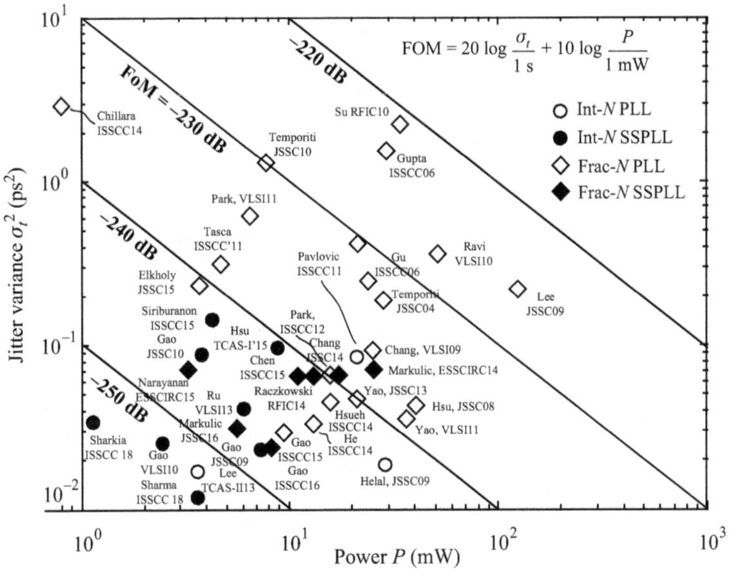

Figure 21.3 FOM of recent PLL designs with integrated LC oscillator

achieve a good FOM with a higher f_{ref} and a lower f_{VCO}, as the former helps to reduce the percentage of static power in the total power and the latter helps to reduce the high-speed related power, approaching closer to the optimum case. In other words, it will be harder to achieve the same FOM with a lower f_{ref} and a higher f_{VCO}. Therefore, while comparing different PLL designs, apart from comparing the FOM, it is also useful to list out the f_{ref} and f_{VCO} for a fair comparison, and give more applause to the design that achieves the same PLL FOM with a higher f_{VCO} and a lower f_{ref}.

21.3 Sub-sampling PLL

Figure 21.4 shows the conceptual and timing diagram of a sub-sampling PD/CP (SSPD/CP) [2]. The high frequency VCO output, assumed to be a sine wave with amplitude A_{VCO} and DC voltage $V_{DC,VCO}$, is sub-sampled by a low-frequency square wave reference clock Ref. When the VCO and Ref are phase aligned and their frequency ratio N is an integer, the sampling point would be the zero crossing (of the falling edge in this example) of the VCO and the sampled voltage V_{sam} has a constant value equal to $V_{DC,VCO}$. When there is phase error, V_{sam} will deviate from $V_{DC,VCO}$. If Ref leads, the sampled voltage would be higher than the $V_{DC,VCO}$. If Ref lags, the sampled voltage would be lower than the $V_{DC,VCO}$. The timing error Δt between VCO and Ref is thus translated to voltage difference between V_{sam} and $V_{DC,VCO}$. A transconductor g_m can be used to convert the voltage difference $V_{sam} - V_{DC,VCO}$ into current, so that a traditional current-driven loop-filter can still be

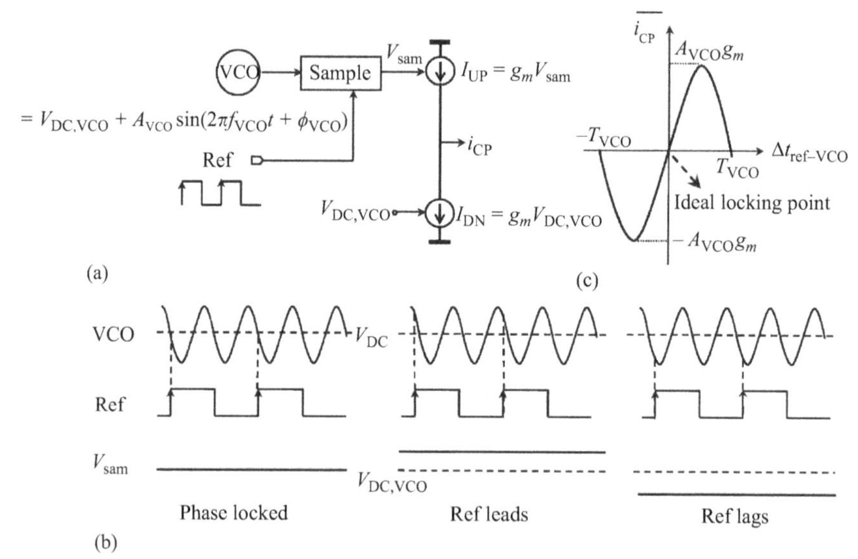

Figure 21.4 Sub-sampling PD/CP: (a) conceptual diagram, (b) timing diagram, and (c) characteristic

used. This g_m can be implemented in a time-continuous way, different from a duty-cycled CP. Here we still call it "CP" to simplify the notation, and implementing the gm as a real CP is actually useful for gain control as we will see later.

In contrast to a traditional CP, the output current is not proportional to $\Delta t/T_{\text{ref}}$, but rather amplitude controlled by the difference of V_{sam} and $V_{\text{DC,VCO}}$. The SSPD/CP transfer characteristic has the same shape as the waveform that is sampled, see Figure 21.4(c). The ideal locking point is the zero crossing where the SSPD/CP gain can be calculated as:

$$\beta_{\text{CP,SS}} = \frac{\Delta \overline{i_{\text{CP}}}}{\Delta \phi_{\text{VCO}}} = \frac{\text{SR}_{\text{sam}}}{2\pi f_{\text{VCO}}} \cdot g_m \tag{21.10}$$

with SR_{sam} the slew rate of the waveform to be sampled at the zero crossing locking point. In the case of LC VCO, $\text{SR}_{\text{sam}} = A_{\text{VCO}} \cdot 2\pi f_{\text{VCO}}$, usually a well-defined value since f_{VCO} is known and VCO amplitude calibration over corners is often performed in practice. We can thus re-write (21.10) as:

$$\beta_{\text{CP,SS}} = A_{\text{VCO}} \cdot g_m = A_{\text{VCO}} \cdot \frac{2I_{\text{CP}}}{V_{\text{gs,eff}}} \tag{21.11}$$

where g_m is assumed to be implemented with a single square-law MOS transistor in saturation and $V_{\text{gs,eff}}$ is the effective gate-source voltage of the transistor. There is no N in (21.11), which means the CP noise in a SSPLL is not multiplied by N^2 when transferred to the output. This is because the phase-detection gain of the SSPD is determined by SR_{sam}. If we fix f_{VCO}, the change in f_{ref} or N only impacts how frequent the sampling would happen but does not impact SR_{sam} and the detection gain.

Viewed from yet another angle, the N factor difference between the SSPLL and the classical PLL can be understood if we look at the PLL as a simple loop-back transceiver. The "signal" is now the VCO phase noise and the function of the PLL loop is to receive this signal, process it and apply it to the VCO to cancel/suppress the VCO phase noise. In a SSPLL, as shown in Figure 21.5(a), the SSPD with an LO (Ref clock) acts as a down converter that directly aliases back the VCO phase noise to around DC. The loop filter acts as a baseband circuit that processes the received signal and applies it to the VCO input. In other words, the SSPLL is analogous to a direct conversion receiver. The SSPD down-converter has no loss but a gain of 1. Therefore, there is no amplification to the SSPD/CP noise. The aliasing of high-frequency VCO noise to low frequency is not a problem because the VCO noise has a steep roll-off. In a classical PLL, the receiver chain consists of a divider and a PFD/CP as shown in Figure 21.5(b). The divider firstly down converts the signal to an intermediate frequency f_{VCO}/N. The PFD/CP together with the Ref clock acts as a second down converter and converts the signal to around DC. In other words, the classical PLL is analogous to a superheterodyne receiver. The divider acts like a very lossy down-converter with a gain of $1/N$, and thus a noise figure of $20 \log N$ even if it has no noise. As a result, any noise in the receiver chain after the divider, like the PFD/CP noise, is amplified by N^2, when referred to the PLL output.

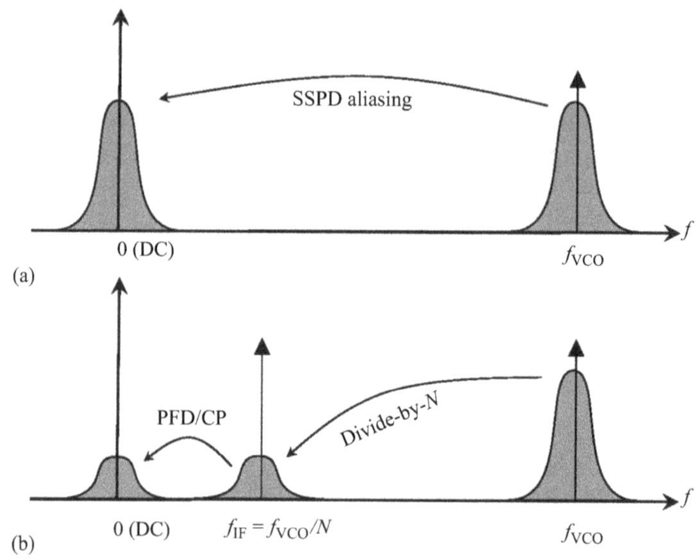

Figure 21.5 PLL's analogy to a loop-back transceiver: (a) SSPLL with a direct conversion receiver and (b) classical PLL with a superheterodyne receiver

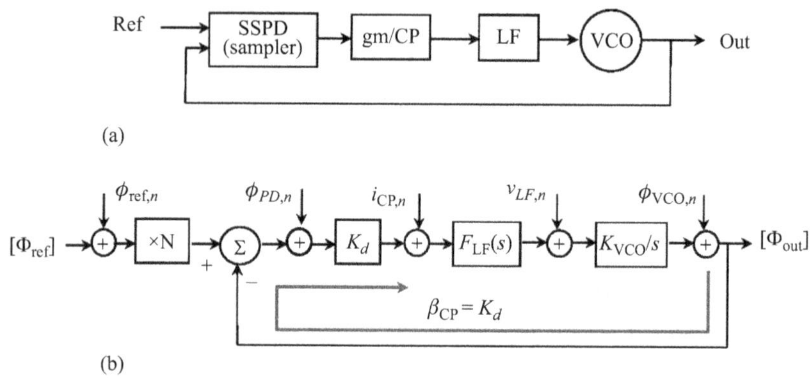

Figure 21.6 SSPLL (a) architecture and (b) phase domain model

The architecture of an SSPLL utilizing the SSPD/CP is shown in Figure 21.6(a). It works without using a divider as soon as the ratio f_{VCO}/f_{ref} is an integer. A linear phase-domain model for the SSPLL is shown in Figure 21.6(b). There is no classical divide-by-N on the feedback path but instead a virtual frequency multiplier "$\times N$" on the reference clock path. This is because the sub-sampling process aliases back the VCO with the closest $N \cdot f_{ref}$ so it works as if the VCO is sampled by a signal with frequency N times higher than Ref. Note that although there is no N^2 factor for

SSPD/CP noise, the reference clock noise in an SSPLL is still multiplied by N^2 when transferred to the output, just as in a classical PLL. This is because the SSPD still relies on the timing of the Ref sampling edge to define the sampling point and a given timing error still corresponds to N times more phase error if we refer it to f_{VCO} instead of f_{ref}.

It is instrumental to compare the β_{CP} between the classical PFD/CP PLL and the SSPLL using (21.3) and (21.11):

$$\frac{\beta_{CP,SS}}{\beta_{CP,PFD}} = N \cdot 4\pi \cdot \frac{A_{VCO}}{V_{gs,eff}} \tag{21.12}$$

The advantage of the SSLL in terms of the detection gain can be more than just a factor of N, as $4\pi \gg 1$ and usually $A_{VCO} > V_{gs,eff}$. Thus, the SSPLL has a much larger β_{CP} than the classical PLL and much more suppression for the CP noise.

Thanks to the high β_{CP}, the CP noise in an SSPLL is greatly suppressed and would have negligible contribution to the total loop noise. In such a case, having an "unnecessarily high" β_{CP} wouldn't further improve the loop noise but does require a large filter capacitor to stabilize the loop. Figure 21.7 shows the SSPD/CP with gain control [2]. Differential VCO outputs and differential sampling are used so the differential zero-crossing can naturally be the locking point. Two switches and a block called "Pulser" are added to the CP. The Pulser generates a pulse Pul, non-overlapping with Ref, and switches on/off I_{UP} and I_{DN} simultaneously for a fraction of time τ_{pul} in each T_{ref}. The mean CP output current and thus $\beta_{CP,SS}$ is reduced. Note that switching on the CP only for a fraction of time also reduces CP noise. The overall $\beta_{CP,SS}$ and the in-band phase noise due to CP can be calculated as [2]:

$$\beta_{CP,SS} = 2A_{VCO} \cdot g_m \cdot \frac{\tau_{pul}}{T_{ref}} \tag{21.13}$$

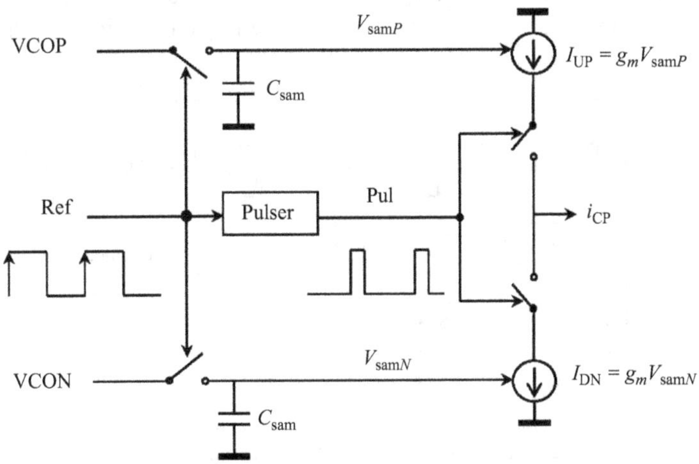

Figure 21.7 Sub-sampling PD/CP with pulse width gain control

$$\mathcal{L}_{\text{in-band,CP,SS}} = \frac{kT\gamma}{A_{\text{VCO}}^2 \cdot g_m} \cdot \frac{T_{\text{ref}}}{\tau_{\text{pul}}} \tag{21.14}$$

By a careful choice of $\tau_{\text{pul}}/T_{\text{ref}}$, the value of $\beta_{\text{CP,SS}}$ will not be "unnecessarily high" but just high enough to keep the CP as a negligible contributor to the total loop noise. Note that there are other ways of reducing $\beta_{\text{CP,SS}}$ as well, for example, reducing the sampled voltage slope, reducing gm by proper sizing/biasing or by adding degeneration. However, the advantage of the Pulser is that it also functions as the slave track-and-hold for the VCO sampling so the SSPD can be a simple switch-and-cap as shown in Figure 21.7. Also by tuning the pulse width, β_{CP} can be tuned in a wide range without affecting the SSPD/CP operation points.

The in-band phase noise due to the simple switch-and-cap SSPD in Figure 21.7 can be derived as [5]:

$$\mathcal{L}_{\text{in-band,SSPD}} = 10 \log \frac{kT}{2C_{\text{sam}} \cdot A_{\text{VCO}}^2 \cdot f_{\text{ref}}} \tag{21.15}$$

Assuming $f_{\text{ref}} = 40$ MHz and $A_{\text{VCO}} = 0.5$ V, a small C_{sam} of 5 fF is enough to keep the SSPD noise at a very low level of −134 dBc/Hz at VCO output. Interestingly, the noise level can be *independent* on f_{VCO} as the scaling of phase noise and the scaling of SSPD gain (defined by VCO slope) with f_{VCO} cancel out.

Figure 21.8 shows the overall architecture of the SSPLL proposed in [2]. The core loop consists of an SSPD/CP with a Pulser, a passive loop filter and a VCO. Since the sub-sampling process cannot distinguish between $N \cdot f_{\text{ref}}$ and other harmonics of f_{ref}, the SSPLL may lock to any integer N as long as $N \cdot f_{\text{ref}}$ fits into the VCO tuning range. A classical PLL with divider is thus used as a frequency-locked-loop (FLL) to ensure correct locking. The key for the FLL is to dominate the loop control when the phase/frequency error is large but avoid adding noise once the phase is locked. This can be achieved, for example, by intentionally adding a large dead zone (DZ) to the FLL PFD/CP (see [2] for the implementation), so that it will

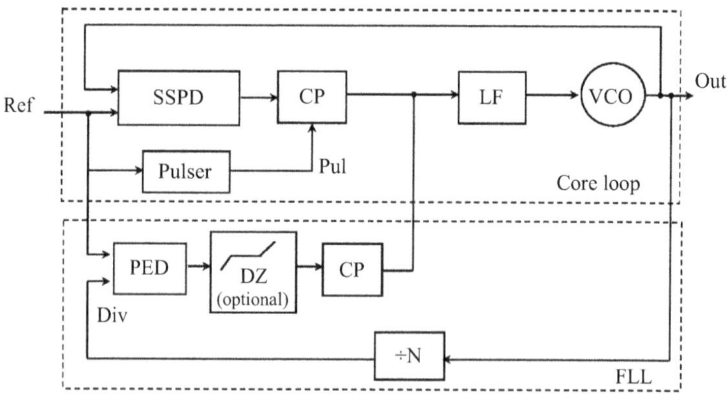

Figure 21.8 Overall SSPLL architecture

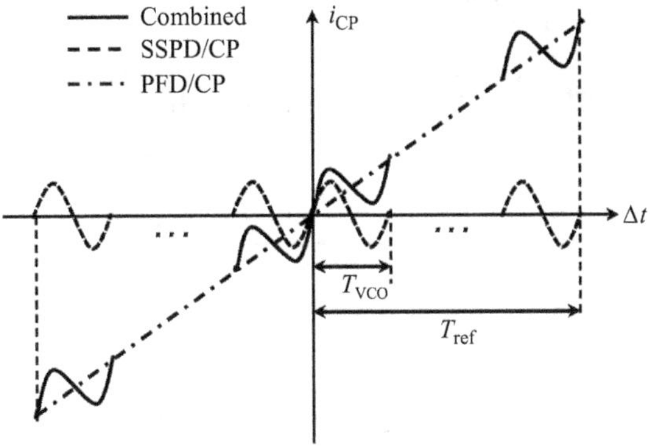

Figure 21.9 Combined SSPD/CP and PFD/CP characteristic

inject no current once the phase error is small and fall within the DZ. However, the work in [6] shows that Figure 21.8 would also work with no DZ because around the locking point $\beta_{CP,SS}$ can be much larger than $\beta_{CP,PFD}$ anyway. The overall characteristic of the combined SSPD/CP and PFD/CP is shown in Figure 21.9. With no DZ, the PFD/CP in the FLL will inject noise even in the locked condition, but its contribution can be small as it is attenuated by $(\beta_{CP,SS} + \beta_{CP,PFD})^2$. After locking, the FLL can be disabled to save power or it can remain on to constantly monitor the phase/frequency error and improve the SSPLL's robustness against disturbances [6].

21.4 Power and spur reduction techniques for sub-sampling PLL

In an SSPLL, the PD and CP noise contributions are low and thus their power can be scaled down progressively. The VCO sampling buffer and Ref buffer for the SSPD could easily become the dominant source of loop noise as well as power. In [2], they respectively account for 30% and 60% of the total loop power.

In most applications, Ref is derived from a low frequency 10s-of-MHz sine wave crystal oscillator (XO). To properly sample the high frequency VCO, Ref needs to have a sharp edge. A Ref buffer converting the sine XO output into a square wave Ref is thus needed. When a CMOS inverter is used as a sine-to-square buffer in the 10s-of-MHz frequency range, a majority of power, as much as 90% [2], could be wasted by the "short-circuit" current due to simultaneous conduction of the NMOS and PMOS during switching. Note that this sine-to-square buffer is needed in a classical PLL too because a three-state PFD uses the Ref to trigger a DFF which also requires the Ref to have a sharp edge.

Figure 21.10 shows a Ref buffer design [5] that can largely eliminate the short-circuit current and drastically reduce the buffer power. The core is an inverter with an NMOS N1 and a PMOS P1. N1 is directly connected to XO while a timing control circuit (TCC) is inserted between P1 and XO. The TCC generates a narrow

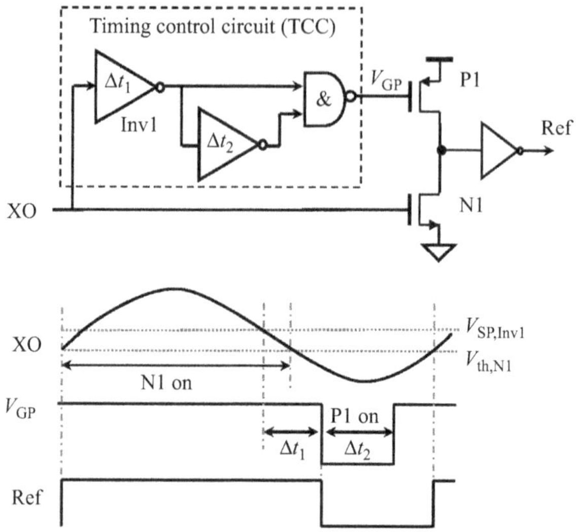

Figure 21.10 Schematic and timing diagram of a low power sine-to-square Ref buffer

pulse V_{GP} from the XO and controls the gate of P1. The delay Δt_1 and Δt_2 are set such that the time when V_{GP} is low (P1 conducts) and the time when XO is high (N1 conducts) are nonoverlapping. Since f_{ref} is low, this timing plan can be easily met. In this way, N1 and P1 will not conduct simultaneously thereby eliminating the short-circuit current. Moreover, for low noise sampling, only the Ref sampling edge (rising edge in this example) needs to be cleaned while the other edge's noise is not relevant. Therefore, N1 can be sized big to maintain low noise, while the P1 and TCC can be sized small to save power. This buffer thus greatly reduces power while maintaining the critical edge's noise performance. It also offers the flexibility of tuning the Ref duty cycle without impacting the critical edge.

One straightforward way of reducing the VCO sampling buffer power is to do buffer-less direct VCO sampling as shown in Figure 21.11. Then the sampling buffer power is simply eliminated. However, the concern is the disturbance of the sampling process to the VCO operation, causing reference spurs. Different spur mechanisms can be identified, namely charge injection, charge sharing, and VCO load/frequency modulation when the sampling switch is turned on/off. The load modulation can be alleviated by adding a complementary switched dummy sampler as shown in Figure 21.11, so that the VCO is always loaded by one C_{sam}. The dummy sampler also helps to cancel the charge injection from the sampling switches to the VCO. The other spur mechanism, the VCO-C_{sam} charge sharing needs more effort to deal with. Charge sharing happens when $V_{sam,on!}$, the voltages on C_{sam} and $V_{VCO,on!}$, the voltage on VCO tank capacitor at the switch-on moment, are not equal. As shown in Figure 21.12, the Ref rising edge defines the switch-off moment where holding starts and voltage is sampled and Ref falling edge defines

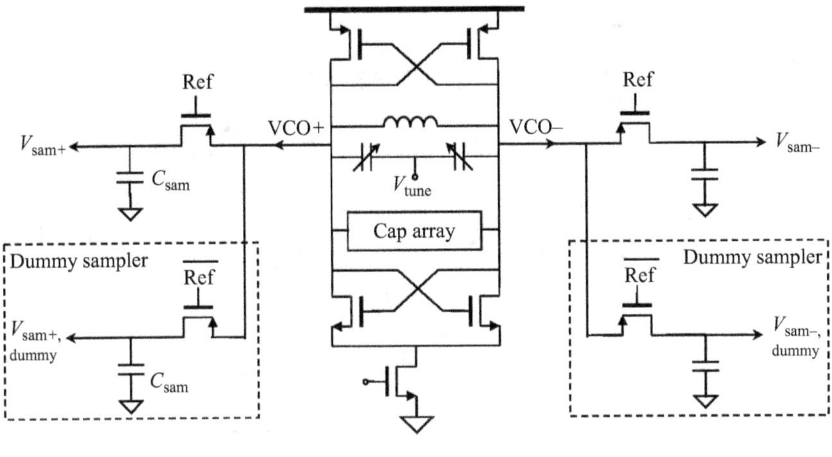

Figure 21.11 Buffer-less direct VCO sampling with dummy sampler

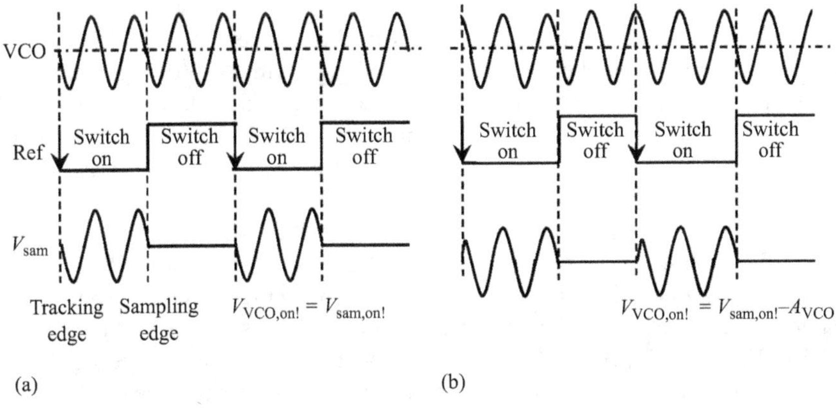

Figure 21.12 Conceptual illustration of (a) the case of minimum charge sharing and (b) the case of maximum charge sharing

the switch-on moment where tracking starts. After the PLL is locked, the sampling edge is phase-locked to a VCO zero-crossing. $V_{sam,on!}$ is well-defined and equal to $V_{VCO,DC}$. In contrast, $V_{VCO,on!}$ depends on the position of the Ref tracking edge which could be anywhere on the VCO waveform.

Apparently, it is desirable to phase lock the tracking edge to VCO zero-crossings as well. Then we will have $V_{VCO,on!} = V_{sam,on!} = V_{VCO,DC}$ and hence no VCO-C_{sam} charge sharing as shown in Figure 21.12(a). To this end, [7] proposed a low spur SSPLL architecture as in Figure 21.13. On top of the SSPLL in Figure 21.8, a sub-sampling DLL (SSDLL) is added which reuses the dummy sampler with $\overline{\text{Ref}}$ as its sampling clock, and reuses the low power buffer in

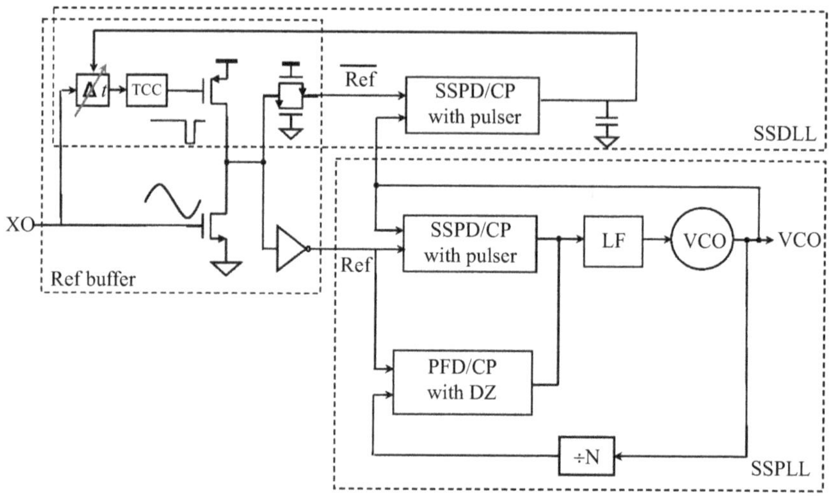

Figure 21.13 Low spur SSPLL architecture

Figure 21.10 to independently tune the Ref tracking edge. Once the loops settle, both the Ref sampling and tracking edges are aligned with the VCO zero-crossings and the condition for no VCO-C_{sam} charge sharing is achieved. Since the SSDLL tuning only affects the timing of Ref tracking edge which is the noncritical edge for the SSPLL, it will neither disturb the SSPLL operation nor add noise to the SSPLL output.

With the aforementioned techniques, the SSPLL design in [5] achieved -125 dBc/Hz in-band phase noise at 200 kHz offset, with $f_{ref} = 55$ MHz and $f_{VCO} = 2.2$ GHz. The loop only consumes 0.7 mW and the reference spur is kept below to be clean 56 dBc. The total integrated PLL jitter is 0.16 ps and total power is 2.5 mW. Its -252 dB FOM is still one of the best as shown in Figure 21.3.

For a particular design, the achievable spur level with buffer-less VCO sampling will be dependent on the matching between the sampler and its dummy and also the ratio between C_{sam} and VCO tank capacitor. When lower reference spur is needed, one can always add isolation buffers between the VCO and the samplers [7]. Note that the loop filter renders no filtering for the spur due to sampling so there is no tradeoff between low (sampling caused) spur and PLL bandwidth. The spur due to CP ripple of course still goes through the loop filter. It is a very important source of spurs in a classical PLL and often caused by UP/DN current mismatch in the three-state PFD/CP. However, for an SSPD/CP, the current mismatch does not add to CP ripple but merely shifts the locking point away from the VCO zero crossing as shown in Figure 21.14(a). The SSPLL can thus achieve low CP ripple without requiring good CP matching. The SSPLL in [7] demonstrated that <-80 dBc reference spur is achievable with a large loop-bandwidth-to-reference-frequency ratio of 1/20 and a simple CP design as shown in Figure 21.14(b).

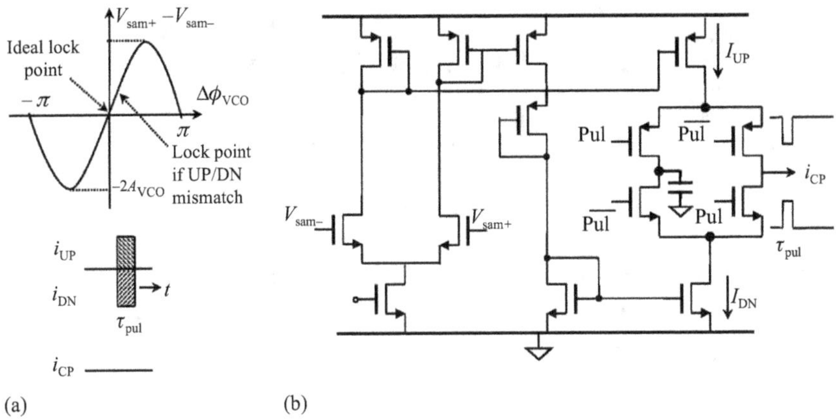

Figure 21.14 (a) Behavior of SSPD/CP with UP/DN mismatch and (b) simple low ripple CP design

21.5 Fractional-*N* sub-sampling PLL

So far the sub-sampling PLL architecture discussed is limited to integer-*N* operation, where the steady-state sampling point is always around the zero crossings. A PLL that is able to work in fractional-*N* mode is apparently more versatile. Fractional-*N* PLL is an essential part of many applications, for example, wireless communication circuitry where the channel spacing is often much smaller than f_{ref}. However, the inherent nature of the sub-sampling phase detector prohibits fractional-*N* operation. If *N* is fractional, it is well known that there will be quantization noise. As shown in Figure 21.15(a), the sampling point of the SSPD can take any value on the VCO waveform even in the locked state. The linearity of the SSPD/CP is then a serious concern. If the sampling point is at the peak or bottom of the sine-wave, the gain would even be zero.

One straight forward method to exploit the SSPLL in fractional-*N* mode is to cascade an integer-*N* SSPLL with a classical fractional-*N* CP PLL [8,9], as shown in Figure 21.16. The classical PLL uses a multimodulus divider (MMDIV) modulated by a delta-sigma modulator (DSM) on the feedback path to perform the fractional-*N* operation. In this way, the overall system is fractional-*N* while the SSPLL is kept in the integer-*N* mode. The cost is the additional area, noise, and power due to the second PLL. However, the fractional-*N* CP PLL design can be more relaxed from the noise perspective because its reference clock is now not the low-frequency crystal oscillator but the output or divided output of the first SSPLL, which can be chosen to be at least a few hundred MHz, much higher than f_{ref}, therefore reducing the fractional-*N* CP PLL's quantization noise and in-band phase noise. Note that the order of SSPLL and fractional-*N* CP PLL can also be swapped and we can put the SSPLL as the second PLL [9]. The advantage of this approach is

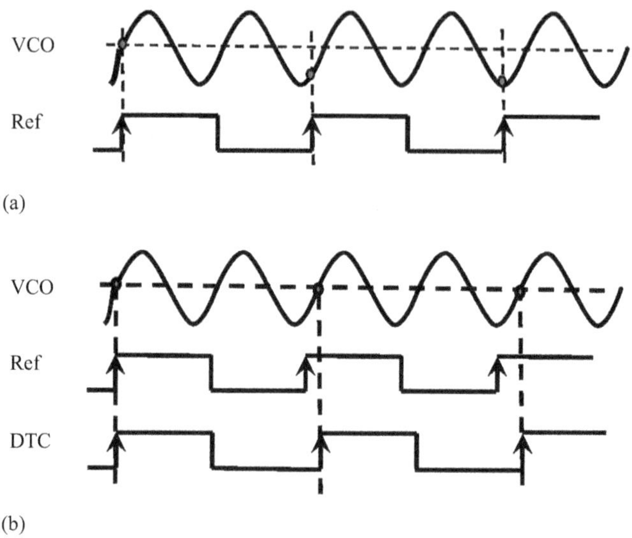

Figure 21.15 (a) Sub-sampling action if N is fractional and (b) fractional-N sub-sampling with the assistance of a DTC

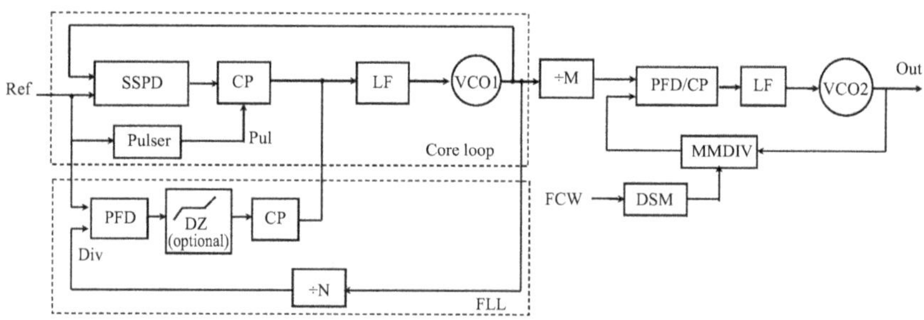

Figure 21.16 Cascading sub-sampling PLL and classical fractional-N CP PLL

that the SSPD/CP can work with a higher f_{ref} than the PFD/CP. In the example of [9], the second stage SSPLL's input is running at 2.8 GHz.

Another alternative is adding a DSM modulated digital-to-time-converter (DTC) to the reference clock path [10–12] to cancel the quantization noise. The key principle is to predict or extract the quantization noise using the DSM and apply it to the DTC to cancel the quantization noise, as shown in Figure 21.15(b). This effectively adds a fractional-N multiplier before the SSPD/CP. As a result, the core SSPLL loop can maintain its integer-N mode operation so that the sampling point would still be around zero crossings. Fractional-N SSPLLs using these concepts have achieved good FOM among fractional-N PLLs as shown in Figure 21.3. The development of very linear DTCs [12,13] should help to further improve the performance.

The third way to achieve fractional-N SSPLL is to add a phase interpolator (PI) on the feedback path [14]. The PI reduces the quantization noise and limits the required linear range on the SSPD/CP. The introduction of an n-bit PI relaxes the dynamic range requirement of the DTC by a factor of $1/2^n$. Given enough number of PI bits, the quantization noise would be small enough and the sampling point would still be around zero crossings in the locked state. Fractional-N SSPLL using this method has also demonstrated excellent performance and achieved a PLL FOM of −246 dB [14]:

$$L_{\text{in-band,TDCQN}} = \frac{(\Delta t_{\text{TDC}})^2}{12} \times \frac{1}{f_{\text{ref}}} \times (2\pi f_{\text{out}})^2$$

21.6 Digital sub-sampling PLL

Digital PLL (DPLL) [15] has been a hot research area for many years. Their advantages include flexible loop bandwidth setting at low cost and small area, utilizing digital calibration to compensate for analog circuit nonidealities, accurately canceling DSM quantization noise in digital domain, and easier porting between technologies. The basic digital PLL architecture is shown in Figure 21.17(a). Compared with the analog counterpart, the PFD/CP is replaced with a time-to-digital converter (TDC), the analog LF with a digital LF and the VCO with a digital-controlled-oscillator (DCO). The TDC is a key component to enable the DPLL. The function of the TDC is to convert timing error into digital code. Figure 21.17(b) shows a basic TDC architecture implemented with buffer delays and DFFs. Similar to

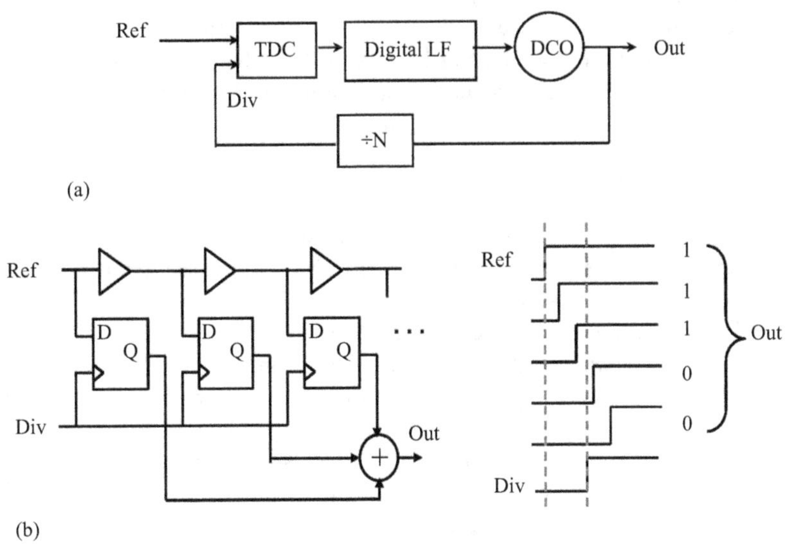

Figure 21.17 (a) Basic digital PLL architecture and (b) basic TDC design

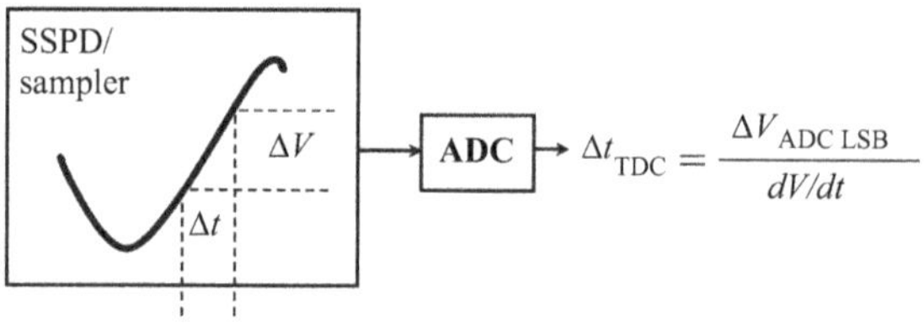

Figure 21.18 The principle of sub-sampling TDC

an ADC, the TDC has limited resolution Δt_{TDC}. It introduces quantization noise and would contribute to PLL in-band phase noise in the following way:

$$\mathcal{L}_{\text{in-band,TDCQN}} = \frac{(\Delta t_{\mathrm{TDC}})^2}{12} \times \frac{1}{f_{\mathrm{ref}}} \times (2\pi f_{\mathrm{DCO}})^2 \tag{21.16}$$

Apparently, the TDC need to have a high resolution or small Δt_{TDC} for low DPLL in-band phase noise. For the basic TDC design in Figure 21.17(b), the TDC resolution is limited by the buffer delay, which is on the order of 10 ps even in technology like 28 nm CMOS. If we use a 40 MHz reference clock and aiming at generating a 5 GHz output, the in-band phase noise according to (21.16) would be worse than -100 dBc/Hz.

The SSPD converts the timing error between the VCO and Ref into a sampled voltage error with a very high gain. One can think that if we digitize the sampled voltage with an ADC, it will effectively become a sub-sampling time-to-digital converter (SSTDC) [16], as shown in Figure 21.18. Simple calculation shows that for a 2 GHz LC VCO with $A_{\mathrm{VCO}} = 0.5$ V and a 10-bit 1-V full scale ADC, the SSTDC resolution would be about 0.16 ps, nearly two orders of magnitude finer than the gate delay. A digital SSPLL with SSTDC has been demonstrated by [17,18,29]. The ADC can also be just 1-bit, like the digital SSPLL design in [19].

21.7 Discussion

One key feature of the SSPD is that it can realize very high phase-detection gain with low power. This is achieved by converting timing/phase error into voltage error by sampling a high slew rate voltage slope, directly from a high-frequency VCO or VCO followed by a slope generator [20]. The waveform to be sampled is not necessary a sine-wave and the sub-sampling technique can also be applied to, for example, ring oscillators [21,22]. In a more advanced process, steeper voltage slopes will be available and the simple switch-cap SSPD will be able to sample faster, thus benefiting from the technology scaling. SSPLLs with SSPD sampling at 10s-of-GHz have been demonstrated in [23] and [24].

In an SSPLL, the SSPD/CP noise is greatly suppressed by the high detection gain so the SSPD/CP design can be simple and very power efficient. The FLL and divider can be turned off in the locked state to save power. Even the VCO sampling buffer power can be eliminated by doing direct VCO sampling. The only remaining loop component is the simple Ref buffer, which is needed in all PLLs. The SSPLL can thus approach "PLL Utopia", where only the Ref path contributes to loop noise and power. Another type of PLL that can approximate this state is the injection-locked PLL (ILPLL). Indeed as shown in Figure 21.3, the current PLL FOM record in literature is −254 dB, held by SSPLLs in [25] and [26], to the best of the authors' knowledge. Another type of PLL that can approximate this state is the injection-locked PLL (ILPLL) [27,28]. Compared with the "open-loop" ILPLL, the SSPLL is more like a traditional PLL with a phase detector and a feedback loop. It can have more robust bandwidth/loop-dynamic control and lower reference spur. However, the ILPLL can achieve a larger filtering bandwidth for VCO noise and can be beneficial, for example, when a ring oscillator is used. Lastly, [27,28] showed that these two techniques can nicely work together with an SSPLL assisting an ILPLL in injection timing control.

References

[1] Gao X., Klumperink E., Geraedts P.J.F., and Nauta B. "Jitter analysis and a benchmarking figure-of-merit for phase-locked loops." *IEEE Trans. Circuits Syst. II*. 2009, vol. 56, no. 2, pp. 117–121.

[2] Gao X., Klumperink E., Bohsali M., and Nauta B. "A low noise sub-sampling PLL in which divider noise is eliminated and PD/CP noise is not multiplied by N^2." *IEEE J. Solid-State Circuits*. 2009, vol. 44, no. 12, pp. 3253–3263.

[3] Kinget P. "Integrated GHz voltage controlled oscillators." In: *Analog Circuit Design: (X)DSL and Other Communication Systems; RF MOST Models; Integrated Filters and Oscillators*. Kluwer, 1999, pp. 353–381.

[4] Vaucher C.S. *Architectures for RF Frequency Synthesizers*. Boston, MA: Kluwer, 2002.

[5] Gao X., Klumperink E., Socci G., Bohsali M., and Nauta B. "A 2.2GHz sub-sampling PLL with 0.16ps_{rms} jitter and -125dBc/Hz in-band phase noise at 700μW loop-components power." *IEEE Symposium on VLSI Circuits*, 2010, pp. 139–140.

[6] Hsu C.-W., Tripurari K., Yu S.-A., and Kinget P.R. "A sub-sampling-assisted phase-frequency detector for low-noise PLLs with robust operation under supply interference." *IEEE Trans. Circuits Syst. I*. 2015, vol. 62, no. 1, pp. 90–99.

[7] Gao X., Klumperink E., Socci G., Bohsali M., and Nauta B. "Spur reduction techniques for phase-locked loops exploiting a sub-sampling phase detector." *IEEE J. Solid-State Circuits*. 2010, vol. 45, no. 9, pp. 1809–1821.

[8] Park D., and Cho S., "A 14.2 mW 2.55-to-3 GHz cascaded PLL with reference injection, 800 MHz delta-sigma modulator and 255fs integrated jitter in 0.13 m CMOS." *IEEE Solid-State Circuits Conference (ISSCC)*, 2012, pp. 344–345.

[9] Yi X., Liang Z., Feng G., Boon C.C., and Meng F. "A 93.4-to-104.8 GHz 57 mW fractional-N cascaded sub-sampling PLL with true in-phase injection-coupled QVCO in 65 nm CMOS." *IEEE Radio Frequency Integrated Circuits Symposium (RFIC)*, 2016, pp. 122–125.

[10] Raczkowski K., Markulic N., Hershberg B., Van Driessche J., and Craninckx J. "A 9.2–12.7 GHz wideband fractional-N subsampling PLL sin 28nm CMOS with 280fs RMS jitter." *IEEE Radio Frequency Integrated Circuits Symposium (RFIC)*, 2014, pp. 89–92.

[11] Chang W.-S., Huang P.-C., and Lee T.-C. "A fractional-N divider-less phase-locked loop with a subsampling phase detector." *IEEE J. Solid-State Circuits*. 2014, vol. 49, no. 12, pp. 2964–2975.

[12] Markulic N., Raczkowski K., Wambacq P., and Craninckx J. "A 10-bit, 550-fs step digital-to-time converter in 28nm CMOS." *IEEE European Solid State Circuits Conference (ESSCIRC)*, 2014, pp. 79–82.

[13] Ru Z., Palattella C., Geraedts P., Klumperink E., and Nauta B. "A high-linearity digital-to-time converter technique: constant-slope charging." *IEEE J. Solid-State Circuits*, 2015, vol. 50, no. 6, pp. 1412–1423.

[14] Narayanan A.T., Katsuragi, M., Kimura, K. *et al.* "A fractional-N sub-sampling PLL using a pipelined phase-interpolator with a FoM of -246dB." *IEEE Radio Frequency Integrated Circuits Symposium (RFIC)*, 2015, pp. 380–383.

[15] R.B. Staszewski, J.L. Wallberg, S. Rezeq, *et al.* "All-Digital PLL and Transmitter for Mobile Phones." *IEEE J. Solid-State Circuits*. 2005, vol. 40, no. 12, pp. 2469–2482.

[16] Gao X. "Low jitter low power phase locked loops using sub-sampling." PhD thesis, University of Twente, 2010.

[17] Siriburanon T., Kondo, S., Kimura, K. *et al.* "A 2.2GHz −242dB-FOM 4.2mW ADC-PLL using digital sub-sampling architecture." *IEEE Solid-State Circuits Conference (ISSCC)*, 2015, pp. 440–441.

[18] Gao X., Burg O., Wang H., *et al.* "A 2.7-4.3GHz 0.16psrms jitter -246.8dB FOM digital fractional-N sampling PLL in 28 nm CMOS." *IEEE International Solid-State Circuits Conference (ISSCC)*, 2016, pp. 174–175.

[19] Ru Z., Geraedts P., Klumperink E., and Nauta B. "A 12GHz 210fs 6mW digital PLL with sub-sampling binary phase detector and voltage-time modulated DCO." *IEEE Symposium on VLSI Circuits*, 2013, pp. 194–195.

[20] Cai D., Fu H., Ren J. *et al.* "A dividerless PLL with low power and low reference spur by aperture-phase detector and phase-to-analog converter." *IEEE Trans. Circuits Syst. I*. 2013, vol. 60, no. 1, pp. 37–50.

[21] Vamvakos S.D., Boecker C., Groen E. *et al.* "A 8.125–15.625 Gb/s SerDes using a sub-sampling ring-oscillator phase-locked loop." *IEEE Custom Integrated Circuits Conference (CICC)*, paper 10.1, 2014.

[22] Sogo K., Toya A., and Kikkawa T. "A ring-VCO-based sub-sampling PLL CMOS circuit with −119 dBc/Hz phase noise and 0.73 ps jitter." *IEEE European Solid State Circuits Conference (ESSCIRC)*, 2012, pp. 253–256.

[23] Yi X., Boon C.C., Sun J., Huang N., and Lim W.M. "A low phase noise 24/77 GHz dual-band sub-sampling PLL for automotive radar applications in 65 nm CMOS technology." *IEEE Asian Solid-State Circuits Conference (A-SSCC)*, 2013, pp. 417–420.

[24] Siriburanon T., Ueno T, Kimura K. *et al.* "A 60-GHz sub-sampling frequency synthesizer using sub-harmonic injection-locked quadrature oscillators." *IEEE Radio Frequency Integrated Circuits Symposium (RFIC)*, 2014, pp. 105–108.

[25] Sharkia A., Mirabbasi S., and Shekhar S. "A 0.01mm2 4.6-to-5.6GHz sub-sampling type-I frequency synthesizer with –254dB FOM." *IEEE International Solid-State Circuits Conference (ISSCC)*, 2018, pp. 256–257.

[26] Sharma J., and Krishnaswamy H. "A 0.01mm^2 4.6-to-5.6GHz sub-sampling type-I frequency synthesizer with –254dB FOM." *IEEE International Solid-State Circuits Conference (ISSCC)*, 2018, pp. 258–259.

[27] Liang C.-F., and Hsiao K.-J. "An injection-locked ring PLL with self-aligned injection window." *IEEE Int. Solid-State Circuits Conference (ISSCC)*, 2011, pp. 90–92.

[28] Lee I.-T., Chen Y-J, Liu S-I. *et al.* "A divider-less sub-harmonically injection-locked PLL with self-adjusted injection timing." *IEEE Int. Solid-State Circuits Conference (ISSCC)*, 2013, pp. 414–415.

[29] Chen Z.-Z., Wang Y-H, Frank Chang M-C *et al.* "Sub-sampling all-digital fractional-N frequency synthesizer with −111dBc/Hz in-band phase noise and an FOM of −242dB." *IEEE Solid-State Circuits Conference (ISSCC)*, paper 14.9, 2015.

Chapter 22

PLLs with nested frequency-locked loop

Taekwang Jang[1], Dong-in Kim[2] and SeongHwan Cho[2]

Phase locked loop (PLL) is a feedback control system that has been used for several decades. As the feedback control loop is quite straight forward, architectural innovations in PLLs have been rather limited. Most PLLs are based on either type-I or type-II control systems and it is usually the circuit implementations that improve their performance. In this chapter, we explore architectural innovations in a PLL, by using additional feedback loop to gain further advantages in performance such as jitter, power consumption, and immunity to supply-noise. In Section 22.1, we look at how a reference-less frequency locked loop (FLL) around the VCO has the potential to improve the phase noise of a VCO. In Section 22.2, we investigate how the FLL-based VCO can be used in a digital PLL for optimal performance in power and noise. Finally, in Section 22.3, we will investigate how this architecture can be used to mitigate jitter due to supply noise.

22.1 VCO noise reduction using reference-less FLL

22.1.1 Basic concept

Voltage-controlled oscillator (VCO) is a fundamental component in many circuits in clock generation, frequency synthesis, and high-speed communication. The key performance parameter of the VCO is its phase noise and improving it while minimizing the associated power and cost has been the focal point of research and development. The most widely used approach to reduce phase noise as well as to stabilize frequency is to employ a phase lock loop (PLL), which periodically calibrates the oscillator using a clean reference frequency that usually has a lower frequency than the VCO. Thus, the PLL cleans the jitter of the VCO at a bandwidth smaller than the reference frequency. Naturally, while such action will reduce the low-frequency components of the jitter (i.e., in-band phase noise of the VCO), high-frequency components beyond the PLL loop bandwidth cannot be suppressed. The stability of PLL occurring from its mixture of discrete and continuous nature

[1]Department of Information Technology and Electrical Engineering (D-ITET), ETH Zürich, Switzerland
[2]School of Electrical Engineering, KAIST (Korea Advanced Institute of Science and Technology), Daejeon, Republic of Korea

also limits the bandwidth typically to below one-tenths of the reference frequency. Hence reducing the VCO phase noise has usually been investigated at the circuit level. In this section, we look at an architectural approach to reduce phase noise, by employing reference-less frequency locked loop around a VCO. By operating the feedback loop at the VCO frequency, the loop bandwidth can be much larger than that of the PLL and thus high-frequency phase noise of the VCO can be suppressed.

A PLL can be considered a feedback system that lowers the VCO phase noise by feeding the absolute jitter (i.e., edge jitter) of the VCO to the VCO itself. To measure the absolute jitter, a low-noise reference clock is often used, which provides a reference time edge that can be compared to the VCO edge using a phase detector (PD). While most schemes use a reference clock for jitter measurement, it is possible to measure the absolute jitter without using a reference clock, but instead using a constant delay as shown in Figure 22.1, where the phase detector (PD) produces the time difference between the VCO output (V) and its delayed version (V_D), and the PD output (e) is accumulated to produce the output (y).

Assuming that the VCO jitter is the only source of noise, the output (y_n) can be expressed as:

$$\begin{aligned} y_n &= \sum_{i=1}^{n} e_i = \sum_{i=1}^{e_i} t_i - (t_{i-1} + t_D) \\ &= t_n - t_0 - nT_D = \tau_n + n(T_0 - T_D) \end{aligned}$$

where τ_i is the absolute jitter of the ith VCO edge, e_i is the time difference between the VCO and its delayed version, T_D is the constant time delay, and T_0 is the period of the VCO. It can be seen that y_n is equal to the absolute jitter τ_i, if the delay is the same as the VCO period (i.e., $T_0 = T_D$). Assuming that $T_0 = T_D$, let's look at how we can use the above scheme to reduce the VCO phase noise. Similar to how a PLL operates, a straight forward approach to reduce the VCO noise is to feed the measured absolute jitter back to the VCO, as shown in Figure 22.2 [1]. When the loop settles and is in lock, the input phase difference must be zero and hence the VCO period is equal to the time delay (T_D). Since operating frequency of this loop is

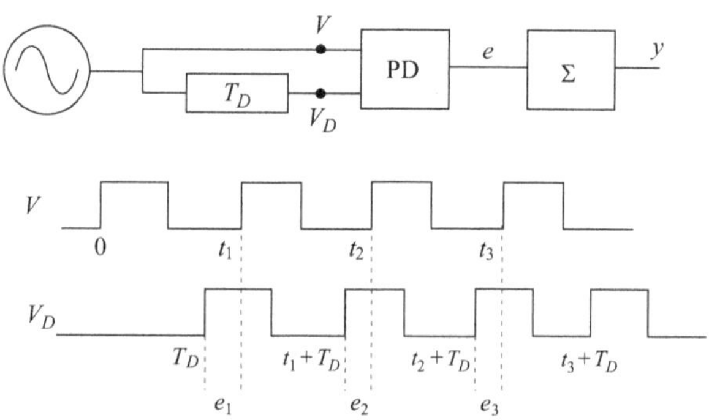

Figure 22.1 Absolute jitter measurement using a constant delay

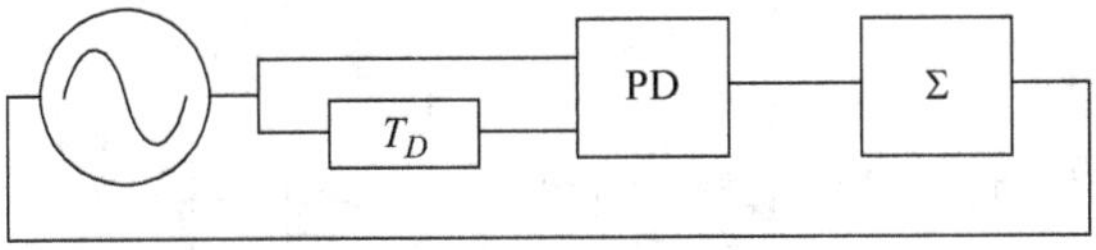

Figure 22.2 A reference-less feedback loop around the VCO

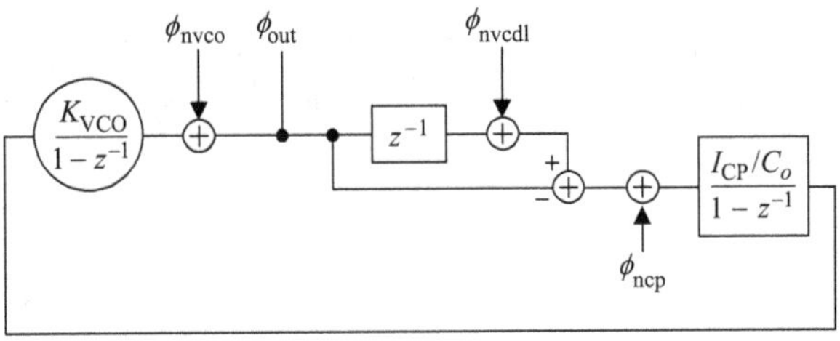

Figure 22.3 Z-domain model of the reference-less feedback loop

determined by the VCO which is much higher than that of typical PLLs that use low reference frequency, the loop bandwidth is much higher and thus the VCO phase noise can be lowered with a high cut-off frequency.

A z-domain block diagram which provides a more accurate model for the delay-line [2] is shown in Figure 22.3 and the resulting noise transfer functions can be expressed as:

$$\frac{\phi_o}{\phi_{nvco}} = \frac{1}{1+\frac{K}{1-z^{-1}}}\frac{\phi_o}{\phi_{nvcdl}} = \frac{\phi_o}{\phi_{ncp}} = \frac{K}{(1-z^{-1})(1-z^{-1}+K)} \text{ where } K = \frac{I_{CP}K_{VCO}}{C_o}$$

For the VCO, the first function yields a highpass filter and thus the loop bandwidth must be increased to reduce the VCO phase noise. Since the operating frequency of the phase detector is the same as the VCO's output frequency, the loop bandwidth can be easily increased without running into stability problems. For the charge-pump and delay-line, the transfer function is effectively an integrator. Thus, care must be taken to reduce the noise from the charge-pump and delay line. It should be noted that VCOs usually dominate the noise compared to the charge-pump or delay lines [3]. Furthermore, since the loop does not contain any divider in the feedback path, the noise from the charge-pump and VCDL is not increased as in the case of conventional PLLs that are used for frequency multiplication.

22.1.2 Implementation and simulation

The feedback loop shown in Figure 22.2 can be implemented as shown in Figure 22.4, where a charge pump and a capacitor is used as the accumulator. Hence the proposed technique can be very useful for integrated PLLs that have

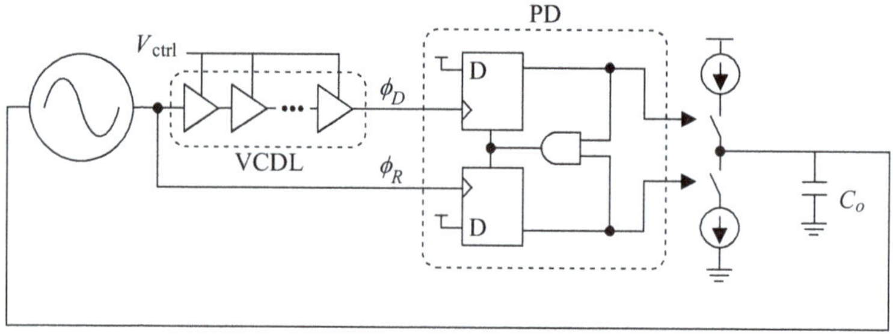

Figure 22.4 Schematic of the reference-less feedback loop

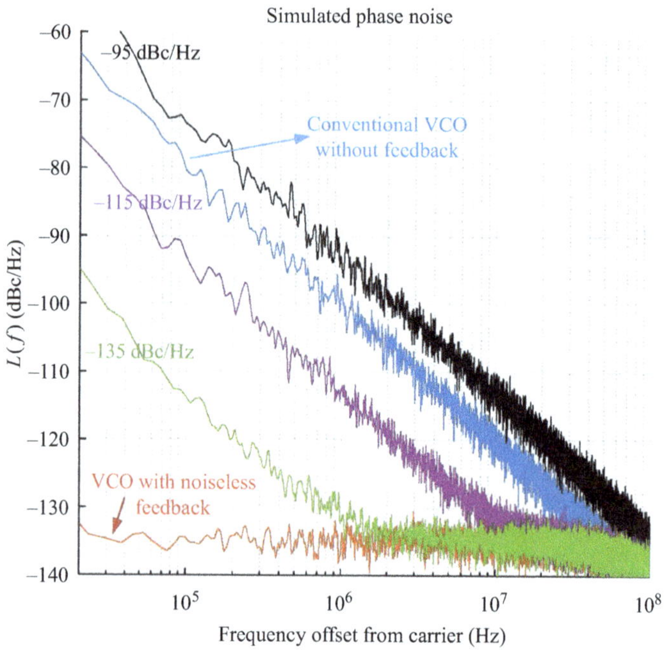

Figure 22.5 Phase noise simulation results

poor VCO phase noise. The schematic shown in Figure 22.4 is simulated using CppSim [4], which is an accurate time-domain behavioral simulator for analog and mixed-signal circuits. For the VCO, the free-running frequency is set as 1 GHz, K_{VCO} as 250 MHz/V, and phase noise as −100 dBc/Hz at 1 MHz offset. Note that $1/f$ noise is not modeled in this simulation. The PD has a reset delay of 0.2 ns, the charge-pump has a current of 100 μA, and the capacitor size is 1 pF. The output phase noise is shown in Figure 22.5, for different charge-pump and VCDL noise of −95, −115, and −135 dBc/Hz, at 1 MHz offset. For comparison, the phase noise of a conventional VCO without any feedback loop is also plotted together with that

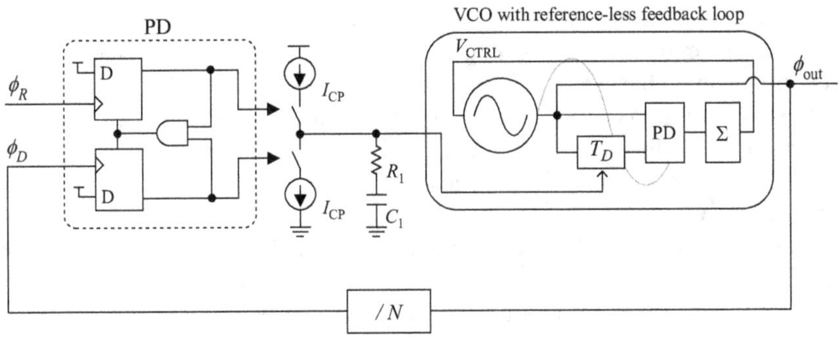

Figure 22.6 PLL using reference-less FLL around VCO

of an ideal FLL which has VCO noise only and does not have any charge-pump or VCDL noise. It can be seen from the ideal FLL that the VCO noise is highpass filtered. For nonideal cases, it can be seen that the noise from the charge-pump and VCDL is integrated, lowpass filtered and may dominate the phase noise such that the feedback loop actually makes the noise worse than a conventional open-loop VCO.

A designer may wonder whether it is worthwhile putting the extra effort to implement the FLL since there are other noise sources that are integrated and the overall noise may become worse than a conventional VCO. It depends on how well the VCDL can be implemented in terms of noise and power. When using CMOS devices for the delay and charge-pump, it is likely that a conventional VCO will be better than the FLL-based VCO as far as the figure-of-merit (FoM) for an oscillator is concerned, assuming they use the same power. However, the FLL-based VCO architecture may be able to achieve phase noise levels at certain frequencies where a conventional VCO cannot. It should also be noted that the architecture of FLL-based VCO can be used in other applications, such as ultra low-noise microwave generation using fiber optics and mode-lock lasers [5].

The reference-less FLL around the VCO can be used in a conventional charge-pump analog PLL as shown in Figure 22.6. It can be seen that the control voltage is fed to the delay cell which indirectly controls the VCO frequency via the FLL. This architecture can relax the conventional trade-offs between VCO noise and in-band noise such as reference noise in clock generators or quantization noise in a fractional-N PLL.

22.2 Frequency-locked loop based on a switched capacitor feedback

All-digital phase-locked loops (ADPLLs) have been widely adopted for their small size, configurability, and portability [6–8]. In order to fully exploit its advanced robustness based on its digital nature, a digitally controlled oscillator (DCO), which is typically implemented with an analogous oscillator combined with voltage, current, or capacitor digital-to-analog converter (DAC), needs to be insensitive to the environmental changes.

As the loop filter is implemented with digital circuits, the gain of DCO (K_{DCO}), is the major variability causing uncertainty in loop dynamics. The gain of DCO, K_{DCO}, needs to be accurately controlled across process, temperature, and supply voltage (PVT) variations.

A DCO is considered the most important building block of a digital PLL for a couple reasons. First, the output phase noise is dominated by DCO noise, especially when a ring oscillator is used for a wide frequency range or a small size. Second, a DCO's gain and frequency range affect the performance of a PLL significantly. Nevertheless, the gain and frequency range vary significantly in accordance with change in process, supply voltage, and temperature (PVT) because its delay relies on the intrinsic parameters of the transistors. The change in the DCO gain results in variation of the PLL loop bandwidth as it moves away from the optimal point. In order to fully exploit the advantages of ADPLL, KDCO needs to be accurately controlled to ensure robust operation under the environmental changes. In addition, a ring oscillator structure that can operate near its theoretical limit [9] is desirable for the optimal noise performance.

In this section, a digital PLL with a nested frequency locked loop (FLL) that linearizes the DCO frequency tuning curve is introduced to provide a stable gain. Therefore, the loop dynamics are insensitive to environmental variations. It further offers a noise reconfiguration capability creating a trade-off between power and noise.

22.2.1 Basic concept

The proposed design adopts a frequency-locked loop (FLL) using a switched capacitor for the frequency feedback [10–16] as introduced in Figure 22.7.

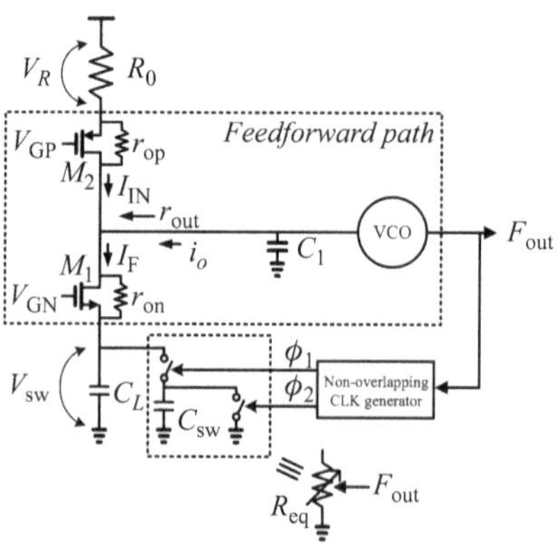

Figure 22.7 Conceptual schematic of the frequency locked loop

The input current, I_{IN}, is generated by applying V_R to a resistor R_0. The feedback current is defined by the following equation:

$$I_F = C_{sw} V_{sw} f_{out} \tag{22.1}$$

where C_{sw}, V_{sw}, and f_{out} are switching capacitance, voltage on C_{sw}, and output frequency, respectively. I_F should be equal to I_{IN} in the steady state assuming a large gain from the frequency detection block. Therefore, the output frequency can be calculated by equalizing the currents from the switched capacitor and the resistor:

$$C_{sw} V_{sw} f_{out} = \frac{V_R}{R_0}$$

$$f_{out} = \frac{V_R}{V_{sw}} \frac{1}{R_0 C_{sw}} \tag{22.2}$$

If an identical voltage is used for both V_R and V_{sw}, the oscillation frequency of the FLL is $1/R_0C_{sw}$. R_0 and C_{sw} can be used to tune the output frequency in the PLL.

Conventional methods of DCO implementation includes a gate voltage control of a current starving transistor [17], a series resistance control to load capacitor [18] or to supply voltage [19], a delay cell size control [20], or a digital current control for a current-controlled oscillator [21], all relying on the transistor's physical characteristics and it is difficult to achieve high linearity or PVT invariance [14]. In contrast, an FLL using switched capacitor can accurately control the output frequency because it is explicitly determined as (22.3) using passive elements [10–16] assuring accurate and linear operation. As a comparison, Figure 22.8 shows a simulated frequency tuning curves of conventional and proposed DCOs.

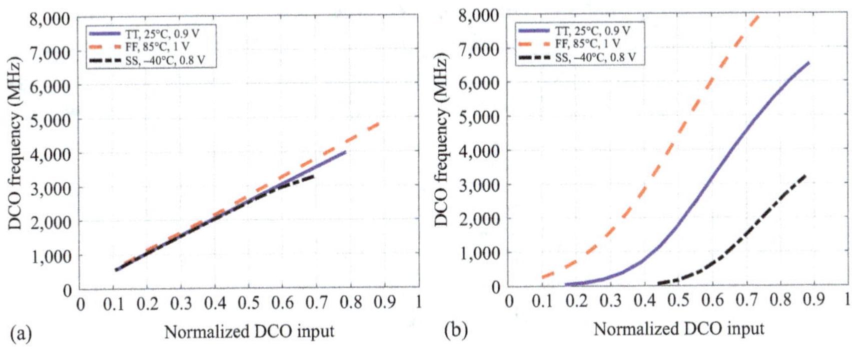

Figure 22.8 Simulated frequency tuning curves. (a) Proposed FLL and (b) a VCO controlled by a digital-to-analog converter

22.2.2 Loop dynamics

The feedforward path of the FLL is from the output current generated by the frequency detection block to the output frequency of the voltage-controlled oscillator (VCO) marked with a dotted line in Figure 22.7. The control voltage (v_{ctrl}) can be calculated as a multiplication of the output current, i_o, and the output impedance of the detection block as shown in the following equation:

$$r_{out} = \left(R_0 + r_{op}\left(1 + g_{mp}R_0\right)\right) \| \left(R_{eq} + r_{on}\left(1 + g_{mn}R_{eq}\right)\right) \\ \simeq r_{op}R_0 g_{mp} \| r_{on}R_{eq}g_{mn} \tag{22.3}$$

where r_{op}, r_{on}, g_{mp}, and g_{mn} are PMOS output resistance, NMOS output resistance, PMOS transconductance, and NMOS transconductance, respectively. Assuming $r_{op} = r_{on} = r_o$, and M_1 and M_2 are in a subthreshold region so that their transconductance is maximized to I_{bias}/mv_T, r_{out} is simplified to $r_oR_0I_{bias}/2mv_T$ where m is $1 + C_d/C_{ox}$ and C_d and C_{ox} are depletion and oxide capacitances, respectively:

$$v_{ctrl} = z_{out}i_o = \frac{r_{out}}{1 + s \cdot r_{out}C_1} i_o \tag{22.4}$$

The output frequency, f_{out}, can be calculated by multiplying the gain of the VCO (K_{VCO}) to (22.5). Therefore, the feedforward transfer function can be expressed as:

$$H_{ff}(s) = \frac{f_{out}}{i_o} = \frac{K_{VCO}r_{out}}{1 + s \cdot r_{out}C_1} \tag{22.5}$$

To analyze the feedback path of the proposed topology, we first need to analyze the transfer function of the switched capacitor converting f_{out} to impedance of switched capacitor, r_{eq}:

$$\frac{\partial r_{eq}}{\partial f_{out}} = \frac{\partial}{\partial f_{out}}\left(\frac{1}{C_{sw}f_{out}}\right) = -\frac{1}{C_{sw}f_{out}^2} \tag{22.6}$$

The change in r_{eq} modulates the gate-to-source voltage of M_1 and results in the change in the output current. Figure 22.9 shows the linearized circuit diagram at the

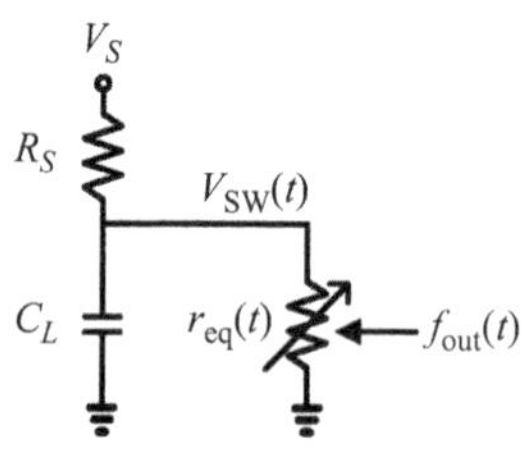

Figure 22.9 Simplified and linearized schematic at the source of M_1 for the analysis of feedback transfer function

source of M_1. R_s is the resistance of M_1 seen at its source ($1/g_{m1}$). C_L is a capacitor connected in parallel to the switched capacitor to lower the ripple magnitude. V_s is a virtual source voltage that provides DC voltage on the switched capacitor. v_{sw} is the voltage on the switched capacitor excluding the sawtooth ripple caused by the switching operation. Then, V_s can be defined as:

$$V_s = \left(1 + \frac{R_s}{R_{\text{eq}}}\right) V_{\text{SW}} \tag{22.7}$$

where V_{sw} is the DC voltage on the switched capacitor. Then, v_{sw} can be found as:

$$V_{\text{sw}}(t) = \frac{r_{\text{eq}}(t)}{R_s + r_{\text{eq}}(t)} V_s \tag{22.8}$$

Therefore, the transfer function from r_{eq} to v_{sw} is determined using the following equation:

$$\frac{\partial v_{\text{sw}}}{\partial r_{\text{eq}}} = \frac{\partial}{\partial r_{\text{eq}}} \left(\frac{r_{\text{eq}}(t)}{R_s + r_{\text{eq}}(t)} V_s \right)_{r_{\text{eq}}(t) = R_{\text{eq}}} = \frac{R_s}{(R_{\text{eq}} + R_s)^2} V_s \tag{22.9}$$

The gain from f_{out} to v_{sw} is calculated by combining (22.7) and (22.10) as shown in the following equation:

$$\begin{aligned} \frac{\partial v_{\text{sw}}}{\partial f_{\text{out}}} &= \frac{\partial v_{\text{sw}}}{\partial r_{\text{eq}}} \frac{\partial r_{\text{eq}}}{\partial f_{\text{out}}} = -\frac{1}{C_{\text{sw}} f_{\text{out}}^2} \frac{R_s V_s}{(R_{\text{eq}} + R_s)^2} \frac{1}{(1 + s/\omega_{p2})} \\ &= -\frac{R_s V_s}{C_{\text{sw}} (1/C_{\text{sw}} + f_{\text{out}} R_s)^2} \frac{1}{(1 + s/\omega_{p2})} \end{aligned} \tag{22.10}$$

ω_{p2} is the secondary pole of the FLL generated by the C_L and parallel resistance of R_s and R_{eq}, $\omega_{p2} = 1/C_L(R_s \| R_{\text{eq}})$.

Finally, the feedback transfer function from f_{out} to i_f is determined by multiplying the transconductance, $-g_{m1}$:

$$H_{\text{fb}}(s) = -g_{m1} \frac{\partial v_{\text{sw}}}{\partial f_{\text{out}}} = \frac{V_s}{C_{\text{sw}} (1/C_{\text{sw}} + f_{\text{out}} R_s)^2 (1 + s/\omega_{p2})} \tag{22.11}$$

Assuming a large feedforward path gain, the closed-loop gain is determined by the feedback factor using the following equation:

$$1/H_{\text{fb}}(0) = \frac{C_{\text{sw}} (1/C_{\text{sw}} + f_{\text{out}} R_s)^2}{V_s} \, [\text{Hz/A}] \tag{22.12}$$

To minimize the gain from the noise current to the output frequency, R_s $(= 1/g_{m1})$ must be minimized. However, given the DC bias current of I_{bias}, there exist an upper bound of the transconductance, I_{bias}/mv_T, when M_1 operates at a subthreshold mode. Therefore, the minimum closed-loop gain can be found as the following:

$$A_{\text{noise}} = \frac{C_{\text{sw}}(1/C_{\text{sw}} + f_{\text{out}} R_s)^2}{V_s} \frac{I_{\text{bias}}}{f_{\text{out}}}$$

$$= \frac{C_{\text{sw}}\left(1/C_{\text{sw}} + R_s/R_{\text{eq}} C_{\text{sw}}\right)^2}{\left(1 + R_s/R_{\text{eq}}\right) V_{\text{sw}}} \frac{I_{\text{bias}}}{f_{\text{out}}}$$

$$= \frac{\left(1 + R_s/R_{\text{eq}}\right)^2}{\left(1 + R_s/R_{\text{eq}}\right) C_{\text{sw}} V_{\text{sw}}} \frac{I_{\text{bias}}}{f_{\text{out}}} = 1 + \frac{R_s}{R_{\text{eq}}} \tag{22.13}$$

$$A_{\text{noise,min}} = 1 + \frac{m v_T / I_{\text{bias}}}{1/f_{\text{out}} C_{\text{sw}}} = 1 + \frac{m v_T}{V_{\text{sw}}} \tag{22.14}$$

A large V_{sw} results in reduced noise gain with the expense of a reduced control voltage range. Therefore, a trade-off needs to be made to provide a small noise gain (e.g., $V_{\text{wsw}} = 10 v_T$ allowing 10% increase of the noise gain) while offering a sufficient control voltage range for the VCO.

22.2.3 Noise analysis

One of the merits of the proposed architecture is a reduction of the VCO noise. The VCO noise is highpass filtered at the FLL bandwidth so that it becomes negligible when calculating the integrated phase noise, as shown in the closed-loop transfer function from the VCO noise to f_{out}, Figure 22.10 (blue, dotted). Therefore, the output noise of the FLL is mostly determined by the frequency detection circuit. It dominates the output noise across most of the frequency range because f_{BW} is large (>100 MHz). Phase noise simulation results of the proposed FLL (blue) and a free funning VCO (red) are plotted in Figure 22.11. The VCO noise is highpass filtered at the FLL bandwidth (208 MHz), and the lower frequency noise is dominated by the detection circuits.

First of all, the current noise generated by the switched capacitor is analyzed to calculate the noise property of the detection circuit. Figure 22.12 shows the transient waveform at the top node of C_{sw}, V_c. The sampling operation happens twice per f_{out} cycle, once for charging it to V_{sw} and once for discharging it to ground. In each sampling process, kT/C noise with bandwidth $1/2f_{\text{out}}$ is generated, so its power spectral density can be written as:

$$S_{v1}(f) = S_{v2}(f) = \frac{2kT}{C_{\text{sw}} f_{\text{out}}} \tag{22.15}$$

where v_1 and v_2 are the sampled voltage on V_c. The amount of charge, $C_{\text{sw}}(v_1(t) - v_2(t))$, is injected at every switching cycle to V_{sw}, so the noise current of the switched capacitor can be calculated using the following equation:

$$i_1(t) = C_{\text{sw}} f_{\text{out}} (v_1(t) - v_2(t)) \tag{22.16}$$

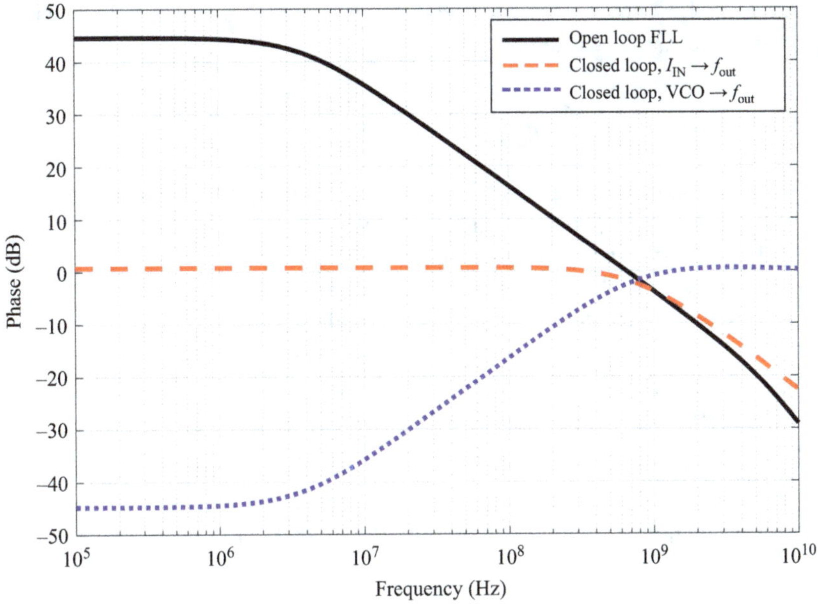

Figure 22.10 Open loop and close loop transfer functions of the frequency locked loop

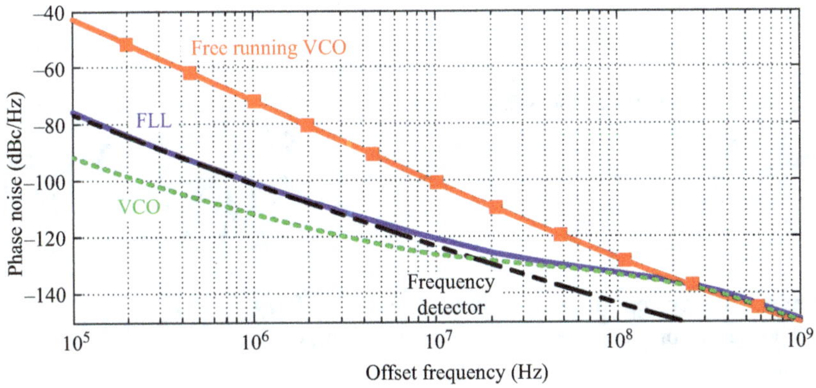

Figure 22.11 Simulated phase noise curves of the free running VCO (red line) and the proposed FLL (blue line), which is composed of noise from VCO and the frequency detection circuit

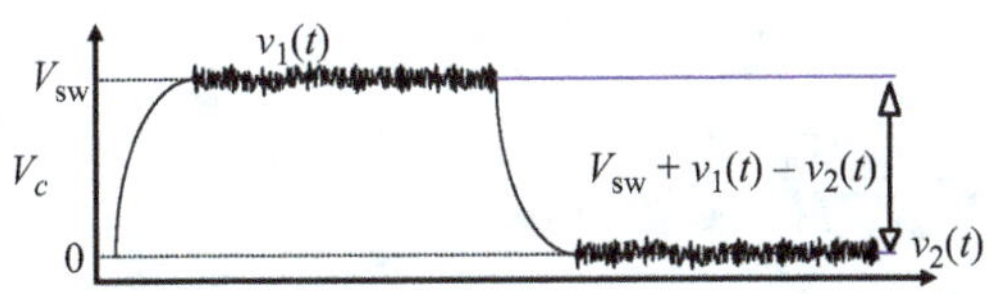

Figure 22.12 Transient waveform at the top node of C_{sw}

As v_1 and v_2 are uncorrelated, the power spectral density of the switched capacitor noise can be written as:

$$\overline{i_1^2} = 4kTf_{\text{out}}C_{\text{SW}} = 4kT\frac{1}{R_{\text{eq}}} \tag{22.17}$$

Therefore, it can be seen that the noise generated by the switched capacitor is equal to the noise of a physical resistor size of R_{eq}.

Current division branches located at the sources of M_1 and M_2 need to be analyzed to calculate the noise transfer functions. For example, the current noise from the switched capacitor, $i_1(t)$, is divided by the impedance ratio of R_{eq} and R_s:

$$i_{\text{out},1} = \frac{R_{\text{eq}}}{R_{\text{eq}} + R_s} i_1 = \alpha i_1 = \frac{1}{A_{\text{noise}}} i_1 \tag{22.18}$$

where $i_{\text{out},1}$ is the output current noise of the detection circuit generated by i_1. As mentioned in (22.15), it is advantageous to increase the size of M_1 to reduce the noise gain. Then, the current division ratio, α, can be simplified as a voltage ratio:

$$\alpha = \frac{\frac{V_{\text{SW}}}{I_{\text{bias}}}}{\frac{V_{\text{SW}}}{I_{\text{bias}}} + \frac{mv_T}{I_{\text{bias}}}} = \frac{V_{\text{SW}}}{V_{\text{SW}} + mv_T} \tag{22.19}$$

The current noise generated by M_1, i_2, is degenerated by R_{eq} and only a small fraction of i_2 contributes to the output as the following:

$$i_{\text{out},2} = (1 - \alpha)i_2 \tag{22.20}$$

Similarly, the current noise from R_0 and M_2 can be calculated, as described the following equations:

$$i_{\text{out},3} = \beta i_3 \tag{22.21}$$

$$i_{\text{out},4} = (1 - \beta)i_4 \tag{22.22}$$

where β is the current division ratio at the source of M_2, $V_R/(V_R + mv_T)$.

The output phase noise generated by $i_{1\text{–}4}$ can be expressed by multiplying the closed-loop gain:

$$L_{C_{\text{SW}}}(f) = \frac{4kT}{I_{\text{bias}}V_{\text{SW}}}\frac{f_{\text{out}}^2}{f^2} \tag{22.23}$$

$$L_{M_1}(f) = \left(\frac{1-\alpha}{\alpha}\right)^2 \frac{4kT\gamma_n g_{m1}}{I_{\text{bias}}^2}\frac{f_{\text{out}}^2}{f^2} \tag{22.24}$$

$$L_{R_0}(f) = \frac{\beta^2}{\alpha^2}\frac{4kT}{I_{\text{bias}}V_R}\frac{f_{\text{out}}^2}{f^2} \tag{22.25}$$

$$L_{M_2}(f) = \left(\frac{1-\beta}{\alpha}\right)^2 \frac{4kT\gamma_p g_{m1}}{I_{\text{bias}}^2} \frac{f_{\text{out}}^2}{f^2} \tag{22.26}$$

By rewriting α and β in (22.28)–(22.31) using (22.23) and (22.27) and by summing all of the noise sources, the overall phase noise can be found as follows:

$$L(f) = \frac{4kT}{I_{\text{bias}}} \left(\frac{1}{V_{\text{sw}}} \left(1 + \frac{\gamma_n m v_T}{V_{\text{sw}}} \right) + A_{\text{noise}}^2 \frac{V_R + \gamma_p m v_T}{(V_R + m v_T)^2} \right) \frac{f_{\text{out}}^2}{f^2} \tag{22.27}$$

The discussion at this point shows that the output noise improves with a larger V_{sw}. A_{noise} is a function of V_{sw} as well, so increasing V_{sw} helps lower the noise not only from M_1 and C_{sw} but also from M_2 and R_0. A smaller overdrive voltage of M_1 helps reduce the noise generated by M_1. Therefore, it is advantageous to maximize the transconductance of M_1 so it operates in the subthreshold region. The latter part of (22.32) also shows increasing V_R helps reduce the output noise generated by M_1 and R_0. Note that the overall phase noise has I_{bias} in the denominator, so the DCO noise can be reconfigured by tuning the bias current while keeping the bias conditions, V_{sw}, V_r, V_{gs1}, and V_{gs2} constant. For instance, the FLL is configured in a low noise mode with high I_{bias} to prioritize noise performance. On the other hand, the FLL is configured in a low current mode when power consumption is more important, sacrificing its noise performance.

For finding an optimal biasing condition of the frequency detector, M_1 and M_2 need to be set at subthreshold mode while maximizing V_{sw} and V_R. Here, an optimal voltage allocation on V_{sw} and V_R is introduced given a fixed supply voltage. First, the voltage assigned to them is defined as V_B as follows:

$$V_B = V_{\text{sw}} + V_R = V_{\text{DD}} - V_{c,\text{range}} - 4v_T \tag{22.28}$$

$V_{c,\text{range}}$ is the input voltage range of the VCO for generating the target frequency under the PVT variation. $4v_T$ is subtracted from V_{DD} as well to give sufficient V_{ds} to M_1 and M_2. Then, (22.32) can be rewritten by substituting V_R with V_B–V_{sw}:

$$L(f) = \frac{4kT}{I_{\text{bias}}} \left(\frac{1}{V_{\text{sw}}} \left(1 + \frac{\gamma_n m v_T}{V_{\text{sw}}} \right) + \left(1 + \frac{m v_T}{V_{\text{sw}}} \right)^2 \frac{V_B - V_{\text{sw}} + \gamma_p m v_T}{(V_B - V_{\text{sw}} + m v_T)^2} \right) \frac{f_{\text{out}}^2}{f^2} \tag{22.29}$$

The phase noise at 10 MHz offset is plotted in Figure 22.13 when T, I_{bias}, γ_n, γ_p, v_T, m, and f_{out} are 300 K, 3 mA, 2/3, 2/3, 26 mV, 1.2 and 2 GHz, respectively. It can be seen that the optimal noise performance is achieved when V_R is close to V_{sw}. The numerical solutions $V_{\text{sw}} = 0.271$ and $V_R = 0.229$ are found when $V_B = 0.5$ after differentiating (22.34) and setting it equal to 0. The reason for V_{sw} is weighted slightly more than V_R is that A_{noise} affects both the noise from C_{sw} and V_R as shown in (22.27).

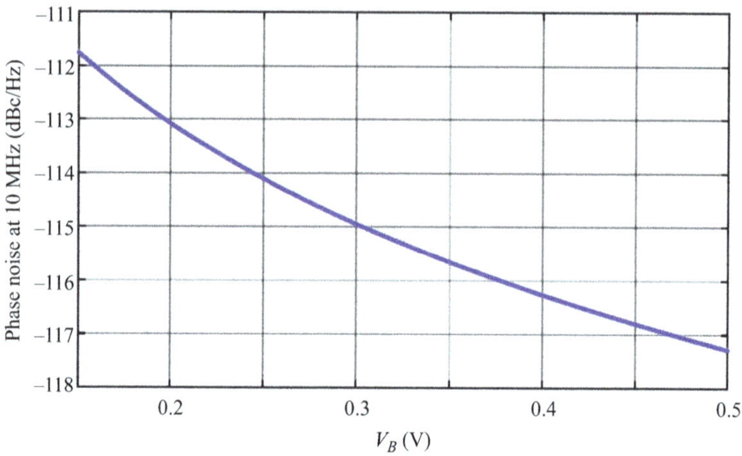

Figure 22.13 Simulated phase noise at 10-MHz offset frequency while varying voltage allocation, V_B

Equation (22.34) can be simplified assuming equal V_R and V_{sw} to provide intuitive understanding on the theoretical limit of the proposed FLL:

$$L_{opt}(f) \simeq \frac{4kT}{I_{bias}}\left(\frac{1}{V_{sw}}\left(1+\frac{\gamma_n m v_T}{V_{sw}}\right)+\frac{1}{V_R}\left(1+\frac{\gamma_p m v_T}{V_R}\right)\right)\frac{f_{out}^2}{f^2}$$
$$= \frac{8kT}{I_{bias}V_{sw}}\left(1+\frac{\gamma_n m v_T}{V_{sw}}\right)\frac{f_{out}^2}{f^2} \tag{22.30}$$

Finally, the optimal figure-of-merit (FoM) of the proposed oscillator can be found as the following:

$$\text{FoM}_{opt} = 10\log\frac{8kT}{I_{bias}V_{sw}}\left(1+\frac{\gamma_n m v_T}{V_{sw}}\right)\frac{f_{out}^2}{f^2}\times\frac{I_{bias}V_{DD}}{1\text{ mW}}\times\frac{f^2}{f_{out}^2}$$
$$= 10\log\frac{8kTV_{DD}}{10^{-3}V_{sw}}\left(1+\frac{\gamma_n m v_T}{V_{sw}}\right) \tag{22.31}$$

It can be seen that FoM of the proposed oscillator only depends on the ratio between V_{DD} and V_{sw} assuming $V_{sw} \gg \gamma_n m v_T$. Then, the theoretical minimum FoM is found when $V_{sw} = V_R = V_{DD}/2$ allowing zero voltage for V_c, range, V_{ds1}, and V_{ds2}:

$$\text{FoM}_{min} = 10\log\frac{8kTV_{DD}}{10^{-3}V_{DD}/2} = 10\log\frac{16kT}{10^{-3}} \approx -161.79 \tag{22.32}$$

The theoretical limit of the proposed oscillator is approximately 3.3 dB worse than that of a conventional ring oscillator [9]:

$$\mathrm{FoM}_{\mathrm{min,ring}} = 10\log\frac{7.33kT}{10^{-3}} \approx -165.2 \tag{22.33}$$

However, the proposed structure provides accurate and linear frequency tenability compared with the conventional ring oscillator whose speed relies on the transistor speed. Furthermore, it offers a noise reconfiguration capability making it more advantageous in various applications.

22.2.4 Circuit implementation

A schematic of the circuit level implementation is shown in Figure 22.14. The gate voltages of M_1 and M_2 are generated using replica cells of the main branch. The amplifiers are designed to consume less than 500 nW so that the noise generated by the amplifiers is located at a very low offset frequency range and filtered by the PLL. The width of M_1 and M_2 are controlled with R_0 and C_{sw} so that V_{gs1} and V_{gs2} remain constant.

C_L needs to be placed in parallel with C_{sw} to minimize the voltage ripple caused by the switching operation. Assuming negligible switch on-resistance, V_{sw} drops abruptly when C_{sw} is connected from and to V_{sw}. Then, V_{sw} increases linearly with time until the next discharging operation. The ripple magnitude is determined by the capacitance ratio between C_L and C_{sw}. C_L should be an order of magnitude larger compared with C_{sw} to minimize the current variation induced by the fluctuation of V_{gs1}. However, a large C_L causes an area penalty and impairs the stability of the FLL by reducing the secondary pole frequency determined by the total capacitance on C_{sw} and $(R_{\mathrm{eq}}\|1/g_{m1})$.

In this study, a multiphase feedback from the VCO is adopted so that the effective switching frequency of C_{sw} becomes N_{phase} times higher, where N_{phase} is

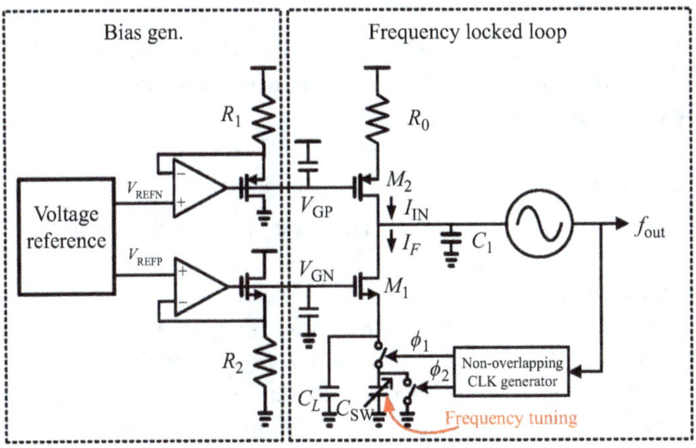

Figure 22.14 Schematic of the frequency locked loop with bias voltage generation

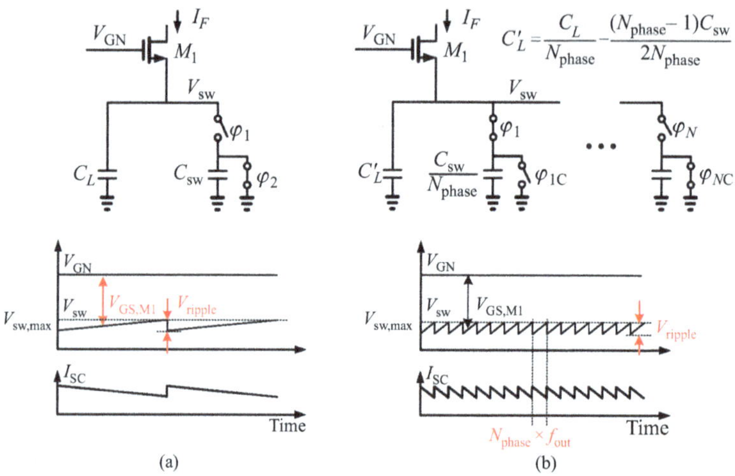

Figure 22.15 Comparison of the voltage ripple caused by the switching operation with (a) a single-phase feedback and (b) multiphase feedback

the number of VCO phases as shown in Figure 22.15. Then, the total capacitance connected in parallel with the switching capacitance can be reduced by the factor N_{phase}, and it helps provide a higher f_{BW} and larger phase margin. Compared with the single-phase feedback scheme, the unit switching capacitance is N_{phase} time smaller requiring a smaller C_L with the same factor. In addition, among other C_{sw} units, nearly half of them are connected to V_{sw} serving as parallel capacitors. Therefore, the size of the additional parallel capacitance is $C_L/N_{phase} - (N_{phase} - 1) C_{sw}/2N_{phase}$. In this design a 5 pF capacitor is connected by using 8-phase feedback which is only 7% of the parallel capacitance required for a single-phase feedback case.

22.3 A supply noise insensitive PLL with a rail-to-rail swing ring oscillator and a wideband noise suppression loop

22.3.1 Motivation

A ring oscillator based PLL is widely used as a clock generator in many applications as it does not require a large area. However, one of the critical drawbacks of ring oscillator-based PLL is that its jitter is sensitive to power supply noise. In system-on-chips (SoCs), the main source of jitter is supply noise and not thermal or $1/f$ noise. To mitigate this issue, many previous works have been suggested, which can be categorized into supply voltage regulation for the oscillator [22,23] or noise cancellation with calibration [18,24,25]. A common solution to regulate supply noise is to use a low dropout (LDO) regulator, which shields the oscillator from the supply noise and prevents it from reaching the supply of the oscillator. However, it requires a larger bandwidth and a large decoupling capacitor. In [22], a replica compensation approach has been proposed to solve this issue by making use of a local replica feedback loop. This topology allows a high-bandwidth amplifier without excessive power consumption resulting in high supply rejection. In [23], a digitally controlled resistor (DCR) is used between the supply and the oscillator for regulation. A digital background-calibration logic searches for a resistance that provides supply-insensitive voltage to the oscillator.

Apart from supply noise regulation, several cancellation techniques have also been introduced [18,24,25], which uses circuits with supply sensitivities that are equal but opposite in magnitude. In [18], a digital background calibration technique is added to the supply noise cancellation circuitry to overcome PVT variations. In [24], a constant-Gm bias circuit is implemented, and the generated current is fed to a current controlled oscillator (CCO), so that the CCO experiences a constant supply current regardless of supply noise. In [25], negative supply sensitivity is added to the output node of the delay cells to ground using additional transistors.

While all these schemes reduce the impact of supply noise, they all reduce the supply headroom of the ring oscillator, and thereby limits the phase noise performance as well as the maximum achievable output frequency. Note that obtaining high enough output frequency is important for low-voltage applications or for processes with slow transistors such as DRAM of flash memory. Thus, reduction in supply headroom for circuits that already suffer from low supply is critical. In addition, supply-noise reduction schemes that use calibration require long settling time or re-calibration under environment changes, which is not desirable in practice. In this section, a supply noise insensitive PLL is presented that features a rail-to-rail ring oscillator that does not have any supply headroom loss [26]. It employs a wideband noise suppression loop (NSL) operating at VCO frequency that results in a wide bandwidth noise filtering and does not require any calibration or long settling time. A prototype PLL implemented in 65 nm CMOS achieves an average spur suppression of 30 dB around the PLL loop bandwidth while consuming 2.73 mW at 3.2 GHz output.

22.3.2 Wide bandwidth noise suppression loop (NSL)

The principle of the wide bandwidth NSL is shown in Figure 22.16(a), which is basically the same reference-less FLL that we discussed earlier. The circuit is composed of a VCO, a delay cell with a delay of T_D, a phase detector (PD), and a charge-pump (CP). The negative feedback loop operates such that the phase error between the VCO and its delayed output is zero. Thus, the loop will lock when the VCO frequency is equal to $1/T_D$. Regarding the jitter at the output of the VCO due to supply noise, it is high-pass filtered by the loop as shown in Figure 22.16(b) and hence the NSL can reduce the supply noise of the VCO. However, noise from the PD, CP, and delay cell is integrated and low-pass filtered as shown in Figure 22.16(b) and thus, the supply noise on these circuits will not be suppressed. Hence, it is important that PD, CP, and T_D are immune from supply noise. Fortunately, it should be noted that VCO is the most supply sensitive circuit that

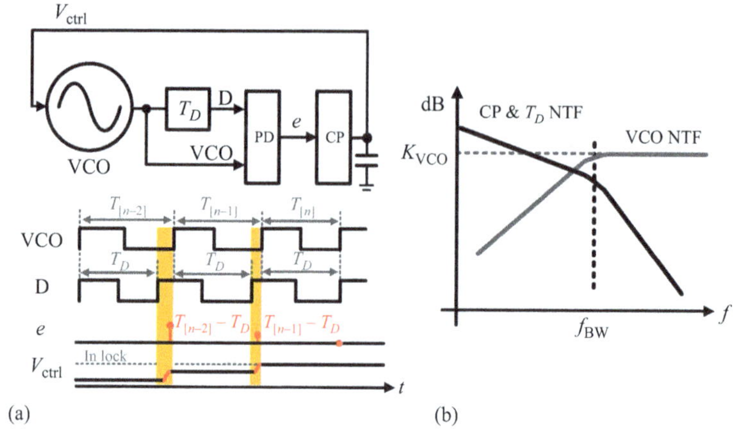

Figure 22.16 (a) Principle of the NSL and (b) its noise transfer function

dominates the supply-induced jitter of a PLL and other circuits do not contribute much to supply induced jitter [27]. Another important aspect of the NSL is that its filtering bandwidth is very large as the loop operates at VCO frequency. Furthermore, since the loop does not have any divider in the feedback path, noise from PD, CP, and delay cell is not multiplied to the output.

22.3.3 PLL with noise suppressed VCO

To implement a complete supply insensitive clock generator, the VCO with NSL is used inside a digital bang-bang PLL as shown in Figure 22.17. It can be seen that the output of the loop filter indirectly controls the VCO frequency via tuning the voltage-controlled delay cell (VCDC), which is unlike conventional PLLs whose loop filter output directly controls the VCO. In the NSL, the supply noise induced jitter is measured by the PD and fed to the VCO every VCO period, which results in a wide bandwidth supply noise filtering. As the supply noise at PD, charge-pump and the delay cell can seriously degrade the output jitter, a low drop-out voltage regulator (LDO) is used for these circuits. It should be noted that the implementation of the FLL is somewhat different from what is shown in Sections 22.1 and 22.2. In Figure 22.2, the time delay of the delay cell (T_D) is compared to the VCO period. Note that this delay is several hundreds of picoseconds for VCOs whose frequencies are in the few GHz range. Since implementing such a lengthy delay element increases power in the advanced CMOS process, only a fraction of the VCO period is used for comparison, which reduces the power consumption of the delay cell as well as the LDO. This is established by comparing the delay of the VDCD with that of the inverter within the VCO. A sub-sampling PD (SSPD) is used to detect the phase error caused by supply noise, as it has a large gain. As the PLL has another loop inside it, stability may be a concern. However, stability can be guaranteed by making the loop bandwidth of the NSL much larger than that of

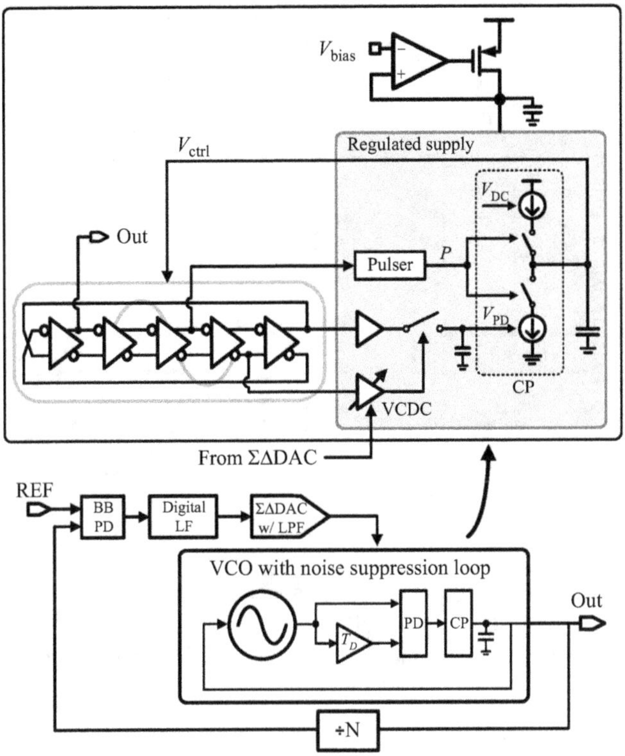

Figure 22.17 Overall architecture of the PLL

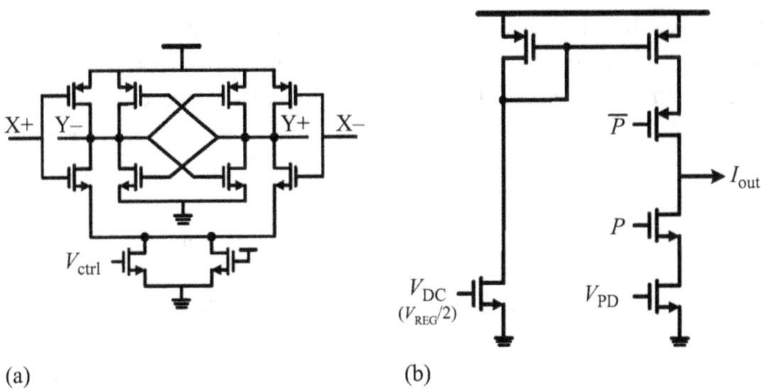

Figure 22.18 Schematic of (a) delay cell of VCO and (b) a charge-pump

the main PLL. This is quite easy to do as the NSL operates at the VCO frequency. In [26], the NSL has a bandwidth of about 300 MHz, while the main PLL loop has about 1 MHz of bandwidth. The schematic of the VCO, its delay cell, and the CP are shown in Figure 22.18. The VCO exhibits a rail-to-rail supply swing and does

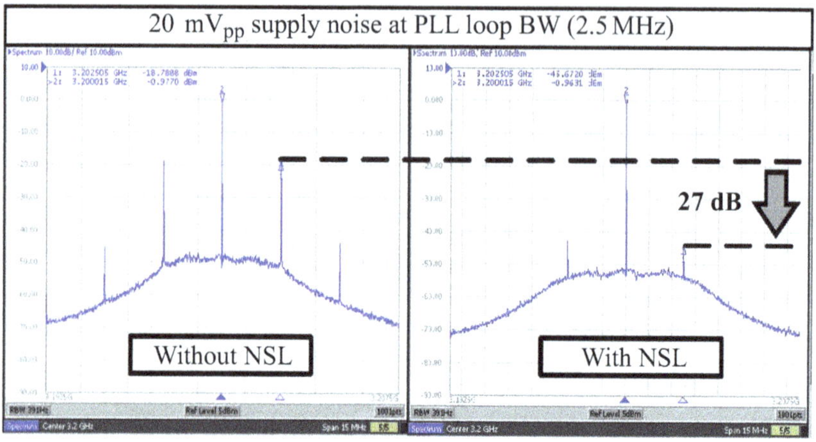

Figure 22.19 Measured output spectrum without and with noise suppression loop

not suffer from any headroom loss. The capacitor of the NSL is only 50 f_F. The supply regulated circuits (i.e., delay cell, PD, CP) and the regulator consume 450 μW and 0.004 mm^2 which are 15% and 8.5% of the total power and area, respectively.

22.3.4 Experimental results

The DPLL with NSL was fabricated in a 65 nm CMOS process and the die micrograph is shown in Figure 22.22. It occupies an active area of 0.047 mm^2 and consumes 2.73 mW from 1 V supply when the output frequency is 3.2 GHz. To validate the NSL, the chip was measured while applying a 20 mV peak-to-peak sinusoidal signal at the supply. The frequency of the sinusoid is 2.5 MHz which is approximately equal to the loop bandwidth of the PLL, which is the most vulnerable frequency. The results are shown in Figure 22.19, where it can be seen that the spur is reduced from −18.7 to −45.7 dBc when the NSL is enabled. The spur and rms jitter also measured over different noise frequencies as shown in Figure 22.20. It can be observed that the supply noise-induced spur is improved by an average of about 30 dB. It can also be seen that rms jitter is also reduced and is independent of the noise applied. The chip was also tested by applying 20 mV$_{rms}$ white noise to the supply. The output phase noise is shown in Figure 22.21, where it can be seen that the jitter is reduced from 14.1 ps to 7.8 ps which is close to 7.5 ps when there is no supply noise. The measured jitter histogram with and without the white noise is shown in Figure 22.22, which shows similar results to the phase noise plot. The performance of the DPLL [26] is summarized and compared with other supply noise insensitive PLLs in Table 22.1.

It can be observed that the supply noise suppression technique is very effective in reducing the effect of supply noise without headroom reduction and without using any calibration.

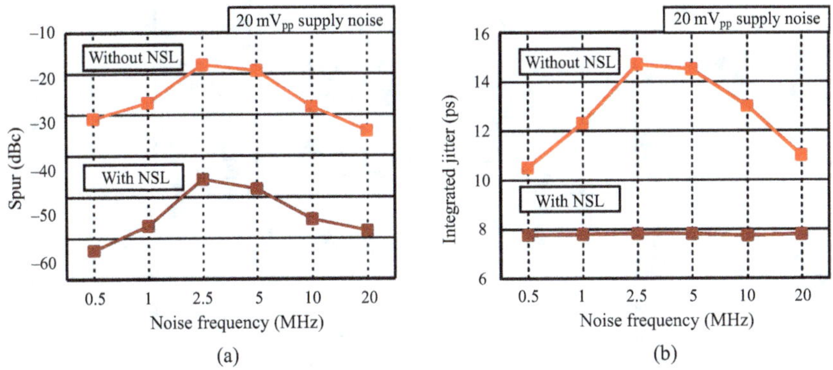

Figure 22.20 Measured (a) supply noise spurs and (b) integrated jitter versus noise frequency without and with supply noise suppression loop

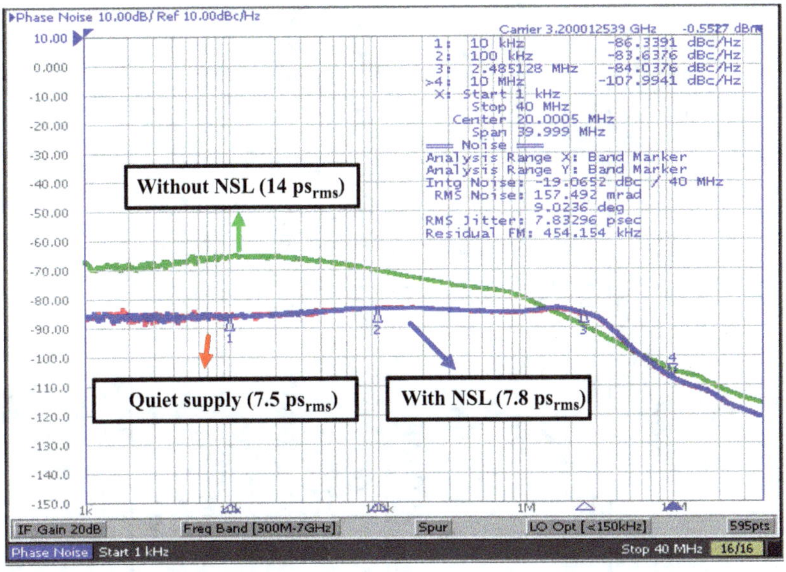

Figure 22.21 Measured phase noise with 20 mV$_{rms}$ white noise on the supply

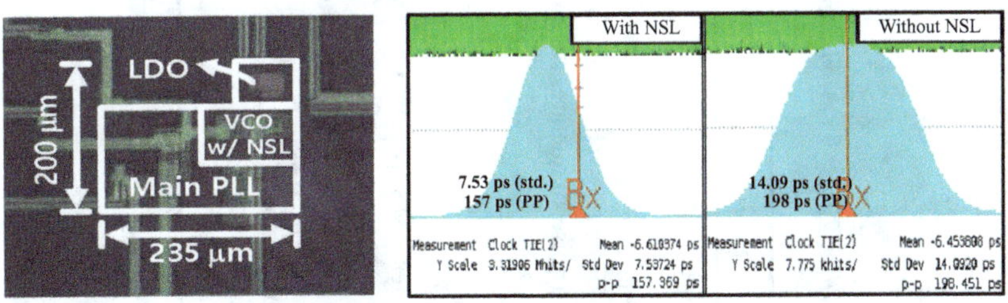

Figure 22.22 Chip photo and measured histograms with 20 mV$_{rms}$ white noise

Table 22.1 Performance summary and comparison

	[18]	[24]	[25]	[27]	[26]
Process [nm]	130	40	40	65	65
Ref. [MHz]	625	26	200	50	50
Output frequency [GHz]	2.5	2.418	3.2	3.2	3.2
Spur or PSNR* improvement [dBc] (noise frequency, Amp)	24* (1 MHz, 30 mV_{pp}) 32* (10 MHz, 30 mV_{pp}) 23* (100 MHz, 30 mV_{pp})	30 (1 MHz, 1 mV_{pp}) 18 (5 MHz, 1 mV_{pp}) 8 (20 MHz, 1 mV_{pp})	11 (10 kHz, 50 mV_{pp}) 10 (100 kHz, 50 mV_{pp}) 6 (10 MHz, 50 mV_{pp})	15 (1 MHz, 50 mV_{pp}) 30 (10 MHz, 50 mV_{pp}) 20 (20 MHz, 50 mV_{pp})	32 (500 kHz, 20 mV_{pp}) 27.5 (2.5 MHz, 20 mV_{pp}) 25.2 (20 MHz, 20 mV_{pp})
Jitter RMS [ps] with and without supply noise	4.8/4.6 (20 mV_{rms} white noise)	3.29/N.A (1 MHz, 1 mV_{pp})	3.85/3.54 (100 kHz. 50 mV_{pp})	1.36/1.31 (30 mV_{pp} white noise)	7.8/7.5 (20 mV_{rms} white noise)
Power [mW]	3.1	6.4	2.915	3.12	2.73
Area [mm^2]	0.08	0.013	0.0216	0.093	0.047
Headroom loss	YES	YES	YES	YES	NO
Calibration	YES	YES	YES	YES	NO

22.4 Conclusions

In this chapter, a VCO noise reduction technique by using reference-less FLL has been reviewed. First, we investigated the operation principles of the noise reduction FLL and how it could potentially reduce the high-frequency phase noise, but not necessarily improve the figure-of-merit of an oscillator. Next, we looked at how the FLL can be implemented in practice for a digital PLL using switched capacitor circuit. The aim was to linearize the DCO frequency tuning curve for a stable gain so that the loop dynamics are insensitive to environmental variations. It also provided a reconfigurable performance in noise and power. Lastly, a supply noise suppression technique was described using the FLL which provides wide bandwidth jitter reduction.

References

[1] Cho S.H., "A self noise cancelling technique for voltage-controlled oscillators." Electron. Lett. 2008, vol. 44, no. 25, pp. 1436–1437.

[2] Lee M.-J. E., Dally W.J., Greer T., *et al.* "Jitter transfer characteristics of delay-locked loops: theories and design techniques." IEEE J. Solid-State Circuits. 2003, vol. 38, no. 4, pp. 614–621.

[3] Maneatis J. G., Kim J., McClatchie I., Maxey J., and Shankaradas M., "Self-biased high-bandwidth low-jitter 1-to-4096 multiplier clock generator PLL." IEEE J. Solid-State Circuits. 2003, vol. 38, no. 11, pp. 1795–1803.

[4] Perrott M. H. CppSim. [Online]. Available: http://www.cppsim.com.

[5] Jung K., Shin J., and Kim J., "Ultralow phase noise microwave generation from mode-locked Er-fiber lasers with subfemtosecond integrated timing jitter." IEEE Photon. J. 2013, vol. 5, no. 3.

[6] August N., Lee H.J., Vandepas M., and Parker R., "A TDC-less ADPLL with 200-to-3200 MHz range and 3 mW power dissipation for mobile SoC clocking in 22 nm CMOS." IEEE International Solid-State Circuits Conference, 2012, pp. 246–248.

[7] Tierno J., Rylyakov A., Friedman D., *et al.* "A DPLL-based per core variable frequency clock generator for an eight-core POWER7TM microprocessor." Symposium on VLSI Circuits, 2010, pp. 85–86.

[8] Rylyakov A., Tierno J., Ainspan H., *et al.* "Bang-bang digital PLLs at 11 and 20 GHz with sub-200 fs integrated jitter for high-speed serial communication applications." IEEE International Solid-State Circuits Conference, 2009, pp. 94–95.

[9] Navid R., Lee T. H., and Dutton R. W. "Minimum achievable phase noise of RC oscillators." IEEE J. Solid-State Circuits. 2005, vol. 40, no. 3, pp. 630–637.

[10] Abidi A. A. "Linearization of voltage-controlled oscillators using switched capacitor feedback." IEEE J. Solid-State Circuits. 1987, vol. 22, no. 3, pp. 494–496.

[11] Ueno K., Asai T., and Amemiya Y. "A 30-MHz, 90-ppm/°C fully-integrated clock reference generator with frequency-locked loop." Proceedings of ESSCIRC, 2009, pp. 392–395.

[12] Jang T., Choi M., Jeong S., Bang S., Sylvester D., and Blaauw D., "A 4.7 nW 13.8 ppm/°C self-biased wakeup timer using a switched-resistor scheme." IEEE International Solid-State Circuits Conference, 2016, pp. 102–103.

[13] Choi M., Jang T., Bang S., Shi Y., Blaauw D., and Sylvester D., "A 110 nW resistive frequency locked on-chip oscillator with 34.3 ppm/°C temperature stability for system-on-chip designs." IEEE J. Solid-State Circuits. 2016, vol. 51, no. 9, pp. 2106–2118.

[14] Youssef M., Zolfaghari A., Darabi H., and Abidi A., "A low-power wide-band polar transmitter for 3G applications." IEEE International Solid-State Circuits Conference, 2011, pp. 378–380.

[15] Sasidhar N., Inti R., and Hanumolu P. K. "Low-noise self-referenced CMOS oscillator." Electron. Lett. 2009, vol. 45, no. 18, pp. 920–921.

[16] Jang T., Jeong S., Jeon D., Choo K. D., Sylvester D., and Blaauw D., "A 2.5 ps 0.8-to-3.2 GHz bang-bang phase- and frequency-detector-based all-digital PLL with noise self-adjustment." IEEE International Solid-State Circuits Conference, 2017, pp. 148–149.

[17] Shen K. Y. J., Farooq S. F. S., Fan Y., *et al.* "A 0.17-to-3.5 mW 0.15-to-5 GHz SoC PLL with 15 dB built-in supply noise rejection and self-bandwidth control in 14 nm CMOS." IEEE International Solid-State Circuits Conference, 2016, pp. 330–331.

[18] Elshazly A., Inti R., Yin W., Young B., and Hanumolu P. K., "A 0.4-to-3 GHz digital PLL with supply-noise cancellation using deterministic background calibration." IEEE International Solid-State Circuits Conference, 2011, pp. 92–94.

[19] Chen Z. Z., Wang Y. H., Shin J., *et al.* "Sub-sampling all-digital fractional-N frequency synthesizer with -111 dBc/Hz in-band phase noise and an FOM of −242 dB." IEEE International Solid-State Circuits Conference, 2015, pp. 376–378.

[20] Kim W., Park J., Park H., and Jeong D. K., "Layout synthesis and loop parameter optimization of a low-jitter all-digital pixel clock generator." IEEE J. Solid-State Circuits. 2014, vol. 49, no. 3, pp. 657–672.

[21] Anand T., Talegaonkar M., Elshazly A., Young B., and Hanumolu P. K., "A 2.5 GHz 2.2 mW/25 μW on/off-state power 2 ps_{rms}-long-term-jitter digital clock multiplier with 3-reference-cycles power-on time." IEEE International Solid-State Circuits Conference, 2013, pp. 256–257.

[22] Alon E., Kim J., Pamarti S., Chang K., and Horowitz M., "Replica compensated linear regulators for supply-regulated phase-locked loops." IEEE Journal of Solid-State Circuits. 2006, vol. 41, no. 2, pp. 413–424.

[23] Huang Y.-C., Liang C.-F., Huang H.-S., and Wang P.-Y., "A 2.4 GHz ADPLL with digital-regulated supply-noise-insensitive and temperature-self-compensated ring DCO." IEEE International Solid-State Circuits Conference, 2014, pp. 270–271.

[24] Yeh C.-W., Hsieh C.-E., and Liu S.-I., “A 3.2 GHz digital phase-locked loop with background supply-noise cancellation.” IEEE International Solid-State Circuits Conference, 2016, pp. 332–333.

[25] Liu J., Jang T., Lee Y., *et al.* “A 0.012 mm^2 3.1 mW bang-bang digital fractional-N PLL with a power-supply-noise cancellation technique and a walking-one-phase-selection fractional frequency divider.” IEEE International Solid-State Circuits Conference, 2014, pp. 268–269.

[26] Kim D., and Cho S. “A supply noise insensitive PLL with a rail-to-rail swing ring oscillator and a wideband noise suppression loop.” Symposium on VLSI Circuits, 2017, pp. 180–181.

[27] Jo Y., Kim H., and Cho S. “A 3.2-GHz supply noise-insensitive PLL using a gate-voltage-boosted source-follower regulator and residual noise cancellation.” IEEE Trans. Very Large Scale Integr. Syst. 2018, vol. 26, no. 10, pp. 2170–2174.

Chapter 23

Time amplified charge pump PLL

Ping-Ying Wang[1]

This chapter introduces a time-amplified charge-pump PLL (TAPLL) which uses a low-noise time amplifier (LNTA) to amplify time difference between a reference clock and a feedback clock all digitally. TAPLL can achieve low in-band noise because the input-referred noise of the loop is divided by the gain of LNTA. Moreover, the bandwidth of PLL can be extended with noiseless circuits to achieve low in-band noise which is one of important keys to implement a wide-band PLL.

This chapter is arranged as follows: First, the noise of a charge pump PLL is discussed, and then design of TAPLL is disclosed followed by the implementations of LNTA and a self-regulated ring oscillator. Finally, a ring oscillator-based TAPLL is measured to demonstrate the superiority of the architecture.

23.1 Noise of a digital charge-pump PLL

A conventional digital charge pump PLL [1] is shown in Figure 23.1. The time difference between a reference clock and a feedback clock is measured by the phase frequency detector (PFD), which consists of the two flip flops with the feedback signal to reset the flip flops. The difference of the pulse width between the outputs of the flip flops equals to the timing error. A charge pump injects current into the loop filter to generate voltage which is proportional to the width of pulse, so the voltage on the loop filter is proportional to the timing error. The oscillator is called voltage-controlled oscillator (VCO), which frequency is proportional to the control voltage linearly with the coefficient K_{VCO}. However, the VCO has two degrees of freedom, the one is the frequency and the other is phases. The frequency adjustment is accumulated in the capacitor and the phase is adjusted by the transient voltage change which is equal to the current of the charge pump multiplied by the resistance. This is the reason why a conventional loop filter is implemented with a noisy resistor and not a noiseless capacitor only.

Recently, a lots of alternatives of PLL [2–8] were developed to replace a digital charge pump PLL by taking advantages of advanced process technologies, the digital charge pump PLL [9–13] is still widely used in industries because it has

[1]CMOS-Crystal Technology, Hsin Chu, Taiwan

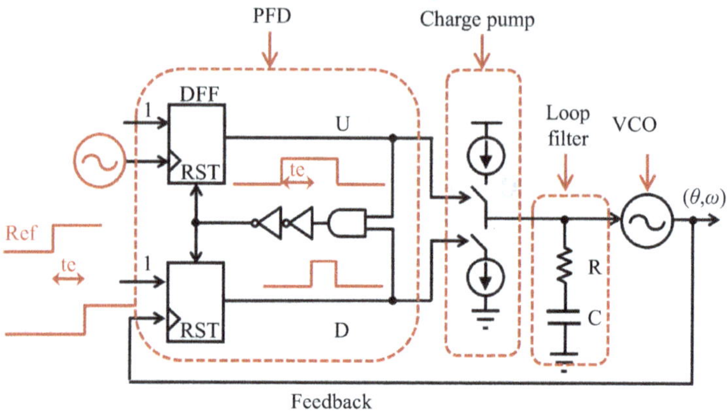

Figure 23.1 Building blocks of a conventional digital charge pump PLL

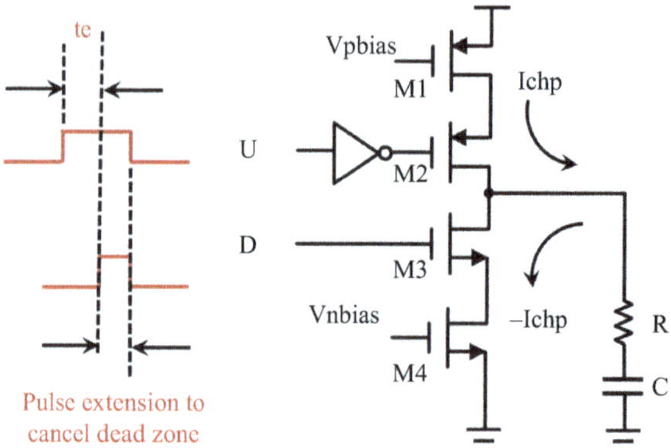

Figure 23.2 Schematic of a simple charge pump

benefits of digital detection of phase difference by using simple logic gates and high-resolution pulse-to-analog voltage generation by the charge pump with passive components. Compared to a time-to-digital converter and digital-to-time converter which induces quantization errors in all digital PLLs [2–8], the charge pump with passive loop filter is easier to achieve subpicosecond resolution and multiple-GHz speed to track phase and frequency of a reference clock.

23.1.1 Noise induced by dead-zone cancellation

However, the bandwidth of the charge pump is still limited by the switching speed of the devices M2/M3 as shown in Figure 23.2. If the width of the pulse to control the switches is smaller than $1/f_{\text{cut-off}}$ of devices M2/M3, the switches cannot be fully turned on, it means that the loop cannot response a small timing error correctly, which is called dead zone. Here, $f_{\text{cut-off}}$ is the cut-off frequency of devices.

To overcome the dead zone, the width of the pulse generated by PFD is extended to maintain the minimum width which is much large than $1/f_{\text{cut-off}}$ of devices. When the timing error is zero, PFD will generate two equal pulses, the charge pump charges and discharges the loop filter simultaneously, as a result, there is no change on the voltage of the loop filter while the minimum width of the pulse is maintained. Accordingly, the loop can response to the timing error correctly within dead zone.

Although the extension of the pulse eliminates the dead zone, it also induces noises. This is because the noise of charging and discharging current cannot be canceled. The noise is generated when the two switches are turned on simultaneously within dead zone to induce variation of the voltage on the loop filter. The noise is proportional to the square root of charge pump's current and minimum with of the pulse extension.

23.1.2 Noise induced by static phase error

In principle, the timing error should be zero if the loop is locked; in practice, there is a static timing error because of mismatch of the current source and the frequency modulation due to the reference clock's noise and VCO's noise. Unfortunately, the static timing error is not constant, it is varied with noise of the current source of charge pump, frequency-modulated noise from VCO, and reference clock.

23.1.3 Noise transfer function of a charge pump

The current of the charge pump and its noise are injected into the loop filter simultaneously. It is impossible to distinguish noises from the current of the charge pump. As a result, there is no way to filter charge pump noise within loop bandwidth by PLL itself. In other words, the noise transfer function of the charge pump is proportional to a transfer function of input signal which is a low-pass filter.

23.2 Time amplified PLL

Since the noise of the charge pump cannot be filtered by the loop and the noise is increased with the current of the charge pump. The performance of a wideband PLL is limited by the noise of the charge pump if loop bandwidth is extended by increasing the current of the charge pump. Accordingly, TAPLL has two advantages: the one is to suppress the noise of the charge pump, and the other is to extend the loop bandwidth without need to increase the current of the charge pump.

Figure 23.3 shows the blocks diagram of the proposed PLL. The time difference between a feedback clock and a reference clock is amplified by LNTA. As a result, input-referred noise of the PLL is suppressed because it is divided by the gain of LNTA, which is 15 in the design. At the same time, the dynamic range of the charge pump current is extended by 15 times, which represents that LNTA suppresses noise of the charge pump and extends the loop bandwidth to filter the noise of VCO simultaneously.

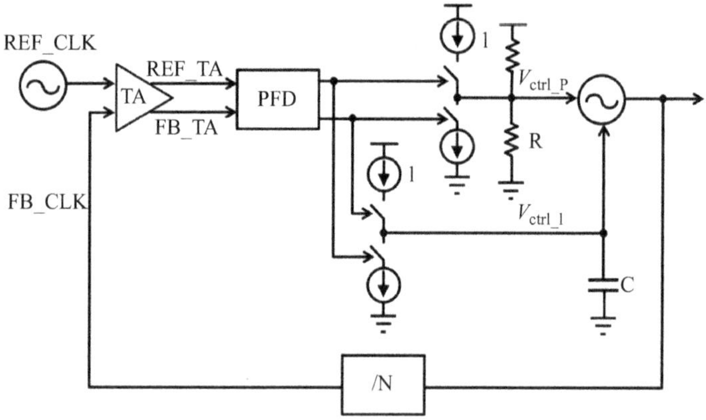

Figure 23.3 Building blocks of TAPLL

23.3 Low-noise time amplifier

LNTA is a pseudo-differential circuit implemented by inverters with push–pull operation. An inverter INV C is AC coupled to INV1 and INV2 by two capacitors C1 and C2 and, two switches connected to the capacitors are controlled by the feedback FB_CLK as shown in Figure 23.4. When the switches are turned on by the feedback clock, the single-ended REF_CLK and a reference voltage V_{ref} are sampled into capacitors as shown in Figure 23.5. When the switches are turned off by the feedback clock, the input of INV C is switched to supply voltage so the capacitors are discharged to the ground. Due to the time difference θ_i between REF_CLK and FB_CLK, the voltages sampled by the capacitors on nodes SA1 and SA2 are different, the discharging time crossing the threshold of the inverters INV1 and INV2 is also different. Furthermore, the time difference θ_o between outputs REF_TA and FB_TA is enlarged due to the slow discharging process as shown in Figure 23.6, which equals to:

$$G_{\text{TA}} * \theta_i \tag{23.1}$$

where G_{TA} is gain of TA:

$$G_{\text{TA}} = \theta_o/\theta_i \tag{23.2}$$

which equals to the slew rate of REF_CLK divided by the slew rate of the discharge on SA1 and SA2.

In this design, slew rate of the discharge is 15 times slower than that of REF_CLK because NMOS in inverter C for the discharge is cascaded with a resistor R to result in a large time constant as shown in Figure 23.4. A tapered slew rate is needed to amplify a time difference in a conventional TA [14], which induces noise. However, in LNTA, the noise of the first stage including resistance

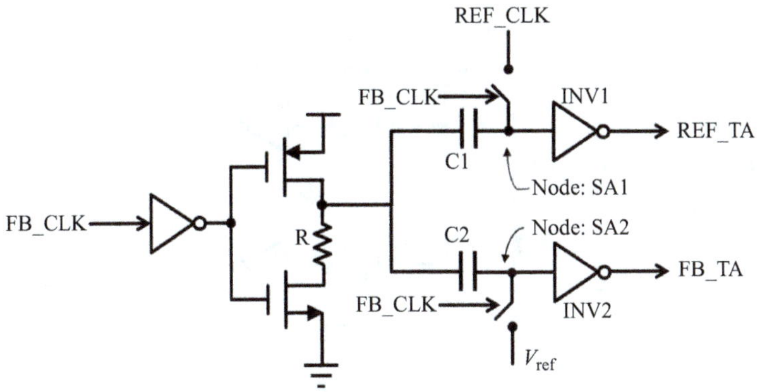

Figure 23.4 Schematic of TA

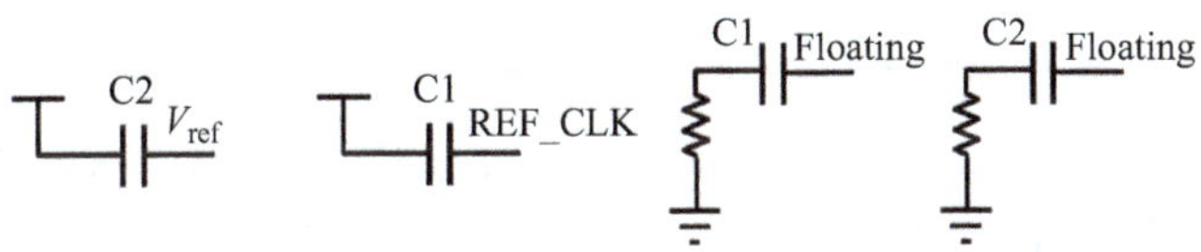

Figure 23.5 Two phases of the capacitors

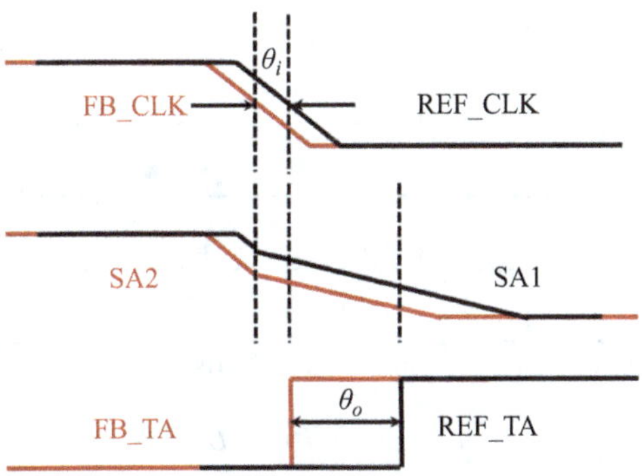

Figure 23.6 Timing diagram

of R and INV C is common mode for the pseudo-differential stages, so the noise is reduced.

The input-referred noise of LNTA is measured by using the configuration of a comparator as shown in Figure 23.7. The bang-bang PD for the test is used to indicate the probability density of the comparator. By fitting the result to a normal distribution, the input-referred noise is 320 μV as shown in Figure 23.8.

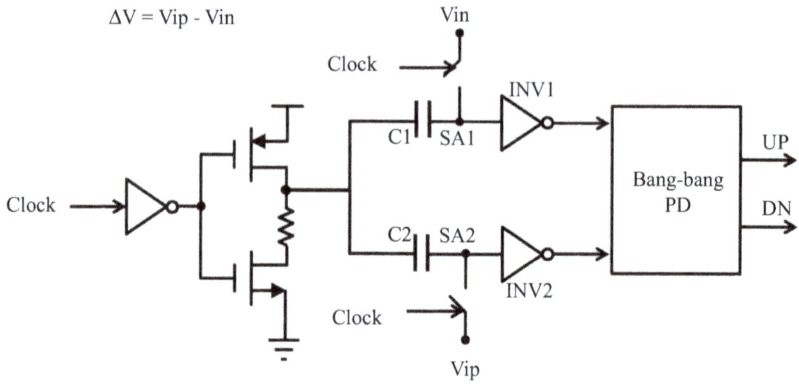

Figure 23.7 Test bench to measure input-referred noise of TA

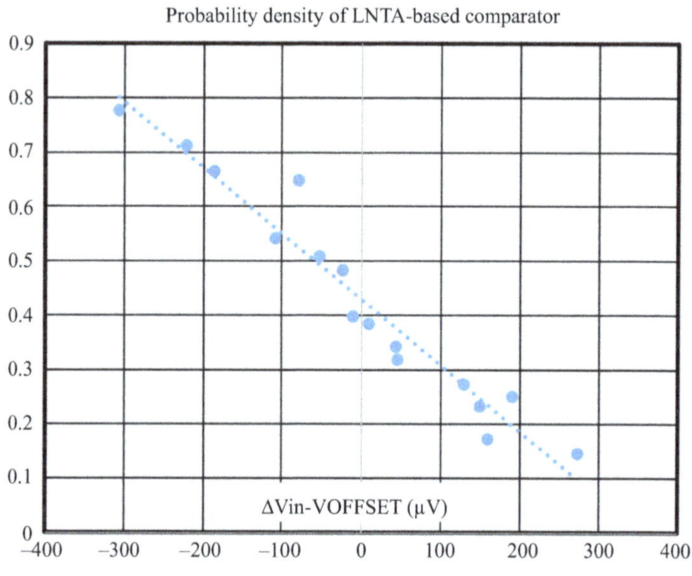

Figure 23.8 Measured probability of LNTA-based comparator

23.4 Self-regulated and self-temperature compensation ring oscillator

Ring oscillator consists of inverters with feedback which is sensitive to the variation of supply voltage and temperature [15]. To improve power supply rejection ratio and reduce temperature coefficient of the ring oscillator for a low supply voltage, a self-regulated and self-temperature compensation circuits are proposed.

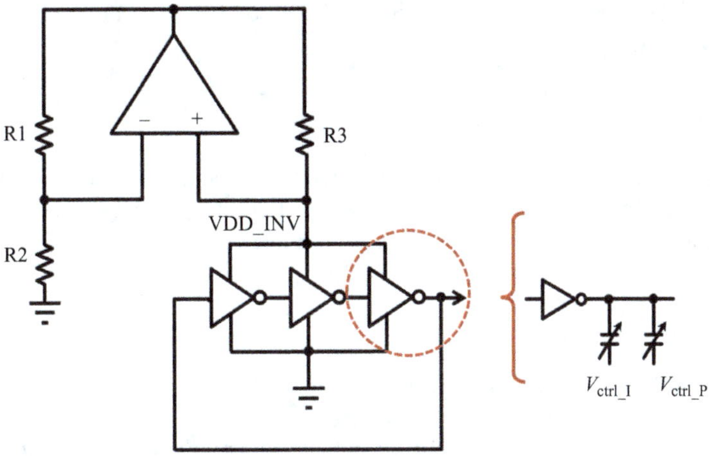

Figure 23.9 Schematic of the self-regulated ring oscillator

Figure 23.9 shows the blocks diagram of the proposed ring oscillator. The drain voltage of inverters VDD_INV is connected to the positive feedback of the operational amplifier, so the drain-source voltage of the inverters is reduced if 1/gm is lower than resistor R3, where gm is transconductance of the ring oscillator. The operational amplifier adjusts the drain voltage of the ring oscillator continuously until 1/gm of the ring oscillator equals to:

$$\mathrm{R2/R1} * \mathrm{R3} \tag{23.3}$$

The period of the ring oscillator is proportional to:

$$C/\mathrm{gm} \tag{23.4}$$

where C is loadings of the inverters. So, the period of the oscillator is proportional to:

$$C * \mathrm{R2/R1} * \mathrm{R3} \tag{23.5}$$

which is independent of a supply voltage. Moreover, the temperature dependence of the period is suppressed because the temperature coefficient of resistors is much smaller than the transconductance of the inverters.

23.5 Ring oscillator-based TAPLL

Ring oscillator is advantageous in wide operation frequency and tuning range as compared to an LC oscillator. However, its spot phase noise is much worse than that of LC oscillators and thus making it impractical to be used in applications for low phase noise such as RF communication. One of the solutions to take advantage of ring oscillator is using a wide-band PLL [16–19] to suppress noise within the

PLL loop including noise of ring oscillator. The loop bandwidth of a charge pump PLL is proportional to:

$$K_{\mathrm{VCO}} * I_{\mathrm{CHP}} * R \tag{23.6}$$

Unfortunately, to increase loop bandwidth by enlarging the magnitude of parameters including VCO gain K_{VCO}, current of charge pump I_{CHP} and resistance of the loop filter R will also increase in-band noise.

Whereas, the loop bandwidth of TAPLL is:

$$G_A * K_{\mathrm{VCO}} * I_{\mathrm{CHP}} * R \tag{23.7}$$

which means the loop bandwidth can be extended by gain of TA without the need to increase the current of the charge pump.

Accordingly, a wide-band TAPLL can have lower in-band noise compared to a conventional wide-band PLL. Moreover, the low-noise time amplifier induces less noise than that of a reference clock, in-band noise of TAPLL will not limited by TAPLL itself.

To demonstrate the analysis and strength of TAPLL, a ring oscillator-based TAPLL is fabricated which achieves lower phase noise than state-of-the-art ring oscillator PLLs [20,21] and comparable performance to LC oscillator [20–22].

23.6 Measurements of ring TAPLL

Figure 23.10 demonstrates that the loop bandwidth can be extended by using TA and the wide-band PLL has less in-band noise. The upper curve is TAPLL, when TA is disabled and the lower curve is a normal TAPLL. The design parameters of the two curves including VCO gain, current of the charge pump, and the loop filter are the same.

Figure 23.11 demonstrates that wide-band PLL by using TA has less noise compared to a conventional wide-band PLL. The upper curve is the phase noise of PLL which bandwidth is extended by increasing the current of the charge pump. The lower curve is the phase noise of PLL with the proposed TA. The phase noise at 1.31 GHz output is reduced 20 dB from 1 K to 10 MHz. It demonstrates that the noise of the loop is suppressed by TA to achieve −120 dBc/Hz at 100 KHz frequency offset and 126 dBc/Hz from 1 MHz to the loop bandwidth which is 10 MHz. In this case, the operation frequency of PFD is 40.9 MHz.

The phase noise at 1.96 GHz is shown in Figure 23.12. It demonstrates that the proposed PLL can achieve −121.9 dBc/Hz at 100 KHz frequency offset, and noise is below −126 dBc/Hz from 1 M to 10 MHz frequency offset. In this case, the operation frequency of PFD is 61.4 MHz.

One of the popular figure of merit (FoM) to benchmark PLLs is:

$$\mathrm{FoM} = 10 * \log((\sigma_t/1\ \mathrm{s})^2 * P_{\mathrm{PLL}}/1\ \mathrm{mW}) \tag{23.8}$$

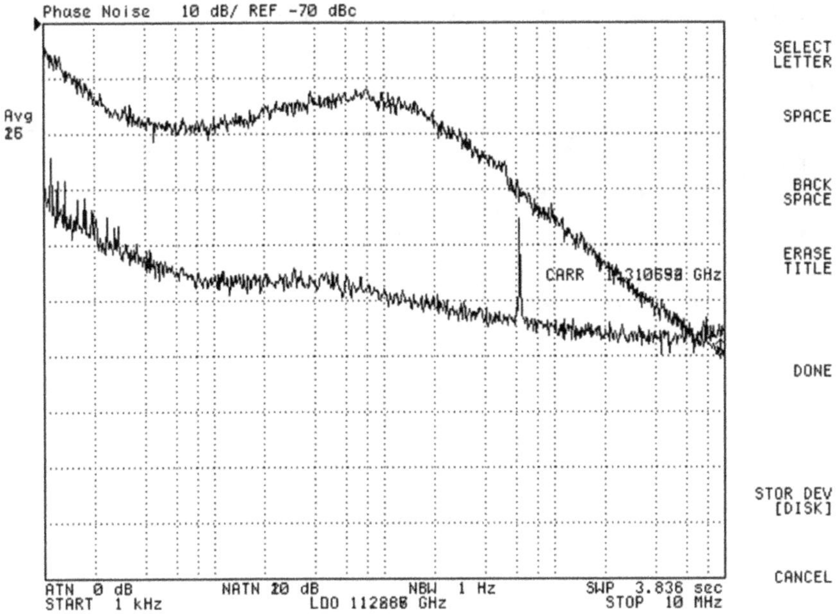

Figure 23.10 Measured phase noise for different loop bandwidth

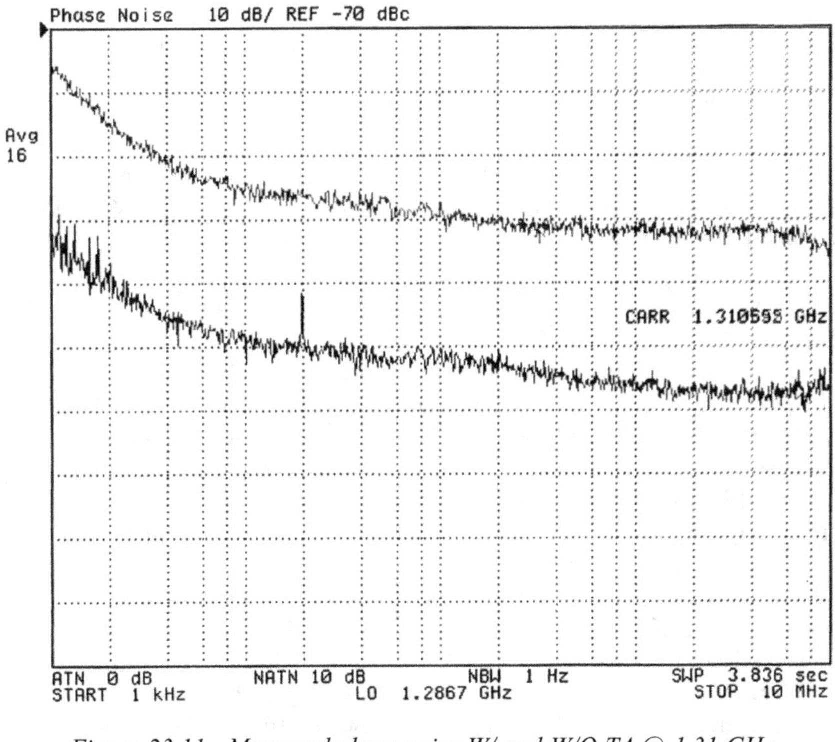

Figure 23.11 Measured phase noise W/ and W/O TA @ 1.31 GHz

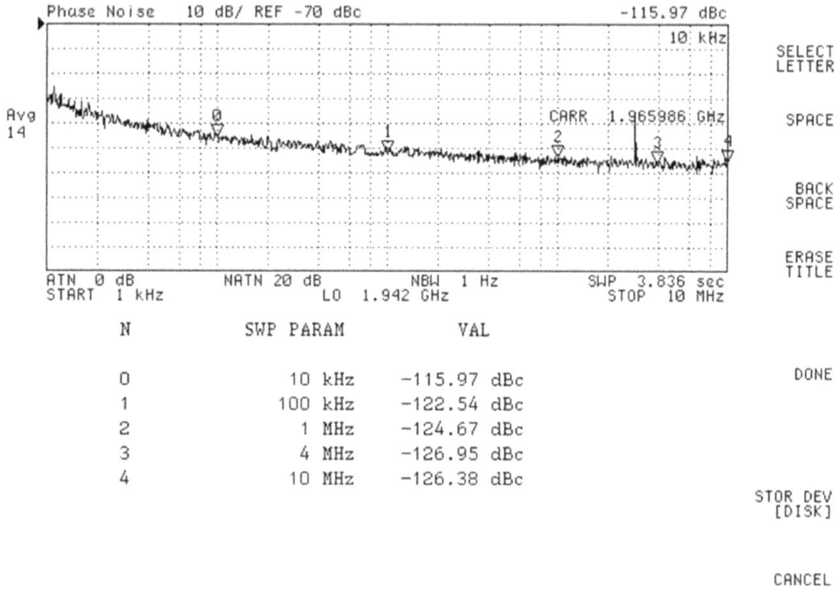

Figure 23.12 Measured phase noise @ 1.965 GHz

where σ_t is one sigma of the jitters, P_{PLL} is the power of PLL. The basic idea of the FoM is that the jitter is independent of the frequency of the reference clock and loop bandwidth. So, we can compare the performance of various DPLLs without need to know frequency of the reference clock and loop bandwidth. However, it is not true for most PLLs. In actuality, jitter can be improved by using the higher frequency of the reference clock or loop bandwidth. Accordingly, a better FoM for benchmark should be:

$$\text{FoM} = 10 * \log((F_{\text{ref}}/1\ \text{MHz}) * (\sigma_t/1\ \text{s})^2 * P_{\text{PLL}}/1\ \text{mW}) \tag{23.9}$$

where F_{ref} is the frequency of the reference clock.

Table 23.1 is the summary of performance and comparison. The in-band phase noise is 10 dB better than ring oscillator PLLs [20,21] if the output is normalized to the same frequency. Also, the in-band phase noise is still 8 dB better than ring oscillator PLLs [20,21] if phase noise is normalized to the same frequency of output, frequency of reference clock and power. The PLL also achieves the same phase noise compared to LC oscillator [22,25,26] if the noise is normalized to the same frequency of output while frequency range of our design is about ten times wider and area is about four times smaller than [22,25,26].

Figure 23.13 is the die micrograph. The chip is fabricated by 45 nm process, which occupies 0.08 mm^2 area. It consumes 20.1 mA current from a 1.2 V supply when it operates at 1.96 GHz. The VCO consumes 16 mA while other circuits including buffer of reference clock consuming 4 mA current.

Table 23.1 Comparisons of phase noise performance

	ISSCC 18 [22]	ISSCC 18 [21]	ISSCC 17 [20]	ISSCC 17 [23]	ISSCC 17 [24]	TAPLL
Oscillator technology	LC	LC	LC	Ring	Ring	Ring
PLL architecture	BB-PLL W/divider	RS-PLL divider-less	DTCD PLL W/divider	SS-PLL divider-less	S/H-PLL W/divider	TAPLL W/divider
Frequency of phase detection	52 MHz	51 MHz	26 MHz	192 MHz	22.6 MHz	61.4 MHz
Output frequency	3.84 GHz	2.55 GHz	2.69 GHz	2.3 GHz	2.42 GHz	1.965 GHz
Phase noise @ 100 KHz	−106 dBc/Hz	−120 dBc/Hz	−113 dBc/Hz	−106 dBc/Hz	−97 dBc/Hz	−122 dBc/Hz
Phase noise @ 1 MHz	−121 dBc/Hz	−125 dBc/Hz	−122 dBc/Hz	−114 dBc/Hz	−104 dBc/Hz	−126 dBc/Hz
Integrated phase noise	0.183 ps (0.01–30 M)	0.11 ps (0.01–100 M)	0.154 ps (0.01–10 M)	0.72 ps (0.01–100 M)	1.5 ps (0.01–50 M)	0.2 ps (0.01–10 M)
Technology area	65 nm 0.61 nm^2	65 nm 0.36 mm^2	14 nm 0.257 mm^2	65 nm 0.0049 mm^2	45 nm 0.096 mm^2	45 nm 0.08 mm^2
Reference clock spur	NA	−67 dBc	−87 dBc	−37 dBc	−60 dBc	−50 dBc
Supply voltage	NA	1.2/0.5	NA	1.2	1.0	1.2
Power	5.28 mW	3.7 mW	13.4 mW	4.59 mW	10 mW	24.2 mW
OSC temperature coefficient	NA*	NA*	NA*	NA**	NA**	0.1%/C

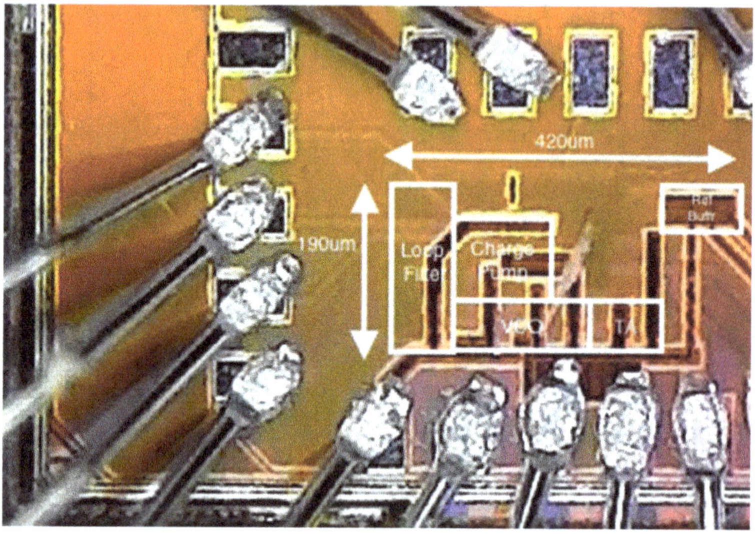

Figure 23.13 Micrograph of the chip in 45 nm CMOS (0.08 mm^2)

References

[1] Gardner F. "Charge-pump phase-lock loop." *IEEE Trans. Commun.* 1980, pp. 1849–1858.

[2] Dunning J., Garcia, G., Lundberg, J. *et al.* "An all-digital phase-locked loop with 50-cycle lock time suitable for high-performance microprocessors." *JSSC*. 1995, pp. 412–422.

[3] Staszewski R.B., Wallberg, J., Rezeq, S. *et al.* "All-digital PLL and GSM/EDGE transmitter in 90 nm CMOS." *ISSCC*. 2005, paper 17.5.

[4] Wang P.-Y., Kao, H.-W., Lin, Y.-Y. *et al.* "DLL-based clock recovery in a PRML channel." *ISSCC*. 2005, paper 31.2.

[5] Wang P.-Y., Yang M.-T., Chen S.-P., Lin M.-H., and Yang J.-B., "RTL-based clock recovery architecture with all-digital duty-cycle correction." *ISSCC*. 2007, paper 14.1.

[6] Chang H.-H. Wang P.Y., Conan Zhan J.-H., and Hsieh B.-Y. "A fractional spur-free ADPLL with loop-gain calibration and phase-noise cancellation for GSM/GPRS/EDGE." *ISSCC*. 2008, paper 10.1.

[7] Hsu C.-M., Straayer M.Z., and Perrot M.H. A low-noise wide-BW 3.6-GHz digital ΔΣ fractional-N frequency synthesizer with a noise-shaping time-to-digital converter and quantization noise cancellation." *JSSC*. 2008, pp. 2776–2786.

[8] Wang P.-Y., Conan Zhan J.-H., Chang H.-H., and Chang H.-M. S. "A digital intensive fractional-N PLL and all-digital self-calibration scheme." *JSSC*. 2009, pp. 2182–2192.

[9] Maneatis J.G. "Low-Jitter process-independent DLL and PLL based on self-biased technique." *JSSC*. 1996, pp. 1723–1732.

[10] Maneatis J.G. "Self-biased, high-bandwidth, low-jitter, 1-to-4096 multiplier clock-generator PLL." *ISSCC*. 2003, paper 24.2.

[11] Riley T., Copeland, M.A., and Kwasniewski T.A. "Delta-sigma modulation in fractional-N frequency synthesis." *JSSC*. 1993, pp. 553–559.

[12] Wang P.-Y., Chen S.-P., and Chen P. "Timing orthogonal capacitance multiplication technique for PLL." *VLSI Symposium*. 2007, paper 16.1.

[13] Wang C.-H., Wang P.-Y., Ke L.-W. *et al*. "A direct digital frequency modulation PLL with all digital on-line self-calibration for quad-band GSM/GPRS transmitter." *VLSI Symposium*. 2009, paper 18.3.

[14] Lee M., and Abidi A.A. "A 9b, 1.25 ps resolution coarse-fine time-to-digital converter in 90 nm CMOS that amplifies a time residue." *VLSI Symposium*. 2007, paper 16.2.

[15] Huang Y.-C., Liang C.-F., Huang H.-S., and Wang P.-Y. "A 2.4 GHz ADPLL with digital-regulated supply-noise-insensitive and temperature-self-compensated ring DCO." *ISSCC*. 2014, paper 15.3.

[16] Liang C.F., and Wang P.Y. "A wideband fractional-N ring PLL using a near-ground pre-distorted switched-capacitor loop filter." *ISSCC*. 2015, paper 10.8.

[17] Swaminathan A., Wang K.J., and Galton I. "A wide-bandwidth 2.4 GHz ISM-band fractional-N PLL with adaptive phase-noise cancellation." *JSSC*. 2007, pp. 2639–2650.

[18] Kong L., and Razavi B. "A 2.4 GHz 4 mW inductor-less RF synthesizer." *ISSCC*. 2015, paper 25.7.

[19] Elkholy A., Talegaonkar M., Anand T., and Hanumolu P.K. "A 6.75-to-8.25 GHz 2.25 mW 190 fs rms integrated-jitter PVT-insensitive injection-locked clock multiplier using all-digital continuous frequency-tracking loop in 65 nm CMOS." *ISSCC*. 2015, paper 10.7.

[20] Yao C.-W., Loke W.-F., and Ni R. "A 14 nm fractional-N digital PLL with 0.14 ps rms jitter and −78 dBc fractional spur for cellular RFICs." *ISSCC*. 2017, pp. 422–423.

[21] Sharma J., and Krishnaswamy H. "A dividerless reference-sampling RF PLL with −253.5 dB jitter FOM and <−67 dBc reference spurs." *ISSCC*. 2018, pp. 258–259.

[22] Bertulessi L., Grimaldi L., Cherniak D., Samori C., and Levantino S. "A low-phase-noise digital bang-bang PLL with lock over a wide lock range." *ISSCC*. 2018, pp. 252–253.

[23] Chuang J., and Krishnaswamy K. "A 0.0049 mm^2 2.3 GHz sub-sampling ring-oscillator PLL with time-based loop filter achieving −236.2 dB jitter-FOM." *ISSCC*. 2017, pp. 328–329.

[24] Long Kong L., and Razavi B. "A 2.4 GHz RF fractional-N synthesizer with $0.25f_{REF}$." *ISSCC*. 2017, pp. 330–331.

Chapter 24

Multiplying DLLs

Shiheng Yang[1], Jun Yin[1], Pui-In Mak[1], and Rui P. Martins[1,2]

The modern SoC chip usually requires multiple clock generators that can achieve low jitter performance over a wide frequency range with the compact area and low power consumption. The multiplying delay-locked-loop (MDLL) employing a ring voltage-controlled oscillator (VCO) is a promising solution to fulfill all these requirements. This chapter focuses on the design of ring-VCO-based MDLL that can achieve low-jitter and spur performance across a wide frequency tuning range in the presence of process-voltage-temperature variations.

24.1 Introduction

The ring VCO (RVCO) has been widely used for the clock-generation circuits due to its wide frequency tuning range, tiny area, and freedom from magnetic pulling when compared with the LC VCO. Unfortunately, the poor phase noise of the RVCO limits the overall jitter performance if a conventional charge-pump based type-II phase-locked loop (PLL) is employed to stabilize the RVCO frequency. It is because the bandwidth (BW) of a type-II PLL is typically smaller than one-tenth of the reference frequency (f_{REF}) to maintain the loop stability, which limits phase noise suppression of the RVCO.

To break the BW limitation, we can directly inject the reference clock (REF) into the VCO to replace the noisy VCO output edge (OUT) every REF period [1] as shown in Figure 24.1(a). Since the VCO jitter is cleaned much faster, the BW of the VCO noise transfer function can be extended to one-fourth of the f_{REF}, aiding to more effective suppression of the VCO phase noise. Assuming the $1/f^3$ phase noise of the VCO is ignored for simplicity, the REF injection can reduce the in-band phase noise by more than 8 dB compared with the conventional type-II PLL as shown in Figure 24.1(b). However, if the VCO frequency f_{VCO} deviates from the targeted frequency $f_{LOCK} = N \times f_{REF}$, the 1st to $(N-1)$th output periods within one

[1]State Key Laboratory of Analog and Mixed-Signal VLSI, Institute of Microelectronics and Faculty of Science and Technology, University of Macau, Macao, China
[2]On leave from the Instituto Superior Técnico, Universidade de Lisboa, Portugal

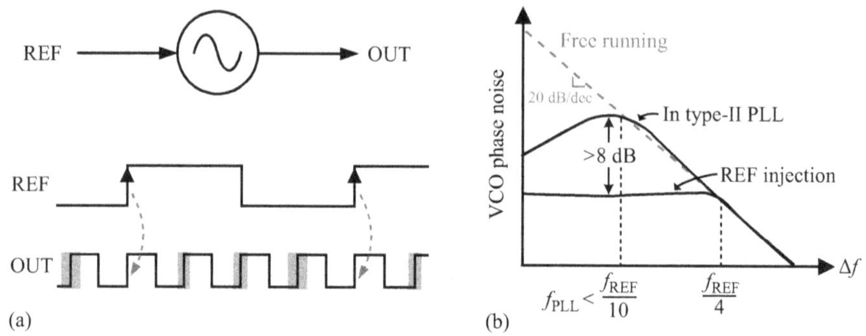

Figure 24.1 (a) Concept of the VCO with REF injection and (b) comparison of the VCO phase noises in a type-II PLL and with REF injection

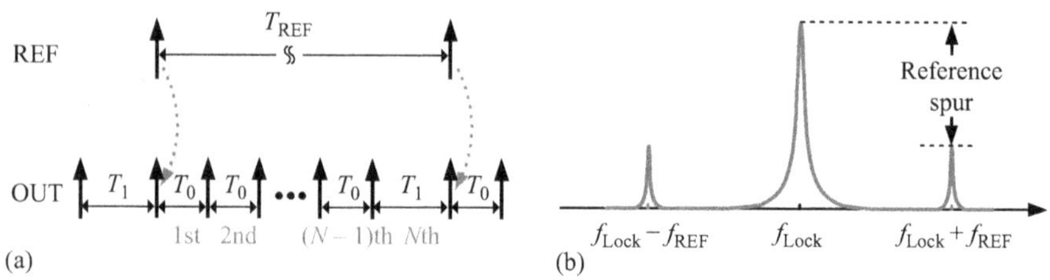

Figure 24.2 (a) Illustration of the static phase offset at the output of the VCO with REF injection and (b) the reference spur induced by the static phase offset

reference cycle will be equal to $T_0 = 1/f_{VCO}$ while the last period will be abruptly changed to $T_1 = T_{REF} - (N - 1)T_0$ due to the edge replacement as shown in Figure 24.2(a). This static phase offset (SPO), that is, $T_1 - T_0$, would cause a deterministic jitter or a reference spur in the output [Figure 24.2(b)]. The relationship between the reference spur and VCO frequency deviation $f_{DEV} = |f_{VCO} - f_{LOCK}|$ can be estimated based on Fourier analysis [2]:

$$\text{Reference spur} \approx -20\log\left(\frac{|T_1 - T_0|}{T_{LOCK}}\right) = -20\log\left(\frac{f_{DEV}}{f_{REF}}\right) \tag{24.1}$$

At $f_{LOCK} = 2$ GHz and $f_{REF} = 100$ MHz, a 0.5% deviation of the f_{VCO}, that is, $f_{DEV} = 10$ MHz, would induce a large reference spur of –20 dBc. If $|f_{DEV}| > f_{REF}$, the VCO can be even locked to a wrong frequency, for example, $(N - 1) \times f_{REF}$ or $(N + 1) \times f_{REF}$. Thus it is necessary to calibrate the VCO frequency to guarantee $f_{VCO} = f_{LOCK}$. The calibration loop or the frequency-tracking loop is favored to operate continuously at the background to track the temperature and supply variation when the clock generator is working.

An effective way to track the VCO frequency is to use a MDLL [2–9]. Figure 24.3(a) shows a typical architecture of the MDLL. The multiplexer (MUX)

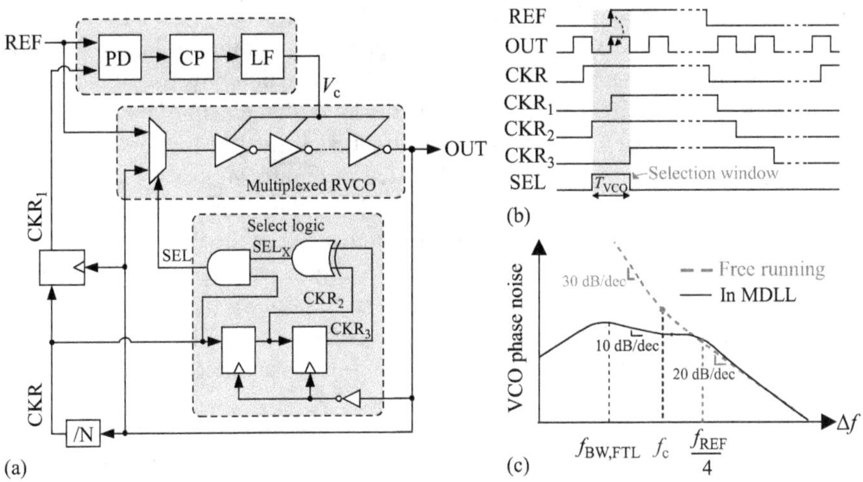

Figure 24.3 (a) Block diagram, (b) timing diagram, and (c) VCO phase noise of a typical MDLL

in the multiplexed RVCO (MRVCO) is controlled by the SEL signal to select the rising edge of REF to replace the noisy edge in the VCO output. At other time, the MUX just let the OUT signal to pass so that the MRVCO operates like a typical RVCO. To generate an SEL signal with a time window aligned with the rising edge of REF, the select logic uses the two consecutive falling edges of the OUT signal to sample the divider output CKR through two D-flip-flops (DFFs) which creates the low-frequency signals CKR_2 and CKR_3. Here CKR_3 delays CKR_2 by one VCO cycle (T_{VCO}). Then the selection window SEL_X is generated by performing the XOR operation for CKR_2 and CKR_3. To avoid the selection of REF when its falling edge comes, we perform the AND operation for SEL_X and CKR to make sure the selection window only appears at the rising edge of the REF. Here the frequency-tracking loop will force the alignment between the rising edges of the REF and CKR_1 through tuning the frequency of the MRVCO. Since the CKR_1 is generated by sampling the CKR with the rising edge of the OUT, its rising edge is always located in the middle of the selection window if the delay of the divider (i.e., from OUT to CKR) is shorter than $T_{VCO}/2$ as shown in Figure 24.3(b). As a result, the selection window is automatically aligned with the rising edge of the REF with a sufficient margin of $\pm T_{VCO}/2$. If the divider delay cannot be kept shorter than $T_{VCO}/2$ in old CMOS technology, a delay can be inserted after the divider to ensure CKR if firstly sampled by the falling edge of OUT to guarantee the alignment of the REF and the selection window. The frequency-tracking loop is implemented by using the conventional charge-pump-based PLL, which will calibrate the VCO frequency in the background.

In ultra-scaled CMOS technology, the $1/f^3$ phase noise corner frequency (f_c) of an RVCO can be as high as several megahertz if short-channel transistors are employed to reduce the power consumption [10]. Unfortunately, the REF injection can only provide a 20 dB/dec phase-noise suppression within the injection bandwidth. Thus the phase noise of the VCO with REF injection will rise at a slope of 10 dB/dec at the offset frequency lower than the f_C. With the help of the frequency-tracking loop, the phase noise at the offset frequency smaller than its bandwidth ($f_{BW,FTL}$) can be further suppressed, as depicted in Figure 24.3(c).

In practice, the delay between OUT and CKR_1, the mismatches of the phase detector and charge pump, and the delay mismatch between the PD and MUX paths will cause an SPO between the REF and OUT edges, leading to a deterministic jitter and reference spur. According to (24.1), an SPO of 10 ps can cause a large reference spur of −34 dBc when the MDLL operates at 2 GHz. Thus it will be necessary to explore novel frequency-tracking loop architecture to minimize the SPO.

In Section 24.2, we will introduce several recently published frequency-tracking loops that aid to reduce the SPO and reference spur. In Section 24.3, we will discuss the MRVCO design that enables the MDLL to maintain a constant jitter performance over a wide frequency range. The phase noise analysis for the MDLL will be presented in Section 24.4. Finally, we will compare the performance of the state-of-the-art MDLLs in Section 24.5.

24.2 Advanced frequency-tracking loop

To eliminate the mismatches of the phase detector and charge pump in the frequency-tracking loop, [5] proposes a digital frequency-tracking loop that employs a 1-bit bang-bang phase detector (BBPD) to extract the phase difference between REF and OUT (Figure 24.4). By increasing the transistor sizes and reducing the rise time of both REF and OUT, the phase offset of the BBPD can be reduced. The Monte Carlo simulation shows the standard deviation of the phase offset can be lower than 0.23 ps with a rise time of 30 ps for REF and OUT. Also by directly comparing the VCO output with REF, the DFF delay in the conventional MDLL [Figure 24.3(a)] is also eliminated. Since the VCO output and REF are

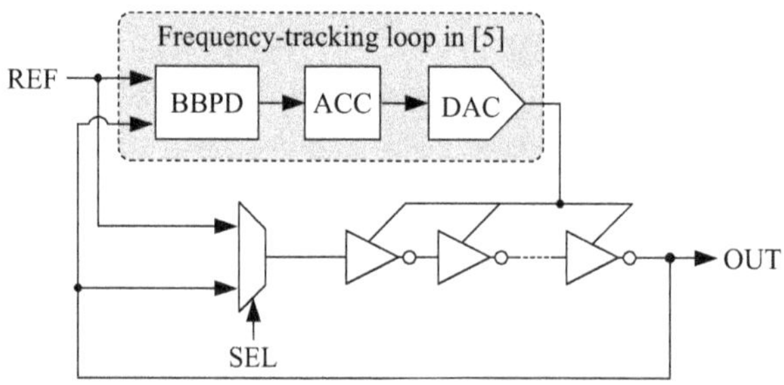

Figure 24.4 Simplified block diagram of the MDLL using a BBPD-based frequency-tracking loop

directly compared, the frequency-tracking loop cannot guarantee the frequency lock, that is, $f_{VCO} = N \times f_{REF}$. The frequency-tracking loop will settle as long as f_{VCO} is integer multiple of f_{REF}, for example, $(N-1) \times f_{REF}$ or $(N+1) \times f_{REF}$. Thus an additional frequency-locked loop or coarse frequency-tracking loop is needed. The frequency-locked loop is usually implemented by counting the number of the VCO cycles within each reference period, and we do not include it in the simplified MDLL block diagram in Figure 24.4 for simplicity. Although the phase offset in the frequency-tracking loop is minimized, the delay mismatch between the BBPD and MUX paths still exists similar to the conventional design which limits the SPO detection accuracy.

The methods to compare the phases between REF and OUT would inevitably introduce the phase offset and path mismatch when implemented by the real circuits. To avoid these phase offsets, we can consider directly extracting the SPO information from the VCO output without comparing it with the REF clock. Since the SPO is defined as $T_1 - T_0$, it can be obtained by directly comparing the periods of VCO output. For example, the periods of the $(N-1)$th and Nth cycle of the VCO output in Figure 24.2(a) already contain the information of T_0 and T_1. If we can precisely measure the periods of these two cycles, the SPO can be obtained. In [2], a gated-ring-oscillator (GRO) based high-resolution time-to-digital converter (TDC) is employed to measure the VCO output periods as shown in Figure 24.5(a). At first, a pulse generator converts VCO cycles with time lengths of T_1 and T_0 into pulses with different time duration every REF cycle. Then the pulse signal PS is fed into the TDC. The GRO-based TDC is enabled only when the PS is high which quantizes the T_0 and T_1. The correlator subtracts T_1 and T_0 in the digital domain, and its digital output Corr represents the SPO as shown in Figure 24.5(b). The accuracy of the obtained SPO directly depends on the TDC resolution. To achieve an SPO less than 0.5 ps for a 2 GHz MDLL, a 10-bit TDC is needed which demands a high power consumption. The MDLL using the TDC-based frequency-tracking for T_0/T_1 comparison also needs an additional frequency-locked loop which is not shown in the simplified block diagram here.

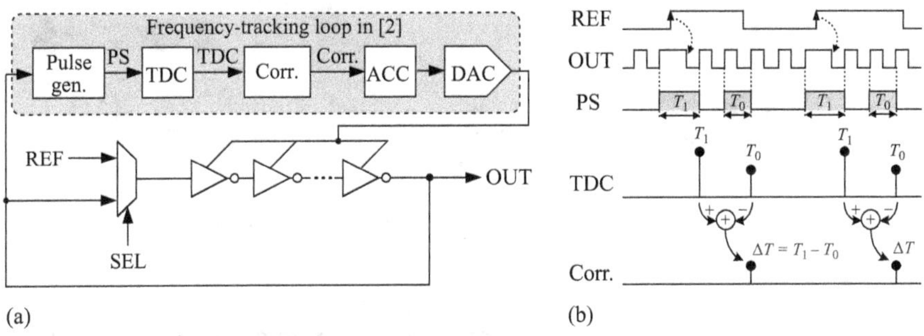

Figure 24.5 (a) Simplified block diagram and (b) timing diagram of the MDLL using a TDC-based frequency-tracking loop for T_1/T_0 comparison

To reduce the power consumption, [6] converts the problem to compare T_1 and T_0 from the time domain to the voltage domain. As shown in Figure 24.6(a), the time lengths of T_1 and T_0 are converted to the voltage signals V_1 and V_2 through the time-to-voltage (T2V) converter at first. Instead of directly subtracting V_1 and V_2 to get a precise SPO, a low-power voltage comparator is employed to only extract the sign of $V_1 - V_2$. The comparator output PWC_{OUT} is accumulated to adjust the frequency of the RVCO. Here both the voltage offsets between the two T2V converters and from the voltage comparator would degrade the SPO detection accuracy [Figure 24.6(b)] which needs to be carefully calibrated. By continuously sending two pulses with the same time length of T_0 to the frequency-tracking loop, that is, $T_{IN1} = T_{IN2} = T_0$, the offset between the upper and down paths can be stored in V_{CAL} by accumulating the PWC_{OUT}. In the frequency-tracking mode to compare V_1 and V_2, we can subtract V_{CAL} from V_2 to nullify the effect of the offset voltage as shown in Figure 24.6(b). Although resolving the T_1 and T_0 comparison in the analog domain benefits for the power consumption, the analog circuits in the pulse width comparator and the digital-to-analog converter (DAC) for offset calibration occupy a substantial amount of area (0.015 mm^2).

To achieve a compact area by taking advantage of the CMOS technology scaling down, [7] proposes a power-efficient frequency-tracking loop architecture to resolve the T_1 and T_0 comparison directly in the digital domain. Since the adjacent edges of OUT that carry the information of T_0 and T_1 do not appear simultaneously, we can store the time length of T_0 (T_1) first, and then compare it with T_1 (T_0). Unlike the voltage information that can be easily stored in a capacitor, storing time information is more difficult. Using a high-resolution TDC to digitize

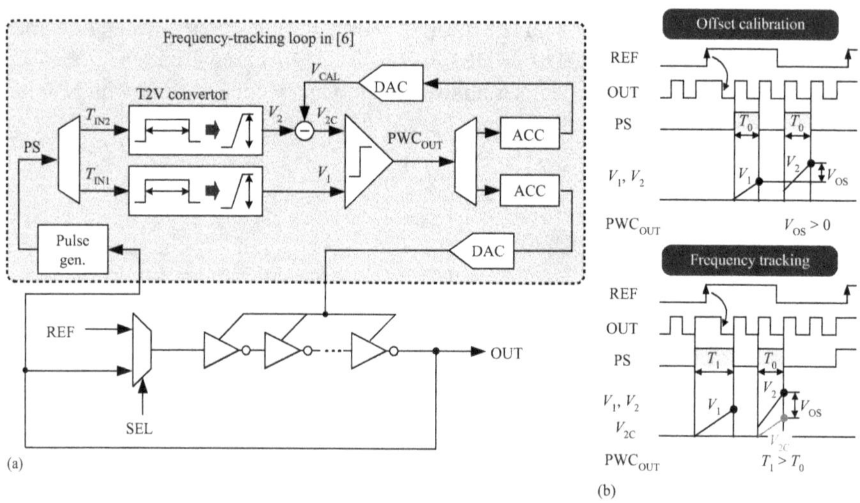

Figure 24.6 (a) Simplified block diagram and (b) timing diagram of the MDLL using an analog pulse-width-comparator-based frequency-tracking loop

T_0 and T_1 directly [2] is power inefficient since the TDC has to cover a large input range which will only be used during the frequency-locking process. After the MDLL is settled, the TDC output will be stable with a small variation.

To improve the power efficiency, we can consider a digital-controlled delay line (DCDL)-based successive approximation (SA) scheme to digitize and store a certain time length [Figure 24.7(a)]. Here, we use the DCDL with a delay time of T_{DCDL} to delay the early edge E_1 and the BBPD, and the feedback loop will automatically adjust the T_{DCDL} until the two edges E_{1d} and E_2 are close enough. Assuming the time intervals between the two input edges E_1 and E_2 is T_0, the settled DCDL control code DTW will jump between the two adjacent values D_1 and D_2, and the corresponding time error Δt_e between T_{DCDL} and T_0 will be smaller than one DCDL step t_q, as shown in Figure 24.7(a). Either the control code DTW_1 or $DTW_1 + 1$ can be stored as DTW_{T0} which sets T_{DCDL} close to T_0 with a maximum quantization error of t_q. The proposed SA scheme allows significant power savings by replacing the high-resolution TDC with a low-power DCDL and BBPD but at the cost of a longer DCDL calibration time. To speed up the comparison of T_0 and T_1 once the DTW_{T0} has been found, we do not repeat the same procedure to find DTW_{T1} and compare it with DTW_{T0}. Instead, we set the DCDL control code to DTW_{T0} and send the input edges E_1 and E_2 with a delay of T_1 to the time-interval comparator as exhibited in Figure 24.7(b). As a result, T_0 and T_1 can be directly compared by the BBPD with one single comparison.

Figure 24.8 exhibits the MDLL block diagram with the frequency-tracking loop employing the DCDL-based SA scheme. The adjacent edge selector extracts two edges E_1 and E_2 with a delay T_0 or T_1 and will be discussed in detail later.

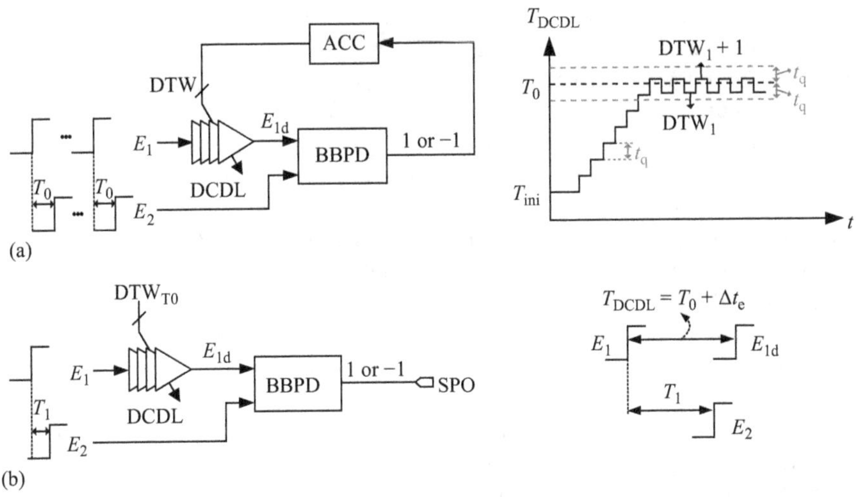

Figure 24.7 Concept of the DCDL-based SA time-length digitization and comparison scheme: (a) The DCDL calibration mode and (b) the T_0/T_1 comparison mode

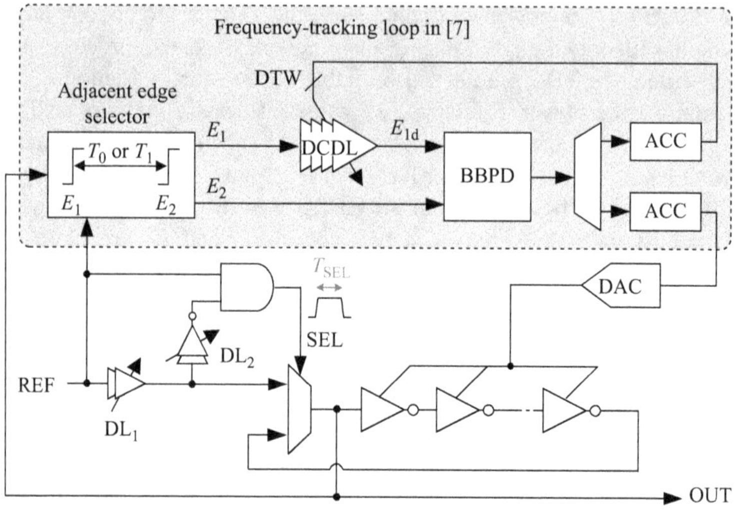

Figure 24.8 Simplified block diagram of the MDLL using an offset-free DCDL-based frequency-tracking loop

In the first phase of calibration as shown in Figure 24.9(a), the delay of the DCDL is continuously adjusted to T_0 every REF cycle according to the BBPD output. While in the second phase of frequency tracking (Figure 24.9(b)), the SA scheme will perform T_0 and T_1 comparison, and the accumulated comparison results are used to correct the VCO frequency. Since both T_0 and T_1 will change after the adjustment of the VCO frequency, the DCDL calibration and the T_0/T_1 comparison will be scheduled in an interleaved manner every REF cycle until the RVCO frequency is locked ($T_0 = T_1$). After that, the loop will stay in the frequency-tracking phase to continuously correct the RVCO frequency error induced by the voltage and temperature variations.

To avoid the use of power-hungry frequency divider, the select window SEL here is generated directly through the REF (Figure 24.8). The rising edge of REF is placed in the middle of the pulse SEL by adjusting the delay lines DL_1 and DL_2. The window length of SEL has to satisfy the condition $T_{LOCK}/N < T_{SEL} < T_{LOCK}$. Here the upper limit is determined by the condition not blocking the transition of the edges in the RVCO right before and after the REF injection while the lower limit is chosen to avoid the glitch caused by the competition between the RVCO edge generated by its former stage and the injected REF edge.

Figure 24.10 shows the block diagram of the time-interval comparator that extracts the adjacent two rising edges from the OUT through DFF_1 and DFF_2. At first, the VCO output is resampled by the falling edge of REF to generate the CKR signal. In the DCDL calibration mode, the CKR is directly utilized to control the reset RST of DFF_1 and DFF_2 to produce E_1 and E_2 with a delay of T_0. Since the CKR signal has a rising edge aligned with the falling edge of OUT, using CKR to reset DFF_1 and DFF_2 can guarantee enough time-margin from the reset port to the

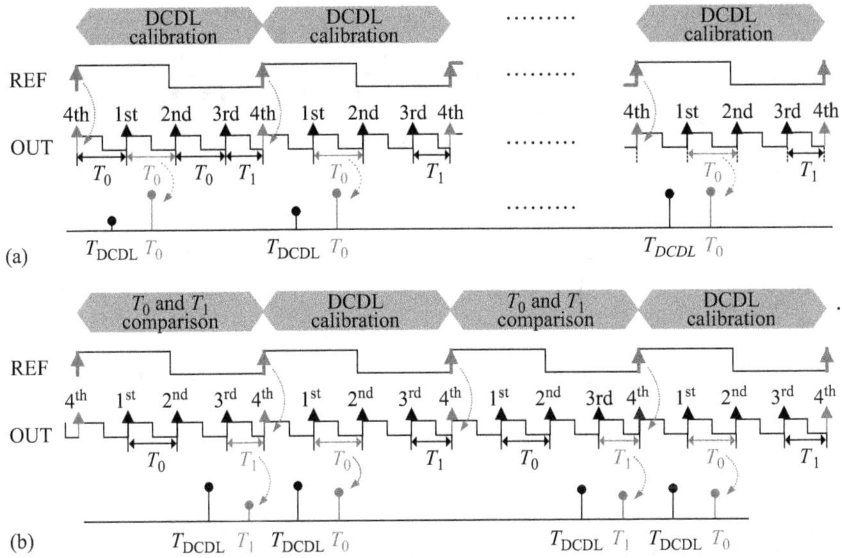

Figure 24.9 Workflow of the offset-free DCDL-based frequency-tracking loop ($N = 4$) in (a) the calibration phase and (b) the frequency tracking phase

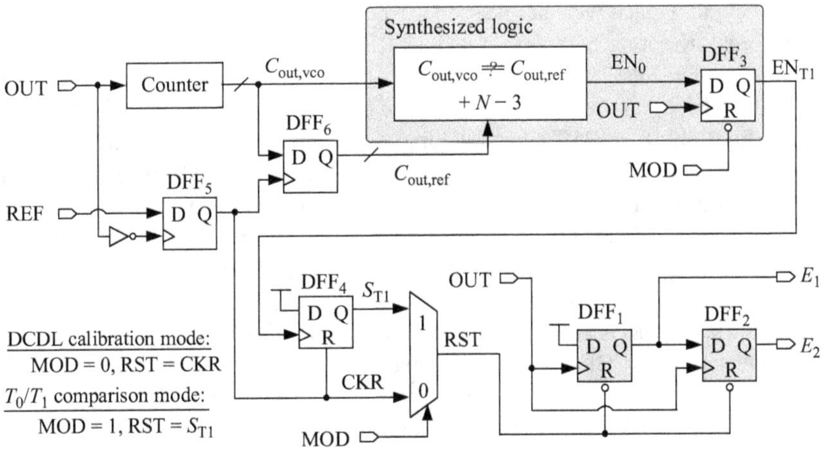

Figure 24.10 Block diagram of the adjacent edge selector

clock port of DFFs to avoid the metastability issues. Therefore, the two rising edges of OUT occurring right after the rising edge of REF can be properly selected as shown in Figure 24.11(a). In the T_0/T_1 comparison mode, the $(N - 1)$th and Nth rising edges of OUT occurring after the rising edge of REF must be precisely

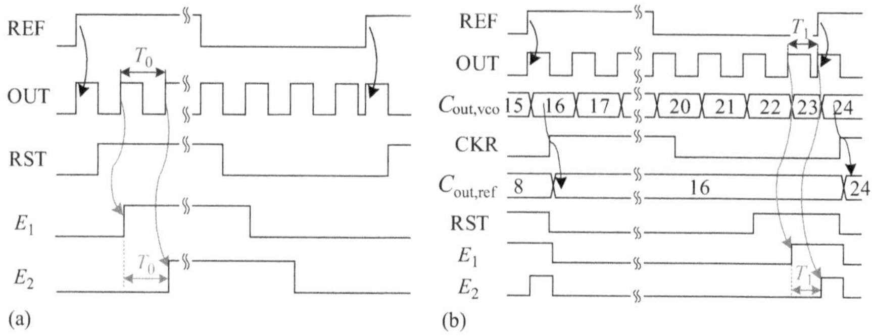

Figure 24.11 Timing diagram of the adjacent edge selector to select E_1 and E_2 with a delay of (a) T_0 and (b) T_1 at $N = 8$

selected to produce E_1 and E_2 with a delay of T_1. This can be obtained with the aid of a counter that continuously computes the rising edges of OUT (Figure 24.10). By comparing the counter output $C_{out,vco}$ with the ($C_{out,ref} + N - 3$), the reset signal of DFF_1 and DFF_2 (RST $= S_{T1}$) will be released right after the ($N - 2$)th rising edge of OUT, to ensure the correct selection of the ($N - 1$)th and Nth rising edges of OUT as shown in Figure 24.11(b). To avoid metastability of DFF_1 and DFF_2, the delay from the rising edge of OUT to the rising edge of RST must be shorter than one VCO cycle. Hence we can insert a DFF_3 clocked by OUT after the comparison logic to break the long path from OUT to RST (Figure 24.10). Besides, the comparison logic and DFF_3 are synthesized separately from other digital logics operating at the REF clock frequency to optimize the delay. We use the DFF_4 to extend the time duration of the RST signal staying at the high voltage level since the DFF_3 output (EN_{T1}) can only stay at the high voltage level for one VCO period. The reset signals of DFF_3 (active-low) and DFF_4 (active-high) controlled by MOD and CKR are properly released several VCO periods before the rising edge of EN_0 and EN_{T1} [Figure 24.11(b)]. According to the post-layout simulation, the delay times from the rising edge of OUT to the rising edge of EN_0 and to the rising edge of RST (RST $= S_{T1}$) are 198 ps and 114 ps, respectively, which ensures enough margin to avoid metastability in DFF_3 and $DFF_{1,2}$ even at $f_{VCO} = 3.2$ GHz. Since we utilize only the rising edges of $DFF_{1,2}$ to generate E_1 and E_2, the unbalanced rising and falling times of the DFF's output would not affect the SPO detection. The reset pins of DFF_5 and DFF_6 (not shown in Figure 24.10) is connected to the global reset signal for the digital circuits, which will be released when the frequency-tracking loop starts to work.

The clock-to-output-delay mismatch between DFF_1 and DFF_2 (ΔT_{DFF}) and the offset of BBPD (ΔT_{BBPD}) may cause a deterministic comparison error. In the DCDL calibration mode, the actual DCDL delay can be given by:

$$T_{DCDL} = T_0 + \Delta t_e + \Delta T_{DFF} + \Delta T_{BBPD} \quad (24.2)$$

Since we reuse the same DFF_1, DFF_2, and BBPD in the T_0/T_1 comparison mode, the net time difference detected by the time-interval comparator is:

$$\Delta T_{\text{actual}} = T_1 + \Delta T_{\text{DFF}} + \Delta T_{\text{BBPD}} - T_{\text{DCDL}} = T_1 - T_0 - \Delta t_e \qquad (24.3)$$

With the block sharing, we eliminate the effects of ΔT_{DFF} and ΔT_{BBPD}; thus the accuracy of T_0/T_1 comparison is only limited by the quantization error t_q of the DCDL.

We can estimate the effect of the DCDL quantization error t_q on the reference spur based on (24.1). Since the frequency-tracking loop ensures $|T_1 - T_0| \leq t_q$ if the quantization error from the DAC and the VCO is negligibly small. The largest f_{DEV} occurs when $|T_1 - T_0| = t_q$. When the MDLL is properly locked, we have $(N - 1) \times T_0 + T_1 = N \times T_{\text{LOCK}}$. If we assume that $f_{\text{VCO}} < f_{\text{LOCK}}$, that is, $T_0 - T_1 = t_q$, the VCO frequency can be obtained as $f_{\text{VCO}} = 1/T_0 = N/(N \times T_{\text{LOCK}} + t_q)$. Hence, the largest frequency deviation can be expressed as $f_{\text{DEV}} \approx f_{\text{LOCK}}^2 \times t_q/N$. By considering the quantization error from the DAC and the VCO, the maximum frequency deviation will further increase to $f_{\text{DEV}} \approx f_{\text{LOCK}}^2 \times t_q/N + K_{\text{DAC}} \times K_{\text{VCO}}$, where K_{DAC} and K_{VCO} are the DAC step and the VCO gain, respectively. Putting the f_{DEV} expression into (24.1), the worst-case reference spur due to the quantization effect from DCDL and from DAC and VCO can be estimated as:

$$\text{Reference spur} \approx 20 \log\left(f_{\text{LOCK}} \times t_q + \frac{K_{\text{DAC}}K_{\text{VCO}}}{f_{\text{REF}}}\right) \qquad (24.4)$$

For example, if we are choosing $t_q \approx 350$ fs and $K_{\text{DAC}}K_{\text{VCO}} \approx 210$ kHz, the worst-case reference spur level is estimated to be −53.5 dBc according to (24.4) when $f_{\text{LOCK}} = 3$ GHz and $f_{\text{REF}} = 200$ MHz. For small t_q and $K_{\text{DAC}}K_{\text{VCO}}$, the VCO phase noise will further contribute to the randomization of the reference spur.

24.3 Wideband multiplexed RVCO and frequency tuning scheme

Assuming the power consumption of a typical M-stage RVCO is dominated by the switching power, that is, the power loss on the loading capacitor of each delay stage, its oscillating frequency can be estimated by:

$$f_{\text{VCO}} = \frac{2I_{\text{DC}}}{M \times C_L \times V_{\text{DD}}} \qquad (24.5)$$

where I_{DC} is the total current consumption, and C_L is the load capacitance of each delay stage. Equation (24.5) indicates f_{VCO} can be changed by either tuning the current I_{DC} [2–6,8,11–14] or the load capacitor C_L [7,9].

Ideally, the MDLL is expected to achieve a constant low jitter performance across the frequency tuning range. The RMS jitter σ_{RMS} of the MDLL can be obtained by integrating the output phase noise $\mathcal{L}_{\text{MDLL}}$ from offset frequencies f_1 to f_2:

$$\sigma_{\text{RMS}} = \frac{\sqrt{2\int_{f_1}^{f_2} \mathcal{L}_{\text{MDLL}}(\Delta f)d(\Delta f)}}{2\pi f_{\text{LOCK}}} \tag{24.6}$$

Assuming the output phase noise of the MDLL is dominated by the VCO phase noise, $\mathcal{L}_{\text{MDLL}}$ is only determined by the phase noise $\mathcal{L}_{\text{VCO}}(\Delta f)$ of the free-running VCO and the bandwidth f_{BW} caused by the REF injection. If, for simplicity, we only consider the $1/f^2$ noise of the VCO as shown in Figure 24.12(a), the RMS jitter of the MDLL when the VCO frequency is properly locked ($f_{\text{VCO}} = f_{\text{LOCK}}$) can be re-expressed as:

$$\sigma_{\text{RMS}} = \frac{\sqrt{2(f_{\text{BW}} - f_1)\mathcal{L}_{\text{VCO}}(f_{\text{BW}}) + 2\int_{f_{\text{BW}}}^{f_2} \mathcal{L}_{\text{VCO}}(\Delta f)d(\Delta f)}}{2\pi f_{\text{VCO}}} \tag{24.7}$$

Since f_{BW} is determined by f_{REF} and can be kept constant as the output frequency changes, the normalized phase noise $\mathcal{L}_{\text{VCO}}(\Delta f)/f_{\text{VCO}}^2$ must be kept constant in order to maintain a constant σ_{RMS} when f_{VCO} changes. According to [15], the thermal-noise-induced phase noise of an inverter-based RVCO can be expressed as:

$$\frac{\mathcal{L}_{\text{VCO}}(\Delta f)}{f_{\text{VCO}}^2} \approx \frac{8\Upsilon}{3\eta} \cdot \frac{kT}{I_{\text{DC}} V_{\text{ov}}} \cdot \frac{1}{\Delta f^2} \tag{24.8}$$

where V_{ov} and γ are the overdrive voltage and the channel noise factor of MOS transistors, respectively, and η is a proportional constant. According to (24.8), the normalized phase noise will increase as I_{DC} reduces, which suggests the current tuning method suffers from a degraded normalized phase noise at low frequency. Interestingly, the normalized phase noise does not depend on C_{L} according to

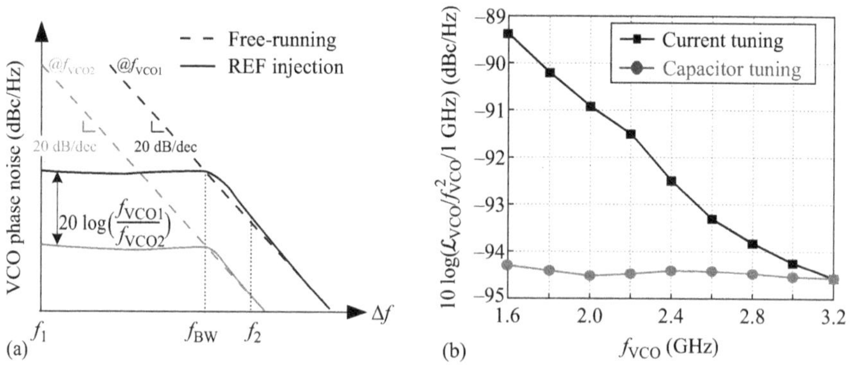

Figure 24.12 (a) Illustration of phase noise scaling with frequency to keep a constant jitter performance and (b) simulated normalized RVCO phase noises using the current-tuning and the capacitor tuning schemes

(24.8). Thus the capacitor tuning method can preserve the normalized phase noise constant over the frequency tuning range. This conclusion is also supported by the simulation results that even including the $1/f^3$ phase noise as shown in Figure 24.12(b). We can see that the RVCO using the capacitor tuning method can achieve a quite constant normalized phase noise of −94.4 dBc/Hz at 1 MHz offset frequency when its frequency is changed from 1.6 to 3.2 GHz. While for the RVCO, its normalized phase noise of the RVCO degrades by 5 dB at the lowest frequency of 1.6 GHz, which corresponds to an 80% RMS jitter increase of the MDLL.

The limited frequency tuning range of the capacitor tuning scheme can be compensated by using multiple RVCOs with little area penalty especially when using the advanced CMOS technology. In [7], two 5-stage varactor-tuned MRVCOs are employed to double the frequency tuning range. Since the varactor-tuned MRVCO features a full swing output, a low-power inverter-based MUX (~20 μW) is used to select the desired output. The un-used MRVCO is turned off to save power. Figure 24.13 shows the schematic of the pseudo-differential inverter-based delay cell used in both MRVCOs. A 3-bit binary-weighted accumulated-MOS (A-MOS) varactor bank is employed for coarse frequency tuning controlled by the frequency-locked loop while a small varactor controlled by the DAC is utilized for frequency tracking. By choosing different transistor and varactor sizes for the delay cells, the two MRVCOs together can cover two frequency bands from 1.55 to 2.47 GHz and 2.35 to 3.35 GHz, with a 120-MHz overlap. They consume ~1.45 mA over the frequency range of 1.55 to 3.35 GHz and exhibit open-loop phase noise of −85 dBc/Hz at 1 MHz offset and 3-GHz carrier frequency. The inverter-based ring RVCOs usually suffer from supply sensitivity which is ~6,500 MHz/V at a 3 GHz carrier in this work. In practice, we need a low-dropout (LDO) regulator to stabilize its supply voltage.

To attain a wide frequency tuning range while keeping a small frequency step is not trivial. In [7], each RVCO aims to cover a frequency range of ~1 GHz. Even with the help of a 3-bit binary coarse-tuning varactor bank, a 10-bit DAC is required to cover a 240-MHz frequency range with a 240-kHz frequency step. To ensure a monotonic transfer function after the layout as well as save the die size, the DAC is implemented with two thermometer-coded tuning banks which are controlled by OTW[10:6] (medium bank) and OTW[5:0] (fine bank), respectively,

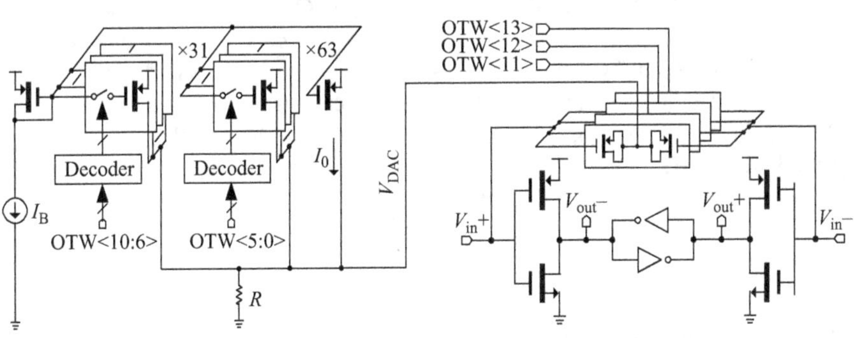

Figure 24.13 Schematic of the DAC (left) and the delay cell of MRVCO (right)

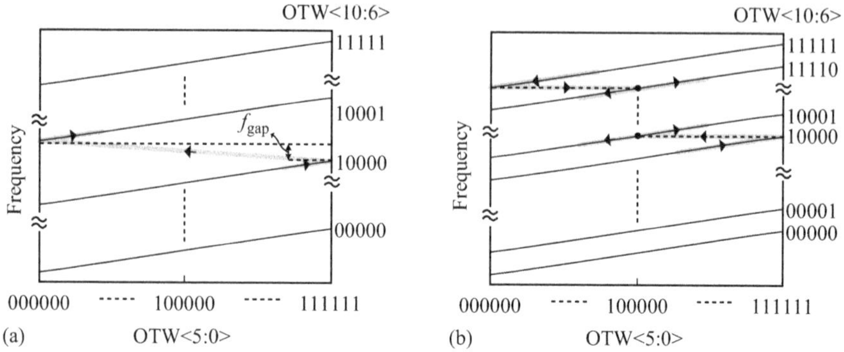

Figure 24.14 Illustration of the band switching using the (a) overlapped and (b) nonoverlapped frequency tuning curves

as shown in Figure 24.13. As a conventional method, we should design the unit transistor size in the medium band to be 32 times of the unit transistor size in the fine bank to guarantee the linearity of the DAC transfer function. However, the frequency step f_{gap} of the DAC would vary a lot due to the process variation when OTW drifts at the boundary between the two tuning banks, for example, from OTW<10:6> = 5'b10000 to OTW<10:6> = 5'b10001, as shown in Figure 24.14(a). If the unit transistor size in the medium band is smaller than 31 times of the unit transistor size in the fine bank, the DAC transfer function will become even non-monotonic. If the OTW of the MDLL happens to settle at this boundary, the jitter and spur performance of the MDLL may suffer from a significant variation from chip to chip due to the changed DAC gain. To guarantee a stable performance, we can design overlapped frequency tuning curves to achieve a seamless band switching control. As shown in Figure 24.14(b), the whole frequency range of the fine bank is designed to cover around twice of one frequency step of the medium bank. When OTW<5:0> is full of logic 1 and needs to be further increased, OTW[10:6] will be raised by 1, and OTW<5:0> will restart to change from the middle point, that is, OTW<5:0> = 6'b100000 instead of the minimum value, that is, OTW [5:0] = 6'b000000. By doing so, we can avoid the OTW to come back and forth across the boundary between the medium and fine banks, which guarantees a stable DAC gain even if OTW drifts at the boundary between the two tuning banks. A similar band switching control at the boundary can also be applied for the coarse (OTW[13:11]) bank design. It is worth mentioning that the seamless band switching method can be applied to other PLLs that requires a large frequency tuning range with a small frequency resolution.

24.4 Phase noise analysis

This section aims to introduce a general method to analyze the phase noise contribution from the VCO, frequency-tracking loop and the reference clock based on the analytical model. We will use the MDLL in [7] (Figure 24.8) as an example to show how to develop a noise model to calculate the noise transfer function (NTF) for each noise source, which helps to analyze the noise contribution from individual building block.

Assuming the DCDL resolution is fine enough, a generalized phase noise model for the MDLL in [7] can be derived in Figure 24.15. For the reference injection part, we adopted the model from [1] and [13] by setting the phase realigning factor $\beta = 1$ for the MDLL:

$$H_{UP} = N \cdot e^{-\frac{j\omega T_{REF}}{2}} \cdot \frac{\sin\left(\frac{\omega T_{REF}}{2}\right)}{\frac{\omega T_{REF}}{2}} \tag{24.9}$$

$$H_{RL1} = 1 - e^{-\frac{j\omega T_{REF}}{2}} \cdot \frac{\sin\left(\frac{\omega T_{REF}}{2}\right)}{\frac{\omega T_{REF}}{2}} \tag{24.10}$$

$$H_{RL2} = 1 - e^{-j\omega T_{REF}} \cdot e^{-\frac{j\omega T_{REF}}{2}} \cdot \frac{\sin\left(\frac{\omega T_{REF}}{2}\right)}{\frac{\omega T_{REF}}{2}} \tag{24.11}$$

Here, H_{UP} represents the up-conversion of the reference noise to the output, and H_{RL1} and H_{RL2} represent the effect of the phase-realignment by the REF injection [1]. Since $\theta_{VCO,FR}$ represents the phase of the free-running VCO within the current RFE cycle, the phase information $\theta_{MDLL,NOI}$ of the MDLL at the time when the next REF edge comes (but does not inject yet) can be obtained by adding the $\theta_{VCO,FR}$ and the phase of the current REF edge. Thus the difference between T_0 and T_1, that is, θ_{SPO}, can be found by subtracting $\theta_{MDLL,NOI}$ from θ_{OUT}. Since we perform the T_0/T_1 comparison every two REF cycles in this design, the θ_{SPO} is scaled down by $2N$ times when referred to the REF frequency.

Assuming DCDL step and $K_{DAC}K_{VCO}$ are small enough, implying that the random input noise of the BBPD dominates over the limit cycle, the BBPD can be

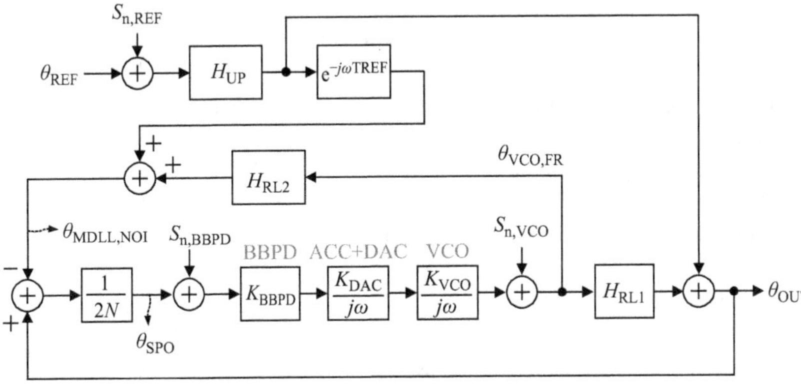

Figure 24.15 Phase noise model of the MDLL using an offset-free DCDL-based frequency-tracking loop

modeled as a linear gain $K_{\mathrm{BBPD}} = \sqrt{2}/(\sqrt{\pi}\cdot\sigma_{\Delta t})$ plus an additive quantization noise $S_{\mathrm{n,BBPD}}$ [16,17], where $\sigma_{\Delta t}$ is the standard deviation of the input time difference which can be well approximated by the RMS jitter of the MDLL output. In case the VCO noise dominates, the PSD of the BBPD input-referred quantization noise can be written as:

$$S_{\mathrm{n,BBPD}} \approx \frac{9\pi^2}{8}\cdot\frac{f_{\mathrm{BW}}^2}{f_{\mathrm{BW}}\cdot\ln\left(1+\frac{f_{\mathrm{ref}}}{2f_{\mathrm{BW}}}\right)}\cdot\frac{\sigma_{\Delta t}^2}{1+\frac{f}{f_{\mathrm{BW}}}} \tag{24.12}$$

The accumulator and DAC together are modeled as a continuous-time integrator $K_{\mathrm{DAC}}/j\omega$ to avoid the conversion between the continuous and discrete-time domains. According to Figure 24.15, the open-loop gain G_{OL} is obtained by breaking the loop at the input of BBPD and the NTF from the REF noise $S_{\mathrm{n,REF}}$, the VCO noise $S_{\mathrm{n,VCO}}$ and the BBPD quantization noise $S_{\mathrm{n,BBPD}}$ to the θ_{OUT} can be calculated as:

$$G_{\mathrm{OL}} = K_0\cdot[H_{\mathrm{RL1}} - H_{\mathrm{RL2}}] \tag{24.13}$$

$$\mathrm{NTF}_{\mathrm{REF}} = H_{\mathrm{UP}}\cdot\frac{\left(1+H_{\mathrm{RL2}}K_0 - H_{\mathrm{RL1}}K_0\cdot e^{-j\omega T_{\mathrm{REF}}}\right)}{1-G_{\mathrm{OL}}} \tag{24.14}$$

$$\mathrm{NTF}_{\mathrm{VCO}} = \frac{H_{\mathrm{RL1}}}{1-G_{\mathrm{OL}}} \tag{24.15}$$

$$\mathrm{NTF}_{\mathrm{BBPD}} = \frac{2H_{\mathrm{RL1}}K_0N}{1-G_{\mathrm{OL}}} \tag{24.16}$$

where K_0 is given by:

$$K_0 = \frac{K_{\mathrm{BBPD}}K_{\mathrm{DAC}}K_{\mathrm{VCO}}}{2N(j\omega)^2} \tag{24.17}$$

Figure 24.16 compares the calculated output phase noise using (24.14)–(24.17) with the results from the behavioral simulation using MATLAB®. Here we consider only the noise contribution from the VCO and BBPD and extract $\sigma_{\Delta t}$ from the behavioral simulation to estimate K_{BBPD} and $S_{\mathrm{n,BBPD}}$ for the calculation. The free-running VCO has a phase noise of −85 dBc/Hz at 1 MHz offset and a $1/f^3$ phase noise corner of 10 MHz. The calculated and simulated overall MDLL phase noises match well at different $K_{\mathrm{DAC}}K_{\mathrm{VCO}}$ gains. The small deviation of the in-band phase noise mainly comes from the approximation of BBPD noise using (24.12). Since $K_{\mathrm{DAC}}K_{\mathrm{VCO}}$ directly affects G_{OL} and thus the FTL bandwidth, a small $K_{\mathrm{BBPD}}K_{\mathrm{DAC}}$ is preferred to reduce the FTL bandwidth, helping to suppress BBPD noise at large offset frequency. As illustrated in Figure 24.17(a), the jitter improvement will saturate when $K_{\mathrm{DAC}}K_{\mathrm{VCO}} < 100$ kHz since the increase of the VCO noise nullifies the reduction of the BBPD noise. Thus $K_{\mathrm{BBPD}}K_{\mathrm{DAC}} \approx 240$ kHz is chosen in [7] to keep a small DAC area while still maintain a low RMS jitter.

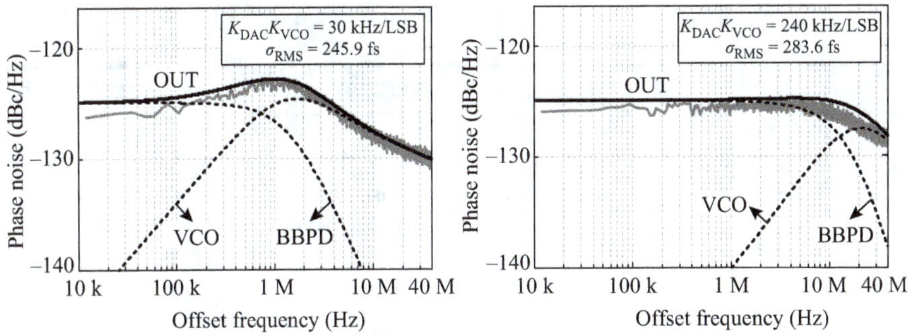

Figure 24.16 Comparison between the simulated and calculated phase noise of the MDLL using an offset-free DCDL-based frequency-tracking loop at 3 GHz and N = 15

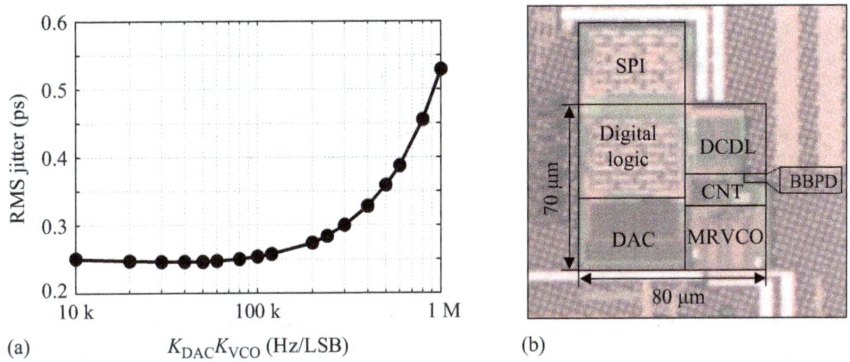

Figure 24.17 (a) Simulated RMS jitter versus $K_{DAC}K_{VCO}$ at 3 GHz and N = 15 and (b) chip micrograph of the MDLL using an offset-free DCDL-based frequency-tracking loop

24.5 Performance comparison

Table 24.1 compares the performances of the state-of-the-art RVCO-based MDLLs from [5] to [7]. The FoM_r [18] is included to count the effect of different REF clock frequencies on the jitter and power performances. Fabricated in 28-nm CMOS, the MDLL in [7] using the offset-free DCDL-based frequency-tracking loop occupies the smallest area of 0.0056 mm^2, which is comparable to the area of a PAD as shown in Figure 24.17(b). Benefitting from the low-power SA time-length digitization and comparison scheme, its frequency-tracking loop achieves the lowest power of 0.3 mW. It also measures constant low jitter performance of ~292 fs over a wide frequency tuning range from 1.5 to 3.5 GHz thanks to the varactor-based dual-MRVCO scheme. The large reference spur of −44 dBc is limited by the coupling from REF input to the MRVCO supplies through the bonding wire and

Table 24.1 Performance comparison of the state-of-the-art RVCO-based MDLLs

	JSSC'2013 [5]	**JSSC'2017 [6]**	**JSSC'2019 [7]**
CMOS technology	130 nm	28 nm	28 nm
Frequency-tracking loop (FTL) architecture	Digital FTL with BBPD	Analog FTL with pulse-width comparison	Digital FTL with offset-free pulse-width comparison
Frequency tuning scheme	Current	Current	Varactor
Supply voltage (V)	1.1	1.1	0.8
Frequency range (GHz)	0.8–2.0	2.4	1.55–3.35
Output frequency (GHz)	1.5	2.4	3.0
REF frequency (MHz)	375	75	200
Multiplication ratio	4	32	15
Reference spur (dBc)	−55.6	−51.4	−44
Integrated jitter (ps) (bandwidth)	0.4 (10 k–100 MHz)	0.7 (1 k–40 MHz)	0.292 (10 k–40 MHz)
Total power (mW)	0.89	1.51	1.45
FTL power (mW)	0.64	0.43	0.3
FoM (dB)[1]	−248.5	−241.3	−249.1
FoM_r (dB)[2]	−245.7	−245.6	−249.1
Active area (mm^2)	0.25	0.024	0.0056

[1]$FoM = 10\log[(\sigma_{rms}/1\ s)^2(P_{DC}/1\ mW)]$.
[2]Normalized to $f_{REF} = 200$ MHz, $FoM_r = 10\log[(\sigma_{rms}/1\ s)^2(P_{DC}/1\ mW)(f_{REF}/200\ MHz)]$.

ESD power rails which can be improved by better PAD arrangement and upsizing the REF buffer to reduce the REF input power. If we reduce the input power of the REF clock, the reference spur can be improved to −52 dBc but at the cost of degraded phase noise performance due to the undersized REF buffer. The ultra-compact area, low power consumption, and low jitter performance facilitate the MDLL in [7] to be flexibly integrated into an SoC system.

References

[1] Ye S., Jansson L., and Galton I. "A multiple-crystal interface PLL with VCO realignment to reduce phase noise." *IEEE J. Solid-State Circuits*. 2002, vol. 37, no. 12, pp. 1795–1803.

[2] Helal B., Straayer M., Wei G.-Y., and Perrott M. "A highly digital MDLL-based clock multiplier that leverages a self-scrambling time-to-digital converter to achieve subpicosecond jitter performance." *IEEE J. Solid-State Circuits*. 2008, vol. 43, no. 4, pp. 855–863.

[3] Gierkink S. "Low-spur, low-phase-noise clock multiplier based on a combination of PLL and recirculating DLL with dual-pulse ring oscillator and self-correcting charge pump." *IEEE J. Solid-State Circuits*. 2008, vol. 43, no. 12, pp. 2967–2976.

[4] Ali T., Hafez A., Drost R., Ho R., and Yang C.-K. "A 4.6 GHz MDLL with −46 dBc reference spur and aperture position tuning." *ISSCC Dig. Tech. Papers*. 2003, pp. 248–249.

[5] Elshazly A., Inti R., Young B., and Hanumolu P.K. "Clock multiplication techniques using digital multiplying delay-locked loops." *IEEE J. Solid-State Circuits*. 2013, vol. 48, no. 6, pp. 1416–1428.

[6] Kim H., Kim Y., Kim T., Ko H.-J., and Cho S. "A 2.4-GHz 1.5-mW digital multiplying delay-locked loop using pulsewidth comparator and double injection technique." *IEEE J. Solid-State Circuits*. 2017, vol. 52, no. 11, pp. 2934–2946.

[7] Yang S., Yin J., Mak P.-I., and Martins P.R. "A 0.0056 mm^2 all-digital MDLL using edge re-extraction, dual-ring VCOs and a 0.3 mW block-sharing frequency-tracking loop achieving 292 fs_{rms} Jitter and −249 dB FOM." *IEEE J. Solid-State Circuits*. 2019, vol. 54, no. 1, pp. 88–98.

[8] Levantino S., Marucci G., Marzin G., Fenaroli A., Samori C., and Lacaita A. "A 1.7 GHz fractional-N frequency synthesizer based on a multiplying delay-locked loop." *IEEE J. Solid-State Circuits*. 2015, vol. 50, no. 11, pp. 2678–2691.

[9] Kundu S., Kim B., and Kim C. "A 0.2–1.45-GHz subsampling fractional-N digital MDLL with zero-offset aperture PD-based spur cancellation and in situ static phase offset detection." *IEEE J. Solid-State Circuits*. 2017, vol. 52, no. 3, pp. 799–811.

[10] Yin J., Mak P.-I., Maloberti F., and Martins R.P. "A time-interleaved ring-VCO with reduced $1/f^3$ phase noise corner, extended tuning range and inherent divided output." *IEEE J. Solid-State Circuits*. 2016, vol. 51, no. 12, pp. 2979–2991.

[11] Park P., Park J., Park H., and Cho S. "An all-digital clock generator using a fractionally injection-locked oscillator in 65 nm CMOS." *ISSCC Dig. Tech. Papers*. 2012, pp. 336–337.

[12] Musa A., Deng W., Siriburanon T., Miyahara M., Okada K., and Matsuzawa A. "A compact, low-power and low-jitter dual-loop injection locked PLL using all-digital PVT calibration." *IEEE J. Solid-State Circuits*. 2014, vol. 49, no. 1, pp. 50–60.

[13] Choi S., Yoo S., Lim Y., and Choi J. "A PVT-robust and low-jitter ring-VCO-based injection-locked clock multiplier with a continuous frequency-tracking loop using a replica-delay cell and a dual-edge phase detector." *IEEE J. Solid-State Circuits*. 2016, vol. 51, no. 8, pp. 1878–1889.

[14] Coombs D., Elkholy A., Nandwana R., Elmallah A., and Hanumolu P. "8.6 A 2.5-to-5.75 GHz 5 mW 0.3 psrms-jitter cascaded ring-based digital injection-locked clock multiplier in 65 nm CMOS." *ISSCC Dig. Tech. Papers*. 2017, pp. 152–153.

[15] Hajimiri A., Limotyrakis S., and Lee T.H. "Jitter and phase noise in ring oscillators." *IEEE J. Solid-State Circuits*. 1999, vol. 34, no. 6, pp. 790–804.

[16] Da Dalt N. "Linearized analysis of a digital bang-bang PLL and its validity limits applied to jitter transfer and jitter generation." *IEEE Trans. Circuits Syst. I, Reg. Papers*. 2008, vol. 55, no. 11, pp. 3663–3675.

[17] Zanuso M., Tasca D., Levantino S., Donadel A., Samori C., and Lacaita A.L. "Noise analysis and minimization in bang-bang digital PLL." *IEEE Trans. Circuits Syst. II, Expr. Briefs*. 2009, vol. 56, no. 11, pp. 835–839.

[18] Zhu J., Nandwana R., Shu G., Elkholy A., Kim S., and Hanumolu P. "A 0.0021 mm^2 1.82 mW 2.2 GHz PLL using time-based integral control in 65 nm CMOS." *IEEE J. Solid-State Circuits*. 2017, vol. 52, no. 1, pp. 8–20.

Chapter 25
Wideband PLLs

Long Kong[1] *and Behzad Razavi*[1]

RF synthesis has generally shied away from ring oscillators due to their much more severe phase noise-power trade-offs than those of inductor-capacitor (LC) topologies. Today's multiband, multimode radios, however, require a number of synthesizers and can greatly benefit from compact, flexible implementations afforded by ring oscillators.

This chapter proposes several phase-locked loop (PLL) architectures that can achieve a wide loop bandwidth, thus suppressing the voltage-controlled oscillator (VCO) phase noise and allowing the use of a ring topology. The concepts are demonstrated for both integer-N and fractional-N RF synthesis. Most of the concepts introduced here are applicable to other PLL and oscillator topologies as well.

This chapter is organized as follows. In Section 25.1, we describe an integer-N synthesizer that achieves a phase noise of −114 dBc/Hz at 10-MHz offset with a spur level of −65 dBc [1]. In Section 25.2, we introduce a new fractional-N synthesizer architecture that achieves a loop bandwidth (BW) as large as $f_{REF}/4$ [2].

25.1 Wideband integer-N RF synthesizer

The need for LC oscillators in RF synthesis has been solidified by various studies revealing that the white-noise-induced phase noise of ring oscillators trades primarily with the power consumption [3,4] and is relatively independent of the number of stages. However, ring oscillators do present compelling advantages: (1) they occupy a smaller area and can be readily placed within a transceiver layout with less serious concerns regarding proximity effects, (2) they entail much less coupling to and from other circuits, (3) they achieve a wider tuning range and can be multiplexed to cover decades of frequencies, and (4) they readily generate multiple phases.

That the phase noise of ring oscillators is difficult to improve at the circuit level forces us to higher levels of abstraction. For example, we can consider the phase noise reduction in the synthesizer architecture.

[1]Department of Electrical Engineering, University of California, Los Angeles, CA, USA

25.1.1 PLL bandwidth limitations

The loop bandwidth is generally constrained by three factors: (1) the PLL reference frequency, f_{REF}, for example, in a mobile phone environment, only a crystal oscillator around 20 MHz is available; (2) the PLL stability limit, often called "Gardner's Limit," and generally accepted to be around $f_{REF}/10$ for type-II topologies; and (3) the ripple amplitude on the VCO control line and hence the output spur level. In the presence of charge pump (CP) nonidealities, the loop bandwidth is reduced to typically $f_{REF}/20$ or less to reduce the reference spurs.

It is helpful to briefly review the different bandwidths encountered in PLL analysis: (a) the input-output transfer function has a certain 3-dB bandwidth, which we call the "PLL bandwidth," f_{BW}, in this chapter; (b) the loop transmission has a unity-gain bandwidth, f_{UGB}; and (c) the VCO noise transfer function, a high-pass response, also has a 3-dB bandwidth, $f_{n,VCO}$. For example, a type-II PLL with a charge pump current I_{CP}, a loop filter capacitor C_1, a VCO gain K_{VCO}, a divide ratio N, and a unity damping factor provides $2\pi f_{BW} \approx 2.5\omega_n$, $2\pi f_{UGB} \approx 2.1\omega_n$, and $2\pi f_{n,VCO} \approx 1.55\omega_n$, where ω_n is the "natural frequency" of the second-order system. For example, if f_{UGB} is chosen in the range of $f_{REF}/20$ to $f_{REF}/10$, then:

$$0.037 f_{REF} < f_{n,VCO} < 0.074 f_{REF} \tag{25.1}$$

*25.1.2 Proposed integer-*N *wideband PLL*

Our approach to suppressing the VCO phase noise is to develop a PLL topology that avoids Gardner's limit and, if necessary, deal with the ripple on the control voltage by additional techniques. Since type-I PLL has only one integrator in the loop, in theory, its loop bandwidth can even exceed $f_{REF}/2$. However, type-I architecture suffers from its own limitations. First, the reference spur becomes unacceptably high at such a wide loop bandwidth. Second, it has a limited capture range. Our proposed architecture will address these issues.

25.1.2.1 Type-I PLL with sampling filter

In a manner similar to charge-pump PLLs [5,6], we can replace the continuous-time filter in a type-I PLL with a discrete-time implementation, aiming to isolate the VCO from the large XOR jumps. As depicted in Figure 25.1(a), we wish to select the timing between the main input and V_F such that S_0 turns on only during a "settled" level. Unfortunately, this is not possible because V_X still jumps between 0 and V_{DD}. Figure 25.1(b) shows the loop's waveforms in the locked condition, indicating that V_X varies at $2f_{REF}$ if the input has a 50 percent duty cycle. When V_F goes high, V_{cont} attempts to track V_X, reaching a certain level, V_1, that is necessary for the VCO to operate at Nf_{REF}*. That is, the loop adjusts the phase error, $\Delta\phi$, until the V_{cont} transient yields a value of V_1 at the end of one T_{REF}.

The above technique does provide a constant voltage, V_1, for the VCO while S_0 is off. We therefore wish to modify the circuit so that the VCO does not sense the

*This is true only if the ripple is small.

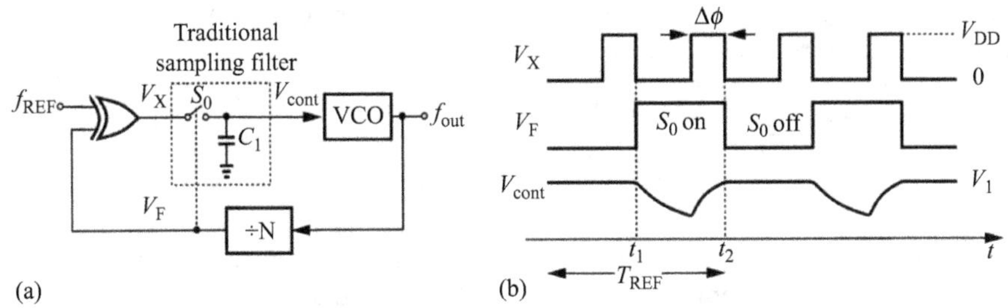

Figure 25.1 *(a) Type-I PLL with traditional sampling filter and (b) time-domain operation*

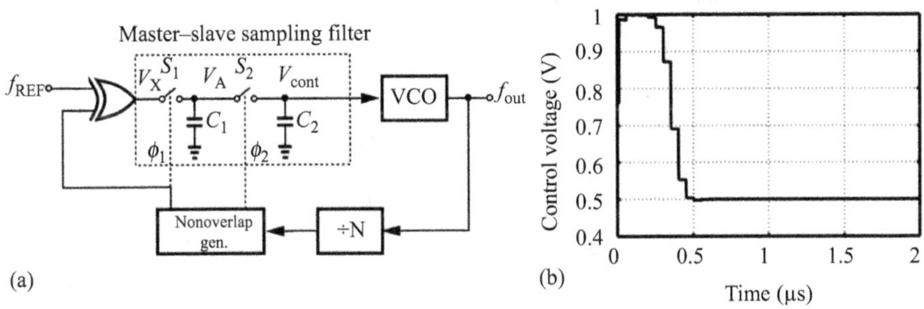

Figure 25.2 *Proposed PLL (a) architecture and (b) settling behavior*

transient from t_1 to t_2. This is accomplished by inserting one more sampling network in the VCO control path (Figure 25.2(a)), with the two now operating in a master-slave manner. The divider output is converted to two nonoverlapping phases, preventing direct feedthrough from V_X to V_{cont}. We expect to observe a large ripple on C_1, similar to that in Figure 25.1(b), but a small ripple on C_2. As an example, Figure 25.2(b) shows the transient behavior with $C_1 = 16$ pF, $C_2 = 1$ pF, $K_{VCO} = 280$ MHz/V, and $N = 120$. The loop bandwidth is about 9 MHz and the loop settles in roughly 10 input cycles.

The PLL architecture employing the master-slave sampling filter (MSSF) displays several interesting and useful properties. As explained below, compared to the traditional type-I PLL, its capture range is much wider, and, in comparison to type-II PLLs, it achieves a loop bandwidth close to $f_{REF}/2$, settles faster, and avoids the difficulties in low-voltage CP design.

25.1.2.2 MSSF transfer function

As explained above, the periodic voltage jumps at V_X in Figure 25.2(a) do not reach V_{cont}, implying that the master-slave filter response has notches at the reference frequency and its harmonics. We examine this response in detail here.

As a continuous-time (CT) approximation, we can say that C_1 switches between V_X and V_{cont} periodically, thus acting as a series resistor, R_{eq}, equal to $1/(f_{CK}C_1)$, where f_{CK} denotes the sampling frequency and is equal to f_{REF} when the loop reaches the locked condition. In other words, the filter resembles a first-order section having a response given by:

$$H(s) = \frac{1}{1 + R_{eq}C_2 s} = \frac{1}{1 + \dfrac{C_2}{C_1 f_{CK}} s} \tag{25.2}$$

Note that this response accounts for charge sharing between C_1 and C_2, but fails to predict the harmonic notches. It is also a crude approximation if the PLL bandwidth approaches $f_{REF}/2$.

A more accurate transfer function is obtained if we consider the MSSF as a zero-order hold (ZOH) circuit. It converts a CT input to a discrete-time output. If $C_2 \ll C_1$ so that charge sharing between the two capacitors can be neglected, then the ZOH output can be expressed as [7]:

$$Y(f) = e^{-j2\pi f T_{CK}/2} \frac{\sin(\pi f T_{CK})}{\pi f T_{CK}} \sum_{n=-\infty}^{\infty} X\left(f - \frac{n}{T_{CK}}\right) \tag{25.3}$$

For the output component of interest, $n = 0$ and

$$Y_0(f) = e^{-j2\pi f T_{CK}/2} \frac{\sin(\pi f T_{CK})}{\pi f T_{CK}} X(f) \tag{25.4}$$

This result, of course, predicts the notches at the harmonic of f_{CK} but disregards charge sharing.

Even though operating as a master-slave storage circuit, the proposed filter exhibits a delay of $T_{CK}/2$, rather than T_{CK}, in the PLL environment. This is because the XOR produces the phase error information twice per cycle. Illustrated in Figure 25.3, this effect can be seen by displacing the f_{REF} edges by a small amount,

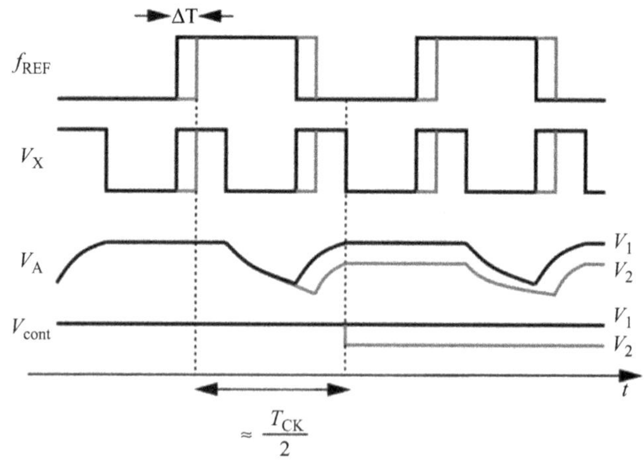

Figure 25.3 XOR and MSSF time-domain waveforms

ΔT, and observing that V_X inherits this change from both the rising edge and the falling edge of V_{in}. Consequently, V_A changes in about $T_{CK}/2$ seconds and is frozen thereafter. If the MSSF delay were as long as T_{CK}, the PLL would become unstable for a unity-gain bandwidth of $f_{REF}/4$.

Equation (25.4) is a reasonable MSSF model for our analysis and design efforts described below, especially because we will select C_2 much less than C_1, thus minimizing charge sharing and improving the ZOH approximation. However, a more accurate model, obtained empirically, is as follows:

$$H_{MSSF}(j\omega) = \frac{1}{1 + \frac{C_2}{C_1 f_{CK}} j\omega} e^{-j\pi f T_{CK}} \frac{\sin(\pi f T_{CK})}{\pi f T_{CK}}. \tag{25.5}$$

Plotted in Figure 25.4 are the magnitude and phase of the MSSF transfer function as predicted by (25.5) and as obtained from transient circuit simulations. Here, $C_1 = 16$ pF, $C_2 = 1$ pF, and $f_{REF} = 20$ MHz. We observe good agreement between the two. In this example, the filter has a 3-dB bandwidth of 7.4 MHz, at which the phase shift reaches $-75°$. To minimize this phase shift (which affects the loop stability), we typically choose $C_1 \gg C_2$ and reduce the contribution of the first fraction in (25.5).

The deep notches in Figure 25.4 distinguish the MSSF from continuous-time filters. These notches suppress the harmonic components generated by the XOR

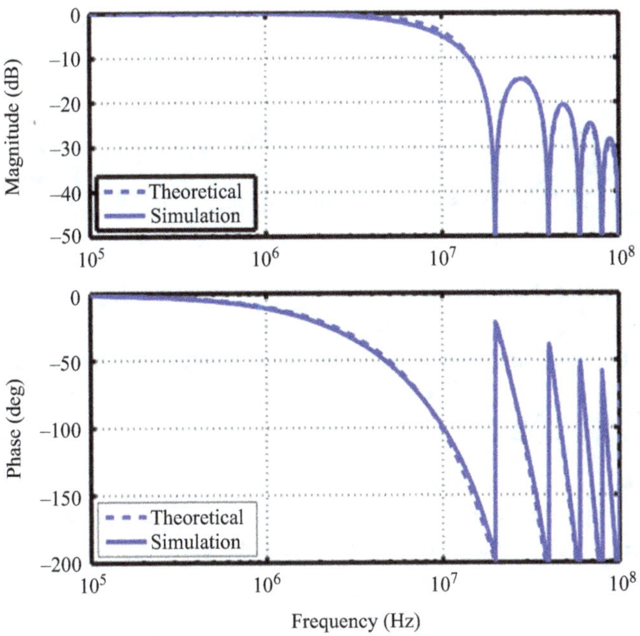

Figure 25.4 Magnitude and phase response of MSSF

gate in Figure 25.2(a), thereby easing the trade-off between the loop bandwidth and the ripple amplitude. Nonetheless, second-order effects do create some ripple and are addressed in Section 25.1.4.

25.1.2.3 Stability considerations

While greatly suppressing the ripple, our proposed PLL is not unconditionally stable. In this section, we deal with this point.

From (25.5), the loop transmission of the topology shown in Figure 25.2(a) can be expressed as:

$$H(j\omega) = \frac{K_{PD}K_{VCO}}{N} \times \frac{1}{j\omega} \times \frac{1}{1 + \dfrac{C_2}{C_1 f_{REF}} j\omega} e^{-j\pi f T_{REF}} \frac{\sin(\pi f T_{REF})}{\pi f T_{REF}} \tag{25.6}$$

where K_{PD} is the phase detector (PD) gain (chosen approximately equal to 2.2 V/rad so as to provide the desired bandwidth). We have approximated the MSSF sampling rate by f_{REF}. To determine the phase margin, we must examine $\angle H(j\omega)$ at the unity-gain bandwidth, f_{UGB}, that is, the frequency at which $|H(j\omega)|$ drops to unity. To this end, we make two approximations: (1) as explained in Section 25.1.2.2, $C_1 \gg C_2$ and hence the fraction $1/[1 + C_2 j\omega/(C_1 f_{REF})]$ contributes negligibly to $\angle H$ and $|H|$, and (2) predicting that $f_{UGB} < f_{REF}/2$, we also neglect the effect of the sinc on $|H|$. It follows that $|H(j\omega)| \approx K_{PD}K_{VCO}/(N\omega)$ and $2\pi f_{UGB} \approx K_{PD}K_{VCO}/N$. The phase contains a $-90°$ contribution by the VCO and $-\pi f T_{REF}$ by the MSSF, $\angle H(j\omega) \approx -\pi/2 - \pi f T_{REF}$. The phase margin, $\pi + \angle H(j2\pi f_{UGB})$, is thus equal to:

$$\mathrm{PM} = \frac{\pi}{2} - \pi f_{UGB} T_{REF} = \frac{\pi}{2} - \frac{K_{PD}K_{VCO}}{2N} T_{REF} \tag{25.7}$$

Equation (25.7) imposes an upper bound of $f_{REF}/2$ on f_{UGB}. The phase margin reaches about 45° for $f_{UGB} = f_{REF}/4$.

25.1.2.4 Closed-loop behavior

As mentioned in Section 25.1.1, the closed-loop input-output bandwidth and the VCO noise transfer bandwidth are of interest. For the former, we have:

$$\frac{\varnothing_{out}}{\varnothing_{in}}(j\omega) = \frac{NH(j\omega)}{1 + H(j\omega)}. \tag{25.8}$$

With the approximation stipulated in Section 25.1.2.3, $H(j\omega) \approx [K_{PD}K_{VCO}/(Nj\omega)] \exp(-j\pi f T_{REF})$, the 3-dB bandwidth is obtained as:

$$2\pi f_{BW} \approx 2\sqrt{3} f_{REF} \sqrt{\frac{\alpha - 1 + \sqrt{(\alpha - 1)^2 + \alpha^3/6}}{\alpha}} \tag{25.9}$$

where $\alpha = 2\pi f_{UGB}/f_{REF}$. Recall from (25.7) that $f_{REF}/4 < f_{UGB} < f_{REF}/2$ for $45° >$ PM > 0, that is, $\pi/2 < \alpha < \pi$. For this range of α, we have:

$$0.55 f_{REF} < f_{BW} < 0.71 f_{REF} \tag{25.10}$$

The key point here is that the closed-loop bandwidth can reach $f_{REF}/2$ with a reasonable phase margin.

The wide bandwidth of the proposed PLL naturally translates to a fast lock transient, for example, about 10 input cycles as shown in Figure 25.2(b).

For the VCO noise transfer, we have $\phi_{out}/\phi_{VCO} = (1 + H)^{-1}$. The 3-dB bandwidth is obtained as:

$$2\pi f_{n,VCO} \approx 2\sqrt{3} f_{REF} \sqrt{\frac{\alpha + 1 - \sqrt{(\alpha + 1)^2 - \alpha^3/6}}{\alpha}} \tag{25.11}$$

which, for $\pi/2 < \alpha < \pi$, falls in the range of:

$$0.16 f_{REF} < f_{n,VCO} < 0.26 f_{REF} \tag{25.12}$$

For a fair comparison, we consider only the lower bound and note that, with respect to the two limits prescribed by (25.1), we have improved the VCO noise suppression bandwidth by about a factor of 2.2 to 4.3. In our synthesizer design, $f_{n,VCO} \approx 0.17 f_{REF}$ for a PM of around 42°.

25.1.2.5 Acquisition range

The MSSF-based PLL provides a much wider acquisition range than the traditional type-I architecture. Fundamentally, this is because the MSSF in Figure 25.2(a) is clocked by the feedback signal, thus behaving differently from the continuous-time filter during the acquisition process. In order to formulate the acquisition range, we construct the open-loop configuration shown in Figure 25.5(a), assuming that the VCO operates at a frequency of f_1. We follow the "beat" component generated by the XOR gate, $f_1/N - f_{REF}$, through the sampling filter and consider two cases. First, suppose the sampling process satisfies the Nyquist rate, that is, $|f_1/N - f_{REF}| < f_1/(2N)$ and hence $(2/3) f_{REF} < f_1/N < 2 f_{REF}$. In this case, the beat component passes through as a "baseband" signal, providing a nearly rail-to-rail voltage swing to the VCO. Figure 25.5(b) plots the simulated control voltage in such a scenario; the VCO is heavily modulated at a rate of $f_1/N - f_{REF}$, producing a strong sideband at the divider output located at $f_1/N - (f_1/N - f_{REF}) = f_{REF}$.[†] In the closed-loop configuration, this sideband yields a dc component at the XOR output, leading to acquisition. The above inequality can be referred to the output as $(2/3) N f_{REF} < f_1 < 2 N f_{REF}$. The loop therefore locks for an initial frequency between 2/3 and 2 times the final value. For example, if the VCO tuning range is from 1.6 GHz to 4.8 GHz, then the loop can always lock to 2.4 GHz. The second case arises if the beat experiences aliasing, that is, if f_1/N falls outside the acquisition range. The MSSF

[†]The MSSF sampling clock too carries sidebands, but they negligibly contribute to the MSSF output.

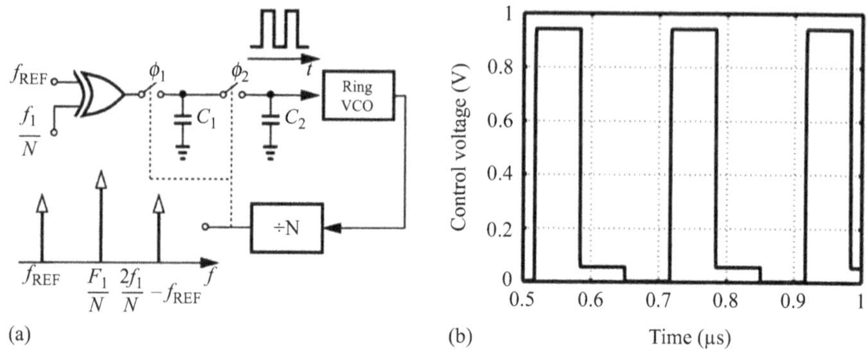

Figure 25.5 (a) Proposed PLL in open-loop configuration and (b) simulated control voltage waveform

output now contains a component at $f_1/N - |f_1/N - f_{REF}|$, which does not lead to lock. Since the free-running VCO range lies well in the acquisition range, no frequency acquisition aid is necessary in our prototype. Circuit simulations confirm these predictions.

25.1.3 Phase noise considerations

The phase noise of the proposed PLL arises from three building blocks, namely, the VCO, the XOR gate, and the sampling filter. We wish to design the VCO according to the overall phase noise specification and reduce to negligible levels the XOR and filter contributions.

25.1.3.1 V3CO phase noise

The VCO is designed as a three-stage inverter-based ring oscillator. Depicted in Figure 25.6, the circuit employs MOS varactors for fine control and banks of switchable capacitors for coarse control. To achieve low flicker-noise-induced phase noise, we choose $W/L = 36$ μm/0.28 μm for both PMOS and NMOS devices in each inverter. The varactors have a W/L of 26 μm/0.2 μm, providing a tuning range of about 200 MHz, and the capacitor banks consist of twelve 25-fF units, offering a range from 2 GHz to 3 GHz. The circuit draws 3.1 mW from a 1-V supply at 2.4 GHz and exhibits a phase noise of −96 dBc/Hz at 1-MHz offset.

Three aspects of the VCO design merit remarks. First, simulations suggest that, among various ring oscillator tuning techniques, varactors cause the least degradation in phase noise as the frequency is varied for a given power consumption. In a starved-inverter topology, for example, the starving transistors themselves contribute significant phase noise as the frequency is decreased. Second, as with other inverter-based rings reported in prior work, the VCO suffers from supply sensitivity. In practice, such VCOs are fed from a low-dropout (LDO) regulator. In our prototype, we have used two separate supply pins for the analog and digital sections. Third, the three node waveforms within the ring can be combined to generate quadrature phases.

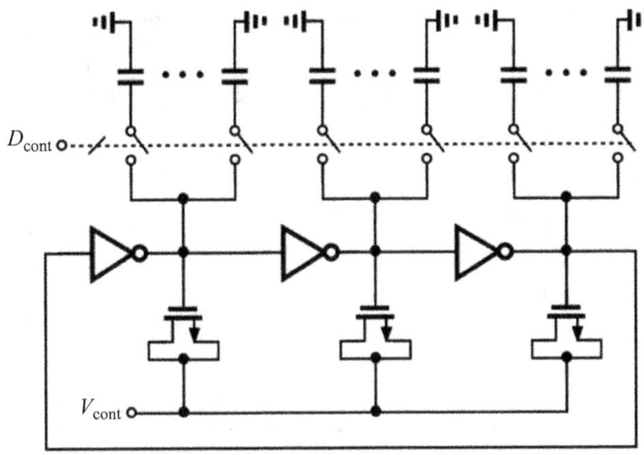

Figure 25.6 VCO implementation

A full-size inverter sensing one node and a half-size inverter sensing another can merge their output nodes, generating ±90° or 180° from 120° phases.

The shaping of the VCO phase noise deserves a note as well. Unlike type-II PLLs, a type-I PLL cannot force flicker-noise-induced phase noise to zero at zero frequency. Nevertheless, by virtue of its large bandwidth, the proposed PLL still displays a smaller integrated phase noise that a type-II architecture would. Figure 25.7 plots the simulated free-running phase noise of the above VCO and the shaping that it experiences in the two loops. (Here, the charge-pump PLL is assumed to have a loop bandwidth of $f_{REF}/20$,[‡] while the type-I PLL is based on our architecture with a bandwidth of $f_{REF}/2$.) Our design leads to an integrated phase noise of 0.35° from 100 kHz to 15 MHz, and the type-II loop to 1.14° for the same range.[§] In practice, the charge pump flicker noise makes this comparison more favorable towards the proposed PLL.

25.1.3.2 PD and MSSF phase noise

In order to minimize the contribution of the PD/MSSF cascade in Figure 25.2(a), we take several measures. First, the XOR incorporates PMOS and NMOS devices with $W/L = 32\ \mu m/80$ nm and $16\ \mu m/80$ nm, respectively, achieving a phase noise of −171 dBc/Hz at 5-MHz offset while consuming 86 μW at 20 MHz. This leads to an in-band phase noise at the PLL output equal to -171 dBc/Hz $+ 20\log(N) =$ -129 dBc/Hz. Second, since S_1 carries large transient currents and can potentially generate high flicker noise, we choose $W/L = 20\ \mu m/100$ nm for this device. Third, the kT/C noise associated with S_2 and C_2 is reduced by selecting $C_2 = 1$ pF ($C_1 = 16$ pF contributes negligibly). This kT/C noise translates to in-band phase

[‡]For a reference spur level below −60 dBc, reported type-II PLLs have a bandwidth of no more than $f_{REF}/20$; hence this choice for a fair comparison with our architecture.

[§]The peaking in our phase noise can be reduced by choosing a smaller bandwidth. In general, some optimization is necessary at this stage.

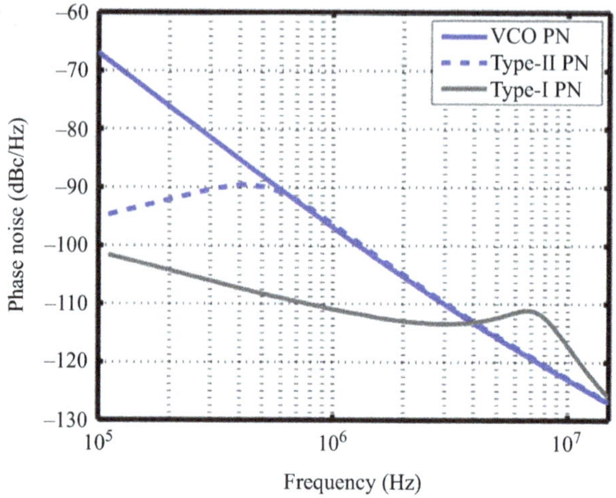

Figure 25.7 VCO phase noise in free-running mode and in type-II and proposed PLLs

noise at the PLL output according to $S_{out,MSSF} \approx [kT/(2C_2)]f_{REF}^{-1}N^2/K_{PD}^2$, where the factor of 2 accounts for the fact that C_2 appears in parallel with C_1 ($\gg C_2$) for about half of the input period. With $C_2 = 1$ pF and $K_{PD} = 2.2$ V/rad, we have $S_{out,MSSF} = -126$ dBc/Hz.

The foregoing study also prescribes a design procedure: we first pick the value of C_2 for negligible phase noise contribution and then choose C_1 to be 10 to 20 times larger. Finally, we size S_1 and the XOR devices for negligible noise as well.

25.1.4 Spur reduction

Despite the transfer notches introduced by the sampling filter, we observe sidebands on the order of −50 to −55 dBc at the VCO output. This phenomenon arises from three mechanisms. First, the large VCO varactors ($W/L = 26$ μm/0.2 μm) draw a significant gate leakage current (~80 nA) from C_2 in Figure 25.2(a), causing a 2-mV droop in each cycle. Second, the leakage, charge injection, and clock feedthrough of S_2 produce another 1 mV_{pp} of ripple. Third, in the presence of ground bond wires, the bounce on the bottom plate of C_1 persists after S_2 turns on, disturbing the control line periodically.

We propose the use of "harmonic" traps to suppress the ripple with little compromise in the bandwidth. Applicable to any PLL architecture and illustrated in Figure 25.8, the idea is to add one or more series resonant branches in parallel with the control line, forming a low impedance to ground at f_{REF}, $2f_{REF}$, etc.

Harmonic traps entail three issues. First, active implementations ultimately present a trade-off between the trap impedance and the power consumption, potentially unable to fight the MSSF output impedance. For this reason, R_S (≈ 2.5 kΩ) is inserted

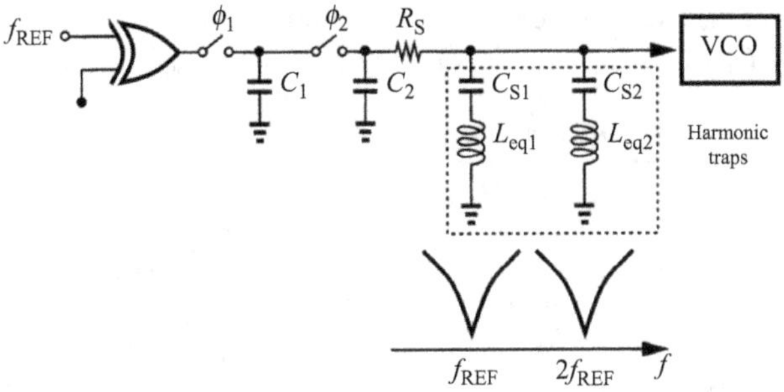

Figure 25.8 Concept of harmonic traps

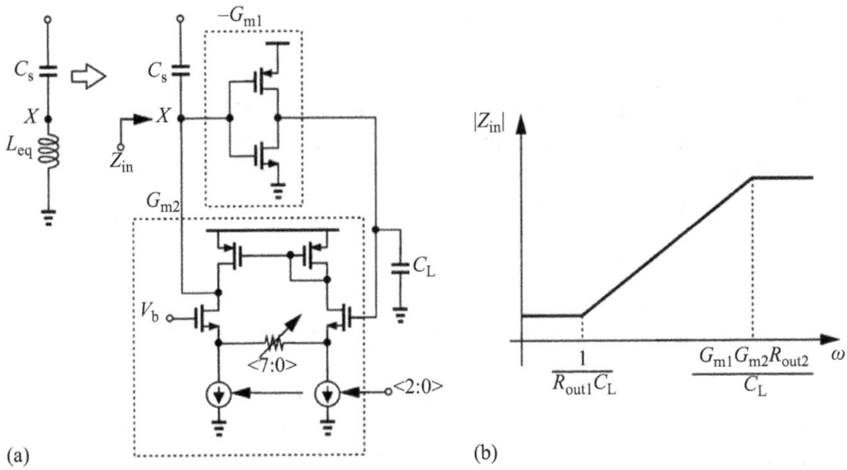

Figure 25.9 (a) Harmonic trap implementation and (b) magnitude of gyrator input impedance

in Figure 25.8. Second, the traps must have a sufficiently high Q so as to contribute negligible phase shift and noise for $f \leq f_{REF}/2$. Third, the traps' resonance frequencies must be calibrated with adequate resolution to deal with PVT variations.

25.1.4.1 Harmonic trap design

Each trap consists of a capacitor in series with an active inductor, obtained by gyrating another capacitor. As shown in Figure 25.9(a), G_{m1} and G_{m2} constitute a gyrator, transforming C_L to $Z_{in} = L_{eq}s = C_L s/(G_{m1}G_{m2})$ if their output impedances are assumed infinite. For example, the f_{REF} trap uses $G_{m1} = 0.92$ mS, $G_{m2} = 54$ μS and $C_L = 3.5$ pF, creating $L_{eq} = 70$ μH. The degeneration resistance and the bias currents within G_{m2} are programmable. The power dissipation is 170 μW.

For design purposes, we need a more accurate expression for Z_{in}. If the output impedances of G_{m1} and G_{m2} are denoted by R_{out1} and R_{out2}, respectively, then:

$$Z_{in} = \frac{(R_{out1}C_L s + 1)R_{out2}}{(R_{out1}C_L s + G_{m1}R_{out1}G_{m2}R_{out2} + 1)}. \tag{25.13}$$

Plotted in Figure 25.9(b), $|Z_{in}|$ reveals an inductive behavior for $(R_{out1}C_L)^{-1} < \omega < G_{m1}G_{m2}R_{out2}/C_L$. It is important that the trap resonance frequency, ω_{res}, lie well between the zero and pole frequencies so that Z_{in} approaches a pure inductor. We therefore view $(R_{out1}C_L)^{-1} << \omega_{res} << G_{m1}G_{m2}R_{out2}/C_L$ as a guideline for choosing C_L. If Z_{in} is rewritten as $[C_L s/(G_{m1}G_{m2}) + (G_{m1}G_{m2}R_{out1})^{-1}]||R_{out2}$, then we recognize that the inductance sees a series resistance equal to $(G_{m1}G_{m2}R_{out1})^{-1}$ and a parallel resistance equal to R_{out2}. Since R_{out2} is sufficiently large in our design, the quality factor is approximately equal to $R_{out1}C_L\omega$, about 15 at $f_{REF} = 20$ MHz. To cover PVT variations, the trap frequency has a programmable range of ±30 percent around its nominal value with a resolution of 0.6 MHz. Circuit simulations indicate that the traps negligibly affect the loop settling time.

The noise contribution is formulated by modeling the gyrator noise by a current source and subjecting it to a (high-pass) transfer function to V_{cont} and another from V_{cont} to ϕ_{out}. Calculations and simulations predict a phase noise of –131 dBc/Hz at 5-MHz offset resulting from the traps.

25.1.4.2 Notch calibration

To calibrate the traps, we must seek an error whose value reaches a minimum as the notch frequency reaches the desired value, for example, f_{REF}. The control ripple amplitude is one such error. But we must also measure this error with reasonable fidelity as, towards the end of calibration, it becomes very small. In our design, for example, a spur level of –60 dBc at the output is equivalent to a ripple amplitude of about 0.28 mV_{pp}.

We employ a Δ modulator as a compact, low-power ADC to measure the ripple waveform and reconstruct it in the digital domain. Shown in Figure 25.10(a), a traditional Δ modulator consists of a comparator and a low-pass feedback network,

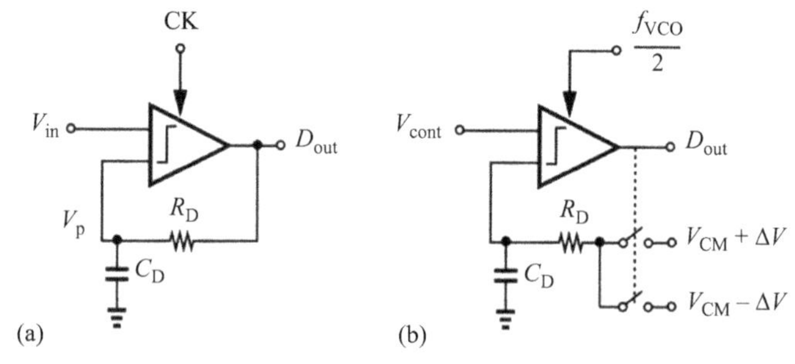

Figure 25.10 (a) Traditional and (b) proposed Δ modulator architectures

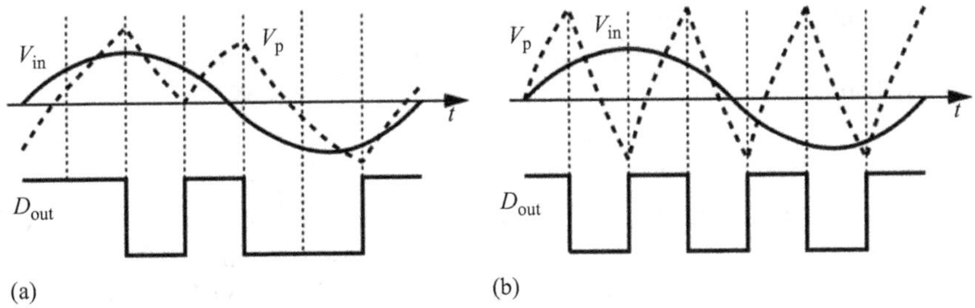

Figure 25.11 Effect of R_DC_D on Δ modulator operation: (a) long time constant and (b) short time constant

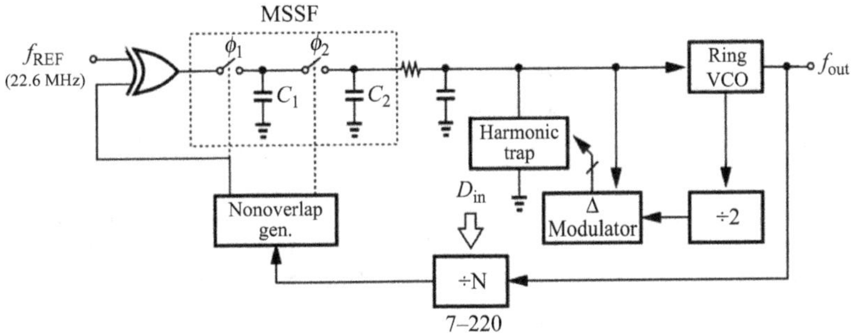

Figure 25.12 Final proposed architecture of the synthesizer

forcing V_p to track V_{in}. As a result, the running average of the pulsewidth-modulated output also tracks V_{in}, provided that R_DC_D is sufficiently long (Figure 25.11(a)). Otherwise, the input peaks do not exceed the peaks of V_P, causing failure (Figure 25.11(b)).

Similarly, the Δ modulator of Figure 25.10(a) fails for small or slow input swings; it simply generates a periodic output at half of the clock frequency if the input peaks do not exceed the peaks of V_p. It can be proved that the sensitivity is given by $V_{DD}\{1 - \exp[-T_{CK}/(2R_DC_D)]\}$, where the comparator output is assumed to swing between 0 and V_{DD}. For example, a sensitivity of 0.28 mV_{pp} with $f_{CK} \approx$ 1.2 GHz translates to $R_DC_D = 3$ μs, demanding very large values for R_D and C_D. To resolve this issue, we modify the architecture as depicted in Figure 25.10(b), where the comparator is clocked at $f_{VCO}/2$ and its output drives a 1-bit DAC with a much smaller swing, $\pm\Delta V$. A ΔV of 25 mV, for example, allows a 20-fold reduction in the R_DC_D product. In this design, we have $R_D = 50$ kΩ, $C_D = 4$ pF, and a StrongArm comparator consuming 80 μW. Since the ripple amplitude is a convex function of the trap resonance frequency, a background calibration is implemented to decide when the calibration has reached a minimum.

The overall synthesizer architecture is shown in Figure 25.12. The feedback divider provides $N = 7$–220, but only the range from 120 to 124 is used for the 2.4-GHz band.

*25.1.5 Experimental results of integer-*N *synthesizer*

The integer-N synthesizer has been fabricated in the TSMC 45-nm digital CMOS technology. As shown in Figure 25.13, the die measures 100 μm × 150 μm. Tested with a 1-V supply, the synthesizer operates from 2 GHz to 3 GHz and consumes 4 mW at 2.4 GHz. The 22.6-MHz input reference is produced by a low-noise crystal oscillator (hence the departure from 20 MHz) and the output is measured by an Agilent spectrum analyzer. The Δ modulator output is sent off-chip and processed in MATLAB®, and the control codes are written back to the chip through a serial bus.

Upon power-up, the PLL locks with the harmonic traps off and then the traps are turned on and calibrated. The initial calibration takes approximately 400 input cycles, but for subsequent frequency changes (initiated by a modulus change), the calibration settings remain constant because the notch frequencies do not depend on the output frequency. The output spectra of the synthesizer with the harmonic traps off and on are plotted in Figure 25.14. The first-order spur falls from −47 dBc to −65 dBc, and the second-order spur falls from −55 dBc to −68.5 dBc. The measured phase noise is shown in Figure 25.15. The in-band phase noise reaches −114 dBc/Hz. Integrated from 1 kHz to 200 MHz, the integrated jitter is equal to 0.97 ps_{rms}, which satisfies the IEEE 802.11 b/g standard.

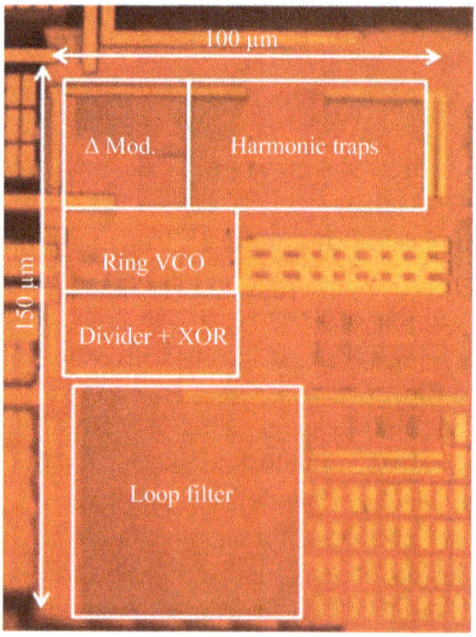

Figure 25.13 Die micrograph of integer-N synthesizer

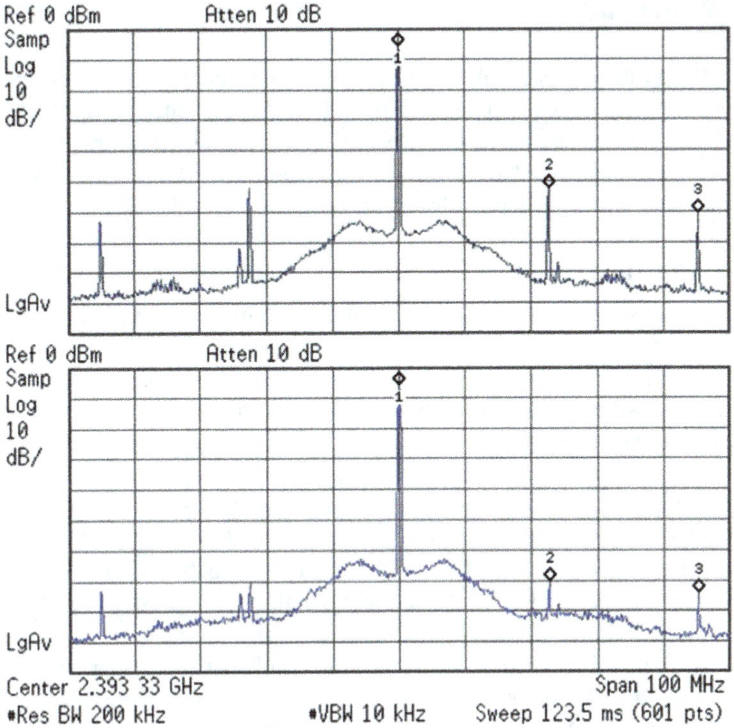

Figure 25.14 Measured output spectrum with harmonic traps turned off (top) and turned on (bottom)

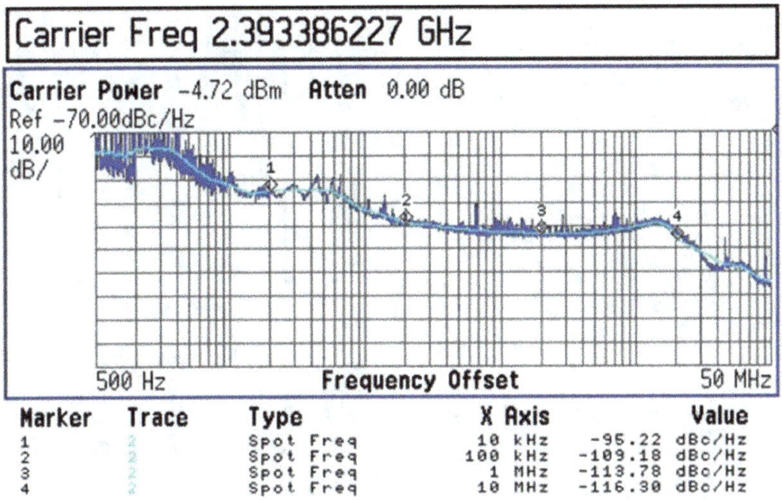

Figure 25.15 Measured phase noise of integer-N synthesizer

25.2 Wideband fractional-N RF synthesizer

Fractional-N synthesis was conceived to use a high reference frequency and avoid the trade-off between the loop bandwidth and the channel spacing [8,9]. However, the use of ΣΔ modulators and the problem of their noise peaking have worked against this premise. To solve this problem, we propose a fractional-N synthesizer architecture that achieves a BW close to $f_{REF}/4$.

25.2.1 Background

Section 25.1 has introduced an integer-N architecture with a BW close to $f_{REF}/2$. We can readily add fractional-N operation to this system [Figure 25.16(a)], but the ΣΔ modulator noise peak at $f_{REF}/2$ would cause high phase noise. To quantify this trade-off, we assume a third-order MASH ΣΔ topology, integrate the synthesizer output phase noise from a zero offset to f_{REF}, and plot the result as a function of the loop bandwidth (Figure 25.16(b)). It can be seen that for the integrated noise to remain negligible (e.g., <0.45°), the bandwidth cannot exceed $0.0007f_{REF}$.[¶] On the other hand, for sufficient VCO phase noise attenuation, we must increase the bandwidth by about a factor of 360. This contention between the ΣΔ and VCO contributions can be avoided only if the ΣΔ noise is reduced independently by another 50 dB.

In order to overcome the bandwidth-noise trade-off, a number of techniques have been introduced. For example, feedforward cancellation by means of a digital-to-analog converter (DAC) can reduce the ΣΔ noise by a maximum of 15 to 33 dB [10]. This concept has been extended to the time domain where a digital-to-time converter (DTC) is utilized to cancel the quantization noise [11], but gain error and nonlinearity issues encountered in DAC-based solutions exist here as well. Another

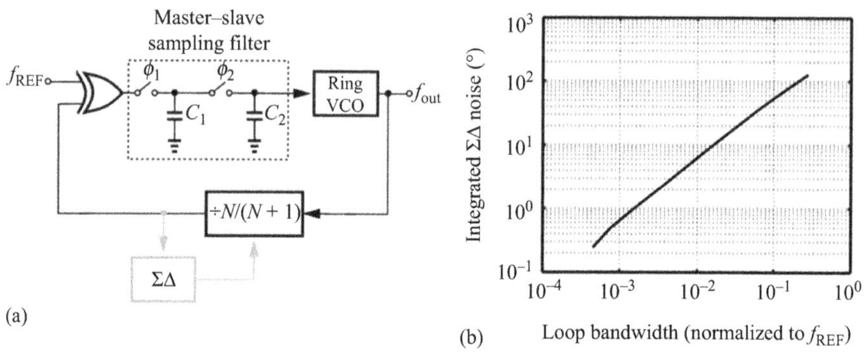

Figure 25.16 (a) Fractional-N synthesizer with MSSF and (b) integrated noise of a third-order ΣΔ modulator at PLL output versus loop bandwidth

[¶]While not reflected here, higher-order loop filters can be used as the loop bandwidth becomes very small, allowing values greater than $0.0007f_{REF}$.

approach is to raise the ΣΔ oversampling ratio and hence the reference frequency. This concept leads to cascaded phase-locked loop (PLL) topologies, wherein an integer-N loop multiplies the main reference and applies the result to a fractional-N loop [12]. The use of two VCOs, however, is generally cumbersome and prone to unwanted couplings. We should also remark that two VCOs operating at high frequencies in a cascade always present a less favorable noise-power trade-off than does a single VCO.

An alternative method is to filter—rather than simply cancel—the ΣΔ noise before it reaches the VCO. Such a filter must, of course, not limit the loop bandwidth, an interesting dilemma. For example, [13] realizes an equivalent FIR filtering action on the ΣΔ noise through the use of multiple paths, each consisting of a feedback divider, a phase/frequency detector (PFD), and a charge pump. In addition to a high power consumption, this architecture also demands linear charge pumps as they carry the unfiltered ΣΔ noise and perform the FIR summing function at their *outputs*, unless a large number of CPs are used to benefit from averaging.

In this section, we introduce a single-loop fractional-N synthesizer that performs the filtering action between the feedback divider and the phase detector. A mostly digital 35-tap FIR structure with arbitrary coefficients suppresses the ΣΔ noise at $f_{REF}/2$ by 62 dB and affords a loop bandwidth of around $f_{REF}/4$.

25.2.2 Proposed noise suppression method

25.2.2.1 Basic idea

We wish to interpose a low-pass filter (LPF) between the feedback divider and the PD so as to remove the phase fluctuations arising from the ΣΔ action (Figure 25.17). From another perspective, the filter receives a pulsewidth-modulated signal from the divider and must produce a "clean," periodic signal having a frequency equal to f_{REF}. If the average divide ratio is equal to $N + \alpha$, where α is the fractional part, then the VCO frequency is given by $(N + \alpha)f_{REF}$.

Operating on a phase (rather than a voltage) quantity, the filter can be more efficiently realized in the digital domain (Figure 25.18), but, unlike standard digital FIR filters, it must process a 1-bit input. We thus face three questions. (1) How do we configure the filter to operate on phase quantities? (2) How do we choose the

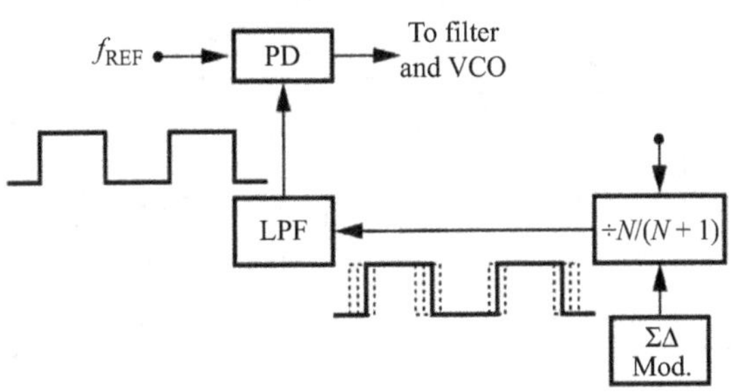

Figure 25.17 Suppression of ΣΔ quantization noise by a low-pass filter

clock frequency of the filter? (3) How do we implement the FIR tap coefficients, which may require a word length of, say, 10 bits, for a 1-bit digital signal?

Let us begin with the first question and note that the divider output contains phase jumps equal to integer multiples of the VCO period, T_{VCO}. Conceptually, we surmise that if delayed copies of the divider output are created and properly summed, then the result can have less jitter. We must therefore return to Figure 25.18 and determine how many delayed copies and with what weighting factors should be added. This is equivalent to defining the necessary frequency response of the FIR filter (Section 25.2.2.3).

It is important to note that mismatches among the α coefficients simply alter the transfer function but do not cause nonlinearity, a critical advantage over DAC-based solutions.

We can consider voltage-domain implementation of the coefficients and the summation in Figure 25.18, but we would face a serious difficulty: the residual ΣΔ noise in the summer output would be subjected to the PD's limiting action, experiencing folding to baseband. According to simulations, such an arrangement would raise the baseband noise by 30 dB. It is thus necessary to seek another method of scaling and combining the copies.

The next issue relates to the choice of the clock frequency and the FIR unit delay, that is, z^{-1} in Figure 25.18. Since the ΣΔ noise reaches a peak at $f_{REF}/2$ and since the plot in Figure 25.16(b) dictates a filter bandwidth well below this value, we expect that the first null of $|H|$ in Figure 25.18, f_1, must be a small fraction of f_{REF}. That is, a unit delay on the order of $T_{REF} = 1/f_{REF}$ is necessary. For example, 35 such units provide zeros at integer multiples of $f_{REF}/35$.

Unfortunately, the units cannot be clocked by f_{REF}: even if a single flipflop clocked by f_{REF} follows the divider, the flipflop's output carries no feedback information, prohibiting lock. Our proposed solution clocks the units by the VCO output and includes enough flipflops in Figure 25.18 to produce a total unit delay equal to T_{REF}. One naturally asks whether, instead, we can use any other low-frequency clock available in the PLL for this purpose so as to reduce the power consumption. We recognize that even if $f_{VCO}/2$ is utilized, the first flipflop's output has a time resolution

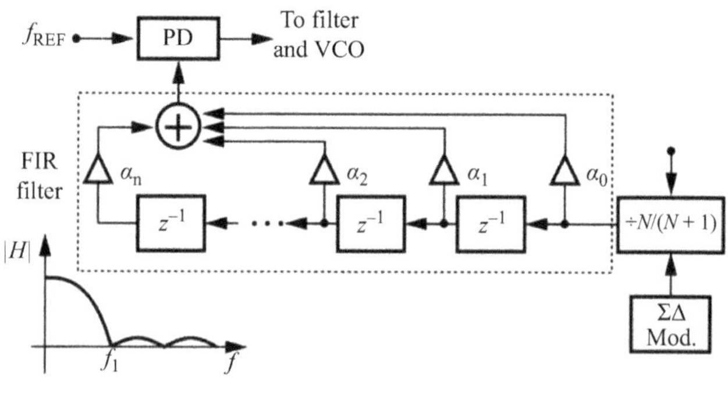

Figure 25.18 Conceptual diagram of the proposed idea

of $2T_{VCO}$ whereas its input exhibits phase jumps equal to $\pm T_{VCO}$, $\pm 2T_{VCO}$, etc. This phase truncation would fold considerable $\Sigma\Delta$ noise to baseband.

The architecture of Figure 25.18 offers two advantages over DAC feedforward cancellation. First, it is inherently a linear system and hence free from noise folding. (The loop imperfections are studied in Section 25.2.4.1.) Second, its cancellation characteristics can be more precisely defined than the latter because it does not rely on matching between a DAC and a charge pump.

We should point out another property of this architecture. Since all of the flipflops in the delay line are clocked by the VCO, any perturbation in the VCO phase readily appears at the end of the delay line; that is, the FIR structure does not filter the VCO output and hence does not limit the PLL bandwidth [12].

25.2.2.2 FIR implementation

We now deal with the implementation of the coefficients and the summation in Figure 25.18. As depicted in Figure 25.19(a), we swap the order of summation and phase detection, that is, we first compute the phase difference between the output of each delay stage and the reference and then combine the PDs' outputs with proper weighting factors.

Sensing the entire $\Sigma\Delta$-induced phase jumps, each PD must be linear enough not to fold noise to baseband. An XOR gate is a suitable candidate. According to transistor-level simulations, the rise in the baseband noise observed at the XOR output is negligible. The pulses generated by the PDs in Figure 25.19(a) represent $\phi_{REF} - \phi_0, \phi_{REF} - \phi_1, \ldots, \phi_{REF} - \phi_n$, where $\phi_1, \ldots, \phi_n$ denote the delay line output phases. Scaling the amplitudes of the pulses by the coefficients and adding the results, we obtain a quantity proportional to $\alpha_0(\phi_{REF} - \phi_0) + \alpha_1(\phi_{REF} - \phi_1) + \cdots + \alpha_n(\phi_{REF} - \phi_n) = (\alpha_0 + \alpha_1 + \cdots + \alpha_n)\phi_{REF} - (\alpha_0\phi_0 + \alpha_1\phi_1 + \cdots + \alpha_n\phi_n)$. The first sum models a constant pulsewidth, that is, a static phase, and the second can be made nearly constant by proper choice of $\alpha_0, \ldots, \alpha_n$. In other words, the tailoring of the FIR response in the frequency domain corresponds to minimizing the variation of this sum in the time domain. To visualize this point, we plot in Figure 25.20 the eye

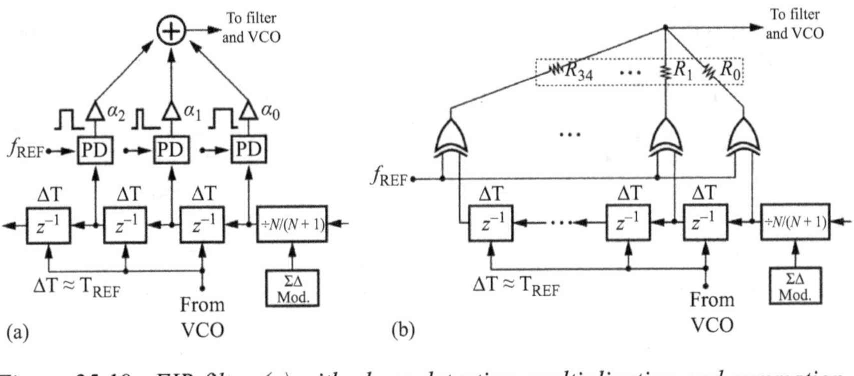

Figure 25.19 FIR filter (a) with phase detection, multiplication and summation and (b) implementation (z^{-1} consists of approximately N flipflops)

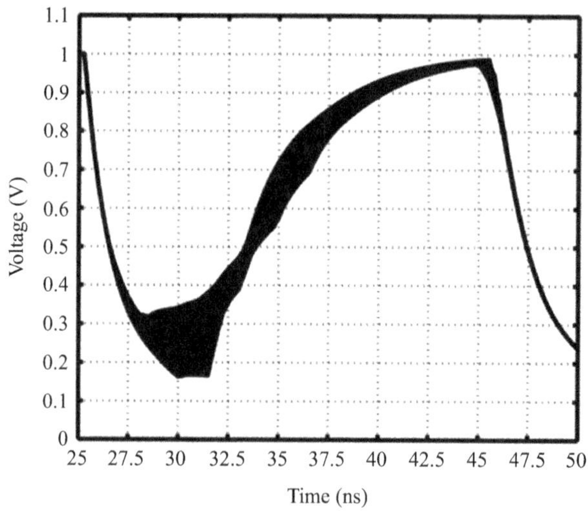

Figure 25.20 Eye diagram of filtered output

diagram produced by a certain filter response (Section 25.2.2.3). As explained in Section 25.2.2.3, only the falling edges matter here and they exhibit a jitter of 7 ps_{pp}. The divider output, on the other hand, produces about 1.68 ns_{pp} of jitter.

The actual implementation is shown in Figure 25.19(b), where resistors perform the addition in the voltage domain, and $\alpha_j = (R_0||R_1|| \ldots ||R_{34})/R_j$. The inverse dependence upon R_j means that we must first invert the coefficients and compute a convenient unit resistor value (Section 25.2.2.3).

25.2.2.3 Filter frequency response

The length and the coefficients of the FIR filter must be chosen so as to create a frequency response that sufficiently attenuates the ΣΔ noise. As a guideline, the plot of Figure 25.16(b) suggests that a suppression of more than 50 dB is necessary if a loop bandwidth near $f_{REF}/4$ is desired. Thus, the FIR design should be tailored such that the area under the *product* of its response and the ΣΔ noise spectrum is small enough. Of course, since the delay elements are clocked at the VCO frequency, we wish to minimize their number.

Simulations can readily rule out certain FIR designs and point to more appropriate choices. If the FIR has 11 equal taps, we recognize that the rejection at $f_{REF}/2$ (= 10 MHz) is equal to 20 log 11 because the transfer function, $(z^{-10} + \cdots + z^{-1} + 1)/11$, assumes a value of 1 at $f = 0$ and 1/11 at $f = f_{REF}/2$. With a 1-1-1 MASH ΣΔ modulator clocked at $f_{REF} = 20$ MHz, the resulting phase noise spectrum at the synthesizer output has an area as high as −17.2 dBc from $f = 0$ to $f = f_{REF}$. Moreover, the product has a local peak around $f = f_{REF}/4.4$ because the FIR filter's 3-dB bandwidth (≈ 0.8 MHz) is excessively large.

The foregoing observations reveal that a greater number of taps must be chosen, both to achieve a higher rejection in the vicinity of $f_{REF}/2$ and to reduce the bandwidth (and hence avoid the local peak). For example, a 35-tap realization with equal coefficients yields a rejection of 20 log 35 ≈ 31 dB at $f_{REF}/2$ and a bandwidth of approximately $f_{REF}/79$. However, even such a design produces an integrated phase noise of 3.5°.

In order to alleviate the situation, we can further increase the number of taps. To achieve a 50-dB reduction in integrated noise, we require 285 taps if the coefficients are equal, which translates to tens of thousands of flipflops. Alternatively, we can select the individual coefficients so as to tailor the response to our needs, a flexibility offered by the resistive combining scheme proposed in Figure 25.19(b). For example, a 35-tap Kaiser-Bessel design [14] provides a 3-dB bandwidth of 400 kHz and a rejection of 65 dB at $f_{REF}/2$. More importantly, the closed-loop output spectrum exhibits an integrated phase noise of 0.38°. The coefficients are given by $\alpha_{34} = \alpha_0 = 0.0011$, $\alpha_{33} = \alpha_1 = 0.0024$, ..., $\alpha_{17} = 0.0575$.

For the ΣΔ noise spectrum assumed here, a 35-tap Kaiser-Bessel filter does not necessarily provide the optimum response. In fact, it is possible to adjust the coefficients manually so as to lower the bandwidth but at the cost of rejection at $f_{REF}/2$ and vice versa. It may even be possible to achieve a slightly lower integrated phase noise through such adjustments. However, another important factor is the realizability of the coefficients, that is, the "resistor spread" in Figure 25.19(b) as well as the resolution necessary in implementing the resistor values. In the example described above, the spread is equal to $\alpha_{17}/\alpha_{34} = 52.3$, and the coefficient values are such that a unit resistor of 400 Ω can realize them with reasonable accuracy.

For the Kaiser-Bessel response, we have the following resistor values: $R_0 = R_{34} = 40$ kΩ, $R_1 = R_{33} = 18.5$ kΩ, ..., $R_{17} = 0.783$ kΩ. We must decide on a unit resistor, R_u, that can yield these values with acceptable accuracy. We choose $R_u = 400$ Ω and use various parallel and series combinations to form $R_0, \ldots, R_{34}$, inevitably incurring some error. The actual filter response has a rejection of 62 dB at $f_{REF}/2$.

25.2.2.4 Effect of nonidealities

The XORs and the resistors in Figure 25.19(b) suffer from two imperfections that can affect the performance. We first consider the effect of resistor mismatches. While introducing no nonlinearity, these mismatches do alter the transfer function, particularly the rejection at $f_{REF}/2$. Fortunately, it is much simpler to obtain a given level of matching between resistors than between transistors (as would be necessary in DACs, charge pumps, and phase interpolators).

In a typical analog layout, resistor mismatches can be readily maintained below about 1 percent. We thus perform Monte Carlo simulations to determine how much the rejection varies. The simulations assume that each unit resistor in the array of $R_0, \ldots, R_{34}$ can vary according to a Gaussian distribution with $\sigma = 1$ percent. We also compute the mismatches of the XOR output resistances from Cadence Monte Carlo simulations and combine the results with the unit resistor mismatches. Figure 25.21 plots the departure of the rejection from 62 dB. The distribution's asymmetry is due to the inverse dependence of the coefficients upon their corresponding resistor values. Here, the samples that fall below 0 dB have a

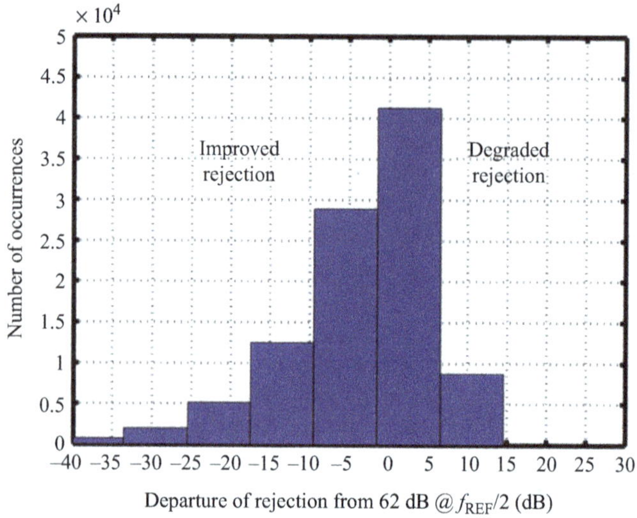

Figure 25.21 *Monte Carlo results showing departure of rejection from 62 dB at $f_{REF}/2$*

greater rejection. Of the 100,000 samples, only about 5 percent experience a degradation of 10 dB or higher. (If even a tighter distribution is desired, a careful layout can yield resistor mismatches less than 0.5 percent, as is practical in A/D and D/A converter design.)

The other imperfection relates to the finite output resistance of the XOR gates as it appears in series with the actual resistors. This resistance must be small enough to introduce negligible mismatches. The XOR design employs large transistors ($W/L = 4$ μm/40 nm) for low flicker noise, and its output resistance is still about 200 Ω, but it has negligible impact on the response.

25.2.3 Proposed synthesizer architecture

In this section, we apply the ΣΔ noise suppression method described above to the synthesizer architecture introduced in Figure 25.16(a). Shown in Figure 25.22, the fractional-N loop consists of the FIR filter, including the XOR PDs and the resistive combiner, a master-slave sampling filter, a ring VCO, a multimodulus feedback divider, and a MASH 1-1-1 ΣΔ modulator. To reduce the ripple on the oscillator's control voltage, a harmonic trap establishes notches at f_{REF} and $2f_{REF}$, and a Δ modulator sets the notch frequencies by measuring and minimizing the ripple. It is demonstrated in Section 25.1 that this architecture can achieve a bandwidth close to $f_{REF}/2$ in the integer-N mode. Figure 25.23 plots the synthesizer ΣΔ-induced phase noise for BW $= f_{REF}/2$ and $f_{REF}/4$. Despite the suppression provided by the FIR action, the output exhibits 7 dB of peaking in the former case. The bandwidth is thus reduced to $f_{REF}/4$ by lowering the PD gain, K_{PD}. (The sampling filter provides notches at mf_{REF} and only a slight attenuation below $f_{REF}/2$.)

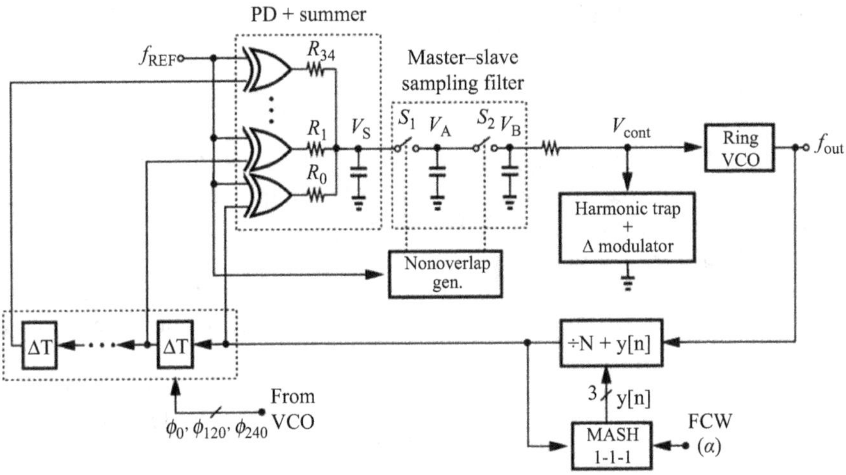

Figure 25.22 Proposed synthesizer architecture

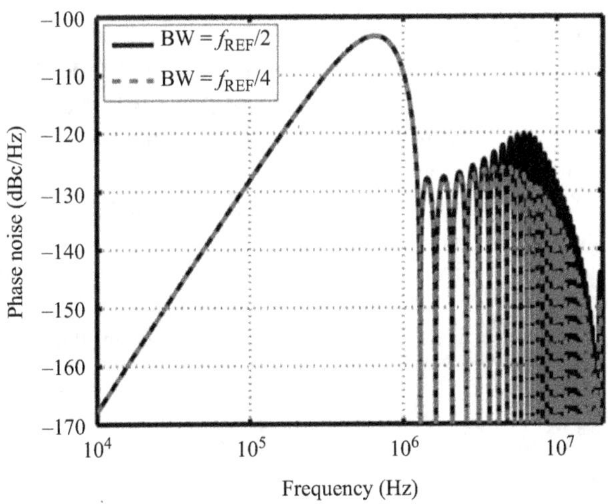

Figure 25.23 Synthesizer ΣΔ-induced phase noise for different loop bandwidths

The MSSF in Figure 25.22 is clocked by nonoverlapping phases derived from f_{REF}, a point of contrast to the integer-N architecture in Section 25.1, where these phases are obtained from the feedback signal. This choice is made here to ensure that the MSSF clock itself is free from the ΣΔ noise and hence does not corrupt the sampled signal.

The use of the MSSF considerably widens the frequency capture range. Following the analysis in Section 25.1.2.5, we can express the range for the present design as:

$$|f_{REF} - f_{FB}| < \frac{f_{REF}}{2} \tag{25.14}$$

where f_{FB} denotes the feedback signal frequency. Thus:

$$\frac{f_{REF}}{2} < f_{FB} < \frac{3f_{REF}}{2} \tag{25.15}$$

Translated to the VCO output, this range is around 1.2 GHz to 3.6 GHz, much wider than our VCO's analog tuning range (320 MHz). In actual measurements, we first turn on only the tap with $\alpha_{17} = 0.0575$ to reach lock and then the remaining taps.

The VCO in Figure 25.22 is realized as a three-stage ring topology (Section 25.1.3.1). The performance can be summarized as follows: overall tuning range: 2.31–3.05 GHz, $K_{VCO} \approx 400$ MHz/V, power consumption: 3 mW, phase noise at 5-MHz offset: -115.5 dBc/Hz.

The MASH 1-1-1 ΣΔ modulator in Figure 25.22 employs a word length of 16 bits for a frequency resolution of 345 Hz at 2.4 GHz. The feedback divider modulus changes in unity steps from 103 to 110 by the modulator. The FIR filter is clocked by three VCO phases for the reasons explained in Section 25.2.4.3.

25.2.4 Design considerations

25.2.4.1 Nonlinearity issues

While reduced considerably by the FIR filter in Figure 25.22, the ΣΔ noise can still experience folding if the MSSF and the VCO are excessively nonlinear. In order to formulate the MSSF nonlinearity, we consider the simplified cascade shown in Figure 25.24(a), where the XOR and R_{eq} ($\approx 45\ \Omega$) represent the parallel combination of the 35 XORs and their summing resistors in Figure 25.22. Switches S_1 and S_2 are controlled by nonoverlapping phases derived from f_{REF}, and the effect of C_0 is ignored for now. Recall from Section 25.2.2.3 that the resistive summer output closely resembles a periodic waveform toggling between zero and V_{DD} (as in an integer-N architecture), with some residual ΣΔ noise on the transitions. This

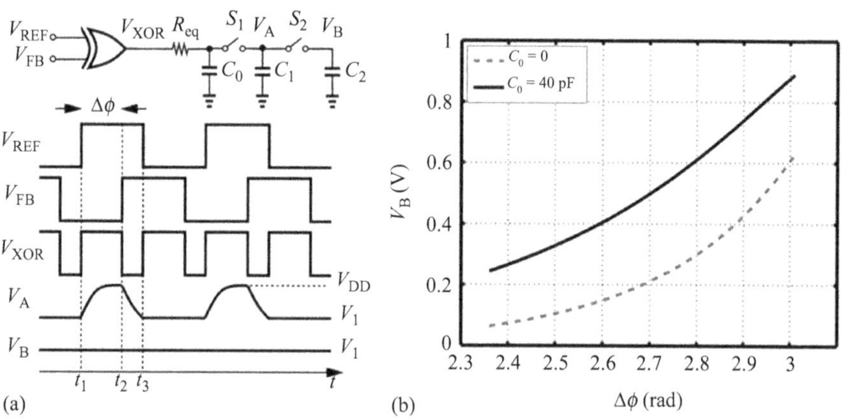

Figure 25.24 (a) XOR PD driving MSSF and (b) control voltage versus static phase error

waveform must therefore be so filtered as to yield an acceptably small ripple on the VCO control line, a task performed by the MSSF in Figure 25.22. (As shown in Section 25.1.2.2, the master-slave sampling action yields zeros in the transfer function located at mf_{REF}.) The operation of the cascade is illustrated by the waveforms in Figure 25.24(a). Switch S_1 turns on as V_{REF} goes high (at $t = t_1$), allowing C_1 to charge to V_{DD}. At $t = t_2$, the PLL feedback signal, V_{FB}, rises, V_{XOR} falls, and C_1 begins to discharge. The phase difference between the falling edge of V_{REF} and the rising edge of V_{FB} is set by the synthesizer such that the discharge lowers V_{A} to the value required by the VCO, V_1. Finally, S_1 turns off and S_2 turns on, impressing V_1 on C_2. In the steady state, the nominal ripple amplitude is zero.

We wish to compute an equivalent PD gain for this cascade and examine its nonlinear behavior. Including the effect of C_0 (= 40 pF) (and neglecting droop due to leakage currents), we write the denominator of the transfer function from V_{XOR} to V_{A} as $R_{\mathrm{eq}}R_{\mathrm{on}}C_0C_1s^2 + (C_0R_{\mathrm{eq}} + C_1R_{\mathrm{on}} + C_1R_{\mathrm{eq}})s + 1$, where R_{on} is the switch resistance, and calculate the poles. The falling transition of V_{A} can be expressed as:

$$V_{\mathrm{A}}(t) = -\frac{p_2}{p_1 - p_2}V_{\mathrm{DD}}e^{p_1 t} + \frac{p_1}{p_1 - p_2}V_{\mathrm{DD}}e^{p_2 t} \tag{25.16}$$

where p_1 and p_2 denote the poles. To calculate V_1 in Figure 25.24(a), we set t to $[(\pi - \Delta\phi)/(2\pi)]T_{\mathrm{REF}}$:

$$V_1 = \frac{V_{\mathrm{DD}}}{p_1 - p_2}\left[-p_2 \exp\frac{p_1(\pi - \Delta\varphi)T_{\mathrm{REF}}}{2\pi} + p_1 \exp\frac{p_2(\pi - \Delta\varphi)T_{\mathrm{REF}}}{2\pi}\right]. \tag{25.17}$$

This PD characteristic is plotted in Figure 25.24(b) for the values chosen in our work. Also shown is the case of $C_0 = 0$, exhibiting a higher nonlinearity. We can prove that, with $C_0 = 40$ pF, the XOR/MSSF cascade introduces negligible noise folding because the residual ΣΔ noise at the summer output is already small. Also, the width of S_1 in Figure 25.22 is programmable so as to vary R_{on} with the output frequency and hence maintain a relatively constant K_{PD}.

The PD gain-nonlinearity trade-off translates to one between the loop bandwidth and noise folding. Based on simulations, the optimum values can be chosen so as to minimize the VCO phase noise and yet avoid significant folding.

25.2.4.2 Delay line implementation

The FIR filter in Figure 25.22 consists of 34 unit delays equal to $T_{\mathrm{REF}} = 1/f_{\mathrm{REF}} \approx$ 50 ns. In other words, the total delay presented to the divider output is 1.7 μs, a value extremely difficult to realize by asynchronous stages such as inverters. We instead opt for flipflops driven by the VCO, requiring about 4,000 of them. The FIR filter data path operates at a low frequency, drawing negligible power. The clock path, on the other hand, is driven by the VCO and consumes a power given by $f_{\mathrm{VCO}}C_{\mathrm{CK}}V_{\mathrm{DD}}^2$, where C_{CK} denotes the total capacitance in the clock path of the flipflops. The true single-phase clock (TSPC) topology [15] proves useful here as it employs only two clocked transistors per flipflop. Shown in Figure 25.25 is the resettable design, consuming 1.5 μW at 2.4 GHz. The overall delay line thus

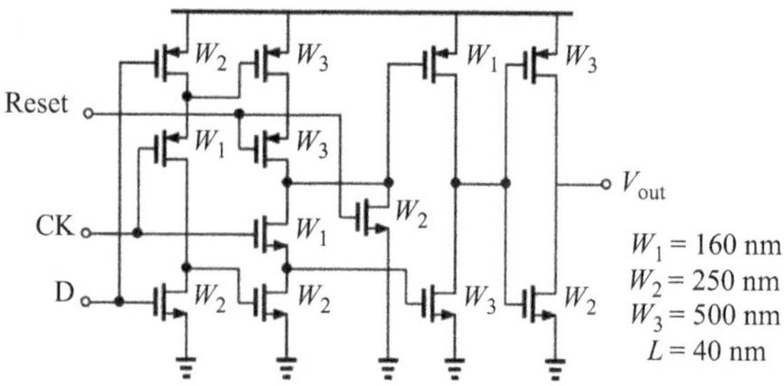

Figure 25.25 TSPC flipflop

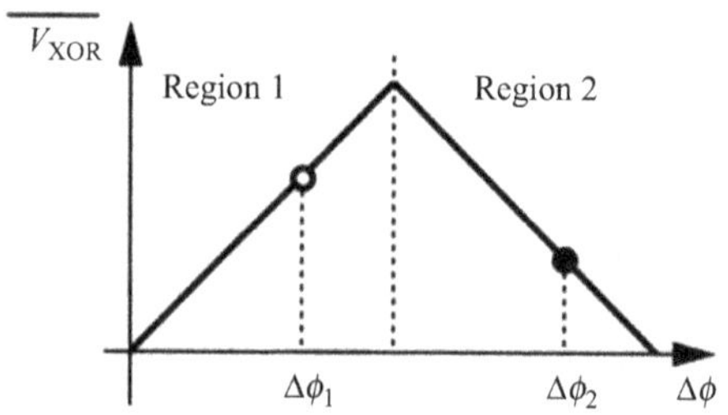

Figure 25.26 Nonmonotonic XOR PD characteristic

dissipates 5.6 mW in 40-nm technology. This amount is expected to fall in more advanced nodes, making the proposed approach more power efficient. For example, according to simulations in 16-nm technology, the power reduces to 1.5 mW at 2.4 GHz, and the capacitance to 0.75 pF.

25.2.4.3 PD nonmonotonicity

XOR gates operating as phase detectors exhibit a nonmonotonic behavior if the phase error grows sufficiently large. As illustrated in Figure 25.26, the PD gain changes sign as the phase error goes from $\Delta\phi_1$ to $\Delta\phi_2$. We observe that this nonmonotonicity folds high frequency noise if the $\Sigma\Delta$-induced phase fluctuations force $\Delta\phi$ to cross from Region 1 to Region 2. Indeed, with as many as 34 unit delays added to the phase of the feedback signal, some XORs in Figure 25.22 can reside near this boundary and experience random polarity reversals. (Recall that each XOR senses the reference and one delayed signal from the delay line). Noise folding can also be seen if we approximate the XOR characteristic by a parabola and recognize that the $\Sigma\Delta$ noise is subjected to such nonlinearity.

To investigate this effect, we recall from Section 25.2.2.1 that the FIR unit delay, ΔT, is nominally equal to T_{REF}. However, the N flipflops within each unit are clocked by the VCO, whose frequency may not be an integer multiple of f_{REF}. That is,

$$\Delta T = NT_{VCO} = \frac{N}{N+\alpha} T_{REF} \approx \left(1 - \frac{\alpha}{N}\right) T_{REF} \tag{25.18}$$

Since the delay departs from T_{REF} by $(\alpha/N)T_{REF}$, it appears as a phase offset and continues to accumulate through the delay line, producing a maximum error of $34(\alpha/N)T_{REF}$ at the end. We can identify three groups of XORs: some fully operate in Region 1, some fully in Region 2, and some randomly cross the boundary, causing noise folding. But even the second group presents an issue: the negative gains carried by these XORs translate to negative FIR taps, distorting the Kaiser-Bessel response, and also partially cancel the average positive gains provided by the first group, reducing the loop gain. For these reasons, we employ a method that confines all of the XORs to Region 1 for any value of α.

To control and limit the amount of phase offset accumulation, we wish to adjust the FIR unit delay for different values of the $\Sigma\Delta$ fractional control word, α. Let us assume that the unit delay is changed from NT_{VCO} to $(N + k)T_{VCO}$, allowing (25.18) to be written as:

$$\Delta T = (N + k)T_{VCO} = \frac{N + k}{N + \alpha} T_{REF} \approx \left(1 + \frac{k - \alpha}{N}\right) T_{REF} \tag{25.19}$$

We recognize that, ideally, k is chosen equal to α so as to minimize the accumulation. But the synchronous delay line offers a resolution of T_{VCO} and hence a minimum k equal to 1, whereas $\alpha < 1$. Fortunately, the three-stage ring VCO provides three phases, improving the resolution to $T_{VCO}/3$. Therefore, we can limit the maximum phase offset to $T_{VCO}/6$ per unit. The accumulated delay reaches $34 \times (69\text{ ps}) \approx 2.36$ ns. To this value, we must add the maximum $\Sigma\Delta$-induced phase fluctuation, about $2T_{VCO} = 0.83$ ns. Now we ensure that, with a departure equal to 2.36 ns + 0.83 ns = 3.19 ns, the XORs remain in Region 1.

25.2.4.4 Overall phase noise

With the $\Sigma\Delta$ noise suppressed by the FIR filter, the synthesizer output phase noise arises primarily from the VCO. The $f_{REF}/4$ bandwidth relaxes the VCO design, allowing a phase noise value as high as −115.5 dBc/Hz at 5-MHz offset. The overall simulated synthesizer phase noise along with the main contributors is plotted in Figure 25.27. The overall integrated jitter is equal to 1.44 ps_{rms}, and despite a peak of −105 dBc/Hz, the $\Sigma\Delta$ noise translates to only 0.44 ps_{rms}.

It is possible to further increase the loop bandwidth but at the cost of higher reference spurs. Also, the reduction in the total jitter begins to diminish for bandwidths exceeding around 5 or 6 MHz.

*25.2.5 Experimental results of fractional-*N *synthesizer*

The fractional-N synthesizer has been fabricated in TSMC's 45-nm digital CMOS technology and characterized with a 22.6-MHz crystal oscillator acting as the reference. Operating from a 1-V supply,[‖] the prototype draws 10 mW, with 3 mW

[‖]The VCO and harmonic trap use a 1-V analog supply while the remaining blocks share a 1-V digital supply.

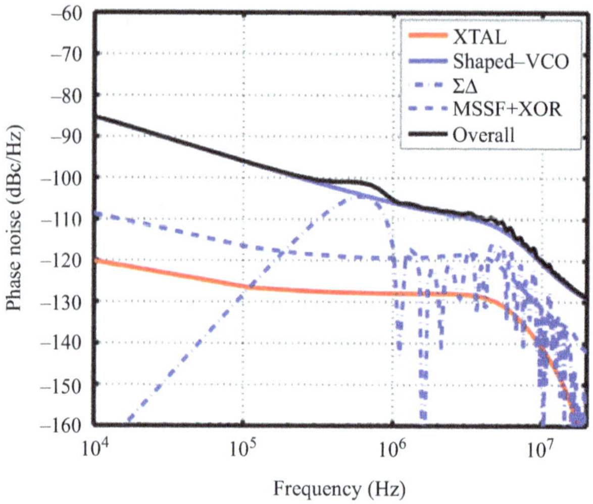

Figure 25.27 Overall simulated PLL phase noise and main contributors

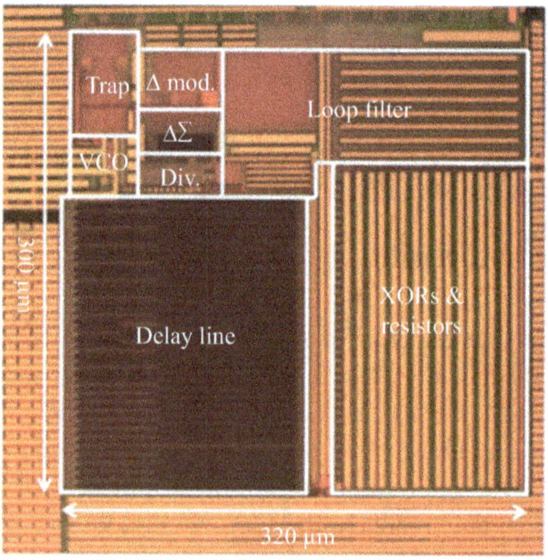

Figure 25.28 Die micrograph of fractional-N synthesizer

dissipated in the VCO, 5.6 mW in the delay line, 0.4 mW in the divider and 0.8 mW in the XOR PDs and MSSF. Shown in Figure 25.28 is the die photo, whose active area measures 300 μm × 320 μm.

Figure 25.29 plots the measured phase noise in the integer-N mode, serving as a "baseline" for quantifying various contributions. In this mode, the MSSF, VCO, and crystal reference phase noise components are present. The in-band noise is

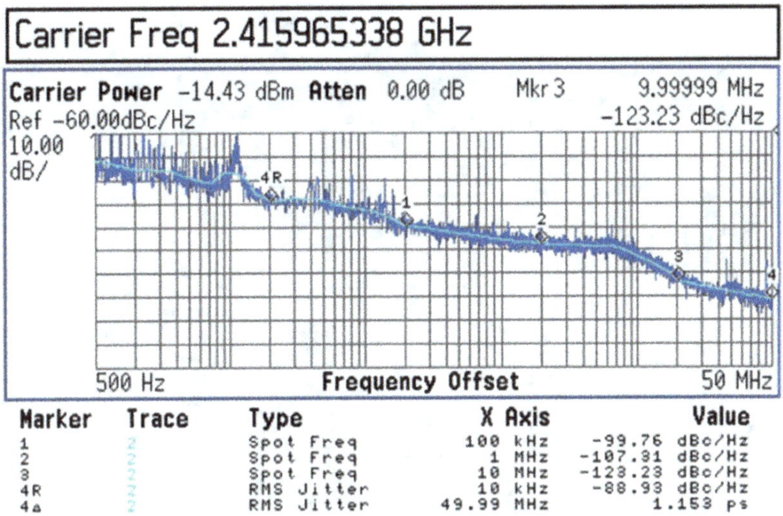

Figure 25.29 Measured phase noise in integer-N mode

about −107 dBc/Hz with a bandwidth of around 5.6 MHz. Integrated from 10-kHz to 50-MHz offset, the rms jitter amounts to 1.15 ps.

Figure 25.30(a) shows the measured output spectra in the fractional-*N* mode in two cases: only two taps of the FIR filter are enabled (top) and all 35 taps are enabled (bottom). The divide ratio is equal to $106 + 2^{-15} + 2^{-16}$. In the former case, the FIR filter provides some attenuation beyond $f_{REF}/2$ ($\approx$ 11 MHz), but the close-in phase noise remains very high. In the latter, on the other hand, the $\Sigma\Delta$ noise falls below the VCO contribution. The reference spurs are at −60 dBc. Figure 25.30(b) plots the measured output phase noise in the fractional-*N* mode for the above two configurations. In the former, the phase noise remains as high as −76 dBc/Hz up to 10-MHz offset. We also observe heavy noise folding due to the MSSF nonlinearity. In the latter, the phase noise at 10-MHz offset falls by 45 dB, reaching −121.4 dBc/Hz, which corresponds to 62 dB of reduction compared to the case without an FIR filter. The integrated rms jitter drops from 100 ps to 1.5 ps. Compared to the integer-*N* mode, the phase noise degrades by 3 dB at 1-MHz offset and 1.8 dB at 10-MHz offset.

Figure 25.31 plots the measured fractional spur levels as a function of the frequency offset. The in-band level is around −41 dBc and the magnitude approximately follows the synthesizer response. These spurs are attributed to the supply line shared between the FIR filter and the XOR PDs. Simulations confirm that by separating these two supplies, the spur level will drop by 13 dB. We note that such in-band spurs negligibly degrade the signals in WiFi applications as they simply cause a small fraction of one OFDM sub-channel to fall atop another. The spurs fall to −60 dBc at the edge of the RF channel.

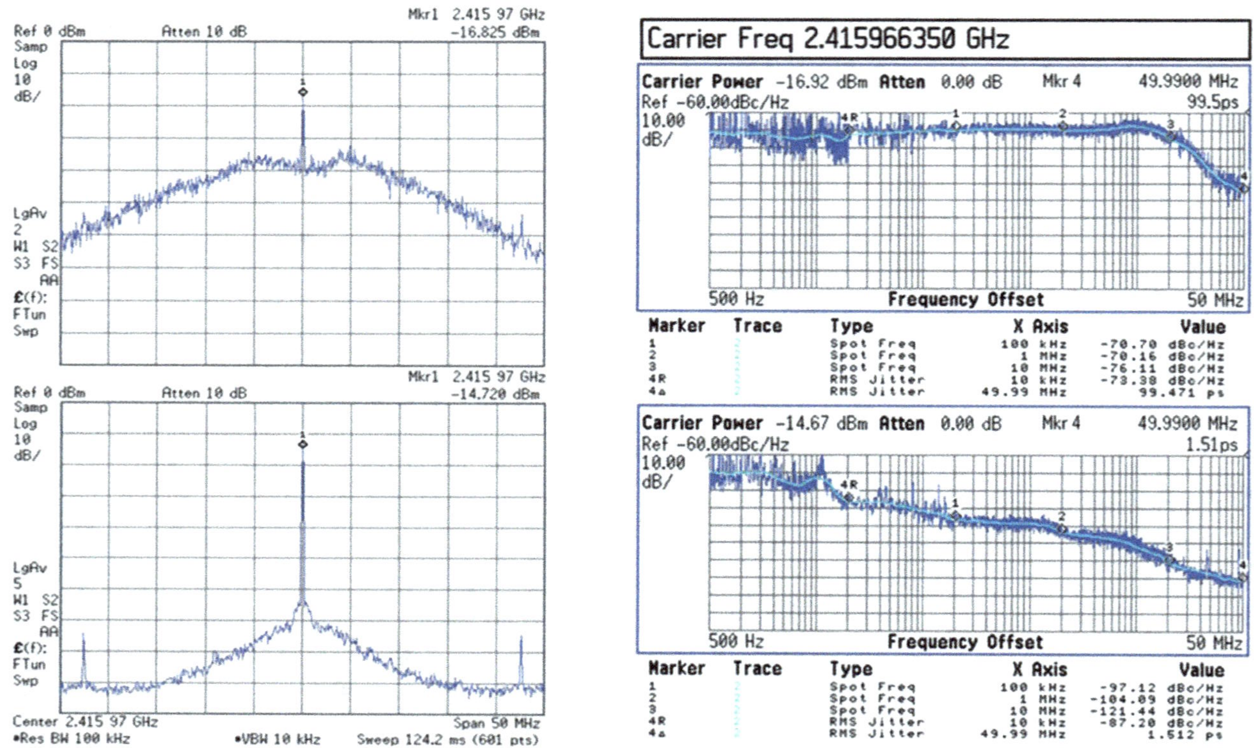

(a) (b)

Figure 25.30 *(a) Measured spectra in fractional-N mode with 2-tap FIR (top) and 35-tap FIR (bottom) and (b) measured phase noise in for these two configurations*

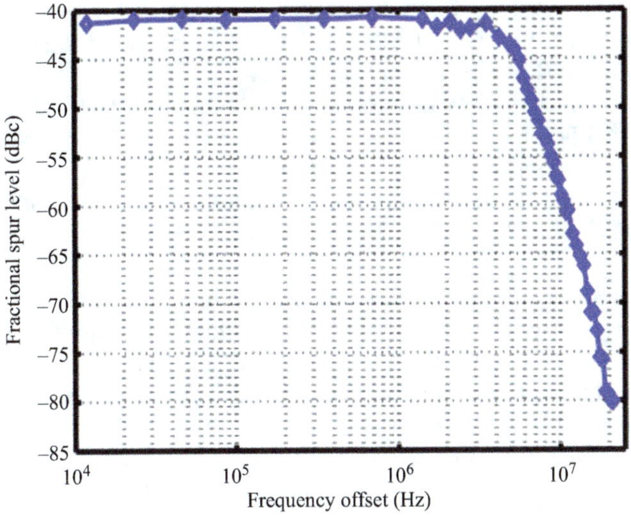

Figure 25.31 Measured fractional spur level versus frequency offset

References

[1] Kong L., and Razavi B. "A 2.4-GHz 4-mW integer-N RF synthesizer." *IEEE J. Solid-State Circuits*. 2016, vol. 51, no. 3, pp. 626–635.

[2] Kong L., and Razavi B. "A 2.4-GHz RF fractional-N synthesizer with BW = $0.25f_{REF}$." *IEEE J. Solid-State Circuits*. 2018, vol. 53, no. 6, pp. 1707–1718.

[3] Abidi A.A. "Phase noise and jitter in CMOS ring oscillators." *IEEE J. Solid-State Circuits*. 2006, vol. 41, no. 8, pp. 1803–1816.

[4] Homayoun A., and Razavi B. "Relation between delay line phase noise and ring oscillator phase noise." *IEEE J. Solid-State Circuits*. 2014, vol. 49, no. 2, pp. 384–391.

[5] Meninger S.E., and Perrott M.H. "A 1-MHz bandwidth 3.6 GHz 0.18-µm CMOS fractional-N synthesizer utilizing a hybrid PFD/DAC structure for reduced broadband phase noise." *IEEE J. Solid-State Circuits*. 2006, vol. 41, no. 4, pp. 966–980.

[6] Wang K.J., Swaminathan A., and Galton I. "Spurious tone suppression techniques applied to a wide-bandwidth 2.4 GHz fractional-N PLL." *IEEE J. Solid-State Circuits*. 2008, vol. 43, no. 12, pp. 2787–2797.

[7] Razavi B. *Principles of Data Conversion System Design*, IEEE Press, 1995.

[8] Gillette G. "The digiphase synthesizer." *Frequency Technology*, pp. 25–29, 1969.

[9] Gibbs J., and Temple R. "Frequency domain yields its data to phase-locked synthesizer." *Electronics*, pp. 107–111, 1978.

[10] Pamarti S., Jansson L., and Galton I. "A wideband 2.4-GHz delta-sigma fractional-N PLL with 1-Mb/s in-loop modulation." *IEEE J. Solid-State Circuits*. 2004, vol. 39, no. 1, pp. 49–61.
[11] Gao X., Burg O., Wang H., *et al.* "A 2.7-to-4.3GHz, 0.16ps_{rms}-jitter, -246.8dB-FOM, digital fractional-N sampling PLL in 28nm CMOS." *ISSCC Dig. Tech. Papers*, 2016, pp. 174–175.
[12] Kong L., and Razavi B. "A 2.4-GHz 6.4-mW fractional-N inductorless RF synthesizer." *IEEE J. Solid-State Circuits*. 2017, vol. 52, no. 8, pp. 2117–2127.
[13] Yu X., Sun Y., Zhang L., Rhee W., and Wang Z. "An FIR-embedded noise filtering method for ΔΣ fractional-N PLL clock generators." *IEEE J. Solid-State Circuits*. 2009, vol. 44, no. 9, pp. 2426–2436.
[14] Kaiser J.F. "Nonrecursive digital filter design using the I_0-sinh window function." *Proc. IEEE Inter. Symp. Circuits Syst.*, 1974, pp. 20–23.
[15] Yuan J., and Svensson C. "High-speed CMOS circuit technique." *IEEE J. Solid-State Circuits*. 1989, vol. 24, no. 1, pp. 62–70.

Index

www.ingramcontent.com/pod-product-compliance
Lightning Source LLC
LaVergne TN
LVHW011536200926
848484LV00001B/271